T0189974

Springer Monographs in Mathematics

More information about this series at http://www.springer.com/series/3733

J.W.P. Hirschfeld • J.A. Thas

General Galois Geometries

J.W.P. Hirschfeld
Department of Mathematics
University of Sussex
Brighton, UK

J.A. Thas
Department of Mathematics
Ghent University
Gent, Belgium

Springer Monographs in Mathematics
ISBN 978-1-4471-7391-5   ISBN 978-1-4471-6790-7   (eBook)
DOI 10.1007/978-1-4471-6790-7

Mathematics Subject Classification (2010): 51E20, 51E12, 51E14, 51E22, 51E23, 94B05, 05B25, 14H50, 14L35, 14M15, 51A05, 51A30, 51A45, 51A50, 51D20, 51E21

Springer London Heidelberg New York Dordrecht
© Springer-Verlag London 2016
Softcover reprint of the hardcover 1st edition 2016

Printed on acid-free paper

Springer-Verlag London Ltd. is part of Springer Science+Business Media (www.springer.com)

For Lucca and Indigo

JWPH

For Sam, Jo and Maya

JAT

Contents

Contents

Preface

This book is the second edition of the third and last volume of a treatise on projective spaces over a finite field, also known as Galois geometries. The first volume, *Projective Geometries over Finite Fields* (1979, 1998), with the second edition referred to as PGOFF2, consists of Parts I to III and contains Chapters 1 to 14 and Appendices I and II. The second volume, *Finite Projective Spaces of Three Dimensions* (1985), referred to as FPSOTD, consists of Part IV and contains Chapters 15 to 21 and Appendices III to V. The present volume comprises Part V and, in its first edition, contains Chapters 22 to 27 and Appendices VI and VII. In this edition, the chapters are numbered from 1 to 7. The scheme of the treatise is indicated by the titles of the parts:

Part I   Introduction
Part II   Elementary general properties
Part III The line and the plane
Part IV $PG(3, q)$
Part V $PG(n, q)$

There are three themes within the book: (a) properties of algebraic varieties over a finite field; (b) the determination of various constants arising from the combinatorics of Galois spaces such as the maximum number of points of a subset under certain linear independence conditions; (c) the identification in Galois spaces of various incidence structures. Many of the results on theme (a) could be equally well stated over an arbitrary field. However, over a finite field, counting arguments come more into play. A significant number of theorems count certain sets and establish the existence of combinatorial structures. Most of Chapters 1 to 4 is on theme (a), whereas Chapter 5 is on theme (c) and Chapter 6 is for the most part on (b). Chapter 7 is on themes (a) and (b).

Chapter 1 on quadrics develops their properties and gives one way of characterising them.

Chapter 2 on Hermitian varieties similarly develops their properties and characterises them in the course of describing all sets of type $(1, r, q + 1)$. This chapter is the one on algebraic varieties that differs most from the classical case, as Hermitian manifolds over the complex numbers are not algebraic varieties.

Chapter 3 on Grassmann varieties and Chapter 4 on Veronese and Segre varieties most closely follow a classical model in the description of their properties. Although most of the characterisations of the Veronesean of quadrics and its projections resemble classical theorems over the complex numbers, the characterisation of Grassmannians is quite different. This is because the Grassmannian characterisation is in terms of an incidence structure, a topic which was studied over the real and complex numbers only for the entire projective space rather than any substructure, whereas the Veronesean and its projections are studied as subsets of $PG(n, q)$ in terms of sections by subspaces. Chapter 4 also contains a section on Hermitian Veroneseans; this section contains no proofs.

Chapter 5 begins with polar spaces, thereby unifying the subjects of Chapters 1 and 2, and it goes on to consider the special case of generalised quadrangles and structures which are natural developments. In this chapter, not every theorem is proved; in particular, no proofs are given for most of the characterisations of generalised quadrangles

Chapter 6 generalises to an arbitrary dimension results of Chapters 18 and 21 from the previous volume: an upper bound is found for the size of a k-cap and the maximum size of a k-arc is found under some restrictions on n and q; the corresponding arcs are generally normal rational curves.

Chapter 7 begins with ovoids and spreads of finite classical polar spaces, which are then generalised to m-systems. Applications to maximal arcs, translation planes, strongly regular graphs, linear codes, generalised quadrangles and semi-partial geometries are given. This is the only chapter without proofs.

The book is conceived as a work of reference and does not have any exercises. However, each individual chapter is suitable for a course of lectures.

Apart from Chapter 5 and the short Chapter 7, complete proofs are given for nearly all results. The last section of each chapter contains all references as well as remarks both on the chapter itself and on related aspects that are not covered.

This volume may be considered as developing over finite fields aspects of the three volumes of Hodge and Pedoe [183, 184, 185], particularly regarding quadrics and Grassmannians. Burau [60] is also an appropriate analogy for quadrics, Grassmannians, Veroneseans and Segre varieties.

Compared to the first edition, this edition contains a considerable amount of new material. In Chapter 4, the characterisation of quadric Veroneseans has been completely rewritten; there is also a section on Hermitian Veroneseans. Chapters 5 and 6 are updated, and contain several new and better proofs. Chapter 7 is new up to the section on ovoids and spreads of finite classical polar spaces, and covers much new material but without proofs.

Status of the subject

Apart from being an interesting and exciting area in combinatorics with beautiful results, Galois geometries have many applications to coding theory, algebraic geometry, incidence geometry, design theory, graph theory, cryptology and group theory. As an example, the theory of linear maximum distance separable codes (MDS codes) is equivalent to the theory of arcs in $PG(n, q)$; so all results of Sections 6.5 to 6.7 can be expressed in terms of linear MDS codes.

Finite projective geometry is essential for finite algebraic geometry, and finite algebraic curves are used to construct interesting classes of codes, the Goppa codes, now also known as algebraic-geometry codes. Many interesting incidence structures and graphs are constructed from finite Hermitian varieties, finite quadrics, finite Grassmannians and finite normal rational curves. Further, most of the objects studied in this book have an interesting group; the classical groups and other finite simple groups appear in this way.

Currently there are several international journals on combinatorics and geometry publishing a large number of papers on Galois geometries; for example, *Ars Combinatoria, Combinatorica, Designs, Codes and Cryptography, European Journal of Combinatorics, Finite Fields and their Applications, Journal of Algebraic Combinatorics, Journal of Combinatorial Designs, Journal of Combinatorial Theory Series A, Journal of Geometry*, and the conference series *Annals of Discrete Mathematics*.

Finite vector spaces and hence also finite projective spaces are of great importance for theoretical computer science. So, in many syllabuses of a computer science degree, there is a course on discrete mathematics with a section on combinatorial structures.

Related topics

There are some interesting topics either not covered or only touched upon in the three volumes. In the *Handbook of Incidence Geometry* [55], edited by Buekenhout, surveys of several of these topics are given. Recent surveys are contained in *Current Research Topics in Galois Geometry* [298], edited by Storme and De Beule.

Finite non-Desarguesian planes are not discussed in the treatise. For references see the chapters in the Handbook [55] on 'Projective planes' by Beutelspacher and 'Translation planes' by Kallaher. See also the book *Foundations of Translation Planes* [28] by Biliotti, Jha and Johnson and the *Handbook of Finite Translation Planes* [187], by Johnson, Jha and Biliotti.

Spreads and partial spreads in $PG(n, q)$ are considered in Chapter 4 of PGOFF2, in Chapter 17 of FPSOTD, and in Chapter 7 here. For blocking sets, only the plane case is considered in Chapter 13 of PGOFF2. For the theory of spreads, partial spreads and blocking sets in n dimensions, see Sections 7 and 8 of the chapter 'Projective geometry over a finite field' by Thas in [55], as well as Chapter 2 by De Beule, Klein and Metsch and Chapter 3 by Blokhuis, Sziklai and Szőnyi in [298].

Flocks of quadrics in $PG(3, q)$ are key objects for the constructions of some new classes of translation planes and generalised quadrangles. They also have other applications. For literature on flocks, see Chapter 7 by Thas in [55], and the books *Translation Generalized Quadrangles* [352] by Thas, K. Thas and Van Maldeghem and *Finite Generalized Quadrangles* [260] by Payne and Thas.

Ovals and ovoids can be generalised by replacing their points with m-dimensional subspaces. These have connections to generalised quadrangles, projective planes, circle geometries, flocks, and other structures; see the last two books.

In Chapter 5, the finite classical generalised quadrangles are considered. Generalised quadrangles are the polar spaces of rank 2, the point of view of Chapter 5, but are also the generalised n-gons with $n = 4$. Generalised 6-gons or hexagons appear in Chapter 1. Standard works on generalised n-gons are the books *Generalized Polygons* [391] by Van Maldeghem and *Moufang Polygons* [385] by Tits and Weiss.

Although null polarities are mentioned in Chapter 7, they are not discussed in detail, nor are pseudo-polarities; references are given there.

The book contains only a few group-theoretical results; also theorems on graphs and designs are rare. Apart from the Handbook of Incidence Geometry, the books of Dembowski [116], Beth, Jungnickel and Lenz [18, 20, 19], Brouwer, Cohen and Neumaier [41], Cameron and van Lint [65], Hughes and Piper [186], Assmus and Key [4] may be consulted. Much material is contained in the *Handbook of Combinatorial Designs* [73], edited by Colbourn and Dinitz.

Reference works on point-line incidence structures and diagram geometries are *Diagram Geometries* [253] by Pasini, *Points and Lines* [288] by Shult, *Diagram Geometry* [56] by Buekenhout and Cohen.

Codes are considered only in Section 2 of Chapter 6. For an introduction, see Hill [170] or van Lint [388]. For further results and geometrical connections, see Cameron and van Lint [65], MacWilliams and Sloane [222], Peterson and Weldon [264], Tonchev [386]. For an introduction to Goppa's algebraic-geometry codes, see Pretzel [267], van Lint and van der Geer [389], Goppa [144], Moreno [240], Niederreiter and Xing [242], Hirschfeld, Korchmáros and Torres [176].

For a range of other topics, see the *Handbook of Finite Fields* [241], edited by Mullen and Panario.

Acknowledgements

Initial versions of Chapters 4 and 7 were typed by Sonia Surmont, of Chapter 5 by Annelies Baeyens, and of Chapter 6 by Erin Pichanick. We are extremely grateful to them for their diligence.

The authors also thank the University of Sussex and Ghent University. Both institutions have supported the writing of this book

The first author is grateful to Professor T.G. Room and Professor W.L. Edge, the supervisors of his MSc and PhD, who started him on this subject.

The second author is also very grateful to Professor J. Bilo, who initiated him into the mysteries of classical geometry thereby providing him with the necessary background to appreciate the beauty of Galois geometry.

Brighton and Ghent   JWPH
25 August 2015   JAT

Terminology

The set $V(n+1, K)$ is $(n+1)$-*dimensional vector space over the field* K and is taken to be the set of vectors $X = (x_0, \ldots, x_n)$, $x_i \in K$. Correspondingly, $\mathrm{PG}(n, K)$ is n-*dimensional projective space* over K and is the set of elements, called *points*, $\mathbf{P}(x)$ with $x \in V(n+1, K) \backslash \{0\}$. When $K = \mathrm{GF}(q) = \mathbf{F}_q$, the finite field of q elements, also called the Galois field of q elements, then $V(n+1, K)$ is written $V(n+1, q)$ and $\mathrm{PG}(n, K)$ is written $\mathrm{PG}(n, q)$. The *order* of $\mathrm{PG}(n, q)$ is q. The number of points in $\mathrm{PG}(n, q)$ is

$$\theta(n) = \frac{q^{n+1} - 1}{q - 1}.$$

A *projectivity*, or *projective transformation*, from S_1 to S_2, with S_1, S_2 both n-dimensional projective spaces over $\mathbf{F}_q$, is a mapping $\mathfrak{T} : S_1 \to S_2$ such that $\mathbf{P}(x)\mathfrak{T} = \mathbf{P}(xT)$ for all vectors $x \neq 0$ and some non-singular $(n+1) \times (n+1)$ matrix T. The group of projectivities from $\mathrm{PG}(n, q)$ to itself is denoted $\mathrm{PGL}(n+1, q)$. A *collineation* from S_1 to S_2 is a mapping $\mathfrak{T} : S_1 \to S_2$ preserving the incidence of points and lines. The *Fundamental Theorem of Projective Geometry* states that $\mathbf{P}(x)\mathfrak{T} = \mathbf{P}(x^\sigma T)$ with σ an automorphism of $\mathbf{F}_q$. Mostly, the properties considered are invariant under $\mathrm{PGL}(n+1, q)$.

A *reciprocity* of $\mathrm{PG}(n, q)$ is a collineation $\mathfrak{T}$ from $\mathrm{PG}(n, q)$ to its dual space; if $\mathfrak{T}$ is a projectivity, then the reciprocity is a *correlation* of $\mathrm{PG}(n, q)$.

A *subspace of dimension* r in $\mathrm{PG}(n, q)$ is a $\mathrm{PG}(r, q)$ and is written Π_r; this notation is used both specifically and generically. Then Π_{-1} is the empty set, Π_0 is a point, Π_1 is a *line*, Π_2 is a *plane*, Π_3 is a *solid*, Π_{n-1} is a *hyperplane*. Also, $\boldsymbol{\pi}(u)$, with $u = (u_0, \ldots, u_n)$, with not all u_i zero, denotes the hyperplane whose points $\mathbf{P}(x_0, \ldots, x_n)$ satisfy the equation

$$u_0 x_0 + \cdots + u_n x_n = 0.$$

A subspace written π_r can have any dimension. In $\mathrm{PG}(n, q)$, the vertices of the simplex of reference are denoted $\mathbf{U}_0, \mathbf{U}_1, \ldots, \mathbf{U}_n$, where $\mathbf{U}_i$ has 1 in the $(i+1)$-th coordinate place and zeros elsewhere, and $\mathbf{U}$ is the unit point. Dually, $\mathbf{u}_0, \mathbf{u}_1, \ldots, \mathbf{u}_n$

are the hyperplane faces of the simplex of reference and $\mathbf{u}$ is the unit hyperplane. The set of all r-spaces in $\mathrm{PG}(n, q)$ is written $\mathrm{PG}^{(r)}(n, q)$.

If two subspaces $\mathcal{S}, \mathcal{S}'$ intersect in a point P, this will generally be written

$$\mathcal{S} \cap \mathcal{S}' = P.$$

For any matrix $M = (m_{ij})$, the *transpose* $M^* = (m'_{ij})$ has $m'_{ij} = m_{ji}$.

The ring $\Gamma = \mathbf{F}_q[X_0, \ldots, X_n]$ is the ring of polynomials in the indeterminates $X_0, \ldots, X_n$ over $\mathbf{F}_q$. For $F_1, \ldots, F_r$ non-zero forms, or homogeneous polynomials, in Γ, the *variety*

$$\mathbf{V}(F_1, \ldots, F_r) = \{\mathbf{P}(x) \in \mathrm{PG}(n, q) \mid F_1(x) = \cdots = F_r(x) = 0\}.$$

So the hyperplane $\pi(u)$ is also written as

$$\mathbf{V}(u_0 X_0 + \cdots + u_n X_n).$$

The term 'variety' here is the set of rational points of a variety in the sense of algebraic geometry. A variety $\mathbf{V}(F)$ is called a *hypersurface*. A hypersurface in $\mathrm{PG}(2, q)$ is a *plane algebraic curve*; a hypersurface in $\mathrm{PG}(3, q)$ is a *surface*. If the hypersurfaces $\mathcal{F}_1$ and $\mathcal{F}_2$ are projectively equivalent, then write $\mathcal{F}_1 \sim \mathcal{F}_2$.

In keeping with the terminology of Chapter 8 of PGOFF2, in $\mathrm{PG}(2, q)$ an *oval* is a $(q + 1)$-arc for q odd and a $(q + 2)$-arc for q even. Other authors use *hyperoval* or *complete oval* in the latter case.

Occasionally, (r, s) denotes the greatest common divisor of r and s.

For more detailed explanation of the foregoing, see Chapter 2 of PGOFF2.

PART V

$\mathrm{PG}(n, q)$

1

Quadrics

1.1 Canonical forms

Quadrics were introduced in Chapter 5 of PGOFF2. The properties of quadrics on
a line were developed in Chapter 6 and in a plane in Chapter 7. The properties of
quadrics in three dimensions were developed in Chapters 15 and 16 of FPSOTD.
Quadrics in five dimensions were also considered in Chapters 15, 17 and 20. First
the essential definitions are recalled. Let $F \in \mathbf{F}_q[X_0, \ldots, X_n]$, where

$$F = \sum_{i=0}^{n} a_i X_i^2 + \sum_{i<j} a_{ij} X_i X_j,$$

be a quadratic form which is *non-degenerate*; that is, F is not reducible to a form
in fewer than $n + 1$ variables by a linear transformation. The variety $\mathbf{V}(F)$ is a
non-singular quadric. Under projectivities of $\mathrm{PG}(n, q)$ there are one or two distinct
non-singular quadrics $\mathcal{Q} = \mathbf{V}(F)$ according as n is even or odd. Equivalently, the
projective linear group $\mathrm{PGL}(n+1, q)$ acting on all non-singular quadrics in $\mathrm{PG}(n, q)$
has one or two orbits as n is even or odd. Throughout the chapter, the notation $\mathcal{Q}_n$ is
used for non-singular quadrics and $\mathcal{W}_n$ for general quadrics.

For n even, $\mathcal{Q}_n \sim \mathcal{P}_n$, where

$$\mathcal{P}_n = \mathbf{V}(X_0^2 + X_1 X_2 + \cdots + X_{n-1} X_n), \quad \textit{parabolic}.$$

For n odd, $\mathcal{Q}_n \sim \mathcal{H}_n$ or $\mathcal{E}_n$, where

$$\mathcal{H}_n = \mathbf{V}(X_0 X_1 + X_2 X_3 + \cdots + X_{n-1} X_n), \quad \textit{hyperbolic};$$
$$\mathcal{E}_n = \mathbf{V}(f(X_0, X_1) + X_2 X_3 + \cdots + X_{n-1} X_n), \quad \textit{elliptic};$$

here f is irreducible over $\mathbf{F}_q$.

In each of the three cases, write $\mathcal{Q}_n = \mathbf{V}(F_n)$, where F_n is the corresponding
quadratic form.

For the method to reduce F to canonical form, see Section 5.1 of PGOFF2.

© Springer-Verlag London 2016
J.W.P. Hirschfeld, J.A. Thas, *General Galois Geometries*, Springer Monographs
in Mathematics, DOI 10.1007/978-1-4471-6790-7_1

Suppose now that the form F may be degenerate. Then the quadric $\mathcal{W}_n = \mathbf{V}(F)$ may be singular and is a *cone* $\Pi_k \mathcal{Q}_s$, the join of the *vertex* Π_k to a non-singular quadric $\mathcal{Q}_s$ in the subspace Π_s with $\Pi_s \cap \Pi_k = \Pi_{-1}$ and $k + s = n - 1$.

If F is reduced to canonical form F_s, then

$$\mathcal{W}_n = \Pi_{n-s-1} \mathcal{Q}_s = \mathbf{V}(F_s),$$

where the vertex $\Pi_k = \Pi_{n-s-1} = \mathbf{V}(X_0, \ldots, X_s)$ is the space of singular points and $\mathcal{Q}_s$ is non-singular in $\Pi_s = \mathbf{V}(X_{s+1}, \ldots, X_n)$.

When $k = -1$ then $\mathcal{W}_n = \Pi_{-1} \mathcal{Q}_n = \mathcal{Q}_n$ and is non-singular.

Lemma 1.1. *The number of projectively distinct quadrics in* $\mathrm{PG}(n, q)$ *is*

$$\tfrac{1}{2}[3n + 1 + (n + 1, 2)],$$

of which $(n + 1, 2)$ *are non-singular and*

$$\tfrac{1}{2}[3n + 1 - (n + 1, 2)]$$

are singular.

Proof. Each quadric may be written as $\Pi_{n-r-1} \mathcal{Q}_r$ for $r \in \overline{\mathbf{N}}_n$. For n even, there is one non-singular quadric for each $r = 0, 2, \ldots, n$ and two non-singular quadrics for each $r = 1, 3, 5, \ldots, n - 1$. Hence the total number of quadrics is

$$\tfrac{1}{2}(n + 2) + 2 \cdot \tfrac{1}{2} n = \tfrac{1}{2}(3n + 2).$$

For n odd, there are two quadrics for each $r = 1, 3, 5, \ldots, n$ and again one for each $r = 0, 2, \ldots, n - 1$. Hence the total number is

$$2 \cdot \tfrac{1}{2}(n + 1) + \tfrac{1}{2}(n + 1) = \tfrac{1}{2}(3n + 3). \qquad \square$$

For $n \leq 5$ the quadrics in $\mathrm{PG}(n, q)$ are now described and listed in Table 1.1: for $n = 5$, only the non-singular quadrics are listed.

Table 1.1. Quadrics for low dimensions

$\mathrm{PG}(1, q)$	$\mathcal{H}_1 = \mathbf{V}(X_0 X_1)$ is two points $\mathbf{U}_0$, $\mathbf{U}_1$
	$\mathcal{E}_1 = \mathbf{V}(f(X_0, X_1))$ is empty
	$\Pi_0 \mathcal{P}_0 = \mathbf{V}(X_0^2)$ is a single point, the join of $\Pi_0 = \mathbf{U}_1$ to the empty quadric $\mathcal{P}_0$ in $\mathbf{u}_1$
$\mathrm{PG}(2, q)$	$\mathcal{P}_2 = \mathbf{V}(X_0^2 + X_1 X_2)$ is a *conic*, comprising $q + 1$ points, no three of which are collinear
	$\Pi_0 \mathcal{H}_1 = \mathbf{V}(X_0 X_1)$ is a line pair $\mathbf{u}_0$, $\mathbf{u}_1$
	$\Pi_0 \mathcal{E}_1 = \mathbf{V}(f(X_0, X_1))$ is a single point $\mathbf{U}_2$
	$\Pi_1 \mathcal{P}_0 = \mathbf{V}(X_0^2)$ is a single line $\mathbf{u}_0$

PG(3, q) $\mathcal{H}_3 = \mathbf{V}(X_0 X_1 + X_2 X_3)$ consists of $(q+1)^2$ points on
            $2(q+1)$ lines, two through each point
            $\mathcal{E}_3 = \mathbf{V}(f(X_0, X_1) + X_2 X_3)$ consists of $q^2 + 1$ points, no
            three of which are collinear
            $\Pi_0 \mathcal{P}_2 = \mathbf{V}(X_0^2 + X_1 X_2)$ is a *cone*, comprising the join of a point
            to a conic
            $\Pi_1 \mathcal{H}_1 = \mathbf{V}(X_0 X_1)$ is a plane pair $\mathbf{u}_0$, $\mathbf{u}_1$
            $\Pi_1 \mathcal{E}_1 = \mathbf{V}(f(X_0, X_1))$ is a single line $\mathbf{U}_2 \mathbf{U}_3$
            $\Pi_2 \mathcal{P}_0 = \mathbf{V}(X_0^2)$ is a single plane $\mathbf{u}_0$

PG(4, q) $\mathcal{P}_4 = \mathbf{V}(X_0^2 + X_1 X_2 + X_3 X_4)$ consists of $(q+1)(q^2+1)$
            points on $(q+1)(q^2+1)$ lines with $q+1$ lines
            through each point
            $\Pi_0 \mathcal{H}_3 = \mathbf{V}(X_0 X_1 + X_2 X_3)$ is a cone comprising the join
            of a point to the hyperbolic surface $\mathcal{H}_3$, that is,
            $q(q+1)^2 + 1$ points in $2(q+1)$ concurrent planes
            $\Pi_0 \mathcal{E}_3 = \mathbf{V}(f(X_0, X_1) + X_2 X_3)$ is a cone comprising the join
            of a point to the elliptic surface $\mathcal{E}_3$, that is,
            $q(q^2+1) + 1$ points on $q^2 + 1$ concurrent lines
            $\Pi_1 \mathcal{P}_2 = \mathbf{V}(X_0^2 + X_1 X_2)$ is the join of a line to a conic,
            and so consists of $q+1$ planes through a line,
            no three in a solid
            $\Pi_2 \mathcal{H}_1 = \mathbf{V}(X_0 X_1)$ is a pair of solids $\mathbf{u}_0$, $\mathbf{u}_1$
            $\Pi_2 \mathcal{E}_1 = \mathbf{V}(f(X_0, X_1))$ is a single plane
            $\Pi_3 \mathcal{P}_0 = \mathbf{V}(X_0^2)$ is a single solid $\mathbf{u}_0$

PG(5, q) $\mathcal{H}_5 = \mathbf{V}(X_0 X_1 + X_2 X_3 + X_4 X_5)$ consists of
            $(q^2+1)(q^2+q+1)$ points on $2(q+1)(q^2+1)$
            planes with $2(q+1)$ planes through a point
            $\mathcal{E}_5 = \mathbf{V}(f(X_0, X_1) + X_2 X_3 + X_4 X_5)$ consists of
            $(q+1)(q^3+1)$ points on $(q^2+1)(q^3+1)$
            lines with $q^2 + 1$ lines through a point

___

The properties of the singular quadrics follow inductively from the properties of non-singular quadrics in lower dimension. So, for the most part, it is reasonable to concentrate on the properties of non-singular quadrics.

1.2 Invariants

In the next theorem, two invariants are given. One, Δ, decides whether the quadric $\mathcal{W}_n$ is singular or not; the other, α, decides whether $\mathcal{W}_n$ is hyperbolic or elliptic in the odd-dimensional non-singular case. In these invariants, factors $1/2$ and $1/4$ appear. This means that, even in the characteristic two case, when the rest of the formula is evaluated in general, the factors 2 and 4 that appear must be cancelled. The invariant Δ is usually called the *discriminant* and α the *Arf invariant*.

First, some results on quadratic equations over $\mathbf{F}_q$ are summarised. As always, $\mathbf{F}_q{}^* = \mathbf{F}_q\backslash\{0\}$; see Sections 1.4, 1.8 of PGOFF2 for more details. Define

$$C(t) = \begin{cases} \frac{1}{2}(1 - t^{(q-1)/2}), & t \in \mathbf{F}_q{}^*, \ q \text{ odd}; \\ t + t^2 + \cdots + t^{2^{h-1}}, & t \in \mathbf{F}_q, \ q = 2^h. \end{cases}$$

Then, for q odd,

$$\begin{aligned} \mathcal{T}_0 &= \{c \in \mathbf{F}_q \mid x^2 - c = 0 \quad \text{has two solutions}\} \\ &= \{t \in \mathbf{F}_q{}^* \mid C(t) = 0\}; \end{aligned}$$

$$\begin{aligned} \mathcal{T}_1 &= \{c \in \mathbf{F}_q \mid x^2 - c = 0 \quad \text{has no solutions}\} \\ &= \{t \in \mathbf{F}_q{}^* \mid C(t) = 1\}. \end{aligned}$$

For $q = 2^h$, the elements of $\mathcal{T}_i$ have *trace* i:

$$\begin{aligned} \mathcal{T}_0 &= \{c \in \mathbf{F}_q \mid x^2 + x + c = 0 \quad \text{has two solutions}\} \\ &= \{t \in \mathbf{F}_q \mid C(t) = 0\}; \end{aligned}$$

$$\begin{aligned} \mathcal{T}_1 &= \{c \in \mathbf{F}_q \mid x^2 + x + c = 0 \quad \text{has no solutions}\} \\ &= \{t \in \mathbf{F}_q \mid C(t) = 1\}. \end{aligned}$$

Also, for q odd, $|\mathcal{T}_0| = |\mathcal{T}_1| = \frac{1}{2}(q - 1)$; for q even, $|\mathcal{T}_0| = |\mathcal{T}_1| = \frac{1}{2}q$.

Another way of phrasing the above is to consider the group homomorphisms

$$\mathbf{F}_q{}^* \overset{\mu}{\to} \mathbf{F}_q{}^* \overset{\rho}{\to} \mathbf{Z}_2 \quad \text{for } q \text{ odd,}$$
$$\mathbf{F}_q \overset{\sigma}{\to} \mathbf{F}_q \overset{\rho}{\to} \mathbf{Z}_2 \quad \text{for } q \text{ even,}$$

where $\mathbf{F}_q$ is regarded as the additive group and $\mathbf{F}_q{}^*$ the multiplicative group of the field, with

$$t\mu = t^2, \quad t\sigma = t + t^2, \quad t\rho = C(t).$$

Then $\mu\rho = 0$, $\sigma\rho = 0$, $\ker \rho = \mathcal{T}_0$.

As before, let $\mathcal{W}_n = \mathbf{V}(F)$ with

$$F = \sum_{i=0}^{n} a_i X_i^2 + \sum_{i<j} a_{ij} X_i X_j.$$

Define $A = (a_{ij})$, where $a_{ii} = 2a_i$, $a_{ji} = a_{ij}$ for $i < j$.

Let $B = (b_{ij})$, where $b_{ii} = 0$, $b_{ji} = -b_{ij} = -a_{ij}$ for $i < j$. Then, with $X = (X_0, X_1, \ldots, X_n)$ and X^* the transpose of X,

$$F = \tfrac{1}{2}XAX^* = \tfrac{1}{2}X(A + B)X^*.$$

When q is even, the formulas for Δ and α in the next theorem should be interpreted as follows. If, in A and B, the terms a_i and a_{ij} are replaced by indeterminates Z_i and Z_{ij}, and Δ and α are evaluated as rational functions over $\mathbf{Z}$, then Z_i and Z_{ij} can be specialised to a_i and a_{ij} to give the result. In the lemma following the theorem, α is obtained for small dimensions.

Theorem 1.2. (i) *The quadric $\mathcal{W}_n$ is singular or not according as Δ is zero or not, where*

$$\Delta = \begin{cases} \frac{1}{2}|A|, & n \text{ even}; \\ |A|, & n \text{ odd}. \end{cases}$$

(ii) *For n odd, the non-singular quadric $\mathcal{Q}_n$ is hyperbolic or elliptic according as $\alpha \in \mathcal{T}_0$ or $\mathcal{T}_1$, where*

$$\alpha = \begin{cases} (-1)^{(n+1)/2}|A|, & q \text{ odd}; \\ \{|B| - (-1)^{(n+1)/2}|A|\}/\{4|B|\}, & q \text{ even}. \end{cases}$$

Proof. With $x = (x_0, x_1, \ldots, x_n) \in \mathrm{PG}(n, q)$, under a projectivity $x \mapsto xT^{-1}$,

$$\begin{aligned} \mathbf{V}(\tfrac{1}{2}XAX^*) &= \mathbf{V}(\tfrac{1}{2}X(A+B)X^*) \\ &\mapsto \mathbf{V}(\tfrac{1}{2}XTAT^*X^*) = \mathbf{V}(\tfrac{1}{2}XT(A+B)T^*X^*). \end{aligned}$$

So, in (i), $\Delta \mapsto \Delta|T|^2$; thus, both Δ and $\Delta|T|^2$ are zero or neither is. In (ii), for q odd, $\alpha \mapsto \alpha|T|^2$; thus, α and $\alpha|T|^2$ are both squares or both non-squares. For q even, it may be shown that $\alpha \mapsto \alpha + t + t^2$, $t \in \mathbf{F}_q$, under the projectivity. Hence the invariance of the conditions has been established.

It now suffices to examine the invariants for the canonical forms:

$$\begin{aligned} \Pi_{n-s-1}\mathcal{P}_s &= \mathbf{V}(X_0^2 + X_1X_2 + \cdots + X_{s-1}X_s), \\ \Pi_{n-s-1}\mathcal{H}_s &= \mathbf{V}(X_0X_1 + X_2X_3 + \cdots + X_{s-1}X_s), \\ \Pi_{n-s-1}\mathcal{E}_s &= \mathbf{V}(X_0^2 + X_0X_1 + dX_1^2 + X_2X_3 + \cdots + X_{s-1}X_s), \end{aligned}$$

with $X^2 + X + d$ irreducible over $\mathbf{F}_q$. This gives the following values for $|A|$ and $|B|$:

$$\Pi_{n-s-1}\mathcal{P}_s : \quad |A| = \begin{cases} (-1)^{n/2}, & s = n; \\ 0, & s < n; \end{cases}$$

$$\Pi_{n-s-1}\mathcal{H}_s : \quad |A| = \begin{cases} (-1)^{(n+1)/2}, & s = n; \\ 0, & s < n; \end{cases}$$
$$|B| = 1, \quad s = n;$$

$$\Pi_{n-s-1}\mathcal{E}_s : \quad |A| = \begin{cases} (1 - 4d)(-1)^{(n+1)/2}, & s = n; \\ 0, & s < n; \end{cases}$$
$$|B| = 1, \quad s = n.$$

So $\Delta \neq 0$ for $s = n$ and $\Delta = 0$ for $s < n$. With $s = n$, the invariant α is given by Table 1.2. Since $X^2 + X + d$ is irreducible, it follows from the table and the formulas for C that $C(\alpha) = 0$ when $\mathcal{W}_n = \mathcal{H}_n$ and $C(\alpha) = 1$ when $\mathcal{W}_n = \mathcal{E}_n$. $\qquad\square$

Table 1.2. Values of the invariant α

$\mathcal{W}_n$	q odd	q even
$\mathcal{H}_n$	1	0
$\mathcal{E}_n$	$1 - 4d$	d

Lemma 1.3. *For q even and $n = 1$ or 3, the invariant α is given, modulo $\mathbf{F}_q \sigma$, as follows:*

(i) $n = 1$, $\alpha = a_0 a_1 / a_{01}^2$;

(ii) $n = 3$,

$$\alpha = \frac{\sum a_i a_{jk} a_{jl} a_{kl} + \sum a_i a_j a_{kl}^2 + (\prod a_{ij}) \sum (a_{kl} a_{mn})^{-1}}{(\sum a_{ij} a_{kl})^2},$$

where the summands in the numerator have four, six, and three terms respectively and that of the denominator also has three terms.

Proof. (i) In this case,

$$A = \begin{bmatrix} 2a_0 & a_{01} \\ a_{01} & 2a_1 \end{bmatrix}, \qquad B = \begin{bmatrix} 0 & a_{01} \\ -a_{01} & 0 \end{bmatrix}.$$

So $|A| = 4a_0 a_1 - a_{01}^2$, $|B| = a_{01}^2$,

$$\alpha = \frac{|B| + |A|}{4|B|} = \frac{a_0 a_1}{a_{01}^2}.$$

(ii) Here,

$$A = \begin{bmatrix} 2a_0 & a_{01} & a_{02} & a_{03} \\ a_{01} & 2a_1 & a_{12} & a_{13} \\ a_{02} & a_{12} & 2a_2 & a_{23} \\ a_{03} & a_{13} & a_{23} & 2a_3 \end{bmatrix}, \qquad B = \begin{bmatrix} 0 & a_{01} & a_{02} & a_{03} \\ -a_{01} & 0 & a_{12} & a_{13} \\ -a_{02} & -a_{12} & 0 & a_{23} \\ -a_{03} & -a_{13} & -a_{23} & 0 \end{bmatrix},$$

and, with φ the Pfaffian of the matrix B,

$$\varphi = a_{01} a_{23} - a_{02} a_{13} + a_{03} a_{12},$$
$$|B| = \varphi^2,$$
$$|A| = a_{01}^2 a_{23}^2 + a_{02}^2 a_{13}^2 + a_{03}^2 a_{12}^2$$
$$- 2a_{01} a_{23} a_{02} a_{13} - 2a_{01} a_{23} a_{03} a_{12} - 2a_{02} a_{13} a_{03} a_{12}$$
$$+ 4 \sum a_i a_{jk} a_{jl} a_{kl} - 4 \sum a_i a_j a_{kl}^2 + 16 \, a_0 a_1 a_2 a_3.$$

So, in $\mathbf{Z}(\{a_i, a_{ij}\})$,

$$\alpha = \frac{|B| - |A|}{4|B|} = \frac{\sum a_i a_j a_{kl}^2 - \sum a_i a_{jk} a_{jl} a_{kl} - 4 a_{0} a_{1} a_{2} a_{3} + a_{01} a_{23} a_{03} a_{12}}{\varphi^2}.$$

Hence, over $\mathbf{F}_q$,

$$\alpha = \frac{\sum a_i a_j a_{kl}^2 + \sum a_i a_{jk} a_{jl} a_{kl} + a_{01} a_{23} a_{03} a_{12}}{\varphi^2}.$$

However,

$$\frac{a_{01} a_{23} a_{02} a_{13} + a_{02} a_{13} a_{03} a_{12}}{\varphi^2} = \frac{a_{02} a_{13}}{\varphi} + \left(\frac{a_{02} a_{13}}{\varphi} \right)^2.$$

So, modulo $\mathbf{F}_q \sigma = \{ t + t^2 \mid t \in \mathbf{F}_q \}$,

$$\alpha = \frac{\sum a_i a_{jk} a_{jl} a_{kl} + \sum a_i a_j a_{kl}^2 + (\prod a_{ij}) \sum (a_{kl} a_{mn})^{-1}}{\varphi^2}. \qquad \square$$

1.3 Tangency and polarity

Consider the non-singular quadric $\mathcal{Q}_n = \mathbf{V}(F)$. With $P \neq Q$, let $P = \mathbf{P}(A)$ and $Q = \mathbf{P}(B)$, where $A = (a_0, \ldots, a_n)$ and $B = (b_0, \ldots, b_n)$. Then

$$F(A + tB) = F(A) + tG(A, B) + t^2 F(B), \qquad (1.1)$$

where

$$G(A, B) = F(A + B) - F(A) - F(B).$$

Definition 1.4. The line ℓ is a *tangent* to $\mathcal{Q}_n$ if $|\ell \cap \mathcal{Q}_n| = 1$.

Lemma 1.5. *Let* $P = \mathbf{P}(A) \in \mathcal{Q}_n$.

(i) *If* $Q \notin \mathcal{Q}_n$, *then* $G(A, B) = 0 \iff PQ$ *is a tangent to* $\mathcal{Q}_n$.
(ii) *If* $Q \in \mathcal{Q}_n$, *then* $G(A, B) = 0 \iff PQ \subset \mathcal{Q}_n$.
(iii) $G(A, B) \neq 0 \iff |PQ \cap \mathcal{Q}_n| = 2$.

Proof. Since $P \in \mathcal{Q}_n$, equation (1.1) becomes

$$F(A + tB) = tG(A, B) + t^2 F(B).$$

The point $\mathbf{P}(A + tB) \in \mathcal{Q}_n$ if and only if

$$0 = tG(A, B) + t^2 F(B). \qquad (1.2)$$

The solution $t = 0$ of (1.2) corresponds to P. Parts (i), (ii), and (iii) now follow. $\square$

Corollary 1.6. *For* q *even, if one of* P *and* Q *is not on* $\mathcal{Q}_n$, *then* PQ *is a tangent if and only if* $G(A, B) = 0$.

Proof. When $F(A + tB) = 0$, equation (1.1) becomes

$$0 = F(A) + tG(A, B) + t^2 F(B). \tag{1.3}$$

If $G(A, B) = 0$, then this becomes

$$0 = F(A) + t^2 F(B), \tag{1.4}$$

which has just one solution.

Conversely, if (1.3) has just one solution, the coefficient of t must be zero.   □

Definition 1.7. A point $\mathbf{P}(A)$ is a *nucleus* of $\mathcal{Q}_n$ if $G(A, B) = 0$ for all points $\mathbf{P}(B)$.

Corollary 1.8. (i) *The quadric $\mathcal{Q}_n$ has a nucleus if and only if q and n are both even.*

(ii) *For q even, $\mathcal{P}_n$ in canonical form has precisely one nucleus $N = \mathbf{U}_0$.*

Proof. This follows immediately from the forms for $G(A, B)$.   □

Remark 1.9. It should be noted that (ii) applies in the case $n = 0$. The empty quadric $\mathcal{P}_0$ has the point $\mathbf{U}_0$ as nucleus.

Definition 1.10. If $G(A, B) = 0$, the points $P = \mathbf{P}(A)$ and $Q = \mathbf{P}(B)$ are *conjugate*. If P is not a nucleus, then, with

$$G(A, X) = F(A + X) - F(A) - F(X), \tag{1.5}$$

the hyperplane $\mathbf{V}(G(A, X))$ is the *polar hyperplane of P*. When $P \in \mathcal{Q}_n$, then $\mathbf{V}(G(A, X))$ is the *tangent hyperplane to $\mathcal{Q}_n$ at P* and is denoted $T_P = T_P(\mathcal{Q}_n)$. If P is the nucleus of $\mathcal{Q}_n$, then $\mathbf{V}(G(A, X)) = \Pi_n$.

Theorem 1.11. (i) $T_P(\mathcal{Q}_n)$ *comprises the points on the tangents to $\mathcal{Q}_n$ at P and the lines on $\mathcal{Q}_n$ through P.*

(ii) $T_P(\mathcal{Q}_n)$ *contains any subspace Π_m such that $P \in \Pi_m \subset \mathcal{Q}_n$.*

Proof. (i) This follows from Lemma 1.5 (i) and (ii).

(ii) Since every line through P of $\mathcal{Q}_n$ lies in the tangent space and every point of Π_m lies on such a line, so $\Pi_m \subset T_P(\mathcal{Q}_n)$.   □

Lemma 1.5 and Corollary 1.6 also hold for general quadrics. In particular, when $\mathcal{W}_n = \Pi_k \mathcal{Q}_t = \mathbf{V}(F)$ is an arbitrary quadric and $P \in \mathcal{W}_n$, the *tangent space to $\mathcal{W}_n$ at P*, denoted $T_P(\mathcal{W}_n)$, is $\mathbf{V}(G(A, X))$ with $G(A, X)$ as in (1.5).

Corollary 1.12. (i) $T_P(\mathcal{W}_n)$ *contains the vertex Π_k.*

(ii) *If $P \in \Pi_k$, then $T_P(\mathcal{W}_n) = \Pi_n$, the whole space.*

It should be noted that if the notation $F_{(i)} = \partial F / \partial X_i$ is adopted so that $F_{(i)}(A)$ is the partial derivative of F with respect to X_i evaluated at A, then

$$G(A, X) = \sum F_{(i)}(A) X_i.$$

Thus the tangent space to a quadric as defined here coincides with that for an arbitrary primal; see Section 2.7 of PGOFF2. A nucleus $\mathbf{P}(A)$ of $\mathcal{Q}_n$ can also be defined as a point at which $F_{(i)}(A) = 0$, all i.

With the canonical forms of Section 1.1 and

$$f(X_0, X_1) = X_0^2 + X_0 X_1 + d X_1^2,$$

the linear form $G(A, X)$ is as follows:

$$\mathcal{Q}_n = \mathcal{P}_n, \quad G(A, X) = 2a_0 X_0 + (a_1 X_2 + a_2 X_1) +$$
$$\cdots + (a_n X_{n-1} + a_{n-1} X_n);$$
$$\mathcal{Q}_n = \mathcal{H}_n, \quad G(A, X) = (a_1 X_0 + a_0 X_1) + (a_3 X_2 + a_2 X_3) +$$
$$\cdots + (a_n X_{n-1} + a_{n-1} X_n);$$
$$\mathcal{Q}_n = \mathcal{E}_n, \quad G(A, X) = (2a_0 + a_1)X_0 + (a_0 + 2da_1)X_1 + (a_3 X_2 + a_2 X_3) +$$
$$\cdots + (a_n X_{n-1} + a_{n-1} X_n).$$

Lemma 1.13. *Let $\mathcal{Q}_n$ be a non-singular quadric.*

(i) *If q and n are not both even, the correspondence*

$$\mathbf{P}(A) \longleftrightarrow \mathbf{V}(G(A, X))$$

is a polarity. For q odd, the set of self-polar points is $\mathcal{Q}_n$. For q even, the polarity is null and every point in $\mathrm{PG}(n, q)$ is self-polar.

(ii) *If q and n are both even, the tangent hyperplanes to $\mathcal{P}_n$ are concurrent at the nucleus $N = \mathbf{U}_0$.*

Even though the points and tangent hyperplanes of $\mathcal{P}_n$ are not related by a polarity for q even, the following lemma plus Theorem 1.11 (ii) are strong enough to prove facts about $\mathcal{P}_n$ for q even which follow from the polar theory for all other $\mathcal{Q}_n$.

Lemma 1.14. *The tangent hyperplanes at $r + 1$ independent points of a Π_r lying on $\mathcal{Q}_n$ are themselves independent.*

Proof. When $(q, n) \not\equiv 0 \pmod 2$, this follows from Lemma 1.13 (i). This leaves the case that q is even and $\mathcal{Q}_n = \mathcal{P}_n$. With $A = (a_0, \dots, a_n)$ and $P = \mathbf{P}(A)$, then $T_P(\mathcal{P}_n) = \mathbf{V}(G(A, X)) = \mathbf{V}(A^\tau X^*)$, where

$$A^\tau = (0, a_2, a_1, a_4, a_3, \dots, a_n, a_{n-1}).$$

Suppose that the points $P = \mathbf{P}(A_i)$, $i = 0, \dots, r$, span Π_r on $\mathcal{P}_n$; that is, they are independent. If the corresponding tangent hyperplanes $\mathbf{V}(A^\tau X^*)$ are dependent, so are the $r + 1$ points $\mathbf{P}(A_i^\tau)$. Hence, under the projectivity fixing $\mathcal{P}_n$ given by

$$x_0 \mapsto x_0, \quad x_{2j-1} \leftrightarrow x_{2j}, \; j = 1, 2, \ldots, \tfrac{1}{2}n,$$

the points $\mathbf{P}(A_i^\sigma)$, where $A^\sigma = (0, a_1, a_2, a_3, a_4, \ldots, a_n)$, are dependent and hence lie in a Π_{r-1}. But $\mathbf{P}(A_i^\sigma)$ is the projection of $P = \mathbf{P}(A)$ from the nucleus $\mathbf{U}_0$ of $\mathcal{P}_n$ onto $\mathbf{u}_0$. Since the $r+1$ points $\mathbf{P}(A_i^\sigma)$ lie in Π_{r-1}, so the $r+1$ points $\mathbf{P}(A_i)$ lie in the r-space $\mathbf{U}_0\Pi_{r-1}$; this follows from the fact that $\mathbf{P}(A_i^\sigma)$ lies on $\mathbf{U}_0\mathbf{P}(A_i)$. However, by hypothesis, the $r+1$ points lie in a Π_r on $\mathcal{P}_n$ and are independent. As Π_r cannot be $\mathbf{U}_0\Pi_{r-1}$, a contradiction is obtained. $\qquad\square$

Definition 1.15. If $\Pi_r \subset \mathcal{Q}_n$, the *tangent space at* Π_r is the intersection of the tangent hyperplanes at $r+1$ independent points $P_0, \ldots, P_r$ of Π_r; in symbols,

$$T_{\Pi_r}(\mathcal{Q}_n) = \bigcap T_{P_i}(\mathcal{Q}_n).$$

Corollary 1.16. (i) *The tangent space of Π_r on $\mathcal{Q}_n$ is a Π_{n-r-1} containing Π_r.*
(ii) $\mathcal{Q}_n \supset \Pi_r \supset \Pi_s \Rightarrow T_{\Pi_s}(\mathcal{Q}_n) \supset \Pi_r$.

Proof. (i) By the lemma, the tangent space is the intersection of $r+1$ independent hyperplanes and so is a Π_{n-r-1}. By Theorem 1.11 (ii), it contains Π_r.
(ii) This follows from Theorem 1.11 (i). $\qquad\square$

In the subsequent results, Π'_{n-r-1} is the tangent space of Π_r on $\mathcal{Q}_n$ and Π'_{n-m-1} is the polar space of Π_m for the cases in which a polarity occurs.
If $\Pi_r \subset \mathcal{W}_n$, the *tangent space*

$$T_{\Pi_r}(\mathcal{W}_n) = \bigcap_{P \in \Pi_r} T_P(\mathcal{W}_n).$$

Corollary 1.17. *If $\mathcal{W}_n = \Pi_k \mathcal{Q}_t$ and $\Pi_r \subset \mathcal{W}_n$ so that $\Pi_r = \Pi_s \Pi_e$ with $\Pi_s \subset \mathcal{Q}_t$ and $\Pi_e \subset \Pi_k$, then*

$$T_{\Pi_r}(\mathcal{W}_n) = T_{\Pi_s}(\mathcal{Q}_t)\Pi_k$$

and has dimension $t - s + k$.

A tangent line meets $\mathcal{Q}_n$ precisely in a point. Now consider what happens in general when a subspace Π_m meets $\mathcal{Q}_n$ in a subspace Π_r that is not the whole of Π_m.

Lemma 1.18. *Suppose $\Pi_m \not\subset \mathcal{Q}_n$ and $\Pi_m \cap \mathcal{Q}_n = \Pi_r$. Then the following hold:*

(i) $\Pi_m \subset T_P(\mathcal{Q}_n)$, *for all P in Π_r, whence $\Pi_m \subset \Pi'_{n-r-1}$;*
(ii) *either* (a) $m = r+1$ *and* $\Pi_m \cap \mathcal{Q}_n = \Pi_{m-1}P_0$
    or (b) $m = r+2$ *and* $\Pi_m \cap \mathcal{Q}_n = \Pi_{m-2}\mathcal{E}_1$;
(iii) *when q is odd,* $\Pi_m \cap \Pi'_{n-m-1} = \Pi_r$ *and* $\Pi_m\Pi'_{n-m-1} = \Pi'_{n-r-1}$;
(iv) *when q is even with n odd and $m = r+1$, then* $\Pi_m \subset \Pi'_{n-m-1}$;
(v) *when q is even with n odd and $m = r+2$, then* $\Pi_m \cap \Pi'_{n-m-1} = \Pi_r$ *and* $\Pi_m\Pi'_{n-m-1} = \Pi'_{n-r-1}$.

Proof. (i) If $P \in \Pi_r$, then any line l through P in Π_m either lies in Π_r and so is on $\mathcal{Q}_n$ or meets Π_r and so $\mathcal{Q}_n$ in the single point P. Thus l is a tangent through P or a line of $\mathcal{Q}_n$ through P; hence $l \subset T_P(\mathcal{Q}_n)$.

(ii) A quadric $\Pi_{n-s-1}\mathcal{Q}_s$ is a Π_r if and only if $\mathcal{Q}_s$ is empty; that is, $\mathcal{Q}_s = \mathcal{P}_0$ or $\mathcal{E}_1$. Hence, if $\Pi_m \cap \mathcal{Q}_n = \Pi_r$, then either $\Pi_r = \Pi_{m-1}\mathcal{P}_0$ or $\Pi_{m-2}\mathcal{E}_1$. Then $\Pi_{m-1}\mathcal{P}_0$ is considered as a repeated Π_{m-1} and $\Pi_{m-2}\mathcal{E}_1$ as a Π_{m-2} which is the intersection of two Π_{m-1} lying over $\mathbf{F}_{q^2}$.

(iii) Π'_{n-m-1} is the intersection of the polar hyperplanes of points of Π_m. So every point of $\Pi_m \cap \Pi'_{n-m-1}$ lies in its own polar hyperplane, which implies that $\Pi_m \cap \Pi'_{n-m-1} \subset \Pi_r$. Since $\Pi_m \subset \Pi'_{n-r-1}$ by (i), so $\Pi'_{n-m-1} \supset \Pi_r$. But, by hypothesis, $\Pi_m \supset \Pi_r$; so $\Pi_r \subset \Pi_m \cap \Pi'_{n-m-1}$. Thus $\Pi_m \cap \Pi'_{n-m-1} = \Pi_r$ and so $\Pi_m\Pi'_{n-m-1} = \Pi'_{n-r-1}$.

(iv) Since q is even, the polarity defined by $\mathcal{Q}_n$ is a null polarity. Thus, if two particular points on a line l are conjugate, then any two points on l are conjugate and l is self-polar: that is, l lies in its polar space. Hence the self-polar lines are the lines of $\mathcal{Q}_n$ and the tangents to $\mathcal{Q}_n$; see Corollary 1.6.

Now, if $r = m - 1$, then every line in Π_m through a point P of $\Pi_m \backslash \Pi_r$ meets Π_r and is tangent to $\mathcal{Q}_n$. So P is conjugate to every point of Π_m and consequently $\Pi_m \subset \Pi'_{n-m-1}$.

(v) Now, with $r = m - 2$, let $P \in \Pi_m \backslash \Pi_r$ and let $Q \in \Pi_m \backslash \{P\}$. Then PQ is self-polar if and only if $PQ \cap \Pi_r = \Pi_0$. However, through P, there is a line of Π_m missing Π_r. But R in Π_m is in Π'_{n-m-1} if and only if R is conjugate to every point Q of Π_m. Hence $\Pi_m \cap \Pi'_{n-m-1} = \Pi_r$ and $\Pi_m\Pi'_{n-m-1} = \Pi'_{n-r-1}$. □

Lemma 1.19. *If $\Pi_m \not\subset \mathcal{Q}_n$, then the following are equivalent:*

(i) Π_r *is the largest subspace on $\mathcal{Q}_n$ such that $\Pi_m \subset T_P(\mathcal{Q}_n)$ for all $P \in \Pi_r$;*
(ii) Π_r *is the largest subspace on $\mathcal{Q}_n$ such that $\Pi_m \subset \Pi'_{n-r-1}$;*
(iii) Π_r *is the singular space of $\Pi_m \cap \mathcal{Q}_n$.*

Proof. P is in the singular space of $\Pi_m \cap \mathcal{Q}_n$

$\Longleftrightarrow$ every line through P in Π_m is a tangent or line of $\mathcal{Q}_n$

$\Longleftrightarrow \Pi_m \subset T_P(\mathcal{Q}_n)$.

Hence Π_r is the singular space of $\Pi_m \cap \mathcal{Q}_n$ if and only if

$$\Pi_m \subset \bigcap_{P \in \Pi_r} T_P(\mathcal{Q}_n) \subset \Pi'_{n-r-1}. \qquad □$$

When the conditions of this lemma hold, then Π_m *touches* $\mathcal{Q}_n$ along Π_r.

Corollary 1.20. *In the case that n and q are not both even, suppose that Π_m and its polar space Π'_{n-m-1} are not contained in $\mathcal{Q}_n$. If $\Pi_m \cap \mathcal{Q}_n = \Pi_k\mathcal{Q}_t$, then the space Π'_{n-m-1} satisfies the following:*

(i) $\Pi'_{n-m-1} \cap \mathcal{Q}_n$ *has singular space Π_k;*
(ii) $\Pi_m \cap \Pi'_{n-m-1} \cap \mathcal{Q}_n = \Pi_k$;

(iii)
$$\Pi_m \cap \Pi'_{n-m-1} = \begin{cases} \Pi_k & \text{when } q \text{ is odd, or } q \text{ is even with } t \text{ odd}; \\ \Pi_k N & \text{when } q \text{ is even with } t \text{ even}, \end{cases}$$

where N is the nucleus of $\mathcal{Q}_t$.

Proof. The hypothesis means that the set of points P in $\Pi_m \cap \mathcal{Q}_n$ such that $|l \cap \mathcal{Q}_n|$ is 1 or $q+1$ for every line l through P in Π_m is Π_k. Any such line l lies in $T_P(\mathcal{Q}_n)$ but lies in Π'_{n-m-1} only if it lies in the polar hyperplane of every point Q in Π_m. This occurs only if l lies in Π_k or does not lie in Π_k and $\Pi_m l \cap \Pi_t$ lies in the polar hyperplane of every point Q in Π_t, where $\mathcal{Q}_t \subset \Pi_t$. These two possibilities give the respective cases for q even in (iii). $\square$

Lemma 1.21. *Suppose that* $\Pi_m \not\subset \mathcal{Q}_n$. *Let* $\Pi_m \cap \mathcal{Q}_n = \Pi_k \mathcal{Q}_t$ *and let* Π_d *be any subspace on* $\Pi_k \mathcal{Q}_t$ *containing* Π_k. *Then*

$$\Pi_m \Pi'_{n-d-1} = \Pi'_{n-k-1},$$

where Π'_{n-d-1} *is the tangent space to* $\mathcal{Q}_n$ *at* Π_d *and* Π'_{n-k-1} *is the tangent space at* Π_k.

Proof. Since $\Pi_d \supset \Pi_k$, so $\Pi'_{n-d-1} \subset \Pi'_{n-k-1}$. By Lemma 1.19, $\Pi_m \subset \Pi'_{n-k-1}$. So $\Pi_m \Pi'_{n-d-1} = \Pi'_{n-k-1}$.

To prove the converse, consider two cases.

(a) $\Pi_k \mathcal{Q}_t$ *spans* Π_m

The space $\Pi_m \cap \Pi'_{n-d-1}$ is the tangent space of $\Pi_k \mathcal{Q}_t$ at Π_d. Since $\Pi_d \supset \Pi_k$, the dimension of $\Pi_m \cap \Pi'_{n-d-1}$ is $t-(d-k-1)+k = m-d+k$ by Corollary 1.17. Hence the dimension of $\Pi_m \Pi'_{n-d-1}$ is

$$m + (n-d-1) - (m-d+k) = n-k-1.$$

So $\Pi_m \Pi'_{n-d-1} = \Pi'_{n-k-1}$.

(b) $\Pi_k \mathcal{Q}_t$ *does not span* Π_m

Then, as in Lemma 1.18, $\Pi_k \mathcal{Q}_t = \Pi_k$. So $\Pi_k = \Pi_d$ and $\Pi_m \subset \Pi'_{n-d-1}$. The result follows. $\square$

Two quadrics have the *same character* if they are both parabolic, both hyperbolic or both elliptic. An absolute definition of character is given in Section 1.4.

Lemma 1.22. *For* $n \geq 2$, *the tangent hyperplane* T_P *at a point* P *of* $\mathcal{Q}_n$ *meets* $\mathcal{Q}_n$ *in a cone* $P\mathcal{Q}_{n-2}$, *where* $\mathcal{Q}_n$ *and* $\mathcal{Q}_{n-2}$ *have the same character.*

Proof. Choose $P = \mathbf{U}_n$. Also, let $\mathbf{U}_{n-1}$ be on $\mathcal{Q}_n$ and let $\mathbf{U}_0, \mathbf{U}_1, \ldots, \mathbf{U}_{n-2}$ be in T_P. So $T_P = \mathbf{u}_{n-1}$ and

$$\mathcal{Q}_n = \mathbf{V}(F(X_0, \ldots, X_{n-1}) + X_{n-1}X_n),$$

where F contains no term in X_{n-1}^2. If $F = \sum_{i \leq j} X_i X_j$, substitute X_n for

$$a_{0,n-1}X_0 + a_{1,n-1}X_1 + \cdots + a_{n-2,n-1}X_{n-2} + X_n.$$

So

$$\mathcal{Q}_n = \mathbf{V}(F(X_0, \ldots, X_{n-2}, 0) + X_{n-1}X_n) \tag{1.6}$$

and

$$\mathcal{Q}_n \cap T_P = \mathbf{V}(X_{n-1}, F(X_0, \ldots, X_{n-2}, 0)) = P\mathcal{Q}_{n-2},$$

where

$$\mathcal{Q}_{n-2} = \mathbf{V}(X_{n-1}, X_n, F(X_0, \ldots, X_{n-2}, 0)). \tag{1.7}$$

Reference to the canonical forms in Section 1.1 shows that $\mathcal{Q}_n$ and $\mathcal{Q}_{n-2}$ have the same character since the quadratic forms in (1.6) and (1.7) which define them differ by $X_{n-1}X_n$. □

1.4 Generators

A subspace of maximum dimension on a quadric $\mathcal{W}_n$ is a *generator*; its dimension $g = g(\mathcal{W}_n)$ is the *projective index* of $\mathcal{W}_n$. The more classical *Witt index* is $g + 1$; this is not used here.

Definition 1.23. For $\mathcal{Q}_n$, the *character* $w = w(\mathcal{Q}_n)$ is defined as follows:

$$w = 2g - n + 3. \tag{1.8}$$

Lemma 1.24. *The character of non-singular quadrics is as follows*:

	$\mathcal{Q}_n$	$\mathcal{P}_n$	$\mathcal{H}_n$	$\mathcal{E}_n$
w		1	2	0

This lemma justifies the names parabolic, hyperbolic and elliptic for the respective quadrics. Sometimes it is convenient to invert (1.8) to give

$$g = \tfrac{1}{2}(n - 3 + w). \tag{1.9}$$

Lemma 1.25. (i) *For $\mathcal{Q}_n$ and $\mathcal{Q}_{n-2}$, non-singular quadrics of the same character,*

$$g(\mathcal{Q}_n) = g(\mathcal{Q}_{n-2}) + 1.$$

(ii)

	$\mathcal{Q}_n$	$\mathcal{P}_n$	$\mathcal{H}_n$	$\mathcal{E}_n$
g		$\tfrac{1}{2}(n-2)$	$\tfrac{1}{2}(n-1)$	$\tfrac{1}{2}(n-3)$

(iii) *Any subspace on $\mathcal{Q}_n$ lies in a generator.*

Proof. (i) This follows directly from Lemma 1.22 and Theorem 1.11 (ii).

(ii) This follows from (i) and the knowledge of $g(\mathcal{Q}_n)$ for low n.

(iii) Induction on n and a similar argument to (i) gives the result.   □

Lemma 1.26. *A generator of* $\mathcal{W}_n = \Pi_k \mathcal{Q}_t$ *is the join of the vertex* Π_k *to a generator of* $\mathcal{Q}_t$.

Now the character $w = w(\mathcal{W}_n)$ of an arbitrary quadric $\mathcal{W}_n = \Pi_k \mathcal{Q}_t$ is defined. Recall that

$k = $ dimension of singular space of $\mathcal{W}_n$,

$n = $ dimension of space in which $\mathcal{W}_n$ is defined by a quadratic form,

$g = $ projective index of $\mathcal{W}_n$.

Define

$$w = 2g - k - n + 2. \tag{1.10}$$

This agrees with (1.8) in the non-singular case, when $k = -1$.

Lemma 1.27. *For a quadric* $\mathcal{W}_n$, *the constants* g *and* w *are as follows*:

$\mathcal{W}_n$	$\Pi_{n-t-1}\mathcal{P}_t$	$\Pi_{n-t-1}\mathcal{H}_t$	$\Pi_{n-t-1}\mathcal{E}_t$
g	$n - \frac{1}{2}(t+2)$	$n - \frac{1}{2}(t+1)$	$n - \frac{1}{2}(t+3)$
w	1	2	0

Note 1.28. (i) The character of $\mathcal{W}_n = \Pi_k \mathcal{Q}_t$ is the same as for the base $\mathcal{Q}_t$.

(ii) When $g = k = n$, then $w = 2$. So it is consistent to write $\Pi_n = \Pi_n \mathcal{H}_{-1}$ and include the whole space Π_n as the quadric $\mathbf{V}(0)$. This becomes relevant when sections of a quadric by a subspace are considered.

Corollary 1.29. *A quadric* $\mathcal{W}_n = \Pi_{n-t-1}\mathcal{Q}_t$ *of character* w *has projective index*

$$g = n - \tfrac{1}{2}(t + 3 - w). \tag{1.11}$$

Lemma 1.30. *If* $\Pi_m \subset \mathcal{Q}_n$ *and* Π'_{n-m-1} *is the tangent space of* Π_m, *then*

$$\Pi'_{n-m-1} \cap \mathcal{Q}_n = \Pi_m \mathcal{Q}'_{n-2m-2},$$

where $\mathcal{Q}'_{n-2m-2}$ *has the same character as* $\mathcal{Q}_n$.

Proof. For $P \in \Pi_m$, every line through P in Π'_{n-m-1} is a tangent or a line of $\mathcal{Q}_n$. So Π_m lies in the singular space of $\Pi'_{n-m-1} \cap \mathcal{Q}_n$. It must be shown that the singular space is no bigger.

Suppose $\Pi_m = \mathbf{U}_0 \mathbf{U}_1 \cdots \mathbf{U}_m$. Then $\mathcal{Q}_n = \mathbf{V}(F)$ with

$$F = X_0 f_0 + \cdots + X_m f_m + g,$$

where each f_i and g are forms in $X_{m+1}, \ldots, X_n$. Since $\mathcal{Q}_n$ is non-singular, the forms $f_0, \ldots, f_m$ are linearly independent. Hence, by a change of coordinates,

$$F = X_0 X_{m+1} + \cdots + X_m X_{2m+1} + g'(X_{m+1}, \ldots, X_n).$$

The non-singularity of F considered as a form in $X_0, \ldots, X_n$ is equivalent to the non-singularity of $G = g'(0, \ldots, 0, X_{2m+2}, \ldots, X_n)$ considered as a form in $X_{2m+2}, \ldots, X_n$. Thus, in $\Pi'_{n-m-1} = \mathbf{V}(X_{m+1}, \ldots, X_{2m+1})$, the equation of $\Pi'_{n-m-1} \cap \mathcal{Q}_n$ is $G = 0$.

It follows that $\Pi'_{n-m-1} \cap \mathcal{Q}_n = \Pi_m \mathcal{Q}'_{n-2m-2}$. Any Π_r lying on $\mathcal{Q}_n$ and containing Π_m lies in Π'_{n-m-1} by Corollary 1.16 (ii). Hence $g(\mathcal{Q}_n) = g(\Pi_m \mathcal{Q}'_{n-2m-2})$. So, if w and w' are the characters of $\mathcal{Q}_n$ and $\mathcal{Q}'_{n-2m-2}$, then

$$\tfrac{1}{2}(n - 3 + w) = n - m - 1 - \tfrac{1}{2}(n - 2m - 2 + 3 - w'),$$

whence $w = w'$. $\qquad\qquad\qquad\qquad\qquad\qquad\qquad\qquad\qquad\qquad\square$

For $m = 0$, the result was given by Lemma 1.22.

Now some numerical properties of the generators are considered before the whole system is described. Let $\mathcal{G} = \mathcal{G}(\mathcal{Q}_n)$ be the set of generators of $\mathcal{Q}_n$.

Notation 1.31. (i) $\rho(d, n; w) = |\{\Pi_g \in \mathcal{G} \mid \Pi_g \text{ contains a fixed } \Pi_d \}|$.
(ii) $\lambda(d, n; w) = |\{\Pi_g \in \mathcal{G} \mid \Pi_g \text{ meets a fixed generator in some } \Pi_d \}|$.
(iii) $\mu(c) = \mu(c, n; w) = |\{\Pi_g \in \mathcal{G} \mid \Pi_g \text{ meets a fixed generator in a fixed } \Pi_{g-c} \}|$.
(iv) $\kappa(n; w) = |\mathcal{G}|$.

In the subsequent results, the following numerical notation is frequently used.

Notation 1.32.

$$[r, s]_+ = \begin{cases} (q^r + 1)(q^{r+1} + 1) \cdots (q^s + 1) & \text{for } s \geq r, \\ 1 & \text{for } s < r; \end{cases}$$

$$[r, s]_- = \begin{cases} (q^r - 1)(q^{r+1} - 1) \cdots (q^s - 1) & \text{for } s \geq r, \\ 1 & \text{for } s < r. \end{cases}$$

Theorem 1.33.

$$\kappa(n; w) = [2 - w, \tfrac{1}{2}(n - w + 1)]_+ = [2 - w, g + 2 - w]_+$$

$$= \begin{cases} [2, s + 1]_+ & \text{for } \mathcal{E}_{2s+1}, \\ [0, s]_+ & \text{for } \mathcal{H}_{2s+1}, \\ [1, s]_+ & \text{for } \mathcal{P}_{2s}. \end{cases}$$

Proof. The set $\{(P, \Pi_g) \mid P \in \Pi_g \in \mathcal{G}\}$ is counted in two ways. By Theorem 1.11 and Lemma 1.22, the set $\{(P_0, \Pi_g) \mid P_0 \in \Pi_g \in \mathcal{G}\}$ for a fixed point P_0 has size $\kappa(n - 2; w)$. Hence

$$\kappa(n-2;w)\,|\mathcal{Q}_n| = \kappa(n;w)\,|\Pi_g|\,.$$

However, from Section 5.2 of PGOFF2,

$$|\mathcal{Q}_n| = \begin{cases} (q^{s+1}+1)(q^s-1)/(q-1) & \text{for } \mathcal{E}_{2s+1}, \\ (q^s+1)(q^{s+1}-1)/(q-1) & \text{for } \mathcal{H}_{2s+1}, \\ (q^{2s}-1)/(q-1) & \text{for } \mathcal{P}_{2s}. \end{cases}$$

The result then follows by induction.   □

Theorem 1.34.

$$\rho(d,n;w) = [2-w, \tfrac{1}{2}(n-1-2d-w)]_+.$$

Proof. By Corollary 1.16, the tangent space Π'_{n-d-1} at Π_d to $\mathcal{Q}_n$ contains all generators through Π_d. By Lemma 1.30,

$$\Pi'_{n-d-1} \cap \mathcal{Q}_n = \Pi_d \mathcal{Q}_{n-2d-2},$$

and has the same character w as $\mathcal{Q}_n$. Hence each generator of $\mathcal{Q}_n$ through Π_d is the join of Π_d to a generator of $\mathcal{Q}_{n-2d-2}$ and conversely. So

$$\rho(d,n;w) = \kappa(n-2d-2;w).$$ □

It may be noted that, when $d=g$,

$$\rho(d,n;w) = [2-w, 1-w]_+ = 1,$$

confirming that the only generator containing a given Π_g is Π_g itself.

Lemma 1.35.
$$\mu(c,n;w) = q^{c(c+3-2w)/2}.$$

Proof. The only generator meeting Π_g in Π_g is Π_g itself; hence $\mu(0) = 1$. Now proceed by induction on c and assume the formula true for all values less than c.

The number of generators meeting Π_g in at least the fixed space Π_{g-c} is

$$\rho(g-c,n;w) = [2-w, c+1-w]_+\,.$$

So, to find $\mu(c)$, subtract from $\rho(g-c,n;w)$ the number of generators meeting Π_g in a $(g-i)$-space containing Π_{g-c} for all i such that $0 \le i < c$. Hence

$$\mu(c) = \rho(g-c,n;w) - \sum_{i=0}^{c-1} \mu(i)\chi(g-c, g-i; g, q),$$

where

$$\chi(g-c, g-i; g, q) = \text{number of } \Pi_{g-i} \text{ through } \Pi_{g-c} \text{ in } \Pi_g$$
$$= [c-i+1, c]_-/[1, i]_-\,,$$

as in PGOFF2, Section 3.1. Hence

$$\mu(c) = [2 - w, c + 1 - w]_+ - \sum_{i=0}^{c-1} q^{i(i+3-2w)/2}[c - i + 1, c]_-/[1, i]_- ,$$

which gives the result after some manipulation.   □

Lemma 1.36. *For* $-1 \leq d \leq g$,

$$\lambda(d, n; w) = q^{c(c+3-2w)/2}[g - d + 1, g + 1]_-/[1, d + 1]_- ,$$

where $c = g - d$, $g = \frac{1}{2}(n - 3 + w)$.

Proof. For $0 \leq d \leq g$,

$$\lambda(d, n; w) = \text{(number of generators meeting } \Pi_g \text{ in a given } \Pi_d)$$
$$\times \text{(number of } \Pi_d \text{ in } \Pi_g)$$
$$= \mu(c)\,\phi(d; g, q).$$

From Section 3.1 of PGOFF2,

$$\phi(d; g, q) = [g - d + 1, g + 1]_-/[1, d + 1]_- .$$

For $d = -1$,

$$\lambda(-1, n; w) = \kappa(n; w) - \sum_{i=0}^{g} \lambda(i, n; w)$$
$$= q^{(g+1)(g+4-2w)/2}$$
$$= \mu(g + 1). \qquad \square$$

In Section 16.3 of FPSOTD, the theory of stereographic projection of a quadric and an ovaloid of $PG(3, q)$ was explained. Here the *stereographic projection* of a non-singular quadric $\mathcal{Q}_n$ onto a hyperplane from a point P_0 on the quadric is considered. Precisely the same argument applies to a variety of degree $d > 2$ if P_0 is taken to be a point of multiplicity $d - 1$.

Let P_0 be any point of $\mathcal{Q}_n$ and let Π_{n-1} be a fixed hyperplane not containing P_0. Let $\mathcal{V} = T_{P_0}(\mathcal{Q}_n)$ be the tangent hyperplane at P_0, let $\mathcal{W} = \mathcal{V} \cap \mathcal{Q}_n$ be the tangent cone, let $\mathcal{V}' = \Pi_{n-1} \cap \mathcal{V}$, and $\mathcal{W}' = \Pi_{n-1} \cap \mathcal{W}$. For example, when $\mathcal{Q}_n = \mathcal{H}_3$, then Π_{n-1} is a plane, $\mathcal{V}$ is a plane meeting $\mathcal{H}_3$ in a line pair $\mathcal{W}$, and $\mathcal{V}'$ is a line meeting $\mathcal{H}_3$ in a point pair $\mathcal{W}'$.

For $P \in \mathcal{Q}_n \backslash \{P_0\}$, define $P' = P_0 P \cap \Pi_{n-1}$. This gives the correspondence

$$P \mapsto P', \qquad P_0 \mapsto \mathcal{V}'.$$

Analytically, let $P_0 = \mathbf{U}_0$, $\Pi_{n-1} = \mathbf{u}_0$,

$$\mathcal{Q}_n = \mathbf{V}(X_0 F_1(X_1, \ldots, X_n) + F_2(X_1, \ldots, X_n)), \qquad (1.12)$$

where $\deg F_i = i$. Then $\mathcal{V} = \mathbf{V}(F_1)$ and $\mathcal{W} = \mathbf{V}(F_1, F_2)$. If $P = \mathbf{P}(a_0, \ldots, a_n)$, then $P' = \mathbf{P}(0, a_1, \ldots, a_n)$. Conversely, if $P' = \mathbf{P}(0, a_1, \ldots, a_n)$, then, from (1.12), $P = \mathbf{P}(a_0, \ldots, a_n)$ with $a_0 = -F_2(a_1, \ldots, a_n)/F_1(a_1, \ldots, a_n)$.

Now the effect of stereographic projection on the generators of $\mathcal{Q}_n$ is described. Let Π_g be a generator of $\mathcal{Q}_n$ and let $P_0 \in \mathcal{Q}_n \backslash \Pi_g$. Let Π_{n-1} be a fixed hyperplane not containing P_0; then Π_g projects from P_0 to $\Pi'_g = P_0 \Pi_g \cap \Pi_{n-1}$. Now Π_g does not lie on $\mathcal{V} = T_{P_0}(\mathcal{Q}_n)$ as otherwise $P_0 \Pi_g$ would lie on $\mathcal{Q}_n$. So Π_g meets $\mathcal{V}$ in a Π_{g-1} lying in $\mathcal{V}$ and $\mathcal{Q}_n$ and so in $\mathcal{W} = \mathcal{V} \cap \mathcal{Q}_n$. Hence the projection of Π_g onto Π_{n-1} from P_0 is a Π'_g not lying in $\mathcal{V}' = \Pi_{n-1} \cap \mathcal{V}$ but meeting $\mathcal{W}' = \Pi_{n-1} \cap \mathcal{W}$ in a Π'_{g-1}.

Conversely, a space Π'_g of Π_{n-1} not on $\mathcal{V}'$ but containing a space Π'_{g-1} of $\mathcal{W}'$ is joined to P_0 by a Π_{g+1}, which contains the generator $P_0 \Pi'_{g-1}$ of $\mathcal{Q}_n$; the space Π_{g+1} meets $\mathcal{Q}_n$ residually in a Π_g, not containing P_0, which projects from P_0 to the space Π'_g. However, $\mathcal{V}'$ is a Π_{n-2} in Π_{n-1} and $\mathcal{W}'$ is a quadric in $\mathcal{V}'$. From Lemma 1.22, $\mathcal{W}' = P_0 \mathcal{Q}_{n-2}$, where $\mathcal{Q}_{n-2}$ has the same character as $\mathcal{Q}_n$; hence $\mathcal{W}'$ is a $\mathcal{Q}_{n-2}$.

The existence of a partition of the generators of $\mathcal{H}_n$ into two sets is now established. Given two generators Π_g and $\bar{\Pi}_g$, define them to be *equivalent* if their intersection Π_t has its dimension t of the same parity as g. It is shown that this relation is an equivalence relation. Trivially, the relation is reflexive and symmetric. Stereographic projection is used to show the transitivity. The key lemma follows.

Lemma 1.37. *If two generators α_1 and α_2 of $\mathcal{H}_{2g+1}$ intersect in Π_{g-1}, then a third generator α_3 intersects α_1 and α_2 in spaces whose dimensions have different parity.*

Proof. Since $\alpha_1 \cap \alpha_2 = \Pi_{g-1}$, so $\alpha_1 \alpha_2 = \Pi_{g+1}$ and $\Pi_{g+1} \cap \mathcal{H}_{2g+1} = \Pi_{g-1} \mathcal{H}_1$, which consists of the pair $\{\alpha_1, \alpha_2\}$. By Lemma 1.19, Π_{g+1} touches $\mathcal{H}_{2g+1}$ along Π_{g-1}. Let $\Pi_{g+1} \cap \alpha_3 = \Pi_m$; then $m \geq 0$, since $2g + 1$ is the dimension of the ambient space. Now, either (a) Π_m lies in exactly one of α_1, α_2 or (b) Π_m lies in Π_{g-1}. In case (b), the polar space Π'_{2g-m} of Π_m contains both Π_{g+1} and α_3. So $\Pi_{g+1} \cap \alpha_3 = \Pi_k$, where $k \geq g + (g+1) - (2g-m) = m+1 > m$, a contradiction. So (a) holds.

Suppose therefore that $\Pi_m \subset \alpha_1$, whence $\alpha_1 \cap \alpha_3 = \Pi_m$. Since Π_{g-1} and Π_m are both contained in α_1, so $\Pi_{g-1} \cap \Pi_m = \Pi_l$, with $l = (g-1) + m - g = m - 1$. Hence $\alpha_2 \cap \alpha_3 = \Pi_{m-1}$. $\qquad\square$

Lemma 1.38. *If the generators α_1 and α_2 of $\mathcal{H}_n$ with intersection Π_t are projected from a point P_0 in $\mathcal{H}_n \backslash (\alpha_1 \cup \alpha_2)$ to spaces α'_1 and α'_2 containing the generators β'_1 and β'_2 of $\mathcal{W}' = \mathcal{H}_{n-2}$ with $\beta'_1 \cap \beta'_2 = \Pi'_{s-1}$, then t and s have the same parity.*

Proof. Take the point P_0 and project stereographically onto Π_{n-1}. Let β'_1 and β'_2 be the spaces in which α'_1 and α'_2 meet $\mathcal{W}' = \Pi_{n-1} \cap T_{P_0}(\mathcal{H}_n) \cap \mathcal{H}_n$.

As in the above description of stereographic projection, $P_0 \alpha'_1 \cap \mathcal{H}_n = \alpha_1 + P_0 \beta'_1$ and $P_0 \alpha'_2 \cap \mathcal{H}_n = \alpha_2 + P_0 \beta'_2$. Now $\alpha_1 \cap \alpha_2 = \Pi_t$ and $P_0 \beta'_1 \cap P_0 \beta'_2 = P_0 \Pi'_{s-1}$, which are of respective dimensions t and s. Both these are of different parity to $\dim(\alpha_1 \cap P_0\beta_2')$, by Lemma 1.37, when the two triples of generators $(\alpha_2, P_0\beta_2', \alpha_1)$ and $(\alpha_1, P_0\beta_1', P_0\beta_2')$ are considered. So s and t have the same parity.   □

Theorem 1.39. *When $\mathcal{Q}_n = \mathcal{H}_n$, the relation on the generators is an equivalence relation with two equivalence classes.*

Proof. It remains to prove that the relation is transitive. Let $\Pi_g^{(1)}, \Pi_g^{(2)}, \Pi_g^{(3)}$ be generators on $\mathcal{H}_n$, with $\Pi_g^{(1)}$ equivalent to $\Pi_g^{(2)}$ and $\Pi_g^{(2)}$ equivalent to $\Pi_g^{(3)}$. There exists a point not on the generators, and so projection may be used. At the i-th stage, $\Pi_g^{(j)}$ corresponds to $\Pi_{g-i}^{(j)} \subset \mathcal{H}_{n-2i}$ and the parity of $(g-i) - \dim(\Pi_{g-i}^{(1)} \cap \Pi_{g-i}^{(2)})$ is the same for all i, by Lemma 1.38; the parity of $(g-i) - \dim(\Pi_{g-i}^{(2)} \cap \Pi_{g-i}^{(3)})$ is also the same for all i.

Successive projection gives three lines l_1, l_2, l_3 on $\mathcal{H}_3$. As $\Pi_g^{(1)}$ is equivalent to $\Pi_g^{(2)}$, so $g - \dim(\Pi_g^{(1)} \cap \Pi_g^{(2)})$ is even; therefore $1 - \dim(l_1 \cap l_2)$ is even. Thus l_1 and l_2 are the same line or are skew. Similarly, l_2 and l_3 are the same or skew. So l_1, l_2, l_3 belong to the same regulus of $\mathcal{H}_3$. Hence the dimension of $l_1 \cap l_3$ is 1 or -1. Thus $1 - \dim(l_1 \cap l_3)$ is even, and so is $g - \dim(\Pi_g^{(1)} \cap \Pi_g^{(3)})$. Hence $\Pi_g^{(1)}$ is equivalent to $\Pi_g^{(3)}$.

From Lemma 1.37 it follows that there are exactly two equivalence classes.   □

Each equivalence class is called a *system of generators*.

Corollary 1.40. *Let Π_g and $\bar{\Pi}_g$ be distinct generators of $\mathcal{H}_n$, with $n = 2g+1$. Their possible intersections are as follows:*

(1) $n - 4s + 1,\ g = 2o,$
$$\dim(\Pi_g \cap \bar{\Pi}_g) = \begin{cases} 0, 2, 4, \ldots, 2s - 2 & \text{same system,} \\ 1, 3, \ldots, 2s - 1 & \text{different systems;} \end{cases}$$

(ii) $n = 4s + 3,\ g = 2s + 1,$
$$\dim(\Pi_g \cap \bar{\Pi}_g) = \begin{cases} -1, 1, 3, \ldots, 2s - 1 & \text{same system,} \\ 0, 2, 4, \ldots, 2s & \text{different systems.} \end{cases}$$

For dimensions up to 9 of hyperbolic quadrics, Table 1.3 gives all the dimensions of intersections of distinct generators that occur.

1.5 Numbers of subspaces on a quadric

Let $N(m; n, w)$ be the number of subspaces Π_m on the quadric $\mathcal{Q}_n$ of character w. In Section 1.4, the number of generators was determined; that is,
$$\kappa(n; w) = N(g; n, w),$$
where $g = \frac{1}{2}(n + w - 3)$. Also, write $N(\Pi_m, \mathcal{W}_n)$ for the number of m-spaces on the quadric $\mathcal{W}_n$; so
$$N(m; n, w) = N(\Pi_m, \mathcal{Q}_n).$$

Table 1.3. Intersection of generators

$\mathcal{Q}_n$	Dimension of generator	Same system	Different systems
$\mathcal{H}_1$	0	$-$	-1
$\mathcal{H}_3$	1	-1	0
$\mathcal{H}_5$	2	0	$-1, 1$
$\mathcal{H}_7$	3	$-1, 1$	$0, 2$
$\mathcal{H}_9$	4	$0, 2$	$-1, 1, 3$

Theorem 1.41. (i) *m-spaces on a general quadric*

$$N(m; n, w) = \left[\tfrac{1}{2}(n + 1 - w) - m, \tfrac{1}{2}(n + 1 - w)\right]_+$$
$$\times \left[\tfrac{1}{2}(n - 1 + w) - m, \tfrac{1}{2}(n - 1 + w)\right]_- / [1, m + 1]_- \quad (1.13)$$
$$= [g + 2 - w - m, g + 2 - w)]_+$$
$$\times [g + 1 - m, g + 1]_- / [1, m + 1]_- . \quad (1.14)$$

(ii) *m-spaces on particular quadrics*

$$N(m; 2s - 1, 0) = N(\Pi_m, \mathcal{E}_{2s-1})$$
$$= [s - m, s]_+[s - 1 - m, s - 1]_- / [1, m + 1]_- ; \quad (1.15)$$
$$N(m; 2s - 1, 2) = N(\Pi_m, \mathcal{H}_{2s-1})$$
$$= [s - 1 - m, s - 1]_+[s - m, s]_- / [1, m + 1]_- ; \quad (1.16)$$
$$N(m; 2s, 1) = N(\Pi_m, \mathcal{P}_{2s})$$
$$= [s - m, s]_+[s - m, s]_- / [1, m + 1]_- . \quad (1.17)$$

(iii) *Points on a general and particular quadrics*

$$N(0; n, w) = (q^{(n+1-w)/2} + 1)(q^{(n-1+w)/2} - 1)/(q - 1) \quad (1.18)$$
$$= (q^n - 1)/(q - 1) + (w - 1)q^{(n-1)/2} ; \quad (1.19)$$
$$N(\Pi_0, \mathcal{E}_{2s-1}) = (q^s + 1)(q^{s-1} - 1)/(q - 1) ; \quad (1.20)$$
$$N(\Pi_0, \mathcal{H}_{2s-1}) = (q^{s-1} + 1)(q^s - 1)/(q - 1) ; \quad (1.21)$$
$$N(\Pi_0, \mathcal{P}_{2s}) = (q^s + 1)(q^s - 1)/(q - 1) . \quad (1.22)$$

(iv) *Generators on a general and particular quadrics*

$$\kappa(n; w) = [2 - w, g + 2 - w]_+ \quad (1.23)$$
$$= \left[2 - w; \tfrac{1}{2}(n + 1 - w)\right]_+ ; \quad (1.24)$$
$$\kappa(2s - 1; 0) = [2, s]_+ ; \quad (1.25)$$
$$\kappa(2s - 1; 2) = [0, s - 1]_+ ; \quad (1.26)$$
$$\kappa(2s; 1) = [1, s]_+ . \quad (1.27)$$

Proof. First the number of points on $\mathcal{Q}_n$ is calculated. By Lemma 1.22,

$$N(0; n, w) = q^{n-1} + 1 + qN(0; n - 2, w) \,. \tag{1.28}$$

However, for dimensions 0 and 1,

$$N(0; 1, 0) = N(0; 0, 1) = 0, \quad N(0; 1, 2) = 2 \,;$$

induction gives (1.18) and (1.19).

Now, by Theorem 1.11 (ii), if $P \in \Pi_m \subset \mathcal{Q}_n$, then $\Pi_m \subset T_P(\mathcal{Q}_n) \cap \mathcal{Q}_n$, which by Lemma 1.22 is $P\mathcal{Q}_{n-2}$ of the same character as $\mathcal{Q}_n$. So Π_m meets $\mathcal{Q}_{n-2}$ in a Π_{m-1} and, conversely, every Π_{m-1} on $\mathcal{Q}_{n-2}$ determines a Π_m on $\mathcal{Q}_n$ through P. Hence

$$N(m; n, w) = N(0; n, w)N(m - 1; n - 2, w)/\theta(m) \,, \tag{1.29}$$

where $\theta(m) = (q^{m+1} - 1)/(q - 1)$. Induction and (1.18) give the result. $\qquad\square$

Corollary 1.42.

$$N(\Pi_0, \Pi_{n-t-1}\mathcal{Q}_t) = (q^n - 1)/(q - 1) + (w - 1)q^{(2n-t-1)/2} \,.$$

Proof. The joins of two points of the base to the vertex give $(n - t)$-spaces intersecting in the vertex Π_{n-t-1}. Hence

$$N(\Pi_0, \Pi_{n-t-1}\mathcal{Q}_t) = N(\Pi_0, \mathcal{Q}_t)(\theta(n - t) - \theta(n - t - 1)) + \theta(n - t - 1) \,,$$

which gives the answer. $\qquad\square$

1.6 The orthogonal groups

The group of projectivities of $\mathrm{PG}(n, q)$ is $\mathrm{PGL}(n+1, q)$. Let $G(\mathcal{Q}_n)$ be the subgroup of $\mathrm{PGL}(n+1, q)$ fixing the form defining $\mathcal{Q}_n$ up to a scalar multiple. This is actually the same as the group fixing $\mathcal{Q}_n$ provided that $\mathcal{Q}_n \neq \mathcal{E}_1$. The group $G(\mathcal{Q}_n)$ is called *orthogonal* and is also denoted by $\mathrm{PGO}(n + 1, q)$ for $\mathcal{P}_n$, by $\mathrm{PGO}_1(n + 1, q)$ for $\mathcal{H}_n$, and by $\mathrm{PGO}_-(n + 1, q)$ for $\mathcal{E}_n$.

Also, let $\mathcal{N}(\mathcal{Q}_n)$ be the set of all quadrics in $\mathrm{PG}(n, q)$ projectively equivalent to $\mathcal{Q}_n$; that is, $\mathcal{N}(\mathcal{Q}_n)$ is the orbit of $\mathcal{Q}_n$ under the action of $\mathrm{PGL}(n + 1, q)$. Again $\mathcal{E}_1$ is a special case and is considered here as a pair of conjugate points in $\mathrm{PG}(1, q^2)$. First, $|G(\mathcal{Q}_n)|$ and $|\mathcal{N}(\mathcal{Q}_n)|$ are calculated.

Lemma 1.43. *The number of quadrics $\mathcal{Q}_n$ in $\mathrm{PG}(n, q)$, $n \geq 2$, containing a given $\Pi_0 \mathcal{Q}_{n-2}$ as a tangent cone is $q^n(q - 1)$.*

Proof. Let $\Pi_0 \mathcal{Q}_{n-2}$ have vertex $\Pi_0 = \mathbf{U}_{n-1}$ and base

$$\mathcal{Q}_{n-2} = \mathbf{V}(X_n, X_{n-1}, f_2(X_0, \ldots, X_{n-2})) \,.$$

So the hyperplane containing $\Pi_0 \mathcal{Q}_{n-2}$ is $\mathbf{u}_n$. Any quadric containing $\mathbf{U}_{n-1}$ and $\mathcal{Q}_{n-2}$ has the form

$$\mathcal{Q} = \mathbf{V}(X_n(a_0 X_0 + \cdots + a_n X_n) + X_{n-1}(b_0 X_0 + \cdots + b_{n-2} X_{n-2}) + f_2).$$

The tangent hyperplane to $\mathcal{Q}$ at $\mathbf{U}_{n-1}$ is

$$\mathbf{V}(a_{n-1} X_n + b_0 X_0 + \cdots + b_{n-2} X_{n-2}).$$

Since this is $\mathbf{u}_n$, so

$$b_0 = \cdots = b_{n-2} = 0, \quad a_{n-1} \neq 0.$$

Thus

$$\mathcal{Q} = \mathbf{V}(X_n(a_0 X_0 + \cdots + a_{n-1} X_{n-1} + a_n X_n) + f_2),$$

which is non-singular since $a_{n-1} \neq 0$. Therefore a_{n-1} may be chosen in $q - 1$ ways and every other a_i in q ways, giving $q^n(q - 1)$ possibilities for $\mathcal{Q} = \mathbf{V}(F)$. This argument relies on the fact that, if $\mathcal{Q} \neq \Pi_{n-2}\mathcal{E}_1$, then it uniquely defines the form F up to a scalar multiple. $\qquad\square$

In this proof, when $\mathcal{Q}_n = \mathcal{E}_3$, then $\Pi_0 \mathcal{Q}_{n-2} = \Pi_0 \mathcal{E}_1$ is a pair of conjugate intersecting lines in the quadratic extension.

Theorem 1.44. (i) *The values of* $|G(\mathcal{Q}_n)|$ *are as follows*:

$$|G(\mathcal{P}_n)| = q^{n^2/4} \prod_{i=1}^{n/2} (q^{2i} - 1); \tag{1.30}$$

$$|G(\mathcal{H}_n)| = 2q^{(n^2-1)/4}(q^{(n+1)/2} - 1) \prod_{i=1}^{(n-1)/2} (q^{2i} - 1); \tag{1.31}$$

$$|G(\mathcal{E}_n)| = 2q^{(n^2-1)/4}(q^{(n+1)/2} + 1) \prod_{i=1}^{(n-1)/2} (q^{2i} - 1). \tag{1.32}$$

(ii) *The values of* $|\mathcal{N}(\mathcal{Q}_n)|$ *are as follows*:

$$|\mathcal{N}(\mathcal{P}_n)| = q^{n(n+2)/4} \prod_{i=1}^{n/2} (q^{2i+1} - 1); \tag{1.33}$$

$$|\mathcal{N}(\mathcal{H}_n)| = \tfrac{1}{2} q^{(n+1)^2/4}(q^{(n+1)/2} + 1) \prod_{i=1}^{(n-1)/2} (q^{2i+1} - 1); \tag{1.34}$$

$$|\mathcal{N}(\mathcal{E}_n)| = \tfrac{1}{2} q^{(n+1)^2/4}(q^{(n+1)/2} - 1) \prod_{i=1}^{(n-1)/2} (q^{2i+1} - 1). \tag{1.35}$$

Proof. First, $|\mathcal{N}(\mathcal{Q}_n)|$ is calculated by counting the set

$$\{(\mathcal{Q}_n, S) \mid \mathcal{Q}_n \text{ a non-singular quadric, } S \text{ a tangent cone of } \mathcal{Q}_n\}$$

in two ways. Let M be the number of cones $\Pi_0 \mathcal{Q}_{n-2}$ in $\mathrm{PG}(n, q)$ for a fixed character w. Then

$$|\mathcal{N}(\mathcal{Q}_n)| \, N(0; n, w) = M q^n (q - 1).$$

However,

$$M = \text{number of } \Pi_{n-1} \text{ in } \mathrm{PG}(n, q)$$
$$\times \text{number of } \Pi_0 \text{ in } \Pi_{n-1}$$
$$\times \text{number of } \mathcal{Q}_{n-2} \text{ in a fixed } \Pi_{n-2} \text{ of } \Pi_{n-1}$$
$$= \theta(n) \, \theta(n - 1) \, |\mathcal{N}(\mathcal{Q}_{n-2})|.$$

Thus

$$|\mathcal{N}(\mathcal{Q}_n)| = \theta(n) \, \theta(n - 1) \, |\mathcal{N}(\mathcal{Q}_{n-2})| \, q^n (q - 1) / N(0; n, w)$$
$$= \frac{(q^{n+1} - 1)(q^n - 1)q^n \, |\mathcal{N}(\mathcal{Q}_{n-2})|}{(q^{(n+1-w)/2} + 1)(q^{(n-1+w)/2} - 1)}.$$

Since $|\mathcal{N}(\mathcal{P}_0)| = 1$, $|\mathcal{N}(\mathcal{E}_1)| = \frac{1}{2}q(q - 1)$, $|\mathcal{N}(\mathcal{H}_1)| = \frac{1}{2}q(q + 1)$, induction now gives $|\mathcal{N}(\mathcal{Q}_n)|$.

Finally,

$$|\mathrm{PGL}(n + 1, q)| = |G(\mathcal{Q}_n)| \, |\mathcal{N}(\mathcal{Q}_n)|$$

in each case, where

$$|\mathrm{PGL}(n + 1, q)| = q^{n(n+1)/2} \prod_{i=2}^{n+1} (q^i - 1).$$

$\square$

For the orders of groups associated to these orthogonal groups, see Appendix I of PGOFF2 or Appendix III of FPSOTD.

Consider the following involutory transformations which fix $\mathcal{Q}_n$. Let Q be any point of $\mathrm{PG}(n, q) \backslash \mathcal{Q}_n$ with the only restriction that, for q and n even, Q is not the nucleus of $\mathcal{Q}_n$. Let $\mu_Q : \mathcal{Q}_n \to \mathcal{Q}_n$ be defined as follows. For $P \in \mathcal{Q}_n$,

$$P\mu_Q = P \qquad \text{if } PQ \text{ is a tangent to } \mathcal{Q}_n;$$
$$P\mu_Q = P' \qquad \text{if } PQ \text{ meets } \mathcal{Q}_n \text{ again at } P'.$$

Lemma 1.45. *The mapping μ_Q can be extended to an element of $G(\mathcal{Q}_n)$.*

Proof. Let $\mathcal{Q}_n = V(F)$, $Q = \mathbf{P}(B) \notin \mathcal{Q}_n$, $P = \mathbf{P}(A) \in \mathcal{Q}_n$. If $P' \in PQ$, then $P' = \mathbf{P}(A + tB)$. If $P' \in \mathcal{Q}_n$, then, as in (1.3),

$$F(A + tB) = F(A) + tG(A, B) + t^2 F(B) = 0.$$

Since $P \in \mathcal{Q}_n$, so $F(A) = 0$. Hence

$$tG(A, B) + t^2 F(B) = 0.$$

The solution $t = 0$ corresponds to P and the solution $t = -G(A, B)/F(B)$ corresponds to P'. Hence $P' = \mathbf{P}(A - G(A, B)B/F(B))$, which is the same point as P when $G(A, B) = 0$, that is, when P lies in the polar hyperplane of Q. In any case, μ_Q is given by

$$\mathbf{P}(x) \mapsto \mathbf{P}\left(x - \frac{G(x, B)}{F(B)}B\right).$$

Thus μ_Q can be extended to an element of $G(\mathcal{Q}_n)$. $\qquad\square$

Since the identity is the only element of $G(\mathcal{Q}_n)$ which fixes Q and all points of $\mathcal{Q}_n$, the extension of μ_Q is necessarily unique. This extension is a perspectivity with centre Q; the axis contains all points P of $\mathcal{Q}_n$ for which PQ is tangent to $\mathcal{Q}_n$.

The extension of μ_Q is also denoted μ_Q.

Theorem 1.46. *The group $G(\mathcal{Q}_n)$ acts transitively on $\mathcal{Q}_n$.*

Proof. Let P, P' be any two points of $\mathcal{Q}_n$. If $PP' \not\subset \mathcal{Q}_n$, let Q be any point on $PP' \backslash \{P, P'\}$. Then μ_Q maps P to P'.

If $PP' \subset \mathcal{Q}_n$, choose P'' such that neither PP'' nor $P'P''$ lies on $\mathcal{Q}_n$. The point P'' exists, since otherwise $\mathcal{Q}_n$ would be singular. Now choose Q on $PP''\backslash\{P, P''\}$ and R on $P'P''\backslash\{P', P''\}$. Then $P\mu_Q\mu_R = P''\mu_R = P''$. $\qquad\square$

Notation 1.47. Let the quadric $\mathcal{Q}_n$ have character w. Then

(i) $\mathcal{S}(m, t, v; n, w)$ is the set of m-spaces Π_m in $PG(n, q)$ with $m \neq n$ such that $\Pi_m \cap \mathcal{Q}_n$ is of type $\Pi_{m-t-1}\mathcal{Q}_t$ where $\mathcal{Q}_t$ has character v;

(ii) $N(m, t, v; n, w) = N(\Pi_{m-t-1}\mathcal{Q}_t, \mathcal{Q}_n) = |\mathcal{S}(m, t, v; n, w)|$.

In Section 1.8, $N(m, t, v; n, w)$ is determined and, in particular, it is shown when it is zero, that is, when $\mathcal{S}(m, t, v; n, w)$ is empty. Here the number of orbits of $\mathcal{S}(m, t, v; n, w)$ under the action of $G(\mathcal{Q}_n)$ is given.

First consider $PG(1, q)$. The quadric $\mathcal{H}_1$ consists of two points and the group $G(\mathcal{H}_1) = PGO_+(2, q)$ has order $2(q - 1)$. As $PGL(2, q)$ acts triply transitively on $PG(1, q)$, as in Section 6.1 of PGOFF2, there is a projectivity fixing both points of $\mathcal{H}_1$ and moving P_1 to P_2, where P_1 and P_2 are any points off $\mathcal{H}_1$. So $PGO_+(2, q)$ acts transitively on the points off $\mathcal{H}_1$.

The quadric $\mathcal{E}_1$ is empty in $PG(1, q)$ but consists of two conjugate points on $PG(1, q^2)$. So, if $\mathcal{E}_1 = \mathbf{V}(F)$ with

$$F = X^2 - bX + c = (X - \alpha)(X - \alpha^q),$$

then, in non-homogeneous coordinates, the projectivity $\mathfrak{T} : t \mapsto t'$ of $PG(1, q^2)$, given by

$$tt'\{e + e' - (\alpha + \alpha^q)\} - (t + t')\{ee' - \alpha^{q+1}\}$$
$$+\{(\alpha + \alpha^q)ee' - \alpha^{q+1}(e + e')\} = 0,$$

is an involution with pairs (α, α^q) and (e, e'). It therefore fixes $\mathcal{E}_1$ and takes e to e'. Thus $\mathrm{PGO}_-(2, q)$ acts transitively on the points of $\mathrm{PG}(1, q)$ off $\mathcal{E}_1$. The group $G(\mathcal{E}_1) = \mathrm{PGO}_-(2, q)$ has order $2(q + 1)$.

Next the conic $\mathcal{P}_2$ is examined. Let

$$\begin{aligned}
\mathcal{O}_1 &= \mathcal{S}(0, -1, 2; 2, 1) = \{\text{points on } \mathcal{P}_2\}, \\
\mathcal{O}_2 &= \mathcal{S}(0, 0, 1; 2, 1) = \{\text{points off } \mathcal{P}_2\}, \\
\mathcal{O}_3 &= \mathcal{S}(1, 0, 1; 2, 1) = \{\text{tangents to } \mathcal{P}_2\}, \\
\mathcal{O}_4 &= \mathcal{S}(1, 1, 2; 2, 1) = \{\text{bisecants of } \mathcal{P}_2\}, \\
\mathcal{O}_5 &= \mathcal{S}(1, 1, 0; 2, 1) = \{\text{external lines of } \mathcal{P}_2\}.
\end{aligned}$$

Theorem 1.46 says that $G(\mathcal{P}_2)$ acts transitively on $\mathcal{O}_3$. In fact, $G(\mathcal{P}_2)$ acts triply transitively on $\mathcal{O}_1$ and $\mathcal{O}_3$, by Corollary 7.15 of PGOFF2. Recall that, for q odd,

$$\mathcal{O}_2 = \mathcal{O}_2^+ \cup \mathcal{O}_2^-,$$

where

$$\mathcal{O}_2^+ = \{\text{external points of } \mathcal{P}_2\}, \quad \mathcal{O}_2^- = \{\text{internal points of } \mathcal{P}_2\};$$

here, a point Q off $\mathcal{P}_2$ is external or internal according as it lies on two or no tangents of $\mathcal{P}_2$, Section 8.2 of PGOFF2. For q even,

$$\mathcal{O}_2 - \{N\} \sqcup \mathcal{O}_2'$$

where N is the nucleus, the meet of all the tangents, and each point of $\mathcal{O}_2'$ lies on precisely one tangent.

Lemma 1.48. (i) $G(\mathcal{P}_2)$ *acts transitively on* $\mathcal{O}_4$ *and* $\mathcal{O}_5$.
(ii) $G(\mathcal{P}_2)$ *has two orbits on* $\mathcal{O}_2$, *namely* $\mathcal{O}_2^+$ *and* $\mathcal{O}_2^-$ *for* q *odd, and* $\{N\}$ *and* $\mathcal{O}_2'$ *for* q *even.*

Proof. (a) For q odd, consider the action of $G(\mathcal{P}_2)$ on $\mathcal{O}_2$, the points off

$$\mathcal{P}_2 = \mathbf{V}(X_0^2 + X_1 X_2).$$

Since each point of $\mathcal{O}_2^+$ is the intersection of two tangents, so $G(\mathcal{P}_2)$ is transitive on $\mathcal{O}_2^+$, the external points. By the polarity, it is transitive on $\mathcal{O}_4$, the bisecants.

Any external line contains an external point. Therefore, to show the transitivity of $G(\mathcal{P}_2)$ on $\mathcal{O}_5$ and, by the polarity, on $\mathcal{O}_2^-$, it suffices to show the transitivity on the external lines through a particular external point. Let $\mathbf{U}_0$ be this point. Then the line $l(t) = \mathbf{V}(X_1 + tX_2)$ is a bisecant or an external line as t is a non-zero square or a non-square.

The projectivity $\mathfrak{T}_c$ given by

$$\mathbf{P}(x_0, x_1, x_2)\mathfrak{T}_c = \mathbf{P}(cx_0, x_1, c^2 x_2)$$

fixes $\mathcal{P}_2$ and transforms $l(t)$ to $l(t/c^2)$. So there is an element of the group transforming any bisecant through $\mathbf{U}_0$ to any other bisecant through $\mathbf{U}_0$ and any external line through $\mathbf{U}_0$ to any other external line through $\mathbf{U}_0$.

(b) For q even, $G(\mathcal{P}_2)$ is similarly transitive on $\mathcal{O}_4$. Since any point of $\mathcal{O}_2'$ is the meet of a tangent and a bisecant, the triple transitivity of $G(\mathcal{P}_2)$ on $\mathcal{O}_1$ ensures the transitivity on $\mathcal{O}_2'$.

To show the transitivity of $G(\mathcal{P}_2)$ on $\mathcal{O}_5$, it suffices to consider the external lines through a particular point of $\mathcal{O}_2'$. Let $Q = \mathbf{P}(1, 0, 1)$ with $\mathcal{P}_2$ as above. Then the line $l(t) = \mathbf{V}(X_0 + tX_1 + X_2)$ contains Q and meets $\mathcal{P}_2$ at $\mathbf{P}(x_0, x_1, x_2)$, where $t^2 x_1^2 + x_1 x_2 + x_2^2 = 0$. So $l(t)$ is a bisecant or an external line according as t^2 and so t is in $\mathcal{T}_0$ or $\mathcal{T}_1$, that is, has trace 0 or 1, Section 1.2. Now the projectivity $\mathfrak{T}_b$, given by

$$\mathbf{P}(x_0, x_1, x_2)\mathfrak{T}_b = \mathbf{P}(x_0 + bx_1, x_1, b^2 x_1 + x_2),$$

fixes Q and $\mathcal{P}_2$ and transforms $l(t)$ to $l(t + b + b^2)$. As $t + b + b^2$ has the same trace as t, any external line through Q can be transformed to any other.   □

For a section $\Pi_{m-t-1}\mathcal{Q}_t$ of $\mathcal{Q}_n$, let $T = n + t - 2m$.

Theorem 1.49. *For given* m, t, v, n, w, *the set* $\mathcal{S}(m, t, v; n, w)$ *acted on by* $G(\mathcal{Q}_n)$ *is either empty or has*

(i) *one orbit when* (a) n *is odd or* (b) n *is even and* t *is odd or* (c) n *is even,* t *is even and* $T = 0$;

(ii) *two orbits when* n *is even,* t *is even and* $T > 0$.

Proof. (1) $t = m$ *with* $(v, w) \neq (1, 1)$

First assume that $m \geq 2$.

Let $\Pi_m^{(i)} \cap \mathcal{Q}_n = \mathcal{W}_m^{(i)}$ with $\Pi_m^{(i)} \in \mathcal{S}$ for $i = 1, 2$ and let $P \in \mathcal{W}_m^{(1)} \cap \mathcal{W}_m^{(2)}$. Project $\mathcal{Q}_n$ from P onto a hyperplane Π_{n-1} not containing P, as in Section 1.4.

Then $\mathcal{Q}_n$ determines a quadric $\mathcal{W}' = \mathcal{Q}_{n-2}$ in Π_{n-2}, and $\mathcal{W}_m^{(1)}$ and $\mathcal{W}_m^{(2)}$ give quadrics $\mathcal{R}_1$ and $\mathcal{R}_2$ of the same type but in dimension $m - 2$. By induction there is a projectivity $\mathfrak{T}$ of Π_{n-2} fixing $\mathcal{Q}_{n-2}$ and mapping $\mathcal{R}_1$ to $\mathcal{R}_2$. Let

$$\Pi_m^{(i)} \cap \Pi_{n-1} = \Pi_{m-1}^{(i)}, \quad i = 1, 2.$$

Extend $\mathfrak{T}$ to Π_{n-1}, and let $\Pi_{m-1}^{(1)}\mathfrak{T} = \Pi_{m-1}^{(1a)}$. In Π_{n-1} there is an elation $\mathfrak{T}'$ with axis Π_{n-2} mapping $\Pi_{m-1}^{(1a)}$ to $\Pi_{m-1}^{(2)}$. Hence $\mathfrak{T}\mathfrak{T}'$ maps $\mathcal{R}_1$ to $\mathcal{R}_2$, $\Pi_{m-1}^{(1)}$ to $\Pi_{m-1}^{(2)}$, and $\mathcal{Q}_{n-2}$ to itself.

Taking $P = \mathbf{U}_n$ with tangent hyperplane $T_P = \mathbf{u}_{n-1}$, the quadric $\mathcal{Q}_n = \mathbf{V}(F)$ with

$$F = f(X_0, \ldots, X_{n-1}) + X_{n-1}X_n.$$

By a linear transformation,

$$F = g(X_0, \ldots, X_{n-2}) + X_{n-1}X_n.$$

Then $\mathcal{Q}_{n-2} = \mathbf{V}(g, X_{n-1}, X_n)$. So $\mathfrak{T}\mathfrak{T}'$ is given by a linear transformation on $X_0, \ldots, X_{n-1}$ taking g to λg and X_{n-1} to $\lambda' X_{n-1}$. Extend $\mathfrak{T}\mathfrak{T}'$ to the whole space by letting $X_n \mapsto (\lambda/\lambda') X_n$. This gives a projectivity $\mathfrak{S}$ fixing P and $\mathcal{Q}_n$, and mapping $\mathcal{W}_m^{(1)}$ to $\mathcal{W}_m^{(2)}$.

Next, let $\mathcal{W}_m^{(1)} \cap \mathcal{W}_m^{(2)} = \emptyset$. Since $G(\mathcal{Q}_n)$ acts transitively on $\mathcal{Q}_n$, there exists $\mathfrak{S}_1$ in $G(\mathcal{Q}_n)$ for which $\mathcal{W}_m^{(1)}\mathfrak{S}_1 = \mathcal{W}_m^{(3)}$ meets $\mathcal{W}_m^{(2)}$. Then application of the preceding argument gives an element $\mathfrak{S}_2$ of $G(\mathcal{Q}_n)$ with $\mathcal{W}_m^{(3)}\mathfrak{S}_2 = \mathcal{W}_m^{(2)}$. Hence $\mathfrak{S}_1\mathfrak{S}_2$ is the required element of $G(\mathcal{Q}_n)$ taking $\mathcal{W}_m^{(1)}$ to $\mathcal{W}_m^{(2)}$.

Since induction was used, the small cases have still to be considered.

First assume that $m = t = 0$. Then the section is a point off the quadric $\mathcal{Q}_n$ with n odd. For $n = 1$, the group $G(\mathcal{Q}_1)$ acts transitively on the set of all points off $\mathcal{Q}_1$, as discussed after Theorem 1.46. So let $n \geq 3$. Assume that P_1 and P_2 are points off $\mathcal{Q}_n$ and let α_1 and α_2 be their polar hyperplanes. Then $\alpha_1 \cap \mathcal{Q}_n$ and $\alpha_2 \cap \mathcal{Q}_n$ are non-singular quadrics $\mathcal{P}_{n-1}^{(1)}$ and $\mathcal{P}_{n-1}^{(2)}$ as $n - 1$ is even. It is sufficient to show that there is an element $\mathfrak{T}$ in $G(\mathcal{Q}_n)$ with $\mathcal{P}_{n-1}^{(1)}\mathfrak{T} = \mathcal{P}_{n-1}^{(2)}$. By induction, as in a previous argument, this reduces to the case $n = 1$ and $m = t = 0$.

Now let $m = t = 1$; then n is even. For $n = 2$ there is nothing to prove, due to Lemma 1.48. So let $n \geq 4$. If $\mathcal{W}_1^{(1)}$ and $\mathcal{W}_1^{(2)}$ are hyperbolic and meet at P, then by projecting from P and applying a previous argument it follows that $G(\mathcal{Q}_n)$ contains an element which maps $\mathcal{W}_1^{(1)}$ onto $\mathcal{W}_1^{(2)}$. If $\mathcal{W}_1^{(1)} \cap \mathcal{W}_1^{(2)} = \emptyset$ then proceed as in the case $\mathcal{W}_m^{(1)} \cap \mathcal{W}_m^{(2)} = \emptyset$.

Finally, assume that $\mathcal{W}_1^{(1)}$ and $\mathcal{W}_1^{(2)}$ are elliptic, and let

$$\mathcal{W}_1^{(1)} = \{P_1, P_1'\}, \quad \mathcal{W}_1^{(2)} = \{P_2, P_2'\}$$

with P_i, P_i' conjugate in a quadratic extension of $\mathbf{F}_q$. Let P_i'' be a point of $P_i P_i'$ off $\mathcal{Q}_n$, $i = 1, 2$. From the case $m = t = 0$, there is an element $\mathfrak{T}$ in $G(\mathcal{Q}_n)$ which maps P_1'' to P_2''. Let

$$P_1\mathfrak{T} = R_1, \quad P_1'\mathfrak{T} = R_1', \quad R_1 P_2' \cap R_1' P_2 = Q, \quad R_1 P_2 \cap R_1' P_2' = Q'.$$

If Q and Q' arc on $\mathcal{Q}_n$, then the plane containing R_1, R_1', P_2, P_2' is on $\mathcal{Q}_n$; so P_2'' is on $\mathcal{Q}_n$, a contradiction. Assume therefore that Q is not on $\mathcal{Q}_n$. Then $R_1\mu_Q = P_2'$ and $R_1'\mu_Q = P_2$. Thus $\mathfrak{T}\mu_Q$ maps $\mathcal{W}_1^{(1)}$ to $\mathcal{W}_1^{(2)}$.

(2) $t = m$ with $w = v = 1$

First assume $m \geq 2$.

Let $\mathcal{S}_P$ be the set of all elements of $\mathcal{S}$ containing the point P, and let G_P be the subgroup of $G(\mathcal{Q}_n)$ fixing P. By projection of $\mathcal{Q}_n$ from P onto a Π_{n-1} not containing P and by using induction on m as in (1), G_P has two orbits on $\mathcal{S}_P$.

Suppose that $\mathcal{W}_m^{(1)}$ and $\mathcal{W}_m^{(2)}$ are in $\mathcal{S}_P$ and also in one orbit $\mathcal{O}$ of $G(\mathcal{Q}_n)$; then $\mathcal{W}_m^{(1)}\mathfrak{T} = \mathcal{W}_m^{(2)}$ for some $\mathfrak{T}$ in $G(\mathcal{Q}_n)$. Let $P\mathfrak{T} = Q$. Now it is shown that there exists $\mathfrak{T}'$ in $G(\mathcal{Q}_n)$ fixing $\mathcal{W}_m^{(2)}$ and mapping Q to P. If PQ is not a line of $\mathcal{W}_m^{(2)}$ then μ_R, with R on PQ but not on $\mathcal{W}_m^{(2)}$, fixes $\mathcal{Q}_n$, fixes $\mathcal{W}_m^{(2)}$ and maps Q to P.

Let PQ be a line of $\mathcal{W}_m^{(2)}$. Consider a point R on $\mathcal{W}_m^{(2)}$ such that neither PR nor QR is on $\mathcal{W}_m^{(2)}$. Further, let A be a point on PR but not on $\mathcal{W}_m^{(2)}$ and let B be a point on QR but not on $\mathcal{W}_m^{(2)}$. Then $\mu_A \mu_B$ fixes both $\mathcal{Q}_n$ and $\mathcal{W}_m^{(2)}$, and maps P to Q. Hence $\mathcal{W}_m^{(2)}$ is in the orbit $\mathcal{O}_P$ of $\mathcal{W}_m^{(1)}$ under G_P. Since $G(\mathcal{Q}_n)$ acts transitively on $\mathcal{Q}_n$ it follows that the number of orbits $\mathcal{O}$ of $\mathcal{S}$ under $G(\mathcal{Q}_n)$ is the number of orbits $\mathcal{O}_P$ of $\mathcal{S}_P$ under G_P.

Since induction was used, the small cases still need to be considered. So assume that $m = t = 0$; then the section is a point off $\mathcal{Q}_n$ and n is even.

First consider q odd. Let P be a point off $\mathcal{Q}_n$, let Π_{n-1} be its polar hyperplane, and let $\mathcal{Q}_n \cap \Pi_{n-1} = \mathcal{Q}_{n-1}$. By (1) and Lemma 1.48, $G(\mathcal{Q}_n)$ has two orbits on the set of all sections $\mathcal{Q}_{n-1}$. Hence $G(\mathcal{Q}_n)$ has two orbits on the set of all points off $\mathcal{Q}_n$.

Now suppose that q is even. By Lemma 1.48 it may be assumed that $n \geq 4$. One orbit consists of a single point, the nucleus N. So take distinct points P_1 and P_2 not in $\mathcal{Q}_n \cup \{N\}$. Let $\mathcal{C}_i$ be a conic on $\mathcal{Q}_n$ with nucleus P_i, $i = 1, 2$. It suffices to show that there exists $\mathfrak{T}$ in $G(\mathcal{Q}_n)$ with $\mathcal{C}_1 \mathfrak{T} = \mathcal{C}_2$. Now $\mathcal{C}_1$ and $\mathcal{C}_2$ may be chosen in such a way that $P \in \mathcal{C}_1 \cap \mathcal{C}_2$. Project $\mathcal{Q}_n$ from P onto a hyperplane Π_{n-1} not containing P. In Π_{n-1} this gives a $\mathcal{Q}_{n-2}$ with nucleus N'. The tangents to $\mathcal{C}_1$ and $\mathcal{C}_2$ at P meet Π_{n-1} in points P_1' and P_2' distinct from N'. By induction on n, the group $G(\mathcal{Q}_{n-2})$ contains an element $\mathfrak{T}'$ with $P_1' \mathfrak{T}' = P_2'$. As in (1), $\mathfrak{T}'$ can be extended to an element of $G(\mathcal{Q}_n)$ that fixes both $\mathcal{Q}_n$ and P, and maps $\mathcal{C}_1$ to $\mathcal{C}_2$. Hence this extension maps P_1 to P_2. The smallest case, where $n = 2$ and $m = t = 0$, is contained in Lemma 1.48.

(3) $-1 \leq t \leq m - 1$

First let $t = -1$. Then $\Pi_m^{(1)}, \Pi_m^{(2)} \subset \mathcal{Q}_n$. For $m = 0$ the result is contained in Theorem 1.46. So assume that $m > 0$. Let $P \in \Pi_m^{(1)} \cap \Pi_m^{(2)}$. By projection of $\mathcal{Q}_n$ onto a hyperplane not containing P and using induction on m as in (1), there is an element $\mathfrak{T}$ of $G(\mathcal{Q}_n)$ that fixes P and maps $\Pi_m^{(1)}$ to $\Pi_m^{(2)}$.

Now assume that $\Pi_m^{(1)} \cap \Pi_m^{(2)} = \Pi_{-1}$. If $P_1 \in \Pi_m^{(1)}$, then there is a point P_2 in $\Pi_m^{(2)} \cap T_{P_1}(\mathcal{Q}_n)$ since $m > 0$. Take $\Pi_m^{(3)} \subset \mathcal{Q}_n$ with the line $P_1 P_2$ in $\Pi_m^{(3)}$. Then there exist $\mathfrak{T}_1$ and $\mathfrak{T}_2$ in $G(\mathcal{Q}_n)$ such that $\mathfrak{T}_1$ maps $\Pi_m^{(1)}$ to $\Pi_m^{(3)}$ and $\mathfrak{T}_2$ maps $\Pi_m^{(2)}$ to $\Pi_m^{(3)}$. Hence $\mathfrak{T}_1 \mathfrak{T}_2^{-1}$ maps $\Pi_m^{(1)}$ to $\Pi_m^{(2)}$.

Next, let $t \geq 0$. Define $\mathcal{S}_{m,t}$ to be the set of all elements of $\mathcal{S}$ containing the subspace Π_{m-t-1} as vertex of a section of $\mathcal{Q}_n$ by an m-space, and let $G_{m,t}$ be the subgroup of $G(\mathcal{Q}_n)$ fixing Π_{m-t-1}. Suppose that $G_{m,t}$ has M orbits on $\mathcal{S}_{m,t}$. Project $\mathcal{Q}_n$ from Π_{m-t-1} onto Π_{n-m+t} with $\Pi_{m-t-1} \cap \Pi_{n-m+t} = \emptyset$. Then $\mathcal{Q}_n$ determines a quadric $\mathcal{Q}_{n-2(m-t)}$ of Π_{n-m+t}, whose equations are now determined.

Let $P_0, \ldots, P_{m-t-1}$ be linearly independent points of Π_{m-t-1}. Take $P_i = \mathbf{U}_i$ and the tangent hyperplane $T_{P_i}(\mathcal{Q}_n) = \mathbf{u}_{m-t+i}$ for $i = 0, 1, \ldots, m - t - 1$; then $\mathcal{Q}_n = \mathbf{V}(F)$ with

$$F = f(X_{2m-2t}, \ldots, X_n) + a_0 X_0 X_{m-t} + \cdots + a_{m-t-1} X_{m-t-1} X_{2m-2t-1}.$$

This gives

$$\mathcal{Q}_{n-2(m-t)} = \mathbf{V}(f, X_0, \ldots, X_{2m-2t-1}).$$

Let $\mathcal{Q}_t^{(1)}$ and $\mathcal{Q}_t^{(2)}$ be sections of $\mathcal{Q}_{n-2(m-t)}$ by subspaces $\Pi_t^{(1)}$ and $\Pi_t^{(2)}$ of $\Pi_{n-2(m-t)}$, where $\mathcal{Q}_t^{(1)}$ and $\mathcal{Q}_t^{(2)}$ are of character v and belong to the same orbit of $G(\mathcal{Q}_{n-2(m-t)})$. Then an element $\mathfrak{T}$ mapping $\mathcal{Q}_t^{(1)}$ to $\mathcal{Q}_t^{(2)}$ is given by a linear transformation on $X_{2m-2t}, \ldots, X_n$ taking f to λf. Extending $\mathfrak{T}$ by the transformation

$$X_i \mapsto X_i, \qquad i = 1, \ldots, m-t-1,$$
$$X_i \mapsto \lambda X_i, \qquad i = m-t, \ldots, 2m-2t-1$$

gives a projectivity fixing Π_{m-t-1} and $\mathcal{Q}_n$ as well as mapping $\Pi_{m-t-1}\mathcal{Q}_t^{(1)}$ to $\Pi_{m-t-1}\mathcal{Q}_t^{(2)}$. If M' is the number of orbits of $G(\mathcal{Q}_{n-2(m-t)})$ on the set of all sections $\mathcal{Q}_t$ of character v of $\mathcal{Q}_{n-2(m-t)}$, then it follows that $M \le M'$. But, by definition, $M \ge M'$, and so $M = M'$.

Suppose that $\mathcal{W}_m^{(1)}$ and $\mathcal{W}_m^{(2)}$ are in $\mathcal{S}_{m,t}$ and also in one orbit $\mathcal{O}$ of $G(\mathcal{Q}_n)$. Since Π_{m-t-1} is the vertex of $\mathcal{W}_m^{(1)}$ and $\mathcal{W}_m^{(2)}$, every element $\mathfrak{T}$ of $G(\mathcal{Q}_n)$ mapping $\mathcal{W}_m^{(1)}$ to $\mathcal{W}_m^{(2)}$ fixes Π_{m-t-1} so is in $G_{m,t}$. Hence $\mathcal{W}_m^{(2)}$ is in the orbit $\mathcal{O}_{m,t}$ of $\mathcal{W}_m^{(1)}$ under $G_{m,t}$. Since $G(\mathcal{Q}_n)$ acts transitively on the set of all $(m-t-1)$-spaces on $\mathcal{Q}_n$, it follows that the number of orbits $\mathcal{O}$ of $\mathcal{S}$ under $G(\mathcal{Q}_n)$ is the same as the number M of orbits $\mathcal{O}_{m,t}$ of $\mathcal{S}_{m,t}$ under $G_{m,t}$. Hence the number of orbits of $\mathcal{S}$ under $G(\mathcal{Q}_n)$ is the number of orbits of $G(\mathcal{Q}_{n-2(m-t)})$ on the set of all sections $\mathcal{Q}_t$ of character v of $\mathcal{Q}_{n-2(m-t)}$.

When $n = 2m - t$, then $\mathcal{Q}_{n-2(m-t)} = \mathcal{Q}_t$ and so $\mathcal{S}$ has just one orbit under $G(\mathcal{Q}_n)$. When $n \ne 2m - t$, and so $n - 2(m - t) > t$, then the number of orbits of $G(\mathcal{Q}_{n-2(m-t)})$ on the set of all sections of character v was calculated in (1) and (2): one orbit when (i) n is odd or (ii) n is even and t is odd; two orbits when n is even and t is even. $\square$

1.7 The polarity reconsidered

Now the complete generalisations of Lemmas 1.22 and 1.30 are given, and sections of $\mathcal{Q}_n$ by subspaces Π and Π' which are polar under the polarity of $\mathcal{Q}_n$ are described.

Lemma 1.50. *Let $\mathcal{Q}_n$ have character w and projective index g, and let a section $\Pi_{m-t-1}\mathcal{Q}_t$ have character v and projective index f. Then, with $T = n + t - 2m$,*

(i) $T \ge v - w$;
(ii) $T + w - v$ *is even.*

Proof. From (1.9) and (1.11),

$$g = \tfrac{1}{2}(n - 3 + w), \quad f = m - \tfrac{1}{2}(t + 3 - v).$$

So $g - f = \tfrac{1}{2}(T + w - v)$ and (ii) follows. However, $g \ge f$ since $\Pi_{m-t-1}\mathcal{Q}_t$ lies on $\mathcal{Q}_n$; hence (i) is obtained. $\square$

Table 1.4. Polar sections

q	$\mathcal{Q}_n$	$\Pi_k \mathcal{V}$	$\Pi_k \mathcal{V}'$
All	$\mathcal{H}_n$	$\Pi_{m-t-1}\mathcal{H}_t$	$\Pi_{m-t-1}\mathcal{H}_{T-1}$
		$\Pi_{m-t-1}\mathcal{E}_t$	$\Pi_{m-t-1}\mathcal{E}_{T-1}$
		$\Pi_{m-t-1}\mathcal{P}_t$	$\Pi_{m-t-1}\mathcal{P}_{T-1}$
All	$\mathcal{E}_n$	$\Pi_{m-t-1}\mathcal{H}_t$	$\Pi_{m-t-1}\mathcal{E}_{T-1}$
		$\Pi_{m-t-1}\mathcal{E}_t$	$\Pi_{m-t-1}\mathcal{H}_{T-1}$
		$\Pi_{m-t-1}\mathcal{P}_t$	$\Pi_{m-t-1}\mathcal{P}_{T-1}$
Odd	$\mathcal{P}_n$	$\Pi_{m-t-1}\mathcal{H}_t$	$\Pi_{m-t-1}\mathcal{P}_{T-1}$
		$\Pi_{m-t-1}\mathcal{E}_t$	$\Pi_{m-t-1}\mathcal{P}_{T-1}$
		$\Pi_{m-t-1}\mathcal{P}_t$	$\begin{cases}\Pi_{m-t-1}\mathcal{H}_{T-1}\\ \Pi_{m-t-1}\mathcal{E}_{T-1}\end{cases}$

Theorem 1.51. *Let $\Pi = \Pi_m$ have polar space $\Pi' = \Pi_{n-m-1}$ with respect to $\mathcal{Q}_n$. Then the possibilities for $\Pi \cap \mathcal{Q}_n = \Pi_k \mathcal{V}$ and $\Pi' \cap \mathcal{Q}_n = \Pi_k \mathcal{V}'$ are listed in Table 1.4.*

Proof. By Corollary 1.20 and Lemma 1.30, $\Pi' \cap \mathcal{Q}_n$ has the same singular space Π_k as $\Pi \cap \mathcal{Q}_n$. So, if $\Pi' \cap \mathcal{Q}_n = \Pi_{m-t-1}\mathcal{Q}_s$, then

$$s = (n-m-1) - (m-t-1) - 1 = n+t-2m-1 = T-1.$$

Now it suffices to look at particular cases of each type of subspace using the standard equations, as, for a given trio m, t, v, there are at most two orbits by Lemma 1.48. $\square$

Corollary 1.52. *In the theorem, let $\mathcal{Q}_n$, $\mathcal{V}$, $\mathcal{V}'$ have respective characters w, v, v' and respective projective indices g, f, f'. Then*

(i) $v' = |2 - w - v|$ *unless* $w = v = 1$, *in which case* $v' = 0$ *or* 2;
(ii) $f + f' - g = k - 1 + \frac{1}{2}(v + v' - w)$.

Proof. (i) This follows from the theorem.
(ii) From (1.8) and (1.10),

$$w = 2g - n + 3,$$
$$v = 2f - k - m + 2,$$
$$v' = 2f' - k - (n-m-1) + 2 = 2f' - n + m - k + 3.$$

Elimination of m and n gives the formula. $\square$

To add something comparable to Theorem 1.51 for $\mathcal{P}_n$ with q even, the following result is available; it only repeats a particular case of Lemma 1.30.

Lemma 1.53. *Let q be even. Then, in the notation of Theorem* 1.51, *with tangent space replacing polar space, the result is as follows:*

Quadric	Section	Tangent section
$\mathcal{Q}_n$	$\Pi_k \mathcal{V}$	$\Pi_k \mathcal{V}'$
$\mathcal{P}_n$	$\Pi_m \mathcal{H}_{-1}$	$\Pi_m \mathcal{P}_{n-2m-2}$

The next result gives more information on the tangency properties when $\mathcal{Q}_n$ does not have a polarity.

Theorem 1.54. *With q even, let N be the nucleus of the parabolic quadric*

$$\mathcal{P}_n = \mathbf{V}(X_0^2 + X_1 X_2 + \cdots + X_{n-1} X_n).$$

(i) *Every section of $\mathcal{P}_n$ through N is parabolic.*
(ii) *There is a bijection between m-spaces Π_m through N of $\mathcal{P}_n$ and $(m-1)$-spaces Π_{m-1} in $\mathbf{u}_0$:*

$$\Pi_m \mapsto \Pi_{m-1} = \Pi_{m-1} \cap \mathbf{u}_0, \quad \Pi_{m-1} \mapsto \Pi_m = \mathbf{U}_0 \Pi_{m-1}.$$

Here, $\Pi_m \cap \mathcal{P}_n$ is a $\Pi_{m-t-1} \mathcal{P}_t$ with Π_m containing N if and only if, with $\mathcal{H}_{n-1} = \mathbf{u}_0 \cap \mathcal{P}_n$, the intersection $\Pi_{m-1} \cap \mathcal{H}_{n-1}$ is one of

$$\Pi_{m-t-1} \mathcal{H}_{t-1}, \quad \Pi_{m-t-1} \mathcal{E}_{t-1}, \quad \Pi_{m-t-2} \mathcal{P}_t.$$

Proof. If $\Pi_m \cap \mathcal{P}_n = \Pi_{m-t-1} \mathcal{P}_t$ and $\Pi_{m-1} = \mathbf{u}_0 \cap \Pi_m$, then Π_{m-1} meets $\mathcal{H}_{n-1}$ in a section $\Pi_{m-s-1} \mathcal{Q}_{s-1}$, which is reduced to canonical form by a projectivity μ of $PG(n, q)$ fixing $\mathbf{u}_0$ and $\mathbf{U}_0$. Hence μ has the matrix

$$\begin{bmatrix} 1 & Z \\ Z^* & M \end{bmatrix},$$

where $Z = (0, 0, \ldots, 0)$ and M is an $n \times n$ matrix with no further restriction. So the possible canonical forms for $\mathcal{Q}_{s-1}$ and correspondingly $\mathcal{P}_t$ are given in Table 1.5. This proves (ii) and so, *a fortiori*, (i). $\square$

1.8 Sections of non-singular quadrics

As in Section 1.6, let $\mathcal{S}(m, t, v; n, w)$ be the set of m-spaces Π_m such that $\Pi_m \cap \mathcal{Q}_n$ is of type $\Pi_{m-t-1} \mathcal{Q}_t$, where $\mathcal{Q}_n$ and $\mathcal{Q}_t$ have respective characters w and v. Also,

$$|\mathcal{S}(m, t, v; n, w)| = N(m, t, v; n, w) = N(\Pi_{m-t-1} \mathcal{Q}_t, \mathcal{Q}_n)$$

and this number will be calculated. As special cases, the formula gives

Table 1.5. Sections through the nucleus

Type	Form for $\mathcal{Q}_{s-1}$	$\Pi_{m-s-1}\mathcal{Q}_{s-1}$
1	$X_1X_2 + \cdots + X_{s-1}X_s$	$\Pi_{m-s-1}\mathcal{H}_{s-1}$
2	$f(X_1, X_2) + X_3X_4 + \cdots + X_{s-1}X_s$	$\Pi_{m-s-1}\mathcal{E}_{s-1}$
3	$X_1^2 + X_2X_3 + \cdots + X_{s-1}X_s$	$\Pi_{m-s-1}\mathcal{P}_{s-1}$

Type	Form for $\mathcal{P}_t$	$\Pi_{m-t-1}\mathcal{P}_t$
1	$X_0^2 + X_1X_2 + \cdots + X_{s-1}X_s$	$\Pi_{m-s-1}\mathcal{P}_s$
2	$X_0^2 + f(X_1, X_2) + X_3X_4 + \cdots + X_{s-1}X_s$	$\Pi_{m-s-1}\mathcal{P}_s$
3	$(X_0 + X_1)^2 + X_2X_3 + \cdots + X_{s-1}X_s$	$\Pi_{m-s}\mathcal{P}_{s-1}$

(1) the number of Π_m lying on $\mathcal{Q}_n$, and here $\Pi_m \cap \mathcal{Q}_n = \Pi_m\mathcal{H}_{-1}$;
(2) the number of points Π_0 not on $\mathcal{Q}_n$, and here $\Pi_m \cap \mathcal{Q}_n = \Pi_{-1}\mathcal{P}_0$.

The formula gives the size of the orbits when $G(\mathcal{Q}_n)$ operates on the lattice of subspaces of $\mathrm{PG}(n,q)$, apart from the case $w = v = 1$. This fact is contained in Theorem 1.49: when $(w, v) \neq (1, 1)$, the sections of $\mathcal{Q}_n$ for a triple (m, t, v) form a single orbit under $G(\mathcal{Q}_n)$; when $(w, v) = (1, 1)$, the sections for a given pair (m, t) form one or two orbits, and the size of these orbits will also be determined.

To obtain the general result, some special cases are first required. These are subsumed in the general result. In Section 1.5, the number

$$N(m; n, w) = N(m, -1, 2; n, w),$$

which is the number of m-spaces on $\mathcal{Q}_n$ of character w, was determined.

The next special cases required are the numbers of bisecants, tangent lines and skew lines to $\mathcal{Q}_n$; the total number of lines on $\mathcal{Q}_n$ has already been determined in Section 1.5.

Lemma 1.55.

(i)
$$N(\Pi_1, \mathcal{Q}_n) = [\tfrac{1}{2}(n - 1 - w), \tfrac{1}{2}(n + 1 - w)]_+$$
$$\times [\tfrac{1}{2}(n - 3 + w), \tfrac{1}{2}(n - 1 + w)]_- / [1, 2]_- ; \quad (1.36)$$

(ii)
$$N(\Pi_0\mathcal{P}_0, \mathcal{Q}_n) = \{(q^n - 1)/(q - 1) + (w - 1)q^{(n-1)/2}\}$$
$$\times \{q^{n-2} - (w - 1)q^{(n-3)/2}\} ; \quad (1.37)$$

(iii)
$$N(\mathcal{H}_1, \mathcal{Q}_n) = \tfrac{1}{2}q^{n-1}\{(q^n - 1)/(q - 1) + (w - 1)q^{(n-1)/2}\} ; \quad (1.38)$$

(iv)
$$N(\mathcal{E}_1, \mathcal{Q}_n) = \tfrac{1}{2}q^{n-1}(q^{\{n+(w-1)^2\}/2} - w^2 + w + 1)$$
$$\times (q^{\{n-(w-1)^2\}/2} + w^2 - 3w + 1)/(q + 1). \quad (1.39)$$

Proof. The number of lines on $\mathcal{Q}_n$ through a point P is $N(0; n-2, w)$, Lemma 1.22. Hence the number of tangents through P is

$$\theta(n-2) - N(0; n-2, w)$$

and the total number of tangents is

$$N(0; n, w)\{\theta(n-2) - N(0; n-2, w)\};$$

this gives (ii).

The number of bisecants through a point P of $\mathcal{Q}_n$ is

$$\theta(n-1) - \theta(n-2) = q^{n-1};$$

hence

$$N(\mathcal{H}_1, \mathcal{Q}_n) = N(0; n, w)\, q^{n-1}/2.$$

The total number of lines in $PG(n, q)$ is, from Theorem 3.1 of PGOFF2,

$$\phi(1; n, q) = [n, n+1]_- / [1, 2]_-. \tag{1.40}$$

Hence (iv) is obtained from the formula

$$N(\mathcal{E}_1, \mathcal{Q}_n) = \phi(1; n, q) - N(\Pi_1, \mathcal{Q}_n) - N(\Pi_0 \mathcal{P}_0, \mathcal{Q}_n) - N(\mathcal{H}_1, \mathcal{Q}_n). \qquad \square$$

Corollary 1.56.

(i)
$$N(1, 1, 0; n, 0) = N(\mathcal{E}_1, \mathcal{E}_n)$$
$$= \tfrac{1}{2} q^{n-1}(q^{(n+1)/2} + 1)(q^{(n-1)/2} + 1)/(q+1); \tag{1.41}$$

(ii)
$$N(1, 1, 0; n, 2) = N(\mathcal{E}_1, \mathcal{H}_n)$$
$$= \tfrac{1}{2} q^{n-1}(q^{(n+1)/2} - 1)(q^{(n-1)/2} - 1)/(q+1); \tag{1.42}$$

(iii)
$$N(1, 1, 0; n, 1) = N(\mathcal{E}_1, \mathcal{P}_n)$$
$$= \tfrac{1}{2} q^{n-1}(q^n - 1)/(q+1). \tag{1.43}$$

Theorem 1.57. *The number of sections $\Pi_{m-t-1}\mathcal{Q}_t$ of character v on $\mathcal{Q}_n$ of character w is*

$$N(m, t, v; n, w)$$
$$= q^{\{T[t+1+vw(2-v)(2-w)] - v(2-v)(w-1)^2\}/2}$$
$$\times [\tfrac{1}{2}\{T + v + (1 + 3v - 2v^2)w - v(2-v)w^2\}, \tfrac{1}{2}(n+1-w)]_+$$
$$\times [\tfrac{1}{2}\{T + 2 - v - (1 - 5v + 2v^2)w - v(2-v)w^2\}, \tfrac{1}{2}(n-1+w)]_-$$
$$\div \{[v(2-v), \tfrac{1}{2}(t+1-v)]_+ [1, \tfrac{1}{2}(t-1+v)]_- [1, m-t]_-\}, \tag{1.44}$$

where $T = n + t - 2m$.

Proof. Consider the spaces Π_d lying on $\mathcal{Q}_n$ and the spaces Π_m meeting $\mathcal{Q}_n$ in a quadric $\Pi_k \mathcal{Q}_t$ of projective index d and character v. Let N_0 be the number of such Π_m through a Π_d. Also, let $N'(\Pi_d, \Pi_k \mathcal{Q}_t)$ be the number of Π_d on $\Pi_k \mathcal{Q}_t$ containing the vertex Π_k. Then, counting pairs (Π_d, Π_m) gives

$$N(\Pi_k \mathcal{Q}_t, \mathcal{Q}_n) = N_0 \, N(\Pi_d, \mathcal{Q}_n) / N'(\Pi_d, \Pi_k \mathcal{Q}_t) \,. \tag{1.45}$$

Here $m = k + t + 1$. From (1.10),

$$v = 2d - (m - t - 1) - m + 2 \,,$$

whence

$$d = m - \tfrac{1}{2}(t - v + 3) \,. \tag{1.46}$$

To find N_0, consider all $\mathcal{W}_m = \Pi_k \mathcal{Q}_t$ on $\mathcal{Q}_n$ through a particular Π_d. Let Π'_{n-d-1} be the tangent space of Π_d with respect to $\mathcal{Q}_n$. Then, by Lemma 1.30,

$$\Pi'_{n-d-1} \cap \mathcal{Q}_n = \Pi_d \mathcal{Q}_{n-2d-2},$$

which has the same character w and projective index g as $\mathcal{Q}_n$.

The required spaces Π_m must satisfy the following properties:

(a) $\mathcal{W}_m$ has projective index at most d;
(b) Π_m touches $\mathcal{Q}_n$ along Π_k.

For (a), it is necessary and sufficient that Π_m contains none of the $(d + 1)$-spaces through Π_d on $\Pi_d \mathcal{Q}_{n-2d-2}$. If (a) is assumed, then it is necessary and sufficient for (b) that $\Pi_m \Pi'_{n-d-1} = \Pi'_{n-k-1}$, using Lemma 1.21. This is equivalent to

$$\Pi_m \cap \Pi'_{n-d-1} = \Pi_r,$$

where

$$r = m + (n - d - 1) - (n - k - 1) = m - d + k \,. \tag{1.47}$$

The space Π_r contains Π_d and, to satisfy (a), $\Pi_r \cap \Pi_d \mathcal{Q}_{n-2d-2} = \Pi_d$. If $r > d$, so $\Pi_r \cap \mathcal{Q}_{n-2d-2}$ is either $\mathcal{P}_0$ or $\mathcal{E}_1$; hence

$$d \le r \le d + 2. \tag{1.48}$$

Three cases are distinguished according as $\mathcal{W}_m$ is parabolic, hyperbolic or elliptic; that is $v = 1, 2$ or 0, where $v = 2d - k - m + 2$, as in (1.10).

(1) $\mathcal{W}_m$ *parabolic*

Since $2 + 2d - k - m = 1$ and $r = m - d + k$, so $r = d + 1$. Therefore, by (a), Π_r is any one of the Π_{d+1} which lie in Π'_{n-d-1} and contain Π_d without being on $\Pi_d \mathcal{Q}_{n-2d-2}$. So the number of Π_r is, in the notation of Section 3.1 of PGOFF2,

$$\chi(d, d+1; n - d - 1, q) - N(\Pi_0, \mathcal{Q}_{n-2d-2})$$
$$= \theta(n - 2d - 2) - N(\Pi_0, \mathcal{Q}_{n-2d-2}) = N_1 \,. \tag{1.49}$$

The spaces Π_m are those m-spaces containing such a $\Pi_r = \Pi_{d+1}$ with the condition that $\Pi_m \cap \Pi'_{n-d-1} = \Pi_r$. If Π_r is fixed, the number of these Π_m is

$$
\begin{aligned}
\psi_{12}(d+1, n-d-1, m; n, q) \\
= q^{(m-d-1)(n-2d-2)}[2d+3-m, d+1]_-/[1, m-d-1]_- \\
= N_2,
\end{aligned}
\tag{1.50}
$$

by Theorem 3.3 of PGOFF2. Thus

$$
\begin{aligned}
N_0 &= N_1 \, N_2 \\
&= q^{\{T(t+1)-(w-1)^2\}/2}(q^{\{T+(w-1)^2\}/2} - w + 1 \\
&\qquad \times [m-t+1, m - \tfrac{1}{2}t]_-/[1, \tfrac{1}{2}t]_- \, ,
\end{aligned}
\tag{1.51}
$$

where N_1 has been evaluated using (1.18) and d has been eliminated by (1.46). Here $n - 2d - 2 = n - 2m + t - v + 3 - 2 = T$.

(2) $\mathcal{W}_m$ hyperbolic

Since $2 + 2d - k - m = 2$ and $r = m - d + k$, so $r = d$ and $\Pi_r = \Pi_d$. Thus the spaces Π_m are those m-spaces such that $\Pi_d = \Pi_m \cap \Pi'_{n-d-1}$. By Theorem 3.3 of PGOFF2, their number is

$$
\begin{aligned}
\psi_{12}(d, n-d-1, m; n, q) \\
= q^{(m-d)(n-2d-1)}[2d+2-m, d+1]_-/[1, m-d]_- \\
= q^{T(t+1)/2}[m-t+1, m - \tfrac{1}{2}(t-1)]_-/[1, \tfrac{1}{2}(t+1)]_- \\
= N_0,
\end{aligned}
\tag{1.52}
$$

using (1.46) with $v = 2$.

(3) $\mathcal{W}_m$ elliptic

Since $2 + 2d - k - m = 0$ and $r = m - d + k$, so $r = d + 2$. Then, from (a), Π_r is one of the spaces Π_{d+2} in Π'_{n-d-1} through the vertex Π_d of $\Pi_d \mathcal{Q}_{n-2d-2}$ but not containing any point of the base $\mathcal{Q}_{n-2d-2}$. So Π_{d+2} is the join of Π_d and a line external to $\mathcal{Q}_{n-2d-2}$. Thus the number of Π_r is the number of lines of $PG(n - 2d - 2, q)$ meeting $\mathcal{Q}_{n-2d-2}$ in some $\mathcal{E}_1$, that is, $N(\mathcal{E}_1, \mathcal{Q}_{n-2d-2})$. Now, (1.46) with $v = 0$ gives $d = m - \tfrac{1}{2}(t+3)$, whence

$$
n - 2d - 2 = n - 2m + (t+3) - 2 = T + 1.
$$

So, from (1.39),

$$
\begin{aligned}
N(\mathcal{E}_1, \mathcal{Q}_{n-2d-2}) &= \tfrac{1}{2}q^T(q^{\{T+1+(w-1)^2\}/2} - w^2 + w + 1) \\
&\qquad \times (q^{\{T+1-(w-1)^2\}/2} + w^2 - 3w + 1)/(q+1) \\
&= N_1 \, .
\end{aligned}
\tag{1.53}
$$

The spaces Π_m are those m-spaces meeting Π'_{n-d-1} in such a $\Pi_r = \Pi_{d+2}$. If Π_r is fixed, the number of Π_m is, by Theorem 3.3 of PGOFF2,

$$
\begin{aligned}
\psi_{12}(d+2, n-d-1, m; n, q) \\
&= q^{(m-d-2)(n-2d-3)}[2d+4-m, d+1]_-/[1, m-d-2]_- \\
&= q^{T(t-1)/2}[m-t+1, m-\tfrac{1}{2}(t+1)]_-/[1, \tfrac{1}{2}(t-1)]_- \\
&= N_2.
\end{aligned}
\tag{1.54}
$$

Thus

$$
\begin{aligned}
N_0 &= N_1 N_2 \\
&= q^{T(t+1)/2}[m-t+1, m-\tfrac{1}{2}(t+1)]_-\, N_3 \\
&\quad \div \{(q+1)[1, \tfrac{1}{2}(t-1)]_-\},
\end{aligned}
\tag{1.55}
$$

where

$$
N_3 = \begin{cases}
[\tfrac{1}{2}T, \tfrac{1}{2}T+1]_+ & \text{for } w = 0, \\
[\tfrac{1}{2}T, \tfrac{1}{2}T+1]_- & \text{for } w = 2, \\
[T, T+1]_- & \text{for } w = 1.
\end{cases}
$$

This completes the calculation of N_0 in the three cases $v = 1, 2, 0$.

To apply the formula (1.45), the numbers $N(\Pi_d, \mathcal{Q}_n)$ and $N'(\Pi_d, \Pi_k \mathcal{Q}_t)$ are required. By (1.46) and the definition,

$$
\begin{aligned}
N(\Pi_d, \mathcal{Q}_n) &= N(m - \tfrac{1}{2}(t-v+3); n, w), \\
N'(\Pi_d, \Pi_k \mathcal{Q}_t) &= N(\Pi_{d-k-1}, \mathcal{Q}_t) = N(\tfrac{1}{2}(t+v-3); t, v).
\end{aligned}
$$

Thus, using (1.13) in Theorem 1.41,

$$
\begin{aligned}
N(\Pi_d, \mathcal{Q}_n)/N'(\Pi_d, \Pi_k \mathcal{Q}_t) \\
&= [\tfrac{1}{2}(T+4-w-v), \tfrac{1}{2}(n+1-w)]_+ [\tfrac{1}{2}(T+2+w-v), \tfrac{1}{2}(n-1+w)]_- \\
&\quad \times [1, \tfrac{1}{2}(t-1+v)]_- \\
&\quad \div \{[2-v, \tfrac{1}{2}(t+1-v)]_+ [1, \tfrac{1}{2}(t-1+v)]_- [1, m-\tfrac{1}{2}(t+1-v)]_-\} \\
&= N_4.
\end{aligned}
\tag{1.56}
$$

So (1.45) becomes

$$
N(\Pi_k \mathcal{Q}_t, \mathcal{Q}_n) = N(m, t, v; n, w) = N_0 N_4.
$$

Thus, from (1.51) and (1.56) with $v = 1$,

$$
\begin{aligned}
N(m, t, 1; n, w) &= q^{\{T(t+1+2w-w^2)-(w-1)^2\}/2} \\
&\quad \times [\tfrac{1}{2}(T+1+2w-w^2), \tfrac{1}{2}(n+1-w)]_+ \\
&\quad \times [\tfrac{1}{2}(T+1+2w-w^2), \tfrac{1}{2}(n-1+w)]_- \\
&\quad \div \{[1, \tfrac{1}{2}t]_+ [1, \tfrac{1}{2}t]_- [1, m-t]_-\}.
\end{aligned}
\tag{1.57}
$$

From (1.52) and (1.56) with $v = 2$,

$$N(m, t, 2; n, w) = q^{T(t+1)/2}$$
$$\times [\tfrac{1}{2}(T + 2 - w), \tfrac{1}{2}(n + 1 - w)]_+ [\tfrac{1}{2}(T + w), \tfrac{1}{2}(n - 1 + w)]_-$$
$$\div \{[0, \tfrac{1}{2}(t - 1)]_+ [1, \tfrac{1}{2}(t + 1)]_- [1, m - t]_-\}. \tag{1.58}$$

From (1.55) and (1.56) with $v = 0$,

$$N(m, t, 0; n, w) = q^{T(t+1)/2}$$
$$\times [\tfrac{1}{2}(T + w), \tfrac{1}{2}(n + 1 - w)]_+ [\tfrac{1}{2}(T + 2 - w), \tfrac{1}{2}(n - 1 + w)]_-$$
$$\div \{[0, \tfrac{1}{2}(t + 1)]_+ [1, \tfrac{1}{2}(t - 1)]_- [1, m - t]_-\}. \tag{1.59}$$

The substitution of $v = 1, 2, 0$ in the 'big formula' (1.44) gives (1.57), (1.58), (1.59) respectively. Thus (1.44) is established. $\square$

It may be noted that all the special cases of (1.44) required for its proof are immediately retrievable from the general formula.

Example 1.58. The number of conics on a quadric $\mathcal{P}_n$, n even:

$$v = w = 1, \quad m = t = 2, \quad T = n - 2,$$
$$N(\mathcal{P}_2, \mathcal{P}_n) = q^{2(n-2)} [\tfrac{1}{2}n, \tfrac{1}{2}n]_+ [\tfrac{1}{2}n, \tfrac{1}{2}n]_-$$
$$\div \{[1, 1]_+ [1, 1]_- [1, 0]_-\}$$
$$= q^{2(n-2)} (q^n - 1)/(q^2 - 1).$$

Now consider, under what conditions on the parameters m, l, v, n, w and the invariant $T = n + t - 2m$, the quadric $\mathcal{Q}_n$ of character w has a section $\Pi_{m-t-1}\mathcal{Q}_t$ of character v. First, the properties of the parameters that are contained within their definition are listed.

Property 1.59. (a) $n - w$ is odd;
(b) $t - v$ is odd;
(c) $T + w - v$ is even;
(d) $n > m \geq 0$;
(e) $m \geq t \geq 1 - v$;
(f) $n \geq (w - 1)^2$.

In Lemma 1.50, it was also shown that $T \geq v - w$.

Theorem 1.60. *Subject to the conditions* (a)–(f) *of Property* 1.59, *a quadric* $\mathcal{Q}_n$ *of character* w *has a section* $\Pi_{m-t-1}\mathcal{Q}_t$ *of character* v *if and only if*

$$T \geq |w - v|.$$

Table 1.6. Values of $f(v, w)$

	w	0	1	2
v				
0		-2	-1	0
1		-1	-2	-1
2		0	-1	-2

Proof. From Notation 1.32, the integer $[r, s]_- = 0$ if and only if $r = 0$. So, from (1.44), the number $N(m, t, v; n, w) > 0$ if and only if

$$T + 2 - v - (1 - 5v + 2v^2)w - v(2 - v)w^2 > 0 ;$$

that is, $T > f(v, w)$ where $f(v, w)$ is given in Table 1.6.

Since $T + w - v$ is even, the minimum value $g(v, w)$ of T is given as follows:

$$g(v, w) = \begin{cases} f(v, w) + 1 & \text{if } f(v, w) + 1 - (w - v) \text{ is even;} \\ f(v, w) + 2 & \text{if } f(v, w) + 1 - (w - v) \text{ is odd.} \end{cases}$$

Hence $g(v, w)$ is given by Table 1.7. Thus $g(v, w) = |w - v|$. □

Table 1.7. Values of $g(v, w)$

	w	0	1	2
v				
0		0	1	2
1		1	0	1
2		2	1	0

Corollary 1.61. *The quadric $\mathcal{Q}_n$ of character w has a section $\Pi_{m-t-1}\mathcal{Q}_t$ of character v if and only if*

(a) $n - w$ *is odd and* $n \geq (w - 1)^2$;
(b) $t - v$ *is odd*;
(c) $n > m \geq t \geq \max(2m - n + |w - v|, 1 - v)$.

Corollary 1.62. *The quadric $\mathcal{Q}_n$ of character w has a section $\Pi_k\mathcal{Q}_t$ of character v with $k = m - t - 1$ if and only if*

(a) $n - w$ *is odd and* $n \geq (w - 1)^2$;
(b) $t - v$ *is odd*;
(c) $1 - v \leq t < n - k - 1$;
(d) $-2 \leq 2k \leq n - t - |w - v| - 2$.

Theorem 1.63. *For a given* $\mathcal{Q}_n = \mathcal{H}_n, \mathcal{E}_n, \mathcal{P}_n$ *the number of projectively distinct pairs* $(\Pi_m, \Pi_{m-t-1}\mathcal{Q}_t)$ *where* $\Pi_m \cap \mathcal{Q}_n = \Pi_{m-t-1}\mathcal{Q}_t$ *is as follows:*

Type of section	$\mathcal{H}_n$	$\mathcal{E}_n$	$\mathcal{P}_n$
Hyperbolic	$\frac{1}{8}(n^2-1)+n$	$\frac{1}{8}(n-1)(n+5)$	$\frac{1}{8}n(n+6)$
Elliptic	$\frac{1}{8}(n^2-1)$	$\frac{1}{8}(n-1)(n+5)$	$\frac{1}{8}n(n+2)$
Parabolic	$\frac{1}{8}(n+1)(n+3)$	$\frac{1}{8}(n+1)(n+3)$	$\frac{1}{8}n(n+6)$
Total	$\frac{1}{8}(3n+1)(n+1)+n$	$\frac{1}{8}(3n+7)(n-1)+n$	$\frac{1}{8}n(3n+14)$

Proof. For each m such that $0 \le m \le n-1$, Corollary 1.61 permits a count of the values of t for which a section $\Pi_{m-t-1}\mathcal{Q}_t$ of character v exists. □

By way of example, the different sections by an m-space for the three quadrics $\mathcal{E}_5$, $\mathcal{P}_6$, $\mathcal{H}_7$ are listed in Table 1.8.

Since the sections of non-singular quadrics have been determined, it is possible to say precisely what are the sections of a singular quadric.

Theorem 1.64. *If, for a fixed t, the non-singular quadric $\mathcal{Q}_n$ has a section $\Pi_i \mathcal{Q}_t$ of character v, then $\Pi_k \mathcal{Q}_n$ has a section $\Pi_j \mathcal{Q}_t$ of character v for all j in the range $i \le j \le i+k+1$.*

Proof. A section of $\Pi_k \mathcal{Q}_n$ is the join of a section Π_s of Π_k to a section $\Pi_i \mathcal{Q}_t$ of $\mathcal{Q}_n$; this join is $\Pi_{s+1+i}\mathcal{Q}_t$, where s may vary from -1 to k. □

1.9 Parabolic sections of parabolic quadrics

Part of Theorem 1.49 is that, in the operation of $G(\mathcal{Q}_n)$ on the set $\mathcal{S}(m,t,v;n,w)$ of m-spaces Π_m meeting $\mathcal{Q}_n$ of character w in a section $\Pi_{m-t-1}\mathcal{Q}_t$ of character v, the only case that two orbits may occur is when $v = w = 1$. The geometrical explanation of this phenomenon is different for q odd and q even. From (1.44),

$$N(m,t,1;n,1) = q^{T(t+2)/2}[\tfrac{1}{2}(T+2),\tfrac{1}{2}n]_+[\tfrac{1}{2}(T+2),\tfrac{1}{2}n]_-$$
$$\div\{[1,\tfrac{1}{2}t]_+[1,\tfrac{1}{2}t]_-[1,m-t]_-\}, \tag{1.60}$$

where, as before,

$$T = n+t-2m.$$

This is the number of Π_m such that

$$\Pi_m \cap \mathcal{P}_n = \Pi_{m-t-1}\mathcal{P}_t. \tag{1.61}$$

In this section only such Π_m are considered.

Table 1.8. Sections in dimensions $5, 6, 7$

	Hyperbolic	Elliptic	Parabolic
$\mathcal{E}_5$			
$m = 4$		$\Pi_0 \mathcal{E}_3$	$\mathcal{P}_4$
$m = 3$	$\mathcal{H}_3$	$\mathcal{E}_3, \Pi_1 \mathcal{E}_1$	$\Pi_0 \mathcal{P}_2$
$m = 2$	$\Pi_0 \mathcal{H}_1$	$\Pi_0 \mathcal{E}_1$	$\mathcal{P}_2, \Pi_1 \mathcal{P}_0$
$m = 1$	$\mathcal{H}_1, \Pi_1$	$\mathcal{E}_1$	$\Pi_0 \mathcal{P}_0$
$m = 0$	Π_0		$\mathcal{P}_0$
$\mathcal{P}_6$			
$m = 5$	$\mathcal{H}_5$	$\mathcal{E}_5$	$\Pi_0 \mathcal{P}_4$
$m = 4$	$\Pi_0 \mathcal{H}_3$	$\Pi_0 \mathcal{E}_3$	$\mathcal{P}_4, \Pi_1 \mathcal{P}_2$
$m = 3$	$\mathcal{H}_3, \Pi_1 \mathcal{H}_1$	$\mathcal{E}_3, \Pi_1 \mathcal{E}_1$	$\Pi_0 \mathcal{P}_2, \Pi_2 \mathcal{P}_0$
$m = 2$	$\Pi_0 \mathcal{H}_1, \Pi_2$	$\Pi_0 \mathcal{E}_1$	$\mathcal{P}_2, \Pi_1 \mathcal{P}_0$
$m = 1$	$\mathcal{H}_1, \Pi_1$	$\mathcal{E}_1$	$\Pi_0 \mathcal{P}_0$
$m = 0$	Π_0		$\mathcal{P}_0$
$\mathcal{H}_7$			
$m = 6$	$\Pi_0 \mathcal{H}_5$		$\mathcal{P}_6$
$m = 5$	$\mathcal{H}_5, \Pi_1 \mathcal{H}_3$	$\mathcal{E}_5$	$\Pi_0 \mathcal{P}_4$
$m = 4$	$\Pi_0 \mathcal{H}_3, \Pi_2 \mathcal{H}_1$	$\Pi_0 \mathcal{E}_3$	$\mathcal{P}_4, \Pi_1 \mathcal{P}_2$
$m = 3$	$\mathcal{H}_3, \Pi_1 \mathcal{H}_1, \Pi_3$	$\mathcal{E}_3, \Pi_1 \mathcal{E}_1$	$\Pi_0 \mathcal{P}_2, \Pi_2 \mathcal{P}_0$
$m = 2$	$\Pi_0 \mathcal{H}_1, \Pi_2$	$\Pi_0 \mathcal{E}_1$	$\mathcal{P}_2, \Pi_1 \mathcal{P}_0$
$m = 1$	$\mathcal{H}_1, \Pi_1$	$\mathcal{E}_1$	$\Pi_0 \mathcal{P}_0$
$m = 0$	Π_0		$\mathcal{P}_0$

For q odd, as in Theorem 1.51, the polar Π'_{n-m-1} of Π_m meets $\mathcal{P}_n$ in either an elliptic or a hyperbolic section. If $\Pi'_{n-m-1} \cap \mathcal{P}_n = \Pi_{m-t-1} \mathcal{E}_{T-1}$, then Π_m is *internal*; if $\Pi'_{n-m-1} \cap \mathcal{P}_n = \Pi_{m-t-1} \mathcal{H}_{T-1}$, then Π_m is *external*. This conforms with the notion of internal and external points of a conic as in Section 8.2 of PGOFF2. Accordingly, write $N_-(m,t,1;n,1)$ for the number of internal Π_m and $N_+(m,t,1;n,1)$ for the number of external Π_m such that (1.61) holds. Hence

$$N_-(m,t,1;n,1) + N_+(m,t,1;n,1) = N(m,t,1;n,1). \qquad (1.62)$$

Theorem 1.65.

(i)
$$N_-(m,t,1;n,1) = q^{T(t+1)/2}[\tfrac{1}{2}(T+2), \tfrac{1}{2}n]_+ [\tfrac{1}{2}T, \tfrac{1}{2}n]_-$$
$$\div \{[0, \tfrac{1}{2}t]_+ [1, \tfrac{1}{2}t]_- [1, m-t]_-\}. \qquad (1.63)$$

(ii)
$$N_+(m,t,1;n,1) = q^{T(t+1)/2}[\tfrac{1}{2}T, \tfrac{1}{2}n]_+ [\tfrac{1}{2}(T+2), \tfrac{1}{2}n]_-$$
$$\div \{[0, \tfrac{1}{2}t]_+ [1, \tfrac{1}{2}t]_- [1, m-t]_-\}. \qquad (1.64)$$

Proof. By definition,

$$N_-(m, t, 1; n, 1) = N(n - m - 1, T - 1, 0; n, 1),$$
$$N_+(m, t, 1; n, 1) = N(n - m - 1, T - 1, 2; n, 1).$$

Application of (1.44) and some manipulation give the required answers, for which (1.60) and (1.62) provide a check. □

Corollary 1.66. *For q odd, the set $\mathcal{S}(m, t, 1; n, 1)$ has one or two orbits under $G(\mathcal{P}_n)$ according as $T = 0$ or $T > 0$. In the former case, the sections of the orbit are all external and have polar sections $\Pi'_{n-m-1} \cap \mathcal{P}_n = \Pi_r$, where $r = n - m - 1 = m - t - 1$.*

Proof. This follows from Theorem 1.49, (1.63) and (1.64). When $T = 0$,

$$\Pi'_{n-m-1} \cap \mathcal{P}_n = \Pi_{m-t-1} \mathcal{H}_{T-1} = \Pi_r.$$ □

For q even, a space Π_m such that (1.61) holds either does or does not contain the nucleus N of $\mathcal{P}_n$. If Π_m does contain N, it is called *nuclear*; if Π_m does not contain N, it is called *non-nuclear*. Accordingly, write $N_0(m, t, 1; n, 1)$ for the number of nuclear Π_m and $N_1(m, t, 1; n, 1)$ for the number of non-nuclear Π_m, both such that $\Pi_m \cap \mathcal{P}_n = \Pi_{m-t-1} \mathcal{P}_t$. So

$$N_0(m, t, 1; n, 1) + N_1(m, t, 1; n, 1) = N(m, t, 1; n, 1). \tag{1.65}$$

Theorem 1.67.

(i)
$$N_0(m, t, 1; n, 1) = q^{tT/2} [\tfrac{1}{2}(T + 2), \tfrac{1}{2}n]_+ [\tfrac{1}{2}(T + 2), \tfrac{1}{2}n]_- \\ \div \{[1, \tfrac{1}{2}t]_+ [1, \tfrac{1}{2}t]_- [1, m - t]_- \}. \tag{1.66}$$

(ii)
$$N_1(m, t, 1; n, 1) = q^{tT/2} [\tfrac{1}{2}T, \tfrac{1}{2}n]_+ [\tfrac{1}{2}T, \tfrac{1}{2}n]_- \\ \div \{[1, \tfrac{1}{2}t]_+ [1, \tfrac{1}{2}t]_- [1, m - t]_- \}. \tag{1.67}$$

Proof. From Theorem 1.54 (ii),

$$N_0(m, t, 1; n, 1) = N(\Pi_{m-t-1} \mathcal{H}_{t-1}, \mathcal{H}_{n-1}) + N(\Pi_{m-t-1} \mathcal{E}_{t-1}, \mathcal{H}_{n-1}) \\ + N(\Pi_{m-t-2} \mathcal{P}_t, \mathcal{H}_{n-1}) \\ = N(m - 1, t - 1, 2; n - 1, 2) + N(m - 1, t - 1, 0; n - 1, 2) \\ + N(m - 1, t, 1; n - 1, 2).$$

Applying (1.44) gives (1.66); then (1.65) gives (1.67). □

Corollary 1.68. *For q even, the set $\mathcal{S}(m, t, 1; n, 1)$ has one or two orbits under $G(\mathcal{P}_n)$ according as $T = 0$ or $T > 0$. In the former case, the sections of the orbit are all nuclear and each is the section of $\mathcal{P}_n$ by the tangent space of a Π_{m-t-1} lying on $\mathcal{P}_n$.*

Proof. This follows from Theorem 1.49, (1.66) and (1.67). In the case that $T = 0$, then $n - 2(m - t - 1) - 2 = t$. So, by Lemma 1.53, the tangent space of a Π_{m-t-1} on $\mathcal{P}_n$ meets $\mathcal{P}_n$ in a section $\Pi_{m-t-1}\mathcal{P}_t$. $\qquad\qquad\square$

The corollaries to Theorems 1.65 and 1.67 for q odd and even can be combined as follows.

Theorem 1.69. *The set $\mathcal{S}(m, t, 1; n, 1)$ has one or two orbits under $G(\mathcal{P}_n)$ according as $T = 0$ or $T > 0$. In the former case, each element of the orbit is the section of $\mathcal{P}_n$ by the tangent space at a Π_{m-t-1} lying on $\mathcal{P}_n$.*

Example 1.70. *Orbits of parabolic sections of $\mathcal{P}_4$.* The five types of parabolic section of $\mathcal{P}_4$ are as follows:

(a) $\mathcal{P}_0$, $m = 0$, $T = 4$, a point off $\mathcal{P}_4$;
(b) $\Pi_0\mathcal{P}_0$, $m = 1$, $T = 2$, a tangent line meeting $\mathcal{P}_4$ in a point;
(c) $\Pi_1\mathcal{P}_0$, $m = 2$, $T = 0$, a plane meeting $\mathcal{P}_4$ in a line;
(d) $\mathcal{P}_2$, $m = 2$, $T = 2$, a plane meeting $\mathcal{P}_4$ in a conic;
(e) $\Pi_0\mathcal{P}_2$, $m = 3$, $T = 0$, a tangent solid meeting $\mathcal{P}_4$ in a cone.

In both cases (c) and (e), for q odd or even, there is a single orbit of $(q + 1)(q^2 + 1)$ elements. In case (a), for q odd, there are $\frac{1}{2}q^2(q^2 - 1)$ internal points and $\frac{1}{2}q^2(q^2 + 1)$ external points; for q even, there is the nucleus and $q^4 - 1$ non-nuclear points. In case (b), for q odd, there are $\frac{1}{2}q(q^4 - 1)$ internal tangents and $\frac{1}{2}q(q + 1)^2(q^2 + 1)$ external tangents; for q even, there are $(q + 1)(q^2 + 1)$ nuclear tangents and $(q + 1)(q^4 - 1)$ non-nuclear tangents. In case (d), for q odd, there are $\frac{1}{2}q^3(q - 1)(q^2 + 1)$ internal conics and $\frac{1}{2}q^3(q+1)(q^2+1)$ external conics; for q even, there are $q^2(q^2+1)$ nuclear conics and $q^2(q^4 - 1)$ non-nuclear conics.

1.10 The characterisation of quadrics

In this section, non-singular quadrics in $\Sigma = \mathrm{PG}(n, q)$ are characterised purely in terms of their intersections with lines of Σ. The characterisation also applies to infinite fields with only a slight rewording. First, a number of definitions are required.

Definition 1.71. (1) A set $\mathcal{K}$ in Σ is of *type* $(r_1, r_2, \ldots, r_s)$ if

$$|l \cap \mathcal{K}| \in \{r_1, r_2, \ldots, r_s\}$$

for all lines l.
(2) Let $\mathcal{K}$ be a set of type $(0, 1, 2, q + 1)$. A line meeting $\mathcal{K}$ in i points is an *i-secant*. The alternative terms for a 0-secant, 1-secant, 2-secant, and $(q + 1)$-secant are *external line*, *unisecant* or *tangent*, *bisecant*, and *line on $\mathcal{K}$* or *line of $\mathcal{K}$*. Some authors use 'tangent' to mean 1-secant or $(q+1)$-secant, but this usage is avoided here. However, in the context of this section, it is convenient to have a single term for this idea. So a *B-line* is defined to be a 1-secant or a $(q + 1)$-secant. ,

(3) The set $\mathcal{K}$ is *quadratic* if
  (a) $\mathcal{K}$ is of type $(0, 1, 2, q + 1)$;
  (b) for each P in $\mathcal{K}$, the union of B-lines through P together with P form the
    tangent space $T_P = T_P(\mathcal{K})$, which is either a hyperplane or Σ itself.
   In (b), if P is not specifically included in $T_P(\mathcal{K})$, there is a difficulty when $n = 1$
   and $\mathcal{K}$ consists of two points.
(4) The point P of the quadratic set $\mathcal{K}$ is *singular* if $T_P(\mathcal{K}) = \Sigma$; in other words,
   there is no bisecant through P. If $\mathcal{K}$ has a singular point it is *singular*.
(5) If a quadratic set does not contain a line, it is an *ovoid*.
(6) A *perspectivity* of Σ is a projectivity fixing all lines through a certain point P_0,
   the *centre*. A quadratic set is *perspective* if there is a non-identity perspectivity
   with centre Q fixing $\mathcal{K}$ for every Q in Σ not in $\mathcal{K} \cup \bigcap T_P$. Every non-singular
   quadric in Σ is perspective, by Lemma 1.45.

Lemma 1.72. *If $\mathcal{K}$ is a quadratic set and Π_s is a subspace of Σ, then $\mathcal{K}' = \mathcal{K} \cap \Pi_s$ is a quadratic set in Π_s for which $T_P(\mathcal{K}') = T_P(\mathcal{K}) \cap \Pi_s$, where P is any point of $\mathcal{K}'$.*

Proof. If l is a line of Σ, then it meets Π_s in 0, 1 or $q + 1$ points. Hence the possibilities for $|(l \cap \Pi_s) \cap \mathcal{K}'|$ are given in Table 1.9.

Table 1.9. Intersection numbers for quadratic sets

$\lvert l \cap \Pi_s \rvert$	$\lvert l \cap \mathcal{K} \rvert$	0	1	2	$q+1$
0		0	0	0	0
1		0	0, 1	0, 1	1
$q+1$		0	1	2	$q+1$

So $\mathcal{K}'$ is a quadratic set. The other part follows similarly.   □

Corollary 1.73. *If Π is a hyperplane and $\mathcal{K}$ is non-singular, then $\Pi \cap \mathcal{K}$ has a singular point P if and only if $\Pi = T_P(\mathcal{K})$.*

Theorem 1.74. *In $\Sigma = \mathrm{PG}(n, q)$, $n \geq 2$, a set $\mathcal{K}$ is a quadratic set if and only if each plane section is a quadratic set.*

Proof. One implication is included in Lemma 1.72. Suppose therefore that every plane section of $\mathcal{K}$ is quadratic. Let l be a line and π a plane containing l. Since $\mathcal{K} \cap l = (\mathcal{K} \cap \pi) \cap l$, it follows that $\mathcal{K}$ is of type $(0, 1, 2, q + 1)$. Further, l is a B-line of $\mathcal{K}$ if and only if l is a B-line of $\mathcal{K} \cap \pi$ for every plane π containing l.

For each $P \in \mathcal{K}$, let the union of the B-lines through P be T_P. If l_1 and l_2 are two of these B-lines, let $\pi = l_1 l_2$; then $\pi \cap \mathcal{K}$ is a quadratic set whose tangent hyperplane at P is π. So π is contained in T_P, and T_P is a subspace. As each plane containing P has a line in T_P, so T_P is a hyperplane or the whole of Σ.   □

Some properties of singular quadratic sets are now developed.

Theorem 1.75. *The set of singular points of a quadratic set is a subspace.*

Proof. Let P, Q be distinct singular points of the quadratic set $\mathcal{K}$ and let R be any point of the line PQ. As PQ is a B-line at P, so $PQ \subset \mathcal{K}$ and $R \in \mathcal{K}$. Let l be any line through R. It must be shown that l is a B-line. So take $R \neq P$ and $l \neq PQ$. If l contains a point S in $\mathcal{K}$ with $S \neq R$, the tangent hyperplane T_S contains P and Q since PS and QS are lines of $\mathcal{K}$. So T_S contains the line PQ and the point R. Hence $RS = l$ is a B-line and lies in $\mathcal{K}$. So R is singular. $\qquad\square$

The next result shows that the theory of quadratic sets is entirely dependent on the theory of non-singular ones. The structure is similar to that of quadrics.

Theorem 1.76. *If $\mathcal{K}$ is a quadratic set, then $\mathcal{K}$ is a cone $\Pi_s \mathcal{K}'$, where Π_s is the subspace of singular points of $\mathcal{K}$ and $\mathcal{K}'$ is a non-singular quadratic set in a subspace Π_{n-s-1} disjoint to Π_s.*

Proof. Let Π_{n-s-1} be any subspace disjoint from Π_s and let $\mathcal{K}' = \mathcal{K} \cap \Pi_{n-s-1}$. If $\mathcal{K}'$ has a singular point P, then $\Pi_{n-s-1} \subset T_P(\mathcal{K})$ by Lemma 1.72. As Π_s also lies in $T_P(\mathcal{K})$ so $T_P(\mathcal{K}) = \Sigma$ and $P \in \Pi_s$, a contradiction. If $Q \in \mathcal{K} \backslash (\Pi_s \cup \Pi_{n-s-1})$, then $Q\Pi_s \cap \Pi_{n-s-1} \neq \emptyset$. So there is a line $P_0 P_1$ with $P_0 \in \Pi_s$, $P_1 \in \Pi_{n-s-1}$, and $Q \in P_0 P_1$. Hence $P_1 \in \mathcal{K}'$ and $Q \in \Pi_s \mathcal{K}'$; therefore $\mathcal{K} \subset \Pi_s \mathcal{K}'$. However, every point of $\Pi_s \mathcal{K}'$ is also in $\mathcal{K}$. $\qquad\square$

Theorem 1.77. *If $\mathcal{K}$ is a quadratic set in $\Sigma = \mathrm{PG}(n, q)$, $n \geq 2$, then every plane section of $\mathcal{K}$ is singular or empty if and only if $\mathcal{K}$ is a subspace or the union of two hyperplanes.*

Proof. In Σ, if $\mathcal{K}$ is a subspace, then, for any plane Π_2, the intersection with $\mathcal{K}$ is one of $\Pi_2, \Pi_1, \Pi_0, \Pi_{-1}$. If $\mathcal{K} = \Pi_{n-1} \cup \Pi'_{n-1}$, then $\mathcal{K} \cap \Pi_2 = \Pi_2$ or $\Pi_1 \cup \Pi'_1$. So, in both cases, a plane section of $\mathcal{K}$ is singular or empty.

To prove the converse, suppose $\mathcal{K}$ is not a subspace. So there are points P and P' with PP' not contained in $\mathcal{K}$. Each plane π through PP' meets $\mathcal{K}$ in two lines l and l' with $P \in l$ and $P' \in l'$. The line l is the only B-line of $\mathcal{K}$ through P in π. Hence the tangent space of $\mathcal{K}$ at P is the hyperplane $\Pi_{n-1} = \bigcup_\pi l$, where the union is taken over all planes π. Similarly, the tangent space of $\mathcal{K}$ at P' is the hyperplane $\Pi'_{n-1} = \bigcup_\pi l'$. Since $\mathcal{K} = \bigcup(l \cup l')$, so $\mathcal{K} = \Pi_{n-1} \cup \Pi'_{n-1}$. $\qquad\square$

Theorem 1.78. *If $\mathcal{K}$ is a quadratic set in Σ other than a subspace, then the smallest subspace containing $\mathcal{K}$ is Σ.*

Proof. The set $\mathcal{K}$ must have at least one non-singular point P, whence T_P is a hyperplane. So every line through P not in T_P contains a second point P' of $\mathcal{K}$. Thus any subspace containing $\mathcal{K}$ contains every line not in T_P and so must be Σ. $\qquad\square$

Definition 1.79. (1) A *generator* of $\mathcal{K}$ is a subspace contained in $\mathcal{K}$ and maximal with respect to inclusion.

(2) Any subspace contained in $\mathcal{K}$ is a *sub-generator* of $\mathcal{K}$.

Theorem 1.80. *Let $\mathcal{K}'$ be a subset of the quadratic set $\mathcal{K}$ such that the line PQ is contained in $\mathcal{K}$ for all P and Q in $\mathcal{K}'$. Then the subspace spanned by $\mathcal{K}'$ is a subgenerator of $\mathcal{K}$.*

Proof. This is by induction on $m = |\mathcal{K}'|$. The result is immediate for $m = 0, 1, 2$. So let $m > 2$, let $P \in \mathcal{K}'$ and let Π_s be the subspace spanned by $\mathcal{K}' \backslash \{P\}$. By the induction hypothesis, Π_s is a sub-generator of $\mathcal{K}$. Every point Q of the subspace $P\Pi_s$ spanned by $\mathcal{K}'$ lies on a line PP' with P' in Π_s. The tangent hyperplane T_P contains $\mathcal{K}' \backslash \{P\}$ and hence Π_s. So every line PP' is a B-line at P and, since it contains two points of $\mathcal{K}$, lies in $\mathcal{K}$; thus $Q \in \mathcal{K}$. $\qquad\square$

Theorem 1.81. *Let $\mathcal{K}$ be a quadratic set with a sub-generator Π and a point P of $\mathcal{K}$ such that $P\Pi$ is not a sub-generator. Then*

(i) *the union Π_P of the lines of $\mathcal{K}$ through P and a point Q of Π is a sub-generator;*
(ii) $\Pi_P \cap \Pi = T_P \cap \Pi$ *and this subspace is a hyperplane in Π and in Π_P;*
(iii) $\dim \Pi_P = \dim \Pi$;
(iv) *if Π is a generator of $\mathcal{K}$, so is Π_P.*

Proof. First, $\Pi_P = P\Pi \cap T_P$. So Π_P is a subspace and $\Pi_P \cap \Pi = T_P \cap \Pi$. Further, Π_P is a sub-generator by Theorem 1.80, where $\mathcal{K}' = (\Pi_P \cap \Pi) \cup \{P\}$ in the notation used there. As Π_P cannot contain Π, so $\Pi_P \cap \Pi$ is a hyperplane of Π since T_P is a hyperplane of Σ. Hence $P(\Pi_P \cap \Pi) = \Pi_P$ is a hyperplane of $P\Pi$ and $\Pi_P \cap \Pi$ is a hyperplane of Π_P. This proves (i) and (ii); part (iii) now follows.

Suppose Π_P is a generator and Π is not. Then Π is properly contained in a generator Π'. Hence, by (ii), $\Pi'_P \cap \Pi'$ is a hyperplane in Π'. So there is a line of $\mathcal{K}$ through P not in Π_P, whence Π_P is not a generator. Thus, if Π_P is a generator, so is Π. The converse result (iv) follows by symmetry. For, let Q be a point of $\Pi \backslash \Pi_P$; then $Q\Pi_P$ is not a sub-generator, unless $P\Pi$ is. As $(\Pi_P)_Q = \Pi$, the result is proved. $\qquad\square$

Theorem 1.82. *The generators of a quadratic set $\mathcal{K}$ all have the same dimension.*

Proof. Suppose Π and Π' are generators of dimensions r and r' with $r < r'$. Let $B = \{P_0, \ldots, P_r\}$ be a generating set for Π. For each P_i, let Π'_i be the union of lines of $\mathcal{K}$ joining P_i to a point of Π'. Then $\Pi'_i \cap \Pi'$ is either a hyperplane of Π' or Π' itself, by Theorem 1.81. As $r < r'$, the intersection of the Π'_i is non-empty since its codimension in Π' is at most r. If P is in this intersection, then $P\Pi$ is a sub-generator by Theorem 1.80 and so Π is not a generator. $\qquad\square$

As for quadrics, the dimension g of a generator of $\mathcal{K}$ is called the *projective index* of $\mathcal{K}$. It should be noted that this is one less than the *Witt index*.

Lemma 1.83. *If Π is a sub-generator of a non-singular quadratic set $\mathcal{K}$, then there exists a generator disjoint from Π.*

Proof. If Π' is a generator meeting Π and $\dim(\Pi' \cap \Pi) = j$, it suffices to show that there exists a generator Π'' such that $\dim(\Pi'' \cap \Pi) = j - 1$. There exists a point P in $\mathcal{K}$ such that $P(\Pi' \cap \Pi)$ is not a sub-generator; otherwise, every point of $\Pi' \cap \Pi$ would be singular for $\mathcal{K}$. Hence $(\Pi' \cap \Pi)_P$ is a sub-generator whose intersection with $\Pi' \cap \Pi$ has dimension $j - 1$, by Theorem 1.81. By the same result, Π'_P is a generator, and contains $(\Pi' \cap \Pi)_P$. Since $P(\Pi' \cap \Pi)$ is not a sub-generator,

$$\Pi' \cap \Pi \not\subset \Pi'_P. \tag{1.68}$$

So

$$(\Pi' \cap \Pi)_P \cap (\Pi' \cap \Pi) = \Pi'_P \cap (\Pi' \cap \Pi). \tag{1.69}$$

It is now shown that

$$\Pi'_P \cap \Pi' \cap \Pi = \Pi'_P \cap \Pi. \tag{1.70}$$

Assume on the contrary that there is a point Q of $\Pi \cap \Pi'_P$ not in Π'. Let Ω be the union of $\Pi' \cap \Pi$, $\Pi'_P \cap \Pi'$, and $\{Q\}$. The join of any two points of Ω is a sub-generator. By Theorem 1.80, Ω spans a sub-generator $\mathcal{K}'$. Since $\mathcal{K}'$ contains the hyperplane $\Pi' \cap \Pi'_P$ of Π'_P and the point Q of Π'_P, so $\Pi'_P \subset \mathcal{K}'$. Since Π'_P is a generator, so $\Pi'_P = \mathcal{K}'$. Hence $\Pi'_P = \mathcal{K}' \supset \Pi' \cap \Pi$, contradicting (1.68). So (1.70) holds. Now, from (1.69),

$$(\Pi' \cap \Pi)_P \cap (\Pi' \cap \Pi) = \Pi'_P \cap \Pi.$$

Since $\dim((\Pi' \cap \Pi)_P \cap (\Pi' \cap \Pi)) = j - 1$, so $\dim(\Pi'_P \cap \Pi) = j - 1$. Thus there is a generator $\Pi'' = \Pi'_P$ with $\dim(\Pi'' \cap \Pi) = j - 1$. $\square$

Corollary 1.84. *If $\mathcal{K}$ is a non-singular quadratic set in $\mathrm{PG}(n, q)$, its projective index g satisfies $2g \leq n - 1$.*

Proof. Let Π and Π' be disjoint generators of $\mathcal{K}$; then $\dim \Pi\Pi' \leq n$. So

$$2g = \dim \Pi + \dim \Pi' = \dim(\Pi \Pi') + \dim(\Pi \cap \Pi') \leq n - 1. \qquad \square$$

Lemma 1.85. *If Π is a sub-generator of the non-singular quadratic set $\mathcal{K}$ and Π is contained in the generator Γ, then there exists a generator Γ' such that $\Gamma \cap \Gamma' = \Pi$.*

Proof. This is by induction on $\dim \Pi = m$. If $m = -1$, the result is that of the previous lemma. Suppose now that the property is satisfied for dimension $m - 1$, with $m \geq 0$. Let Π' be a hyperplane of Π and let P be a point of $\Pi \backslash \Pi'$. There exists a generator Γ_1 such that $\Gamma \cap \Gamma_1 = \Pi'$ since $\dim \Pi' = m - 1$. Then $\Gamma' = (\Gamma_1)_P$ is the required generator. $\square$

Theorem 1.86. *Every sub-generator of a non-singular quadratic set $\mathcal{K}$ is the intersection of the generators containing it.*

Proof. This follows from the preceding lemma. $\square$

Lemma 1.87. *If $\mathcal{K}$ is a quadratic set in $\mathrm{PG}(n, q)$, $n \geq 2$, which is not a subspace, then any collineation σ fixing every point of $\mathcal{K}$ is the identity.*

Proof. Since $\mathcal{K}$ is not a subspace, it contains non-singular points P and Q. Hence σ fixes every line through P other than a tangent line. It follows that σ also fixes these tangent lines. So σ is a perspectivity with centre P. Similarly, it is a perspectivity with centre Q. If R is any point not on the line PQ, the lines RP and RQ are fixed. So R is fixed and σ is the identity. $\qquad\square$

Lemma 1.88. *If $\mathcal{K}$ is a quadratic set in $\Sigma = \mathrm{PG}(n, q)$, $n \geq 2$, and P in $\Sigma \backslash \mathcal{K}$ is a point not lying on all tangent hyperplanes, then there is at most one perspectivity other than the identity with centre P which fixes $\mathcal{K}$.*

Proof. Since P exists, $\mathcal{K}$ is not a subspace. Let σ and σ' be perspectivities with centre P fixing $\mathcal{K}$. Let $P_1 P_2$ be a bisecant of $\mathcal{K}$ through P with $P_1, P_2 \in \mathcal{K}$. First, let $P_i \sigma' = P_i$, $i = 1, 2$. Then σ' fixes both T_{P_1} and T_{P_2}, whence σ' is the identity, a contradiction. So $P_1 \sigma' = P_2$ and $P_2 \sigma' = P_1$. Analogously, $P_1 \sigma = P_2$, $P_2 \sigma = P_1$. So $\sigma^{-1}\sigma'$ is a perspectivity with centre P fixing $\mathcal{K}$, P_1 and P_2. Again, $\sigma^{-1}\sigma'$ is the identity, implying that $\sigma = \sigma'$. $\qquad\square$

Theorem 1.89. *If $\mathcal{K}$ is a quadratic set other than a subspace and every plane section of $\mathcal{K}$ is the empty set, a point, a line or a conic, then $\mathcal{K}$ is a quadric.*

Proof. This is by induction on the dimension n. The result is in the hypotheses when $n = 1$ or 2. So let $n \geq 3$, and let $P'Q' \cap \mathcal{K} = \{P', Q'\}$. A hyperplane Π through $P'Q'$ therefore does not meet $\mathcal{K}$ in a subspace. So, by induction, $\Pi \cap \mathcal{K}$ is a quadric.

If all points of $\mathcal{K}$ outside Π were singular, they would be contained in a subspace Π'. So $\mathcal{K} \subset \Pi \cup \Pi'$, whence $\mathcal{K}$ has bisecants through each point of $(\mathcal{K} \cap \Pi')\backslash\Pi$, a contradiction. So $\mathcal{K}$ has a non-singular point P not in Π. Choose coordinates such that (i) $P = \mathbf{U}_0$, (ii) $\Pi = \mathbf{u}_0$, (iii) $T_P = \mathbf{u}_n$. Then

$$\Pi \cap \mathcal{K} = \mathbf{V}\left(X_0, \ \sum_1^n{}' a_{ij} X_i X_j\right).$$

The summation sign $\sum'$ indicates summation over all i and j with $i \leq j$, whereas $\sum''$ indicates summation with $i < j$.

Consider the pencil of quadrics $\mathcal{F}_t$, $t \neq 0$, where

$$\mathcal{F}_t = \mathbf{V}\left(t X_0 X_n + \sum_1^n{}' a_{ij} X_i X_j\right).$$

Each $\mathcal{F}_t$ contains $\Pi \cap \mathcal{K}$, passes through P, and has $T_P(\mathcal{K})$ as the tangent hyperplane at P.

Now, let Q be a point of $\Pi \cap \mathcal{K}$ not lying in T_P; suppose $Q = \mathbf{P}(0, c_1, \ldots, c_n)$, $c_n \neq 0$. Not all points of $\Pi \cap \mathcal{K}$ outside T_P are singular for $\Pi \cap \mathcal{K}$. So, let Q be non-singular for $\Pi \cap \mathcal{K}$; then

$$T_Q(\mathcal{F}_t) = \mathbf{V}\left(t c_n X_0 + 2\sum_1^n a_{ii} c_i X_i + \sum_1^n{}'' a_{ij}(c_i X_j + c_j X_i)\right).$$

However,

$$T_Q(\mathcal{K}) = \mathbf{V}\left(\lambda X_0 + 2\sum_1^n a_{ii}c_i X_i + \sum_1^n {}'' a_{ij}(c_i X_j + c_j X_i)\right), \quad \text{with } \lambda \neq 0.$$

So there exists some t, say $t = b$, such that $T_Q(\mathcal{F}_b) = T_Q(\mathcal{K})$. It is now shown that $\mathcal{K} = \mathcal{F}_b$.

Let R be a point of $\mathcal{K}$ other than P and not in Π. The points P, Q, R are not collinear; so the plane $\alpha = PQR$ meets $\mathcal{K}$ in a conic $\mathcal{C}$ and $\mathcal{F}_b$ in a conic $\mathcal{C}'$. Let $l = \alpha \cap \Pi$; then P and R do not belong to the line l, although Q does. There are two cases.

(a) l *contains a point Q' in $\mathcal{K}$ with $Q' \neq Q$.* Then $\mathcal{C}$ and $\mathcal{C}'$ have the points P, Q, Q' in common as well as the tangents at P and Q, by Lemma 1.72; thus $\mathcal{C} = \mathcal{C}'$. Let A be any point not in the union of the planes α through PQ meeting Π in a tangent l to $\mathcal{K}$, and so in particular outside the hyperplane $PT_Q(\Pi \cap \mathcal{K})$, which is the union of the planes β through PQ meeting Π in a B-line to $\mathcal{K}$. Then A belongs to $\mathcal{K}$ if and only if it belongs to $\mathcal{F}_b$.

(b) l *is a tangent to $\mathcal{K}$.* Then RP belongs neither to $\mathcal{K}$ nor to $\mathcal{F}_b$. Let S be a point of $\mathcal{K}$ not in the hyperplane $PT_Q(\Pi \cap \mathcal{K})$. The plane PRS meets $\mathcal{K}$ and $\mathcal{F}_b$ in conics $\mathcal{D}$ and $\mathcal{D}'$ which coincide outside the line PR. Since PR is contained in neither $\mathcal{D}$ nor $\mathcal{D}'$, so $\mathcal{D} = \mathcal{D}'$; hence $R \in \mathcal{F}_b$.

Thus $\mathcal{K} \subset \mathcal{F}_b$. Similarly $\mathcal{F}_b \subset \mathcal{K}$. This concludes the proof. $\qquad\square$

Theorem 1.90. *In* $\mathrm{PG}(n, q)$, *a perspective quadratic set $\mathcal{K}$ which is not a subspace is a quadric.*

Proof. If the set is the union of two subspaces, neither of which is contained in the other, then it follows, for example from Theorem 1.77, that $\mathcal{K}$ is the union of two distinct hyperplanes and so is a quadric. From now on, assume that this is not the case.

(a) $n = 2$. Here $\mathcal{K}$ must be non-singular and non-empty. So it has no three points collinear and is therefore a $(q + 1)$-arc. Let $\mathbf{U}_0, \mathbf{U}_1$ be points of $\mathcal{K}$ and let $\mathbf{U}_0\mathbf{U}_2$ and $\mathbf{U}_1\mathbf{U}_2$ be the tangents at $\mathbf{U}_0$ and $\mathbf{U}_1$. Every point of $\mathbf{u}_2$ not in $\mathcal{K}$ is of the form $\mathbf{P}(1, m, 0)$ with $m \neq 0$. There exists a non-identity perspectivity σ with centre $\mathbf{P}(1, m, 0)$ fixing $\mathcal{K}$ and $\mathbf{U}_2$ but interchanging $\mathbf{U}_0$ and $\mathbf{U}_1$. Hence the axis of σ is $\mathbf{V}(mX_0 + X_1)$; if $p = 2$, the axis contains $\mathbf{P}(1, m, 0)$ and, if $p \neq 2$, the axis passes through the harmonic conjugate $\mathbf{P}(1, -m, 0)$ of $\mathbf{P}(1, m, 0)$ with respect to $\mathbf{U}_0$ and $\mathbf{U}_1$. Thus $\mathbf{U}$ is mapped to $\mathbf{P}(1, m^2, -m)$, whence $\mathcal{K} = \mathbf{V}(X_0 X_1 - X_2^2)$.

(b) $n > 2$. Let Π be a plane such that $\Pi \cap \mathcal{K}$ is not a subspace. If $\Pi \cap \mathcal{K}$ is non-singular, it is a conic by (a). If $\Pi \cap \mathcal{K}$ is singular, it is a line pair and therefore a quadric. Hence, by Theorem 1.89, $\mathcal{K}$ is a quadric. $\qquad\square$

To reach the final characterisation of quadrics, more properties of perspective quadratic sets are required. For the next theorem, let P, Q be distinct points of a quadratic set $\mathcal{K}$, and define $S_{PQ} = T_P \cap T_Q$. For the remainder of this section, $\mathcal{K}$ is always non-singular.

Lemma 1.91. (i) S_{PQ} has codimension two in Σ.
(ii) If A, B, C are collinear points of $\mathcal{K}$, then $S_{AB} = S_{AC}$.

Proof. (i) If $T_P = T_Q$, then $PQ \subset \mathcal{K}$. It is first shown that if R is any other point on PQ, then $T_P = T_R$. Let $A \in T_P \backslash PQ$; then either AP and AQ are both tangents or both lines of $\mathcal{K}$.

(a) Suppose AP, AQ are tangents. The plane APR lies in T_P. If AR is a bisecant, then it meets $\mathcal{K}$ in another point R'. So PR' and QR' are lines of $\mathcal{K}$. Hence any line l through R other than AR meets PR' and QR' in a point of $\mathcal{K}$; so $l \subset \mathcal{K}$. Thus the plane APR lies in $\mathcal{K}$. So AR is not a bisecant and is hence a tangent.

(b) Suppose AP, AQ are lines of $\mathcal{K}$. Again, any line l through R other than AR contains distinct points of $\mathcal{K}$ on AP and AQ, and so lies on $\mathcal{K}$. Hence the plane APQ lies in $\mathcal{K}$. So AR is a line of $\mathcal{K}$.

From (a) and (b), it follows that $T_P \subset T_R$. Since $\mathcal{K}$ is non-singular, $T_P = T_R$ for all R on PQ. Let $A \notin T_P \cup \mathcal{K}$. Then AR is a bisecant for every R on PQ; so AP, AQ, AR all meet $\mathcal{K}$ again at the respective points P_1, Q_1, R_1.

If P_1, Q_1, R_1 are not collinear, then each side of the triangle $P_1 Q_1 R_1$ meets the line PQ in a point of $\mathcal{K}$. So these sides are lines of $\mathcal{K}$ and the plane of the triangle lies on $\mathcal{K}$; thus A lies on $\mathcal{K}$. So P_1, Q_1, R_1 are collinear and R_1 lies on $P_1 Q_1$ for every R on PQ.

Let $PQ \cap P_1 Q_1 = S$. Then AS is a tangent to $\mathcal{K}$. Thus $A \in T_S$ and $T_S = \Sigma$. So $\mathcal{K}$ is singular, a contradiction. Hence $T_P \neq T_Q$ and so S_{PQ} has codimension two.

(ii) If $D \in S_{AB}$ and $D \notin \mathcal{K}$, then AD and BD are tangents. If DC is not a tangent, it meets $\mathcal{K}$ again at C'. The plane $\alpha = ABD$ lies in S_{AB}, hence $C' \in S_{AB}$, and AC' and BC' lie in $\mathcal{K}$. This gives enough points of α on $\mathcal{K}$ to mean that α lies in $\mathcal{K}$, a contradiction. If $D \in \mathcal{K} \backslash AB$, then $ABD \subset \mathcal{K}$. In both cases, $D \in T_C$. □

Lemma 1.92. If PQ is a bisecant of $\mathcal{K}$, then S_{PQ} is spanned by $S_{PQ} \cap \mathcal{K}$ providing $S_{PQ} \cap \mathcal{K}$ is non-empty.

Proof. The set $S_{PQ} \cap \mathcal{K}$ is non-singular, for a singular point R would be such that T_R contains S_{PQ} as well as P and Q; so R would be singular for $\mathcal{K}$, since S_{PQ} and P span S_P. Now Theorem 1.78 gives the result. □

Lemma 1.93. If σ is a perspectivity fixing $\mathcal{K}$ with centre R not in $\mathcal{K}$ and $P\sigma = Q$, where $P, Q \in \mathcal{K}$ and $P \neq Q$, then the axis of σ contains S_{PQ}.

Proof. Since $P\sigma = Q$, so $T_P \sigma = T_Q$. Hence $T_P \cap T_Q = S_{PQ}$ is in the axis of σ. □

Let AB be a bisecant of $\mathcal{K}$, with $A, B \in \mathcal{K}$, and let $P \in AB \backslash \mathcal{K}$. Let l be a line of $\mathcal{K}$ through A; then $Bl \not\subset \mathcal{K}$. By Theorem 1.81, there exists a unique E on l such that $BE \subset \mathcal{K}$. Let C be a point of $l \backslash \{A, E\}$ and let $D = PC \cap BE$. See Figure 1.1.

Lemma 1.94. With A, B, C, D, E as in Figure 1.1,

(i) $S_{AB} \neq S_{BC}$;
(ii) $S_{AB} S_{CD}$ is a hyperplane.

Fig. 1.1. Points on a quadratic set

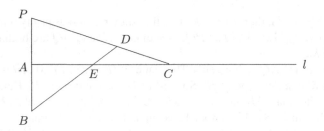

Proof. (i) If $S_{AB} = S_{BC}$, then $S_{AB} = S_{AC} = S_{CD}$ and T_B contains A.

(ii) $S_{AB} \cap S_{CD} = T_A \cap T_B \cap T_C \cap T_D$

$$= T_A \cap T_C \cap T_B \cap T_E, \quad \text{since } T_B \cap T_D = T_E \cap T_B,$$

$$= T_A \cap T_B \cap T_C, \quad \text{since } T_C \cap T_E = T_A \cap T_C,$$

using Lemma 1.91 (ii) twice. By (i), $\dim(T_A \cap T_B \cap T_C) = n - 3$. So $S_{AB}S_{CD}$ is a hyperplane. $\qquad\square$

Lemma 1.95. *Let σ be the perspectivity with centre P, axis $S_{AB}S_{CD}$, and such that $A\sigma = B$. Then σ fixes $\mathcal{K}$.*

Proof. First it is shown that $A, B \notin S_{AB}S_{CD}$ in order to guarantee the existence of σ. Suppose that $A \in S_{AB}S_{CD}$. As $A \notin S_{AB}$, so $AS_{AB} = T_A = S_{AB}S_{CD}$. Hence $S_{CD} \subset T_A$ and so $S_{AC} = S_{CD}$. By Lemma 1.91 (ii), $S_{AC} \subset T_E$; so $S_{CD} \subset T_E$. Hence $S_{ED} = S_{CD}$. Again by Lemma 1.91 (ii), $S_{ED} \subset T_B$ and so $S_{CD} \subset T_B$. Consequently, S_{CD} is contained in T_C, T_D, T_A, T_B. Hence $S_{AB} = S_{BC} = S_{CD}$, contradicting Lemma 1.94 (i). Analogously, $B \notin S_{AB}S_{CD}$.

Since $E \in S_{AB} \cap S_{CD}$ so E is on the axis. The perspectivity σ depends only on $\mathcal{K}, P, A, B, C$; so write

$$\sigma = \sigma(\mathcal{K}, P, A, B, C).$$

Now, σ maps every line of $\mathcal{K}$ through A to a line through B and every line of $\mathcal{K}$ through C to a line through D. Hence

$$T_A\sigma = (AS_{AB})\sigma = BS_{AB} = T_B;$$

analogously, $T_C\sigma = T_D$.

Choose Q in $\mathcal{K}$ not in the plane ABC and consider the following two possibilities.

(a) *QE is not a line of $\mathcal{K}$*

Let $l' = QF$, with $F \in AC$, be on $\mathcal{K}$, and consider the solid $\Pi_3 = ABCQ$. It can be shown that $\Pi_3 \cap \mathcal{K}$ is non-singular. It follows that every point of $\Pi_3 \cap \mathcal{K}$ lies on at most two lines of $\Pi_3 \cap \mathcal{K}$. For any point R of $\Pi_3 \cap \mathcal{K}$ such that RE is not a line of $\mathcal{K}$, the tangent hyperplane T_R intersects AE and BE in points A' and B';

then RA' and RB' are lines of $\mathcal{K}$. If $R' \in \Pi_3 \cap \mathcal{K}$ is such that $R'E$ is a line of $\mathcal{K}$, then choose E' such that neither $E'E$ nor $E'R'$ are lines of $\mathcal{K}$. Then the previous argument can be applied with R' for R and E' for E. Thus each point of $\Pi_3 \cap \mathcal{K}$ lies on exactly two lines of $\Pi_3 \cap \mathcal{K}$. Hence $\Pi_3 \cap \mathcal{K}$ is a hyperbolic quadric, by Theorem 16.2.6 of FPSOTD.

In Π_3 there exists a perspectivity σ' with centre P and axis $\Pi_3 \cap S_{AB}S_{CD}$ fixing $\Pi_3 \cap \mathcal{K}$ and such that $A\sigma' = B$. Hence $\sigma|_{\Pi_3} = \sigma'$ and $Q\sigma \in \mathcal{K}$. It also follows that $B\sigma = A$, and so $T_B\sigma = T_A$. Since E is non-singular, such a point Q always exists.

(b) *QE is a line of $\mathcal{K}$*

If QA and QB are lines of $\mathcal{K}$, then $Q \in S_{AB}$ and so $Q\sigma = Q$. Thus it may be assumed that QA is not on $\mathcal{K}$. It may also be assumed that $Q \notin S_{AB}S_{CD}$, since otherwise $Q\sigma = Q$. There exists a point E' in $S_{AB} \cup S_{CD}$ such that $E' \in \mathcal{K}$ and QE' is a bisecant of $\mathcal{K}$. Otherwise, T_Q contains S_{AB} and S_{CD}, by Lemma 1.92, as $S_{AB} \cap \mathcal{K}$ and $S_{CD} \cap \mathcal{K}$ are not empty since they both contain E; this is excluded by the fact that Q is neither singular nor in $S_{AB}S_{CD}$.

Let $E' \in S_{AB}$. Through Q there is a line QA' in $\mathcal{K}$ meeting AE' in A'. Then $A' \ne E', A$. Now consider the perspectivity $\sigma_1 = \sigma(\mathcal{K}, P, A, B, A')$; so $A'\sigma_1 = B'$ where $B' = PA' \cap E'B$. Applying the result of the previous paragraph to σ_1 shows that $Q\sigma_1 \in \mathcal{K}$. As σ and σ_1 both interchange T_A and T_B, the map $\sigma\sigma_1^{-1}$ is the identity, whence $\sigma = \sigma_1$. □

Theorem 1.96. *In* $\mathrm{PG}(n, q)$, *a non-singular quadratic set containing a line is perspective.*

Proof. This merely restates the previous lemma in a different form. □

Theorem 1.97. *In* $\mathrm{PG}(n, q)$, *a non-singular quadratic set $\mathcal{K}$ is a quadric or an ovoid. If $\mathcal{K}$ is an ovoid, then it is one of the following:*

 (i) *a $(q+1)$-arc in* $\mathrm{PG}(2, q)$;
 (ii) *an ovaloid of* $\mathrm{PG}(3, q)$, $q > 2$;
(iii) *an elliptic quadric in* $\mathrm{PG}(3, 2)$.

Proof. The first part follows from Theorems 1.90 and 1.96. The second part follows from Theorem 5.55. □

1.11 Further characterisations of quadrics

In Section 1.10 two restrictions are put on a subset $\mathcal{K}$ of $\Sigma = \mathrm{PG}(n, q)$:

(a) $\mathcal{K}$ is of type $(0, 1, 2, q+1)$;
(b) $\mathcal{K}$ has a tangent space at each point P, which is either a hyperplane or Σ itself.

In this section, condition (b) is dropped. In what follows, condition (a) still holds. So the definitions of Section 1.10 still apply for a set $\mathcal{K}$ is of type $(0, 1, 2, q+1)$. A point P of $\mathcal{K}$ is *singular* if there is no bisecant of $\mathcal{K}$ through it. The set of singular points of $\mathcal{K}$ is the *singular space of $\mathcal{K}$*, and $\mathcal{K}$ is *singular* or *non-singular* according as it has singular points or not.

Lemma 1.98. *The singular space S of K is a subspace of Σ.*

Proof. Let $P, Q \in S$; then it must be shown that any other point R of PQ is also singular. Since P is singular, so $PQ \subset K$. Let $A \in K \backslash PQ$. Then the three lines AP, AQ, PQ lie on K and hence the plane APQ also lies on K; so $AR \subset K$. Hence R is singular. $\qquad\qquad\qquad\qquad\qquad\qquad\qquad\qquad\qquad\qquad\qquad\qquad\qquad\square$

Theorem 1.99. *If K is a set of type $(0, 1, 2, q+1)$ in Σ with singular space Π_r, then K is a cone $\Pi_r K'$, where K' is a non-singular set of type $(0, 1, 2, q+1)$ in a subspace Π_{n-r-1} skew to Π_r.*

Proof. Let Π_{n-r-1} be any space of dimension $n - r - 1$ skew to the singular space Π_r and let $K' = \Pi_{n-r-1} \cap K$. Now take A in Π_r and A' in K'; then $AA' \subset K$. For any point P of $K \backslash (\Pi_r \cup K')$, the line $AP \subset K$. The space $P\Pi_r$ meets Π_{n-r-1} in a point B; so $B \in K'$. Hence K is the cone $\Pi_r K'$.

It must still be shown that K' is non-singular; it is a set of type $(0, 1, 2, q + 1)$ from its definition. Suppose K' has a singular point Q. Let $P \in K \backslash \Pi_r$ and also let $B = P\Pi_r \cap \Pi_{n-r-1}$. If $B = Q$, then $PQ \subset K$. If $B \neq Q$, then, with $A = BP \cap \Pi_r$, the planes APQ and ABQ coincide. However, $BQ \subset K$ and A is singular; so $ABQ \subset K$. Thus $PQ \subset K$ and Q is a singular point of K, which is a contradiction. $\qquad\qquad\qquad\qquad\qquad\qquad\qquad\qquad\qquad\qquad\qquad\qquad\qquad\qquad\qquad\square$

Theorem 1.100. *A k-set of type $(0, 1, 2, q+1)$ in $\mathrm{PG}(n, q)$, with $n \geq 3$, $q > 2$, and such that $\theta(n) > k \geq \theta(n-1)$, is one of the following:*

(a) $\Pi_{n-1} \cup \Pi_r'$ *for some* $r = -1, 0, 1, \ldots, n-1$;
(b) $\Pi_t P_{n-t-1}$ *for some* $t = -1, 1, \ldots, n-3$ *when n is even and equally some* $t = 0, 2, \ldots, n-3$ *when n is odd*;
(c) $\Pi_t H_{n-t-1}$ *for some* $t = -1, 1, \ldots, n-4$ *when n is odd and equally some* $t = 0, 2, \ldots, n-4$ *when n is even*;
(d) $W_n \cup \Pi_r$, *where W_n is one of the quadrics (b) and $\Pi_r \subset \Pi_t N$ with N the nucleus of a base P_{n-t-1}, and q is even*;
(e) $\Pi_{n-3} K' \cup \Pi_r$, *where K' is a $(q+1)$-arc in a Π_2 skew to Π_{n-3} and $\Pi_r \subset \Pi_{n-3} N$ with N the nucleus of K', and q is even*.

Theorem 1.101. *In $\mathrm{PG}(n, q)$, with $q > 3$, $n \geq 4$, a k-set of type $(0, 1, 2, q+1)$ with $k = \theta(n-1) - q^{g+1}$, where g is the largest dimension of a subspace in K, is one of the following:*

(a) $\Pi_t \mathcal{E}_{n-t-1}$ *for some* $t = -1, 1, \ldots, n-4$ *when n is odd and similarly some* $t = 0, 2, \ldots, n-4$ *when n is even*;
(b) $\Pi_{n-4} K'$, *where K' is a (q^2+1)-cap in $\mathrm{PG}(3, q)$ skew to Π_{n-4} and q is even*;
(c) Π_{n-2}.

The last theorem has the following improvement.

Theorem 1.102. *In $\mathrm{PG}(n, q)$, $q > 3$, $n \geq 3$, q odd, a k-set of type $(0, 1, 2, q+1)$ with $\theta(n) > k > \theta(n-1) - q^{n-2} + q^{n-3}$ is either a quadric or $\Pi_{n-1} \cup \Pi_r'$ for some $r = -1, 0, 1, \ldots, n-1$.*

From the previous three results, the following result for the non-singular case is immediate.

Theorem 1.103. *In* $\mathrm{PG}(n, q)$ *with* $n \geq 4$, $q > 2$, *let* $\mathcal{K}$ *be a non-singular* k-*set of type* $(0, 1, 2, q + 1)$.

(i) *If* $\theta(n) > k \geq \theta(n - 1)$, *then one of the following holds:*
    (a) $k = \theta(n - 1)$, *n is even and* $\mathcal{K} = \mathcal{P}_n$;
    (b) $k = \theta(n - 1) + q^{(n-1)/2}$, *n is odd and* $\mathcal{K} = \mathcal{H}_n$;
    (c) $k = \theta(n - 1) + 1$ *and* $\mathcal{K} = \Pi_t \mathcal{P}_{n-t-1} \cup \{N\}$ *or* $\mathcal{K} = \Pi_{n-3}\mathcal{K}' \cup \{N\}$,
    where $\mathcal{K}'$ *is a* $(q + 1)$-*arc in a* Π_2 *skew to* Π_{n-3}, *and* N *is the nucleus of*
    $\mathcal{P}_{n-t-1}$ *and* $\mathcal{K}'$ *in the two cases.*
(ii) *If* $k = \theta(n - 1) - q^{g+1}$, *where* g *is the largest dimension of a subspace in* $\mathcal{K}$,
    then n *is odd,* $g + 1 = \frac{1}{2}(n - 1)$, *and* $\mathcal{K} = \mathcal{E}_n$.
(iii) *If* q *is odd,* $q > 3$ *and* $\theta(n) > k > \theta(n - 1) - q^{n-2} + q^{n-3}$, *then*
    (a) *for* n *even,* $\mathcal{K} = \mathcal{P}_n$;
    (b) *for* n *odd,* $\mathcal{K} = \mathcal{H}_n$ *or* $\mathcal{E}_n$.

1.12 The Principle of Triality

On the hyperbolic quadric $\mathcal{H}_7$, known sometimes as the *triality quadric* or the *Study quadric* after the discoverer of the principle, consider the two systems of generators $\mathcal{A}$ and $\mathcal{B}$. From Theorem 1.33,

$$|\mathcal{A}| = |\mathcal{B}| = \tfrac{1}{2}\kappa(7; 2) = \tfrac{1}{2}[0, 3]_+ = (q + 1)(q^2 + 1)(q^3 + 1).$$

From Theorem 1.41 (iii),

$$|\mathcal{H}_7| = N(0; 7, 2) = (q^3 + 1)(q^4 - 1)/(q - 1) = \tfrac{1}{2}\kappa(7; 2).$$

Then it can be shown that the solids in $\mathcal{A}$ correspond to the points of a quadric $\mathcal{H}_7'$ of $\mathrm{PG}'(7, q)$ such that solids in $\mathcal{A}$ containing a given line are mapped to the points of a line of $\mathcal{H}_7'$; similarly, the solids in $\mathcal{B}$ correspond to the points of a quadric $\mathcal{H}_7''$ of $\mathrm{PG}''(7, q)$. A *triality* is a permutation τ of $\mathcal{H}_7 \cup \mathcal{A} \cup \mathcal{B}$ such that

$$\mathcal{H}_7\tau = \mathcal{A}, \quad \mathcal{A}\tau = \mathcal{B}, \quad \mathcal{B}\tau = \mathcal{H}_7,$$

and such that incidence is preserved, where *incidence* is defined as follows:

(1) a point is incident with a solid if it lies in the solid;
(2) two points are incident if their join lies on $\mathcal{H}_7$;
(3) two solids of the same system are incident if they meet in a line;
(4) two solids of different systems are incident if they meet in a plane.

Then

$$\mathcal{H}_7 \overset{\tau}{\to} \mathcal{A} \overset{\tau}{\to} \mathcal{B} \overset{\tau}{\to} \mathcal{H}_7$$

is induced by collineations

$$\text{PG}(7,q) \overset{\sigma_1}{\to} \text{PG}'(7,q) \overset{\sigma_2}{\to} \text{PG}''(7,q) \overset{\sigma_3}{\to} \text{PG}(7,q)$$

such that

$$\mathcal{H}_7 \sigma_1 = \mathcal{H}_7', \quad \mathcal{H}_7' \sigma_2 = \mathcal{H}_7'', \quad \mathcal{H}_7'' \sigma_3 = \mathcal{H}_7.$$

The map τ is called a triality by analogy with a duality of a projective space.

Example 1.104. To give an explicit example, a *trilinear correspondence* is introduced. Write points as follows:

in $\text{PG}(7,q)$ as $\mathbf{P}(x)$ with $x = (x_0, x_1, \ldots, x_7)$;
in $\text{PG}'(7,q)$ as $\mathbf{P}(y)$ with $y = (y_0, y_1, \ldots, y_7)$;
in $\text{PG}''(7,q)$ as $\mathbf{P}(z)$ with $z = (z_0, z_1, \ldots, z_7)$.

Let

$$\begin{aligned}
\mathcal{H}_7 &= \mathbf{V}(X_0 X_4 + X_1 X_5 + X_2 X_6 + X_3 X_7), \\
\mathcal{H}_7' &= \mathbf{V}(Y_0 Y_4 + Y_1 Y_5 + Y_2 Y_6 + Y_3 Y_7), \\
\mathcal{H}_7'' &= \mathbf{V}(Z_0 Z_4 + Z_1 Z_5 + Z_2 Z_6 + Z_3 Z_7).
\end{aligned}$$

Consider the following trilinear form:

$$\begin{aligned}
T(X,Y,Z) = {} & \begin{vmatrix} X_0 & X_1 & X_2 \\ Y_0 & Y_1 & Y_2 \\ Z_0 & Z_1 & Z_2 \end{vmatrix} + \begin{vmatrix} X_4 & X_5 & X_6 \\ Y_4 & Y_5 & Y_6 \\ Z_4 & Z_5 & Z_6 \end{vmatrix} \\
& + X_3(Z_0 Y_4 + Z_1 Y_5 + Z_2 Y_6) + X_7(Y_0 Z_4 + Y_1 Z_5 + Y_2 Z_6) \\
& + Y_3(X_0 Z_4 + X_1 Z_5 + X_2 Z_6) + Y_7(Z_0 X_4 + Z_1 X_5 + Z_2 X_6) \\
& + Z_3(Y_0 X_4 + Y_1 X_5 + Y_2 X_6) + Z_7(X_0 Y_4 + X_1 Y_5 + X_2 Y_6) \\
& - X_3 Y_3 Z_3 - X_7 Y_7 Z_7.
\end{aligned}$$

A pair $(\mathbf{P}(x), \mathbf{P}(y))$ of points in $\mathcal{H}_7 \times \mathcal{H}_7'$ represents an incident pair in $\mathcal{H}_7 \times \mathcal{A}$ if and only if the linear form $T(x, y, Z)$ is identically zero in Z; the similar condition applies for a cyclic permutation of x, y, z.

For example, to find the solid of $\mathcal{A}$ that corresponds to the point $\mathbf{P}(y)$ of $\mathcal{H}_7'$ when $y = (1, 0, \ldots, 0)$, put this value of y in $T(X, y, Z)$ and require that the coefficients of all Z_i, $i \in \{0, 1, \ldots, 7\}$ vanish. This gives the solid $\mathbf{V}(X_1, X_2, X_4, X_7)$.

The collineations $\sigma_1, \sigma_2, \sigma_3$, with

$$\begin{aligned}
&\text{PG}(7,q)\sigma_1 = \text{PG}'(7,q), \ \text{PG}'(7,q)\sigma_2 = \text{PG}''(7,q), \ \text{PG}''(7,q)\sigma_3 = \text{PG}(7,q), \\
&(s_0, s_1, \ldots, s_7)\sigma_i = (s_0, s_1, \ldots, s_7), \quad i = 1, 2, 3,
\end{aligned}$$

define a triality τ. Explicit calculations are possible using the form $T(X, Y, Z)$.

1.13 Generalised hexagons

Definition 1.105. A *generalised hexagon* is an incidence structure $\mathcal{S} = (\mathcal{P}, \mathcal{B}, \mathrm{I})$ in which $\mathcal{P}$ and $\mathcal{B}$ are non-empty, disjoint sets of *points* and *lines*, and for which I is a symmetric point-line incidence relation satisfying the following properties.

(1) Each point is incident with $1 + t$ lines, with $t \geq 1$, and two distinct points are incident with at most one line.

(2) Each line is incident with $1 + s$ points, with $s \geq 1$, and two distinct lines are incident with at most one point.

(3) $\mathcal{S}$ contains no ordinary k-gon, as a subgeometry, for $2 \leq k < 6$.

(4) Any two elements of $\mathcal{P} \cup \mathcal{B}$ are contained in some ordinary 6-gon, again as a subgeometry.

The integers s and t are the *parameters* of the generalised hexagon, and $\mathcal{S}$ has *order* (s, t); if $s = t$, then $\mathcal{S}$ has *order* s. There is a point-line duality for a generalised hexagon of order (s, t) for which in any statement 'point' and 'line' are interchanged, and s and t are interchanged.

Theorem 1.106. *The order of the known finite generalised hexagons is one of the following:*

(a) $(s, 1)$ *with* $s \geq 1$;

(b) $(1, t)$ *with* $t \geq 1$;

(c) (q, q) *for* q *any prime power*;

(d) (q, q^3), *and dually* (q^3, q), *for* q *any prime power.*

Example 1.107. Let $\mathcal{S}' = (\mathcal{P}', \mathcal{B}', \mathrm{I}')$ be any finite projective plane of order s. Define

(a) $\mathcal{P} = \mathcal{P}' \cup \mathcal{B}'$;

(b) $\mathcal{B} = \{(P', l') \in \mathcal{P}' \times \mathcal{B}' \mid P' \mathrm{I}' l'\}$;

(c) $P \mathrm{I} l$ with $P \in \mathcal{P}$ and $l \in \mathcal{B}$ if and only if $P \in l$.

Then $\mathcal{S} = (\mathcal{P}, \mathcal{B}, \mathrm{I})$ is a generalised hexagon of order $(s, 1)$. Conversely, any generalised hexagon with $t = 1$ is of this type.

Trialities and generalised hexagons are closely related. Let $\mathcal{H}_7$ be the Study quadric and let $\mathcal{P}_6$ be a parabolic quadric on it. If τ is a triality and l is a line of $\mathcal{P}_6$, then τ maps the points of l onto the $q + 1$ generators of $\mathcal{A}$ through a line l' of $\mathcal{H}_7$. Let $\mathcal{B}$ be the set of all lines l' on $\mathcal{P}_6$. Then the incidence structure $(\mathcal{P}_6, \mathcal{B}, \mathrm{I})$, with I the natural incidence, is a generalised hexagon of order q.

Up to duality and isomorphism, no other generalised hexagon of order q, with $q \neq 1$, is known. Also, all known generalised hexagons of order (q, q^3), and dually (q^3, q), arise from trialities.

1.14 Notes and references

Sections 1.1–1.9

These sections continue the material in Chapter 5 of PGOFF2. They are based on Hirschfeld [174], although many details come from Segre [276]. In particular, the proof of Theorem 1.57 is based on Segre's treatment, although the formula (1.44), due to Hirschfeld [173], is an amalgamation of several formulas established by matrix-theoretic methods by Dai and Feng [81] and by Feng and Dai [132]. See also De Bruyn [84].

The nature of the transitivity of the orthogonal group, as expounded in Theorem 1.49, seems more difficult in the projective space than in the vector space case. See Higman [165], Artin [3], Dieudonné [122] for the vector space case, and Dye [124] for the connection between the two cases. The size of the stabiliser in the orthogonal group of an isotropic subspace in the vector space case has been calculated by Derr [117].

Section 1.10

This section follows Buekenhout [54].

Section 1.11

The complication of further characterisations, even when the set $\mathcal{K}$ is non-singular is illustrated by the work of Lefèvre [200]. Theorems 1.100 and 1.101 are due to Tallini [302, 303] and Theorem 1.102 to Lefèvre-Percsy [202].

For characterisations of quadrics by intersection numbers, see Schillewaert [270] and De Winter and Schillewaert [105].

Sections 1.12–1.13

See Hirschfeld and Thas [180], Tits [379], Van Maldeghem [391].

2

Hermitian varieties

2.1 Introduction

In $\mathbf{F}_q = \mathrm{GF}(q)$, q square, the map $x \mapsto x^{\sqrt{q}} = \bar{x}$ is an involutory automorphism. For a matrix $A = (a_{ij})$, write $\bar{A} = (\bar{a}_{ij})$. Then a *Hermitian form* F is an element of $\mathbf{F}_q[X_0, X_1, X_2, \ldots, X_n]$ such that

$$F = XH\bar{X}^*,$$

where $X = (X_0, X_1, X_2, \ldots, X_n)$, $\bar{H}^* = H$ and $H \neq 0$. As in Section 5.1 of PGOFF2, F can be reduced by a non-singular linear transformation to the canonical form

$$F_r = X_0\bar{X}_0 + X_1\bar{X}_1 + \cdots + X_r\bar{X}_r.$$

The variety $\mathbf{V}(F_r)$ in $\mathrm{PG}(n, q)$ is a *Hermitian variety*, which is non-singular when $r = n$. The Hermitian variety $\mathbf{V}(F_r)$ is written $\mathcal{U}_r$ or $\mathcal{U}_{r,q}$; that is,

$$\mathcal{U}_r = \mathbf{V}(X_0\bar{X}_0 + X_1\bar{X}_1 + \cdots + X_r\bar{X}_r). \tag{2.1}$$

Similarly to quadrics, $\mathbf{V}(F_r) = \Pi_{n-r-1}\mathcal{U}_r$, where $\mathcal{U}_r$ is the non-singular Hermitian variety in the r-space $\mathbf{U}_0\mathbf{U}_1 \cdots \mathbf{U}_r = \mathbf{V}(X_{r+1}, \ldots, X_n)$ and

$$\Pi_{n-r-1} = \mathbf{V}(X_0, \ldots, X_r) = \mathbf{U}_{r+1}\mathbf{U}_{r+2}\cdots\mathbf{U}_n;$$

that is, the points of $\Pi_{n-r-1}\mathcal{U}_r$ comprise all the points of the lines joining any point of Π_{n-r-1} to any point of $\mathcal{U}_r$. As for quadrics, Π_{n-r-1} is the vertex of the cone and $\mathcal{U}_r$ a base.

Theorem 2.1. *There are $n + 1$ projectively distinct Hermitian varieties in $\mathrm{PG}(n, q)$.*

Proof. From above, there is one variety $\Pi_{n-r-1}\mathcal{U}_r$ for each r in $\{0, 1, \ldots, n\}$. □

Lemma 2.2. *A section of a Hermitian variety is still a Hermitian variety.*

© Springer-Verlag London 2016
J.W.P. Hirschfeld, J.A. Thas, *General Galois Geometries*, Springer Monographs in Mathematics, DOI 10.1007/978-1-4471-6790-7_2

Proof. If $\mathcal{U} = \mathbf{V}(F)$ is Hermitian with F in $\mathbf{F}_q[X_0, X_1, \ldots, X_n]$, then a section by a hyperplane $\mathbf{V}(L)$, where $L = X_0 - a_1X_1 - \cdots - a_nX_n$, is $\mathcal{U}' = \mathbf{V}(L, G)$, where

$$G(X_1, \ldots, X_n) = F(a_1X_1 + \cdots + a_nX_n, X_1, \ldots, X_n).$$

As G is a Hermitian form, so $\mathcal{U}'$ is a Hermitian variety in $\mathbf{V}(L)$. Thus the result follows by induction. $\qquad\qquad\square$

The behaviour of low-dimensional varieties gives a feeling for higher dimensional ones. In Table 2.1, all types up to five dimensions are described.

2.2 Tangency and polarity

The notation $\mathcal{U}_n$ or $\mathcal{U}_{n,q}$ is also used for any non-singular Hermitian variety of $\mathrm{PG}(n, q)$. So consider $\mathcal{U}_n = \mathbf{V}(F)$. Let $P = \mathbf{P}(A)$ and $Q = \mathbf{P}(B)$, where $A = (a_0, \ldots, a_n)$ and $B = (b_0, \ldots, b_n)$. If $F(X) = XH\bar{X}^*$, then

$$F(A + tB) = (A + tB)H(\bar{A} + \bar{t}\bar{B})^*$$
$$= AH\bar{A}^* + tBH\bar{A}^* + \bar{t}AH\bar{B}^* + t\bar{t}BH\bar{B}^*.$$

So $\mathbf{P}(A + tB)$ lies on $\mathcal{U}_n$ if and only if

$$0 = F(A) + tG(A, B) + \bar{t}\overline{G(A, B)} + t\bar{t}F(B), \qquad (2.2)$$

where $G(A, B) = BH\bar{A}^*$.

The line l is a *tangent* to $\mathcal{U}_n$ if $|l \cap \mathcal{U}_n| = 1$.

Lemma 2.3. *Let $P = \mathbf{P}(A)$ and $Q = \mathbf{P}(B)$.*

(i) *If $F(A)F(B) - G(A, B)\overline{G(A, B)} \neq 0$, then $|PQ \cap \mathcal{U}_n| = \sqrt{q} + 1$.*
(ii) *If $F(A)F(B) - G(A, B)\overline{G(A, B)} = 0$, then*
    (a) *$F(A) = F(B) = G(A, B) = 0 \Rightarrow PQ \subset \mathcal{U}_n$;*
    (b) *otherwise $|PQ \cap \mathcal{U}_n| = 1$.*
(iii) *Suppose $P \in \mathcal{U}_n$.*
    (a) *If $Q \notin \mathcal{U}_n$, then PQ is a tangent to $\mathcal{U}_n$ if and only if $G(A, B) = 0$.*
    (b) *If $Q \in \mathcal{U}_n$, then PQ is a line of $\mathcal{U}_n$ if and only if $G(A, B) = 0$.*

Proof. (i), (ii) With $F(A) = a$, $F(B) = b$, $G(A, B) = \delta$, (2.2) becomes

$$bt\bar{t} + \bar{\delta}\bar{t} + \delta t + a = 0; \qquad (2.3)$$

here $\bar{b} = b$, $\bar{a} = a$.

First, let $a = b = 0$. If $\delta = 0$, then $PQ \subset \mathcal{U}_n$. Now, let $\delta \neq 0$. Then (2.3) has $t = 0$ and $t = \infty$ as solutions corresponding to P and Q on $\mathcal{U}_n$. Every other t satisfies

$$t^{\sqrt{q}-1} = -\delta^{-(\sqrt{q}-1)}, \qquad (2.4)$$

Table 2.1. Hermitian varieties in $\mathrm{PG}(n, q)$ for $n \leq 5$

$\mathrm{PG}(0, q)$ $\mathcal{U}_0 = \mathbf{V}(X_0 \bar{X}_0)$ is empty

$\mathrm{PG}(1, q)$ $\mathcal{U}_1 = \mathbf{V}(X_0 \bar{X}_0 + X_1 \bar{X}_1)$ is $\sqrt{q} + 1$ points forming a subline $\mathrm{PG}(1, \sqrt{q})$
$\qquad\qquad \Pi_0 \mathcal{U}_0 = \mathbf{V}(X_0 \bar{X}_0)$ is the single point $\mathbf{U}_1$

$\mathrm{PG}(2, q)$ $\mathcal{U}_2 = \mathbf{V}(X_0 \bar{X}_0 + X_1 \bar{X}_1 + X_2 \bar{X}_2)$ is a Hermitian arc (unital) comprising
$\qquad\qquad$ $q\sqrt{q} + 1$ points; through each point of $\mathcal{U}_2$ there is a unique line
$\qquad\qquad$ meeting $\mathcal{U}_2$ in a $\Pi_0 \mathcal{U}_0$, whereas all other lines meet $\mathcal{U}_2$ in a $\mathcal{U}_1$
$\qquad\qquad \Pi_0 \mathcal{U}_1 = \mathbf{V}(X_0 \bar{X}_0 + X_1 \bar{X}_1)$ comprises $\sqrt{q} + 1$ lines concurrent at $\mathbf{U}_2$
$\qquad\qquad \Pi_1 \mathcal{U}_0 = \mathbf{V}(X_0 \bar{X}_0)$ is the single line $\mathbf{u}_0$

$\mathrm{PG}(3, q)$ $\mathcal{U}_3 = \mathbf{V}(X_0 \bar{X}_0 + X_1 \bar{X}_1 + X_2 \bar{X}_2 + X_3 \bar{X}_3)$ comprises $(q+1)(q\sqrt{q}+1)$
$\qquad\qquad$ points on $(\sqrt{q}+1)(q\sqrt{q}+1)$ lines; there are as many plane sections
$\qquad\qquad$ $\Pi_0 \mathcal{U}_1$ as points and the remaining plane sections are of type $\mathcal{U}_2$
$\qquad\qquad \Pi_0 \mathcal{U}_2 = \mathbf{V}(X_0 \bar{X}_0 + X_1 \bar{X}_1 + X_2 \bar{X}_2)$ is a cone comprising the join of the
$\qquad\qquad$ vertex $\mathbf{U}_3$ to a Hermitian curve
$\qquad\qquad \Pi_1 \mathcal{U}_1 = \mathbf{V}(X_0 \bar{X}_0 + X_1 \bar{X}_1)$ is $\sqrt{q} + 1$ collinear planes
$\qquad\qquad \Pi_2 \mathcal{U}_0 = \mathbf{V}(X_0 \bar{X}_0)$ is the plane $\mathbf{u}_0$

$\mathrm{PG}(4, q)$ $\mathcal{U}_4 = \mathbf{V}(\sum_{i=0}^{4} X_i \bar{X}_i)$ comprises $(q+1)(q^2\sqrt{q}+1)$ points on
$\qquad\qquad$ $(q\sqrt{q}+1)(q^2\sqrt{q}+1)$ lines with $q\sqrt{q}+1$ lines through each point
$\qquad\qquad \Pi_0 \mathcal{U}_3 = \mathbf{V}(\sum_{i=0}^{3} X_i \bar{X}_i)$ is a cone with vertex $\mathbf{U}_4$ and base a Hermitian
$\qquad\qquad$ surface $\mathcal{U}_3$; its generators are planes
$\qquad\qquad \Pi_1 \mathcal{U}_2 = \mathbf{V}(X_0 \bar{X}_0 + X_1 \bar{X}_1 + X_2 \bar{X}_2)$ is a cone with vertex the line $\mathbf{U}_3 \mathbf{U}_4$
$\qquad\qquad$ and base a Hermitian curve; its generators are planes
$\qquad\qquad \Pi_2 \mathcal{U}_1 = \mathbf{V}(X_0 \bar{X}_0 + X_1 \bar{X}_1)$ comprises $\sqrt{q} + 1$ solids through the plane
$\qquad\qquad$ $\mathbf{U}_2 \mathbf{U}_3 \mathbf{U}_4$
$\qquad\qquad \Pi_3 \mathcal{U}_0 = \mathbf{V}(X_0 \bar{X}_0)$ is the solid $\mathbf{u}_0$

$\mathrm{PG}(5, q)$ $\mathcal{U}_5 = \mathbf{V}(\sum_{i=0}^{5} X_i \bar{X}_i)$ comprises $(q^2 + q + 1)(q^2\sqrt{q}+1)$ points on
$\qquad\qquad$ $(\sqrt{q}+1)(q\sqrt{q}+1)(q^2\sqrt{q}+1)$ planes
$\qquad\qquad \Pi_0 \mathcal{U}_4 = \mathbf{V}(\sum_{i=0}^{4} X_i \bar{X}_i)$ is a cone with vertex $\mathbf{U}_5$ and base $\mathcal{U}_4$; its generators
$\qquad\qquad$ are planes
$\qquad\qquad \Pi_1 \mathcal{U}_3 = \mathbf{V}(\sum_{i=0}^{3} X_i \bar{X}_i)$ is a cone with vertex the line $\mathbf{U}_4 \mathbf{U}_5$ and base the
$\qquad\qquad$ surface $\mathcal{U}_3$; its generators are solids
$\qquad\qquad \Pi_2 \mathcal{U}_2 = \mathbf{V}(X_0 \bar{X}_0 + X_1 \bar{X}_1 + X_2 \bar{X}_2)$ comprises $q\sqrt{q} + 1$ solids through the
$\qquad\qquad$ plane $\mathbf{U}_3 \mathbf{U}_4 \mathbf{U}_5$
$\qquad\qquad \Pi_3 \mathcal{U}_1 = \mathbf{V}(X_0 \bar{X}_0 + X_1 \bar{X}_1)$ is $\sqrt{q} + 1$ hyperplanes through the solid $\mathbf{u}_0 \cap \mathbf{u}_1$
$\qquad\qquad \Pi_4 \mathcal{U}_0 = \mathbf{V}(X_0 \bar{X}_0)$ is the hyperplane $\mathbf{u}_0$

---

which has $\sqrt{q} - 1$ solutions by PGOFF2, Section 1.5(v). Thus $|PQ \cap \mathcal{U}_n| = \sqrt{q} + 1$.

If $b = 0$, $a \neq 0$, then the substitution $t \mapsto t^{-1}$ gives an equation of the form (2.3) with $b \neq 0$.

Finally, if $b \neq 0$, the substitution $t \mapsto t - \bar{\delta}/b$ transforms (2.3) to

$$t\bar{t} + (ab - \delta\bar{\delta})/b^2 = 0. \tag{2.5}$$

If $ab - \delta\bar{\delta} = 0$, then $|PQ \cap \mathcal{U}_n| = 1$; if $ab - \delta\bar{\delta} \neq 0$, then, again by PGOFF2, Section 1.5(v), (2.5) has $\sqrt{q} + 1$ solutions, which means that $|PQ \cap \mathcal{U}_n| = \sqrt{q} + 1$.

(iii) This follows immediately from (i) and (ii) with $F(A) = 0$. $\square$

When $G(A, B) = 0$, the points $P = \mathbf{P}(A)$ and $Q = \mathbf{P}(B)$ are *conjugate*. With $X = (X_0, X_1, \ldots, X_n)$, the hyperplane $\mathbf{V}(G(X, A))$ is the *polar hyperplane* of P and is denoted $P\mathfrak{U}$, the image of P under the Hermitian polarity $\mathfrak{U}$ with matrix H. If $\Pi_r = P_0 P_1 \cdots P_r$, then the polar space $\Pi_{n-r-1} = \Pi_r \mathfrak{U}$ of Π_r is

$$P_0 \mathfrak{U} \cap P_1 \mathfrak{U} \cap \cdots \cap P_r \mathfrak{U}.$$

This is independent of the choice of $P_0, P_1, \ldots, P_r$ in Π_r in the sense that, if

$$\Pi_r = P_0 P_1 \cdots P_r = Q_0 Q_1 \cdots Q_r,$$

then $P_0 \mathfrak{U} \cap P_1 \mathfrak{U} \cap \cdots \cap P_r \mathfrak{U} = Q_0 \mathfrak{U} \cap Q_1 \mathfrak{U} \cap \cdots \cap Q_r \mathfrak{U}$. Two spaces are *conjugate* if they are contained in polar spaces.

When $P \in \mathcal{U}_n$, the polar hyperplane of P is also the *tangent hyperplane at P* and written T_P or $T_P(\mathcal{U}_n)$. Similarly, if $\Pi_r = P_0 P_1 \cdots P_r \subset \mathcal{U}_n$, then the *tangent space at Π_r* is $T_{\Pi_r} = \Pi_{n-r-1} = \bigcap_i T_{P_i}$.

For $\mathcal{U}_n$ in canonical form, that is, $\mathcal{U}_n = \mathbf{V}(X_0 \bar{X}_0 + X_1 \bar{X}_1 + \cdots + X_n \bar{X}_n)$,

$$G(X, A) = \sum_{i=0}^{n} \bar{a}_i X_i. \tag{2.6}$$

Lemma 2.4. *If $\Pi_r \subset \mathcal{U}_n$, then the tangent space at Π_r is a Π_{n-r-1} which contains Π_r and any Π_s on $\mathcal{U}_n$ through Π_r.*

Proof. This follows from Lemma 2.3 and the definition of tangent space. $\square$

It should be observed that the polarity of $\mathcal{U}_n$ lies behind its reduction to the canonical form (2.1). For, let $\mathcal{U}$ be any non-singular Hermitian variety with polarity $\mathfrak{U}$ in $\Pi_n = \mathrm{PG}(n, q)$. Choose P_0 in $\Pi_n \backslash \mathcal{U}$ and let $\pi_0 = P_0 \mathfrak{U}$. Now, choose P_1 in $\pi_0 \backslash \mathcal{U}$ and let $\pi_1 = P_1 \mathfrak{U}$. Continue this process and choose $P_0, P_1, \ldots, P_n$ so that $P_0, P_1, \ldots, P_i$ span an i-space and so that P_i lies in $(\pi_0 \cap \pi_1 \cap \cdots \cap \pi_{i-1}) \backslash \mathcal{U}$. Take $\mathbf{U}_i = P_i$, $i = 0, 1, \ldots, n$, as the vertices of the simplex of reference. Hence

$$\mathcal{U} = \mathbf{V}(c_0 X_0 \bar{X}_0 + c_1 X_1 \bar{X}_1 + \cdots + c_n X_n \bar{X}_n).$$

Now, a suitable choice of the unit point or, equivalently, the projective transformation $x_i' = d_i x_i$, $i = 0, 1, \ldots, n$, with $d_i \bar{d}_i = c_i$, gives the required form

$$\mathcal{U} = \mathbf{V}(X_0 \bar{X}_0 + X_1 \bar{X}_1 + \cdots + X_n \bar{X}_n),$$

where the dashes have been omitted.

When considering spaces lying on $\mathcal{U}$, another canonical form is useful. Since $\mathcal{U}_n$ is projectively unique, it can also be written as $\mathbf{V}(G_n)$, where

$$G_n(X_0, \ldots, X_n)$$
$$= \begin{cases} X_0\bar{X}_0 + (X_1\bar{X}_2 + X_2\bar{X}_1) + \cdots + (X_{n-1}\bar{X}_n + X_n\bar{X}_{n-1}), & n \text{ even,} \\ (X_0\bar{X}_1 + X_1\bar{X}_0) + \cdots + (X_{n-1}\bar{X}_n + X_n\bar{X}_{n-1}), & n \text{ odd.} \end{cases} \quad (2.7)$$

Lemma 2.5. *The tangent hyperplane at a point P of $\mathcal{U}_n$ is a cone $\Pi_0 \mathcal{U}_{n-2}$.*

Proof. Let $P = \mathbf{U}_0$ and choose $\mathbf{U}_1 \in \mathcal{U}_n$, but not in the tangent prime T_P. Choose $\mathbf{U}_2, \ldots, \mathbf{U}_n$ in $T_P \cap T_{\mathbf{U}_1}$. Then, by a suitable choice of unit point, $\mathcal{U}_n = \mathbf{V}(F)$, where

$$F = X_0\bar{X}_1 + X_1\bar{X}_0 + F'(X_2, \ldots, X_n).$$

Then F' can be reduced to canonical form so that

$$F = X_0\bar{X}_1 + X_1\bar{X}_0 + X_2\bar{X}_2 + \cdots + X_n\bar{X}_n.$$

So

$$\begin{aligned} \mathcal{U}_n \cap T_P &= \mathbf{V}(X_1, X_2\bar{X}_2 + \cdots + X_n\bar{X}_n) \\ &= \{\mathbf{P}(\lambda, 0, x_2, \ldots, x_n) \mid x_2\bar{x}_2 + \cdots + x_n\bar{x}_n = 0\} \\ &= \mathbf{U}_0 \mathcal{U}_{n-2}. \end{aligned}$$ □

Notation 2.6. Let $\mu_n = |\mathcal{U}_n|$.

Corollary 2.7. *The tangent hyperplane at P meets $\mathcal{U}_n$ in $q\mu_{n-2} + 1$ points.*

Theorem 2.8. *The number of points on $\mathcal{U}_n$ is*

$$\mu_n = \{(\sqrt{q})^{n+1} + (-1)^n\}\{(\sqrt{q})^n - (-1)^n\}/(q-1) \quad (2.8)$$
$$= \theta(n-1) + \{q^n - (-\sqrt{q})^n\}/(\sqrt{q}+1). \quad (2.9)$$

Proof. Any line through a point P of $\mathcal{U}_n$ not in T_P meets $\mathcal{U}_n$ in $\sqrt{q}$ other points. Hence

$$\begin{aligned} \mu_n &= \sqrt{q}\{\theta(n-1) - \theta(n-2)\} + q\mu_{n-2} + 1 \\ &= q^{n-1}\sqrt{q} + 1 + q\mu_{n-2}. \end{aligned}$$

Put $\alpha_n = \mu_n/(\sqrt{q})^n$. Then

$$\alpha_n = (\sqrt{q})^{n-1} + (1/\sqrt{q})^n + \alpha_{n-2}.$$

Separate calculations for n even and odd give the desired result. □

Now let $\mu_n^{(t)} = |\Pi_t \mathcal{U}_{n-t-1}|$; thus $\mu_n = \mu_n^{(-1)}$.

Corollary 2.9. $\mu_n^{(t)} = \theta(n-1) + \{q^n - (-\sqrt{q})^{n+t+1}\}/(\sqrt{q}+1).$ $\quad (2.10)$

Proof. Any two points of the base $\mathcal{U}_{n-t-1}$ are joined to the vertex Π_t by two Π_{t+1} having precisely the vertex in common. Therefore

$$\mu_n^{(t)} = \theta(t) + \{\theta(t+1) - \theta(t)\}\mu_{n-t-1}$$

and the result follows. $\qquad\square$

Let $\mathcal{N}(\mathcal{U}_n)$ be the set of non-singular Hermitian varieties in $\mathrm{PG}(n, q)$.

Lemma 2.10. *The number of Hermitian varieties $\mathcal{U}_n$ containing a given $\Pi_0 \mathcal{U}_{n-2}$ as tangent cone is*

$$(q-1)q^{n-1}\sqrt{q}\,.$$

Proof. Let $\Pi_0 = \mathbf{U}_{n-1}$ and $\mathcal{U}_{n-2} = \mathbf{V}(X_n, X_{n-1}, F_{n-2})$, where

$$F_{n-2} = X_0\bar{X}_0 + \cdots + X_{n-2}\bar{X}_{n-2}.$$

The hyperplane containing $\Pi_0 \mathcal{U}_{n-2}$ is $\mathbf{u}_n$. Any Hermitian variety containing $\mathbf{U}_{n-1}$ and $\mathcal{U}_{n-2}$ has the form

$$\mathbf{V}(X_n\bar{f} + \bar{X}_n f + X_{n-1}\bar{g} + \bar{X}_{n-1}g + F_{n-2}),$$

where f is linear in $X_0, \ldots, X_n$ and g is linear in $X_0, \ldots, X_{n-2}$. Since the tangent hyperplane at $\mathbf{U}_{n-1}$ is $\mathbf{u}_n$, so the only term involving $\bar{X}_{n-1}$ is $cX_n\bar{X}_{n-1}$ with $c \neq 0$ and hence the only term involving X_{n-1} is $\bar{c}X_{n-1}\bar{X}_n$; so $g = 0$.

The non-singularity of $\mathcal{U}_n$ is equivalent to $c \neq 0$. Thus the number of choices for the form f and thus for $\mathcal{U}_n$ is $(q-1)q^{n-1}\sqrt{q}$. This argument relies on the fact that, if $\mathcal{U}_n = \mathbf{V}(F)$, then $\mathcal{U}_n$ defines F uniquely up to a scalar multiple. $\qquad\square$

Lemma 2.11. *The number of cones $\Pi_0 \mathcal{U}_{n-2}$ in $\mathrm{PG}(n, q)$ is*

$$\theta(n)\theta(n-1)\,|\mathcal{N}(\mathcal{U}_{n-2})|\,.$$

Proof.   The number of cones

$$= \text{number of } \Pi_{n-1} \text{ in } \mathrm{PG}(n, q) \times \text{number of } \Pi_0 \text{ in } \Pi_{n-1}$$
$$\times \text{ number of } \mathcal{U}_{n-2} \text{ in a fixed } \Pi_{n-2} \text{ of } \Pi_{n-1}. \qquad\square$$

Notation 2.12. Let

$$\overline{[r, s]} = \begin{cases} \prod_{i=r}^{i=s} \{(\sqrt{q})^i - (-1)^i\} & \text{for } r \leq s, \\ 1 & \text{for } r > s. \end{cases}$$

Theorem 2.13. *The total number of Hermitian varieties in $\mathrm{PG}(n, q)$ is*

$$|\mathcal{N}(\mathcal{U}_n)| = q^{n(n+1)/4}[2, n+1]_- / \overline{[2, n+1]}.$$

Proof. Count $\{(\mathcal{U}_n, \mathcal{S}) \mid \mathcal{U}_n$ a Hermitian variety, $\mathcal{S}$ a tangent cone of $\mathcal{U}_n\}$ in two ways:

$$|\mathcal{N}(\mathcal{U}_n)|\, \mu_n = \theta(n)\theta(n-1)\,|\mathcal{N}(\mathcal{U}_{n-2})|\,(q-1)q^{n-1}\sqrt{q}.$$

Hence

$$|\mathcal{N}(\mathcal{U}_n)| = \frac{\theta(n)\theta(n-1)(q-1)q^{n-1}\sqrt{q}}{\mu_n}\,|\mathcal{N}(\mathcal{U}_{n-2})|.$$

Repetition of this recurrence relation gives the result, after a separate calculation for n even and odd; it is only required that $|\mathcal{N}(\mathcal{U}_0)| = 1$ and $|\mathcal{N}(\mathcal{U}_1)| = \sqrt{q}(q+1)$, from PGOFF2, Section 6.2. $\square$

The group $G = G(\mathcal{U}_n)$, the *unitary* group, is the group of projectivities fixing $\mathcal{U}_n$.

Corollary 2.14.

$$|G| = |G(\mathcal{U}_n)| = |\mathrm{PGU}(n+1, q)| = q^{n(n+1)/4}\overline{[2, n+1]}.$$

Proof. All $\mathcal{U}_n$ are projectively equivalent; in other words, $\mathcal{N}(\mathcal{U}_n)$ is a single orbit under $\mathrm{PGL}(n+1, q)$. So

$$|G| = |\mathrm{PGL}(n+1, q)|/|\mathcal{N}(\mathcal{U}_n)|.$$

As $|\mathrm{PGL}(n+1, q)| = q^{n(n+1)/2}[2, n+1]_-$, the result follows. $\square$

Lemma 2.15. *If* $\Pi_m \cap \mathcal{U}_n = \Pi_v \mathcal{U}_s$, *then the polar space* Π'_{n-m-1} *of* Π_m *satisfies the following:*

(i) $\Pi'_{n-m-1} \cap \mathcal{U}_n$ *also has singular space* Π_v;
(ii) $\Pi_m \cap \Pi'_{n-m-1} = \Pi_v$.

Proof. The point P is in Π_v if and only if every line l through P in Π_m is either a tangent or lies in $\mathcal{U}_n$. This is precisely the condition for P to be conjugate to every point of Π_m; hence $\Pi_v \subset \Pi'_{n-m-1}$. Since a point P in Π_v is conjugate to every point of Π'_{n-m-1}, so Π_v belongs to the singular space Π'_w of $\Pi'_{n-m-1} \cap \mathcal{U}_n$. Similarly, $\Pi'_w \subset \Pi_v$ and consequently $\Pi'_w = \Pi_v$.

Since every point P of $\Pi_m \cap \Pi'_{n-m-1}$ is conjugate to all points of $\Pi_m \cup \Pi'_{n-m-1}$, it belongs to the singular space Π_v both of $\Pi_m \cap \mathcal{U}_n$ and of $\Pi'_{n-m-1} \cap \mathcal{U}_n$. Hence $\Pi_v = \Pi_m \cap \Pi'_{n-m-1}$. $\square$

Corollary 2.16. *If* $\Pi_m \cap \mathcal{U}_n = \Pi_v \mathcal{U}_s$, *then* $\Pi'_{n-m-1} \cap \mathcal{U}_n = \Pi_v \mathcal{U}_{T-1}$, *where* Π_m *and* Π'_{n-m-1} *are polar spaces, and* $T = n - 2m + s$.

2.3 Generators and sub-generators

The following definitions are similar to those for quadrics.

Definition 2.17. (1) A *generator* of $\mathcal{U}_n$ is a subspace of maximum dimension lying on $\mathcal{U}_n$.
(2) A *sub-generator* of $\mathcal{U}_n$ is any subspace lying on $\mathcal{U}_n$.
(3) The dimension of a generator is the *projective index* of $\mathcal{U}_n$; as for quadrics, it is denoted by $g = g(\mathcal{U}_n)$.

Lemma 2.18. *The projective index* $g(\mathcal{U}_n) = \lfloor \frac{1}{2}(n-1) \rfloor$.

Proof. If Π_g is a generator, then by Lemma 2.4 it lies in its own tangent space Π'_{n-g-1}. So $g \leq n - g - 1$ and hence $g \leq \frac{1}{2}(n-1)$.

When n is even, $\mathbf{V}(X_0, X_1, X_3, \ldots, X_{n-1})$ is a space of dimension $\frac{1}{2}n - 1$ on

$$\mathbf{V}(X_0 \bar{X}_0 + (X_1 \bar{X}_2 + X_2 \bar{X}_1) + \cdots + (X_{n-1}\bar{X}_n + X_n \bar{X}_{n-1}));$$

when n is odd, $\mathbf{V}(X_0, X_2, X_4, \ldots, X_{n-1})$ is a space of dimension $\frac{1}{2}(n-1)$ on

$$\mathbf{V}((X_0 \bar{X}_1 + X_1 \bar{X}_0) + \cdots + (X_{n-1}\bar{X}_n + X_n \bar{X}_{n-1})).$$

So the upper bound is achieved. □

Now the number of sub-generators of a given dimension is calculated.

Theorem 2.19. *The number of* Π_r *on* $\mathcal{U}_n$ *is*

$$\nu_{r,n} = N(\Pi_r, \mathcal{U}_n) = \overline{[n-2r, n+1]}/[1, r+1]_- . \tag{2.11}$$

Proof. The following set is counted in two ways:

$$\{(\Pi_r, \Pi_{r-1}) \mid \Pi_r \subset \mathcal{U}_n, \Pi_{r-1} \subset \Pi_r\}.$$

Hence

$$\nu_{r,n}\, \theta(r) = \nu_{r-1,n}\, M, \tag{2.12}$$

where M is the number of Π_r on $\mathcal{U}_n$ through a given Π_{r-1}.

To calculate M, consider the polar Π'_{n-r} of Π_{r-1}. It contains Π_{r-1} and meets $\mathcal{U}_n$ in $\Pi_{r-1}\mathcal{U}_{n-2r}$ with the same Π_{r-1} as vertex, by Lemma 2.15. So take a Π_{n-2r} in Π'_{n-r} skew to Π_{r-1}. It meets $\mathcal{U}_n$ in a $\mathcal{U}_{n-2r}$ whose points joined to Π_{r-1} form the Π_r of $\mathcal{U}_n$ containing Π_{r-1}. Thus $M = \mu_{n-2r}$.

Now, (2.12) becomes

$$\nu_{r,n}\, \theta(r) = \nu_{r-1,n}\, \mu_{n-2r}. \tag{2.13}$$

Iteration gives that

$$\nu_{r,n} = \frac{\mu_{n-2r}\mu_{n-2(r-1)} \cdots \mu_n}{\theta(r)\theta(r-1) \cdots \theta(0)}.$$

Since $\theta(i) = (q^{i+1} - 1)/(q-1)$ and μ_j is given by (2.8), the result follows. □

2.4 Sections of $\mathcal{U}_n$

If $\Pi_m \subset \mathrm{PG}(n, q)$, then from Theorem 2.1 there are $m + 1$ types of Hermitian varieties in Π_m, namely $\Pi_{m-s-1}\mathcal{U}_s$ for $s = 0, \ldots, m$. However, when the intersection of Π_m with $\mathcal{U}_n$ is considered, it may also occur that Π_m lies entirely on $\mathcal{U}_n$. Suppose therefore that

$$\Pi_m \cap \mathcal{U}_n = \Pi_v \mathcal{U}_s ; \tag{2.14}$$

then

$$s + v = m - 1. \tag{2.15}$$

The section is non-singular when $v = -1$, and Π_m lies entirely on $\mathcal{U}_n$ when $s = -1$. The parameters n, m, v, s satisfy the following:

$$-1 \le v \le m, \tag{2.16}$$
$$-1 \le s \le m, \tag{2.17}$$
$$0 \le m \le n - 1. \tag{2.18}$$

The question now is to determine for what values of v and s there is a section as in (2.14).

Lemma 2.20. *There is a section* $\Pi_m \cap \mathcal{U}_n = \Pi_v \mathcal{U}_s$ *if and only if*

$$T \ge 0, \tag{2.19}$$

where

$$T = n - 2m + s. \tag{2.20}$$

Proof. When $m \le \lfloor \frac{1}{2}(n - 1) \rfloor$, the condition $T \ge 0$ means that $m \ge s \ge -1$, so that (2.19) is equivalent to (2.17); then it must be shown that $\Pi_v \mathcal{U}_s$ exists for each s in $I = \{-1, 0, 1, \ldots, m\}$ or, equivalently, for each v in I. When $m > \lfloor \frac{1}{2}(n - 1) \rfloor$, condition (2.19) means that $m \ge s \ge 2m - n$ or, equivalently, $-1 \le v \le n - m - 1$. Thus, if the result is established for $m \le \lfloor \frac{1}{2}(n - 1) \rfloor$, the polarity gives the result for $m > \lfloor \frac{1}{2}(n - 1) \rfloor$.

Now let $\Pi_m = \mathbf{U}_0 \mathbf{U}_1 \cdots \mathbf{U}_m = \mathbf{V}(X_{m+1}, \ldots, X_n)$ with $m \le \lfloor \frac{1}{2}(n - 1) \rfloor$. For $-1 \le s \le m$, write

$$K_s = \sum_{i=0}^{s} X_i \bar{X}_i + \sum_{j=s+1}^{m} (X_j \bar{X}_{2m+1-j} + X_{2m+1-j}\bar{X}_j) + \sum_{i=m+1}^{n} X_i \bar{X}_i .$$

Then $\mathbf{V}(K_s)$ is non-singular and

$$\mathbf{V}(K_s) \cap \Pi_m = \mathbf{V}\left(\sum_{i=0}^{s} X_i \bar{X}_i, X_{m+1}, \ldots, X_n \right)$$

$$= \Pi_v \mathcal{U}_s.$$

$\square$

Corollary 2.21. *For fixed m and n, the number of projectively inequivalent sections $\Pi_m \cap \mathcal{U}_n$ of $\mathcal{U}_n$ is*

$$m + 2 \quad \text{when } m \le \lfloor \tfrac{1}{2}(n-1) \rfloor,$$
$$n - m + 1 \text{ when } m > \lfloor \tfrac{1}{2}(n-1) \rfloor.$$

The next result looks at the orbits of subspaces Π_m under the action of the unitary group $G(\mathcal{U}_n) = \mathrm{PGU}(n+1, q)$.

Theorem 2.22. (i) *Two subspaces are in the same orbit of $G(\mathcal{U}_n)$ if and only if they have the same parameters m and s, where m is their dimension and $\mathcal{U}_s$ is the type of base of their intersection with $\mathcal{U}_n$.*
(ii) *If there is a projectivity $\mathfrak{T} : \Pi_m \to \Pi'_m$ such that $(\Pi_m \cap \mathcal{U}_n)\mathfrak{T} = \Pi'_m \cap \mathcal{U}_n$, then it can be extended to an element of $G(\mathcal{U}_n)$.*

Proof. The variety $\mathcal{U}_n$ and the space Π_m are reduced simultaneously to canonical form. Consider the section $\Pi_v \mathcal{U}_s$, where $\mathcal{U}_s \subset \Pi_s$; thus Π_v and Π_s are skew. The polar of Π_m is Π'_t with $t = n - m - 1$.

By Lemma 2.15, Π'_t contains Π_v. Choose in Π'_t a space Π'_w skew to Π_v with $\Pi'_w \Pi_v = \Pi'_t$; so

$$w = t - v - 1 = n - m - v - 2.$$

Since $w \ge -1$, (2.19) is satisfied. By construction, Π'_w is conjugate to Π_m. The polar Π_d of Π'_w has dimension

$$d = n - w - 1 = m + v + 1.$$

Also Π_d contains Π_m. Choose in Π_d a space Π'_v skew to Π_m with $\Pi'_v \Pi_m = \Pi_d$ in a way that is specified below.

The set of points P of Π_m such that P is conjugate to every point of Π_m is Π_v, and Π_v is the same set with respect to Π'_t. Hence, when $w \ne -1$, the space Π'_w meets $\mathcal{U}_n$ in a non-singular Hermitian variety. So Π'_w and its polar space Π_d are skew, as any intersection would be singular for such a Hermitian variety, Lemma 2.15. Thus $\Pi'_w \Pi_d = \mathrm{PG}(n, q)$ and $\Pi_d \cap \mathcal{U}_n$ is non-singular.

Take in Π_d the polar space Π_{2v+1} of Π_s with respect to $\Pi_d \cap \mathcal{U}_n$. In Π_{2v+1} take Π'_v skew to Π_v with $\Pi'_v \cap \mathcal{U}_n$ non-singular. Then, with $S_m = \Pi'_v \Pi_s$, the set $S_m \cap \Pi_v$ is empty and $S_m \cap \mathcal{U}_n$ is non-singular. Also, in S_m, the polar of Π_s with respect to $S_m \cap \mathcal{U}_n$ is the v-space Π'_v, which is skew to both Π_v and Π_s. Thus $\Pi'_v \cap \mathcal{U}_n$ is non-singular and the polarity $\mathfrak{U}$ of $\mathcal{U}_n$ induces a reciprocity, as in PGOFF2, Section 2.1, between Π'_v and Π_v.

Thus there are four mutually skew spaces

$$\Pi_s, \; \Pi_v, \; \Pi'_w, \; \Pi'_v.$$

By construction, the spaces Π_s, Π'_w, Π'_v are skew and mutually conjugate. Their join is a space Π_e with

$$e = s + w + v + 2 = m + w + 1,$$

and $\Pi_e \cap \mathcal{U}_n$ is non-singular. For, if P is singular on $\Pi_e \cap \mathcal{U}_n$, then P is conjugate to all points of Π'_w and to all points of S_m; so $P \in \Pi'_w \cap S_m$, a contradiction.

As in the derivation of the canonical form following Lemma 2.4, in the three spaces Π_s, Π'_w, Π'_v it is possible to choose $s+1, w+1, v+1$ points so that they form $e+1$ independent points in Π_e with any two of these points conjugate.

The reciprocity induced between Π'_v and Π_v by $\mathfrak{U}$ transforms the $v+1$ points of Π'_v, which may be considered as vertices of a simplex, to $v+1$ faces of a simplex in Π_v; hence the vertices of this simplex in Π_v form $v+1$ independent points. These $v+1$ and the above $e+1$ give $n+1$ independent points. Thus, in a suitable order, these points are the vertices of the simplex of reference of a coordinate system. Correspondingly,

$$\Pi_m = \Pi_v \Pi_s = \mathbf{V}(X_{m+1}, X_{m+2}, \dots, X_n).$$

Table 2.2. Equation for $\mathcal{U}_n$

	Π_s	Π_v	Π'_w	Π'_v
Π_s	H_1	0	0	0
Π_v	0	0	0	H_4
Π'_w	0	0	H_2	0
Π'_v	0	$\bar{H}_4$	0	H_3

Then $\mathcal{U}_n = \mathbf{V}(XH\bar{X}^*)$ with $\bar{H}^* = H$, where H is the matrix of Table 2.2; here, H_1, H_2, H_3, H_4 are square diagonal submatrices, the three matrices H_1, H_2, H_3 have all their elements in $\mathbf{F}_{\sqrt{q}}$, and the elements of $\bar{H}_4$ are the conjugates of the elements of H_4. The zero submatrices indicate the conjugacy of the spaces bordering the matrix H. A suitable choice of the unit point allows H_i to be reduced to the identity matrix of the appropriate order. Thus the simultaneous reduction of $\mathcal{U}_n$ and $\Pi_m \cap \mathcal{U}_n$ to canonical form has been achieved.

The first part of the theorem now follows since the canonical forms are determined by the integers n, m, v.

To prove the second part, note that, in the reduction to canonical form, the $m+1$ reference points chosen in Π_m consist of $v+1$ in Π_v and $s+1$ in Π_s, every two of which are conjugate. Now, it is shown that, for any choice of $v+1$ independent points $P_0, \dots, P_v$ in Π_v, there exists Π'_v in Π_{2v+1} and independent points $P'_0, \dots, P'_v$ in Π'_v such that $P'_i \notin \mathcal{U}_n$, $\Pi_v \cap \Pi'_v = \emptyset$, $\Pi'_v \cap \mathcal{U}_n$ is non-singular, $P'_0, \dots, P'_v$ are mutually conjugate, and P_i is conjugate to P'_j for $i \neq j$. Therefore it is sufficient to show that, for any $v+1$ independent points $Q_0, \dots, Q_v$ in Π_v, there is a projectivity $\mathfrak{S}$ fixing $\Pi_{2v+1} \cap \mathcal{U}_n$ and Π_v, and with $P_i\mathfrak{S} = Q_i$, $i = 0, 1, \dots, v$. Choose a space Π''_v on $\Pi_{2v+1} \cap \mathcal{U}_n$ skew to Π_v and choose independent points $P''_0, \dots, P''_v$ in Π''_v with P_i

conjugate to P_j'', $i \neq j$. With respect to the reference points $P_0, \ldots, P_v, P_0'', \ldots, P_v''$ and a suitable unit point, $\Pi_{2v+1} \cap \mathcal{U}_n$ is represented by the canonical form G_{2v+1} as in (2.7). This proves the existence of the projectivity $\mathfrak{S}$.

Since the projectivity $\mathfrak{T}$ from Π_m to Π_m' transforms the chosen $(m+1)$-tuple in Π_m to such an $(m+1)$-tuple in Π_m', by the preceding paragraph the reference points and the unit point may be chosen in a new system of coordinates $(x_0', x_1', \ldots, x_n')$ so that

$$\Pi_m' = \mathbf{V}(X_{m+1}', \ldots, X_n')$$

and so that $\mathfrak{T}$ has equations

$$x_0' : x_1' : \cdots : x_m' = c_0 x_0 : c_1 x_1 : \cdots : c_m x_m \,;$$

here, $\mathcal{U}_n$ preserves its equation in these coordinates apart from changing X_i to X_i'. Since $\Pi_m \cap \mathcal{U}_n$ and $\Pi_m' \cap \mathcal{U}_n$ have the same equations, apart from changing X_i to X_i', so

$$c_0 \bar{c}_0 = c_1 \bar{c}_1 = \cdots = c_m \bar{c}_m \,.$$

The projectivity $\mathfrak{T}$ has an extension to $\mathrm{PG}(n, q)$ with equations

$$x_i' = \begin{cases} c_i x_i, & i = 0, \ldots, m; \\ c_0 x_i, & i = m+1, \ldots, m+w+1; \\ c_{i+m-n} x_i, & i = m+w+2, \ldots, n; \end{cases}$$

this is an element of $G(\mathcal{U}_n)$. Thus the theorem is established. $\qquad\square$

Now, the size of the orbits in the previous theorem must be determined. Let $N(\Pi_v \mathcal{U}_{m-v-1}, \mathcal{U}_n)$ be the number of m-spaces Π_m meeting $\mathcal{U}_n$ in a section of type $\Pi_v \mathcal{U}_{m-v-1}$. This number was determined in Theorem 2.19 in the case that $v = m$, that is when Π_m lies on $\mathcal{U}_n$.

Theorem 2.23. *The number of sections $\Pi_v \mathcal{U}_{m-v-1}$ of $\mathcal{U}_n$ is*

$$N(\Pi_v \mathcal{U}_{m-v-1}, \mathcal{U}_n) = q^{T(s+1)/2} \overline{[s+2, n+1]} / \{ \overline{[1, T]}[1, v+1]_- \},$$

where

$$s + v = m - 1, \tag{2.15}$$

$$T = n - 2m + s. \tag{2.20}$$

Proof. The required number $N(\Pi_v \mathcal{U}_s, \mathcal{U}_n)$ is equal to the product of $N(\Pi_v, \mathcal{U}_n)$ and the number of sections $\Pi_v \mathcal{U}_s$ with a given Π_v.

So, let Π_v be given and let Π_{n-v-1} be skew to Π_v with $\Pi_{n-v-1} \cap \mathcal{U}_n$ non-singular; put $\Pi_{n-v-1} \cap \mathcal{U}_n = \mathcal{U}_{n-v-1}$. If Π_{n-v-1}' is the tangent space of $\mathcal{U}_n$ at Π_v, then $\Pi_{n-v-1}' \cap \mathcal{U}_{n-v-1}$ is a non-singular $\mathcal{U}_{n-2v-2}$. Each section $\Pi_v \mathcal{U}_s$ is contained in Π_{n-v-1}', by Lemma 2.4, and $\Pi_v \mathcal{U}_s \cap \mathcal{U}_{n-2v-2}$ is a non-singular $\mathcal{U}_s'$. Conversely, each non-singular $\mathcal{U}_s'$ on $\mathcal{U}_{n-2v-2}$ defines a section $\Pi_v \mathcal{U}_s'$ of the prescribed type. Hence the number of sections $\Pi_v \mathcal{U}_s$ with given Π_v is $N(\Pi_{-1} \mathcal{U}_s, \mathcal{U}_{n-2v-2})$.

Now, $N(\Pi_{-1}\mathcal{U}_s, \mathcal{U}_r) = \rho_{s,r}$ is calculated by counting the following set in two ways:

$$\{(\Pi_s, \Pi_{s+1}) \mid \Pi_s \subset \Pi_{s+1},\ \Pi_s \cap \mathcal{U}_r \text{ non-singular},\ \Pi_{s+1} \cap \mathcal{U}_r \text{ non-singular}\}.$$

For a given Π_s, the number of such pairs (Π_s, Π_{s+1}) is the number of points of Π'_{r-s-1} not on $\Pi'_{r-s-1} \cap \mathcal{U}_r$, with Π'_{r-s-1} the polar space of Π_r. Hence

$$\rho_{s,r}\{\theta(r-s-1) - \mu_{r-s-1}\} = \rho_{s+1,r}\{\theta(s+1) - \mu_{s+1}\}.$$

Since $\rho_{r-1,r}$ is the number of non-tangent hyperplanes to $\mathcal{U}_r$ in $\mathrm{PG}(r, q)$, so

$$\rho_{r-1,r} = \theta(r) - \mu_r.$$

Hence

$$\rho_{s,r} = \frac{\{\theta(s+1) - \mu_{s+1}\}\{\theta(s+2) - \mu_{s+2}\} \cdots \{\theta(r) - \mu_r\}}{\{\theta(1) - \mu_1\}\{\theta(2) - \mu_2\} \cdots \{\theta(r-s-1) - \mu_{r-s-1}\}}.$$

By (2.9),

$$
\begin{aligned}
\rho_{s,r} &= \frac{\{q^{s+1}\sqrt{q} + (-\sqrt{q})^{s+1}\} \cdots \{q^r\sqrt{q} + (-\sqrt{q})^r\}}{\{\sqrt{q}+1\}\{q\sqrt{q} + (-\sqrt{q})\} \cdots \{q^{r-s-1}\sqrt{q} + (-\sqrt{q})^{r-s-1}\}} \\
&= q^{(s+1)(r-s)/2}\frac{\{(\sqrt{q})^{s+2} - (-1)^{s+2}\} \cdots \{(\sqrt{q})^{r+1} - (-1)^{r+1}\}}{\{\sqrt{q}+1\}\{(\sqrt{q})^2 - 1\} \cdots \{(\sqrt{q})^{r-s} - (-1)^{r-s}\}} \\
&= q^{(s+1)(r-s)/2}\overline{[s+2, r+1]}/\overline{[1, r-s]}.
\end{aligned}
$$

Hence, for $r = n - 2v - 2$,

$$\rho_{s,n-2v-2} = q^{(s+1)T/2}\overline{[s+2, n-2v-1]}/\overline{[1, T]}.$$

Since $N(\Pi_v, \mathcal{U}_n) = \overline{[n-2v, n+1]}/[1, v+1]_-$ by (2.11), the final result follows. $\square$

2.5 The characterisation of Hermitian varieties

This is a continuation of the treatment in FPSOTD, Section 19.4, where Hermitian surfaces were characterised as subsets of $\mathrm{PG}(3, q)$ meeting every line in $1, r$ or $q+1$ points with some further restrictions. Some definitions and results are recalled. A subset $\mathcal{K}$ of $\mathrm{PG}(n, q)$ is a $k_{r,n,q}$ if r is a fixed integer with $1 \le r \le q$ such that

(1) $|\mathcal{K}| = k$;
(2) $|l \cap \mathcal{K}| = 1, r$ or $q+1$ for each line l;
(3) $|l \cap \mathcal{K}| = r$ for some line l.

A $k_{1,n,q}$ is a hyperplane. A $k_{2,n,2}$ is the complement of a cap with at least one unisecant; the only cap of $\mathrm{PG}(n, 2)$ with no unisecant is the complement of a hyperplane.

From now on, assume that $q > 2$. From FPSOTD, Theorem 19.4.4, there are seven types of plane section $\mathcal{K}'$, with $|\mathcal{K}'| = k'$, of such a $\mathcal{K}$:

I. a Hermitian arc (unital), that is, a set of type $(1, \sqrt{q} + 1)$ with $r = \sqrt{q} + 1$ and $k' = q\sqrt{q} + 1$;

II. a subplane $PG(2, \sqrt{q})$, that is, a set of type $(1, \sqrt{q} + 1)$ with $r = \sqrt{q} + 1$ and $k' = q + \sqrt{q} + 1$;

III. a set of type $(0, r - 1)$ plus an external line, whence $k' = (r - 1)q + r$ and $r - 1 \mid q$;

IV. the complement of a set of type $(0, q + 1 - r)$, whence $k' = r(q + 1)$ and $(q + 1 - r) \mid q$;

V. the union of r concurrent lines, whence $k' = rq + 1$;

VI. a single line, $k' = q + 1$;

VII. a plane, $k' = q^2 + q + 1$.

Definition 2.24. (1) A point P of $\mathcal{K}$ is *singular* if every line through P is either a unisecant or a line of $\mathcal{K}$.

(2) The set $\mathcal{K}$ is *singular* or *non-singular* according as it has singular points or not.

In $PG(3, q)$, the fundamental result is contained in FPSOTD, Theorem 19.5.13 and Section 19.6.

Theorem 2.25. *Let* $\mathcal{K}$ *be a* $k_{r,3,q}$ *in* $\Pi_3 = PG(3, q)$.

(i) *When* $r = 1$, *then* $\mathcal{K}$ *is a plane.*

(ii) *When* $r = 2$, *then* $\mathcal{K}$ *is one of the following*:

   (a) $\Pi_2 \cup \Pi_0$,

   (b) $\Pi_2 \cup \Pi_1$,

   (c) $\Pi_2 \cup \Pi_2'$.

(iii) *When* $r = q$, *then* $\mathcal{K}$ *is one of the following*:

   (a) $(\Pi_3 \backslash \Pi_2) \cup \Pi_1$ *with* $\Pi_1 \subset \Pi_2$,

   (b) $(\Pi_3 \backslash \Pi_1) \cup \Pi_0$ *with* $\Pi_0 \subset \Pi_1$,

   (c) $\Pi_3 \backslash \Pi_0$.

(iv) *When* $3 \leq r \leq q - 1$, *then one of the following occurs.*

   (a) *If* $\mathcal{K}$ *is singular, then* $\mathcal{K}$ *is* r *planes through a line or a cone* $\Pi_0 \mathcal{K}'$ *with base* $\mathcal{K}'$ *a set of type* I, II, III *or* IV *as above.*

   (b) *If* $\mathcal{K}$ *is non-singular, then*

      (1) *for* q *odd,* $r = \sqrt{q} + 1$ *and* $\mathcal{K} = \mathcal{U}_{3,q}$;

      (2) *for* q *even and* $q > 4$, *either* $r = \sqrt{q} + 1$ *and* $\mathcal{K} = \mathcal{U}_{3,q}$ *or* $r = \frac{1}{2}q + 1$ *and* $\mathcal{K} = \mathcal{R}_3$, *the projection of a quadric* $\mathcal{P}_4$;

      (3) *for* $q = 4$, $\mathcal{K} = \mathcal{U}_{3,4}$ *or* $\mathcal{K} = \mathcal{R}_3$ *or* $\mathcal{K}$ *contains sections of type* II.

A similar result is true in $PG(n, q)$. First, singular sets are considered. As for $n = 3$, the study of singular sets $k_{r,n,q}$ reduces to that of non-singular ones.

Lemma 2.26. *The singular points of a* $k_{r,n,q}$ *form a subspace.*

The subspace of singular points of $\mathcal{K}$ is the *singular space* of $\mathcal{K}$.

Theorem 2.27. *If $\mathcal{K}$ is a singular $k_{r,n,q}$ with singular space Π_d, then one of the following holds:*

(a) $d = n - 1$ *and $\mathcal{K}$ is a hyperplane;*
(b) $d = n - 2$ *and $\mathcal{K}$ is r hyperplanes through Π_d, with $r > 1$;*
(c) $d \leq n - 3$ *and $\mathcal{K} = \Pi_d \mathcal{K}'$, where $\mathcal{K}'$ is a non-singular $k_{r,n-d-1,q}$.*

Now the cases $r = 2$ and $r = q$ are considered.

Theorem 2.28. *If $\mathcal{K}$ is a $k_{2,n,q}$, then $\mathcal{K} = \Pi_{n-1} \cup \Pi_i'$ for some Π_i' not contained in Π_{n-1}.*

Proof. (a) $n = 2$. Suppose $\mathcal{K}$ contains no line. Then $\mathcal{K}$ is a k-arc meeting every line of the plane, a contradiction. So $\mathcal{K}$ contains a line Π_1. If $\mathcal{K} \backslash \Pi_1$ contains a 3-arc then $\mathcal{K}$ is the whole plane. Hence it follows that $\mathcal{K} = \Pi_1 \cup \Pi_1'$ or $\Pi_1 \cup \Pi_0$.

(b) $n > 2$.

(i) $\mathcal{K}$ *is non-singular*

The proof proceeds by induction on n. Since not every plane is contained in $\mathcal{K}$, some plane meets $\mathcal{K}$ in a line or a $k_{2,2,q}'$. Hence $\mathcal{K}$ has a unisecant l with point of contact Q.

Let Π_{n-1} be a hyperplane of $\mathrm{PG}(n,q)$ not containing Q and let $\mathcal{K}'$ be the projection of $\mathcal{K} \backslash \{Q\}$ from Q onto Π_{n-1}. If l' is any line of Π_{n-1} then, by (a),

$$Ql' \cap \mathcal{K} = \Pi_2, \ \Pi_1 \cup \Pi_1', \ \Pi_1 \cup \Pi_0, \ \text{or} \ \Pi_1.$$

Hence $|l' \cap \mathcal{K}'| = 1, 2$, or $q + 1$. Let $l \cap \Pi_{n-1} = \{R\}$; then at least one line m through R in Π_{n-1} meets $\mathcal{K}'$ in two points. For, otherwise, every plane through l meets $\mathcal{K}$ in a line and so Q is singular, a contradiction. So $\mathcal{K}'$ is a $k_{2,n-1,q}'$. Hence, by the induction hypothesis, $\mathcal{K}' = \Pi_{n-2} \cup \Pi_d'$, for $0 \leq d \leq n - 2$. Consequently, $\mathcal{K} \subset Q\Pi_{n-2} \cup Q\Pi_d'$.

(1) $d < n - 2$. If $S \in Q\Pi_{n-2} \backslash \mathcal{K}$, then there is a line through S with no point in $\mathcal{K}$, a contradiction; so $Q\Pi_{n-2} \subset \mathcal{K}$. If P_1 and P_2 are points of $\mathcal{K} \backslash Q\Pi_{n-2}$, the line $P_1 P_2$ meets $Q\Pi_{n-2}$ and so contains a third point of $\mathcal{K}$; therefore $P_1 P_2 \subset \mathcal{K}$. Hence, if $|\mathcal{K} \backslash Q\Pi_{n-2}| \geq 2$, then $\mathcal{K} \backslash Q\Pi_{n-2}$ is an affine subspace of the affine space $\mathrm{PG}(n,q) \backslash Q\Pi_{n-2}$ since $q > 2$. In this case, $\mathcal{K} = Q\Pi_{n-2} \cup \Pi_t''$ with $t \geq 1$; since $\mathcal{K}$ is non-singular, $Q\Pi_{n-2} \cap \Pi_t'' = \emptyset$, a contradiction. Therefore $\mathcal{K} = \Pi_{n-1} \cup \Pi_0$ with Π_0 not in Π_{n-1}.

(2) $d = n - 2$. If $Q\Pi_{n-2} \not\subset \mathcal{K}$ and $Q\Pi_{n-2}' \not\subset \mathcal{K}$, then there is a line with no point in $\mathcal{K}$, a contradiction. So, let $Q\Pi_{n-2} \subset \mathcal{K}$. Now, proceed exactly as in (1).

(ii) $\mathcal{K}$ *is singular*

The result follows from Theorem 2.27. □

Now the case $r = q$ is considered.

Theorem 2.29. *If $\mathcal{K}$ is a $k_{q,n,q}$ in $\Pi_n = \mathrm{PG}(n,q)$, $q \geq 3$, then $\mathcal{K} = (\Pi_n \backslash \Pi_i) \cup \Pi_{i-1}$ with $\Pi_{i-1} \subset \Pi_i$ for $0 \leq i \leq n - 1$.*

Proof. Let Q be a point of $\Pi_n \backslash \mathcal{K}$. If l_1 and l_2 are two unisecants through Q with points of contact P_1 and P_2, the line $P_1 P_2$ either belongs to $\mathcal{K}$ or is a q-secant. Since $q \geq 3$, there is another point P_3 of $\mathcal{K}$ on $P_1 P_2$. Each line l of the plane $\pi = l_1 l_2$ through P_3 other than $P_1 P_2$ and $Q P_3$ is a unisecant of $\mathcal{K}$ since $l \cap l_1$ and $l \cap l_2$ are not in $\mathcal{K}$. It follows that every line of π through P_1 other than $P_1 P_2$ is also a unisecant and that all points of $\mathcal{K}$ in π lie on $P_1 P_2$. As $\mathcal{K}$ has no external lines, so $\mathcal{K} \cap \pi = P_1 P_2$. Thus it has been shown that, for any two unisecants through Q, all the lines of the pencil determined by these two are unisecants and that the plane of the pencil meets $\mathcal{K}$ in a line. Hence the unisecants through Q generate a Π_i meeting $\mathcal{K}$ in a Π_{i-1}. Since every line through Q not in this Π_i is a q-secant, it follows that $\mathcal{K}$ consists of the points of Π_{i-1} plus the points not in Π_i. $\qquad\square$

The previous two theorems mean that the rest of the characterisation can be restricted to

$$3 \leq r \leq q - 1. \tag{2.21}$$

Lemma 2.30. *If $\mathcal{K}$ is a $k_{r,n,q}$ with $3 \leq r \leq q - 1$ such that $\mathcal{K}$ has a section $\mathcal{K}'$ by a plane π containing a triangle of lines of $\mathcal{K}$ with $\pi \not\subset \mathcal{K}$, then one of the following occurs:*

(a) $\mathcal{K} = \Pi_{n-3}\mathcal{K}'$, *where $\mathcal{K}'$ is a section of type IV;*
(b) $q = 2^h$, $r = \frac{1}{2}q + 1$, *and the singular space of $\mathcal{K}$ has dimension at most $n - 4$.*

Proof. Since π is not contained in $\mathcal{K}$, the section $\mathcal{K}'$ must be of type IV; that is, $\mathcal{K}'$ is the complement in π of a maximal arc of type $(0, q + 1 - r)$. So, by Theorem 12.7 of PGOFF2, $q \equiv 0 \pmod{q - r + 1}$; hence, with $q = p^h$,

$$p^h - r + 1 = p^m \tag{2.22}$$

for some m with $0 < m < h$. Thus

$$r - 1 = p^m(p^{h-m} - 1). \tag{2.23}$$

This gives two possibilities:

(1) $r - 1 \neq p^m$, in which case every plane section of $\mathcal{K}$ is of type IV, V, VI or VII;
(2) $r - 1 = p^m$ and $p^{h-m} - 1 = 1$, whence $p = 2$, $m = h - 1$, and $r = \frac{1}{2}q + 1$.

In case (1), let Π_3 be a solid through π. By the Corollary to Lemma 19.4.7 of FPSOTD, there is a unisecant l_1 of $\Pi_3 \cap \mathcal{K}$ and so of $\mathcal{K}$ with point of contact P. Let l be any line through P other than l_1 and let $\pi_1 = l l_1$. The section $\pi_1 \cap \mathcal{K}$ cannot be of type IV or VII since they do not have unisecants. So $\pi_1 \cap \mathcal{K}$ is of type V or VI, that is, r lines of a pencil or a single line; in both these cases, P is singular. So every solid through π meets the singular space S of $\mathcal{K}$, while S does not meet π since $\pi \cap \mathcal{K}$ is non-singular. Hence $S = \Pi_{n-3}$ and $\mathcal{K} = \Pi_{n-3}\mathcal{K}'$.

In case (2), it may happen that (a) occurs. If it does not, then the dimension of S is at most $n - 4$. $\qquad\square$

This section continues the investigation under the hypothesis that a $k_{r,n,q}$ has no section of type IV. The case when such a section occurs is investigated in Section 2.6.

Lemma 2.31. *Let $\mathcal{K}$ be a $k_{r,n,q}$ with $3 \leq r \leq q - 1$ such that* (a) $\mathcal{K}$ *contains a hyperplane* Π_{n-1} *with* $\mathcal{K} \neq \Pi_{n-1}$, (b) $\mathcal{K}$ *has no plane section of type* IV. *Then $\mathcal{K}$ is the union of r hyperplanes of a pencil or $\mathcal{K} = \Pi_{n-3}\mathcal{K}'$ with $\mathcal{K}'$ a plane section of type* III.

Proof. Let π be a plane in Π_{n-1} and let P be a point of $\mathcal{K}\backslash\Pi_{n-1}$. First it is shown that the solid $P\pi$ contains a line l of $\mathcal{K}$ not in π. Suppose otherwise and consider the set $\mathcal{K}' = (\mathcal{K} \cap P\pi)\backslash\pi$. It is a k'-set of type $(0, r-1)$ in $P\pi$ with $2 \leq r - 1 \leq q - 1$. By Lemma 19.4.7 of FPSOTD, such a set $\mathcal{K}'$ does not exist.

Let $P' = l \cap \pi$, let l' be any line through P' other than l and let $\pi' = ll'$. The plane π' meets Π_{n-1} in a line $m \neq l$. So π' has the two lines l and m in $\mathcal{K}$ in a section of type V or VII; in the former case, the r lines contain P'. Hence l' is either a unisecant or a line of $\mathcal{K}$; so P' is singular. Thus $\mathcal{K}$ is singular and its singular space S is necessarily in Π_{n-1}. Also every plane in Π_{n-1} meets S. Thus the dimension of S is at least $n - 3$. Now Theorem 2.27 gives the result. □

For $n = 3$, the characterisation of Hermitian surfaces was completed in Section 19.5 of FPSOTD.

Definition 2.32. A subset $\mathcal{K}$ of $\mathrm{PG}(n, q)$ is *regular* if

(1) $\mathcal{K}$ is a $k_{r,n,q}$;
(2) $3 \leq r \leq q - 1$;
(3) $\mathcal{K}$ has no plane section of type IV.

Lemma 2.33. *If $\mathcal{K}$ is a regular $k_{r,n,q}$ with $n \geq 4$, then $\mathcal{K}$ cannot have plane sections both of type* I *and of type* II.

Proof. Let π be a plane meeting $\mathcal{K}$ in a section $\mathcal{K}'$ of type I. Then, from Theorem 2.25, a solid through π meets $\mathcal{K}$ either in a cone $P\mathcal{K}'$ or in a Hermitian surface $\mathcal{U}_{3,q}$. If there are M solids of the latter type, then there are $\theta(n-3) - M$ of the former. Hence

$$k = q\sqrt{q} + 1 + Mq(q\sqrt{q} + 1) + \{\theta(n-3) - M\}\{(q-1)(q\sqrt{q}+1)+1\}$$
$$= q\sqrt{q}(q^{n-2} + M) + \theta(n-2). \qquad (2.24)$$

If π_0 is a plane meeting $\mathcal{K}$ in a plane section $\mathcal{K}_0$ if type II, then every solid through π_0 meets $\mathcal{K}$ in a cone $P\mathcal{K}_0$, by Theorem 2.25. Hence

$$k = q + \sqrt{q} + 1 + \theta(n-3)\{(q-1)(q+\sqrt{q}+1)+1\}$$
$$= \theta(n-1) + q^{n-2}\sqrt{q}. \qquad (2.25)$$

Equating (2.24) and (2.25) gives that

$$M = q^{n-3}(1 + \sqrt{q} - q) < 0,$$

a contradiction. □

Lemma 2.34. *If $\mathcal{K}$ is a regular $k_{r,n,q}$ with $n \geq 4$, then $\mathcal{K}$ cannot have plane sections both of type II and of type III.*

Proof. If π is a plane meeting $\mathcal{K}$ in a section $\mathcal{K}'$ of type III, then again by Theorem 2.25 any solid through π meets $\mathcal{K}$ in a cone $P\mathcal{K}'$. Hence

$$k = (r-1)q + r + \theta(n-3)\{(q-1)[(r-1)q + r] + 1\}$$
$$= (r-1)q^{n-1} + rq^{n-2} + \theta(n-3). \qquad (2.26)$$

If there is a section of type III and of type II, then $r = \sqrt{q} + 1$. So substituting this in (2.26) and equating it to (2.25) gives $q = 1$, a contradiction. $\square$

Lemma 2.35. *If $\mathcal{K}$ is a regular $k_{r,n,q}$ with $n \geq 4$, and $\mathcal{K}$ has plane sections of type I and type III, then q^{n-3} solids through a plane of type I meet $\mathcal{K}$ in a Hermitian surface.*

Proof. Equating (2.24) and (2.26) with $r = \sqrt{q} + 1$ implies that $M = q^{n-3}$. $\square$

Theorem 2.36. *If $\mathcal{K}$ is a regular $k_{r,n,q}$ with $n \geq 3$ and if there is a plane π meeting $\mathcal{K}$ in a section $\mathcal{K}'$ of type II, then $\mathcal{K} = \Pi_{n-3}\mathcal{K}'$.*

Proof. The result is true for $n = 3$ by Theorem 2.25. So let $n \geq 4$ and proceed by induction. Thus every hyperplane through π meets $\mathcal{K}$ in a $k'_{r,n-1,q}$ which is a cone $\Pi_{n-4}\mathcal{K}'$. The theorem will follow if it is shown that the points of any such Π_{n-4} are singular for $\mathcal{K}$. For, considering a second hyperplane through π, it then follows that $\mathcal{K}$ has at least a Π_{n-3} of singular points, and hence exactly a Π_{n-3} of singular points.

Let Π_{n-1} be one such hyperplane and let P be a point of the vertex Π_{n-4}, that is, the singular space of $\Pi_{n-1} \cap \mathcal{K}$. Suppose that P is non-singular for $\mathcal{K}$, and let l be an r-secant of $\mathcal{K}$ through P; necessarily, $r = \sqrt{q} + 1$. The line l cannot belong to Π_{n-1} and so a plane through l meets Π_{n-1} in a line l' other than l, where l' is a line or a unisecant of $\Pi_{n-1} \cap \mathcal{K}$. The number of lines through P in $\Pi_{n-1} \cap \mathcal{K}$ is

$$\{\theta(n-4) - \theta(n-5)\}(q + \sqrt{q} + 1) + \theta(n-5) = q^{n-4}\sqrt{q} + \theta(n-3); \quad (2.27)$$

this is calculated by looking at the number of lines through P in each $(n-3)$-space $\Pi_{n-4}Q$, where Q varies in $\mathcal{K}'$. Thus the number of lines through P in Π_{n-1} but not in $\mathcal{K}$ is

$$q^{n-4}(q^2 - \sqrt{q}). \qquad (2.28)$$

Therefore the number of planes through l meeting Π_{n-1} in a line of $\mathcal{K}$ is given by (2.27) and the number meeting Π_{n-1} in a unisecant of $\mathcal{K}$ by (2.28).

By Lemmas 2.33 and 2.34, each of these planes meeting Π_{n-1} in a line of $\mathcal{K}$ must be a section of type V with $r = \sqrt{q} + 1$ and so has $r(q-1) + 1 = q\sqrt{q} + q - \sqrt{q}$ points in common with $\mathcal{K}\backslash l$. Similarly, the other planes meet $\mathcal{K}$ in a section of type II and so meet $\mathcal{K}\backslash l$ in q points. Using these numbers to find the size of $\mathcal{K}\backslash l$ gives

$$k - (\sqrt{q} + 1) = (q^{n-4}\sqrt{q} + \theta(n-3))(q\sqrt{q} + q - \sqrt{q}) + q^{n-4}(q^2 - \sqrt{q})q,$$

whence

$$k = \theta(n-1) + q^{n-2}\sqrt{q} + q^{n-3}(q-1). \tag{2.29}$$

Comparing this with (2.25) gives $q^{n-3}(q-1) = 0$. This contradiction proves the result. □

Lemma 2.37. *Let $\mathcal{K}$ be a regular $k_{r,n,q}$ with $n \geq 3$ that contains a hyperplane Π_{n-1}. Then $\mathcal{K}$ is one of the following:*

(a) *r hyperplanes of a pencil;*
(b) *$\Pi_{n-3}\mathcal{K}_{III}$, where $\mathcal{K}_{III}$ is a plane section of type III.*

Proof. Let π be any plane in Π_{n-1}, and let P be any point of $\mathcal{K}\backslash\Pi_{n-1}$. The solid $P\pi$ contains a line l of $\mathcal{K}$ not in π and therefore not in Π_{n-1}; for, otherwise, the points of $\mathcal{K} \cap (P\pi\backslash\pi)$ constitute a k'-set $\mathcal{K}'$ with

$$k' = (r-2)(q^2+q+1)+1$$

and of type $(0, r-1)$ in $P\pi$. Such sets do not exist by Lemma 19.4.7 of FPSOTD.

Let $P' = l \cap \pi$ and let l' be any line through P' other than l. The plane $\pi' = ll'$ meets Π_{n-1} in a line m through P'. Thus π' contains the lines l and m of $\mathcal{K}$. Hence π' meets $\mathcal{K}$ in r lines of a pencil with centre P' or lies in $\mathcal{K}$. So l' is either a line of $\mathcal{K}$ or a unisecant at P'. Thus P' is singular. Therefore the singular space Π_d of $\mathcal{K}$ lies in Π_{n-1}, and every plane of Π_{n-1} meets Π_d. Hence $d \geq n-3$ and the result follows from Theorem 2.27. □

Theorem 2.38. *If $\mathcal{K}$ is a regular $k_{r,n,q}$ with $n \geq 4$ and if there is a plane π meeting $\mathcal{K}$ in a section $\mathcal{K}'$ of type III, then $\mathcal{K} = \Pi_{n-3}\mathcal{K}'$.*

Proof. Let $n = 4$, and take two solids Π_3 and Π_3' through π. From Theorem 2.25, $\Pi_3 \cap \mathcal{K} = P\mathcal{K}'$ and $\Pi_3' \cap \mathcal{K} = P'\mathcal{K}'$. The section $\mathcal{K}'$ is an $((r-2)q+r-1; r-1)$-arc plus an external line l. The points P and P' are distinct as are the planes $\alpha = Pl$ and $\alpha' = P'l$. The line PP' is skew to π, since otherwise it would belong to both Π_3 and Π_3', and hence to π. The planes α and α' lie in $\mathcal{K}$. If it is shown that the solid $\alpha\alpha'$ lies in $\mathcal{K}$, then the result follows from Lemma 2.37.

So suppose that the solid $\alpha\alpha'$ is not contained in $\mathcal{K}$. Since $\alpha\alpha'$ contains two planes in $\mathcal{K}$, it meets $\mathcal{K}$ in r planes through the line l, by Theorem 2.25; hence PP' is an r-secant of $\mathcal{K}$.

Any plane β through PP' does not lie in $\mathcal{K}$ and meets π in a point B. If $B \in \mathcal{K}'$, the lines BP and BP' lie in $\mathcal{K}$ and hence $\beta \cap \mathcal{K}$ is r lines of a pencil. If $B \notin \mathcal{K}'$, the lines BP and BP' are unisecants to $\mathcal{K}$ with contacts P and P'; then $\beta \cap \mathcal{K}$ can contain no lines, as otherwise such a line would have to pass through P and P'. So, in this case, $\beta \cap \mathcal{K}$ is of type I by Lemma 2.34 and $r = \sqrt{q}+1$.

The planes β of the first type number $(r-1)q+r$ and each meets $\mathcal{K}$ in $r(q-1)+1$ points off the line PP'. There are $\theta(2) - \{(r-1)q+r\}$ planes β of the second type, each of which contains $q\sqrt{q}+1-r$ points of $\mathcal{K}\backslash PP'$. Thus a count of the points of $\mathcal{K}\backslash PP'$ on the planes through PP' gives

$$k - r = \{(r-1)q + r\}\{r(q-1) + 1\} + \{q^2 + q + 1 - (r-1)q - r\}$$
$$\times \{q\sqrt{q} + 1 - r\};$$

hence, with $r = \sqrt{q} + 1$,

$$k = q\sqrt{q}(q^2 + q + 1) + q + 1. \tag{2.30}$$

However, with $r = \sqrt{q} + 1$ and $n = 4$, the number k is given by (2.26); namely,

$$k = q^3\sqrt{q} + q^2(\sqrt{q} + 1) + q + 1.$$

Thus $q\sqrt{q} = q^2$, a contradiction. This proves the result for $n = 4$.

Now let $n \geq 5$. Each solid χ through π meets $\mathcal{K}$ in a cone $Q\mathcal{K}'$. It suffices to show that Q is always singular. Let m be a line through Q. If it lies in the solid χ, it cannot be an r-secant. Suppose therefore that m is not in χ and so is skew to π. The 4-space $m\pi$ meets $\mathcal{K}$ in a $k'_{r,4,q}$ that is a cone $\Pi_1\mathcal{K}'$, by the previous part of the proof. Also $\chi \cap \Pi_1 = \{Q\}$. Hence Q is singular for $k'_{r,4,q}$; so the line m is not an r-secant of $k'_{r,4,q}$ and therefore not an r-secant of $\mathcal{K}$. Thus Q is singular and the result follows.   □

The previous results allow a summary for sections of a non-singular regular set.

Theorem 2.39. *Let $\mathcal{K}$ be a regular, non-singular $k_{r,n,q}$ and let Π_2 and Π_3 be spaces not contained in $\mathcal{K}$. Then*

(i) $\Pi_2 \cap \mathcal{K}$ *is of type* I, V *or* VI;
(ii) $\Pi_3 \cap \mathcal{K}$ *is a plane, r planes of a pencil, $\Pi_0\mathcal{K}_I$, or $\mathcal{U}_3$, where $\mathcal{K}_I$ is a section of type* I.

Now the study of regular sets $k_{r,n,q}$ is continued for $n \geq 4$ and $q > 4$.

Lemma 2.40. *Let $\mathcal{K}$ be a non-singular, regular $k_{r,n,q}$ with $n \geq 4$ and $q > 4$. Then through any point P of $\mathcal{K}$ there passes a section of type* I, *whence q is a square and $r = \sqrt{q} + 1$.*

Proof. Suppose there is no unisecant through P. Let l_r be an r-secant through P and let π', π'' be distinct planes through l_r. By Theorem 2.39, $\pi' \cap \mathcal{K}$ is r lines through P' and $\pi'' \cap \mathcal{K}$ is r lines through P''; also $\pi'\pi'' \cap \mathcal{K}$ is r planes through $P'P''$. It follows that P' is singular for $\mathcal{K}$, which is a contradiction. If l_1 is a unisecant through P, then, by Theorem 2.39, the plane $\pi = l_1l_r$ meets $\mathcal{K}$ in a section of type I.   □

Definition 2.41. For any point P of a regular, non-singular set $\mathcal{K}$, the *tangent space* T_P is the union of the unisecants and lines of $\mathcal{K}$ through P.

Lemma 2.42. *The tangent space has the following properties:*

(i) T_P *is a hyperplane;*
(ii) *the singular space of $T_P \cap \mathcal{K}$ is $\{P\}$;*
(iii) *if $P \neq Q$ then $T_P \neq T_Q$;*

(iv) *a non-tangent hyperplane meets* $\mathcal{K}$ *in a non-singular* $k'_{r,n-1,q}$.

Proof. Let $\mathcal{L}$ be the set consisting of the unisecants and lines of $\mathcal{K}$ through P. By the previous lemma, there is a plane π through P meeting $\mathcal{K}$ in a Hermitian arc $\mathcal{K}'$. Let l be the unisecant to $\mathcal{K}'$ in π at P. Each of the $\theta(n-3)$ solids Π_3 through π meets $\mathcal{K}$ either in a cone $Q\mathcal{K}'$ or $\mathcal{U}_3$. Let α be the tangent plane at P to $\Pi_3 \cap \mathcal{K}$; here α must contain l. The lines through P in α belong to $\mathcal{L}$ and so are the only lines of $\mathcal{L}$ in Π_3. Also, any line of $\mathcal{L}\backslash\{l\}$ is joined to π by some Π_3 and lies in the tangent plane at P to $\Pi_3 \cap \mathcal{K}$. Since distinct solids through π give distinct tangent planes, the number of lines in $\mathcal{L}$ is $q\theta(n-3) + 1 = \theta(n-2)$.

Now consider the pencil of lines containing two lines l_1 and l_2 of $\mathcal{L}$. The plane $l_1 l_2$ cannot meet $\mathcal{K}$ in a section of type I. Hence $l_1 l_2$ is of type V, VI or VII, by Theorem 2.39. In each case, all the lines of the pencil are in $\mathcal{L}$. Since $|\mathcal{L}| = \theta(n-2)$, the lines of $\mathcal{L}$ must be the set of lines through P in a hyperplane; that is, T_P is a hyperplane.

The set $T_P \cap \mathcal{K}$ is a $k'_{1,n-1,q}$ or a $k'_{r,n-1,q}$ for which P is singular. Suppose that $T_P \cap \mathcal{K}$ has another singular point P'; then every point of PP' is singular. So every point Q of PP' has T_P as tangent hyperplane. Hence every line through Q not in T_P is an r-secant.

Let α be a plane through PP' but not contained in T_P. Either α belongs to $\mathcal{K}$ or meets it in a line or in r lines of a pencil with centre Q_0, a point of PP'. This means that in α there is no r-secant through any point of PP' in the first two cases and through Q_0 in the third case. This contradicts the result of the previous paragraph. Hence P is the only singular point of $T_P \cap \mathcal{K}$. This is (ii). Parts (iii) and (iv) now follow. □

Corollary 2.43. *For any point P in $\mathcal{K}$, the meet $T_P \cap \mathcal{K}$ is a cone $P\mathcal{K}'$, where $\mathcal{K}'$ is a regular, non-singular* $k'_{r,n-2,q}$.

Proof. Let $\Pi_{n-2} \subset T_P$ with $P \notin \Pi_{n-2}$. Then $\mathcal{K}' = \Pi_{n-2} \cap \mathcal{K}$ is non-singular and meets every line of $\mathcal{K}$ through P. Conversely, since P is singular in $T_P \cap \mathcal{K}$, the join QP is a line of $\mathcal{K}$ for every point Q of $\mathcal{K}'$. Hence $T_P \cap \mathcal{K} = P\mathcal{K}'$. □

Theorem 2.44. *Let $\mathcal{K}$ be a non-singular, regular $k_{r,n,q}$ with $n \geq 3$ and $q > 4$. Then, with $\mu_n = |\mathcal{U}_{n,q}|$,*

(i) $k = \mu_n$;   (2.31)

(ii) *every section of type* I *is a Hermitian curve.*

Proof. (i) For $n = 3$, the theorem is part of Theorem 2.25. For $n \geq 4$, let P be a point of $\mathcal{K}$ and T_P the tangent hyperplane to $\mathcal{K}$ at P. Thus $T_P \cap \mathcal{K} = P\mathcal{K}'$, where $\mathcal{K}'$ is a regular, non-singular $k'_{r,n-2,q}$. By the definition of T_P, each of the q^{n-1} lines through P not in T_P is an r-secant of $\mathcal{K}$. Since $r = \sqrt{q} + 1$ by Lemma 2.40,

$$k = q^{n-1}\sqrt{q} + 1 + qk'.$$

This is the same recurrence relation as for μ_n in Theorem 2.8. Since a Hermitian arc has the same number of points as a Hermitian curve and the result is true for $n = 3$, the result is true for all n.

(ii) For any section by a plane π of type I, by (2.24),

$$k = q\sqrt{q}(q^{n-2} + M) + \theta(n - 2),$$

where M is the number of solids through π meeting $\mathcal{K}$ in a $\mathcal{U}_{3,q}$. From (2.31) above, $k = \mu_n$; that is, from (2.9),

$$k = q^{n/2}\{q^{n/2} - (-1)^n\}/(\sqrt{q} + 1) + \theta(n - 1).$$

Hence

$$M = \{q^{n-5/2} - (-1)^n q^{(n-3)/2}\}/(\sqrt{q} + 1).$$

Since $M \geq 1$, the section $\pi \cap \mathcal{K}$ is a Hermitian curve. $\qquad\square$

As in the previous results, let $\mathcal{K}$ be regular and non-singular, with $q > 4$. For $P \in \mathcal{K}$, the meet $T_P \cap \mathcal{K}$ is the *tangent cone at* P. From Corollary 2.43, the tangent cone is $P\mathcal{K}'$ with $\mathcal{K}' \subset \Pi_{n-2}$. From Theorem 2.44 when $n = 4$ and Theorem 2.25 when $n = 5$, the tangent cone is a Hermitian variety.

Consider a Hermitian variety $\mathcal{W}$ in $\mathrm{PG}(n, q)$ with singular space Π_0. If $\Pi_0 = \mathbf{U}_n$ and $\mathcal{W}$ is reduced to canonical form

$$\mathcal{W} = \mathbf{V}(X_0\bar{X}_0 + X_1\bar{X}_1 + \cdots + X_{n-1}\bar{X}_{n-1}),$$

then associate to any point $P = \mathbf{P}(a_0, a_1, \ldots, a_n)$ other than $\mathbf{U}_n$ the hyperplane

$$\Pi_P = \mathbf{V}(X_0\bar{a}_0 + X_1\bar{a}_1 + \cdots + X_{n-1}\bar{a}_{n-1}).$$

As Π_P does not depend on a_n, the hyperplane Π_P is associated to every point other than $\mathbf{U}_n$ on the line $P\mathbf{U}_n$. Hence the hyperplane Π_P is associated to the line $P\mathbf{U}_n$. Conversely, associated to any hyperplane $\Pi = \mathbf{V}(b_0X_0 + \cdots + b_{n-1}X_{n-1})$ is the line

$$l_\Pi = \{\mathbf{P}(\bar{b}_0, \bar{b}_1, \ldots, \bar{b}_{n-1}, t) \mid t \in \mathbf{F}_q \cup \{\infty\}\}.$$

If l_Π is associated to Π, with $\mathbf{U}_n$ in Π, and if Π' is any hyperplane not through $\mathbf{U}_n$, then $\Pi \cap \Pi'$ is the *polar hyperplane of* $l_\Pi \cap \Pi'$ *with respect to* $\mathcal{W} \cap \Pi'$.

Theorem 2.45. *Let $\mathcal{K}$ be a non-singular, regular $k_{r,n,q}$ with $n \geq 4$ and $q > 4$. If every tangent cone of $\mathcal{K}$ is a Hermitian variety, then $\mathcal{K}$ is a non-singular Hermitian variety.*

Proof. Let Π be a non-tangent hyperplane, let P be a point of $\Pi \cap \mathcal{K} = \mathcal{K}_1$ and let $\Gamma = T_P \cap \mathcal{K} = P\mathcal{K}'$. The $(n - 2)$-space $\beta_P = T_P \cap \Pi$ is tangent to $\mathcal{K}_1$ at P and so $\beta_P \cap \mathcal{K}_1$ is a cone $P\mathcal{K}''$, which is the same as $\Gamma \cap \beta_P$. Hence $\Gamma \cap \beta_P$ has only one singular point, namely P.

In the hyperplane T_P let l_P be the line associated with β_P as described before the theorem; it is a 1-secant to $\mathcal{K}$ with point of contact P. To each P in $\mathcal{K}_1$ is associated such a line l_P.

To see this when $n = 4$, consider the case that

$$\mathcal{K} = \mathcal{U}_4 = \mathbf{V}(X_0\bar{X}_0 + X_1\bar{X}_1 + X_2\bar{X}_3 + X_3\bar{X}_2 + X_4\bar{X}_4).$$

Let $\Pi = \mathbf{u}_4$ and $P = \mathbf{U}_3$. Then

$$T_P = \mathbf{u}_2,$$
$$\mathcal{K}_1 = \mathcal{U}_3 = \mathbf{V}(X_0\bar{X}_0 + X_1\bar{X}_1 + X_2\bar{X}_3 + X_3\bar{X}_2, X_4),$$
$$\mathcal{K}' = \mathcal{U}_2 = \mathbf{V}(X_0\bar{X}_0 + X_1\bar{X}_1 + X_4\bar{X}_4, X_2, X_3),$$
$$\beta_P = \mathbf{u}_2 \cap \mathbf{u}_4,$$
$$\mathcal{K}'' = \mathcal{U}_1 = \mathbf{V}(X_0\bar{X}_0 + X_1\bar{X}_1, X_2, X_3, X_4),$$
$$\Gamma = \mathbf{U}_3\mathcal{K}',$$
$$\beta_P \cap \mathcal{K}_1 = \beta_P \cap \Gamma = \mathbf{U}_3\mathcal{K}'' = \mathbf{V}(X_0\bar{X}_0 + X_1\bar{X}_1, X_2, X_4),$$
$$l_P = \mathbf{U}_3\mathbf{U}_4 = \mathbf{V}(X_0, X_1, X_2).$$

It is now shown that the lines l_P are concurrent at a point P_0. To do this it suffices to show that any two lines l_P intersect. So, let P_1, P_2 be points of $\mathcal{K}_1$; for $i = 1, 2$, let T_i be the tangent hyperplane at P_i, and let l_i be the line associated to P_i. To prove that l_1 and l_2 meet, two cases are distinguished: (a) $P_1P_2 \not\subset \mathcal{K}$; (b) $P_1P_2 \subset \mathcal{K}$.

(a) The $(n-2)$-space $T_1 \cap T_2$ contains neither P_1 nor P_2 and so $T_1 \cap T_2 \cap \mathcal{K}$ is a non-singular Hermitian variety. It follows that l_1 and l_2 are the lines joining P_1 and P_2 to the pole P_0 of the $(n-3)$-space $T_1 \cap T_2 \cap \Pi = \beta_1 \cap \beta_2$ with respect to the Hermitian variety $T_1 \cap T_2 \cap \mathcal{K}$.

(b) Both β_1 and β_2 contain P_1P_2. Let α be a plane in Π through P_1P_2 but in neither β_1 nor β_2; it does not lie in $\mathcal{K}$. If $\alpha \cap \mathcal{K} = P_1P_2$, then α would be in both T_1 and T_2, and so in $\beta_1 \cap \beta_2$. Also, if α met $\mathcal{K}$ in r lines of a pencil with centre P_i, it would belong to β_i. Thus α meets $\mathcal{K}$ in r lines of a pencil through a point P of P_1P_2 with $P \neq P_1, P_2$. Let m_1 and m_2 be two of these lines other than P_1P_2; also let $Q_1 \in m_1\backslash\{P\}$, and let $Q_2 \in m_2\backslash\{P\}$ with Q_2 not on P_1Q_1 or P_2Q_1. Then the points P_1, P_2, Q_1, Q_2 of $\mathcal{K}_1$ have no three collinear and none of the lines P_1Q_1, P_1Q_2, P_2Q_1, P_2Q_2, Q_1Q_2 belong to $\mathcal{K}$. Hence, from (a), the lines l_1' and l_2' associated to Q_1 and Q_2 meet in a point P_0; they also must meet l_1 and l_2. Since P_i, Q_1, Q_2 are not collinear, so l_i, l_1', l_2' are not coplanar, for $i = 1, 2$. Therefore l_1 and l_2 also contain P_0.

The point P_0 through which all the lines l_P pass as P varies in $\mathcal{K}_1$ is called the *pole* of Π; it is not in $\mathcal{K} \cup \Pi$.

From Theorem 2.44 (i), the number of lines l_P is

$$k_1 = \mu_{n-1}. \tag{2.32}$$

On the other hand, let N denote the number of unisecants of $\mathcal{K}$ through P_0. Then counting the points of $\mathcal{K}$ on the lines through P_0 gives

$$k = N + r\{\theta(n-1) - N\},$$

where $r = \sqrt{q} + 1$. Hence

$$N = \{(\sqrt{q} + 1)\theta(n - 1) - k\}/\sqrt{q} = k_1,$$

using (2.31) and (2.32). Thus every unisecant through P_0 is a line associated to a point of $\mathcal{K}_1$. Hence $\mathcal{K}_1$ is the set of points of contact of the unisecants through P_0, and these points generate Π, which can now be called the *polar hyperplane of P_0 with respect to $\mathcal{K}$.*

From above, the correspondence which associates to a non-tangent hyperplane Π its pole P_0 is bijective. Let $\mathfrak{S}$ be the bijection from the points to the hyperplanes of $PG(n, q)$ in which each point of $\mathcal{K}$ is mapped to its tangent hyperplane and each point not on $\mathcal{K}$ is mapped to its polar hyperplane.

It must be shown that $\mathfrak{S}$ is involutory; that is, if $\Pi = P\mathfrak{S}$ and $P' \in \Pi$, then the hyperplane $\Pi' = P'\mathfrak{S}$ contains P. Let P and P' be off $\mathcal{K}$, and let l be an r-secant of $\mathcal{K}$ through P' in Π. The r lines PQ for $Q \in l \cap \mathcal{K}$ are unisecants of $\mathcal{K}$. So the plane $\pi = Pl$ meets $\mathcal{K}$ in a Hermitian curve $\mathcal{U}_2$ by Theorems 2.39 and 2.44 (ii). Also, l is the polar of P with respect to $\mathcal{U}_2$, and so the polar l' of P' with respect to $\mathcal{U}_2$ contains P. The line l' contains the r points of contact of the tangents to $\mathcal{U}_2$ through P'. Hence the polar hyperplane Π' of P' contains l' and so P.

The other cases of P and P' are simpler. Thus the mapping $\mathfrak{S}$ is bijective and involutory, and transforms the points of a hyperplane Π into the hyperplanes through the point $P = \Pi\mathfrak{S}^{-1}$, as well as vice versa. So $\mathfrak{S}$ is a polarity for which $\mathcal{K}$ is the set of self-polar points. Hence $\mathcal{K} = \mathcal{U}_n$. $\square$

Theorem 2.46. *If $\mathcal{K}$ is a non-singular, regular $k_{r,n,q}$ with $n \geq 4$ and $q > 4$, then $\mathcal{K}$ is a non-singular Hermitian variety.*

Proof. From the previous result and Theorems 2.25 and 2.44 (ii), the result is true for $n = 4$ and $n = 5$.

Let $n \geq 6$ and proceed by induction. The tangent cone Γ at a point P of $\mathcal{K}$ has base $\mathcal{K}'$, which is a regular, non-singular $k_{r,n-2,q}$. Since $n - 2 \geq 4$, the set $\mathcal{K}'$ is a non-singular Hermitian variety by the induction hypothesis. So Γ is a Hermitian variety and then Theorem 2.45 gives the result. $\square$

2.6 The characterisation of projections of quadrics

In Section 2.5, a description was given of sets $\mathcal{K} = k_{r,n,q}$ with one of the following properties:

(i) $r = 1, 2$ or q;
(ii) $\mathcal{K}$ is singular;
(iii) $\mathcal{K}$ is non-singular with no plane section of type IV, $3 \leq r \leq q - 1$, and $q > 4$.

This leaves the case that $\mathcal{K}$ is non-singular, $3 \leq r \leq q - 1$, and either $q = 4$ or $\mathcal{K}$ has a plane section of type IV.

When $q = 4$, the topic is sets $k_{3,n,4}$, that is, sets of type $(1, 3, 5)$ in $PG(n, 4)$ with at least one 3-secant. As explained in Section 19.6 of FPSOTD, these sets form a vector space over $\mathbf{F}_2$ of dimension $\frac{1}{3}(n+1)(n^2 + 2n + 3)$. In $\mathrm{PG}(3, 4)$ there are seven distinct types of non-singular $k_{3,3,4}$.

For the remainder of this section, although some notice is taken of the case $q = 4$, the main topic is the case that

(a) $\mathcal{K}$ is non-singular and has a plane section of type IV;
(b) $q > 4$, $n \geq 3$, $3 \leq r \leq q - 1$.

It follows from Lemma 2.30 that $q = 2^h$ with $h \geq 2$ and $r = \frac{1}{2}q + 1$.

Theorem 2.47. *Let $\mathcal{K}$ be a non-singular $k_{r,n,q}$ with $3 \leq r \leq q - 1$ having a plane section of type* IV. *Then,*

(i) *if $q > 4$, $\mathcal{K}$ contains exactly one hyperplane Π_{n-1};*
(ii) *if $q = 4$ and $\mathcal{K}$ contains no section of type* I *or* II, *it contains exactly one hyperplane Π_{n-1}.*

Proof. For $n = 3$, the result is contained in Theorems 19.4.8 and 19.4.9 of FPSOTD, which show that $\mathcal{K}$ is a set $\mathcal{K}_1$ that contains precisely one plane. Subsequently, in Theorem 19.4.17, it is shown that $\mathcal{K}_1$ is $\mathcal{R}_3$, which is the projection of a quadric $\mathcal{P}_4$ onto a solid Π_3 from a point other than the nucleus. Now suppose that $n > 3$.

By the Corollary to Theorem 19.4.7, there exists a unisecant l_1 of $\mathcal{K}$ with point of contact P; let l_2 be an r-secant through P. So the plane $\pi = l_1 l_2$ is of type III. Let Π_4 be a 4-dimensional space containing π and consider the solids $\alpha_1, \ldots, \alpha_{q+1}$ in Π_4 containing π. Then $\alpha_i \cap \mathcal{K}$ is either a cone $P_i \mathcal{K}_{\mathrm{III}}$ with vertex P_i and a section $\mathcal{K}_{\mathrm{III}} = \pi \cap \mathcal{K}$ or $\alpha_i \cap \mathcal{K}$ is a set $\mathcal{R}_3^{(i)}$; this follows from Theorem 2.25 for $q > 4$ and from Theorem 19.6.8 of FPSOTD for $q = 4$.

Suppose there are s cones with vertices $P_1, P_2, \ldots, P_s$ and $q + 1 - s$ sets $\mathcal{R}_3^{(1)}, \ldots, \mathcal{R}_3^{(q+1-s)}$. If $\Pi_4 \cap \mathcal{K}$ contains a solid Π_3, then Π_3 contains the line l_0 of $\mathcal{K}$ in π, the plane $l_0 P_i$, and the plane $\Pi_2^{(j)}$ of $\mathcal{K}$ in $\mathcal{R}_3^{(j)}$. It follows that Π_3 is unique and is the union of the $q + 1$ planes $l_0 P_i$, $\Pi_2^{(j)}$.

Now suppose that the $q+1$ planes $l_0 P_i$, $\Pi_2^{(j)}$ do not lie in a solid: they are in any case the only planes in $\Pi_4 \cap \mathcal{K}$ through l_0. Consider the solid Π_3' containing two of the planes; then $\Pi_3' \cap \mathcal{K}$ is a cone with base type IV or V. If the base is of type IV, then $\Pi_3' \cap \mathcal{K}$ contains exactly two of the $q + 1$ planes. Also, any other plane Π_2 through l_0 in Π_3' is of type V. This plane Π_2 and π define a solid Π_3'' for which $\Pi_3'' \cap \mathcal{K}$ is an $\mathcal{R}_3$. Since Π_2 is of type V and since l_0 is in the plane on this $\mathcal{R}_3$, the line l_0 contains the exceptional point Q_0 of $\mathcal{R}_3$. So π is also of type V, a contradiction. Thus $\Pi_3' \cap \mathcal{K}$ consists of $\frac{1}{2}q + 1$ planes through l_0. Hence the solids Π_3' and the sections of type VII through l_0 in such solids form a $2 - (q+1, \frac{1}{2}q+1, 1)$ design, whence the number of such solids is $2(q + 1)/(\frac{1}{2}q + 1)$, which is not an integer. Thus the $q + 1$ planes $l_0 P_i$, $\Pi_2^{(j)}$ are the $q + 1$ planes through l_0 of a solid.

Consider now all the solids α_i, $i = 1, 2, \ldots, N = (q^{n-2} - 1)/(q - 1)$, which pass through π. Then $\alpha_i \cap \mathcal{K}$ is a cone $P_i \mathcal{K}_{\mathrm{III}}$ or a set $\mathcal{R}_3^{(i)}$. Suppose there are t cones with vertices $P_1, P_2, \ldots, P_t$ and $N - t$ sets $\mathcal{R}_3^{(1)}, \ldots, \mathcal{R}_3^{(N-t)}$. If $\mathcal{K}$ contains a hyperplane Π_{n-1}, then Π_{n-1} contains the line l_0, the plane $l_0 P_i$, and the plane

$\Pi_2^{(j)}$ of $\mathcal{K}$ on $\mathcal{R}_3^{(j)}$. It follows that Π_{n-1} is unique and is the union of the N planes $l_0 P_i$, $\Pi_2^{(j)}$, which all contain l_0. However, the solid containing at least two of these planes contains exactly $q + 1$, by the same argument as above for the Π_4 containing π and the two planes. Hence the planes $l_0 P_i$, $\Pi_2^{(j)}$ are all the planes through l_0 of a hyperplane Π_{n-1}. So $\mathcal{K}$ contains exactly one hyperplane Π_{n-1}. $\square$

Corollary 2.48. *If $\mathcal{K}$ is a non-singular $k_{r,n,q}$ with $3 \le r \le q - 1$, then $\mathcal{K}$ contains at most one hyperplane.*

Proof. Suppose $\mathcal{K}$ contains two hyperplanes Π_{n-1}, Π'_{n-1}; they intersect in a Π_{n-2}. Let P be a point of Π_{n-2} and let l be an r-secant of $\mathcal{K}$ through P; also let π be a plane through l such that $\pi \cap \Pi_{n-2} = \{P\}$. Now π contains l, and meets Π_{n-1} and Π'_{n-1} in lines of $\mathcal{K}$; so π is of type IV. For $q = 4$, since any plane contains at least one line of $\mathcal{K}$, no plane section is of type I or II. Hence, by the theorem, $\mathcal{K}$ contains exactly one hyperplane, a contradiction. $\square$

Let $\mathcal{K}$ contain the unique hyperplane Π_{n-1}. As in Theorem 19.4.9 of FPSOTD for $\mathrm{PG}(3, q)$, define $\mathcal{J}$, the *residual of $\mathcal{K}$*, to be

$$\mathcal{J} = (\mathrm{PG}(n, q) \backslash \mathcal{K}) \cup \Pi_{n-1}.$$

This may also be written $\mathcal{J} = \mathcal{K} \triangle \Pi_{n-1}$, where $X \triangle Y$ is the complement of the symmetric difference of the two sets X and Y. For $q = 4$, this operation defines a vector space over $\mathbf{F}_2$ on the sets $\mathcal{K}$ of type $(1, 3, 5)$ as above; see Section 19.6 of FPSOTD.

Corollary 2.49. *If $\mathcal{K}$ is a non-singular $k_{r,n,q}$ with $3 \le r \le q - 1$ and contains a hyperplane Π_{n-1}, then*

  (i) *$\mathcal{K}$ contains a section of type IV and no section of type I or II;*
 (ii) *$\mathcal{J}$ is also a non-singular set of type $(1, \frac{1}{2}q + 1, q + 1)$ containing the same unique hyperplane as $\mathcal{K}$.*

Proof. If $|l \cap \mathcal{K}| = i$ for a line l not in Π_{n-1}, then $|l \cap \mathcal{J}| = q + 2 - i$. So $\mathcal{J}$ is a set of type $(1, q + 2 - r, q + 1)$; note that $3 \le q + 2 - r \le q - 1$. Suppose P is a singular point of $\mathcal{J}$. If $P \notin \Pi_{n-1}$, then any line l through P contains two points of $\mathcal{J}$ and so $|l \cap \mathcal{J}| = q + 1$. Hence $\mathcal{J} = \mathrm{PG}(n, q)$, a contradiction. If $P \in \Pi_{n-1}$, then any line through P contains 1 or $q + 1$ points of $\mathcal{K}$; that is, P is singular for $\mathcal{K}$, a contradiction. Hence $\mathcal{J}$ is a non-singular set of type $(1, q + 2 - r, q + 1)$ containing Π_{n-1}.

By Corollary 2.48, Π_{n-1} is the only hyperplane of $\mathcal{K}$. Any plane section of $\mathcal{K}$ contains at least one line and is consequently not of type I or II. Similarly $\mathcal{J}$ contains no sections of type I or II. If l_1 is a unisecant of $\mathcal{K}$ with point of contact P and if l_r is an r-secant through P, then the plane $l_1 l_r$ is of type III for $\mathcal{K}$ and hence of type IV for $\mathcal{J}$. Consequently $r = \frac{1}{2}q + 1$. In this argument, $\mathcal{K}$ and $\mathcal{J}$ can be interchanged. $\square$

The nature of sets $\mathcal{K}$ containing a hyperplane is now investigated in more detail.

Definition 2.50. (1) A non-empty subset $\mathcal{S}$ of $\mathrm{PG}(d, q)$ is a *projective Shult space* with ambient space $\mathrm{PG}(n, q)$ if

(a) $\mathcal{S}$ spans $\mathrm{PG}(n, q)$;

(b) there is a non-empty subset $\mathcal{L}$ of the set of lines in $\mathcal{S}$ such that, given a point $S \in \mathcal{S}$ and a line $l \in \mathcal{L}$ not containing S, then the line SQ is in $\mathcal{L}$ for exactly one or for all points Q of l.

(2) The Shult space $\mathcal{S}$ is *non-degenerate* if there is no point $A \in \mathcal{S}$ such that $AQ \in \mathcal{L}$ for every point Q of $\mathcal{S} \backslash \{A\}$.

Projective Shult spaces in $\mathrm{PG}(n, q)$ are discussed in Section 5.3, and classified in Theorems 5.51 and 5.52.

Definition 2.51. If $\mathcal{K}$ in $\mathrm{PG}(n, q)$ contains a hyperplane Π_{n-1} and P is any point of $\mathcal{K} \backslash \Pi_{n-1}$, then the *support of P* is

$$\mathcal{S}_P = \{Q \in \Pi_{n-1} \mid PQ \subset \mathcal{K}\}.$$

In Theorem 2.56, it is in fact shown that $\mathcal{S}_P$ is a non-degenerate, projective Shult space in Π_{n-1}, or possibly, when $n = 4$, an elliptic quadric. When $n = 3$, it is proved in Lemma 19.4.16 of FPSOTD that $\mathcal{S}_P$ is a conic.

The number of projectively distinct non-singular quadrics $\mathcal{Q}_n$ in $\mathrm{PG}(n, q)$ is one or two according as n is even or odd. For n even, $\mathcal{Q}_n = \mathcal{P}_n$ and, for n odd, $\mathcal{Q}_n = \mathcal{H}_n$ or $\mathcal{E}_n$. In the respective cases, the character w of $\mathcal{Q}_n$ is 1, 2 or 0; see Section 1.4. Also from (1.19),

$$|\mathcal{Q}_n| = q^{n-1} + q^{n-2} + \cdots + q + 1 + (w - 1)q^{(n-1)/2}.$$

The character $w = 0$ is also assigned to a $(q^2 + 1)$-cap, of which $\mathcal{E}_3$ is an example.

Theorem 2.52. *Let*

(a) $\mathcal{Q}_{n+1}$ *be a non-singular quadric of character w in $PG(n + 1, q)$, q even;*

(b) Q *be a point off $\mathcal{Q}_{n+1}$ other than the nucleus when $\mathcal{Q}_{n+1} = \mathcal{P}_{n+1}$;*

(c) Π_n *be a hyperplane not containing Q.*

Then the projection of $\mathcal{Q}_{n+1}$ from Q onto Π_n is a non singular set $\mathcal{R}_n$ of type $(1, \frac{1}{2}q + 1, q + 1)$ in Π_n containing a hyperplane Π_{n-1} of Π_n with

$$|\mathcal{R}_n| = \tfrac{1}{2}q^n + q^{n-1} + q^{n-2} + \cdots + q + 1 + \tfrac{1}{2}(w - 1)q^{n/2}.$$

Proof. Let l be a line in Π_n. The plane Ql meets $\mathcal{Q}_{n+1}$ in a point, a line, a line pair or a conic. In the case that $Ql \cap \mathcal{Q}_{n+1} = \mathcal{P}_2$, either Q is the nucleus of $\mathcal{P}_2$, in which case the lines joining Q to the points of $\mathcal{P}_2$ are $q + 1$ distinct tangents, or the lines joining Q to $\mathcal{P}_2$ are $\frac{1}{2}q$ bisecants and one tangent.

Let $\mathcal{R}_n = \{P' = PQ \cap \Pi_n \mid P \in \mathcal{Q}_{n+1}\}$; then Table 2.3 is obtained. Thus $\mathcal{R}_n$ is of type $(1, \frac{1}{2}q + 1, q + 1)$. The tangents to $\mathcal{Q}_{n+1}$ through Q meet $\mathcal{Q}_{n+1}$ in $\mathcal{P}_n$, $\mathcal{P}_n$ or $\Pi_0 \mathcal{P}_{n-1}$ as $\mathcal{Q}_{n+1}$ is $\mathcal{H}_{n+1}$, $\mathcal{E}_{n+1}$ or $\mathcal{P}_{n+1}$; they meet Π_n in a hyperplane Π_{n-1}. Since $\theta(n - 1)$ is the number of tangent lines through Q to $\mathcal{Q}_{n+1}$, so

Table 2.3. Intersection types

$Ql \cap Q_{n+1}$	Point	Line	Line pair	Conic
$|l \cap \mathcal{R}_n|$	1	$q+1$	$q+1$	$q+1$ or $\frac{1}{2}q+1$

$$k = |\mathcal{R}_n| = \theta(n-1) + \tfrac{1}{2}(|Q_{n+1}| - \theta(n-1)),$$

whence k is as required.

To show that $\mathcal{R}_n$ is non-singular, it suffices to show that for any point P of Q_{n+1}, joined to Q by a line l there exists a plane π through l meeting Q_{n+1} in a $\mathcal{P}_2$ for which Q is not the nucleus; if P projects to P', then π projects to a $(\frac{1}{2}q+1)$-secant of $\mathcal{R}_n$ through P'.

First, let l be a bisecant of Q_{n+1}. Then every plane through l meets Q_{n+1} in a conic $\mathcal{P}_2$ or a line pair $\Pi_0\mathcal{H}_1$. If there are b_0 of the former and b_1 of the latter,

$$b_0 + b_1 = \theta(n-1),$$
$$(q-1)b_0 + (2q-1)b_1 + 2 = \theta(n) + (w-1)q^{n/2},$$

whence

$$b_0 = q^{n-1} - (w-1)q^{(n-2)/2} > 0.$$

Now let l be a tangent to Q_{n+1} through Q and let s_{n+1} be the number of planes through l meeting Q_{n+1} in a line pair $\Pi_0\mathcal{H}_1$. When $Q_{n+1} = \mathcal{P}_{n+1}$ and the nucleus of $\mathcal{P}_{n+1}$ is on l, then $s_{n+1} = 0$; otherwise,

$$s_{n+1} = \tfrac{1}{2}(|Q_{n-1}| - t_{n-1}),$$

where t_{n-1} is the number of tangent lines to Q_{n-1} through a point P off Q_{n-1} other than the nucleus. It follows that there is a bisecant m through Q such that the plane ml does not meet Q_{n+1} in a line pair. So ml meets Q_{n+1} in a conic.   □

When n is even, $\mathcal{R}_n$ is also denoted $\mathcal{R}_n^+$ or $\mathcal{R}_n^-$ according as it is the projection of $\mathcal{H}_{n+1}$ or $\mathcal{E}_{n+1}$.

A description of $\mathcal{R}_n$ is required which does not depend on projection from higher space. So, let $F(X_0, X_1, \ldots, X_{n-1})$ be a non-degenerate quadratic form over $\mathbf{F}_q$ and let H be an additive subgroup of $\mathbf{F}_q$ of index 2. Let

$$\mathcal{F}_\lambda = \mathbf{V}(F(X_0, X_1, \ldots, X_{n-1}) + \lambda X_n^2)$$

be a quadric in $\mathrm{PG}(n, q)$; here $\mathcal{F}_\infty = \mathbf{V}(X_n^2) = \mathbf{u}_n$. Also it may be assumed that F is one of the following forms:

(i) $X_0^2 + X_1 X_2 + \cdots + X_{n-2}X_{n-1}$, for n odd;
(ii) $X_0 X_1 + X_2 X_3 + \cdots + X_{n-2}X_{n-1}$ or $f(X_0, X_1) + X_2 X_3 + \cdots + X_{n-2}X_{n-1}$ with f irreducible, for n even.

Theorem 2.53. *The set* $\mathcal{R}_n = \bigcup_{\lambda \in H \cup \{\infty\}} \mathcal{F}_\lambda$ *for* $n \geq 2$.

Proof. Consider together the cases that $\mathcal{Q}_{n+1} = \mathcal{P}_{n+1}, \mathcal{H}_{n+1}, \mathcal{E}_{n+1}$. Now write $\mathcal{Q}_{n+1} = \mathbf{V}(G_{n+1})$, so that in each case $G_{n+1} = G_{n-1} + X_n X_{n+1}$ with $n \geq 1$. From the above canonical forms, $G_0 = X_0^2$ in the parabolic case, $G_1 = X_0 X_1$ in the hyperbolic case and $G = f(X_0, X_1)$ in the elliptic case. Let $Q = \mathbf{P}(0, 0, \ldots, 0, 1, 1)$ and consider the pencil of hyperplanes in $\mathrm{PG}(n + 1, q)$ through the subspace $\mathbf{V}(X_n, X_{n+1})$. Let

$$\mathcal{V}_t = \mathbf{V}(tX_n + X_{n+1}) \cap \mathcal{Q}_{n+1}.$$

For $t \neq \infty$,

$$\mathcal{V}_t = \mathbf{V}(G_{n-1} + tX_n^2, tX_n + X_{n+1}).$$

In particular, $\mathcal{V}_0 = \mathbf{V}(X_{n+1}) \cap \mathcal{Q}_{n+1}$, $\mathcal{V}_\infty = \mathbf{V}(X_n) \cap \mathcal{Q}_{n+1}$.

(i) For $\mathcal{Q}_{n+1} = \mathcal{P}_{n+1}$,

$$G_{n-1} + tX_n^2 = X_0^2 + X_1 X_2 + \cdots + X_{n-2} X_{n-1} + tX_n^2.$$

So

$$\mathcal{V}_t = \Pi_0 \mathcal{P}_{n-1}, \quad \text{for all } t.$$

(ii) For $\mathcal{Q}_{n+1} = \mathcal{H}_{n+1}$,

$$G_{n-1} + tX_n^2 = X_0 X_1 + \cdots + X_{n-2} X_{n-1} + tX_n^2.$$

So

$$\mathcal{V}_t = \begin{cases} \mathcal{P}_n & \text{for } t \neq 0, \infty, \\ \Pi_0 \mathcal{H}_{n-1} & \text{for } t = 0, \infty. \end{cases}$$

(iii) For $\mathcal{Q}_{n+1} = \mathcal{E}_{n+1}$,

$$G_{n-1} + tX_n^2 = f(X_0, X_1) + X_2 X_3 + \cdots + X_{n-2} X_{n-1} + tX_n^2.$$

So

$$\mathcal{V}_t = \begin{cases} \mathcal{P}_n & \text{for } t \neq 0, \infty, \\ \Pi_0 \mathcal{E}_{n-1} & \text{for } t = 0, \infty. \end{cases}$$

In each case, the $\mathcal{V}_t$ have a $\mathcal{Q}_{n-1}$ in common, of the same character w as $\mathcal{Q}_{n+1}$.

If $A = \mathbf{P}(a_0, a_1, \ldots, a_{n+1})$ lies in $\mathcal{Q}_{n+1}$, then QA meets $\mathcal{Q}_{n+1}$ again in $A' = \mathbf{P}(a_0, a_1, \ldots, a_{n-1}, a_{n+1}, a_n)$. If A also lies in $\mathcal{V}_t$, then

$$A = \mathbf{P}(a_0, a_1, \ldots, a_n, ta_n), \quad A' = \mathbf{P}(a_0, a_1, \ldots, a_{n-1}, ta_n, a_n),$$

whence A' lies in $\mathcal{V}_{1/t}$. Further, QA is a tangent when $t = 1$, and so the tangents through Q meet $\mathcal{Q}_{n+1}$ in $\mathcal{V}_1$.

Now project $\mathcal{Q}_{n+1}$ from Q to $\mathbf{u}_{n+1} = \mathbf{V}(X_{n+1})$:

$$A = \mathbf{P}(a_0, a_1, \ldots, a_n, a_{n+1}) \mapsto \mathbf{P}(a_0, a_1, \ldots, a_{n-1}, a_n + a_{n+1}, 0).$$

Hence $\mathcal{V}_1 \mapsto \mathbf{V}(X_n, X_{n+1})$ and, for $t \neq 1$,

$$\{\mathcal{V}_t, \mathcal{V}_{1/t}\} \mapsto \mathcal{W}_t = \mathbf{V}\left(G_{n-1} + \frac{t}{t^2 + 1}X_n^2, X_{n+1}\right).$$

For, regarding the projection as $\mathbf{P}(x) \mapsto \mathbf{P}(x')$,

$$x_0' = x_0, \ldots, x_{n-1}' = x_{n-1}, x_n' = x_n + x_{n+1}, x_{n+1}' = 0.$$

So, if $\mathbf{P}(X) \in \mathcal{V}_t$ for $t \neq 1$, then $x_{n+1} = tx_n$, whence $x_n' = (t+1)x_n$. Also, $\mathcal{W}_t$ is the same type of quadric as $\mathcal{V}_t$.

Let $K = \{t/(t^2 + 1) \mid t \in \mathbf{F}_q \backslash \{1\}\}$; then $|K| = \frac{1}{2}q$. It is now shown that K is a group. Write $t = \lambda/\lambda'$; then $t/(t^2 + 1) = \lambda\lambda'/(\lambda^2 + \lambda'^2)$. So

$$\frac{\lambda\lambda'}{\lambda^2 + \lambda'^2} + \frac{\mu\mu'}{\mu^2 + \mu'^2} = \frac{(\lambda\mu + \lambda'\mu')(\lambda\mu' + \lambda'\mu)}{(\lambda\mu + \lambda'\mu')^2 + (\lambda\mu' + \lambda'\mu)^2}.$$

Therefore K is a subgroup of $\mathbf{F}_q$ of index 2. It has thus been shown that

$$\mathcal{R}_n = \bigcup_{\lambda \in K} \mathcal{W}_\lambda \cup \mathbf{V}(X_n, X_{n+1}).$$

Select F as G_{n-1}. By Lemma 19.4.12 of FPSOTD, $K = \beta H$ for some $\beta \in \mathbf{F}_q \backslash \{0\}$. So the projectivity $\mathbf{P}(x) \mapsto \mathbf{P}(x')$, given by

$$x_0' = x_0, \ldots, x_{n-1}' = x_{n-1}, x_n' = \sqrt{\beta}\, x_n,$$

transforms $\mathcal{R}_n$ to the required form.   □

Now a result on the characterisation of elliptic quadrics in $\mathrm{PG}(3, q)$ for q even is established that might have been shown in Chapter 16 or 18 of FPSOTD. A weaker version is required in the subsequent theorem. For q odd, or $q = 4$, a $(q^2 + 1)$-cap is an elliptic quadric, by Theorem 16.1.7 of FPSOTD.

Lemma 2.54. *A $(q^2 + 1)$-cap $\mathcal{K}$ in $\mathrm{PG}(3, q)$ containing $\frac{1}{2}(q^3 - q^2 + 2q)$ conics is an elliptic quadric $\mathcal{E}_3$.*

Proof. From above, it suffices to consider q even with $q \geq 8$. First it is shown that there exist points P and Q on $\mathcal{K}$ such that

(a) through P there are $\frac{1}{2}q^2 + 1$ conics in $\mathcal{K}$;
(b) through both P and Q there are $\frac{1}{2}q + 1$ conics in $\mathcal{K}$.

If there were no point P for which (a) holds, then every point of $\mathcal{K}$ would be on at most $\frac{1}{2}q^2$ conics. Counting the size of the set

$$\{(A, \mathcal{C}) \mid A \in \mathcal{K}, \mathcal{C} \text{ a conic in } \mathcal{K}, A \in \mathcal{C}\}$$

in two ways gives the following:

$$\tfrac{1}{2}(q^3 - q^2 + 2q)(q + 1) \leq \tfrac{1}{2}q^2(q^2 + 1),$$

a contradiction.

If, given P, there is no point Q of $\mathcal{K}$ for which (b) holds, then through P and any one of the q^2 points of $\mathcal{K}\backslash\{P\}$ there can be at most $\frac{1}{2}q$ conics of $\mathcal{K}$. So, a count of the set

$$\{(A,\mathcal{C}) \mid A \in \mathcal{K}\backslash\{P\}, \mathcal{C} \text{ a conic in } \mathcal{K} \text{ containing } A \text{ and } P\}$$

gives

$$(\tfrac{1}{2}q^2 + 1)q \leq \tfrac{1}{2}q \cdot q^2,$$

another contradiction.

Since P and Q satisfy (a) and (b), let T_P and T_Q be the tangents planes to $\mathcal{K}$ at P and Q, let $\mathcal{C}_0, \mathcal{C}_1, \ldots, \mathcal{C}_{q/2}$ be distinct conics on $\mathcal{K}$ through P and Q, and let $\mathcal{C}$ be a conic on $\mathcal{K}$ containing P but not Q. The plane π of $\mathcal{C}$ meets T_P in a line which is tangent at P to at most one $\mathcal{C}_i$, say $\mathcal{C}_0$. Therefore π meets the planes of $\mathcal{C}_1, \mathcal{C}_2, \ldots, \mathcal{C}_{q/2}$ in bisecants of $\mathcal{K}$; if $P_1, P_2, \ldots, P_{q/2}$ denote the points of $\mathcal{K}$ other than P on these bisecants, where $P_i \in \mathcal{C}_i$, then $\mathcal{C}_i \cap \mathcal{C} = \{P, P_i\}$.

Let $\mathcal{Q}$ be the quadric containing $\mathcal{C}_1, \mathcal{C}_2$ and P_3; the nine conditions necessary for $\mathcal{C}_1, \mathcal{C}_2$ and P_3 to lie on a quadric ensure the existence of $\mathcal{Q}$, and it is impossible for two quadrics to meet in two conics plus a point.

Since T_P contains the tangent lines at P to $\mathcal{C}_1$ and $\mathcal{C}_2$, it is also the tangent plane at P to $\mathcal{Q}$; it also contains the tangent line to $\mathcal{C}$ at P. As $\mathcal{Q}$ contains the four points P_1, P_2, P_3, P_4 of $\mathcal{C}$ and, as the tangent plane T_P to $\mathcal{Q}$ at P contains the tangent to $\mathcal{C}$ at P, so $\mathcal{Q}$ contains $\mathcal{C}$. Also, each conic $\mathcal{C}_i$, $i = 3, \ldots, \frac{1}{2}q$, lies on $\mathcal{Q}$, since $\mathcal{Q}$ contains the three points P, Q, P_i of $\mathcal{C}_i$ and the tangent planes T_P and T_Q to $\mathcal{Q}$ at P and Q contain the tangents of $\mathcal{C}_i$ at P and Q.

From (a), there exists a conic $\mathcal{D}$ on $\mathcal{K}$ which contains P but not Q and which does not touch $\mathcal{C}_0$ at P. If $\mathcal{D}$ is substituted for $\mathcal{C}$ in the above argument, then it follows that $\mathcal{C}_0$ lies on any quadric containing $\frac{1}{2}q - 1$ of the conics $\mathcal{C}_1, \mathcal{C}_2, \ldots, \mathcal{C}_{q/2}$; hence $\mathcal{Q}$ also contains $\mathcal{C}_0$.

The number of points in $\mathcal{C}_0 \cup \mathcal{C}_1 \cup \cdots \cup \mathcal{C}_{q/2} \cup \mathcal{C}$ is at least

$$2 + (q-1)(\tfrac{1}{2}q + 1) + (\tfrac{1}{2}q - 1) = \tfrac{1}{2}(q^2 + 2q)$$
$$\geq \tfrac{1}{2}(q^2 + q + 4).$$

Hence, by the Corollary to Lemma 18.1.8 of FPSOTD, $\mathcal{K}$ lies on $\mathcal{Q}$; so $\mathcal{K} = \mathcal{Q}$. $\square$

To prove the next result a further definition is required.

Definition 2.55. A *semi-quadric* in $\mathrm{PG}(n, q)$ is a pair $(\mathcal{P}, \mathcal{L})$ where $\mathcal{P}$ is a set of points and $\mathcal{L}$ is a set of lines of $\mathrm{PG}(n, q)$ such that one of the following holds:

(1) $\mathcal{P}$ is a non-singular quadric $\mathcal{Q}_n$ and $\mathcal{L}$ is the set of lines on $\mathcal{Q}_n$;
(2) $\mathcal{P}$ is a non-singular Hermitian variety $\mathcal{U}_n$ and $\mathcal{L}$ is the set of lines on $\mathcal{U}_n$;
(3) $\mathcal{P} = \mathrm{PG}(n, q)$ and $\mathcal{L}$ is the set of lines of a linear complex.

Theorem 2.56. *In* $\mathrm{PG}(n,q)$ *with* $n \geq 4$, *let* $\mathcal{K}$ *be a non-singular* $k_{r,n,q}$ *still with* $3 \leq r \leq q - 1$ *and containing a hyperplane* Π_{n-1}. *If* P *is a point of* $\mathcal{K}\backslash\Pi_{n-1}$, *then the support* $\mathcal{S}_P$ *of* P *is a non-singular quadric in* Π_{n-1}. *Thus*

$$\mathcal{S}_P = \begin{cases} \mathcal{P}_{n-1} & \text{for } n \text{ odd,} \\ \mathcal{H}_{n-1} \text{ or } \mathcal{E}_{n-1} & \text{for } n \text{ even.} \end{cases}$$

Proof. If l is a line through P, then l contains a point of Π_{n-1}, whence

$$|l \cap \mathcal{K}| = \tfrac{1}{2}q + 1 \text{ or } q + 1.$$

Consider all lines $l_1, l_2, \ldots, l_m$ through P which lie on $\mathcal{K}$, and let $l_i \cap \Pi_{n-1} = S_i$; that is, $\mathcal{S}_P = \{S_i \mid i = 1, 2, \ldots, m\}$. Suppose that three distinct lines l_1, l_2, l_3 lie in a plane Π_2 and consider a solid Π_3 containing Π_2. Since $\Pi_3 \cap \mathcal{K}$ contains a plane in Π_{n-1} as well as the lines l_1, l_2, l_3, it is singular by Lemma 19.4.10 of FPSOTD. In particular, it must be the join of a point to a plane section of type IV, V or VII; that is, $\Pi_3 \cap \mathcal{K}$ consists of either $q + 2$ concurrent planes no three of which have a line in common, or $\tfrac{1}{2}q + 1$ planes through a line or Π_3 itself. In each case, the plane Π_2 lies on $\mathcal{K}$ and hence every line through P in Π_2 is on $\mathcal{K}$. So $\mathcal{S}_P$ is a set of type $(0, 1, 2, q + 1)$ in Π_{n-1}.

Now it is shown that $\mathcal{S}_P$ is a non-degenerate Shult space in some subspace Π_s of Π_{n-1}. Let l be a line of $\mathcal{S}_P$ and S_i a point of $\mathcal{S}_P\backslash\{l\}$. Consider the solid $\Pi_3 = ll_i$, where $l_i = PS_i$. Then $\Pi_3 \cap \mathcal{K}$ contains a plane of Π_{n-1}, the plane Pl and the line l_i skew to l. So $\Pi_3 \cap \mathcal{K}$ is Π_3 or a cone with base of type IV and vertex V on l. In the first case, the plane $S_i l$ is in $\mathcal{S}_P$; in the second case, there is just one line in $\mathcal{S}_P$ through S_i containing a point of l, namely VS_i, since the plane PVS_i lies in $\mathcal{K}$. Thus, if $\mathcal{S}_P$ contains at least one line l, it is a projective Shult space of type $(0, 1, 2, q + 1)$ in some subspace Π_s of Π_{n-1}; if $\mathcal{S}_P$ contains no line, it is a cap.

If Π_3 is a solid containing P, then in each of the cases there is at least one line of $\Pi_3 \cap \mathcal{K}$ through P; so $\mathcal{S}_P$ is non-empty. Suppose $\mathcal{S}_P$ is degenerate with singular point A. Let Q be a point of $\mathcal{K}\backslash\Pi_{n-1}$ other than P, and let Π_3 contain A, P, Q. Then $\Pi_3 \cap \mathcal{S}_P$ consists of lines through A. If $\Pi_3 \cap \mathcal{K}$ is non-singular, then $\Pi_3 \cap \mathcal{S}_P$ is the support of P in $\Pi_3 \cap \mathcal{K}$ and so, by Lemma 19.4.10 of FPSOTD, is a $(q + 1)$-arc, a contradiction. Thus $\Pi_3 \cap \mathcal{K}$ is singular and is therefore the join of a point to a section of type III, IV, V or VII.

(i) If the section is of type III, then $\Pi_3 \cap \mathcal{S}_P = \{A\}$ and A is the vertex of the cone. Hence AQ is on $\mathcal{K}$.

(ii) If the section is of type IV, then $\Pi_3 \cap \mathcal{S}_P$ is a pair of distinct lines which meet in the vertex V of the cone. Since V is the only singular point of $\Pi_3 \cap \mathcal{S}_P$, so $A = V$ and AQ is on $\mathcal{K}$.

(iii) If the section is of type V, then $\Pi_3 \cap \mathcal{S}_P$ is a line through A, namely the line of intersection of the $\tfrac{1}{2}q + 1$ planes of $\Pi_3 \cap \mathcal{K}$. Hence AQ is on $\mathcal{K}$.

(iv) If the section is of type VII, then $\Pi_3 \cap \mathcal{K} = \Pi_3$ and again AQ is on $\mathcal{K}$.

In each case AQ is on $\mathcal{K}$, whence A is a singular point of $\mathcal{K}$, a contradiction.

If $\mathcal{S}_P$ contains no lines, it is a cap. If $\mathcal{S}_P$ has a projective index at least one, then, by Theorem 5.51, it is a semi-quadric in a subspace Π_s of Π_{n-1}. As $\mathcal{S}_P$ is of type $(0, 1, 2, q+1)$, it cannot be a Hermitian variety. The symplectic case is also excluded as every line of Π_{n-1} in $\mathcal{S}_P$ is a line of it considered as a Shult space. Thus $\mathcal{S}_P$ is a cap or a non-singular quadric spanning a subspace Π_s of Π_{n-1}.

Suppose $s < n - 1$. Let l be a line of Π_s with $l \cap \mathcal{S}_P = \emptyset$, and let Π_3 be a solid containing P and l such that $\Pi_3 \cap \Pi_s = l$. Then, in $\Pi_3 \cap \mathcal{K}$, there is no line containing P, a contradiction. Hence $\mathcal{S}_P$ spans Π_{n-1}.

It remains to show that when $\mathcal{S}_P$ is a cap, then $n = 4$ and $\mathcal{S}_P = \mathcal{E}_3$. Let Π_3 be the solid containing P and three points S_1, S_2, S_3 of $\mathcal{S}_P$. As $\Pi_3 \cap \mathcal{K}$ is necessarily non-singular, it follows that it is an $\mathcal{R}_3$ and $\Pi_3 \cap \mathcal{S}_P$ is a conic. So the points of $\mathcal{S}_P$ and the conics $\Pi_3 \cap \mathcal{S}_P$ form a $3 - (m, q + 1, 1)$ design with $m = |\mathcal{S}_P|$. However, if Π_3 is an arbitrary solid through P, S_1, S_2, then $\Pi_3 \cap \mathcal{K}$ is non-singular and $\Pi_3 \cap \mathcal{S}_P$ is again a conic. Thus every plane through two points of $\mathcal{S}_P$ meets it in a conic. The number of planes through a line in Π_{n-1} is $N = (q^{n-2} - 1)/(q - 1)$. Hence

$$m = N(q - 1) + 2 = q^{n-2} + 1.$$

However, the maximum number M of points of a cap in $\mathrm{PG}(d, q)$ with $q > 2$ satisfies $M = q^2 + 1$ for $d = 3$ and $M < q^{d-1} + 1$ for $d > 3$: see Section 6.3; so $n = 4$ and $m = q^2 + 1$. Since every plane of the Π_3 containing $\mathcal{S}_P$ intersects $\mathcal{S}_P$ in a conic or just one point, so $\mathcal{S}_P = \mathcal{E}_3$ by Lemma 2.54.   □

Theorem 2.57. *In* $\mathrm{PG}(n, q)$, *let* $\mathcal{K}$ *be a non-singular* $k_{r,n,q}$ *with* $3 \le r \le q - 1$ *containing a hyperplane* Π_{n-1} *and let* P *be a point of* $\mathcal{K} \backslash \Pi_{n-1}$. *If the support* $\mathcal{S}_P$ *of* P *has character* w, *then*

$$k = \tfrac{1}{2}q^n + q^{n-1} + q^{n-2} + \cdots + q + 1 + \tfrac{1}{2}(w - 1)q^{n/2}.$$

Proof. For $n = 3$, this was proved in Theorem 19.4.9 of FPSOTD. For $d \ge 4$, Theorem 2.56 gives that

$$m = |\mathcal{S}_P| = |\mathcal{Q}_{n-1}| = q^{n-2} + q^{n-3} + \cdots + q + 1 + (w - 1)q^{(n-2)/2}.$$

There are $(q^n - 1)/(q - 1)$ lines through P; of these, m are lines of $\mathcal{K}$ and the remainder are $(\tfrac{1}{2}q + 1)$-secants. Hence

$$
\begin{aligned}
k &= 1 + qm + \tfrac{1}{2}q[(q^n - 1)/(q - 1) - m] \\
&= 1 + \tfrac{1}{2}qm + \tfrac{1}{2}(q^n - 1)/(q - 1) \\
&= 1 + \tfrac{1}{2}[q^{n-1} + q^{n-2} + \cdots + q + 1 + (w - 1)q^{n/2}] \\
&\quad + \tfrac{1}{2}[q^n + q^{n-1} + \cdots + q] \\
&= \tfrac{1}{2}q^n + q^{n-1} + q^{n-2} + \cdots + q + 1 + \tfrac{1}{2}(w - 1)q^{n/2}.
\end{aligned}
$$

□

This theorem shows that k is the same as $|\mathcal{R}_n|$ in Theorem 2.52. The aim now is to show that, if $\mathcal{K}$ is as in the previous two theorems, then $\mathcal{K} = \mathcal{R}_n$. It is necessary to deal separately with the cases of n odd and even.

First a rather curious lemma is required. Consider the pencil $\mathcal{L}$ in $PG(2, q)$, q even, of plane quadrics $\mathcal{F}_\lambda$, where

$$\mathcal{F}_\lambda = \mathbf{V}(X_0^2 + bX_0X_1 + cX_1^2 + \lambda X_2^2);$$

here, λ varies in $\mathbf{F}_q \cup \{\infty\}$ and $X^2 + bX + c$ is irreducible. So $\mathcal{F}_\infty = \mathbf{V}(X_2^2)$ is the line $\mathbf{u}_2$ and $\mathcal{F}_0 = \mathbf{V}(X_0^2 + bX_0X_1 + cX_1^2) = \{\mathbf{U}_2\}$, a point not lying on $\mathcal{F}_\infty$. The other $q - 1$ quadrics $\mathcal{F}_\lambda$ are all conics, no two of which have a point of intersection, since $\mathcal{F}_0 \cap \mathcal{F}_\infty = \emptyset$. If H is an additive subgroup of $\mathbf{F}_q$, it is shown in Theorem 12.12 of PGOFF2 that $\mathcal{K}' = \cup_{\lambda \in H}\mathcal{F}_\lambda$ is a maximal arc. However, implicit in the proof is the following result.

Lemma 2.58. *Let $H \subset \mathbf{F}_q$ with $|H| = \frac{1}{2}q$ and let $\mathcal{K}' = \cup_{\lambda \in H}\mathcal{F}_\lambda$. If there is some line l other than $\mathcal{F}_\infty$ with $l \cap \mathcal{K}' = \emptyset$, then H is an additive group and $\mathcal{K}'$ is a maximal arc of type $(0, \frac{1}{2}q)$.*

Proof. Let $l = V(a_0X_0 + a_1X_1 + X_2)$ with not both a_0 and a_1 zero. Note that any line through $\mathbf{U}_2$ meets every $\mathcal{F}_\lambda$. Let $\lambda \in H$ and so $l \cap \mathcal{F}_\lambda = \emptyset$. However, $\mathbf{P}(x_0, x_1, x_2) \in l \cap \mathcal{F}_\lambda$ when

$$x_0^2 + bx_0x_1 + cx_1^2 + \lambda(a_0^2x_0^2 + a_1^2x_1^2) = 0;$$

that is,

$$x_0^2(1 + \lambda a_0^2) + bx_0x_1 + (c + \lambda a_1^2)x_1^2 = 0,$$

or

$$x^2 + x + d = 0,$$

where

$$x = (1 + \lambda a_0^2)x_0/(bx_1), \qquad d = (1 + \lambda a_0^2)(c + \lambda a_1^2)/b^2.$$

So

$$d = e_0 + e_1\lambda + e_2\lambda^2,$$

where

$$e_0 = c/b^2, \qquad e_1 = (ca_0^2 + a_1^2)/b^2, \qquad e_2 = a_0^2a_1^2/b^2.$$

Also, $e_1 + \sqrt{e_2} = (ca_0^2 + ba_0a_1 + a_1^2)/b^2 \neq 0$, since $X^2 + bX + c$ is irreducible.

Now, $x^2 + x + d = 0$ has two solutions or none in $\mathbf{F}_{2^h}$ as $T(d) = 0$ or 1, where

$$T(d) = d + d^2 + d^4 + \cdots + d^{2^{h-1}}.$$

The trace function T from $\mathbf{F}_{2^h}$ onto $\mathbf{F}_2$ is an additive homomorphism, and satisfies

$$T(u + v) = T(u) + T(v),$$
$$T(u^2) = T(u)^2 = T(u).$$

Also, since $X^2 + bX + c$ is irreducible, so is $X^2 + X + c/b^2$, whence $T(e_0) = 1$. Thus

$$T(d) = T(e_0) + T(e_1\lambda) + T(e_2\lambda^2) = 1 + T(e_1\lambda) + T(\sqrt{e_2}\lambda)$$
$$= 1 + T((e_1 + \sqrt{e_2})\lambda).$$

Now, as $l \cap \mathcal{F}_\lambda = \emptyset$ for each $\lambda \in H$, so $T(d) = 1$ for each $\lambda \in H$. Hence

$$T((e_1 + \sqrt{e_2})\lambda) = 0, \quad \text{or} \quad T(\mu) = 0,$$

where $\mu = (e_1 + \sqrt{e_2})\lambda$. The $\frac{1}{2}q$ solutions μ of this equation form an additive group, the kernel of the function T. So H is also an additive group. The rest of the lemma follows as in Theorem 12.12 of PGOFF2. □

It may be noted that, in this case, $\mathcal{K}'$ is the complement of the dual of a regular oval, where a regular oval is defined to be a conic plus its nucleus.

Theorem 2.59. *In* $\mathrm{PG}(n, q)$, *n odd and* $n \geq 5$, *let* $\mathcal{K}$ *be a non-singular* $k_{r,n,q}$ *with* $3 \leq r \leq q - 1$ *containing a plane section of type IV and, for* $q = 4$, *also no section of type I or II. Then* $\mathcal{K} = \mathcal{R}_n$, *the projection of the quadric* $\mathcal{P}_{n+1}$ *in* $\mathrm{PG}(n + 1, q)$ *onto* $\mathrm{PG}(n, q)$.

Proof. By Theorem 2.53, it must be shown that $\mathcal{K}$ comprises $\frac{1}{2}q + 1$ quadrics of a pencil, one being a hyperplane Π_{n-1} and the others cones $V_i\mathcal{P}_{n-1}$, $i = 1, 2, \ldots, \frac{1}{2}q$. From Theorem 2.47, $\mathcal{K}$ contains a unique hyperplane Π_{n-1}. If P is any point of $\mathcal{K}\backslash\Pi_{n-1}$, then its support $\mathcal{S}_P$ in Π_{n-1} is a quadric $\mathcal{P}_{n-1}$, by Theorem 2.56. Let Q_0 be the nucleus of $\mathcal{P}_{n-1}$; the line PQ_0 is a $(\frac{1}{2}q + 1)$-secant of $\mathcal{K}$. So let P' be any point of $\mathcal{K} \cap PQ_0$ other than P and Q_0, and let S be any point of $\mathcal{S}_P$.

Suppose that $P'S$ is not a line of $\mathcal{K}$. Consider the plane $\pi = PP'S$. It contains the two $(\frac{1}{2}q + 1)$-secants PP' and $P'S$, and the two $(q + 1)$-secants PS and SQ_0; so π is a section of type IV. However, choose a solid Π_3 containing π in such a way that $\Pi_3 \cap \mathcal{P}_{n-1}$ is a conic $\mathcal{P}_2$. Then $\Pi_3 \cap \mathcal{K}$ is an $\mathcal{R}_3$. As π contains the nucleus Q_0 of the conic $\mathcal{P}_2$, it is of type V, from Table 19.5 of FPSOTD; this contradicts that π is of type IV. So $P'S$ is a line of $\mathcal{K}$. Hence $\mathcal{S}_{P'} = \mathcal{S}_P = \mathcal{P}_{n-1}$.

Now it is shown that if, for $S, S' \in \mathcal{P}_{n-1}$, the lines PS and $P'S'$ intersect, then $S = S'$. Suppose that $S \neq S'$ and that $PS \cap P'S' = T$. The plane π containing PS and $P'S'$ contains Q_0 and therefore it is of type V. However, π contains the three lines PS, $P'S'$ and SQ_0, which are not concurrent, a contradiction. Hence $S = S'$.

This means that if P_1 and P_2 are any two points on $PQ_0 \cap \mathcal{K}$ other than Q_0, then the two cones $P_1\mathcal{P}_{n-1}$ and $P_2\mathcal{P}_{n-1}$, where $P_i\mathcal{P}_{n-1}$ comprises the points on all joins P_iQ for Q in $\mathcal{P}_{n-1}$, intersect exactly in $\mathcal{P}_{n-1}$. So, if

$$PQ_0 \cap \mathcal{K} = \{Q_0, P_1, P_2, \ldots, P_{q/2}\},$$

where P is some P_i, say P_1, define

$$\mathcal{K}_0 = \Pi_{n-1} \cup \bigcup_{i=1}^{q/2} P_i\mathcal{P}_{n-1}.$$

Then $\mathcal{K}_0 \subset \mathcal{K}$ and

$$|\mathcal{K}_0| = (q^n - 1)/(q - 1) + \tfrac{1}{2}q + \tfrac{1}{2}q(q - 1)|\mathcal{P}_{n-1}|$$
$$= (q^n - 1)/(q - 1) + \tfrac{1}{2}q + \tfrac{1}{2}q(q - 1)(q^{n-1} - 1)/(q - 1)$$
$$= \tfrac{1}{2}q^n + (q^n - 1)/(q - 1)$$
$$= k,$$

by Theorem 2.57. So $\mathcal{K}_0 = \mathcal{K}$.

Now coordinates are attached to $\mathcal{K}$. Let

$$\Pi_{n-1} = \mathbf{V}(X_n), \quad \mathcal{P}_{n-1} = \mathbf{V}(X_0^2 + X_1 X_2 + \cdots + X_{n-2} X_{n-1}, X_n)$$
$$P_i = \mathbf{P}(\sqrt{\lambda_i}, 0, \ldots, 0, 1)$$

with $\lambda_1 = 0$. Then $Q_0 = U_0$ and

$$P_i \mathcal{P}_{n-1} = \mathbf{V}(X_0^2 + X_1 X_2 + \cdots + X_{n-2} X_{n-1} + \lambda_i X_n^2).$$

Lemma 2.58 can be used to now show that $H = \{\lambda_i \mid i = 1, 2, \ldots, \tfrac{1}{2}q\}$ is an additive group. Select a plane Π_2 meeting Π_{n-1} in a line skew to $\mathcal{P}_{n-1}$, namely,

$$\Pi_2 = \mathbf{V}(bX_0 + cX_1 + X_2, X_3, X_4, \ldots, X_{n-1}),$$

where $X^2 + bX + c$ is irreducible. Then

$$\Pi_2 \cap P_i \mathcal{P}_{n-1} = \mathbf{V}(X_0^2 + bX_0 X_1 + cX_1^2 + \lambda_i X_n^2, bX_0 + cX_1 + X_2, X_3, \ldots, X_{n-1}).$$

So $\Pi_2 \cap \mathcal{K}$ consists of the line $\Pi_2 \cap \Pi_{n-1}$ plus $\tfrac{1}{2}q - 1$ conics of a pencil plus the point P_1, the nucleus of each of the $\tfrac{1}{2}q - 1$ conics. So $\Pi_2 \cap \mathcal{K}$ is of type III containing a unisecant l, which is a 0-secant of $\mathcal{K} \backslash \Pi_{n-1}$ and is not the line $\Pi_2 \cap \Pi_{n-1}$ of the pencil. So, by the lemma, H is an additive group. Thus, by Theorem 2.53, $\mathcal{K} = \mathcal{R}_n$.
□

It remains to consider the case that n is even. Let $\mathcal{K}$ be a non-singular $k_{r,n,q}$ with $3 \le r \le q - 1$ containing a hyperplane Π_{n-1} and let $P \in \mathcal{K} \backslash \Pi_{n-1}$. Then, from Theorem 2.56, $\mathcal{S}_P = \mathcal{H}_{n-1}$ or $\mathcal{E}_{n-1}$. Suppose it is shown when $\mathcal{S}_P = \mathcal{H}_{n-1}$ that $\mathcal{K}$ is projectively unique and so the projection of a quadric $\mathcal{H}_{n+1}$. Now take the other case and let $\mathcal{K}$ be such that $\mathcal{S}_P = \mathcal{E}_{n-1}$; then

$$k = \tfrac{1}{2}q^n + q^{n-1} + q^{n-2} + \cdots + q + 1 - \tfrac{1}{2}q^{n/2}$$

and, with $\mathcal{Q}$ the residual of $\mathcal{K}$,

$$|\mathcal{Q}| = (q^{n+1} - 1)/(q - 1) - k + (q^n - 1)/(q - 1)$$
$$= \tfrac{1}{2}q^n + q^{n-1} + q^{n-2} + \cdots + q + 1 + \tfrac{1}{2}q^{n/2}.$$

So, for $\mathcal{Q}$, the support of a point is an $\mathcal{H}_{n-1}$ and hence $\mathcal{Q}$ is the projection of $\mathcal{H}_{n+1}$. Thus $\mathcal{K}$ is projectively unique and so the projection of $\mathcal{E}_{n+1}$.

Theorem 2.60. *In* $\mathrm{PG}(n, q)$, *n even and* $n \ge 4$, *let* $\mathcal{K}$ *be a non-singular* $k_{r,n,q}$ *with* $3 \le r \le q - 1$ *containing a plane section of type* IV *and, for* $q = 4$, *also no section of type* I *or* II. *Then* $\mathcal{K}$ *is the projection of either a hyperbolic quadric* $\mathcal{H}_{n+1}$ *or an elliptic quadric* $\mathcal{E}_{n+1}$ *in* $\mathrm{PG}(n + 1, q)$ *onto* $\mathrm{PG}(n, q)$.

Proof. By Theorem 2.47, $\mathcal{K}$ contains a unique hyperplane Π_{n-1}. If P is any point of $\mathcal{K}\backslash\Pi_{n-1}$, then its support $\mathcal{S}_P$ in Π_{n-1} is a quadric $\mathcal{H}_{n-1}$ or $\mathcal{E}_{n-1}$, by Theorem 2.56. However, by the remark above, it suffices to consider the case that $\mathcal{S}_P = \mathcal{H}_{n-1}$. Now, by Theorem 2.53, it must be shown that $\mathcal{K}$ comprises $\frac{1}{2}q + 1$ quadrics of a pencil, one of which is the hyperplane Π_{n-1}, another the cone $P\mathcal{H}_{n-1}$ and the remainder parabolic quadrics $\mathcal{P}_n^{(i)}$, $i = 1, 2, \ldots, \frac{1}{2}q - 1$, each of which contains $\mathcal{H}_{n-1}$ and has P as its nucleus.

Each line l through P not joining P to $\mathcal{H}_{n-1}$ meets Π_{n-1} in a point of $\mathcal{K}$ and is therefore a $(\frac{1}{2}q+1)$-secant of $\mathcal{K}$: it contains $\frac{1}{2}q - 1$ points other than P and the point of Π_{n-1}. The quadrics $\mathcal{P}_n^{(i)}$ are now constructed by suitably selecting from every such line l one of the $\frac{1}{2}q - 1$ points for each quadric.

Let $\Pi_{n-1} = \mathbf{u}_n$, let $P = \mathbf{U}_n$ and let

$$\mathcal{S}_P = \mathcal{H}_{n-1} = \mathbf{V}(X_0 X_1 + X_2 X_3 + \cdots + X_{n-2} X_{n-1}, X_n).$$

In Π_{n-1}, take the line $l = \mathbf{V}(X_0 + X_1, X_2 + X_3, X_4, \ldots, X_n)$. Then $l \cap \mathcal{S}_P = \{R\}$, where $R = \mathbf{P}(1, 1, 1, 1, 0, \ldots, 0)$. If l' is any line in Π_{n-1} not on $\mathcal{S}_P$, then in the plane Pl' there are through P two, one or no lines of $\mathcal{K}$ and respectively $q - 1, q$ or $q + 1$ lines $(\frac{1}{2}q + 1)$-secant to $\mathcal{K}$ according as $|l' \cap \mathcal{S}_P| = 2, 1$ or 0; the plane Pl' meets $\mathcal{K}$ correspondingly in a section of type IV, V or III. In particular, Pl meets $\mathcal{K}$ in a section of type V. The line l contains the point $Q = \mathbf{P}(0, 0, 1, 1, 0, \ldots, 0)$ in Π_{n-1}. So

$$PQ \cap \mathcal{K} = \{T_\lambda = \mathbf{P}(0, 0, \lambda, \lambda, 0, \ldots, 0, 1) \mid \lambda \in H \cup \{\infty\}\},$$

where $|H| = \frac{1}{2}q$ and $T_\infty = Q$. Consider, for $\lambda \neq \infty$, the line

$$l_\lambda = RT_\lambda = \{M_{\mu\lambda} = \mathbf{P}(\mu, \mu, \mu + \lambda, \mu + \lambda, 0, \ldots, 0, 1) \mid \mu \in \mathbf{F}_q \cup \{\infty\}\}$$

of $\mathcal{K}$, where $M_{\infty\lambda} = R$. Define, for $\mu \neq \infty$ and $\lambda \neq 0, \infty$, the set

$$\mathcal{N}_{\mu\lambda} = \mathcal{S}_P \cap \mathcal{S}_{M_{\mu\lambda}}.$$

Then, for any point $N \in \mathcal{N}_{\mu\lambda}$, the lines NP, $NM_{\mu\lambda}$ and $\Pi_{n-1} \cap NPM_{\mu\lambda}$ are all lines of $\mathcal{K}$. As $\pi = NPM_{\mu\lambda}$ contains the $(\frac{1}{2}q+1)$-secant $PM_{\mu\lambda}$ and three concurrent lines of $\mathcal{K}$, it follows that $\pi \cap \mathcal{K}$ is of type V and $\Pi_{n-1} \cap \pi$ is a tangent to $\mathcal{S}_P$.

The point $S_{\mu\lambda} = \Pi_{n-1} \cap PM_{\mu\lambda} = \mathbf{P}(\mu, \mu, \mu + \lambda, \mu + \lambda, 0, \ldots, 0)$ lies on l. Then, not only is $NS_{\mu\lambda} = \Pi_{n-1} \cap \pi$ a tangent to $\mathcal{S}_P$, but conversely, if $S_{\mu\lambda}V$ is a tangent to $\mathcal{S}_P$ with point of contact V, then $V \in \mathcal{N}_{\mu\lambda}$. Thus $\mathcal{N}_{\mu\lambda}$ is the set of contact points of the tangents to $\mathcal{S}_P$ from $S_{\mu\lambda}$. Let $\mathcal{M}_{\mu\lambda}$ be the cone with vertex $M_{\mu\lambda}$ and base $\mathcal{N}_{\mu\lambda}$, and let $\Gamma = \bigcup_{\mu \in \mathbf{F}_q} \mathcal{M}_{\mu\lambda}$. Now

$$\mathcal{N}_{\mu\lambda} = \{\mathbf{P}(y_0, y_1, \ldots, y_{n-1}, 0) \mid F = G = 0\},$$

where

$$F = \mu(y_0 + y_1) + (\mu + \lambda)(y_2 + y_3), \quad G = y_0 y_1 + y_2 y_3 + \cdots + y_{n-2} y_{n-1}.$$

Fig. 2.1. The structure of $\mathcal{K}$

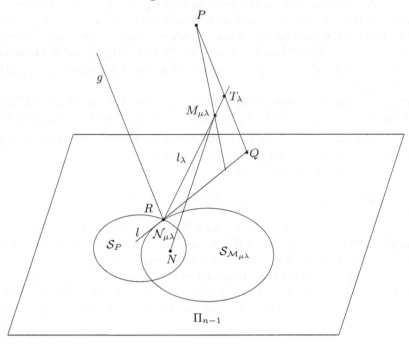

So

$$\mathcal{M}_{\mu\lambda} = \{\mathbf{P}(x_0, x_1, \ldots, x_n) \mid x_0 = \mu + ty_0,\ x_1 = \mu + ty_1,\ x_2 = \mu + \lambda + ty_2,$$
$$x_3 = \mu + \lambda + ty_3,\ x_4 = ty_4,\ \ldots,\ x_{n-1} = ty_{n-1},\ x_n = 1,$$
$$F = G = 0,\ t \in \mathbf{F}_q \cup \{\infty\}\} \,.$$

Elimination of $\mu, t, y_0, y_1, \ldots, y_{n-1}$ from the equations for $\mathcal{M}_{\mu\lambda}$ and homogenisation give $G_\lambda(x_0, x_1, \ldots, x_n) = 0$, where

$$G_\lambda = X_0 X_1 + X_2 X_3 + \cdots + X_{n-2} X_{n-1} + \lambda^2 X_n^2 \,.$$

Thus Γ_λ is a subset of the quadric $\mathcal{Q}_\lambda = \mathbf{V}(G_\lambda)$, where $\lambda \in H \backslash \{0\}$.

Each $\mathcal{N}_{\mu\lambda}$ is a $\mathcal{P}_{n-2}$, whence

$$|\mathcal{N}_{\mu\lambda}| = q^{n-3} + q^{n-4} + \cdots + q + 1 \,.$$

Further,

$$\mathcal{N}_{\mu\lambda} \cap \mathcal{N}_{\rho\lambda} = \mathbf{V}(X_0 + X_1, X_2 + X_3, X_n, X_0 X_1 + X_2 X_3 + \cdots + X_{n-2} X_{n-1})$$
$$= R\mathcal{P}_{n-4}$$

for $\mu \neq \rho$. So, if $\mathcal{N} = \bigcap_{\mu \in \mathbf{F}_q} \mathcal{N}_{\mu\lambda}$, then $\mathcal{N} = R\mathcal{P}_{n-4}$ and

$$|\mathcal{N}| = q^{n-4} + q^{n-5} + \cdots + q + 1.$$

As $\mathcal{N}_{\mu\lambda}$ is a $\mathcal{P}_{n-2}$ and $\mathcal{M}_{\mu\lambda} = M_{\mu\lambda}\mathcal{N}_{\mu\lambda}$, it follows that

$$|\mathcal{N}_{\mu\lambda}| = q^{n-2} + q^{n-3} + \cdots + q + 1.$$

Also, for any point N in $\mathcal{N}\backslash\{R\}$, the plane $RT_\lambda N$ is on $\mathcal{K}$, whence, for $\mu \neq \rho$,

$$T_\lambda \mathcal{N} \subset \mathcal{M}_{\mu\lambda} \cap \mathcal{M}_{\rho\lambda}.$$

Conversely, if $M \in \mathcal{M}_{\mu\lambda} \cap \mathcal{M}_{\rho\lambda}$ for $\mu \neq \rho$ and $M \notin RT_\lambda$, then $MRT_\lambda \cap \Pi_{n-1}$ is a line of $\mathcal{N}$, and so the plane MRT_λ is on $\mathcal{K}$, whence $M \in T_\lambda \mathcal{N}$. This means that $\mathcal{M}_{\mu\lambda} \cap \mathcal{M}_{\rho\lambda} = T_\lambda \mathcal{N}$; this fact is also obtainable from the equations for $\mathcal{M}_{\mu\lambda}$, $\mathcal{M}_{\rho\lambda}$, $T_\lambda \mathcal{N}$. Therefore

$$\mathcal{M} = \bigcap_{\mu \in \mathbf{F}_q} \mathcal{M}_{\mu\lambda} = T_\lambda \mathcal{N} \quad \text{and} \quad |\mathcal{M}| = q^{n-3} + q^{n-4} + \cdots + q + 1.$$

Thus

$$|\Gamma_\lambda| = q(|\mathcal{M}_{\mu\lambda}| - |\mathcal{M}|) + |\mathcal{M}| = q^{n-1} + q^{n-3} + q^{n-4} + \cdots + q + 1.$$

In fact, $\mathcal{Q}_\lambda\backslash\Gamma_\lambda$ consists of q^{n-3} lines through R. So let g be a line of $\mathcal{Q}_\lambda$ through R other than l_λ. If $g \subset \Pi_{n-1}$, then g is on $\mathcal{K}$. Assume therefore that $g \not\subset \Pi_{n-1}$. If gl_λ is a plane of $\mathcal{Q}_\lambda$, then $gl_\lambda \cap \Pi_{n-1}$ is a line of $\mathcal{S}_P$, where $\mathcal{S}_P \subset \mathcal{Q}_\lambda$. It now follows that $gl_\lambda \subset T_\lambda \mathcal{N}$ and so $g \subset \mathcal{K}$.

If the plane gl_λ is not on $\mathcal{Q}_\lambda$, then let Π_3 be a solid containing gl_λ and intersecting $\mathcal{Q}_\lambda$ in a hyperbolic quadric $\mathcal{H}_3$. The latter contains one line other than l_λ of each $\mathcal{M}_{\rho\lambda}$ for $\rho \in \mathbf{F}_q$. These q lines form with g a regulus on $\mathcal{H}_3$. Let g' be a line other than l_λ of the complementary regulus. Since $|g' \cap \mathcal{K}| \geq q$, so g' is a line of $\mathcal{K}$ and $g \cap g'$ is on $\mathcal{K}$, whence g is on $\mathcal{K}$. Hence $\mathcal{Q}_\lambda \subset \mathcal{K}$.

Any two $\mathcal{Q}_\lambda$ intersect in $\mathcal{S}_P$. Thus $\Pi_{n-1} \cup P\mathcal{S}_P \cup \bigcup_{\lambda \in H\backslash\{0\}} \mathcal{Q}_\lambda$ is contained in $\mathcal{K}$ and has the same number of points as $\mathcal{K}$; so this set is $\mathcal{K}$. Therefore

$$\mathcal{K} = \bigcup_{t \in H' \cup \{\infty\}} \mathcal{F}_t,$$

where $\mathcal{F}_t = \mathbf{V}(X_0 X_1 + X_2 X_3 + \cdots + X_{n-2}X_{n-1} + tX_n^2)$, $|H'| = \frac{1}{2}q$ and $0 \in H'$; here $\mathcal{F}_0 = P\mathcal{S}_P$, $\mathcal{F}_\infty = \Pi_{n-1}$.

Now, exactly as in the proof of Theorem 2.59, if a plane section of type III through P is considered, Lemma 2.58 shows that H' is an additive subgroup of $\mathbf{F}_q$; for example, take the plane

$$\Pi_2 = \mathbf{V}(X_0 + X_1, bX_1 + cX_2 + X_3, X_4, \ldots, X_{n-1}),$$

where $X^2 + bX + c$ is irreducible. Then, by Theorem 2.53, $\mathcal{K}$ is the projection of $\mathcal{H}_{n+1}$; that is, $\mathcal{K} = \mathcal{R}_n^+$. $\qquad\square$

Theorem 2.61. *In* $\mathrm{PG}(n, q)$ *with* $n \geq 3$ *and* $q > 4$, *a non-singular* $k_{r,n,q}$ *with* $3 \leq r \leq q - 1$ *is either a non-singular Hermitian variety with* $r = \sqrt{q} + 1$ *or the projection of a non-singular quadric in* $\mathrm{PG}(n + 1, q)$ *with* $r = \frac{1}{2}q + 1$.

Proof. The case $n = 3$ was summarised in Theorem 2.25. When $n \geq 4$ and the set has no plane section of type IV, the result is given by Theorem 2.46; when $n \geq 4$ and the set has a plane section of type IV, the result is given by Theorems 2.59 and 2.60. $\square$

This theorem can be reworded as follows.

Corollary 2.62. *The projectively distinct non-singular sets* $k_{r,n,q}$ *in* $\mathrm{PG}(n, q)$, *with* $n \geq 3$, *with* $3 \leq r \leq q - 1$ *and with* $q > 4$, *are given in Table* 2.4.

Table 2.4. Types of $k_{r,n,q}$

		n odd	n even
$q = p^h$, $p > 2$	h odd	$-$	$-$
	h even	$\mathcal{U}_n$	$\mathcal{U}_n$
$q = 2^h$	h odd	$\mathcal{R}_n$	$\mathcal{R}_n^+$, $\mathcal{R}_n^-$
	h even	$\mathcal{U}_n$, $\mathcal{R}_n$	$\mathcal{U}_n$, $\mathcal{R}_n^+$, $\mathcal{R}_n^-$

2.7 Notes and references

Sections 2.1–2.3

Although the theory of Hermitian forms over finite fields and their associated semi-linear groups is already contained in books such as Jordan [188] and Dickson [119], the first accounts with greater emphasis on the geometry rather than the group theory were given independently by Bose and Chakravarti [37] and in the monumental paper of Segre [279]. These sections follow in style the early sections on quadrics. See also De Bruyn [84].

In Section 19.3 of FPSOTD, regular systems of lines on $\mathcal{U}(3, q^2)$ are considered. The only type that exists is a *hemisystem*; this is a subset $\mathcal{L}$ of the generators of $\mathcal{U}(3, q^2)$ such that through every point of $\mathcal{U}(3, q^2)$ there pass exactly $(q + 1)/2$ lines of $\mathcal{L}$. Segre [279]constructed a hemisystem in the case $q = 3$. In 2005, Cossidente and Penttila [77] constructed hemisystems for all odd q. See also Cossidente and Penttila [78] and Bamberg, Giudici and Royle [12].

Section 2.4

The proof of Theorem 2.22, which is Witt's theorem, follows the treatment of Segre [279]. The fundamental formula of Theorem 2.23 is due to Wan and Yang [396], although they give a different proof.

Sections 2.5–2.6

These sections on the characterisation of Hermitian varieties as well as sets of type $(1, r, q + 1)$ are an amalgamation of Tallini Scafati [308], Hirschfeld and Thas [179, 178], and Glynn [139].

Lemma 2.54 was considerably improved by Brown [45]: a $(q^2 + 1)$-cap $\mathcal{K}$ in $\mathrm{PG}(3, q)$, q even and $q \neq 2$, containing at least one conic is an elliptic quadric $\mathcal{E}_3$.

For other characterisations of Hermitian varieties by intersection numbers, see De Winter and Schillewaert [104] and Schillewaert and Thas [271].

3

Grassmann varieties

3.1 Plücker and Grassmann coordinates

Let Π_r be an r-space in $\mathrm{PG}(n, K)$, $n \geq 3$, $1 \leq r \leq n - 2$, and let

$$\mathbf{P}(x^{(0)}), \mathbf{P}(x^{(1)}), \ldots, \mathbf{P}(x^{(r)}),$$

with $x^{(i)} = (x_0^i, x_1^i, \ldots, x_n^i)$, be $r + 1$ linearly independent points of Π_r. Write

$$T_x = \begin{bmatrix} x_0^0 & x_1^0 & \cdots & x_n^0 \\ x_0^1 & x_1^1 & \cdots & x_n^1 \\ \vdots & \vdots & \cdots & \vdots \\ x_0^r & x_1^r & \cdots & x_n^r \end{bmatrix}.$$

Also, write $(i_0 i_1 \cdots i_r)_x$, or $(i_0 i_1 \cdots i_r)$ if no confusion is possible, for the determinant of order $r + 1$ whose columns are the $(i_0 + 1)$-th, $(i_1 + 1)$-th,...,$(i_r + 1)$-th columns of the matrix T_x with $i_0, i_1, \ldots, i_r \in \{0, 1, \ldots, n\}$. If two of the i_j are interchanged the sign of $(i_0 i_1 \cdots i_r)_x$ changes, and if two of the i_j are equal then $(i_0 i_1 \cdots i_r)_x = 0$. Note also that at least one of the determinants $(i_0 i_1 \cdots i_r)_x$ is not zero.

Lemma 3.1. *Let* $\mathbf{P}(x^{(0)}), \mathbf{P}(x^{(1)}), \ldots, \mathbf{P}(x^{(r)})$ *and* $\mathbf{P}(y^{(0)}), \mathbf{P}(y^{(1)}), \ldots, \mathbf{P}(y^{(r)})$ *be two sets of* $r + 1$ *linearly independent points of the* r-space Π_r *of* $\mathrm{PG}(n, K)$, *where* $n \geq 3$ *and* $1 \leq r \leq n - 2$. *Then* $(i_0 i_1 \cdots i_r)_y = t(i_0 i_1 \cdots i_r)_x$ *for some* $t \in K \backslash \{0\}$ *which is independent of* $i_0, i_1, \ldots, i_r$.

Proof. First, $T_y = TT_x$ with $T = (t_{ij})$ a non-singular $(r + 1) \times (r + 1)$ matrix over K. Hence, if $t = |T| \neq 0$, then $(i_0 i_1 \cdots i_r)_y = t(i_0 i_1 \cdots i_r)_x$ for any $i_0, i_1, \ldots, i_r$ in $\{0, 1, \ldots, n\}$. $\square$

Now choose $\mathbf{c}(n + 1, r + 1)$ ordered $(r + 1)$-tuples $(i_0, i_1, \ldots, i_r)$ such that $i_0, i_1, \ldots, i_r$ are distinct elements of $\{0, 1, \ldots, n\}$ and such that the $\{i_0, i_1, \ldots, i_r\}$

© Springer-Verlag London 2016
J.W.P. Hirschfeld, J.A. Thas, *General Galois Geometries*, Springer Monographs in Mathematics, DOI 10.1007/978-1-4471-6790-7_3

are all subsets of order $r+1$ of $\{0, 1, \ldots, n\}$. Further, order the set $\mathcal{V}$ of these $(r+1)$-tuples.

Reconsider the $r + 1$ linearly independent points $\mathbf{P}(x^{(0)}), \mathbf{P}(x^{(1)}), \ldots, \mathbf{P}(x^{(r)})$ of Π_r. Then a *coordinate vector* of Π_r is

$$L = (l_0, l_1, \ldots, l_{\mathbf{c}(n+1,r+1)-1}),$$

where the l_j are the elements $(i_0 i_1 \cdots i_r)_x$ with $(i_0, i_1, \ldots, i_r)$ in $\mathcal{V}$ in the given order. These elements l_j are the *coordinates* of the r-space Π_r of $\mathrm{PG}(n, K)$. By Lemma 3.1, L is determined by Π_r up to a factor of proportion. Write

$$\Pi_r = \mathbf{\Pi}_r(L).$$

For $n = 3$ and $r = 1$ these coordinates were introduced in Section 15.2 of FPSOTD. In this case, the coordinates are also called *Plücker coordinates*. In the general case they are called *Grassmann coordinates*.

Consider a projectivity ξ of $\mathrm{PG}(n, K)$ with matrix A, and let $\Pi_r \xi = \Pi_r'$ with $\Pi_r = \mathbf{\Pi}_r(L)$ and $\Pi_r' = \mathbf{\Pi}_r(L')$. By a standard matrix manipulation as in Lemma 15.2.8 of FPSOTD,

$$tL' = L\tilde{A},$$

where $t \in K \backslash \{0\}$ and where the elements of $\tilde{A}$ are, up to the sign, minors of order $r + 1$ of the matrix A. Also, $\tilde{A}$ is non-singular.

Next let $\boldsymbol{\pi}(u^{(0)}), \boldsymbol{\pi}(u^{(1)}), \ldots, \boldsymbol{\pi}(u^{(n-r-1)})$ be $n - r$ linearly independent hyperplanes containing the r-space Π_r of $\mathrm{PG}(n, K)$, where $u^{(i)} = (u_0^i, u_1^i, \ldots, u_n^i)$. Write

$$T_u = \begin{bmatrix} u_0^0 & u_1^0 & \cdots & u_n^0 \\ u_0^1 & u_1^1 & \cdots & u_n^1 \\ \vdots & \vdots & \cdots & \vdots \\ u_0^{n-r-1} & u_1^{n-r-1} & \cdots & u_n^{n-r-1} \end{bmatrix}.$$

Also, write $(j_0 j_1 \cdots j_{n-r-1})_u$, or $(j_0 j_1 \cdots j_{n-r-1})$ if no confusion is possible, for the determinant of order $n - r$ whose columns coincide with the $(j_0 + 1)$-th, $(j_1 + 1)$-th,...,$(j_{n-r-1} + 1)$-th columns of the matrix T_u, with $j_0, j_1, \ldots, j_{n-r-1}$ in $\{0, 1, \ldots, n\}$. At least one of the $n - r$ determinants $(j_0 j_1 \cdots j_{n-r-1})$ is not zero.

For each element $(i_0, i_1, \ldots, i_r)$ of $\mathcal{V}$ an $(n - r)$-tuple $(j_0, j_1, \ldots, j_{n-r-1})$ is chosen so that

$$(i_0, i_1, \ldots, i_r, j_0, j_1, \ldots, j_{n-r-1})$$

is an even permutation of $(0, 1, \ldots, n)$. The set of these ordered $(n-r)$-tuples, which number $\mathbf{c}(n + 1, r + 1) = \mathbf{c}(n + 1, n - r)$, is denoted by $\mathcal{W}$. Then the ordering of $\mathcal{V}$ induces an ordering of $\mathcal{W}$.

Consider again the $n - r$ linearly independent hyperplanes containing Π_r. Then a *dual coordinate vector* of Π_r is

$$\hat{L} = (\hat{l}_0, \hat{l}_1, \ldots, \hat{l}_{\mathbf{c}(n+1,r+1)-1}),$$

where the $\hat{l}_i$ are the elements $(j_0 j_1 \cdots j_{n-r-1})_u$ with $(j_0, j_1, \ldots, j_{n-r-1})$ in $\mathcal{W}$ in the given order. By the dual of Lemma 3.1, $\hat{L}$ is determined by Π_r up to a factor of proportion.

Lemma 3.2. *The coordinates* $\rho l_i = \hat{l}_i$ *for* $i = 0, 1, \ldots, \mathbf{c}(n+1, r+1) - 1$; *that is, up to a factor of proportion,* $\hat{L}$ *is* L.

Proof. Let $\mathbf{P}(x^{(0)}), \mathbf{P}(x^{(1)}), \ldots, \mathbf{P}(x^{(r)})$ be $r+1$ linearly independent points of Π_r with $x^{(i)} = (x_0^i, x_1^i, \ldots, x_n^i)$, $i = 0, 1, \ldots, r$. The l_j are of the form $(i_0 i_1 \cdots i_r)_x$. Consider $n - r$ points $\mathbf{P}(x^{(r+1)}), \mathbf{P}(x^{(r+2)}), \ldots, \mathbf{P}(x^{(n)})$ of PG(n, K) such that $\mathbf{P}(x^{(0)}), \mathbf{P}(x^{(1)}), \ldots, \mathbf{P}(x^{(n)})$, with

$$x^{(i)} = (x_0^i, x_1^i, \ldots, x_n^i), \quad i = r+1, r+2, \ldots, n,$$

are linearly independent. As hyperplanes $\pi(u^{(0)}), \pi(u^{(1)}), \ldots, \pi(u^{(n-r-1)})$ choose

$$\Pi_r \mathbf{P}(x^{(r+1)}) \mathbf{P}(x^{(r+2)}) \cdots \mathbf{P}(x^{(r+i-1)}) \mathbf{P}(x^{(r+i+1)}) \cdots \mathbf{P}(x^{(n)}),$$

with $i = 1, 2, \ldots, n - r$. So, for u_j^i take the cofactor of x_j^{r+i+1} in the matrix

$$D = \begin{bmatrix} x_0^0 & x_1^0 & \cdots & x_n^0 \\ x_0^1 & x_1^1 & \cdots & x_n^1 \\ \vdots & \vdots & \cdots & \vdots \\ x_0^n & x_1^n & \cdots & x_n^n \end{bmatrix},$$

$j = 0, 1, \ldots, n$ and $i = 0, 1, \ldots, n - r - 1$. If E is the matrix obtained from D by replacing each element x_j^i by its cofactor, then each minor of order $n - r$ of E is equal to the product of $|D|^{n-r-1}$ and the algebraic complement of the similarly placed minor in the matrix D. Hence

$$(j_0 j_1 \cdots j_{n-r-1})_u = |D|^{n-r-1} (-1)^d (i_0 i_1 \cdots i_r)_x,$$

where

$$\{j_0, j_1, \ldots, j_{n-r-1}\} \cup \{i_0, i_1, \ldots, i_r\} = \{0, 1, \ldots, n\},$$

with $j_0 < j_1 < \cdots < j_{n-r-1}$, $i_0 < i_1 < \cdots < i_r$, and

$$d = i_0 + i_1 + \cdots + i_r + r + 1 + (r+2)(r+1)/2.$$

Now the result follows. ☐

If $n = 2r + 1$ then often the ordering in $\mathcal{W}$ is chosen in such a way that $\rho l_i = \hat{l}_{i+\mathbf{c}(2r+2, r+1)/2}$, where indices are taken modulo $\mathbf{c}(2r + 2, r + 1)$. Hence, if $(i_0 i_1 \cdots i_r)_x$ is a coordinate of Π_r and $(j_0 j_1 \cdots j_r)_u$ is a dual coordinate of Π_r', then their positions differ by $\mathbf{c}(2r + 2, r + 1)/2$, where

$$|\{i_0, i_1, \ldots, i_r\} \cup \{j_0, j_1, \ldots, j_r\}| = 2r + 2.$$

In Lemma 3.3, for $n = 2r+1$ and r even, the following condition is not imposed: for a coordinate $(i_0 i_1 \cdots i_r)_x$ of Π_r and a dual coordinate $(j_0 j_1 \cdots j_r)_u$ of Π_r with $|\{i_0, i_1, \ldots, i_r\} \cup \{j_0, j_1, \ldots, j_r\}| = n+1$, the permutation

$$(i_0, i_1, \ldots, i_r, j_0, j_1, \ldots, j_r)$$

has to be even.

Lemma 3.3. *Let Π_r be an r-space of $\mathrm{PG}(n, K)$ with coordinates $(i_0\, i_1 \cdots i_r)_x$ for $n \geq 3$. Also, let Π_{n-r-1} be an $(n-r-1)$-space with dual coordinates $(i_0\, i_1 \cdots i_r)_u$, where $1 \leq r \leq n-2$. Then $\Pi_r \cap \Pi_{n-r-1} \neq \emptyset$ if and only if*

$$\sum (i_0\, i_1 \cdots i_r)_x (i_0\, i_1 \cdots i_r)_u = 0\,.$$

Proof. Choose $r+1$ linearly independent points $\mathbf{P}(x^{(0)}), \mathbf{P}(x^{(1)}), \ldots, \mathbf{P}(x^{(r)})$ in Π_r and $n-r$ linearly independent points $\mathbf{P}(y^{(0)}), \mathbf{P}(y^{(1)}), \ldots, \mathbf{P}(y^{(n-r-1)})$ in Π_{n-r-1}. Let

$$\Delta = \begin{vmatrix} x_0^0 & x_1^0 & \cdots & x_n^0 \\ x_0^1 & x_1^1 & \cdots & x_n^1 \\ \vdots & \vdots & \cdots & \vdots \\ x_0^r & x_1^r & \cdots & x_n^r \\ y_0^0 & y_1^0 & \cdots & y_n^0 \\ y_0^1 & y_1^1 & \cdots & y_n^1 \\ \vdots & \vdots & \cdots & \vdots \\ y_0^{n-r-1} & y_1^{n-r-1} & \cdots & y_n^{n-r-1} \end{vmatrix}\,.$$

Then $\Pi_r \cap \Pi_{n-r-1} \neq \emptyset$ if and only if $\Delta = 0$. By the Laplace expansion of Δ along the first $r+1$ rows and by Lemma 3.2 the required result is immediate. $\qquad\square$

Corollary 3.4. *Let Π_r be an r-space of $\mathrm{PG}(n, K)$, where $1 \leq r \leq n-2$ and $n \geq 3$, with coordinate vector $L = (l_0, l_1, \ldots)$. If $l_k = (i_0\, i_1 \cdots i_r)_x$ then $l_k = 0$ if and only if $\Pi_r \cap \mathbf{U}_{i_{r+1}} \mathbf{U}_{i_{r+2}} \cdots \mathbf{U}_{i_n} \neq \emptyset$, where $\{i_0, i_1, \ldots, i_n\} = \{0, 1, \ldots, n\}$.*

Proof. Let $\mathbf{U}_{i_{r+1}} \mathbf{U}_{i_{r+2}} \cdots \mathbf{U}_{i_n} = \Pi_{n-r-1}$. As $\Pi_{n-r-1} = \mathbf{V}(X_{i_0}, X_{i_1}, \ldots, X_{i_r})$, the space Π_{n-r-1} has $(j_0 j_1 \cdots j_r)_u \neq 0$ if and only if

$$\{j_0, j_1, \ldots, j_r\} = \{i_0, i_1, \ldots, i_r\}.$$

By Lemma 3.3, $(i_0\, i_1 \cdots i_r)_x = 0$ if and only if $\Pi_r \cap \Pi_{n-r-1} \neq \emptyset$. $\qquad\square$

Lemma 3.5. *Let Π_r be an r-space of $\mathrm{PG}(n, K)$, where $1 \leq r \leq n-2$ and $n \geq 3$, and let $\Pi_r \cap \mathbf{V}(X_{i_1}, X_{i_2}, \ldots, X_{i_r})$ be a point $\mathbf{P}(x)$, where $i_1, \ldots, i_r$ are distinct elements of $\{0, 1, \ldots, n\}$. If $\mathbf{P}(x^{(0)}), \mathbf{P}(x^{(1)}), \ldots, \mathbf{P}(x^{(r)})$ are linearly independent points of Π_r, where $x^{(i)} = (x_0^i, x_1^i, \ldots, x_n^i)$, then up to a factor of proportion*

$$x = ((0\, i_1 \cdots i_r)_x, (1\, i_1 \cdots i_r)_x, \ldots, (n\, i_1 \cdots i_r)_x)\,.$$

Proof. Since $\Pi_r \cap \mathbf{V}(X_{i_1}, X_{i_2}, \ldots, X_{i_r})$ is a point $P = \mathbf{P}(x)$, there is at least one hyperplane $\mathbf{V}(X_i)$ not containing $\mathbf{P}(x)$ for some i in $\{0, 1, \ldots, n\}$. Then, by Corollary 3.4, $(ii_1 i_2 \cdots i_r)_x \neq 0$. Hence the matrix

$$
D = \begin{bmatrix} x_{i_1}^0 & x_{i_2}^0 & \cdots & x_{i_r}^0 \\ x_{i_1}^1 & x_{i_2}^1 & \cdots & x_{i_r}^1 \\ \vdots & \vdots & \cdots & \vdots \\ x_{i_1}^r & x_{i_2}^r & \cdots & x_{i_r}^r \end{bmatrix}
$$

has rank r. If $x = (x_0, x_1, \ldots, x_n)$ is the coordinate vector of P, then there are elements $t_0, \ldots, t_r$ determined up to a factor of proportion and not all zero such that

$$
x_i = t_0 x_i^0 + t_1 x_i^1 + \cdots + t_r x_i^r,
$$

$i = 0, 1, \ldots, n$. Since

$$
t_0 x_{i_s}^0 + t_1 x_{i_s}^1 + \cdots + t_r x_{i_s}^r = 0,
$$

$s = 1, 2, \ldots, r$ and since rank $D = r$, take for the t_j the minors of order r of D with alternating signs. Hence, up to a factor of proportion, x_i is $(i\, i_1\, i_2 \cdots i_r)_x$ for $i = 0, 1, \ldots, n$. $\square$

Lemma 3.6. *Let* $\mathbf{P}(x^{(0)}), \mathbf{P}(x^{(1)}), \ldots, \mathbf{P}(x^{(r)})$, *with* $x^{(i)} = (x_0^i, x_1^i, \ldots, x_n^i)$, *be* $r + 1$ *linearly independent points of the r-space* Π_r *of* $\mathrm{PG}(n, K)$, $1 \leq r \leq n - 2$ *and* $n \geq 3$. *The point* $\mathbf{P}(x)$, *where* $x = (x_0, x_1, \ldots, x_n)$, *is contained in* Π_r *if and only if*

$$
x_{i_0}(i_1\, i_2 \cdots i_{r+1})_x - x_{i_1}(i_0\, i_2 \cdots i_r)_x + \cdots + (-1)^{r+1} x_{i_{r+1}}(i_0\, i_1 \cdots i_r)_x = 0, \tag{3.1}
$$

for each choice of distinct $i_0, i_1, \ldots, i_{r+1}$ *in* $\{0, 1, \ldots, n\}$.

Proof. The point $\mathbf{P}(x)$ is contained in Π_r if and only if there exist elements $t_0, t_1, \ldots, t_r$ in K, not all zero, such that

$$
x = t_0 x^{(0)} + t_1 x^{(1)} + \cdots + t_r x^{(r)}.
$$

Equivalently,

$$
x_i = t_0 x_0^i + t_1 x_1^i + \cdots + t_r x_r^i, \tag{3.2}
$$

$i = 0, 1, \ldots, n$. Since the rank of the $(r + 1) \times (n + 1)$ matrix with elements x_j^i is equal to $r + 1$, the system (3.2) of $n + 1$ linear equations in $r + 1$ unknowns has at least one solution if and only if all minors of order $r + 2$ of the matrix

$$
\begin{bmatrix} x_0 & x_1 & \cdots & x_n \\ x_0^0 & x_1^0 & \cdots & x_n^0 \\ x_0^1 & x_1^1 & \cdots & x_n^1 \\ \vdots & \vdots & \cdots & \vdots \\ x_0^r & x_1^r & \cdots & x_n^r \end{bmatrix}
$$

are zero. Expanding these minors along the first row, the conditions (3.1) are obtained.   □

Theorem 3.7. *For any r-space Π_r of $\mathrm{PG}(n, K)$, $1 \leq r \leq n - 1$ and $n \geq 3$,*

$$
\begin{aligned}
(i_0\, i_1 \cdots i_r)(j_0\, j_1 \cdots j_r) &- (j_0\, i_1 \cdots i_r)(i_0\, j_1 \cdots j_r) \\
&+ (j_1\, i_1 \cdots i_r)(i_0\, j_0\, j_2 \cdots j_r) \\
&- \cdots + (-1)^{r+1}(j_r\, i_1 \cdots i_r)(i_0\, j_0 \cdots j_{r-1}) = 0\,, \quad (3.3)
\end{aligned}
$$

where $i_0, i_1, \ldots, i_r, j_0, j_1, \ldots, j_r$ are arbitrarily chosen elements in $\{0, 1, \ldots, n\}$.

Proof. Assume that $\Pi_r \cap \mathbf{V}(X_{i_1}, X_{i_2}, \ldots, X_{i_r})$, with $i_1, i_2, \ldots, i_r$ distinct, is a point $\mathbf{P}(x)$. By Lemma 3.5, up to a factor of proportion,

$$
x = ((0\, i_1\, i_2 \cdots i_r), (1\, i_1\, i_2 \cdots i_r), \ldots, (n\, i_1\, i_2 \cdots i_r))\,.
$$

Since $\mathbf{P}(x)$ belongs to Π_r, by Lemma 3.6,

$$
\begin{aligned}
(i_0\, i_1 \cdots i_r)(j_0\, j_1 \cdots j_r) &- (j_0\, i_1 \cdots i_r)(i_0\, j_1 \cdots j_r) \\
&+ (j_1\, i_1 \cdots i_r)(i_0\, j_0\, j_2 \cdots j_r) \\
&- \cdots + (-1)^{r+1}(j_r\, i_1 \cdots i_r)(i_0\, j_0 \cdots j_{r-1}) = 0\,, \quad (3.4)
\end{aligned}
$$

for any distinct $i_0, j_0, j_1, \ldots, j_r$.

If $\Pi_r \cap \mathbf{V}(X_{i_1}, X_{i_2}, \ldots, X_{i_r})$, with $i_1, i_2, \ldots, i_r$ distinct, is at least of projective dimension one then, by Corollary 3.4, $(i\, i_1 i_2 \cdots i_r) = 0$ for $i = 0, 1, \ldots, n$. Hence (3.4) is also satisfied in this case.

If $i_1, i_2, \ldots, i_r$ or $i_0, j_0, j_1, \ldots, j_r$ are not all distinct, then (3.4) is trivially satisfied. In conclusion, (3.4) is satisfied whenever $i_0, i_1, \ldots, j_r$ are arbitrarily chosen in $\{0, 1, \ldots, n\}$.   □

By Theorem 3.7, the coordinates of Π_r satisfy a number of *quadratic relations*. These relations (3.3) can also be written as follows:

$$
(i_0\, i_1 \cdots i_r)(j_0\, j_1 \cdots j_r) = \sum_{s=0}^{r} (j_s\, i_1 \cdots i_r)(j_0 \cdots j_{s-1}\, i_0\, j_{s+1} \cdots j_r)\,. \quad (3.5)
$$

In particular, putting $j_2 = i_2, \ldots, j_r = i_r$, (3.5) becomes

$$
\begin{aligned}
(i_0\, i_1 \cdots i_r)(j_0\, j_1\, i_2 \cdots i_r) &= (j_0\, i_1 \cdots i_r)(i_0\, j_1\, i_2 \cdots i_r) \\
&\quad + (j_1\, i_1 \cdots i_r)(j_0\, i_0\, i_2 \cdots i_r)\,. \quad (3.6)
\end{aligned}
$$

These are the *elementary quadratic relations*. When $r = 1$ or $r = n - 2$, it follows that (3.5) and (3.6) are the same. For $r = 1$, (3.6) was also derived in Lemma 15.2.2 of FPSOTD.

Now let $1 < r < n - 2$; in this case, $n \geq 5$. Suppose that $(k_0 k_1 \cdots k_r) \neq 0$ if and only if

$$\{k_0, k_1, \ldots, k_r\} = \{0, 1, 2, 6, 7, \ldots, r+3\} \text{ or } \{3, 4, 5, 6, 7, \ldots, r+3\},$$

where $k_0, k_1, \ldots, k_r \in \{0, 1, \ldots, n\}$. Then the relations (3.6) are satisfied. Since $(0\,1\,2\,6\,7 \cdots r+3)(3\,4\,5\,6\,7 \cdots r+3) \neq 0$, the relations (3.5) are not satisfied. Hence elements $(k_0 k_1 \cdots k_r)$, where

$$(k_0\ k_1\ k_2\ \cdots\ k_r) = -(k_1\ k_0\ k_2\ \cdots\ k_r), \quad (k_0\ k_0\ k_2\ \cdots\ k_r) = 0 \qquad (3.7)$$

and similar relations, which satisfy the elementary quadratic relations and which are not all zero, do not necessarily correspond to some r-space of $PG(n, K)$.

Theorem 3.8. *Let* $1 \leq r \leq n-2$, $n \geq 3$. *If the elements* $(k_0 k_1 \cdots k_r)$, *where* $k_i \in \{0, 1, \ldots, n\}$, *are not all zero and satisfy both* (3.7) *and the quadratic relations* (3.5), *then these elements correspond to exactly one r-space* Π_r *of* $PG(n, K)$.

Proof. Suppose, for example, that $(0\,1 \cdots r) \neq 0$. If there is an r-space Π_r corresponding to the $c(n+1, r+1)$ given elements, then, by Corollary 3.4,

$$\Pi_r \cap V(X_0, X_1, \ldots, X_r) \neq \emptyset.$$

Hence

$$\Pi_r \cap V(X_0, X_1, \ldots, X_{s-1}, X_{s+1}, \ldots, X_r)$$

is a point $P(x^{(s)})$, $s = 0, 1, \ldots, r$. By Lemma 3.5, up to a factor of proportion,

$$x^{(s)} = (x_0^s, x_1^s, \ldots, x_n^s),$$

where $x_i^s = (0\,1 \cdots s-1\,i\,s+1 \cdots r)$, $s = 0, 1, \ldots, r$ and $i = 0, 1, \ldots, n$. Calculating the determinant

$$\Delta = |x_i^s|, \quad i, s = 0, 1, \ldots, r,$$

it follows that $(0\,1 \cdots r)^{r+1} \neq 0$. Hence the points $P(x^{(0)}), P(x^{(1)}), \ldots, P(x^{(r)})$, are linearly independent. Now it is shown that, up to a factor of proportion, the given elements $(k_0 k_1 \cdots k_r)$ are the elements $(k_0\ k_1\ k_2\ \cdots\ k_r)_x$ corresponding to the points $P(x^{(s)})$ of Π_r; more precisely, it is shown that

$$(k_0\ k_1\ k_2\ \cdots\ k_r)_x = (k_0\ k_1\ k_2\ \cdots\ k_r)(0\,1 \cdots r)^r. \qquad (3.8)$$

From above,

$$(0\,1 \cdots r)_x = (0\,1 \cdots r)^{r+1}. \qquad (3.9)$$

Also,

$$(0\,1 \cdots s-1\,i\,s+1 \cdots r)_x = x_i^s(0\,1 \cdots r)^r \qquad (3.10)$$
$$= (0\,1 \cdots s-1\,i\,s+1 \cdots r)(0\,1 \cdots r)^r,$$

$s = 0, 1, \ldots, r$ and $i = 0, 1, \ldots, n$. Now proceed by induction on the number ν of k_j in the set $\{r+1, r+2, \ldots, n\}$. Equation (3.8) is satisfied if $\nu = 0$ or 1. Without loss of generality, it must be shown that

$$(j_1 \, j_2 \, \cdots \, j_\nu \, \nu \, \nu + 1 \, \cdots \, r)_x = (j_1 \, j_2 \, \cdots \, j_\nu \, \nu \, \nu + 1 \, \cdots \, r)(0 1 \cdots r)^r, \qquad (3.11)$$

where $j_1, j_2, \ldots, j_\nu \in \{r+1, r+2, \ldots, n\}$ and $\nu > 1$. Since both the elements $(k_0 \, k_1 \, k_2 \, \cdots \, k_r)$ and $(k_0 \, k_1 \, k_2 \, \cdots \, k_r)_x$ satisfy the quadratic relations (3.5), so

$$(0 1 \cdots r)(j_1 \, j_2 \, \cdots \, j_\nu \, \nu \, \nu + 1 \, \cdots \, r)$$
$$= \sum_{s=1}^{\nu} (j_s \, 1 \, 2 \, \cdots \, r)(j_1 \, j_2 \, \cdots \, j_{s-1} \, 0 \, j_{s+1} \, \cdots \, j_\nu \, \nu \, \nu + 1 \, \cdots \, r)$$
$$+ \sum_{s=\nu}^{r} (s \, 1 \, 2 \, \cdots \, r)(j_1 \, j_2 \, \cdots \, j_\nu \, \nu \, \cdots \, s - 1 \, 0 \, s + 1 \, \cdots \, r)$$
$$= \sum_{s=1}^{\nu} (j_s \, 1 \, 2 \, \cdots \, r)(j_1 \, j_2 \, \cdots \, j_{s-1} \, 0 \, j_{s+1} \, \cdots \, j_\nu \, \nu \, \nu + 1 \, \cdots \, r). \qquad (3.12)$$

Also,

$$(0 1 \cdots r)_x (j_1 \, j_2 \, \cdots \, j_\nu \, \nu \, \nu + 1 \, \cdots \, r)_x$$
$$= \sum_{s=1}^{\nu} (j_s \, 1 \, 2 \, \cdots \, r)_x (j_1 \, j_2 \, \cdots \, j_{s-1} \, 0 \, j_{s+1} \, \cdots \, j_\nu \, \nu \, \nu + 1 \, \cdots \, r)_x. \qquad (3.13)$$

By (3.9) and (3.10),

$$(0 1 \cdots r)_x = (0 1 \cdots r)^{r+1},$$
$$(j_s \, 1 \, 2 \, \cdots \, r)_x = (j_s \, 1 \, 2 \, \cdots \, r)(0 1 \cdots r)^r.$$

By induction,

$$(j_1 \, j_2 \, \cdots \, j_{s-1} \, 0 \, j_{s+1} \, \cdots \, j_\nu \, \nu \, \nu + 1 \, \cdots \, r)_x$$
$$= (j_1 \, j_2 \, \cdots \, j_{s-1} \, 0 \, j_{s+1} \, \cdots \, j_\nu \, \nu \, \nu + 1 \, \cdots \, r)(0 1 \cdots r)^r.$$

Hence (3.13) becomes

$$(0 1 \cdots r)^{r+1} (j_1 \, j_2 \, \cdots \, j_\nu \, \nu \, \nu + 1 \, \cdots \, r)_x \qquad\qquad (3.14)$$
$$= \sum_{s=1}^{\nu} (0 1 \cdots r)^{2r} (j_s \, 1 \, 2 \, \cdots \, r)(j_1 \, j_2 \, \cdots \, j_{s-1} \, 0 \, j_{s+1} \, \cdots \, j_\nu \, \nu \, \nu + 1 \, \cdots \, r).$$

Comparing (3.14) with (3.12), and since $(0 1 \cdots r) \neq 0$, this gives

$$(j_1 \, j_2 \, \cdots \, j_\nu \, \nu \, \nu + 1 \, \cdots \, r)_x = (j_1 \, j_2 \, \cdots \, j_\nu \, \nu \, \nu + 1 \, \cdots \, r)(0 1 \cdots r)^r,$$

which is (3.11), as required. $\square$

Example 3.9. For $n = 4$ and $r = 1$, a coordinate vector of a line l is

$$L = ((01), (02), (03), (04), (12), (13), (14), (23), (24), (34)).$$

The coordinates of l satisfy the following five quadratic relations :

$$(12)(34) + (23)(14) + (31)(24) = 0,$$
$$(02)(34) + (23)(04) + (30)(24) = 0,$$
$$(01)(34) + (13)(04) + (30)(14) = 0,$$
$$(01)(24) + (12)(04) + (20)(14) = 0,$$
$$(01)(23) + (12)(03) + (20)(13) = 0.$$

These relations, though linearly independent, must in fact be equivalent to three conditions only, giving seven independent homogeneous parameters with which to define a line l.

Example 3.10. For $n = 5$ and $r = 2$, the 20 Grassmann coordinates of a plane π are the following:

$$(012), (013), (014), (015), (023), (024), (025), (034), (035), (045),$$
$$(123), (124), (125), (134), (135), (145), (234), (235), (245), (345).$$

These coordinates satisfy the following quadratic relations:

$$(ijk)(uvw) = (ivw)(ujk) + (jvw)(iuk) + (kvw)(iju),$$

with $i, j, k, u, v, w \in \{0, 1, 2, 3, 4, 5\}$.

In particular for $k = w$ the following elementary quadratic relations are obtained:

$$(ijk)(uvk) = (ivk)(ujk) + (jvk)(iuk),$$

with $i, j, k, u, v \in \{0, 1, 2, 3, 4, 5\}$.

It can be shown that, among these 20 Grassmann coordinates there are exactly 35 linearly independent quadratic relations. In fact, these relations are equivalent to 10 conditions only, giving 10 independent homogeneous parameters with which to define a plane π in $\mathrm{PG}(5, K)$.

3.2 Grassmann varieties

Let $\mathrm{PG}^{(r)}(n, K)$ be the set of all r-spaces of $\mathrm{PG}(n, K)$, with $1 \le r \le n - 2$ and $n \ge 3$. If

$$L = (l_0, l_1, \ldots, l_{\mathbf{c}(n+1, r+1)-1})$$

is a coordinate vector of $\Pi_r \in \mathrm{PG}^{(r)}(n, K)$, then $\mathbf{P}(L)$ is a point of the projective space $\mathrm{PG}(N, K)$, with $N = \mathbf{c}(n + 1, r + 1) - 1$.

The mapping which associates $\mathbf{P}(L)$ to Π_r is denoted by $\mathfrak{G}$. The algebraic variety $\mathrm{PG}^{(r)}(n, K)\mathfrak{G}$ of $\mathrm{PG}(N, K)$ is called the *Grassmannian* or the *Grassmann variety* of the r-spaces of $\mathrm{PG}(n, K)$. It is denoted by $\mathcal{G}_{r,n,K}$ or $\mathcal{G}_{r,n}$. In the finite case, when $K = \mathbf{F}_q$, it is also denoted by $\mathcal{G}_{r,n,q}$. By Theorems 3.7 and 3.8, $\mathcal{G}_{r,n}$ is the intersection of the quadrics of $\mathrm{PG}(N, K)$ represented by the equations (3.5). For $r = 1$ and $n = 3$, the dimension $N = 5$ and (3.5) represents only one quadric of $\mathrm{PG}(5, K)$. In this case, $\mathcal{G}_{1,3}$ is the hyperbolic quadric of Section 15.4 of FPSOTD. This quadric $\mathcal{G}_{1,3}$ is also called the *Klein quadric*.

Theorem 3.11. *The Grassmannian* $\mathcal{G}_{r,n,q}$ *has* $\phi(r; n, q)$ *points.*

Proof. $|\mathcal{G}_{r,n,q}| = |\mathrm{PG}^{(r)}(n,q)| = \phi(r; n, q).$ □

Theorem 3.12. *No hyperplane of* $\mathrm{PG}(N, K)$ *contains* $\mathcal{G}_{r,n}.$

Proof. Suppose that the hyperplane $\pi(u)$, where

$$u = (u_0, u_1, \ldots, u_N)$$

contains $\mathcal{G}_{r,n}$. Consider the r-space $\mathbf{U}_{i_0} \mathbf{U}_{i_1} \cdots \mathbf{U}_{i_r}$ of $\mathrm{PG}(n, K)$, where

$$\{i_0, i_1, \ldots, i_r\} \subset \{0, 1, \ldots, n\}.$$

For this r-space, $(k_0 k_1 \cdots k_r) = 0$ whenever $(k_0, k_1, \ldots, k_r) \neq \{i_0, i_1, \ldots, i_r\}$. If l_i is the coordinate of $\mathbf{U}_{i_0} \mathbf{U}_{i_1} \cdots \mathbf{U}_{i_r}$ corresponding to the set $\{i_0, i_1, \ldots, i_r\}$ then $l_i \neq 0$ while all other coordinates of this r-space are zero. Since $\mathcal{G}_{r,n}$ is contained in $\pi(u)$, so $u_i = 0$. Hence all coordinates of u are zero, a contradiction. □

If $\mathbf{V}(F_1), \mathbf{V}(F_2), \ldots$ are the quadrics represented by (3.5), then

$$\mathcal{G}_{r,n,K} = \mathbf{V}_{N,K}(F_1, F_2, \ldots).$$

Also, $\overline{\mathcal{G}}_{r,n,K} = \mathbf{V}_{N,\overline{K}}(F_1, F_2, \ldots)$ is the Grassmannian $\mathcal{G}_{r,n,\overline{K}}$ of the r-spaces of $\mathrm{PG}(n, \overline{K})$.

The following result is stated without proof.

Theorem 3.13. (i) *The algebraic variety* $\mathcal{G}_{r,n}$ *is absolutely irreducible and rational.*
(ii) *All points of* $\overline{\mathcal{G}}_{r,n}$ *are simple.*
(iii) *The dimension of* $\mathcal{G}_{r,n}$ *is* $(r+1)(n-r)$.
(iv) *The order of* $\mathcal{G}_{r,n}$ *is*

$$[(r+1)(n-r)]! \frac{((r))((n-r-1))}{((n))},$$

where $((m)) = 1! \, 2! \cdots m!.$

Hence $\mathcal{G}_{1,3}$ has dimension 4 and order 2, while $\mathcal{G}_{1,4}$ has dimension 6 and order 5, and $\mathcal{G}_{1,n}$ has dimension $2(n-1)$ and order $[2(n-1)]!/\{(n-1)! \, n!\}$. Also $\mathcal{G}_{2,5}$ has dimension 9 and order 42, while $\mathcal{G}_{r,2r+1}$ has dimension $(r+1)^2$ and order

$$[(r+1)^2]! \frac{1! \, 2! \cdots r!}{(r+1)! \, (r+2)! \cdots (2r+1)!}.$$

Theorem 3.14. *The Grassmannians* $\mathcal{G}_{r,n}$ *and* $\mathcal{G}_{n-r-1,n}$ *are projectively equivalent for* $1 \leq r \leq n-2$ *and* $n \geq 3$.

Proof. It may be assumed that $n \neq 2r + 1$. So, suppose that the coordinate of Π_r in $\mathrm{PG}^{(r)}(n, K)$ in position $s + 1$, and the dual coordinate of Π_{n-r-1} in $\mathrm{PG}^{(n-r-1)}(n, K)$ in the same position, correspond to the same ordered $(r + 1)$-tuple

$$(i_0, i_1, \ldots, i_r), \ s = 0, 1, \ldots, \mathbf{c}(n + 1, r + 1).$$

If $\mathbf{P}(x^{(0)}), \mathbf{P}(x^{(1)}), \ldots, \mathbf{P}(x^{(r)})$ are $r + 1$ linearly independent points of Π_r, then the hyperplanes $\boldsymbol{\pi}(x^{(0)}), \boldsymbol{\pi}(x^{(1)}), \ldots, \boldsymbol{\pi}(x^{(r)})$ have a Π_{n-r-1} as intersection. Hence any coordinate vector of Π_r is also a dual coordinate vector of Π_{n-r-1}. By Lemma 3.2, any dual coordinate vector of Π_{n-r-1} is also a coordinate vector of Π_{n-r-1}. Hence $\mathcal{G}_{r,n} = \mathcal{G}_{n-r-1,n}$.

If the assumption on positions at the beginning of the proof is not made, then there is a projectivity of $\mathrm{PG}(N, K)$ of the form $\rho x_i' = \epsilon_i x_j$ which takes $\mathcal{G}_{r,n}$ to $\mathcal{G}_{n-r-1,n}$; here, $\epsilon_i \in \{+1, -1\}$ and $i, j = 0, 1, \ldots, \mathbf{c}(n + 1, r + 1) - 1$. □

Now the case $n = 2r + 1$ is considered in more detail. It is assumed that, for any two coordinates $(i_0\, i_1\, \cdots\, i_r)_x$ and $(j_0\, j_1\, j_2\, \cdots\, j_r)_x$ of Π_r, where

$$|\{i_0, i_1, \ldots, i_r\} \cup \{j_0, j_1, \ldots, j_r\}| = 2r + 2,$$

their positions in the coordinate vector of Π_r differ by $\mathbf{c}(2r + 2, r + 1)/2 = m$; if $l_i = (i_0\, i_1\, \cdots\, i_r)_x$ and $l_{i+m} = (j_0\, j_1\, j_2\, \cdots\, j_r)_x$, $i \in \{0, 1, \ldots, m - 1\}$, then assume that the permutation $(i_0, i_1, \ldots, i_r, j_0, j_1, \ldots, j_r)$ is even. Let $\Pi_r = \mathbf{P}_r(L)$ and $\Pi_r' = \mathbf{P}_r(L')$ be r-spaces of $\mathrm{PG}(2r + 1, K)$, where

$$L = (l_0, l_1, \ldots) \text{ and } L' = (l_0', l_1', \ldots). \tag{3.15}$$

By Lemmas 3.2 and 3.3, $\Pi_r \cap \Pi_r' \neq \emptyset$ if and only if

$$\sum_{i=0}^{m-1} (l_i l_{i+m}' + l_i' l_{i+m}) = 0 \quad \text{for } r \text{ odd},$$

$$\sum_{i=0}^{m-1} (l_i l_{i+m}' - l_i' l_{i+m}) = 0 \quad \text{for } r \text{ even}.$$

When r is odd, then associated to $\mathrm{PG}^{(r)}(2r+1, K)$ is the polarity δ of $\mathrm{PG}(N, K)$ with bilinear form

$$\sum_{i=0}^{m-1} (X_i X_{i+m}' + X_i' X_{i+m}). \tag{3.16}$$

If $K = \mathbf{F}_q$ with q odd, then δ is the orthogonal polarity defined by the hyperbolic quadric

$$\mathbf{V}\left(\sum_{i=0}^{m-1} X_i X_{i+m}\right).$$

For $r = 1$ and $n = 3$ this quadric coincides with the Grassmannian $\mathcal{G}_{1,3}$. If $K = \mathbf{F}_q$ with q even, then δ is a null polarity.

When r is even, then associated to $\mathrm{PG}^{(r)}(2r+1, K)$ is the null polarity δ of $\mathrm{PG}(N, K)$ with bilinear form

$$\sum_{i=0}^{m-1} (X_i X'_{i+m} - X'_i X_{i+m}). \tag{3.17}$$

The polarity δ is the *fundamental polarity* associated to $\mathcal{G}_{r,2r+1}$.

Consider in $\mathrm{PG}(2r+1, K)$ the correlation η represented by

$$\rho x_i = u_i, \quad i = 0, 1, \ldots, 2r+1. \tag{3.18}$$

If $\Pi_r \eta = \Pi'_r$ with $\Pi_r = \mathbf{P}_r(L)$, $\Pi'_r = \mathbf{P}_r(L')$ as in (3.15), then, by Lemma 3.2,

$$\rho l'_i = \begin{cases} l_{i+m}, & i = 0, 1, \ldots, m-1, \\ (-1)^{r+1} l_{i+m}, & i = m, m+1, \ldots, N, \end{cases}$$

where the indices are taken modulo $N+1$. Hence, to the correlation η, there corresponds the following projectivity of $\mathrm{PG}(N, K)$ that leaves $\mathcal{G}_{r,2r+1}$ invariant:

$$\rho x'_i = \begin{cases} x_{i+m}, & i = 0, 1, \ldots, m-1, \\ (-1)^{r+1} x_{i+m}, & i = m, m+1, \ldots, N. \end{cases} \tag{3.19}$$

This projectivity is denoted by ζ.

Now, all subspaces of $\mathrm{PG}(N, K)$ contained in $\mathcal{G}_{r,n}$ are determined. First, the lines on $\mathcal{G}_{r,n}$ are considered.

Definition 3.15. The set of all r-spaces Π_r of $\mathrm{PG}(n, K)$ contained in a given $(r+1)$-space Π_{r+1} and containing a given $(r-1)$-space Π_{r-1} is a *pencil of r-spaces* and is denoted by (Π_{r-1}, Π_{r+1}).

Theorem 3.16. *The image of a pencil of r-spaces under $\mathfrak{G}$ is a line of $\mathcal{G}_{r,n}$, and conversely.*

Proof. Let Π_{r-1} be a given $(r-1)$-space of $\mathrm{PG}(n, K)$ and let Π_{r+1} be a given $(r+1)$-space through Π_{r-1}. Also, let l be a line of Π_{r+1} skew to Π_{r-1} and let $\mathbf{P}(x^{(0)}), \mathbf{P}(x^{(1)}), \ldots, \mathbf{P}(x^{(r-1)})$ be r linearly independent points of Π_{r-1}. Consider distinct elements Π_r^1, Π_r^2 of the pencil (Π_{r-1}, Π_{r+1}). The intersections of Π_r^1 and Π_r^2 with l are denoted by $\mathbf{P}(x_r^1)$ and $\mathbf{P}(x_r^2)$. The coordinates of Π_r^i, $i = 1, 2$, are determined by the matrix

$$T_x^i = \begin{bmatrix} x^{(0)} \\ x^{(1)} \\ \vdots \\ x^{(r-1)} \\ x_r^i \end{bmatrix}.$$

Let Π_r be an arbitrary element of the pencil (Π_{r-1}, Π_{r+1}). A coordinate vector of the point $\Pi_r \cap l$ is of the form

$$t_1 x_r^1 + t_2 x_r^2,$$

$t_1, t_2 \in K$ and not both zero. Conversely, any vector of this type defines one element Π_r of (Π_{r-1}, Π_{r+1}). The coordinates of Π_r are determined by the matrix

$$T_x = \begin{bmatrix} x^{(0)} \\ x^{(1)} \\ \vdots \\ x^{(r-1)} \\ t_1 x_r^1 + t_2 x_r^2 \end{bmatrix}.$$

If

$$L_i = (l_0^i, l_1^i, \ldots), i = 1, 2,$$

is the coordinate vector defined by the matrix T_x^i, then

$$t_1 L_1 + t_2 L_2$$

is the coordinate vector defined by the matrix T_x. Hence $(\Pi_{r-1}, \Pi_{r+1})\mathfrak{G}$ is the line joining the points $\Pi_r^1 \mathfrak{G}$ and $\Pi_r^2 \mathfrak{G}$.

Conversely, let l be a line on $\mathcal{G}_{r,n}$ and let $\mathbf{P}(L_1), \mathbf{P}(L_2)$ be distinct points of l. Suppose that the intersection of $\mathbf{P}(L_1)\mathfrak{G}^{-1} = \Pi_r^1$ and $\mathbf{P}(L_2)\mathfrak{G}^{-1} = \Pi_r^2$ is an $(r-1)$-space Π_{r-1}. Then $\Pi_r^1 \Pi_r^2$ is an $(r+1)$-space Π_{r+1}; so

$$(\Pi_{r-1}, \Pi_{r+1})\mathfrak{G} = \mathbf{P}(L_1)\mathbf{P}(L_2) = l.$$

Now it is shown that $\Pi_r^1 \cap \Pi_r^2$ is an $(r-1)$-space. Let $\Pi_r^1 \cap \Pi_r^2$ be a d-space, with $-1 \leq d \leq r-1$, let $\mathbf{P}(x^{(0)}), \mathbf{P}(x^{(1)}), \ldots, \mathbf{P}(x^{(d)})$ be $d+1$ linearly independent points of Π_d, and also let $\mathbf{P}(x^{(0)}), \mathbf{P}(x^{(1)}), \ldots, \mathbf{P}(x^{(d)}), \mathbf{P}(x_i^{(d+1)}), \ldots, \mathbf{P}(x_i^{(r)})$ be $r+1$ linearly independent points of Π_r^i, for $i = 1, 2$. Let ξ be a projectivity of $PG(n, K)$ with

$$\mathbf{P}(x^{(j)})\xi = \mathbf{U}_j, \quad j = 0, 1, \ldots, d,$$
$$\mathbf{P}(x_1^{(j)})\xi = \mathbf{U}_j, \quad j = d+1, d+2, \ldots, r,$$
$$\mathbf{P}(x_2^{(d+j)})\xi = \mathbf{U}_{r+j}, \quad j = 1, 2, \ldots, r-d.$$

In Section 3.1, it is shown that ξ induces a projectivity $\tilde{\xi}$ of $PG(N, K)$ which leaves $\mathcal{G}_{r,n}$ invariant. Then

$$\mathbf{P}(L_1)\tilde{\xi} = (\mathbf{U}_0 \mathbf{U}_1 \cdots \mathbf{U}_r)\mathfrak{G},$$
$$\mathbf{P}(L_2)\tilde{\xi} = (\mathbf{U}_0 \mathbf{U}_1 \cdots \mathbf{U}_d \mathbf{U}_{r+1} \cdots \mathbf{U}_{2r-d})\mathfrak{G}.$$

The points $\mathbf{P}(L_1)\tilde{\xi}$ and $\mathbf{P}(L_2)\tilde{\xi}$ are on the line $l\tilde{\xi}$ of $\mathcal{G}_{r,n}$. For the space $\mathbf{U}_0 \mathbf{U}_1 \cdots \mathbf{U}_r$, the coordinate $(k_0 \, k_1 \, \cdots \, k_r) \neq 0$ if and only if

$$\{k_0, k_1, \ldots, k_r\} = \{0, 1, \ldots, r\};$$

for the space $\mathbf{U}_0\mathbf{U}_1\cdots\mathbf{U}_d\mathbf{U}_{r+1}\cdots\mathbf{U}_{2r-d}$, the coordinate $(k_0\,k_1\,\cdots\,k_r)\neq 0$ if and only if

$$\{k_0,k_1,\ldots,k_r\}=\{0,1,\ldots,d,r+1,\ldots,2r-d\}.$$

Since each point of $l\tilde{\xi}$ is on $\mathcal{G}_{r,n}$, for any two given elements $t_1,t_2\in K$, not both zero, there is an r-space Π_r with

$$(0\,1\,\cdots\,r)=t_1,\quad (0\,\cdots\,d\,r+1\,\cdots\,2r-d)=t_2,$$

and $(k_0\,k_1\,\cdots\,k_r)=0$ in all other cases. Choose $t_1,t_2\neq 0$. Then

$$(r\,r-1\,\cdots\,0)(0\,1\,\cdots\,d\,r+1\,\cdots\,2r-d)\neq 0.$$

By (3.5),

$$(r\,r-1\,\cdots\,0)(0\,1\,\cdots\,d\,r+1\,\cdots\,2r-d)$$

$$=\sum_{s=0}^{d}(s\,r-1\,\cdots\,0)(0\,\cdots\,s-1\,r\,s+1\,\cdots\,d\,r+1\,\cdots\,2r-d)$$

$$+\sum_{s=r+1}^{2r-d}(s\,r-1\,\cdots\,0)(0\,\cdots\,d\,r+1\,\cdots\,s-1\,r\,s+1\,\cdots\,2r-d)$$

$$=\sum_{s=r+1}^{2r-d}(s\,r-1\,\cdots\,0)(0\,\cdots\,d\,r+1\,\cdots\,s-1\,r\,s+1\,\cdots\,2r-d).$$

Hence $(s\,r-1\,\cdots\,0)\neq 0$ for some $s\in\{r+1,r+2,\ldots,2r-d\}$; that is,

$$\{0,1,\ldots,r-1,s\}=\{0,\ldots,d,r+1,\ldots,2r-d\}.$$

Consequently, $d=r-1$, and so $\Pi_r^1\cap\Pi_r^2$ is an $(r-1)$-space. □

Theorem 3.16 is equivalent to saying that the images $\Pi_r^1\mathfrak{G}$ and $\Pi_r^2\mathfrak{G}$ of two distinct r-spaces Π_r^1 and Π_r^2 of $PG(n,K)$ are on a common line of $\mathcal{G}_{r,n}$ if and only if the intersection $\Pi_r^1\cap\Pi_r^2$ is an $(r-1)$-space or, equivalently, if and only if $\Pi_r^1\Pi_r^2$ is an $(r+1)$-space.

Theorem 3.17. *The number of lines on $\mathcal{G}_{r,n,q}$ is equal to*

$$\prod_{i=3}^{n+1}(q^i-1)\Big/\Big\{\prod_{i=1}^{r}(q^i-1)\prod_{i=1}^{n-r-1}(q^i-1)\Big\}. \tag{3.20}$$

Proof. By Theorem 3.16, the number of lines of $\mathcal{G}_{r,n,q}$ is equal to the number of pencils (Π_{r-1},Π_{r+1}). Hence it is equal to the product of $|PG^{(r-1)}(n,q)|$ and the number of $(r+1)$-spaces containing a given Π_{r-1}. So it is equal to

$$\phi(r-1;n,q)\chi(r-1,r+1;n,q)$$
$$= \frac{[n-r+2,n+1]_-[3,n-r+1]_-}{[1,r]_-[1,n-r-1]_-}$$
$$= \frac{[3,n+1]_-}{[1,r]_-[1,n-r-1]_-},$$

as required.   □

Lemma 3.18. *The number of lines on* $\mathcal{G}_{r,n,q}$ *through one of its points is equal to*

$$\theta(n-r-1)\,\theta(r)\,. \qquad (3.21)$$

Proof. The number of lines of $\mathcal{G}_{r,n,q}$ through a given point P on it is equal to the number of pencils (Π_{r-1},Π_{r+1}) with $\Pi_{r-1}\subset\Pi_r\subset\Pi_{r+1}$ and $\Pi_r=P\mathfrak{G}^{-1}$. Hence this number equals

$$\chi(r,r+1;n,q)\,\phi(r-1;r,q)=\theta(n-r-1)\,\theta(r)\,. \qquad □$$

Lemma 3.19. *A line* l *of* $\mathrm{PG}(N,K)$ *having at least three points in common with* $\mathcal{G}_{r,n}$ *is entirely contained in it.*

Proof. Since $\mathcal{G}_{r,n}$ is the intersection of the quadrics of $\mathrm{PG}(N,K)$ represented by the equations (3.5), so the line l has at least three points in common with each of these quadrics, and hence is contained in all the quadrics. It follows that l is contained in $\mathcal{G}_{r,n}$.   □

An s-space Π_s which is contained in $\mathcal{G}_{r,n}$ but in no $(s+1)$-space Π_{s+1} of $\mathcal{G}_{r,n}$ is a *maximal space* or *maximal subspace* of $\mathcal{G}_{r,n}$. The next result describes all such maximal spaces.

Theorem 3.20. *The variety* $\mathcal{G}_{r,n}$ *contains two systems* $\mathcal{S}_\mathrm{L}$ *and* $\mathcal{S}_\mathrm{G}$ *of maximal spaces:*

(i) $\mathcal{S}_\mathrm{L}$ *consists of the* $(n-r)$-*spaces* Π_{n-r} *with* $\Pi_{n-r}\mathfrak{G}^{-1}$ *the set of all* r-*spaces through a common* Π_{r-1};
(ii) $\mathcal{S}_\mathrm{G}$ *consists of the* $(r+1)$-*spaces* Π_{r+1} *with* $\Pi_{r+1}\mathfrak{G}^{-1}$ *the set of all* r-*spaces contained in a common* Π_{r+1}.

Proof. Let Π_{r-1} be an $(r-1)$-space of $\mathrm{PG}(n,K)$ and let $\mathcal{R}$ be the set of all the r-spaces containing Π_{r-1}. If Π_r^1,Π_r^2 are distinct elements of $\mathcal{R}$, then $\Pi_r^1\cap\Pi_r^2$ is the $(r-1)$-space Π_{r-1}, and so they belong to a pencil of r-spaces which is completely contained in $\mathcal{R}$. Hence, if P_1 and P_2 are distinct points of $\mathcal{R}\mathfrak{G}$, then the line P_1P_2 is contained in $\mathcal{R}\mathfrak{G}$; so $\mathcal{R}\mathfrak{G}$ is a subspace of $\mathrm{PG}(N,K)$.

Let Π_{n-r} be a subspace of $\mathrm{PG}(N,K)$ which is skew to Π_{r-1}. The bijection which maps each element of $\mathcal{R}$ onto its intersection with Π_{n-r} is denoted by δ. Then $\delta^{-1}\mathfrak{G}$ is a bijection of Π_{n-r} onto $\mathcal{R}\mathfrak{G}$ which maps the lines of Π_{n-r} onto the lines of $\mathcal{R}\mathfrak{G}$, and so $\mathcal{R}\mathfrak{G}$ has dimension $n-r$. If $\mathcal{R}\mathfrak{G}$ is properly contained in the subspace π of $\mathcal{G}_{r,n}$, then let $P\in\pi\backslash\mathcal{R}\mathfrak{G}$. The r-space $P\mathfrak{G}^{-1}$ of $\mathrm{PG}(n,K)$ does not contain Π_{r-1} but has an $(r-1)$-space in common with each r-space through Π_{r-1}. This contradiction shows that the $(n-r)$-space $\mathcal{R}\mathfrak{G}$ is a maximal space of $\mathcal{G}_{r,n}$.

Similarly, let Π_{r+1} be an $(r+1)$-space of $\mathrm{PG}(n,K)$ and let $\mathcal{S}$ be the set of all the r-spaces contained in Π_{r+1}. If Π_r^1, Π_r^2 are distinct elements of $\mathcal{S}$, then $\Pi_r^1\Pi_r^2$ is the $(r+1)$-space Π_{r+1}, and so they belong to a pencil of r-spaces which is completely contained in $\mathcal{S}$. Hence, if P_1 and P_2 are distinct points of $\mathcal{S}\mathfrak{G}$, then the line P_1P_2 is contained in $\mathcal{S}\mathfrak{G}$; so $\mathcal{S}\mathfrak{G}$ is a subspace of $\mathrm{PG}(N,K)$. Since the pencils of hyperplanes of Π_{r+1} are mapped by $\mathfrak{G}$ onto the lines of $\mathcal{S}\mathfrak{G}$, so $\mathfrak{G}$ induces a reciprocity from Π_{r+1} to $\mathcal{S}\mathfrak{G}$. Hence $\mathcal{S}\mathfrak{G}$ has dimension $r+1$. If $\mathcal{S}\mathfrak{G}$ is properly contained in the subspace π of $\mathcal{G}_{r,n}$, then let $P \in \pi\backslash\mathcal{S}\mathfrak{G}$. The r-space $P\mathfrak{G}^{-1}$ of $\mathrm{PG}(n,K)$ is not contained in Π_{r+1} but has an $(r-1)$-space in common with each r-space in Π_{r+1}, a contradiction. This shows that the $(r+1)$-space $\mathcal{S}\mathfrak{G}$ is a maximal space of $\mathcal{G}_{r,n}$.

Now consider an arbitrary maximal space π of $\mathcal{G}_{r,n}$; then π has dimension at least 1. If l is a line of π, then each element of the pencil $(\Pi_{r-1}, \Pi_{r+1}) = l\mathfrak{G}^{-1}$ has at least an $(r-1)$-space in common with each element $P\mathfrak{G}^{-1}$, where $P \in \pi$. So $P\mathfrak{G}^{-1}$ contains Π_{r-1} or is contained in Π_{r+1}. Suppose, for at least one point $P' \in \pi\backslash l$, that the r-space $P'\mathfrak{G}^{-1}$ contains Π_{r-1}. For any element $P\mathfrak{G}^{-1}$, with $P \in \pi\backslash(l \cup \{P'\})$, it is known that $P\mathfrak{G}^{-1} \subset \Pi_{r+1}$ or $\Pi_{r-1} \subset P\mathfrak{G}^{-1}$ and that $P\mathfrak{G}^{-1} \cap P'\mathfrak{G}^{-1}$ is an $(r-1)$-space. Hence $\Pi_{r-1} \subset P\mathfrak{G}^{-1}$. So all elements of $\pi\mathfrak{G}^{-1}$ contain Π_{r-1}. Since π is maximal, it is the image of the set of all r-spaces containing Π_{r-1}. If, for at least one point $P' \in \pi\backslash l$, the r-space $P'\mathfrak{G}^{-1}$ is contained in Π_{r+1}, then, analogously, π is the image of the set of all r-spaces contained in Π_{r+1}. $\square$

The system $\mathcal{S}_{\mathrm{L}}$ of maximal spaces of $\mathcal{G}_{r,n}$ corresponding to the $(r-1)$-spaces of $\mathrm{PG}(n,K)$ is called the *Latin system* and its elements the *Latin $(n-r)$-spaces*. The system $\mathcal{S}_{\mathrm{G}}$ corresponding to the $(r+1)$-spaces of $\mathrm{PG}(n,K)$ is called the *Greek system* and its elements the *Greek $(r+1)$-spaces*. Note that the Latin and the Greek spaces have the same dimension if and only if $n = 2r+1$.

Let Π_{r-1} be an $(r-1)$-space of $\mathrm{PG}(n,K)$ and let Π_{n-r} be an $(n-r)$-space skew to Π_{r-1}. In the first part of the proof of Theorem 3.20, it was shown that $\mathfrak{G}$ induces a collineation ξ of Π_{n-r} onto the corresponding maximal space of $\mathcal{G}_{r,n}$. From the first part of the proof of Theorem 3.16, ξ preserves the cross-ratio of any four collinear points of Π_{n-r}. Hence ξ is a projectivity.

Similarly, Let Π_{r+1} be an $(r+1)$-space of $\mathrm{PG}(n,K)$. In the second part of the proof of Theorem 3.20, it was shown that $\mathfrak{G}$ induces a reciprocity ξ of Π_{r+1} onto the corresponding maximal space of $\mathcal{G}_{r,n}$. Again, by Theorem 3.16, ξ preserves the cross-ratio of any four hyperplanes of Π_{r+1} in the same pencil, and hence ξ is a correlation.

Lemma 3.21.   (i) *Any line l of $\mathcal{G}_{r,n}$ is contained in one Latin space and one Greek space.*

(ii) *Any two distinct Latin or two distinct Greek spaces have at most one point in common.*

(iii) *If $\Pi_{n-r} \in \mathcal{S}_{\mathrm{L}}$ and $\Pi_{r+1} \in \mathcal{S}_{\mathrm{G}}$, then $\Pi_{n-r} \cap \Pi_{r+1}$ is the empty set or a line.*

Proof. Let $l\mathfrak{G}^{-1}$ be the pencil (Π_{r-1}, Π_{r+1}). Then the Latin space defined by Π_{r-1} is the only Latin space through l and the Greek space defined by Π_{r+1} is the only Greek space through l.

For $\Pi_{n-r} \in \mathcal{S}_L$ and $\Pi_{r+1} \in \mathcal{S}_G$, let the corresponding spaces of $\mathrm{PG}(n, K)$ be denoted by Π'_{r-1} and Π'_{r+1}. If $\Pi'_{r-1} \subset \Pi'_{r+1}$, then $\Pi_{n-r} \cap \Pi_{r+1}$ is a line; if $\Pi'_{r-1} \not\subset \Pi'_{r+1}$, then $\Pi_{n-r} \cap \Pi_{r+1} = \emptyset$. $\square$

Let $\mathcal{T}_r(\Pi_g, \Pi_h)$ be the set of all r-spaces of $\mathrm{PG}(n, K)$ through the g-space Π_g and contained in the h-space Π_h. In the previous notation, $\mathcal{T}_r(\Pi_{r-1}, \Pi_{r+1})$, where $\Pi_{r-1} \subset \Pi_{r+1}$, is the pencil (Π_{r-1}, Π_{r+1}). Let $\mathcal{G}_{r,n}^{(s)}$ be the set of all s-spaces on $\mathcal{G}_{r,n}$.

Theorem 3.22. (i) *For $0 \le s \le n - r$, the set $\mathcal{T}_r(\Pi_{r-1}, \Pi_{r+s})$, is an element of $\mathcal{G}_{r,n}^{(s)}$; here $\Pi_{r-1} \subset \Pi_{r+s}$.*

(ii) *For $0 \le s \le r+1$, the set $\mathcal{T}_r(\Pi_{r-s}, \Pi_{r+1})$, where $\Pi_{r-s} \subset \Pi_{r+1}$, is an element of $\mathcal{G}_{r,n}^{(s)}$.*

(iii) *Any s-space of $\mathcal{G}_{r,n}$, where $0 \le s \le \max(n - r, r + 1)$, is obtained in one of these ways.*

Proof. (i) It is first shown that $\mathcal{T}_r(\Pi_{r-1}, \Pi_{r+s}) \in \mathcal{G}_{r,n}^{(s)}$ for $0 \le s \le n-r$. If Π_{n-r} is a space skew to Π_{r-1}, then $\Pi_{n-r} \cap \Pi_{r+s} = \Pi_s$. The maximal space of $\mathcal{G}_{r,n}$ defined by Π_{r-1} is denoted by Π'_{n-r}. Now consider the projectivity $\xi : \Pi_{n-r} \to \Pi'_{n-r}$; then $\mathcal{T}_r(\Pi_{r-1}, \Pi_{r+s})\mathfrak{G} = \Pi_s \xi$ is an s-space Π'_s of Π'_{n-r}.

(ii) Similarly, $\mathcal{T}_r(\Pi_{r-s}, \Pi_{r+1}) \in \mathcal{G}_{r,n}^{(s)}$, where $0 \le s \le r + 1$. For, if the maximal space of $\mathcal{G}_{r,n}$ defined by Π_{r+1} is denoted by Π'_{r+1}, consider, as above, the correlation $\xi : \Pi_{r+1} \to \Pi'_{r+1}$; then $\mathcal{T}_r(\Pi_{r-s}, \Pi_{r+1})\mathfrak{G}$ is an s-space Π'_s of Π'_{r+1}.

(iii) Conversely, consider Π_s in $\mathcal{G}_{r,n}^{(s)}$, where $0 \le s \le \max(n-r, r+1)$. The space Π'_s is contained in at least one element of $\mathcal{S}_L \cup \mathcal{S}_G$. So, either $\Pi'_s \subset \Pi'_{n-r}$ where $\Pi'_{n-r} \in \mathcal{S}_L$ or $\Pi'_s \subset \Pi'_{r+1}$ where $\Pi'_{r+1} \in \mathcal{S}_G$. The $(r - 1)$-space corresponding to Π'_{n-r} is denoted by Π_{r-1} and the $(r + 1)$-space corresponding to Π'_{r+1} is denoted by Π_{r+1}.

(a) Let Π_{n-r} be an $(n - r)$-space of $\mathrm{PG}(n, K)$ skew to Π_{r-1}. From the projectivity $\xi : \Pi_{n-r} \to \Pi'_{n-r}$, it follows that $\Pi'_s \xi^{-1}$ is an s-space Π_s of Π_{n-r}. Hence $\Pi'_s \mathfrak{G}^{-1}$ is the set $\mathcal{T}_r(\Pi_{r-1}, \Pi_s\Pi_{r-1} = \Pi_{r+s})$.

(b) From the correlation $\xi : \Pi_{r+1} \to \Pi'_{r+1}$, it follows that $\Pi'_s \xi^{-1}$ is the set of all hyperplanes of Π_{r+1} containing a fixed Π_{r-s}. Hence $\Pi'_s \mathfrak{G}^{-1} = \mathcal{T}_r(\Pi_{r-s}, \Pi_{r+1})$. $\square$

Remark 3.23. By Lemma 3.21, any s-space of $\mathcal{G}_{r,n}$, where $s > 1$, is contained in exactly one element of $\mathcal{S}_L \cup \mathcal{S}_G$.

Theorem 3.24. *For $K = \mathbf{F}_q$, the sizes of $\mathcal{S}_L, \mathcal{S}_G, \mathcal{G}_{r,n}^{(s)}$ are as follows:*

(i) $|\mathcal{S}_L| = \phi(r - 1; n, q)$ *and* $|\mathcal{S}_G| = \phi(r + 1; n, q)$;

(ii) *for $n < 2r + 1$ and $1 < s \le n - r$,*

$$|\mathcal{G}_{r,n}^{(s)}| = \phi(r - 1; n, q)\phi(s; n - r, q) + \phi(r + 1; n, q)\phi(s; r + 1, q);$$

(iii) *for* $n < 2r + 1$ *and* $n - r < s \le r + 1$,

$$|\mathcal{G}_{r,n}^{(s)}| = \phi(r+1; n, q)\phi(s; r+1, q);$$

(iv) *for* $n > 2r + 1$ *and* $1 < s \le r + 1$,

$$|\mathcal{G}_{r,n}^{(s)}| = \phi(r-1; n, q)\phi(s; n-r, q) + \phi(r+1; n, q)\phi(s; r+1, q);$$

(v) *for* $n > 2r + 1$ *and* $r + 1 < s \le n - r$,

$$|\mathcal{G}_{r,n}^{(s)}| = \phi(r-1; n, q)\phi(s; n-r, q);$$

(vi) *for* $n = 2r + 1$ *and* $1 < s \le r + 1$,

$$|\mathcal{G}_{r,n}^{(s)}| = 2\phi(r+1; 2r+1, q)\phi(s; r+1, q);$$

(vii) *for* $n = 2r + 1$,
$$|\mathcal{S}_{\mathrm{L}}| = |\mathcal{S}_{\mathrm{G}}| = \phi(r+1; 2r+1, q).$$

Proof. First,

$$\begin{aligned}
|\mathcal{S}_{\mathrm{L}}| &= \mathrm{PG}^{(r-1)}(n, q) = \phi(r-1; n, q);\\
|\mathcal{S}_{\mathrm{G}}| &= \mathrm{PG}^{(r+1)}(n, q) = \phi(r+1; n, q).
\end{aligned}$$

When $n = 2r + 1$, then $\phi(r - 1; 2r + 1, q) = \phi(r + 1; 2r + 1, q)$; hence (vii) is shown.

Since any s-space of $\mathcal{G}_{r,n}$, with $s > 1$, is contained in exactly one element of $\mathcal{S}_{\mathrm{L}} \cup \mathcal{S}_{\mathrm{G}}$, the number of these s-spaces is equal to the sum of the number of all s-spaces in elements of $\mathcal{S}_{\mathrm{L}}$ and the number of all s-spaces in elements of $\mathcal{S}_{\mathrm{G}}$. This gives (ii)–(vi). $\qquad\square$

Theorem 3.25. *If the parameters* r, s, t, n *satisfy*

$$1 \le r \le n - 2, \quad -1 \le t \le r - 2, \quad r + 2 \le s \le n,$$

and Π_t, Π_s *are subspaces of* $\mathrm{PG}(n, K)$ *with* $\Pi_t \subset \Pi_s$, *then* $\mathcal{T}_r(\Pi_t, \Pi_s)\mathfrak{G}$ *is projectively equivalent both to* $\mathcal{G}_{r-t-1,s-t-1}$ *and to* $\mathcal{G}_{s-r-1,s-t-1}$.

Proof. From Section 3.1 it may be assumed, without loss of generality, that

$$\Pi_s = \mathbf{U}_0\mathbf{U}_1\cdots\mathbf{U}_s, \quad \Pi_t = \mathbf{U}_{s-t}\mathbf{U}_{s-t+1}\cdots\mathbf{U}_s.$$

For any $\Pi_r \in \mathcal{T}_r(\Pi_t, \Pi_s)$ choose $r - t$ linearly independent points

$$\mathbf{P}(x^{(0)}), \mathbf{P}(x^{(1)}), \dots, \mathbf{P}(x^{(r-t-1)})$$

in $\Pi_r \cap \mathbf{U}_0\mathbf{U}_1\cdots\mathbf{U}_{s-t-1}$. Let

$$x^{(i)} = (x_0^i, x_1^i, \dots, x_{s-t-1}^i, 0, \dots, 0),$$

$i = 0, 1, \ldots, r - t - 1$. The coordinates of Π_r are determined by the $(r+1) \times (n+1)$ matrix

$$
\begin{bmatrix}
x_0^0 & x_1^0 & \cdots & x_{s-t-1}^0 & 0 & 0 & \cdots & 0 & 0 & 0 & \cdots & 0 \\
x_0^1 & x_1^1 & \cdots & x_{s-t-1}^1 & 0 & 0 & \cdots & 0 & 0 & 0 & \cdots & 0 \\
\vdots & \vdots & \cdots & \vdots & \vdots & \vdots & \cdots & \vdots & \vdots & \vdots & \cdots & \vdots \\
x_0^{r-t-1} & x_1^{r-t-1} & \cdots & x_{s-t-1}^{r-t-1} & 0 & 0 & \cdots & 0 & 0 & 0 & \cdots & 0 \\
0 & 0 & \cdots & 0 & 1 & 0 & \cdots & 0 & 0 & 0 & \cdots & 0 \\
0 & 0 & \cdots & 0 & 0 & 1 & \cdots & 0 & 0 & 0 & \cdots & 0 \\
\vdots & \vdots & \cdots & \vdots & \vdots & \vdots & \cdots & \vdots & \vdots & \vdots & \cdots & \vdots \\
0 & 0 & \cdots & 0 & 0 & 0 & \cdots & 1 & 0 & 0 & \cdots & 0
\end{bmatrix}.
$$

Now, with $k_0, k_1, \ldots, k_r$ distinct,

$$(k_0\, k_1 \cdots k_r) = 0$$

if $\{s - t, s - t + 1, \ldots, s\} \not\subset \{k_0, k_1, \ldots, k_r\}$ or if $\{k_0, k_1, \ldots, k_r\} \not\subset \{0, 1, \ldots, s\}$. If both

$$\{s - t, s - t + 1, \ldots, s\} \subset \{k_0, k_1, \ldots, k_r\} \subset \{0, 1, \ldots, s\},$$
$$\{l_0, l_1, \ldots, l_{r-t-1}\} = \{k_0, k_1, \ldots, k_r\} \backslash \{s - t, s - t + 1, \ldots, s\},$$

then, up to the sign, $(k_0\, k_1 \cdots k_r)$ is equal to $(l_0\, l_1 \cdots l_{r-t-1})$, where the latter is calculated with respect to the matrix

$$
\begin{bmatrix}
x_0^0 & x_1^0 & \cdots & x_{s-t-1}^0 \\
x_0^1 & x_1^1 & \cdots & x_{s-t-1}^1 \\
\vdots & \vdots & \cdots & \vdots \\
x_0^{r-t-1} & x_1^{r-t-1} & \cdots & x_{s-t-1}^{r-t-1}
\end{bmatrix}.
$$

From these considerations it follows that $\mathcal{T}_r(\Pi_t, \Pi_s)\mathfrak{G}$ is projectively equivalent to the Grassmann variety $\mathcal{G}_{r-t-1,s-t-1}$. By Theorem 3.14, this variety is projectively equivalent to $\mathcal{G}_{s-r-1,s-t-1}$. $\quad\square$

Corollary 3.26. *The image of the set of all r-spaces of $\mathrm{PG}(n, K)$ containing a given t-space Π_t, where $1 \leq r \leq n - 2$ and $-1 \leq t \leq r - 2$, is projectively equivalent both to $\mathcal{G}_{r-t-1,n-t-1}$ and to $\mathcal{G}_{n-r-1,n-t-1}$.*

Proof. This is Theorem 3.25 with $s = n$. $\quad\square$

Corollary 3.27. *The image of the set of all r-spaces of $\mathrm{PG}(n, K)$ contained in a given s-space Π_s, where $1 \leq r \leq n - 2$ and $r + 2 \leq s \leq n$, is projectively equivalent both to $\mathcal{G}_{r,s}$ and to $\mathcal{G}_{s-r-1,s}$.*

Proof. This is Theorem 3.25 with $s = -1$. □

Corollary 3.28. *The Grassmannian $\mathcal{G}_{r,n}$ contains a subvariety projectively equivalent to $\mathcal{G}_{1,n'}$ if and only if $3 \leq n' \leq \max(n - r + 1, r + 2)$.*

Proof. For $t = r - 2$, Theorem 3.25 gives subvarieties projectively equivalent to $\mathcal{G}_{1,s-r+1}$, where $r + 2 \leq s \leq n$; for $s = r + 2$ the theorem also gives subvarieties projectively equivalent to $\mathcal{G}_{1,r-t+1}$, where $-1 \leq t \leq r - 2$.

For $n' > \max(n - r + 1, r + 2)$, the variety $\mathcal{G}_{1,n'}$ contains $(n' - 1)$-spaces. The Latin spaces of $\mathcal{G}_{r,n}$ have dimension $n - r < n' - 1$, and the Greek spaces of $\mathcal{G}_{r,n}$ have dimension $r + 1 < n' - 1$. Hence, in this case, $\mathcal{G}_{r,n}$ cannot contain subvarieties projectively equivalent to $\mathcal{G}_{1,n'}$. □

In the last part of this section all projectivities of $\mathrm{PG}(N, K)$ leaving $\mathcal{G}_{r,n}$ invariant are determined. It is necessary to distinguish between the cases $n = 2r + 1$ and $n \neq 2r + 1$.

Theorem 3.29. *If ξ is a projectivity of $\mathrm{PG}(N, K)$ leaving $\mathcal{G}_{r,n}$ invariant, then ξ also leaves $\mathcal{S}_\mathrm{L}$ and $\mathcal{S}_\mathrm{G}$ invariant or interchanges them. For $n \neq 2r + 1$, they are left invariant.*

Proof. First, ξ maps a maximal space of $\mathcal{G}_{r,n}$ onto a maximal space. For $n \neq 2r + 1$, the Greek and Latin spaces have different dimensions, and so ξ leaves $\mathcal{S}_\mathrm{L}$ and $\mathcal{S}_\mathrm{G}$ invariant.

Now assume that $n = 2r + 1$. Let $\Pi_{r+1} \in \mathcal{S}_\mathrm{L} \cup \mathcal{S}_\mathrm{G}$; for example, $\Pi_{r+1} \in \mathcal{S}_\mathrm{L}$. Choose a space $\Pi'_{r+1} \in \mathcal{S}_\mathrm{L}$ which has exactly one point P in common with Π_{r+1}; then $\Pi_{r+1}\xi \cap \Pi'_{r+1}\xi = P\xi$. By Lemma 3.21, the maximal spaces $\Pi_{r+1}\xi$ and $\Pi'_{r+1}\xi$ both belong to $\mathcal{S}_\mathrm{L}$ or to $\mathcal{S}_\mathrm{G}$.

Next consider any space $\Pi''_{r+1} \in \mathcal{S}_\mathrm{L}$, $\Pi''_{r+1} \neq \Pi_{r+1}$. The $(r - 1)$-spaces of $\mathrm{PG}(2r + 1, K)$ which correspond to Π_{r+1} and Π''_{r+1} are denoted by Π_{r-1} and Π''_{r-1}. There exists a finite number of distinct $(r - 1)$-spaces $\Pi^0_{r-1}, \Pi^1_{r-1}, \ldots, \Pi^k_{r-1}$ such that

$$\Pi_{r-1}\Pi^0_{r-1}, \Pi^0_{r-1}\Pi^1_{r-1}, \ldots, \Pi^{k-1}_{r-1}\Pi^k_{r-1}, \Pi^k_{r-1}\Pi''_{r-1}$$

are r-spaces. Hence to $\Pi^0_{r-1}, \Pi^1_{r-1}, \ldots, \Pi^k_{r-1}$ there correspond Latin spaces

$$\Pi^0_{r+1}, \Pi^1_{r+1}, \ldots, \Pi^k_{r+1}$$

such that

$$\Pi_{r+1} \cap \Pi^0_{r+1}, \Pi^0_{r+1} \cap \Pi^1_{r+1}, \ldots, \Pi^{k-1}_{r+1} \cap \Pi^k_{r+1}, \Pi^k_{r+1} \cap \Pi''_{r+1}$$

are points. By a previous argument, the maximal spaces

$$\Pi_{r+1}\xi, \Pi^0_{r+1}\xi, \Pi^1_{r+1}\xi, \ldots, \Pi^k_{r+1}\xi, \Pi''_{r+1}\xi$$

all belong to $\mathcal{S}_\mathrm{L}$ or all to $\mathcal{S}_\mathrm{G}$. Hence $\Pi_{r+1}\xi$ and $\Pi''_{r+1}\xi$ both belong to $\mathcal{S}_\mathrm{L}$ or both to $\mathcal{S}_\mathrm{G}$. Thus it has been shown that ξ maps the elements of $\mathcal{S}_\mathrm{L}$ either all to $\mathcal{S}_\mathrm{L}$ or all to $\mathcal{S}_\mathrm{G}$. Analogously, ξ maps all elements of $\mathcal{S}_\mathrm{G}$ to $\mathcal{S}_\mathrm{L}$ or to $\mathcal{S}_\mathrm{G}$. Hence ξ leaves both $\mathcal{S}_\mathrm{L}$ and $\mathcal{S}_\mathrm{G}$ invariant or interchanges them. □

From (3.18) and (3.19) in the case $n = 2r+1$ the correlation η of $\mathrm{PG}(2r+1, K)$ represented by

$$\rho x_i = u_i, \quad i = 0, 1, \ldots, 2r+1,$$

induces a projectivity ζ of $\mathrm{PG}(N, K)$ which leaves $\mathcal{G}_{r,2r+1}$ invariant. Then ζ interchanges the systems $\mathcal{S}_{\mathrm{L}}$ and $\mathcal{S}_{\mathrm{G}}$ of $\mathcal{G}_{r,2r+1}$.

Let $G(\mathcal{G}_{r,n})$ be the subgroup of $\mathrm{PGL}(N+1, K)$ leaving $\mathcal{G}_{r,n}$ fixed.

Lemma 3.30. *Let* $\theta : \xi \to \tilde{\xi}$ *map each element* ξ *of* $\mathrm{PGL}(n+1, K)$ *onto the corresponding element* $\tilde{\xi}$ *of* $G(\mathcal{G}_{r,n})$. *Then*

(i) θ *is a monomorphism of* $\mathrm{PGL}(n+1, K)$ *into* $G(\mathcal{G}_{r,n})$;
(ii) *distinct elements of* $G(\mathcal{G}_{r,n})$ *induce distinct permutations of* $\mathcal{G}_{r,n}$.

Proof. First it is shown that the identity mapping of $\mathcal{G}_{r,n}$ is induced only by the identity $\mathfrak{I}$ of $\mathrm{PGL}(N+1, K)$. Let $\mathfrak{I}' \in \mathrm{PGL}(N+1, K)$ fix $\mathcal{G}_{r,n}$ point-wise. Choose distinct points $P, P' \in \mathcal{G}_{r,n}$, and let $P\mathfrak{G}^{-1} = \Pi_r$, $P'\mathfrak{G}^{-1} = \Pi'_r$. There is a finite number of elements $\Pi_r^0, \Pi_r^1, \ldots, \Pi_r^k \in \mathrm{PG}^{(r)}(n, K)$ such that

$$\Pi_r \cap \Pi_r^0, \Pi_r^0 \cap \Pi_r^1, \ldots, \Pi_r^{k-1} \cap \Pi_r^k, \Pi_r^k \cap \Pi'_r$$

are $(r-1)$-spaces. This means that there is a finite number of points $P_0, P_1, \ldots, P_k$ on $\mathcal{G}_{r,n}$ such that $PP_0, P_0P_1, \ldots, P_{k-1}P_k, P_kP'$ are lines of $\mathcal{G}_{r,n}$. Hence all points of the subspace of $\mathrm{PG}(N, K)$ generated by $\mathcal{G}_{r,n}$ are fixed by $\mathfrak{I}'$. By Theorem 3.12, all points of $\mathrm{PG}(N, K)$ are fixed by $\mathfrak{I}'$; hence $\mathfrak{I}'$ is the identity of $\mathrm{PGL}(N+1, K)$.

From the preceding paragraph it follows that, if $\delta, \delta' \in G(\mathcal{G}_{r,n})$ coincide on $\mathcal{G}_{r,n}$, then $\delta = \delta'$.

Let $\xi, \xi' \in \mathrm{PGL}(n+1, K)$, with $\xi \neq \xi'$. Then the mappings induced by ξ and ξ' on $\mathrm{PG}^{(r)}(n, K)$ are distinct; hence $\tilde{\xi} \neq \tilde{\xi}'$. So θ is an injection of $\mathrm{PGL}(n+1, K)$ into $G(\mathcal{G}_{r,n})$. Again, consider elements $\xi, \xi' \in \mathrm{PGL}(n+1, K)$; then $\widetilde{\xi\xi'}$ and $\tilde{\xi}\tilde{\xi}'$ coincide on $\mathcal{G}_{r,n}$. Hence $\widetilde{\xi\xi'} = \tilde{\xi}\tilde{\xi}'$ and θ is a monomorphism of $\mathrm{PGL}(n+1, K)$ into $G(\mathcal{G}_{r,n})$. $\square$

Theorem 3.31. *For* $n \neq 2r+1$, *the mapping* θ *is an isomorphism of* $\mathrm{PGL}(n+1, K)$ *onto* $G(\mathcal{G}_{r,n})$. *For* $n = 2r+1$,

$$G(\mathcal{G}_{r,n}) = \mathrm{PGL}(n+1, K)\theta \cup (\mathrm{PGL}(n+1, K)\theta)\zeta.$$

Proof. First suppose that $n \neq 2r+1$. It must be shown that

$$G(\mathcal{G}_{r,n}) = \mathrm{PGL}(n+1, K)\theta.$$

Consider an element $\delta \in G(\mathcal{G}_{r,n})$. By Theorem 3.29, δ leaves $\mathcal{S}_{\mathrm{L}}$ and $\mathcal{S}_{\mathrm{G}}$ invariant. Consider an $(r+1)$-space $\Lambda_{r+1}^1 \in \mathcal{S}_{\mathrm{G}}$. The $(r+1)$-space $\Lambda_{r+1}^1 \delta = \Lambda_{r+1}^{1'}$ also belongs to $\mathcal{S}_{\mathrm{G}}$. The corresponding $(r+1)$-spaces of $\mathrm{PG}(n, K)$ are denoted by Π_{r+1}^1 and $\Pi_{r+1}^{1'}$. By the remark just preceding Lemma 3.21, $\mathfrak{G}$ induces a correlation of

Π_{r+1}^1 onto Λ_{r+1}^1 and of $\Pi_{r+1}^{1'}$ onto $\Lambda_{r+1}^{1'}$. Hence $\mathfrak{G}\delta\mathfrak{G}^{-1}$ induces a projectivity ξ_1 of Π_{r+1}^1 onto $\Pi_{r+1}^{1'}$.

Next consider an $(r+1)$-space Π_{r+1}^2 of $\mathrm{PG}(n, K)$ for which $\Pi_{r+1}^1 \cap \Pi_{r+1}^2$ is an r-space Π_r. Corresponding to Π_{r+1}^2 is an $(r+1)$-space $\Lambda_{r+1}^2 \in \mathcal{S}_G$. So, let $\Lambda_{r+1}^2\delta = \Lambda_{r+1}^{2'} \in \mathcal{S}_G$ and let $\Pi_{r+1}^{2'}$ be the corresponding $(r+1)$-space of $\mathrm{PG}(n, K)$. Again $\mathfrak{G}\delta\mathfrak{G}^{-1}$ induces a projectivity ξ_2 of Π_{r+1}^2 onto $\Pi_{r+1}^{2'}$.

Since $\Pi_{r+1}^1 \cap \Pi_{r+1}^2$ is an r-space, the intersection $\Lambda_{r+1}^1 \cap \Lambda_{r+1}^2$ is a point P. Hence $\Lambda_{r+1}^{1'} \cap \Lambda_{r+1}^{2'}$ is a point P', whence $\Pi_{r+1}^{1'} \cap \Pi_{r+1}^{2'}$ is an r-space Π_r'. Since $P\delta = P'$, so $\Pi_r\xi_1 = \Pi_r\xi_2 = \Pi_r'$. In Π_r, now choose an $(r-1)$-space Π_{r-1}. If $\Pi_{r+1}^1\Pi_{r+1}^2 = \Pi_{r+2}$, then $\mathcal{T}_r(\Pi_{r-1}, \Pi_{r+2})\mathfrak{G}$ is a plane Π_2. If the lines $\Pi_2 \cap \Lambda_{r+1}^1$ and $\Pi_2 \cap \Lambda_{r+1}^2$ are l_1 and l_2, then $l_1 \cap l_2 = P$. The plane $\Pi_2\delta = \Pi_2'$ intersects Λ_{r+1}^1 and $\Lambda_{r+1}^{2'}$ in the lines $l_1' = l_1\delta$ and $l_2' = l_2\delta$; also $l_1' \cap l_2' = P'$. Now, $l_1'\mathfrak{G}^{-1}$ is the pencil $(\Pi_{r-1}\xi_1, \Pi_{r+1}^{1'})$ and $l_2'\mathfrak{G}^{-1}$ is the pencil $(\Pi_{r-1}\xi_2, \Pi_{r+1}^{2'})$. These two pencils belong to $\Pi_2'\mathfrak{G}^{-1}$; hence, by Theorem 3.22, all elements of the two pencils contain a common $(r-2)$-space and are contained in a common $(r+1)$-space, or contain a common $(r-1)$-space and are contained in a common $(r+2)$-space. Since the elements of the pencils generate $\Pi_{r+1}^{1'}\Pi_{r+1}^{2'}$, which is an $(r+2)$-space, they all contain a common $(r-1)$-space.

Hence $\Pi_{r-1}\xi_1 = \Pi_{r-1}\xi_2 = \Pi_{r-1}'$. Consequently, the actions of ξ_1 and ξ_2 on the set of all hyperplanes of Π_r coincide, which means that ξ_1 and ξ_2 also coincide on all points of Π_r.

Consider distinct $(r+1)$-spaces Π_{r+1}^3 and Π_{r+1}^4 of $\mathrm{PG}(n, K)$. Let their intersection be an s-space Π_s for some $s \in \{-1, 0, 1, \ldots, r\}$. There exists a finite number of distinct $(r+1)$-spaces

$$\Pi_{r+1}^3 = \Pi_{r+1}^5, \Pi_{r+1}^6, \ldots, \Pi_{r+1}^k = \Pi_{r+1}^4$$

in $\mathrm{PG}(n, K)$ such that

$$\Pi_{r+1}^5 \cap \Pi_{r+1}^6, \Pi_{r+1}^6 \cap \Pi_{r+1}^7, \ldots, \Pi_{r+1}^{k-1} \cap \Pi_{r+1}^k$$

are r-spaces which contain Π_s. As in the preceding paragraphs, the spaces

$$\Pi_{r+1}^5, \Pi_{r+1}^6, \ldots, \Pi_{r+1}^k$$

define projectivities $\xi_3 = \xi_5, \xi_6, \ldots, \xi_k = \xi_4$. Also, ξ_i and ξ_{i+1} coincide on all points of $\Pi_{r+1}^i \cap \Pi_{r+1}^{i+1}$, $i = 5, 6, \ldots, k-1$, and consequently coincide on all points of Π_s. Hence ξ_3 and ξ_4 coincide on all points of Π_s.

Define ξ as follows. Let P be an arbitrary point of $\mathrm{PG}(n, K)$. Consider any $(r+1)$-space Π_{r+1}^0 containing P and the corresponding projectivity ξ_0. Then define $P\xi = P\xi_0$. From the above, $P\xi_0$ is independent of the choice of Π_{r+1}^0 through P. If l is a line of $\mathrm{PG}(n, K)$, then choosing Π_{r+1}^0 through l it follows that $l\xi$ is a line of $\mathrm{PG}(n, K)$; if Π_r is an r-space of $\mathrm{PG}(n, K)$, then choosing Π_{r+1}^0 through Π_r gives that $\Pi_r\mathfrak{G}\delta = \Pi_r\xi\mathfrak{G}$. Hence ξ is a projectivity of $\mathrm{PG}(n, K)$ and $\xi\theta = \tilde{\xi} = \delta$. This proves the theorem in the case that $n \neq 2r + 1$.

Finally, suppose that $n = 2r + 1$. From above,

$$\mathrm{PGL}(2r + 2, K)\theta \cup (\mathrm{PGL}(2r + 2, K)\theta)\zeta \subset G(\mathcal{G}_{r,2r+1}).$$

Let $\delta \in G(\mathcal{G}_{r,2r+1})$ leave $\mathcal{S}_{\mathrm{L}}$ and $\mathcal{S}_{\mathrm{G}}$ invariant; then, as in the case $n \neq 2r + 1$, it follows that $\delta \in \mathrm{PGL}(2r + 2, K)\theta$. Now suppose that δ interchanges $\mathcal{S}_{\mathrm{L}}$ and $\mathcal{S}_{\mathrm{G}}$. Then $\delta\zeta^{-1} \in G(\mathcal{G}_{r,2r+1})$ and leaves both $\mathcal{S}_{\mathrm{L}}$ and $\mathcal{S}_{\mathrm{G}}$ invariant. Consequently $\delta\zeta^{-1} \in \mathrm{PGL}(2r + 2, K)\theta$; that is, $\delta \in (\mathrm{PGL}(2r + 2, K)\theta)\zeta$. □

It follows from this proof that $(\mathrm{PGL}(2r + 2, K)\theta)\zeta$ is induced by the set of all correlations of $\mathrm{PG}(2r + 1, K)$.

Corollary 3.32.   (i) *For* $n \neq 2r + 1$,

$$|G(\mathcal{G}_{r,n,q})| = |\mathrm{PGL}(n + 1, q)|.$$

(ii) *For* $n = 2r + 1$,

$$|G(\mathcal{G}_{r,2r+1,q})| = 2\,|\mathrm{PGL}(2r + 2, q)|.$$

Proof. This follows immediately from the theorem.   □

Theorem 3.33. *If* δ *is a permutation of* $\mathcal{G}_{r,n}$ *such that both* δ *and* δ^{-1} *fix* $\mathcal{G}_{r,n}^{(1)}$, *then* δ *can be extended to an element of* $G(\mathcal{G}_{r,n})$.

Proof. The permutations δ and δ^{-1} map each s-space Π_s of $\mathcal{G}_{r,n}$ onto an s-space of $\mathcal{G}_{r,n}$. As in the proof of Theorem 3.29, it can be shown that, for $n \neq 2r + 1$, both δ and δ^{-1} leave each of $\mathcal{S}_{\mathrm{L}}$ and $\mathcal{S}_{\mathrm{G}}$ invariant, whereas, for $n = 2r + 1$, both δ and δ^{-1} leave the pair $\{\mathcal{S}_{\mathrm{L}}, \mathcal{S}_{\mathrm{G}}\}$ invariant. Similarly to the proof of Theorem 3.31, when $n \neq 2r + 1$ the permutation δ naturally defines a projectivity ξ of $\mathrm{PG}(n, K)$; also, δ is the restriction of $\xi\theta = \tilde{\xi}$ to $\mathcal{G}_{r,n}$. Thus δ can be extended to $\tilde{\xi}$. When $n = 2r + 1$, either δ or $\delta\zeta^{-1}$ defines the projectivity ξ and is correspondingly the restriction of $\xi\theta = \tilde{\xi}$ to $\mathcal{G}_{r,n}$; here, δ can be extended to one of $\tilde{\xi}$ and $\tilde{\xi}\zeta$. In both cases, δ can be extended to an element of $G(\mathcal{G}_{r,n})$.   □

3.3 A characterisation of Grassmann varieties

In this section it is always assumed that the objects considered are finite. However, all theorems stated can be generalised to the infinite case. The main goal is to characterise the finite Grassmann varieties in terms of their subspaces.

Let $\mathcal{P}$ be a non-empty set whose elements are called *points*, and let $\mathcal{B}$ be a non-empty set consisting of subsets of $\mathcal{P}$. The elements of $\mathcal{B}$ are called *lines*. The pair $(\mathcal{P}, \mathcal{B})$ is a *partial linear space* (PLS) if the following conditions are satisfied:

(1) any two distinct points in $\mathcal{P}$ belong to at most one line in $\mathcal{B}$;
(2) any line in $\mathcal{B}$ contains at least two points of $\mathcal{P}$;
(3) $\mathcal{B}$ is a *covering* of $\mathcal{P}$; that is, $\mathcal{P}$ is the union of all elements of $\mathcal{B}$.

The points P, P' of $\mathcal{P}$ are *collinear* if there is a line l of $\mathcal{B}$ containing P and P'; in this case, write $P \sim P'$. If P and P' are non-collinear, write $P \nsim P'$. Note that $P \sim P$. If P and P' are distinct points of the line l, then l is also denoted by PP'.

Definition 3.34. (1) If any two points of $\mathcal{P}$ are collinear, then $(\mathcal{P}, \mathcal{B})$ is a *linear space* (LS).

(2) Otherwise, $(\mathcal{P}, \mathcal{B})$ is a *proper partial linear space* (PPLS).

(3) A subset $\mathcal{P}'$ of the PLS $(\mathcal{P}, \mathcal{B})$ is a *subspace* if any two of its points are collinear and the line joining them is completely contained in $\mathcal{P}'$.

(4) The subspace $\mathcal{P}'$ is a *maximal subspace* if it is not properly contained in any subspace of $(\mathcal{P}, \mathcal{B})$.

(5) If each line l of the PLS $(\mathcal{P}, \mathcal{B})$ contains at least three points, the space $(\mathcal{P}, \mathcal{B})$ is *irreducible*.

(6) A PPLS $(\mathcal{P}, \mathcal{B})$ is *connected* if for any two points P, P' there exist points $P_1, P_2, \ldots, P_k$ such that $P \sim P_1 \sim P_2 \sim \cdots \sim P_k \sim P'$.

Consider the Grassmann variety $\mathcal{G}_{r,n}$, $1 \leq r \leq n-2$. Let $\mathcal{P} = \mathcal{G}_{r,n}$ and let $\mathcal{B}$ be the set of all lines of $\mathcal{G}_{r,n}$. Then $(\mathcal{P}, \mathcal{B})$ is a PPLS whose subspaces are the subspaces of $\mathrm{PG}(N, q)$ contained in $\mathcal{G}_{r,n}$ and whose maximal subspaces are the maximal spaces of $\mathcal{G}_{r,n}$. Also, $(\mathcal{P}, \mathcal{B})$ is irreducible and connected; see the proof of Lemma 3.30.

Lemma 3.35. *Let $(\mathcal{P}, \mathcal{B})$ be the PPLS corresponding to the Grassmann variety $\mathcal{G}_{r,n}$. Then $(\mathcal{P}, \mathcal{B})$ satisfies the following conditions.*

(i) *If $P, P', P'' \in \mathcal{P}$ with $P \sim P' \sim P'' \sim P$, then there is a subspace containing these points.*

(ii) *The set of maximal subspaces is partitioned into two families $\mathcal{S}_{\mathrm{L}}$ and $\mathcal{S}_{\mathrm{G}}$ with the following further properties.*

   (a) *If $\pi \in \mathcal{S}_{\mathrm{L}}$ and $\pi' \in \mathcal{S}_{\mathrm{G}}$, then $\pi \cap \pi' = \emptyset$ or $\pi \cap \pi' \in \mathcal{B}$.*

   (b) *For each $l \in \mathcal{B}$ there is a unique $\pi \in \mathcal{S}_{\mathrm{L}}$ and a unique $\pi' \in \mathcal{S}_{\mathrm{G}}$ such that $\pi \cap \pi' = l$.*

   (c) *If π, π', π'' are distinct elements of $\mathcal{S}_{\mathrm{L}}$ for which $\pi \cap \pi', \pi' \cap \pi'', \pi'' \cap \pi$ are distinct points, then any element of $\mathcal{S}_{\mathrm{L}}$ other than π, π', π'' having distinct points in common with π and π' also has a point in common with π''.*

   Similarly, if π, π', π'' are distinct elements of $\mathcal{S}_{\mathrm{G}}$ for which the spaces $\pi \cap \pi', \pi' \cap \pi'', \pi'' \cap \pi$ are distinct points, then any element of $\mathcal{S}_{\mathrm{G}}$ other than π, π', π'' having distinct points in common with π and π' also has a point in common with π''.

(iii) *There exist distinct subspaces $\pi_1, \pi_2, \ldots, \pi_{r+1}$ such that $\pi_1 \in \mathcal{B}, \pi_{r+1} \in \mathcal{S}_{\mathrm{G}}$, $\pi_i \subset \pi_{i+1}$ and such that there is no subspace π other than π_i and π_{i+1} with $\pi_i \subset \pi \subset \pi_{i+1}$.*

   Similarly, there exist distinct subspaces $\pi_1, \pi_2, \ldots, \pi_{n-r}$ such that $\pi_1 \in \mathcal{B}$, $\pi_{n-r} \in \mathcal{S}_{\mathrm{L}}, \pi_i \subset \pi_{i+1}$ and such that there is no subspace π other than π_i and π_{i+1} with $\pi_i \subset \pi \subset \pi_{i+1}$.

Proof. Let P, P', P'' be distinct points of $\mathcal{P}$, with $P \sim P' \sim P'' \sim P$. It may be assumed that P, P', P'' are not on a common line of $\mathcal{B}$. The plane $PP'P''$ has three lines in common with each of the quadrics represented by equations (3.5), and hence is contained in all these quadrics. So the plane $PP'P''$ is contained in $\mathcal{G}_{r,n}$. Therefore P, P', P'' lie in a subspace of $(\mathcal{P}, \mathcal{B})$.

Properties (ii)(a) and (ii)(b) are proved in Theorem 3.20 and Lemma 3.21. Let π, π', π'' be distinct elements of $\mathcal{S}_L$ for which $\pi \cap \pi', \pi' \cap \pi'', \pi'' \cap \pi$ are distinct points. The $(r-1)$-spaces of $\mathrm{PG}(n, q)$ which correspond to π, π', π'' are denoted by $\Pi_{r-1}, \Pi'_{r-1}, \Pi''_{r-1}$. Since $\pi \cap \pi', \pi' \cap \pi'', \pi'' \cap \pi$ are distinct points, the spaces $\Pi_{r-1}, \Pi'_{r-1}, \Pi''_{r-1}$ contain a common $(r-2)$-space Π_{r-2}. If the space $\pi''' \in \mathcal{S}_L \backslash \{\pi, \pi', \pi''\}$ has distinct points in common with π and π', then the $(r-1)$-space of $\mathrm{PG}(n, q)$ corresponding to π''' contains Π_{r-2}. Hence π'' and π''' have a point in common. Similarly, let π, π', π'' be distinct elements of $\mathcal{S}_G$ for which $\pi \cap \pi', \pi' \cap \pi'', \pi'' \cap \pi$ are distinct points. The $(r+1)$-spaces of $\mathrm{PG}(n, q)$ which correspond to π, π', π'' are denoted by $\Pi_{r+1}, \Pi'_{r+1}, \Pi''_{r+1}$. Since $\pi \cap \pi', \pi' \cap \pi'', \pi'' \cap \pi$ are distinct points, the spaces $\Pi_{r+1}, \Pi'_{r+1}, \Pi''_{r+1}$ are contained in a common $(r+2)$-space Π_{r+2}. If the space $\pi''' \in \mathcal{S}_G \backslash \{\pi, \pi', \pi''\}$ has distinct points in common with π and π', then the $(r+1)$-space of $\mathrm{PG}(n, q)$ corresponding to π''' is contained in Π_{r+2}. Hence π'' and π''' have a point in common.

Property (iii) follows immediately. $\square$

Let $(\mathcal{P}, \mathcal{B})$ be a connected irreducible PPLS. It is a *Grassmann space of index r*, with $r \geq 1$, if the following axioms are satisfied.

A1. If $P, P', P'' \in \mathcal{P}$ with $P \sim P' \sim P'' \sim P$, then there is a subspace of $(\mathcal{P}, \mathcal{B})$ containing these points.

A2. The set of maximal subspaces of $(\mathcal{P}, \mathcal{B})$ is partitioned into two families, say $\mathcal{S}$ and $\mathcal{T}$, with the following properties.

  I. If $\pi \in \mathcal{S}$ and $\pi' \in \mathcal{T}$, then $\pi \cap \pi' = \emptyset$ or $\pi \cap \pi' \in \mathcal{B}$.

  II. For each $l \in \mathcal{B}$ there is a unique $\pi \in \mathcal{S}$ and a unique $\pi' \in \mathcal{T}$ such that $l \subset \pi$ and $l \subset \pi'$.

  III. Let π, π', π'' be distinct elements of $\mathcal{S}$ for which $\pi \cap \pi', \pi' \cap \pi'', \pi'' \cap \pi$ are distinct points. Then any element of $\mathcal{S} \backslash \{\pi, \pi', \pi''\}$ having distinct points in common with π and π' also has a point in common with π''.

A3. There exist $r+1$ distinct subspaces π_i such that $\pi_1 \subset \pi_2 \subset \cdots \subset \pi_{r+1}$, with $\pi_1 \in \mathcal{B}$, $\pi_{r+1} \in \mathcal{T}$, and such that there is no subspace π with $\pi_i \subset \pi \subset \pi_{i+1}$ other than $\pi = \pi_i, \pi_{i+1}$ for $i = 1, 2, \ldots, r$.

Lemma 3.36. *Let* $(\mathcal{P}, \mathcal{B})$ *be the* PPLS *corresponding to the Grassmann variety* $\mathcal{G}_{r,n}$. *Then* $(\mathcal{P}, \mathcal{B})$ *is a Grassmann space both of index r and of index* $n - r - 1$.

Proof. Putting $\mathcal{S} = \mathcal{S}_L$ and $\mathcal{T} = \mathcal{S}_G$ in Lemma 3.35 shows that $(\mathcal{P}, \mathcal{B})$ is a Grassmann space of index r; putting $\mathcal{S} = \mathcal{S}_G$ and $\mathcal{T} = \mathcal{S}_L$ shows that $(\mathcal{P}, \mathcal{B})$ is a Grassmann space of index $n - r - 1$. $\square$

Definition 3.37. Let $(\mathcal{P}, \mathcal{B})$ and $(\mathcal{P}', \mathcal{B}')$ be two Grassmann spaces.

(1) A bijection ξ from $\mathcal{P}$ to $\mathcal{P}'$ is an *isomorphism* or *collineation* from $(\mathcal{P}, \mathcal{B})$ to $(\mathcal{P}', \mathcal{B}')$ if $\mathcal{B}'$ is the set of all images of the elements of $\mathcal{B}$ under ξ.

(2) In this case, $(\mathcal{P}, \mathcal{B})$ and $(\mathcal{P}', \mathcal{B}')$ are *isomorphic*.

If ξ is an isomorphism from $(\mathcal{P}, \mathcal{B})$ to $(\mathcal{P}', \mathcal{B}')$, then ξ maps the subspaces and maximal subspaces of $(\mathcal{P}, \mathcal{B})$ onto the subspaces and maximal subspaces of $(\mathcal{P}', \mathcal{B}')$.

Let π and π' be distinct maximal subspaces belonging to the same system $\mathcal{D}$ of maximal subspaces of the Grassmann space $(\mathcal{P}, \mathcal{B})$. If $P \in \pi$, $P' \in \pi'$, $P \neq P'$, then by the connectivity of $(\mathcal{P}, \mathcal{B})$ there exist distinct points $P_1 = P, P_2, \ldots, P_k = P'$ such that $P_1 \sim P_2 \sim \cdots \sim P_k$. Let π_i be the element of $\mathcal{D}$ which contains the line $P_i P_{i+1}$, $i = 1, 2, \ldots, k - 1$. By A2.II, $\pi_i = \pi_{i+1}$ or $\pi_i \cap \pi_{i+1}$ is a point, $i = 1, 2, \ldots, k - 1$. If ξ is an isomorphism from $(\mathcal{P}, \mathcal{B})$ to $(\mathcal{P}', \mathcal{B}')$ then $\pi_i \xi = \pi_{i+1} \xi$ or $\pi_i \xi \cap \pi_{i+1} \xi$ is a point. Hence, by A2.I, $\pi_1 \xi, \pi_2 \xi, \ldots, \pi_k \xi$ belong to the same system $\mathcal{D}'$ of maximal subspaces of $(\mathcal{P}', \mathcal{B}')$. Consequently, $\pi \xi$ and $\pi' \xi$ belong to $\mathcal{D}'$. Therefore ξ maps each system of maximal subspaces of $(\mathcal{P}, \mathcal{B})$ onto a system of maximal subspaces of $(\mathcal{P}', \mathcal{B}')$.

Henceforth, it is assumed that $(\mathcal{P}, \mathcal{B})$ is a Grassmann space of index $r \geq 2$.

Lemma 3.38. (i) *No line in $\mathcal{B}$ is a maximal subspace.*

(ii) *Two distinct elements of the same system of maximal subspaces have at most one point in common.*

Proof. Let $l \in \mathcal{B}$ be a maximal subspace, say $l \in \mathcal{S}$. By A2.II, there is a $\pi \in \mathcal{T}$ which contains l. Since $\mathcal{T} \cap \mathcal{S} = \emptyset$, so l is properly contained in π, and hence l is not maximal, a contradiction. Consequently no element of $\mathcal{B}$ is a maximal subspace.

Suppose that π and π' are distinct elements of $\mathcal{T} \cup \mathcal{S}$ which have distinct points P and P' in common. The line $l = PP'$ belongs to exactly one element of $\mathcal{S}$ and to exactly one element of $\mathcal{T}$. Hence π and π' belong to distinct families. $\square$

Lemma 3.39. *Let $\pi, \pi' \in \mathcal{S}$, $\pi \neq \pi'$, and $P \in \pi \cap \pi'$. If $\pi'' \in \mathcal{T}$, and if $\pi \cap \pi'' = l$ and $\pi' \cap \pi'' = l'$ are lines, then $l \cap l' = P$ and so π'' contains P.*

Proof. By Lemma 3.38, $\pi \cap \pi' = P$; so $l \neq l'$. Let $Q \in l$, $Q' \in l'$, with P, Q, Q' distinct; then $P \sim Q \sim Q' \sim P$. By A1, the points P, Q, Q' are contained in a subspace α_1, and α_1 is contained in a maximal subspace α_2. If $\alpha_2 \in \mathcal{S}$ then, by Lemma 3.38, $\pi = \alpha_2 = \pi'$, a contradiction; so $\alpha_2 \in \mathcal{T}$. Since α_2 contains the distinct points Q, Q' of π'', so $\pi'' = \alpha_2$ by Lemma 3.38; thus $P \in \pi''$. Finally, $P \in \pi \cap \pi'' = l$ and $P \in \pi' \cap \pi'' = l'$. $\square$

Lemma 3.40. *Each element of $\mathcal{T}$ provided with its lines has the structure of the points and lines of a projective space.*

Proof. If $\pi \in \mathcal{T}$, then π with its lines is a linear space. It is sufficient to show that in that linear space the Veblen–Wedderburn axiom holds.

Let l_1 and l_2 be distinct lines of π, with $l_1 \cap l_2 = P$; if l_3 and l_4 are distinct lines of π, each meeting both l_1 and l_2 at points other than P, then it must be shown that l_3 and l_4 meet at a point.

Let $l_1 \cap l_3 = P_1$, $l_2 \cap l_3 = P_2$, $l_1 \cap l_4 = Q_1$, $l_2 \cap l_4 = Q_2$. Through l_i there is exactly one maximal subspace $\alpha_i \in \mathcal{S}$, $i = 1, 2, 3, 4$. If $\alpha_i = \alpha_j$, $i \neq j$, then $\alpha_i \cap \pi$ contains all points of $l_i \cup l_j$, in contradiction to A2.I. Hence the spaces $\alpha_1, \alpha_2, \alpha_3, \alpha_4$ are distinct. By Lemma 3.38,

$$\alpha_1 \cap \alpha_2 = P, \quad \alpha_1 \cap \alpha_3 = P_1, \quad \alpha_2 \cap \alpha_3 = P_2,$$
$$\alpha_1 \cap \alpha_4 = Q_1, \quad \alpha_2 \cap \alpha_4 = Q_4.$$

By A2.III, α_3 and α_4 have a common point Q. Since

$$Q \in \alpha_3 \cap \alpha_4, \quad \pi \in \mathcal{T}, \alpha_3 \cap \pi = l_3, \quad \alpha_4 \cap \pi = l_4,$$

so $l_3 \cap l_4 = Q$, by Lemma 3.39.   □

Any projective space belonging to $\mathcal{T}$ contains projective planes. The set of all these projective planes is denoted by $\mathcal{C}$. Then any element of $\mathcal{C}$ is a subspace of $(\mathcal{P}, \mathcal{B})$, and is contained in exactly one element of $\mathcal{T}$.

Lemma 3.41. *Let*

(a) *π and π' be distinct elements of $\mathcal{T}$ which intersect in the point P;*
(b) *$\alpha_1, \alpha_2, \alpha_3$ be distinct elements of $\mathcal{S}$ containing P;*
(c) *$\alpha_i \cap \pi = l_i \in \mathcal{B}$, $\alpha_i \cap \pi' = l_i' \in \mathcal{B}$, $i = 1, 2, 3$;*
(d) *l_1, l_2, l_3 belong to a common plane in π.*

Then l_1', l_2', l_3' also belong to a common plane in π'.

Proof. The lines l_1, l_2, l_3 are distinct, as are l_1', l_2', l_3'. Let Π_2 be the plane containing l_1, l_2, l_3, and let Π_2' be the plane of π' containing l_1' and l_2'. It must be shown that $l_3' \subset \Pi_2'$.

Let l be a line of Π_2 which does not contain P, and let α be the element of $\mathcal{S}$ which contains l. If $\alpha \cap \pi'$ is not a line, then let l' be any line of Π_2' which does not contain P; if $\alpha \cap \pi'$ is a line, then let l' be any line of Π_2' which is distinct from $\alpha \cap \pi'$ and does not contain P.

Let $l \cap l_i = Q_i$, $i = 1, 2, 3$, and let $l' \cap l_i' = Q_i'$, $i = 1, 2$. The five points $Q_1, Q_2, Q_3, Q_1', Q_2'$ are distinct. Let α' be the element of $\mathcal{S}$ which contains l'. Then $\alpha, \alpha', \alpha_1, \alpha_2, \alpha_3$ are distinct, with

$$\alpha \cap \alpha_i = Q_i, i = 1, 2, 3, \quad \alpha' \cap \alpha_i = Q_i', i = 1, 2.$$

Since $\alpha, \alpha_1, \alpha_2$ meet in pairs in distinct points and since α' has distinct points in common with α_1 and α_2, then A2.III implies that α and α' have a point in common.

Let $\alpha \cap \alpha' = Q$; here, Q is distinct from Q_1 and Q_1'. Consequently, the maximal subspaces $\alpha, \alpha', \alpha_1$ meet in pairs in distinct points, and α_3 meets α and α_1 in the distinct points Q_3 and P. Hence, by A2.III, the intersection of α_3 and α' is a point Q'. Since $\alpha_3 \cap \pi' = l_3'$, $\alpha' \cap \pi' = l'$, $\alpha_3 \cap \alpha' = Q'$, by Lemma 3.39, $l' \cap l_3' = Q'$. So l_3' contains distinct points P and Q' of the plane Π_2'; that is, $l_3' \subset \Pi_2'$.   □

Let $P \in \mathcal{P}$ and $\pi \in \mathcal{C}$, with $P \in \pi$. Then the set consisting of all the elements of $\mathcal{S}$ meeting π at lines through P is denoted by $\mathcal{R}(P, \pi)$.

Lemma 3.42. *If $\pi \in \mathcal{C}$ with $P \in \pi$, then $|\mathcal{R}(P, \pi)| \geq 3$. If $\pi, \pi' \in \mathcal{C}$, with $P \in \pi$ and $P' \in \pi'$, then $|\mathcal{R}(P, \pi) \cap \mathcal{R}(P', \pi')| \geq 2$ implies that $\mathcal{R}(P, \pi) = \mathcal{R}(P', \pi')$.*

Proof. Let $\pi \in \mathcal{C}$, $P \in \mathcal{P}$, where $P \in \pi$. Choose distinct lines l_1, l_2, l_3 in π through P. The elements of $\mathcal{S}$ containing l_1, l_2, l_3 are denoted by $\alpha_1, \alpha_2, \alpha_3$; then $\alpha_1, \alpha_2, \alpha_3$ belong to $\mathcal{R}(P, \pi)$. If $\alpha_i = \alpha_j$, $i \neq j$, then $\pi \cap \alpha$, with α the element of $\mathcal{T}$ containing π, contains all points of $l_i \cup l_j$. This contradicts A2.I. Hence $\alpha_i \neq \alpha_j$ for $i \neq j$, and so $|\mathcal{R}(P, \pi)| \geq 3$.

Next, let $\pi, \pi' \in \mathcal{C}$, $P \in \pi$, $P' \in \pi'$, and $|\mathcal{R}(P, \pi) \cap \mathcal{R}(P', \pi')| \geq 2$. Choose distinct elements α_1 and α_2 in $\mathcal{R}(P, \pi) \cap \mathcal{R}(P', \pi')$. The points P and P' belong to both α_1 and α_2. By Lemma 3.38, $P = P'$. If $\pi = \pi'$, then $\mathcal{R}(P, \pi) = \mathcal{R}(P', \pi')$. So assume that $\pi \neq \pi'$. Let α and α' be the elements of $\mathcal{T}$ which contain π and π'. For at least one of α_1, α_2 the lines $\alpha_i \cap \pi$ and $\alpha_i \cap \pi'$ are distinct, say $\alpha_1 \cap \pi \neq \alpha_1 \cap \pi'$. If $\alpha = \alpha'$ then the distinct lines $\alpha_1 \cap \pi, \alpha_1 \cap \pi'$ belong to $\alpha \cap \alpha_1$, in contradiction to A2.I. Hence $\alpha \neq \alpha'$ and $\alpha \cap \alpha' = P$. Now consider a subspace $\alpha_3 \in \mathcal{R}(P, \pi) \backslash \{\alpha_1, \alpha_2\}$. Since $\alpha_3 \cap \alpha' \neq \emptyset$, so it is a line l'. By Lemma 3.41, the lines $\alpha_1 \cap \alpha', \alpha_2 \cap \alpha', l'$ belong to a common plane. The lines $\alpha_1 \cap \alpha', \alpha_2 \cap \alpha'$ belong to π'; hence $l' \subset \pi'$. This means that $\alpha_3 \in \mathcal{R}(P, \pi')$. Consequently $\mathcal{R}(P, \pi) \subset \mathcal{R}(P, \pi')$. Analogously, $\mathcal{R}(P, \pi') \subset \mathcal{R}(P, \pi)$. So $\mathcal{R}(P, \pi') = \mathcal{R}(P, \pi)$, and the theorem is proved. $\square$

The set whose elements are the subsets $\mathcal{R}(P, \pi)$ of $\mathcal{S}$ is denoted by $\mathcal{R}$.

Lemma 3.43. *The pair $(\mathcal{S}, \mathcal{R})$ is a connected irreducible PLS. Also, two elements in $\mathcal{S}$ are collinear if and only if they have a common point in $\mathcal{P}$.*

Proof. Let π and π' be two distinct elements of $\mathcal{S}$. If $\pi \cap \pi' = P$, then let l be a line in π through P. There is a subspace $\alpha \in \mathcal{T}$ through l meeting π' in at least one point. Hence $\alpha \cap \pi'$ is a line l'. Let Π_2 be the plane of $\mathcal{T}$ containing the lines l and l'. Since $\pi, \pi' \in \mathcal{R}(P, \Pi_2)$, so π and π' are collinear in $(\mathcal{S}, \mathcal{R})$. If $\pi \cap \pi' = \emptyset$, then there is no element in $\mathcal{R}$ through π and π'.

Let $\pi \in \mathcal{S}$ and choose a line l in π. Through l there is a maximal subspace $\pi' \in \mathcal{T}$. Let $P \in l$ and let Π_2 be a plane of π' containing l. Then $\pi \in \mathcal{R}(P, \Pi_2)$, and so $\mathcal{R}$ is a covering of $\mathcal{S}$. By Lemma 3.42, any element of $\mathcal{R}$ contains at least three elements of $\mathcal{S}$, and two distinct elements of $\mathcal{S}$ belong to at most one element of $\mathcal{R}$. Hence $(\mathcal{S}, \mathcal{R})$ is an irreducible PLS.

Let π and π' be distinct elements of $\mathcal{S}$. Choose $P \in \pi$ and $P' \in \pi'$. Since $(\mathcal{P}, \mathcal{B})$ is connected, there are points $P = P_1, P_2, \ldots, P_k = P'$ with $P_1 \sim P_2 \sim \cdots \sim P_k$. Let α_i be the element of $\mathcal{S}$ containing the line $P_i P_{i+1}, i = 1, 2, \ldots, k - 1$. Then $\pi \cap \alpha_1 \neq \emptyset$, $\pi' \cap \alpha_{k-1} \neq \emptyset$, and $\alpha_i \cap \alpha_{i+1} \neq \emptyset$, for $i = 1, 2, \ldots, k - 1$. Hence the three pairs $\{\pi, \alpha_1\}, \{\alpha_i, \alpha_{i+1}\}, \{\pi', \alpha_{k-1}\}$ are collinear in $(\mathcal{S}, \mathcal{R})$. This shows that $(\mathcal{S}, \mathcal{R})$ is connected. $\square$

Let $P \in \mathcal{P}$, and let $\mathcal{S}_P$ be the set of all elements of $\mathcal{S}$ containing P. Then $\mathcal{S}_P$ is a subspace of the PLS $(\mathcal{S}, \mathcal{R})$.

Lemma 3.44. *Let P be a point of P and let π be an element of T through P. If π is the projective space* $\text{PG}(s+1,q)$, *then* $\mathcal{S}_P$ *provided with its lines is isomorphic to the linear space formed by the points and lines of* $\text{PG}(s,q)$.

Proof. Let $\mathcal{L}$ be the set of all lines of π through P. By A2.I, each element of $\mathcal{S}_P$ meets π in a line of $\mathcal{L}$. Let ϕ be the mapping

$$\phi : \pi' \in \mathcal{S}_P \mapsto \pi' \cap \pi \in \mathcal{L};$$

then ϕ is a bijection of $\mathcal{S}_P$ onto $\mathcal{L}$.

First, ϕ^{-1} maps each pencil of lines in $\mathcal{L}$ onto a line of $(\mathcal{S},\mathcal{R})$ which is contained in $\mathcal{S}_P$. Conversely, consider a line of the PLS $(\mathcal{S},\mathcal{R})$ which is contained in $\mathcal{S}_P$. Such a line is of type $\mathcal{R}(P,\Pi_2)$. If α_1 and α_2 are distinct elements of $\mathcal{R}(P,\Pi_2)$, then $\alpha_1 \cap \pi$ and $\alpha_2 \cap \pi$ are distinct lines of π, which determine a plane Π_2'. By Lemma 3.42, $\mathcal{R}(P,\Pi_2) = \mathcal{R}(P,\Pi_2')$. Then $\mathcal{R}(P,\Pi_2)\phi = \mathcal{R}(P,\Pi_2')\phi$ consists of all lines of Π_2' through P. Hence $\mathcal{R}(P,\Pi_2)\phi$ is a pencil of lines in $\mathcal{L}$.

Note that $\mathcal{L}$ provided with its pencils of lines is isomorphic to the structure of points and lines of $\text{PG}(s,q)$. Since ϕ is an isomorphism of the linear space formed by $\mathcal{S}_P$ and its lines onto the linear space formed by $\mathcal{L}$ and its pencils, it has been shown that $\mathcal{S}_P$ provided with its lines is isomorphic to the linear space formed by the points and lines of $\text{PG}(s,q)$. □

Lemma 3.45. *Each π in* T *is an* $(r+1)$-*dimensional projective space over the same field* $\mathbf{F}_q$.

Proof. By A3, there is a maximal subspace α in T which is an $(r+1)$-dimensional projective space over some field $\mathbf{F}_q$. Let $\pi \in \mathcal{T}\backslash\{\alpha\}$, and choose a point $P \in \alpha$ and a point $P' \in \pi$, with $P \neq P'$. Since $(\mathcal{P},\mathcal{B})$ is connected, there are distinct points $P = P_1, P_2, \ldots, P_k = P'$ such that $P_i \sim P_{i+1}$ for $i = 1, 2, \ldots, k-1$. Let α_i be the element of T containing $P_i P_{i+1}$; then α_i and α_{i+1} have a common point P_{i+1}. Also, α and α_1 both contain $P = P_1$, and α_{k-1} and π both contain $P' = P_k$. Suppose that α_1 is the projective space $\text{PG}(r'+1,q')$. By Lemma 3.44, $\mathcal{S}_P$ provided with its lines is isomorphic to the linear space formed by the points and lines both of $\text{PG}(r,q)$ and of $\text{PG}(r',q')$. Hence $r = r'$ and $q = q'$. Repeating the argument, it finally follows that π is $\text{PG}(r+1,q)$. □

Corollary 3.46. *Each* $\mathcal{S}_P$, *provided with its lines, is the linear space formed by the points and lines of* $\text{PG}(r,q)$.

Proof. Let $P \in \mathcal{P}$. Choose a line l containing P, and let π be the element of T containing l. By Lemma 3.45, π is the space $\text{PG}(r+1,q)$. Now, by Lemma 3.44, $\mathcal{S}_P$ is the space $\text{PG}(r,q)$. □

Lemma 3.47. *Let* $\alpha_1,\alpha_2,\alpha_3$ *be distinct elements of* S *which are pairwise collinear in* $(\mathcal{S},\mathcal{R})$. *If* $\alpha_1,\alpha_2,\alpha_3$ *contain a common point P of* P, *then there exists a projective plane over* $\mathbf{F}_q$ *in* $(\mathcal{S},\mathcal{R})$ *through them.*

Proof. By Corollary 3.46, $\mathcal{S}_P$ is the projective space $\mathrm{PG}(r, q)$, which contains $\alpha_1, \alpha_2, \alpha_3$ as points. Hence there is a plane over $\mathbf{F}_q$ in $(\mathcal{S}, \mathcal{R})$ which contains $\alpha_1, \alpha_2, \alpha_3$. □

Lemma 3.48. *Let* $\alpha_1, \alpha_2, \alpha_3$ *be elements of* $\mathcal{S}$ *which are pairwise collinear in* $(\mathcal{S}, \mathcal{R})$, *and suppose as subspaces of* $(\mathcal{P}, \mathcal{B})$ *that they do not contain a common point of* $\mathcal{P}$. *Then there exists a projective plane over* $\mathbf{F}_q$ *in* $(\mathcal{S}, \mathcal{R})$ *which contains* $\alpha_1, \alpha_2, \alpha_3$.

Proof. Since $\alpha_1, \alpha_2, \alpha_3$ are distinct and do not belong to a common line of $(\mathcal{S}, \mathcal{R})$, let $\alpha_2 \cap \alpha_3 = P_1$, $\alpha_3 \cap \alpha_1 = P_2$, $\alpha_1 \cap \alpha_2 = P_3$ in $(\mathcal{P}, \mathcal{B})$. As $\alpha_1, \alpha_2, \alpha_3$ do not contain a common point, so P_1, P_2, P_3 are distinct. By A2.II, the lines $P_1 P_2$, $P_2 P_3$, $P_3 P_1$ of $\mathcal{B}$ are distinct. By A1, there is a subspace of $(\mathcal{P}, \mathcal{B})$ which contains P_1, P_2, P_3, and this subspace contains a line of α_i, $i = 1, 2, 3$; hence it is contained in an element π of $\mathcal{T}$. Consequently P_1, P_2, P_3 generate a projective plane Π_2 of π.

Now it is shown that

$$\tilde{\Pi}_2 = \{\pi' \in \mathcal{S} \mid \pi' \cap \Pi_2 \in \mathcal{B}\}$$

is a projective plane of $(\mathcal{S}, \mathcal{R})$ which is isomorphic to Π_2. Let π'_1, π'_2 be any two distinct elements of $\tilde{\Pi}_2$. They meet Π_2 in two lines l_1, l_2 of $\mathcal{B}$ that are distinct; here, let $l_1 \cap l_2 = P$. Hence π'_1 and π'_2 belong to the line $\mathcal{R}(P, \Pi_2)$ in $\mathcal{R}$; also $\mathcal{R}(P, \Pi_2) \subset \tilde{\Pi}_2$. Now, corresponding to each line l in Π_2 there is an element of $\mathcal{S}$ containing l. This gives an isomorphism from the dual of the plane Π_2 to the linear space formed by the elements of $\tilde{\Pi}_2$ and the lines of $(\mathcal{S}, \mathcal{R})$ contained in $\tilde{\Pi}_2$. Hence $\tilde{\Pi}_2$ is a projective plane isomorphic to the dual of Π_2 and so also to $\mathrm{PG}(2, q)$. □

Lemma 3.49. *Each subspace of* $(\mathcal{S}, \mathcal{R})$ *is a projective space over* $\mathbf{F}_q$.

Proof. This follows immediately from the previous two lemmas. □

Lemma 3.50. *Let* P *be a point in* $\mathcal{P}$ *and let* π *be an element of* $\mathcal{S}$ *which does not contain* P. *Then the set*

$$\mathcal{D} = \{\pi' \in \mathcal{S} \mid P \in \pi', \pi \cap \pi' \neq \emptyset\}$$

is either a line in $\mathcal{R}$ *or the empty set.*

Proof. Assume that through P there are two distinct elements of $\mathcal{S}$, say α_1, α_2, both meeting π in points: $\pi \cap \alpha_1 = P_1$, $\pi \cap \alpha_2 = P_2$. By Lemma 3.38, $P_1 \neq P_2$. Since the points P_1, P_2, P are pairwise collinear, there is an element α in $\mathcal{T}$ containing these points. If Π_2 is the plane of α containing P_1, P_2, P, the line $\mathcal{R}(P, \Pi_2)$ in $(\mathcal{S}, \mathcal{R})$ consists of those elements in $\mathcal{S}$ containing P and any point of the line $P_1 P_2$. Now assume that α_3 in $\mathcal{S}$ contains P, that $\alpha_3 \cap \pi \neq \emptyset$, and that $\alpha_3 \notin \mathcal{R}(P, \Pi_2)$. Let $\pi \cap \alpha_3 = P_3$; then $P_3 \notin P_1 P_2$. Again, P, P_1, P_3 are contained in some plane Π'_2 contained in some element of $\mathcal{T}$. The intersection of the planes Π_2 and Π'_2 is the line $P P_1$. Let α and α' be the elements of $\mathcal{T}$ containing Π_2 and Π'_2. By Lemma 3.38,

$\alpha = \alpha'$. Hence the points P_1, P_2, P_3 belong to α, and so the plane $P_1 P_2 P_3$ belongs to $\pi \cap \alpha$, contradicting A2.I. This proves that $\mathcal{D}$ is a line in $\mathcal{R}$.

Next, assume that through P there is at least one element of $\mathcal{S}$, say π', which has a point Q' in common with π. The element of $\mathcal{T}$ containing the line PQ' has a line l in common with π. Let $Q'' \in l \backslash \{Q'\}$. Then $P \sim Q''$, and the element of $\mathcal{S}$ through PQ'' belongs to $\mathcal{D}$. By the first part of the proof, $\mathcal{D}$ is a line in $\mathcal{R}$. So it has been shown that $\mathcal{D} = \emptyset$ or a line in $\mathcal{R}$. $\square$

Corollary 3.51. *The set $(\mathcal{S}, \mathcal{R})$ is a proper* PLS. *More precisely, if $P \in \mathcal{P}, \pi \in \mathcal{S}$, with $P \notin \pi$, then there is at least one element of $\mathcal{S}_P$ which has no point in common with π.*

Proof. Let $\pi \in \mathcal{S}$ and let P be a point not in π. Since $\mathcal{S}_P$ is the projective space $\mathrm{PG}(r, q)$, with $r \geq 2$, and since $\mathcal{D} = \emptyset$ or a line in $\mathcal{S}_P$, there are elements in $\mathcal{S}_P$ which do not belong to $\mathcal{D}$. Hence there are elements through P which have no point in common with π, which means that there exists at least one pair of non-collinear points in $(\mathcal{S}, \mathcal{R})$. $\square$

Lemma 3.52. *For any $P \in \mathcal{P}$, the set $\mathcal{S}_P$ is a maximal subspace of $(\mathcal{S}, \mathcal{R})$.*

Proof. Suppose that $\mathcal{S}_P$ is not maximal. Then there exists an element $\pi \in \mathcal{S}$, with $\pi \notin \mathcal{S}_P$, such that π is collinear in $(\mathcal{S}, \mathcal{R})$ with each element of $\mathcal{S}_P$. By Lemma 3.43, π has a point in common with each element of $\mathcal{S}_P$. By Corollary 3.51, $\mathcal{S}_P$ contains an element which has no point in common with π, a contradiction. $\square$

The family consisting of the maximal subspaces $\mathcal{S}_P$ of $(\mathcal{S}, \mathcal{R})$ is denoted by $\tilde{\mathcal{T}}$. Note that each element of $\tilde{\mathcal{T}}$ is an r-dimensional projective space over $\mathbf{F}_q$.

Lemma 3.53. *Two distinct maximal subspaces in $\tilde{\mathcal{T}}$ have at most one element of $\mathcal{S}$ in common; that is, $|\mathcal{S}_P \cap \mathcal{S}_Q| \leq 1$ for distinct points $P, Q \in \mathcal{P}$.*

Proof. If $P \sim Q$ and $P \neq Q$, then $\mathcal{S}_P \cap \mathcal{S}_Q$ is the unique element of $\mathcal{S}$ containing the line PQ. If $P \not\sim Q$, there is no element of $\mathcal{S}$ containing both P and Q. $\square$

Lemma 3.54. *Let π, π' be distinct elements of $\mathcal{S}$ containing the point P. If α_1 and α_2 are distinct elements of $\mathcal{S} \backslash \{\pi, \pi'\}$ both meeting π and π' at distinct points, then*

(i) *α_1 and α_2 are collinear in $(\mathcal{S}, \mathcal{R})$;*
(ii) *any element $\alpha_3 \in \mathcal{S}$ belonging to the line $\alpha_1 \alpha_2$ of $(\mathcal{S}, \mathcal{R})$ either meets both π and π' in distinct points or belongs to the line $\pi \pi'$ of $(\mathcal{S}, \mathcal{R})$.*

Proof. By A2.III, the maximal subspaces α_1 and α_2 have a point in common; that is, they are collinear in $(\mathcal{S}, \mathcal{R})$. If $\alpha_1 \cap \alpha_2 = Q$, then $P \neq Q$ since $P \notin \alpha_1 \cup \alpha_2$.

If $Q \in \pi$, then any $\alpha_3 \in \mathcal{S}$ belonging to the line $\alpha_1 \alpha_2$ of $(\mathcal{S}, \mathcal{R})$ has a point in common with π. Hence in $(\mathcal{S}, \mathcal{R})$ any such α_3 is collinear with π.

Now let $Q \notin \pi$. By Lemma 3.50, the set of all elements in $\mathcal{S}$ through Q and having a point in common with π is the line $\alpha_1 \alpha_2$ of $(\mathcal{S}, \mathcal{R})$. Hence each element of the line $\alpha_1 \alpha_2$ contains a point of π.

Similarly, each element of the line $\alpha_1 \alpha_2$ contains a point of π'.

Next, assume that the element $\alpha_3 \in \alpha_1 \alpha_2$ contains P. It must be shown that α_3 belongs to the line $\pi\pi'$ of (S, R). So $\alpha_3 \in S_P$ and $\alpha_3 \cap \alpha_1 = Q$. By Lemma 3.50, the set of all elements in S through P and having a point in common with α_1 is the line $\pi\pi'$ of (S, R). Hence $\alpha_3 \in \pi\pi'$ and the result is proved.   □

Let π, π' be distinct elements of S which are collinear in (S, R). Then π and π' have a common point P. Denote by $S(\pi, \pi')$ the set consisting of all elements in S that either belong to the line $\pi\pi'$ of (S, R) or meet both π and π' at points of $P \backslash \{P\}$.

Lemma 3.55. *The set $S(\pi, \pi')$ is a subspace of (S, R), which properly contains the line $\pi\pi'$ of (S, R).*

Proof. Let $\pi \cap \pi' = P$ and let α be an element of T containing P. By the axiom A2.I, $\pi \cap \alpha = l \in B$ and $\pi' \cap \alpha = l' \in B$. Let $Q \in l \backslash \{P\}$ and let $Q' \in l' \backslash \{P\}$; then $Q \sim Q'$. The space $\pi'' \in S$ which contains the line QQ' belongs to $S(\pi, \pi')$, but not to $\pi\pi'$. Hence $S(\pi, \pi')$ properly contains the line $\pi\pi'$ of (S, R).

It must still be shown that $S(\pi, \pi')$ is a subspace of (S, R). Three cases are considered.

(1) Let α_1, α_2 be distinct elements of $S(\pi, \pi')$ which both belong to the line $\pi\pi'$. Then α_1, α_2 are collinear in (S, R) and the line $\alpha_1 \alpha_2 = \pi\pi'$ is completely contained in $S(\pi, \pi')$.

(2) Let α_1, α_2 be distinct elements of $S(\pi, \pi')$, and suppose that $P \notin \alpha_1$ and $P \notin \alpha_2$. By Lemma 3.54, α_1 and α_2 are collinear in (S, R), and the line $\alpha_1 \alpha_2$ is completely contained in $S(\pi, \pi')$.

(3) Let α_1, α_2 be distinct elements of $S(\pi, \pi')$, where $P \notin \alpha_1$ and $\alpha_2 \in \pi\pi'$. By Lemma 3.50, the spaces α_1 and α_2 have a common point Q. Hence α_1 and α_2 are collinear in (S, R). If $Q \in \pi$, then each $\alpha_3 \in \alpha_1 \alpha_2$ contains a point of π; if $Q \notin \pi$, then, by Lemma 3.50, each $\alpha_3 \in \alpha_1 \alpha_2$ contains a point of π. Similarly, each $\alpha_3 \in \alpha_1 \alpha_2$ contains a point of π'. Since α_2 is the only element of S containing P and Q, so all spaces in $\alpha_1 \alpha_2 \backslash \{\alpha_2\}$ belong to $S(\pi, \pi') \backslash \pi\pi'$. Hence the line $\alpha_1 \alpha_2$ of (S, R) is completely contained in $S(\pi, \pi')$.

From (1), (2), (3), it follows that $S(\pi, \pi')$ is a subspace of (S, R).   □

Lemma 3.56. *Each $S(\pi, \pi')$ is a maximal subspace of (S, R).*

Proof. Let $\pi \cap \pi' = P$, and suppose that $S(\pi, \pi')$ is not maximal. Then there exists an element π'' in S not in $S(\pi, \pi')$ such that in (S, R) the element π'' is collinear with each element of $S(\pi, \pi')$. Hence π'' has a point in common with each element of $S(\pi, \pi')$. If $P \notin \pi''$, then π'' meets both π and π' at points of $P \backslash \{P\}$. Hence $\pi'' \in S(\pi, \pi')$, a contradiction; so $P \in \pi''$.

Let $\alpha_1 \in S(\pi, \pi') \backslash \pi\pi'$. Since π'' contains P and $\pi'' \cap \alpha_1 \neq \emptyset$, by Lemma 3.50, $\pi'' \in \pi\pi'$; hence $\pi'' \in S(\pi, \pi')$, again a contradiction.   □

The family consisting of the maximal subspaces $S(\pi, \pi')$ of (S, R) is denoted by $\tilde{S}$. By Lemma 3.55, each element of $\tilde{S}$ properly contains a line of (S, R).

Lemma 3.57. *Let π and π' be two distinct collinear elements of $(\mathcal{S}, \mathcal{R})$, and suppose that $\pi \cap \pi' = P$. Then $\mathcal{S}_P$ and $\mathcal{S}(\pi, \pi')$ are the only maximal subspaces of $(\mathcal{S}, \mathcal{R})$ containing π and π', and $\mathcal{S}_P \cap \mathcal{S}(\pi, \pi') = \pi\pi'$.*

Proof. First, $\pi\pi'$ is contained in both $\mathcal{S}_P$ and $\mathcal{S}(\pi, \pi')$; also, $\mathcal{S}_P \cap \mathcal{S}(\pi, \pi') = \pi\pi'$. Now, let Π be any subspace of $(\mathcal{S}, \mathcal{R})$ through π and π'. It must be shown that $\Pi \subset \mathcal{S}_P$ or $\Pi \subset \mathcal{S}(\pi, \pi')$.

If each element of Π contains P, then $\Pi \subset \mathcal{S}_P$. So, assume that $\pi'' \in \Pi$ and $P \notin \pi''$. Since π'' is collinear with π and π', it meets π and π' at distinct points; hence $\pi'' \in \mathcal{S}(\pi, \pi')$. Each element π_0 of Π which contains P has a point in common with π'' and hence, by Lemma 3.50 belongs to the line $\pi\pi'$ of $(\mathcal{S}, \mathcal{R})$. This means that $\pi_0 \in \mathcal{S}(\pi, \pi')$, and so $\Pi \subset \mathcal{S}(\pi, \pi')$. $\qquad\square$

Corollary 3.58. *The only maximal subspaces of $(\mathcal{S}, \mathcal{R})$ are the elements of $\tilde{\mathcal{S}} \cup \tilde{\mathcal{T}}$.*

Proof. This is immediate from Lemma 3.57. $\qquad\square$

Remark 3.59. From their definitions, $\tilde{\mathcal{S}} \cap \tilde{\mathcal{T}} = \emptyset$.

Lemma 3.60. *If α_1, α_2 are distinct elements of $\mathcal{S}(\pi, \pi')$, then they have a common point, and $\mathcal{S}(\alpha_1, \alpha_2) = \mathcal{S}(\pi, \pi')$.*

Proof. Since $\alpha_1, \alpha_2 \in \mathcal{S}(\pi, \pi')$, they are collinear in $(\mathcal{S}, \mathcal{R})$; so they have a common point Q in $\mathcal{P}$. By Lemma 3.57, $\mathcal{S}_Q$ and $\mathcal{S}(\alpha_1, \alpha_2)$ are the only maximal subspaces containing the line $\alpha_1\alpha_2$ of $(\mathcal{S}, \mathcal{R})$. Since $\mathcal{S}(\pi, \pi')$ is a subspace containing the line $\alpha_1\alpha_2$, so $\mathcal{S}(\pi, \pi') \subset \mathcal{S}_Q$ or $\mathcal{S}(\pi, \pi') \subset \mathcal{S}(\alpha_1, \alpha_2)$. Since not all elements of $\mathcal{S}(\pi, \pi')$ have a common point, so $\mathcal{S}(\pi, \pi') \subset \mathcal{S}(\alpha_1, \alpha_2)$. Since $\mathcal{S}(\pi, \pi')$ is maximal, it follows that $\mathcal{S}(\alpha_1, \alpha_2) = \mathcal{S}(\pi, \pi')$. $\qquad\square$

Lemma 3.61. *If β, β', β'' are distinct elements in $\tilde{\mathcal{S}}$ with the intersections*

$$\beta \cap \beta' = \{\pi''\}, \ \beta' \cap \beta'' = \{\pi\}, \ \beta'' \cap \beta = \{\pi'\},$$

where $\pi \neq \pi'$, then π, π', π'' have a point of $\mathcal{P}$ in common.

Proof. If $\pi = \pi''$, then also $\pi = \pi'$, a contradiction. Hence $\pi \neq \pi''$ and analogously $\pi' \neq \pi''$. The spaces π, π', π'' are collinear in pairs in $(\mathcal{S}, \mathcal{R})$. By Lemmas 3.47 and 3.48, they are contained in a subspace γ of $(\mathcal{S}, \mathcal{R})$. Denote by β_0 the maximal subspace of $(\mathcal{S}, \mathcal{R})$ which contains γ. Since $\pi, \pi' \in \beta_0 \cap \beta''$, so $\beta_0 \in \tilde{\mathcal{T}}$ by Lemma 3.57. Hence there is some point $P \in \mathcal{P}$ for which $\beta_0 = \mathcal{S}_P$. It follows that π, π', π'' all contain the point P. $\qquad\square$

Lemma 3.62. *The pair $(\mathcal{S}, \mathcal{R})$ is a Grassmann space of index $r - 1$.*

Proof. From Lemma 3.43 and Corollary 3.51, it follows that $(\mathcal{S}, \mathcal{R})$ is a connected irreducible PPLS.

Let π, π', π'' be pairwise collinear elements of $(\mathcal{S}, \mathcal{R})$. By Lemmas 3.47 and 3.48, they are contained in a subspace of $(\mathcal{S}, \mathcal{R})$. This means that A1 is satisfied.

The set of all maximal subspaces of $(\mathcal{S}, \mathcal{R})$ is partitioned into the families $\tilde{\mathcal{S}}$ and $\tilde{\mathcal{T}}$. Consider an element $\beta \in \tilde{\mathcal{T}}$ and an element $\beta' \in \tilde{\mathcal{S}}$. Then $\beta = \mathcal{S}_P$ for some point $P \in \mathcal{P}$. Take an element $\pi \in \mathcal{S}_P \cap \beta'$ and let $\pi' \in \beta' \backslash \{\pi\}$. By Lemma 3.60, $\beta' = \mathcal{S}(\pi, \pi')$. If $\pi \cap \pi' = P$, then Lemma 3.57 implies that $\mathcal{S}_P \cap \mathcal{S}(\pi, \pi') = \pi\pi'$. Now assume that $P \notin \pi'$. By Lemma 3.50, the set of all elements of $\mathcal{S}_P$ which have a point in common with π' is a line l in $\mathcal{R}$. Take π'' in $l \backslash \{\pi\}$; then $\pi'' \in \mathcal{S}(\pi, \pi')$. By Lemma 3.60, $\mathcal{S}(\pi, \pi') = \mathcal{S}(\pi, \pi'')$. Again, by Lemma 3.57, $\mathcal{S}_P \cap \mathcal{S}(\pi, \pi'') = \pi\pi''$. Hence, if $\beta \cap \beta' \neq \emptyset$, then $\beta \cap \beta'$ is a line of $(\mathcal{S}, \mathcal{R})$. This shows that A2.I is satisfied.

Let l be a line of $(\mathcal{S}, \mathcal{R})$. Suppose that $\pi, \pi' \in l$, $\pi \neq \pi'$, and let $\pi \cap \pi' = P$. By Lemma 3.57, $\mathcal{S}_P$ in $\tilde{\mathcal{T}}$ and $\mathcal{S}(\pi, \pi')$ in $\tilde{\mathcal{S}}$ are the only maximal subspaces of $(\mathcal{S}, \mathcal{R})$ containing l. Consequently A2.II is satisfied.

Next, let β, β', β'' be distinct elements in $\tilde{\mathcal{S}}$ for which

$$\beta \cap \beta' = \{\pi''\}, \ \beta' \cap \beta'' = \{\pi\}, \ \beta'' \cap \beta = \{\pi'\}$$

are distinct elements of $\mathcal{S}$. Now consider an element $\beta_0 \in \tilde{\mathcal{S}} \backslash \{\beta, \beta', \beta''\}$ having distinct elements in common with β and β'. Let $\beta_0 \cap \beta = \{\alpha\}$ and $\beta_0 \cap \beta' = \{\alpha'\}$. It is now shown that $\beta_0 \cap \beta'' \neq \emptyset$.

From Lemma 3.61, π, π', π'' contain a common point P in $\mathcal{P}$ and α, α', π'' contain a common point Q in $\mathcal{P}$.

First assume that $P = Q$. Since $\beta \cap \mathcal{S}_P = \emptyset$ or a line of $(\mathcal{S}, \mathcal{R})$, the elements π'', π', α are collinear in $(\mathcal{S}, \mathcal{R})$; similarly, π'', π, α' are collinear in $(\mathcal{S}, \mathcal{R})$. Hence, in the projective space $\mathcal{S}_P$, the plane $\pi\pi'\pi''$ also contains α and α'. Consequently, in $(\mathcal{S}, \mathcal{R})$, the lines $\alpha\alpha'$ and $\pi\pi'$ meet at an element $\alpha'' \in \mathcal{S}$. However, $\alpha\alpha' \subset \beta_0$ and $\pi\pi' \subset \beta''$, so that $\beta_0 \cap \beta'' \neq \emptyset$.

Next, assume that $P \neq Q$, and so $\alpha \neq \pi'$. In $(\mathcal{S}, \mathcal{R})$, the elements α and π' are collinear; so $\alpha \cap \pi' = P'$ for some $P' \in \mathcal{P}$. Similarly, $\alpha' \neq \pi$ and $\alpha' \cap \pi = Q'$. Now, $P \neq Q$ implies that the points P, Q, P', Q' are distinct. In $(\mathcal{P}, \mathcal{B})$, the points P, Q, P' are pairwise collinear; hence they are contained in a maximal subspace τ of $(\mathcal{S}, \mathcal{R})$. Since PQ and PP' are contained in the distinct elements π'' and π' of $\mathcal{S}$, so $\tau \in \mathcal{T}$. Similarly, the points P, Q, Q' are contained in a maximal subspace τ' in $\mathcal{T}$. The subspaces τ and τ' have at least two points in common, and so they coincide. Hence $P, Q, P', Q' \in \mathcal{T}$; so the points P' and Q' are collinear. Let π_0 be the element of $\mathcal{S}$ which contains P' and Q'. Since P' belongs to π_0 and α, and since Q' belongs to π_0 and α', so $\pi_0 \in \mathcal{S}(\alpha, \alpha') = \beta_0$. Since P' belongs to π_0 and π', and Q' belongs to π_0 and π, so $\pi_0 \in \mathcal{S}(\pi, \pi') = \beta''$. Consequently, $\beta_0 \cap \beta'' \neq \emptyset$. Since $\beta_0 \cap \beta'' \neq \emptyset$ in both cases, so A2. III is satisfied.

By Corollary 3.46, each $\mathcal{S}_P$ in $\tilde{\mathcal{T}}$ is an r-dimensional projective space over $\mathbf{F}_q$. This means that, for each $\mathcal{S}_P$ in $\tilde{\mathcal{T}}$, there exist distinct subspaces $\beta_1, \beta_2, \ldots, \beta_r$ of $(\mathcal{S}, \mathcal{R})$ such that β_1 is a line, that $\beta_1 \subset \beta_2 \subset \cdots \subset \beta_r$ and that there is no subspace β with $\beta_i \subsetneq \beta \subsetneq \beta_{i+1}$ for $i = 1, 2, \ldots, r - 1$. This proves that A3 is satisfied.

Thus $(\mathcal{S}, \mathcal{R})$ is a Grassmann space of index $r - 1$. $\qquad \square$

Lemma 3.63. *If $(\mathcal{S}, \mathcal{R})$ is isomorphic to the* PPLS *corresponding to the Grassmann variety $\mathcal{G}_{r-1,n}$, then $(\mathcal{P}, \mathcal{B})$ is isomorphic to the* PPLS *corresponding to the Grassmann variety $\mathcal{G}_{r,n}$.*

Proof. By hypothesis there exists a bijection ξ of $\mathcal{S}$ onto $\mathcal{G}_{r-1,n}$ such that the set of all lines of $\mathcal{G}_{r-1,n}$ consists of all images under ξ of the elements of $\mathcal{R}$. Then ξ maps the maximal subspaces of $(\mathcal{S}, \mathcal{R})$ onto the maximal subspaces of $\mathcal{G}_{r-1,n}$. By the observation preceding Lemma 3.38, ξ maps each of $\tilde{\mathcal{S}}$ and $\tilde{\mathcal{T}}$ onto a system of maximal spaces of $\mathcal{G}_{r-1,n}$.

First let $n \neq 2r-1$. Then the elements of $\mathcal{S}_L$, namely the Latin spaces of $\mathcal{G}_{r-1,n}$, and the elements of $\mathcal{S}_G$, the Greek spaces of $\mathcal{G}_{r-1,n}$, have different dimensions. By Corollary 3.46, the elements of $\tilde{\mathcal{T}}$ have dimension r, which is the dimension of the elements of $\mathcal{S}_G$.

Next, let $n = 2r - 1$. In Section 3.2, it was shown that $G(\mathcal{G}_{r-1,2r-1})$ contains elements interchanging $\mathcal{S}_L$ and $\mathcal{S}_G$. It follows that also in this case it may be assumed that ξ maps the elements of $\tilde{\mathcal{T}}$ onto the elements of $\mathcal{S}_G$ and the elements of $\tilde{\mathcal{S}}$ onto the elements of $\mathcal{S}_L$. As in Section 3.2, the mapping which associates the points of $\mathcal{G}_{r-1,n}$ to the elements of $\mathrm{PG}^{(r-1)}(n,q)$ is denoted by $\mathfrak{G}$.

Let $\Pi_r \in \mathrm{PG}^{(r)}(n,q)$, and let $\mathcal{R}^{r-1}(\Pi_r)$ be the set of all $(r-1)$-dimensional subspaces of Π_r. Then

$$\mathcal{R}^{r-1}(\Pi_r) \subset \mathrm{PG}^{(r-1)}(n,q) \text{ and } \mathcal{R}^{r-1}(\Pi_r)\,\mathfrak{G}\,\xi^{-1} = \mathcal{S}_P \in \mathcal{T}.$$

Now consider the mapping

$$\psi : \mathrm{PG}^{(r)}(n,q) \to \mathcal{P},$$

defined by

$$\Pi_r\psi = P \iff \mathcal{R}^{r-1}(\Pi_r)\,\mathfrak{G}\,\xi^{-1} = \mathcal{S}_P.$$

Then ψ is a bijection of $\mathrm{PG}^{(r)}(n,q)$ onto $\mathcal{P}$.

Let Π_r, Π_r', Π_r'' be elements of $\mathrm{PG}^{(r)}(n,q)$, with at least two of them distinct, and let $\Pi_r\psi = P$, $\Pi_r'\psi = P'$, $\Pi_r''\psi = P''$. Then the following are equivalent:

(a) $\Pi_r \cap \Pi_r' \cap \Pi_r''$ is an element Π_{r-1} of $\mathrm{PG}^{(r-1)}(n,q)$;
(b) $\mathcal{R}^{(r-1)}(\Pi_r) \cap \mathcal{R}'^{(r-1)}(\Pi_r) \cap \mathcal{R}''^{(r-1)}(\Pi_r) = \{\Pi_{r-1}\}$;
(c) $(\mathcal{R}^{(r-1)}(\Pi_r)\,\mathfrak{G}\,\xi^{-1}) \cap (\mathcal{R}'^{(r-1)}(\Pi_r)\,\mathfrak{G}\,\xi^{-1}) \cap (\mathcal{R}''^{(r-1)}(\Pi_r)\,\mathfrak{G}\,\xi^{-1})$

$$= \{\Pi_{r-1}\,\mathfrak{G}\,\xi^{-1}\};$$

(d) $\mathcal{S}_P \cap \mathcal{S}_{P'} \cap \mathcal{S}_{P''} = \{\pi\}$ with $\pi = \Pi_{r-1}\,\mathfrak{G}\,\xi^{-1}$ in $\mathcal{S}$;
(e) $P, P', P'' \in \pi$.

Hence $\Pi_r \cap \Pi_r' \cap \Pi_r''$ is an element Π_{r-1} of $\mathrm{PG}^{(r-1)}(n,q)$ if and only if P, P', P'' belong to a common element of $\mathcal{S}$.

Let Π_r and Π_r' be distinct elements of $\mathrm{PG}^{(r)}(n,q)$, with $\Pi_r\psi = P$, $\Pi_r'\psi = P'$. If $\Pi_r \cap \Pi_r' \in \mathrm{PG}^{(r-1)}(n,q)$, then by the preceding paragraph the points P and P' belong to a common element of $\mathcal{S}$, and hence are collinear in $(\mathcal{P}, \mathcal{B})$. Conversely, if P and P' are collinear, then the line PP' belongs to an element of $\mathcal{S}$, and hence $\Pi_r \cap \Pi_r' \in \mathrm{PG}^{(r-1)}(n,q)$.

Assume again that $\Pi_r \cap \Pi_r' = \Pi_{r-1} \in \mathrm{PG}^{(r-1)}(n,q)$. Then each element of $\mathrm{PG}^{(r)}(n,q)$ through Π_{r-1} belongs to the unique π in $\mathcal{S}$ containing $P = \Pi_r\psi$ and

$P' = \Pi'_r \psi$. Conversely, if $P'' \in \pi$, then $P'' \psi^{-1}$ contains Π_{r-1}. So all elements of $\mathrm{PG}^{(r)}(n, q)$ through Π_{r-1} are mapped by ψ onto all points of π. It follows from the preceding paragraph that each π in $\mathcal{S}$ corresponds in this way to a Π_{r-1} in $\mathrm{PG}^{(r-1)}(n, q)$.

Next, let $\alpha \in \mathcal{T}$ and let P, P', P'' be three points of α which do not belong to a common line of $(\mathcal{P}, \mathcal{B})$. Since P, P', P'' are pairwise collinear, the spaces $P\psi^{-1} = \Pi_r, P'\psi^{-1} = \Pi'_r, P''\psi^{-1} = \Pi''_r$ intersect in pairs at some element of $\mathrm{PG}^{(r-1)}(n, q)$. If Π_r, Π'_r, Π''_r contain a common element of $\mathrm{PG}^{(r-1)}(n, q)$, then P, P', P'' belong to a common π in $\mathcal{S}$. Hence $\alpha \cap \pi$ contains three non-collinear points, a contradiction. Consequently Π_r, Π'_r, Π''_r do not contain a common element Π_{r-1}, whence they all lie in some Π_{r+1} in $\mathrm{PG}^{(r+1)}(n, q)$.

Now, let Q be any point of α, and suppose that P, P', P'' are three points of α that do not belong to a common line of $(\mathcal{P}, \mathcal{B})$. Then the r-spaces $\Pi_r, \Pi'_r, \Lambda_r = Q\psi^{-1}$ belong to a common $(r + 1)$-space; hence $\Lambda_r \subset \Pi_{r+1}$. Conversely, let Λ_r be any r-space contained in Π_{r+1}. Since $\Pi_r = \Lambda_r$ or $\Pi_r \cap \Lambda_r$ is an $(r-1)$-space, the points P and $Q = \Lambda_r \psi$ are collinear.

Similarly Q and P' are collinear as are Q and P''. If $Q \notin \alpha$, then the maximal subspaces of $(\mathcal{P}, \mathcal{B})$ containing $PP'Q$, $P'P''Q$, $PP''Q$ are elements of $\mathcal{S}$, which gives a contradiction; hence $Q \in \alpha$. So it has been shown that all elements of $\mathrm{PG}^{(r)}(n, q)$ contained in Π_{r+1} are mapped by ψ onto all points of α.

Consider now any Π'_{r+1} in $\mathrm{PG}^{(r+1)}(n, q)$. If $\Pi^1_r, \Pi^2_r \subset \Pi'_{r+1}$, with $\Pi^1_r \neq \Pi^2_r$, then $\Pi^1_r \psi = P_1$ and $\Pi^2_r \psi = P_2$ are collinear. The points of the element of $\mathcal{T}$ through the line $P_1 P_2$ are mapped by ψ^{-1} onto the r-spaces belonging to the $(r + 1)$-space Π'_{r+1} which contains Π^1_r and Π^2_r. Hence each $(r+1)$-space of $\mathrm{PG}(n, q)$ corresponds to some element of $\mathcal{T}$.

Next, let l be a line of $(\mathcal{P}, \mathcal{B})$. Then l is contained in a unique element π of $\mathcal{S}$ and a unique element α of $\mathcal{T}$; also $\pi \cap \alpha = l$. So $\pi \psi^{-1} \cap \alpha \psi^{-1} = l \psi^{-1}$. Since $l \psi^{-1} \neq \emptyset$, so it is the pencil (Π_{r-1}, Π_{r+1}) of r-spaces, with Π_{r-1} the $(r - 1)$-space corresponding to π and Π_{r+1} the $(r + 1)$-space corresponding to α. Conversely, consider a pencil (Π_{r-1}, Π_{r+1}) of r-spaces in $\mathrm{PG}(n, q)$. If π is the element of $\mathcal{S}$ corresponding to Π_{r-1} and α is the element of $\mathcal{T}$ corresponding to Π_{r+1}, then the image $(\Pi_{r-1}, \Pi_{r+1})\psi = \pi \cap \alpha$. Since $(\Pi_{r-1}, \Pi_{r+1})\psi \neq \emptyset$, so $\pi \cap \alpha$ is a line l of $(\mathcal{P}, \mathcal{B})$.

It has been shown that ψ is a bijection of $\mathrm{PG}^{(r)}(n, q)$ onto $\mathcal{P}$ such that $\mathcal{B}$ is the set of all images of the pencils of r-spaces of $\mathrm{PG}(n, q)$. In other words, the Grassmann space $(\mathcal{P}, \mathcal{B})$ is isomorphic to the PPLS defined by the Grassmann variety $\mathcal{G}_{r,n}$. □

Theorem 3.64.   (i) *In a Grassmann space $(\mathcal{P}, \mathcal{B})$ of index $r = 1$, any two distinct elements of $\mathcal{S}$ have exactly one point in common.*

(ii) *If $(\mathcal{P}, \mathcal{B})$ is a connected irreducible PPLS satisfying A1, A2.I, A2.II, and*

   A2.III$'$: *any two distinct elements of $\mathcal{S}$ have exactly one point in common,*

then $(\mathcal{P}, \mathcal{B})$ is a Grassmann space of index 1.

Proof. Let $(\mathcal{P}, \mathcal{B})$ be a Grassmann space of index $r = 1$. Then Lemmas 3.38 to 3.45 hold, so that each element of $\mathcal{T}$ is a projective plane over $\mathbf{F}_q$. Let π and π' be distinct elements of $\mathcal{S}$, and suppose that $\pi \cap \pi' = \emptyset$. Let k be the minimum number for which there exist distinct points $P_1, P_2, \dots, P_k$, with $P_1 \in \pi$, $P_k \in \pi'$, and $P_1 \sim P_2 \sim \cdots \sim P_k$. Then $P_2, P_3, \dots, P_{k-1} \notin \pi \cap \pi'$. Note that k exists by the connectivity of $(\mathcal{P}, \mathcal{B})$. Assume that $k > 2$. Let π'' be the element of $\mathcal{S}$ which contains the line $P_2 P_3$, and let α be the element of $\mathcal{T}$ which contains the line $P_1 P_2$. Since $\pi \cap \alpha \neq \emptyset$, so $\pi \cap \alpha$ is a line l of $(\mathcal{P}, \mathcal{B})$. Similarly $\pi'' \cap \alpha$ is a line l''. The space α is a projective plane; hence l and l'' have a common point P. The points $P, P_3, P_4, \dots, P_k$ are distinct, and $P \sim P_3 \sim P_4 \sim \cdots \sim P_k$ with $P \in \pi$. This contradicts the assumption on the minimality of k. Hence $k = 2$, which means that there are points P_1 and P_2, with $P_1 \in \pi$, $P_2 \in \pi'$ and $P_1 \sim P_2$.

Let α' be the element of $\mathcal{T}$ which contains the line $P_1 P_2$. Since $\pi \cap \alpha' \neq \emptyset$, so $\pi \cap \alpha'$ is a line m. Similarly $\pi' \cap \alpha'$ is a line m'. Since α' is a projective plane, the lines m and m' have a point in common, whence $\pi \cap \pi' \neq \emptyset$, a contradiction. Therefore any two distinct elements of $\mathcal{S}$ have exactly one point in common.

Next, let $(\mathcal{P}, \mathcal{B})$ be a connected irreducible PPLS satisfying A1, A2.I, A2.II and A2.III$'$. Then A2.III is trivially satisfied. Lemmas 3.38 to 3.45 are satisfied for a certain $r \geq 1$. Hence $(\mathcal{P}, \mathcal{B})$ is a Grassmann space of index r. If $r \geq 2$, then, by Corollary 3.51, there exist disjoint elements in $\mathcal{S}$, in contradiction to A2.III$'$. This gives the conclusion that $r = 1$. $\square$

Theorem 3.65. *Any Grassmann space of index $r \geq 1$ is isomorphic to the PPLS defined by some Grassmann variety $\mathcal{G}_{r,n}$.*

Proof. Let $(\mathcal{P}, \mathcal{B})$ be a Grassmann space of index r, where $r > 2$. Assume that each Grassmann space of index $r - 1$ is isomorphic to the PPLS defined by some Grassmann variety $\mathcal{G}_{r-1,n}$.

Then, by Lemmas 3.62 and 3.63, the space $(\mathcal{P}, \mathcal{B})$ is isomorphic to the PPLS defined by the Grassmann variety $\mathcal{G}_{r,n}$. So it is only necessary to show that any Grassmann space of index 1 is isomorphic to the PPLS defined by some $\mathcal{G}_{1,n}$.

Let $(\mathcal{P}, \mathcal{B})$ be a Grassmann space of index 1. By Theorem 3.64, any two distinct elements of $\mathcal{S}$ have exactly one point in common. Since Lemmas 3.38 to 3.45 hold, each element of $\mathcal{T}$ is a projective plane over $\mathbf{F}_q$. For any P in $\mathcal{P}$, let $\mathcal{S}_p$ denote the set of all elements of $\mathcal{S}$ through P. Further, let $\tilde{\mathcal{T}} = \{\mathcal{S}_p \mid P \in \mathcal{P}\}$. Consider a point P of $\mathcal{P}$ and a plane α of $\mathcal{T}$ through P. Through each line in α through P, there is a unique element of $\mathcal{S}_p$; by A2.I, $q + 1$ distinct elements of $\mathcal{S}_p$ are obtained in this way. Hence $(\mathcal{S}, \tilde{\mathcal{T}})$ is an irreducible linear space, and it is now shown that it is isomorphic to the linear space formed by the points and lines of some $\mathrm{PG}(n, q)$. It is sufficient to show that the Veblen–Wedderburn axiom holds in $(\mathcal{S}, \tilde{\mathcal{T}})$.

Let π, π', π'' be distinct elements of $\mathcal{S}$, let

$$\pi \cap \pi' = P'', \quad \pi' \cap \pi'' = P, \quad \pi'' \cap \pi = P',$$

and assume that P, P', P'' are distinct. Further, let α and α' be distinct elements of $\mathcal{S} \setminus \{\pi''\}$, let $\alpha \cap \pi'' = P'$, $\alpha' \cap \pi'' = P$, $\alpha \cap \alpha' = Q$, and assume again that P, P', Q

are distinct. It must be shown that P'' and Q belong to a common element of $\mathcal{S}$; that is, $P'' \sim Q$. By A1, the pairwise collinear points P, P', P'' belong to a maximal space π_1, which by A2.II is an element of $\mathcal{T}$. Similarly the pairwise collinear points P, P', Q belong to a maximal space π_2 in $\mathcal{T}$. As the line PP' is contained in both π_1 and π_2, so $\pi_1 = \pi_2$ by A2.II. Hence $P'' \sim Q$, and consequently $(\mathcal{S}, \tilde{\mathcal{T}})$ is isomorphic to the linear space formed by the points and lines of some $PG(n, q)$. Since the partial linear space $(\mathcal{P}, \mathcal{B})$ is proper, there exist disjoint lines in $(\mathcal{S}, \tilde{\mathcal{T}})$, and so $n \geq 3$.

Now consider the mapping

$$\Phi : \tilde{\mathcal{T}} \to \mathcal{P}, \quad \mathcal{S}_P \mapsto P.$$

Then Φ is a bijection of $\tilde{\mathcal{T}}$ onto $\mathcal{P}$. Let δ be a plane in $(\mathcal{S}, \tilde{\mathcal{T}})$ and let π, π', π'' be any three independent points of δ. Further, let

$$\pi \cap \pi' = P'', \quad \pi' \cap \pi'' = P, \quad \pi'' \cap \pi = P'.$$

Then the points P, P', P'' are distinct. Since they are pairwise collinear in $(\mathcal{P}, \mathcal{B})$, there exists a maximal subspace α containing them; by A2.II, this subspace belongs to $\mathcal{T}$. Let β be an element of $\mathcal{S}$ containing a line l of α. Suppose that l is distinct from the line PP'. If $PP' \cap l = Q$, then $\mathcal{S}_Q$ contains π'' and β, and has an element in common with $\mathcal{S}_{P''}$. But $\mathcal{S}_{P''}$ is the line $\pi\pi'$ of $(\mathcal{S}, \tilde{\mathcal{T}})$, and hence β belongs to the plane δ.

Conversely, let β be an element of the plane δ. Suppose, for example, that $\beta \neq \pi''$; so let $\beta \cap \pi'' = Q$. Then the lines $\mathcal{S}_{P''}$ and $\mathcal{S}_Q$ of $(\mathcal{S}, \tilde{\mathcal{T}})$ have an element in common; hence $P'' \sim Q$. Suppose that $Q \notin \alpha$. The pairwise collinear and distinct points P, P'', Q are contained in a maximal subspace α', which belongs to $\mathcal{T}$. Since α and α' share the line PP'', so $\alpha = \alpha'$. Hence $Q \in \alpha$, a contradiction. Consequently $Q \in \alpha$. From A2.I, $\beta \cap \alpha$ is a line of $(\mathcal{P}, \mathcal{B})$. Therefore it has been shown that δ consists of all elements of $\mathcal{S}$ meeting α at a line of $(\mathcal{P}, \mathcal{B})$. As the lines of δ are the elements $\mathcal{S}_P$ with $P \in \alpha$, it follows that, for any plane α'' of $\mathcal{T}$, the set of all spaces in $\mathcal{S}$ having a line in common with α'' is a plane of $(\mathcal{S}, \tilde{\mathcal{T}})$.

Let $\pi \in \mathcal{S}$, let δ be a plane of $(\mathcal{S}, \tilde{\mathcal{T}})$, and assume that $\pi \in \delta$. From the preceding paragraph, δ consists of all elements of $\mathcal{S}$ containing a line of some plane α of $\mathcal{T}$. The lines of the pencil (π, δ) of $(\mathcal{S}, \tilde{\mathcal{T}})$ are the elements $\mathcal{S}_P$ with P in $\pi \cap \alpha = l$; hence $(\pi, \delta)\Phi = l$. Conversely, consider any l' in $\mathcal{B}$. Let $l' \subset \pi' \in \mathcal{S}$ and $l' \subset \alpha' \in \mathcal{T}$. Then $l'\Phi^{-1}$ is the pencil (π', δ') with δ' the plane of $(\mathcal{S}, \tilde{\mathcal{T}})$ which consists of all elements of $\mathcal{S}$ having a line in common with α'.

It has been shown that the Grassmann space $(\mathcal{P}, \mathcal{B})$ of index 1 is isomorphic to the PPLS defined by the Grassmann variety $\mathcal{G}_{1,n}$. $\square$

To end this section, some examples of proper partial linear spaces are given; these show that none of the axioms A2.I, A2.II, A2.III, nor the conditions of connectivity or irreducibility, can be deleted in the characterisation of Grassmann varieties.

Example 3.66. Let $\mathcal{P}$ consist of all 2-subsets of the set $\{1, 2, 3, 4\}$; let $\mathcal{B}$ consist of the elements of the form $\{\{a, b\}, \{a, c\}\}$ with a, b, c distinct. Then $(\mathcal{S}, \mathcal{B})$ is a connected PPLS which satisfies A1, A2.I, A2.II, A2.III. However, it is not irreducible.

Example 3.67. Let $(\mathcal{P}, \mathcal{B})$ and $(\mathcal{P}', \mathcal{B}')$ be Grassmann spaces with $\mathcal{P} \cap \mathcal{P}' = \emptyset$. Then $(\mathcal{P} \cup \mathcal{P}', \mathcal{B} \cup \mathcal{B}')$ is an irreducible PPLS which satisfies A1, A2.I, A2.II, A2.III, but which is not connected.

Example 3.68. Let $(\mathcal{P}, \mathcal{B})$ be a Grassmann space with lines having at least four points. Let $l \in \mathcal{B}$, and let $(l, \mathcal{L})$ be an irreducible linear space with $|\mathcal{L}| > 1$. Then $(\mathcal{P}, (\mathcal{B} \backslash \{l\}) \cup \mathcal{L})$ is a connected irreducible PPLS which satisfies A1, A2.II, A2.III, but not A2.I.

Example 3.69. Let $(\mathcal{P}, \mathcal{B})$ be a Grassmann space. If the elements of $\mathcal{S}$ are projective spaces of dimension s over $\mathbf{F}_q$, then embed one element Π_s of $\mathcal{S}$ in a projective space Π_{s+1} of dimension $s+1$ over $\mathbf{F}_q$ for which $\Pi_{s+1} \cap \mathcal{P} = \Pi_s$. Let $\mathcal{L}$ be the set of all lines of Π_{s+1}. Then $(\mathcal{P} \cup \Pi_{s+1}, \mathcal{B} \cup \mathcal{L})$ is a connected irreducible PPLS which satisfies A1, A2.I, A2.III, but not A2.II.

Example 3.70. Let $(\mathcal{P}, \mathcal{B})$ be a Grassmann space with lines having at least four points. Choose a point P in $\mathcal{P}$ and let $\mathcal{B}' = \{l \backslash \{P\} \mid l \in \mathcal{B}\}$, $\mathcal{P}' = \mathcal{P} \backslash \{P\}$. Then $(\mathcal{P}', \mathcal{B}')$ is a connected irreducible PPLS which satisfies A1, A2.I, A2.II; for neither of the systems of maximal subspaces is A2.III satisfied.

It is not known whether or not A1 can be deleted in the characterisation of Grassmann varieties.

3.4 Embedding of Grassmann spaces

Let $(\mathcal{P}, \mathcal{B})$ be a Grassmann space. If $\mathcal{P}$ is a point set of $\mathrm{PG}(n, q)$ and $\mathcal{B}$ is a line set of $\mathrm{PG}(n, q)$, then $(\mathcal{P}, \mathcal{B})$ is *embedded* in $\mathrm{PG}(n, q)$. Grassmann varieties are examples of embedded Grassmann spaces. In this section all embedded Grassmann spaces are determined.

First it is shown that not every embedded Grassmann space is a Grassmann variety. Consider the Grassmann variety $\mathcal{G}_{1,7}$ of the lines of $\mathrm{PG}(7, q)$. The number of points of $\mathcal{G}_{1,7}$ is a polynomial of degree 12 in q. The number of points on the lines having at least two points in common with $\mathcal{G}_{1,7}$ is a polynomial $a_0 q^{25} + a_1 q^{24} + \cdots + a_{24} q + a_{25}$. By Theorem 3.12, the projective space generated by $\mathcal{G}_{1,7}$ has dimension 28. Hence, for q large enough, $\mathrm{PG}(28, q)$ contains a point P such that each line through it has at most one point in common with $\mathcal{G}_{1,7}$.

Let Π_{27} be a hyperplane of $\mathrm{PG}(28, q)$, which does not contain P. The intersection of the cone $P\mathcal{G}_{1,7}$ with the hyperplane Π_{27} is a variety which, together with the projections of the lines of $\mathcal{G}_{1,7}$ is a Grassmann space embedded in Π_{27}. Assume that this Grassmann space $(\mathcal{P}, \mathcal{B})$ is a Grassmann variety. Since the maximal subspaces of $(\mathcal{P}, \mathcal{B})$ have dimensions 2 and 6, the only candidates for the Grassmann variety are $\mathcal{G}_{1,7}$ and $\mathcal{G}_{5,7}$. By Theorem 3.12, these two projectively equivalent varieties are not contained in a $\mathrm{PG}(27, q)$, a contradiction. So, this is an example of a Grassmann space which is embedded in a projective space, but which is not a Grassmann variety.

To determine all embeddings of Grassmann spaces, it is necessary to introduce *homomorphisms* between projective spaces. Let ψ be the semi-linear transformation of the vector space $V(m + 1, K)$ into the vector space $V(n + 1, K)$ defined by the $m \times n$ matrix T over K and the automorphism σ of K. The kernel of ψ is a subspace of $V(m + 1, K)$ and the image of ψ is a subspace of $V(n + 1, K)$. Also,

$$\dim(\ker \psi) + \dim(\operatorname{im} \psi) = m + 1.$$

Now consider the projective spaces over K defined by $V(m + 1, K)$, $V(n + 1, K)$, the kernel of ψ, and the image of ψ. These spaces are respectively denoted by $\mathrm{PG}(m, K)$, $\mathrm{PG}(n, K)$, $P(\ker \psi)$, and $P(\operatorname{im} \psi)$. Then

$$\dim(P(\ker \psi)) + \dim(P(\operatorname{im} \psi)) = m - 1.$$

The semi-linear transformation ψ induces a mapping ξ from $\mathrm{PG}(m, K) \backslash P(\ker \psi)$ onto $P(\operatorname{im} \psi)$. Such a mapping ξ is a *homomorphism* of $\mathrm{PG}(m, K)$ into $\mathrm{PG}(n, K)$ or onto $P(\operatorname{im} \psi)$. If ψ is bijective, that is, if T is a non-singular $(n + 1) \times (n + 1)$ matrix over K, then ξ is a collineation between projective spaces.

Now consider the Grassmann variety $\mathcal{G}_{r,n}$ of the r-spaces in $\mathrm{PG}(n, q)$. The projective space generated by $\mathcal{G}_{r,n}$ is denoted by $\mathrm{PG}(N, q)$. Suppose that ξ is a homomorphism of $\mathrm{PG}(N, q)$ into $\mathrm{PG}(N', q)$, where $\ker \psi \neq \{0\}$ and with the condition that any line of $\mathrm{PG}(N, q)$ having at least two distinct points in common with $\mathcal{G}_{r,n}$ has no point in common with $P(\ker \psi)$. Then ξ maps the points and lines of $\mathcal{G}_{r,n}$ onto the points and lines of a Grassmann space $(\mathcal{P}, \mathcal{B})$ which is embedded in $P(\operatorname{im} \psi)$. Since the dimension of $P(\operatorname{im} \psi)$ is less than N, an argument of one of the previous paragraphs shows that $(\mathcal{P}, \mathcal{B})$ is not a Grassmann variety. The example given at the beginning of this section is constructed in this way.

It is now shown that any Grassmann space embedded in a projective space can be obtained in the way described above. The proof is given for $\mathbf{F}_q$, but is valid for any field K.

Theorem 3.71. *Let $(\mathcal{P}, \mathcal{B})$ be a Grassmann space of index r that is embedded in a projective space $\mathrm{PG}(s, q)$ and let ξ be an isomorphism from the PPLS defined by $\mathcal{G}_{r,n}$ onto the PPLS $(\mathcal{P}, \mathcal{B})$. Then there is a unique homomorphism ψ from $\mathrm{PG}(N, q)$, the space generated by $\mathcal{G}_{r,n}$, into $\mathrm{PG}(s, q)$ that induces ξ on $\mathcal{G}_{r,n}$.*

Proof. Let $(\mathcal{P}, \mathcal{B})$ be a Grassmann space of index r, with $r \geq 1$, which is embedded in $\mathrm{PG}(s, q)$. It may be assumed that $\mathcal{P}$ generates $\mathrm{PG}(s, q)$. From Section 3.3, there is an isomorphism ξ from the PPLS corresponding to some Grassmann variety $\mathcal{G}_{r,n}$ onto $(\mathcal{P}, \mathcal{B})$. The projective space generated by $\mathcal{G}_{r,n}$ is denoted by $\mathrm{PG}(N, q)$.

It is sufficient to prove the theorem for the pair (r, n), where $1 \leq r \leq n - 2$, under the assumption that it is already established for all pairs $(r', n') \neq (r, n)$, with $r' \leq r$, $n' \leq n$, and $1 \leq r' \leq n' - 2$.

Let π be a hyperplane of $\mathrm{PG}(n, q)$. If $r \leq n - 3$, then, by Corollary 3.27, $\mathfrak{G}$ maps the r-spaces of π onto the points of a subvariety $\mathcal{G}_{r,n-1}$ of $\mathcal{G}_{r,n}$; if $r = n - 2$, then $\mathfrak{G}$ maps the $(n-2)$-spaces of π onto the points of a maximal $(n-1)$-space π' of $\mathcal{G}_{n-2,n}$. For $r \leq n-3$, the images under ξ of the points and lines of $\mathcal{G}_{r,n-1}$ form a Grassmann space $(\mathcal{P}', \mathcal{B}')$ isomorphic to $\mathcal{G}_{r,n-1}$; for $r = n - 2$, a maximal $(n - 1)$-space $\pi' \xi$ of

$(\mathcal{P}, \mathcal{B})$ is obtained. Let $\mathrm{PG}(N', q)$ be the projective space generated by $\mathcal{G}_{r,n-1}$. For $r \leq n - 3$, the induction step gives a unique homomorphism ψ_π from $\mathrm{PG}(N', q)$ into $\mathrm{PG}(s, q)$ which coincides with ξ on $\mathcal{G}_{r,n-1}$; if $r = n - 2$, the restriction of ξ to π' is a collineation ψ_π of π' onto $\pi'\xi$. The semi-linear transformations $\bar{\psi}_\pi$ of $V(N' + 1, q)$ into $V(s + 1, q)$ which correspond to ψ_π are determined by π up to a scalar multiple.

Next, let P be a point of $\mathrm{PG}(n, q)$. When $r \geq 2$, then, by Corollary 3.26, $\mathfrak{G}$ maps the r-spaces through P onto the points of a subvariety $\mathcal{G}_{r-1,n-1}$ of $\mathcal{G}_{r,n}$; when $r = 1$, then $\mathfrak{G}$ maps the lines through P onto the points of a maximal $(n - 1)$-space π'' of $\mathcal{G}_{1,n}$. For $r \geq 2$, the images under ξ of the points and lines of $\mathcal{G}_{r-1,n-1}$ form a Grassmann space $(\mathcal{P}'', \mathcal{B}'')$ isomorphic to $\mathcal{G}_{r-1,n-1}$; for $r = 1$, a maximal $(n - 1)$-space $\pi''\xi$ of $(\mathcal{P}, \mathcal{B})$ is obtained. Let $\mathrm{PG}(N'', q)$ be the projective space generated by $\mathcal{G}_{r-1,n-1}$. For $r \geq 2$, the induction hypothesis implies that there is a unique homomorphism ψ_P from $\mathrm{PG}(N'', q)$ into $\mathrm{PG}(s, q)$ which coincides with ξ on $\mathcal{G}_{r-1,n-1}$; when $r = 1$, the restriction of ξ to π'' is a collineation ψ_P of π'' onto $\pi''\xi$. It may be noted that the semi-linear transformations $\bar{\psi}_P$ of $V(N'' + 1, q)$ into $V(s + 1, q)$ which correspond to ψ_P are determined by P up to a scalar multiple.

It is now shown that ψ_π and ψ_P have the same associated field automorphism. First suppose that $P \in \pi$. The notation of the preceding paragraphs is used. Let W be the image under $\mathfrak{G}$ of the set of all r-spaces of π containing P. If $r \neq 1, n - 2$, then, by Theorem 3.25, $W = \mathcal{G}_{r,n-1} \cap \mathcal{G}_{r-1,n-1}$ is a Grassmann variety $\mathcal{G}_{r-1,n-2}$. If $r = n - 2$ and $n \neq 3$, then $W = \pi' \cap \mathcal{G}_{n-3,n-1}$ is a maximal $(n - 2)$-space α' of $\mathcal{G}_{n-3,n-1}$; if $r = 1$ and $n \neq 3$, then $W = \pi'' \cap \mathcal{G}_{1,n-1}$ is a maximal $(n - 2)$-space α'' of $\mathcal{G}_{1,n-1}$; if $r = 1$ and $n = 3$, then $W = \pi' \cap \pi''$ is a line l.

Let l be a line of W, and let P_1, P_2, P_3, P_4 be any four distinct points of l. Then

$$P_i \psi_P = P_i \psi_\pi = P_i \xi, \quad i = 1, 2, 3, 4.$$

Let $P_i \xi = Q_i$, $i = 1, 2, 3, 4$. Then the cross-ratio $\{Q_1, Q_2; Q_3, Q_4\}$ is equal to $\{P_1, P_2; P_3, P_4\}^\sigma$ and to $\{P_1, P_2; P_3, P_4\}^{\sigma'}$, where σ and σ' are the field automorphisms associated to ψ_P and ψ_π; hence $\sigma = \sigma'$.

Next, let $P \notin \pi$. Consider a point Q and a hyperplane α of $\mathrm{PG}(n, q)$, where $P \in \alpha$, $Q \notin \alpha$, $Q \in \pi$. The field automorphisms associated to ψ_P, ψ_α, ψ_Q, ψ_π are respectively denoted by σ, ρ', ρ, σ'. By a previous argument, $\sigma = \rho' = \rho = \sigma'$; hence $\sigma = \sigma'$.

Now let P and P' be distinct points of $\mathrm{PG}(n, q)$. By considering a hyperplane through P and P', it is seen that the field automorphisms associated to ψ_P and $\psi_{P'}$ coincide. Similarly, with π a hyperplane of $\mathrm{PG}(n, q)$, the field automorphism associated to ψ_π is independent of the choice of π. This common field automorphism, associated to each P and each π, is denoted by σ.

The following notation is required. For $i = 0, 1, \ldots, n$, let $\psi_i = \psi_{\mathbf{U}_i}$ and let $\psi^i = \psi_{\mathbf{u}_i}$. Recall that $\mathbf{U}_i \in \mathbf{u}_j$ for $i \neq j$. For the corresponding semi-linear transformations, write $\bar{\psi}_i$ and $\bar{\psi}^i$. Consider now homomorphisms ψ_i and ψ^j with $i \neq j$. Let W_i^j be the image under $\mathfrak{G}$ of the set of all r-spaces of $\mathbf{u}_j$ containing $\mathbf{U}_i$. On W_i^j the homomorphisms ψ_i and ψ^j coincide with ξ. First, let W_i^j be the Grassmann variety $\mathcal{G}_{r-1,n-2}$ and let $\text{PG}(N^*, q)$ be the projective space generated by $\mathcal{G}_{r-1,n-2}$. Let $\bar{\Phi}_i$ be the restriction of $\bar{\psi}_i$ and $\bar{\Phi}^j$ the restriction of $\bar{\psi}^j$ to $V(N^* + 1, q)$. By induction, the semi-linear transformations $\bar{\Phi}_i$ and $\bar{\psi}^j$ differ only by a scalar multiple. If W_i^j is a projective space Π_m, then again let $\bar{\Phi}_i$ and $\bar{\Phi}^j$ be the restrictions of $\bar{\psi}_i$ and $\bar{\psi}^j$ to $V(m + 1, q)$; again, they differ only by a scalar multiple. In both cases, it is possible to choose $\bar{\psi}_i$ and $\bar{\psi}^j$ in such a way that $\bar{\Phi}_i$ and $\bar{\Phi}^j$ coincide. This process is called *normalisation*.

Fix $\bar{\psi}^n$ and normalise each of $\bar{\psi}_0, \bar{\psi}_1, \ldots, \bar{\psi}_{n-1}$ with respect to it. Next normalise $\bar{\psi}^0$ with respect to $\bar{\psi}_1$, and $\bar{\psi}_n$ with respect to $\bar{\psi}^0$. Then normalise $\bar{\psi}^1, \bar{\psi}^2, \ldots, \bar{\psi}^{n-1}$ with respect to $\bar{\psi}_n$. Now consider $\bar{\psi}_i$ and $\bar{\psi}_j$, $i \neq j$ and $i, j \in \{0, 1, \ldots, n-1\}$. The image under $\mathfrak{G}$ of the set of all r-spaces through $\mathbf{U}_i$ and $\mathbf{U}_j$ is denoted by $\mathcal{W}_{i,j}$; the image under $\mathfrak{G}$ of the set of all r-spaces of $\mathbf{u}_n$ through $\mathbf{U}_i$ and $\mathbf{U}_j$ is denoted by $\mathcal{W}_{i,j}^n$. Let $\text{PG}(M, q)$ be the projective space generated by $\mathcal{W}_{i,j}$ and let $\bar{\mathcal{W}}_{i,j}^n$ be the set of all vectors representing the points of $\mathcal{W}_{i,j}^n$. By previous arguments, the restrictions of $\bar{\psi}_i$ and $\bar{\psi}_j$ to $V(M + 1, q)$ differ only by a scalar multiple. Since $\bar{\psi}_i$ and $\bar{\psi}_j$ coincide on $\bar{\mathcal{W}}_{i,j}^n \subset V(M + 1, q)$ and $\bar{\mathcal{W}}_{i,j}^n$ contains a non-zero vector, it follows that the restrictions of $\bar{\psi}_i$ and $\bar{\psi}_j$ to $V(M + 1, q)$ coincide. Repeating this argument shows that any two elements of $\bar{\psi}_1, \bar{\psi}_2, \ldots, \bar{\psi}_n, \bar{\psi}^1, \bar{\psi}^2, \ldots, \bar{\psi}^n$ coincide on the intersection of their common domain and $\bar{\mathcal{G}}_{r,n}$, the set of all vectors representing the points of $\mathcal{G}_{r,n}$.

Let $V(N + 1, q)$ be the vector space generated by $\bar{\mathcal{G}}_{r,n}$. If E_i is the vector of $V(N + 1, q)$ with one in the $(i + 1)$-th place and zeros elsewhere, then the vectors $E_0, E_1, \ldots, E_N$ are contained in $\bar{\mathcal{G}}_{r,n}$. Consider the vector E_i. It is in the common domain of $r + 1$ semi-linear transformations $\bar{\psi}_{i_0}, \bar{\psi}_{i_1}, \ldots, \bar{\psi}_{i_r}$. Then $\bar{\psi}_{i_0} E_i = \bar{\psi}_{i_1} E_i = \cdots = \bar{\psi}_{i_r} E_i$. Now define as follows a semi-linear transformation $\bar{\psi}$ from $V(N + 1, q)$ into $V(s + 1, q)$ with associated field automorphism σ:

$$\bar{\psi} E_i = \bar{\psi}_{i_j} E_i, \text{ with } i = 0, 1, \ldots, N, \ j = 0, 1, \ldots, r.$$

Consider now the basis $\{E_{i_0}, E_{i_1}, \ldots\}$ of the domain $\bar{\psi}_i$; then $\bar{\psi} E_{i_j} = \bar{\psi}_i E_{i_j}$. Hence $\bar{\psi}$ agrees with $\bar{\psi}_0, \bar{\psi}_1, \ldots, \bar{\psi}_n$. Next consider the basis $\{E_{j_0}, E_{j_1}, \ldots\}$ of the domain of $\bar{\psi}^j$. Then, since $\bar{\psi}^j$ agrees with any $\bar{\psi}_i$, it follows that $\bar{\psi} E_{j_i} = \bar{\psi}^j E_{j_i}$. Hence $\bar{\psi}$ agrees with $\bar{\psi}^0, \bar{\psi}^1, \ldots, \bar{\psi}^n$. Thus the homomorphism ψ from $\text{PG}(N, q)$ into $\text{PG}(s, q)$, which corresponds with $\bar{\psi}$, agrees with $\psi_0, \psi_1, \ldots, \psi_n, \psi^0, \psi^1, \ldots, \psi^n$.

For a point P of $\text{PG}(n, q)$, let its *weight* $w(P)$ be the number of non-zero coordinates, and let the *weight* of a set of linearly independent points be the sum of its members' weights. The *weight* of an r-space of $\text{PG}(n, q)$ is the minimum of all the weights of its linearly independent sets of size $r + 1$; then the smallest possible weight of an r-space is $r + 1$. The weight of a point Q of $\mathcal{G}_{r,n}$ is the weight of the r-space $Q\mathfrak{G}^{-1}$ of $\text{PG}(n, q)$. Let Π_r be an r-space of weight at most $2r + 1$. Such a space must have a point of weight one in one of its independent sets of size $r + 1$ of minimal weight. Hence this space contains one of the points $\mathbf{U}_i$, and so ξ and ψ agree on the corresponding point of $\mathcal{G}_{r,n}$.

It is now shown by induction on the weight that ξ and ψ agree on all the points of $\mathcal{G}_{r,n}$. First, assume that ξ and ψ agree on all points of $\mathcal{G}_{r,n}$ of weight $r + 1$,

$r + 2, \ldots, h - 1$ and let $h \geq 2r + 2$. Let P be a point of $\mathcal{G}_{r,n}$ of weight h. Suppose that $P_0, P_1, \ldots, P_r$ define the r-space $\Pi_r = P\mathfrak{G}^{-1}$ of $\mathrm{PG}(n, q)$. Now it may be assumed that $h = \sum_{i=0}^{r} w(P_i)$. There exist indices i and j, with $i \neq j$, for which $w(P_i) \geq 2$, $w(P_j) \geq 2$; so take P_0, P_1 with $w(P_0) \geq 2$, $w(P_1) \geq 2$ and also let $P_i = \mathbf{P}(X_i)$, $i = 0, 1$. Now choose points $Q_i = \mathbf{P}(Y_i)$, $i = 0, 1, 2, 3$, with $X_0 = Y_0 + Y_1$, $X_1 = Y_2 + Y_3$, and

$$w(Q_0) < w(P_0), \ w(Q_1) < w(P_0), \ w(Q_2) < w(P_1), \ w(Q_3) < w(P_1).$$

Then P_0, Q_0, Q_1 are distinct as are P_1, Q_2, Q_3. Suppose that $Q_0 \in \Pi_r$; then also $Q_1 \in \Pi_r$. At most one of the points Q_0, Q_1 belongs to $\Pi_{r-1} = P_1 P_2 \cdots P_r$; so assume that $Q_1 \notin \Pi_{r-1}$. Then the points $Q_1, P_1, P_2, \ldots, P_r$ define Π_r, and

$$w(Q_1) + \sum_{i=1}^{r} w(P_i) < \sum_{i=0}^{r} w(P_i) = h,$$

a contradiction. Hence $Q_0 \notin \Pi_r$; similarly $Q_1, Q_2, Q_3 \notin \Pi_r$.

Let

$$\Pi_r^1 = Q_0 P_1 P_2 \cdots P_r, \ \ \Pi_r^2 = Q_1 P_1 P_2 \cdots P_r,$$
$$\Pi_r^3 = Q_2 P_0 P_2 \cdots P_r, \ \ \Pi_r^4 = Q_3 P_0 P_2 \cdots P_r.$$

The r-spaces Π_r^1 and Π_r^2 are distinct, as are Π_r^3 and Π_r^4. For $i = 1, 2, 3, 4$, write $\Pi_r^i \mathfrak{G} = A_i$. Since $w(A_i) < h$, so, by induction, ψ and ξ agree on the four A_i. As the r-spaces Π_r^1 and Π_r^2 have an $(r-1)$-space in common, so, by Theorem 3.16, $A_1 A_2$ is a line of $\mathcal{G}_{r,n}$; similarly, so is $A_3 A_4$. Since $\Pi_r^1 \cap \Pi_r^2 \neq \Pi_r^3 \cap \Pi_r^4$, so $A_1 A_2 \neq A_3 A_4$. As Π_r belongs to the pencils defined by both the pairs $\{\Pi_r^1, \Pi_r^2\}$ and $\{\Pi_r^3, \Pi_r^4\}$, so $A_1 A_2 \cap A_3 A_4 = P$. Hence

$$P\psi = (A_1 A_2)\psi \cap (A_3 A_4)\psi = (A_1)\psi(A_2)\psi \cap (A_3)\psi(A_4)\psi$$
$$= (A_1)\xi(A_2)\xi \cap (A_3)\xi(A_4)\xi = (A_1 A_2)\xi \cap (A_3 A_4)\xi$$
$$= P\xi.$$

Therefore ξ and ψ agree on P. This shows that ξ and ψ agree on all points of $\mathcal{G}_{r,n}$.

Finally, it is shown that ψ is uniquely defined by ξ. To do this, let ψ' be a homomorphism from $\mathrm{PG}(N, q)$ to $\mathrm{PG}(s, q)$ which agrees with ξ on $\mathcal{G}_{r,n}$. A corresponding semi-linear transformation is denoted by $\bar{\psi}'$. Then, for the restrictions

$$\bar{\psi}'_0, \bar{\psi}'_1, \ldots, \bar{\psi}'_n, \bar{\psi}'^0, \bar{\psi}'^1, \ldots, \bar{\psi}'^n \qquad (3.22)$$

of $\bar{\psi}'$, the map $\bar{\psi}'^i$ is normalised with respect to $\bar{\psi}'^j$ for all $i \neq j$. From the uniqueness of the homomorphisms $\psi_0, \psi_1, \ldots, \psi_n, \psi^0, \psi^1, \ldots, \psi^n$, it follows that the transformations (3.22) are the transformations

$$\bar{\psi}_0, \bar{\psi}_1, \ldots, \bar{\psi}_n, \bar{\psi}^0, \bar{\psi}^1, \ldots, \bar{\psi}^n$$

up to a common factor of proportion. Hence $\bar{\psi}$ and $\bar{\psi}'$ are equal up to a factor of proportion. Therefore $\psi = \psi'$. $\qquad \square$

3.5 Notes and references

Section 3.1

This is taken from Segre [277].

Section 3.2

For more details on Grassmann varieties, see for example Burau [60], and Hodge and Pedoe [183], where Theorem 3.13 is proved.

Section 3.3

This is taken from Tallini [307], Bichara and Tallini [25, 26]. The final section on the independence of the axioms is due to Bichara and Mazzocca [22].

Let $\Gamma = (\mathcal{P}, \mathcal{L}, \mathcal{B})$ be a finite Buekenhout incidence structure of points, lines and blocks admitting the diagram

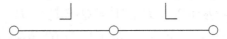

and satisfying the following:

(Sp) blocks are subspaces of $(\mathcal{P}, \mathcal{L})$;
(Sp*) each line is the intersection of any two blocks containing it.

These structures were determined by Sprague [295, 296]. Essentially, this is an alternative proof of the characterisation theorem in Section 3.3. An infinite version of Sprague's theorem is proved in Shult [286]. Shult [288] points out that, in several papers and books, including Section 24.5 of GGG1, these two axioms are overlooked. In Sprague's original paper [296], these axioms are unstated, but are used in the proof. If the axioms are not satisfied, there are counterexamples; see Shult [288].

Characterisations of the Grassmann varieties $\mathcal{G}_{1,n}$ and $\mathcal{G}_{2,n}$, in terms of points and lines only, are given by Lo Re and Olanda [209] and Biondi [29]. In these papers the authors prove that their axioms are equivalent to the axioms A1, A2, A3, and then apply Theorem 3.65.

The Grassmann varieties associated to an affine space are characterised by Bichara and Mazzocca [23, 24]. Other characterisations of Grassmann varieties are discussed in Section 5.10.

Section 3.4

This is taken from Wells [397]. However, Theorem 3.71 was first obtained by Havlicek [158, 159].

4

Veronese and Segre varieties

4.1 Veronese varieties

The *Veronese variety* of all quadrics of $PG(n, K)$, $n \geq 1$, is the variety

$$\mathcal{V} = \{ \mathbf{P}(x_0^2, x_1^2, \ldots, x_n^2, x_0 x_1, \ldots, x_0 x_n, x_1 x_2, \ldots, x_1 x_n, \ldots, x_{n-1} x_n) \mid \\ \mathbf{P}(X) \text{ is a point of } PG(n, K) \}$$

of $PG(N, K)$ with $N = n(n+3)/2$, where $X = (x_0, x_1, \ldots, x_n)$; then $\mathcal{V}$ is a variety of dimension n. It is also called the *Veronesean of quadrics* of $PG(n, K)$, or simply the *quadric Veronesean* of $PG(n, K)$. It can be shown that the quadric Veronesean is absolutely irreducible and non-singular.

Let $PG(N, K)$ consist of all points $\mathbf{P}(Y)$ with

$$Y = (y_{00}, y_{11}, \ldots, y_{nn}, y_{01}, y_{02}, \ldots, y_{0n}, y_{12}, \ldots, y_{1n}, \ldots, y_{n-1,n}).$$

For y_{ij} also write y_{ji}. Then $\mathcal{V}$ belongs to the intersection of the quadrics $\mathbf{V}(F_{ij})$ and $\mathbf{V}(F_{abc})$ with $i, j \in \{0, 1, \ldots, n\}$, $i \neq j$ and $a, b, c \in \{0, 1, \ldots, n\}$, with a, b, c distinct, where

$$F_{ij} = Y_{ij}^2 - Y_{ii} Y_{jj}, \quad F_{abc} = Y_{aa} Y_{bc} - Y_{ab} Y_{ac}.$$

It is now shown that the variety $\mathcal{V}$ is the intersection of these $(n+1)n^2/2$ quadrics.

Lemma 4.1. *The quadric Veronesean $\mathcal{V}$ of $PG(n, K)$ is the intersection of all quadrics $\mathbf{V}(F_{ij})$ and $\mathbf{V}(F_{abc})$.*

Proof. Let $P = \mathbf{P}(Y)$, with

$$Y = (y_{00}, y_{11}, \ldots, y_{n-1,n}),$$

be a point of the intersection of the quadrics $\mathbf{V}(F_{ij})$ and $\mathbf{V}(F_{abc})$. Then

$$(y_{00}, y_{11}, \ldots, y_{nn}) \neq (0, 0, \ldots, 0),$$

© Springer-Verlag London 2016
J.W.P. Hirschfeld, J.A. Thas, *General Galois Geometries*, Springer Monographs in Mathematics, DOI 10.1007/978-1-4471-6790-7_4

since $y_{00} = y_{11} = \cdots = y_{nn} = 0$ and $y_{ij}^2 - y_{ii}y_{jj} = 0$ imply that $y_{ij} = 0$ for all i, j. Suppose, for example, that $y_{00} \neq 0$. Put

$$y_{00} = 1 = x_0, \ y_{01} = x_1, \ y_{02} = x_2, \ \ldots, \ y_{0n} = x_n.$$

Let $i, j \in \{1, 2, \ldots, n\}$ with $i \neq j$. Then, since $y_{00}y_{ij} - y_{0i}y_{0j} = 0$, so $y_{ij} = x_ix_j$. Since $y_{0j}^2 - y_{00}y_{jj} = 0$ for $j \neq 0$, so $y_{jj} = x_j^2$. Hence $y_{ij} = x_ix_j$ for all i, j in $\{0, 1, 2, \ldots, n\}$; that is, P belongs to the quadric Veronesean $\mathcal{V}$. $\quad\square$

Theorem 4.2. *The quadric Veronesean $\mathcal{V}$ of* $\mathrm{PG}(n, K)$ *consists of all points* $\mathbf{P}(Y)$, *with* $Y = (y_{00}, y_{11}, \ldots, y_{n-1,n})$, *of* $\mathrm{PG}(N, K)$ *for which* rank $[y_{ij}] = 1$.

Proof. Let $\mathbf{P}(Y)$ be a point for which rank $[y_{ij}] = 1$. Then $\mathbf{P}(Y)$ belongs to the intersection of the quadrics $\mathbf{V}(F_{ij})$ and $\mathbf{V}(F_{abc})$. By Lemma 4.1, $\mathbf{P}(Y)$ belongs to the quadric Veronesean $\mathcal{V}$.

Conversely, let $\mathbf{P}(Y)$ be a point of the Veronesean $\mathcal{V}$. Then

$$y_{ij}y_{ab} - y_{ib}y_{aj} = x_ix_jx_ax_b - x_ix_bx_ax_j = 0,$$

for all $i, j, a, b \in \{0, 1, \ldots, n\}$. Hence rank $[y_{ij}] = 1$. $\quad\square$

Let $\zeta : \mathrm{PG}(n, K) \to \mathrm{PG}(N, K)$, with $N = n(n+3)/2$ and $n \geq 1$, be defined by

$$\mathbf{P}(x_0, x_1, \ldots, x_n) \mapsto \mathbf{P}(y_{00}, y_{11}, \ldots, y_{n-1,n})$$

with $y_{ij} = x_ix_j$. Then ζ is a bijection of $\mathrm{PG}(n, K)$ onto $\mathcal{V}$. It then follows that the variety $\mathcal{V}$ is rational.

Theorem 4.3. *The quadrics of* $\mathrm{PG}(n, K)$ *are mapped by ζ onto all hyperplane sections of $\mathcal{V}$.*

Proof. Let $\mathbf{V}(F)$, with $F = \sum a_{ij}X_iX_j$, be a quadric $\mathcal{Q}$ of $\mathrm{PG}(n, K)$. Then $\mathcal{Q}\zeta$ consists of all points $\mathbf{P}(Y)$ of $\mathcal{V}$ for which $\sum a_{ij}y_{ij} = 0$; that is, $\mathcal{Q}\zeta$ is a hyperplane section of $\mathcal{V}$.

Conversely, let $\mathcal{H}$ be the intersection of $\mathcal{V}$ and the hyperplane $\mathbf{V}(\sum a_{ij}Y_{ij})$ of $\mathrm{PG}(N, K)$. Then $\mathcal{H}\zeta^{-1}$ consists of all points $\mathbf{P}(X)$ of $\mathrm{PG}(n, K)$ satisfying $\sum a_{ij}x_ix_j = 0$; that is, $\mathcal{H}\zeta^{-1}$ is a quadric of $\mathrm{PG}(n, K)$. $\quad\square$

Theorem 4.3 explains why $\mathcal{V}$ is called the Veronesean of quadrics of $\mathrm{PG}(n, K)$.

Corollary 4.4. *No hyperplane of* $\mathrm{PG}(N, K)$ *contains the quadric Veronesean $\mathcal{V}$.*

Theorem 4.5. *The Veronese variety $\mathcal{V}$ of all quadrics of* $\mathrm{PG}(n, K)$, $n \geq 1$, *has order* 2^n.

Proof. Let $\overline{K}$ be the algebraic closure of K, and let $\overline{\mathcal{V}}$ be the corresponding extension of $\mathcal{V}$. In $\mathrm{PG}(N, \overline{K})$ take the intersection $\Pi_{N-n} \cap \overline{\mathcal{V}}$. Let $\Pi_{N-1}^1, \Pi_{N-1}^2, \ldots, \Pi_{N-1}^n$ be n linearly independent hyperplanes of $\mathrm{PG}(N, \overline{K})$ through Π_{N-n}. Then, writing $\mathcal{H}_i = \overline{\mathcal{V}} \cap \Pi_{N-1}^i$, it follows that

$$|\overline{\mathcal{V}} \cap \Pi_{N-n}| = |\mathcal{H}_1\zeta^{-1} \cap \mathcal{H}_2\zeta^{-1} \cap \cdots \cap \mathcal{H}_n\zeta^{-1}|.$$

Since $\mathcal{H}_1\zeta^{-1}, \mathcal{H}_2\zeta^{-1}, \ldots, \mathcal{H}_n\zeta^{-1}$ are n linearly independent quadrics of $PG(n, \overline{K})$, so $|\overline{\mathcal{V}} \cap \Pi_{N-n}| = 2^n$ for a general Π_{N-n} in $PG(N, \overline{K})$. This means that the quadric Veronesean of $PG(n, K)$ has order 2^n.   □

Notation 4.6. Henceforth, the quadric Veronesean of $PG(n, K)$, $n \geq 1$, is denoted by $\mathcal{V}_n^{2^n}$ or simply $\mathcal{V}_n$. For $n = 1$, the Veronesean $\mathcal{V}_1^2$ is a conic of $PG(2, K)$; referring to the classification of plane quadrics in Section 7.2 of PGOFF2, a conic is a non-singular plane quadric. For $n = 2$, the Veronesean is a surface $\mathcal{V}_2^4$ of order 4 in $PG(5, K)$. For $n = 3$, the Veronesean is a variety $\mathcal{V}_3^8$ of dimension 3 and order 8 of $PG(9, K)$.

Remark 4.7. For $n = 1$ and any four points P_1, P_2, P_3, P_4 on $PG(1, K)$, the cross-ratio $\{P_1, P_2; P_3, P_4\} = \{P_1\zeta, P_2\zeta; P_3\zeta, P_4\zeta\}$.

From now on it is assumed that K is the finite field $\mathbf{F}_q$, although many of the results also hold in the case of a general field.

Theorem 4.8. *The number of points on $\mathcal{V}_n$ is $|\mathcal{V}_n| = \theta(n)$.*

Proof. The variety $\mathcal{V}_n$ is bijectively mapped by ζ^{-1} onto $PG(n, q)$. Hence $\mathcal{V}_n$ has $\theta(n)$ points.   □

Let ξ be a projectivity of $PG(n, q)$. Then ξ defines a permutation ξ' of the quadric Veronesean $\mathcal{V}_n$.

Lemma 4.9. *The permutation ξ' of $\mathcal{V}_n$ is induced by a projectivity $\widetilde{\xi}$ of $PG(N, q)$.*

Proof. Let $\overline{\mathbf{F}}_q$ be the algebraic closure of $\mathbf{F}_q$, and let ζ' be the bijection which maps the quadric $\mathcal{Q} = \mathbf{V}(F)$, with $F = \sum a_{ij}X_iX_j$, of $PG(n, \overline{\mathbf{F}}_q)$ onto the hyperplane $\mathbf{V}(\sum a_{ij}Y_{ij})$ of $PG(N, \overline{\mathbf{F}}_q)$. For any hyperplane Π_{N-1} of $PG(N, \overline{\mathbf{F}}_q)$, let $\Pi_{N-1}\overline{\eta}$ be the hyperplane $\Pi_{N-1}\zeta'^{-1}\overline{\xi}\zeta'$, with $\overline{\xi}$ the extension of ξ to $PG(n, \overline{\mathbf{F}}_q)$. Then $\overline{\eta}$ is a permutation of the set of all hyperplanes of $PG(N, \overline{\mathbf{F}}_q)$. Since $\overline{\xi}$ maps a pencil of quadrics onto a pencil of quadrics, the permutation $\overline{\eta}$ maps a pencil of hyperplanes onto a pencil of hyperplanes. Also, $\overline{\xi}$ leaves invariant the cross-ratio of any four elements of any pencil of quadrics of $PG(N, \overline{\mathbf{F}}_q)$. It follows that $\overline{\eta}$ is a projectivity of the dual space of $PG(N, \overline{\mathbf{F}}_q)$. Let $\overline{\widetilde{\xi}}$ be the corresponding projectivity of $PG(N, \overline{\mathbf{F}}_q)$, with $\widetilde{\xi}$ the projectivity induced by $\overline{\widetilde{\xi}}$ on $PG(N, q)$. Now it follows that $\widetilde{\xi}$ leaves $\mathcal{V}_n$ invariant and induces ξ' on $\mathcal{V}_n$. This proves that ξ' is induced by at least one element $\widetilde{\xi}$ of $PGL(N + 1, q)$.   □

Theorem 4.10. *If Π_s is any s-dimensional subspace of $PG(n, q)$, then $\Pi_s\zeta$ is a quadric Veronesean $\mathcal{V}_s$, which is the complete intersection of $\mathcal{V}_n$ and the space $PG(s(s + 3)/2, q)$ containing $\mathcal{V}_s$.*

Proof. By Lemma 4.9, let Π_s contain the points $\mathbf{P}(E_0), \mathbf{P}(E_1), \ldots, \mathbf{P}(E_s)$, where E_i is the vector with 1 in the $(i+1)$-th place and zeros elsewhere. So $\Pi_s\zeta$ consists of the points $\mathbf{P}(Y)$, where, up to permutation, the coordinates of the vector Y are

$$x_0^2, x_1^2, \ldots, x_s^2, x_0x_1, x_0x_2, \ldots, x_0x_s, \ldots, x_1x_s, \ldots, x_{s-1}x_s, 0, 0, \ldots, 0,$$

where $x_i \in \mathbf{F}_q$ and $(x_0, x_1, \ldots, x_s) \neq (0, 0, \ldots, 0)$. Hence $\Pi_s\zeta$ is a quadric Veronesean Π_s.

The subspace π of $\mathrm{PG}(N, q)$ which contains $\mathcal{V}_s$ is the intersection of the hyperplanes $\mathbf{V}(Y_{ij})$, with i and j not both belonging to $\{0, 1, \ldots, s\}$. Then the intersection of π and $\mathcal{V}_n$ corresponds to the set of all points $\mathbf{P}(x_0, x_1, \ldots, x_n)$ of $\mathrm{PG}(n, q)$ with $x_{s+1} = x_{s+2} = \cdots = x_n = 0$; that is, $\pi \cap \mathcal{V}_n = \mathcal{V}_s$.   □

As a particular case, the lines of $\mathrm{PG}(n, q)$ are mapped onto conics of $\mathcal{V}_n$.

Theorem 4.11. *The quadric Veronesean $\mathcal{V}_n$ is a $\theta(n)$-cap of $\mathrm{PG}(N, q)$, where the dimension $N = n(n+3)/2$.*

Proof. Suppose that P_1, P_2, P_3 are distinct collinear points of $\mathcal{V}_n$. Let π be a plane of $\mathrm{PG}(n, q)$ containing the points $P_1\zeta^{-1}, P_2\zeta^{-1}, P_3\zeta^{-1}$. By Theorem 4.10, $\pi\zeta$ is a quadric Veronesean $\mathcal{V}_2^4$ which is contained in a subspace Π_5 of $\mathrm{PG}(N, q)$. The mapping ζ defines a bijection of the set of all plane quadrics of π containing $P_1\zeta^{-1}, P_2\zeta^{-1}, P_3\zeta^{-1}$ onto the set of all hyperplanes of Π_5 containing P_1, P_2, P_3. There are $q^2 + q + 1$ plane quadrics of π through $P_1\zeta^{-1}, P_2\zeta^{-1}, P_3\zeta^{-1}$, and there are $q^3 + q^2 + q + 1$ hyperplanes of Π_5 through P_1, P_2, P_3. This yields a contradiction; so $\mathcal{V}_n$ is a $\theta(n)$-cap of $\mathrm{PG}(n, q)$.   □

Now a converse of Theorem 4.10 is established.

Theorem 4.12. *For $(q, s) \neq (2, 1)$, any quadric Veronesean $\mathcal{V}_s$ contained in $\mathcal{V}_n$, with $n > 1$, is of the form $\Pi_s\zeta$, with Π_s some s-dimensional subspace of $\mathrm{PG}(n, q)$.*

Proof. First, the theorem is proved for $s = 1$ with $q \geq 3$. So let $\mathcal{C}$ be a conic which is contained in $\mathcal{V}_n$. Let P_1, P_2, P_3 be three distinct points of $\mathcal{C}$, and let π be a plane of $\mathrm{PG}(n, q)$ containing $P_1\zeta^{-1}, P_2\zeta^{-1}, P_3\zeta^{-1}$. By Theorem 4.10, $\pi\zeta$ is a quadric Veronesean $\mathcal{V}_2^4$ in a subspace Π_5 of $\mathrm{PG}(N, q)$. Take the $q^2 + q + 1$ hyperplanes of Π_5 containing $\mathcal{C}$. If π' is any of these hyperplanes, then $(\pi' \cap \mathcal{V}_2^4)\zeta^{-1} = \mathcal{C}'$ is a plane quadric of π which contains the set $\mathcal{C}\zeta^{-1}$ of order $q + 1$. In this way the $q^2 + q + 1$ distinct plane quadrics of π containing $\mathcal{C}\zeta^{-1}$ are obtained. Hence $\mathcal{C}\zeta^{-1}$ is necessarily a line of the plane π. It has therefore been shown that, for any conic $\mathcal{C}$ on $\mathcal{V}_n$, the set $\mathcal{C}\zeta^{-1}$ is a line of $\mathrm{PG}(n, q)$.

Next, the theorem is established for $s = q = 2$. So let $\mathcal{V}_2^4$ be a quadric Veronesean which is contained in $\mathcal{V}_n$. Let P_1, P_2, P_3, P_4 be four distinct points of $\mathcal{V}_2^4$, and let Π_3 be a solid containing $P_1\zeta^{-1}, P_2\zeta^{-1}, P_3\zeta^{-1}, P_4\zeta^{-1}$. By Theorem 4.10, $\Pi_3\zeta$ is a quadric Veronesean $\mathcal{V}_3^8$ in a subspace Π_9 of $\mathrm{PG}(N, 2)$. Consider the 15 hyperplanes of Π_9 containing $\mathcal{V}_2^4$. If π' is any of these hyperplanes, then $(\pi' \cap \mathcal{V}_3^8)\zeta^{-1} = \mathcal{Q}$ is a quadric of Π_3 which contains the set $\mathcal{V}_2^4\zeta^{-1}$ of size 7. In this way, 15 distinct quadrics of Π_3 are obtained, all containing $\mathcal{V}_2^4 \zeta^{-1}$. Hence $\mathcal{V}_2^4 \zeta^{-1}$ is necessarily a plane of the solid Π_3. It has therefore been shown that, for any Veronesean $\mathcal{V}_2^4$ on $\mathcal{V}_n$, the set $\mathcal{V}_2^4 \zeta^{-1}$ is a plane of Π_n.

Consider any quadric Veronesean $\mathcal{V}_s$ contained in $\mathcal{V}_n$, with $q \geq 3$. Let Q_1, Q_2 be distinct points on $\mathcal{V}_s \zeta^{-1} = \mathcal{R}$. The points $Q_1 \zeta$ and $Q_2 \zeta$ are contained in a conic $\mathcal{C}$ of $\mathcal{V}_s$. By the first part of the proof, $\mathcal{C}\zeta^{-1}$ is the line $Q_1 Q_2$. As $\mathcal{C}\zeta^{-1} \subset \mathcal{V}_s \zeta^{-1}$, the line $Q_1 Q_2$ belongs to $\mathcal{R}$. Hence $\mathcal{R}$ is a subspace of $\mathrm{PG}(n, q)$. Since both $|\mathcal{R}|$ and $|\mathcal{V}_s|$ equal $\theta(s)$, so $\mathcal{R}$ is an s-dimensional subspace of $\mathrm{PG}(n, q)$. Thus any quadric Veronesean $\mathcal{V}_s$ contained in $\mathcal{V}_n$ is of the form $\Pi_s \zeta$, with Π_s some s-dimensional subspace of $\mathrm{PG}(n, q)$.

Finally, consider any quadric Veronesean $\mathcal{V}_s$ contained in $\mathcal{V}_n$, with $q = 2$ and $s > 2$. Let Q_1, Q_2 be distinct points on $\mathcal{V}_s \zeta^{-1} = \mathcal{R}$. Further, let $\mathcal{V}_2^4$ be a quadric Veronesean on $\mathcal{V}_s$ containing $Q_1 \zeta$ and $Q_2 \zeta$. By the second part of the proof the set $\mathcal{V}_2^4 \zeta^{-1}$ is a plane of $\mathrm{PG}(n, 2)$. As $\mathcal{V}_2^4 \zeta^{-1} \subset \mathcal{V}_s \zeta^{-1}$ the plane $\mathcal{V}_2^4 \zeta^{-1}$ belongs to $\mathcal{R}$. Consider a second quadric Veronesean $\mathcal{V}_2^{4'}$ on $\mathcal{V}_s$ containing $Q_1 \zeta$ and $Q_2 \zeta$. This yields a second plane $\mathcal{V}_2^{4'} \zeta^{-1}$ belonging to $\mathcal{R}$. So the line $\mathcal{V}_2^4 \zeta^{-1} \cap \mathcal{V}_2^{4'} \zeta^{-1} = Q_1 Q_2$ belongs to $\mathcal{R}$. Hence $\mathcal{R}$ is a subspace of $\mathrm{PG}(n, 2)$. Since $|\mathcal{R}| = |\mathcal{V}_s| = \theta(s)$, so $\mathcal{R}$ is an s-dimensional subspace of $\mathrm{PG}(n, 2)$. Thus any quadric Veronesean $\mathcal{V}_s$ contained in $\mathcal{V}_n$ is of the form $\Pi_s \zeta$, with Π_s some s-dimensional subspace of $\mathrm{PG}(n, 2)$. $\square$

Let $q = 2$. Then any triple of points on $\mathcal{V}_n$ is a conic. As for $n > 1$ there are more triples of points on $\mathcal{V}_n$ than lines in $\mathrm{PG}(n, 2)$, so Theorem 4.12 does not hold for $(q, s) = (2, 1)$ with $n > 1$.

Corollary 4.13. *For $q \neq 2$, any two points of $\mathcal{V}_n$ are contained in a unique conic of $\mathcal{V}_n$.*

Proof. Let P_1, P_2 be distinct points of $\mathcal{V}_n$. Then $P_1 \zeta^{-1}$ and $P_2 \zeta^{-1}$ are contained in a unique line of $\mathrm{PG}(n, q)$; that is, P_1 and P_2 are contained in a unique $\mathcal{V}_1^2$, which is a conic. $\square$

Corollary 4.14. *For $(q, s) \neq (2, 1)$, the quadric $\mathcal{V}_n$ contains $\phi(s; n, q)$ quadric Veroneseans $\mathcal{V}_s$.*

Proof. The number of Veroneseans $\mathcal{V}_s$ on $\mathcal{V}_n$ is the number of s-dimensional subspaces of $\mathrm{PG}(n, q)$. $\square$

Let ξ be a projectivity of $\mathrm{PG}(n, q)$. Then ξ defines a permutation ξ' of $\mathcal{V}_n$ which, by Lemma 4.9, is induced by a projectivity $\widetilde{\xi}$ of $\mathrm{PG}(N, q)$. However, $\xi_1 \neq \xi_2$ implies that $\widetilde{\xi}_1 \neq \widetilde{\xi}_2$. Let $G(\mathcal{V}_n)$ be the subgroup of $\mathrm{PGL}(N + 1, q)$ leaving $\mathcal{V}_n$ fixed.

Theorem 4.15. *Let $n \geq 2$ with $(q, n) \neq (2, 2)$. Then, for any projectivity ξ of $\mathrm{PG}(n, q)$,*

(i) *the corresponding permutation ξ' of $\mathcal{V}_n$ is induced by a unique element $\widetilde{\xi}$ of $G(\mathcal{V}_n)$;*

(ii) *the mapping $\theta : \mathrm{PGL}(n + 1, q) \to G(\mathcal{V}_n)$, given by $\xi \mapsto \widetilde{\xi}$, is an isomorphism.*

Proof. Let $\widetilde{\xi}$ be any element of $G(\mathcal{V}_n)$. The corresponding permutation of $\mathcal{V}_n$ is denoted by ξ'. It is now shown that ξ' corresponds to a projectivity of $\mathrm{PG}(n, q)$.

Let ξ be the permutation of the points of $\mathrm{PG}(n, q)$ which corresponds to ξ'. First, let $q \geq 3$. Since, on $\mathcal{V}_n$, the map ξ' transforms conics to conics, so, on $\mathrm{PG}(n, q)$, the permutation ξ maps lines to lines. By Remark 4.7, ξ leaves the cross-ratio of any four collinear points invariant. By the fundamental theorem of projective geometry, the permutation ξ is a projectivity of $\mathrm{PG}(n, q)$. Next, let $q = 2$, $n > 2$. Since, on $\mathcal{V}_n$, the map ξ' transforms a $\mathcal{V}_2^4$ to another one $\mathcal{V}_2^{4'}$, so, on $\mathrm{PG}(n, 2)$, the permutation ξ maps planes to planes. Hence ξ is a projectivity of $\mathrm{PG}(n, 2)$. Consequently, for $(q, n) \neq (2, 2)$, the map ξ' corresponds to a projectivity of $\mathrm{PG}(n, q)$.

Next, consider any projectivity ξ of $\mathrm{PG}(n, q)$ and also the corresponding permutation ξ' of $\mathcal{V}_n$. Assume that ξ' is induced by the projectivities $\widetilde{\xi}$ and $\widetilde{\xi}'$ of $\mathrm{PG}(N, q)$. It must be shown that $\widetilde{\xi} = \widetilde{\xi}'$.

If $\widetilde{\eta} = \widetilde{\xi}'\widetilde{\xi}^{-1}$, then $\widetilde{\eta}$ induces the identity mapping of $\mathcal{V}_n$. Consider distinct points P_1, P_2 on $\mathcal{V}_n$. If $q = 2$, then $\widetilde{\eta}$ fixes all points of the line $P_1 P_2$.

Now let $q > 2$. The points P_1, P_2 are contained in exactly one conic $\mathcal{C}$ on $\mathcal{V}_n$. Let π be the plane of $\mathcal{C}$; then the projectivity $\widetilde{\eta}$ fixes all points of $\mathcal{C}$, and so fixes each point of π. In particular, $\widetilde{\eta}$ fixes all points of the line $P_1 P_2$. By Corollary 4.4, the Veronesean $\mathcal{V}_n$ generates $\mathrm{PG}(N, q)$. Let $P_1, P_2, \ldots, P_{N+1}$ be linearly independent points on $\mathcal{V}_n$. The projectivity $\widetilde{\eta}$ fixes each point of the line $P_i P_j$, for $i, j = 1, 2, \ldots, N + 1$ and $i \neq j$. It follows that $\widetilde{\eta}$ is the identity mapping of $\mathrm{PG}(N, q)$; so $\widetilde{\xi} = \widetilde{\xi}'$.

Since $\xi \neq \eta$ implies that $\widetilde{\xi} \neq \widetilde{\eta}$, it has been shown that $\theta : \xi \mapsto \widetilde{\xi}$ is an isomorphism of $\mathrm{PGL}(n + 1, q)$ onto $G(\mathcal{V}_n)$.   □

Let $q = n = 2$. Then, for any projectivity ξ of $\mathrm{PG}(2, 2)$, the corresponding permutation ξ' of $\mathcal{V}_2$ is induced by a unique element $\widetilde{\xi}$ of $G(\mathcal{V}_2)$; but the group order $|\mathrm{PGL}(3, 2)| < |G(\mathcal{V}_2)| = 7!$. For $n = 1$, Theorem 4.15 also holds.

Corollary 4.16. $|G(\mathcal{V}_n)| = |\mathrm{PGL}(n + 1, q)|$ *for* $(q, n) \neq (2, 2)$.

In the rest of this chapter, to avoid exceptions, when $q = 2$ a conic of the quadric Veronesean $\mathcal{V}_n$ is by definition any triple of points of $\mathcal{V}_n$ which is the image of a line of $\mathrm{PG}(n, 2)$.

Apart from the conic, the quadric Veronesean which is most studied and characterised is the surface $\mathcal{V}_2^4$ of $\mathrm{PG}(5, q)$. In the second part of this section, some interesting properties of this surface are established, several of which can be generalised to all Veroneseans.

So consider the quadric Veronesean $\mathcal{V}_2^4$. By Corollary 4.14, the variety $\mathcal{V}_2^4$ contains $q^2 + q + 1$ conics and, by Corollary 4.13, any two points of $\mathcal{V}_2^4$ are contained in a unique one of these conics. Since the conics of $\mathcal{V}_2^4$ correspond to the lines of $\mathrm{PG}(2, q)$, any two of these conics have a unique point in common.

To the quadrics of $\mathrm{PG}(2, q)$ there correspond all hyperplane sections of $\mathcal{V}_2^4$. The hyperplane is uniquely determined by the plane quadric if and only if the latter is not a single point. If the quadric $\mathcal{C}$ of $\mathrm{PG}(2, q)$ is a repeated line, then the corresponding hyperplane Π_4 of $\mathrm{PG}(5, q)$ meets $\mathcal{V}_2^4$ in a conic; if $\mathcal{C}$ is two distinct lines, then Π_4

meets $\mathcal{V}_2^4$ in two conics with exactly one point in common; if $\mathcal{C}$ is a conic, then Π_4 meets $\mathcal{V}_2^4$ in a rational quartic curve.

The planes of $\mathrm{PG}(5, q)$ which meet $\mathcal{V}_2^4$ in a conic are called the *conic planes* of $\mathcal{V}_2^4$.

Theorem 4.17. *Any two conic planes* π *and* π' *of* $\mathcal{V}_2^4$ *have exactly one point in common, and this common point belongs to* $\mathcal{V}_2^4$.

Proof. Let $\pi \cap \mathcal{V}_2^4 = \mathcal{C}$ and $\pi' \cap \mathcal{V}_2^4 = \mathcal{C}'$ As $|\mathcal{C} \cap \mathcal{C}'| = 1$, suppose that $\pi \cap \pi'$ is a line and let $\pi \cup \pi'$ be contained in two distinct hyperplanes of $\mathrm{PG}(5, q)$. To these hyperplanes there correspond two distinct quadrics of $\mathrm{PG}(2, q)$, which both contain the distinct lines $\mathcal{C}\zeta^{-1}$ and $\mathcal{C}'\zeta^{-1}$. This contradiction proves the theorem. $\qquad\square$

Theorem 4.18. *The union of the conic planes of* $\mathcal{V}_2^4$ *is the hypersurface* $\mathbf{V}(F) = \mathcal{M}_4^3$ *of order three, where*

$$F = \begin{vmatrix} Y_{00} & Y_{01} & Y_{02} \\ Y_{01} & Y_{11} & Y_{12} \\ Y_{02} & Y_{12} & Y_{22} \end{vmatrix}. \tag{4.1}$$

Proof. Let $l = \mathbf{V}(u_0 X_0 + u_1 X_1 + u_2 X_2)$ be any line of $\mathrm{PG}(2, q)$. Then, by multiplying this form in turn by X_0, X_1, X_2, it follows that the conic $\mathcal{C} = l\zeta$ is the section of $\mathcal{V}_2^4$ by the plane

$$\mathbf{V}(u_0 Y_{00} + u_1 Y_{01} + u_2 Y_{02},\ u_0 Y_{01} + u_1 Y_{11} + u_2 Y_{12},\ u_0 Y_{02} + u_1 Y_{12} + u_2 Y_{22}). \tag{4.2}$$

A point $\mathbf{P}(y_{00}, y_{11}, y_{22}, y_{01}, y_{02}, y_{12})$ belongs to a conic plane, that is, a plane of the form (4.2), if and only if $F(y_{00}, y_{11}, y_{22}, y_{01}, y_{02}, y_{12}) = 0$. $\qquad\square$

Theorem 4.19. *The hypersurface* $\mathcal{M}_4^3$ *has* $(q^2 + q + 1)(q^2 + 1)$ *points.*

Proof. The hypersurface $\mathcal{M}_4^3$ is the union of the $q^2 + q + 1$ conic planes of $\mathcal{V}_2^4$. By Theorem 4.17, any two conic planes have exactly one point in common which belongs to $\mathcal{V}_2^4$. Also, each point of $\mathcal{V}_2^4$ belongs to at least one conic plane. Hence

$$|\mathcal{M}_4^3| = (q^2 + q + 1)q^2 + (q^2 + q + 1) = (q^2 + q + 1)(q^2 + 1). \qquad\square$$

If the characteristic of $\mathbf{F}_q$ is two, then $\mathcal{M}_4^3 = \mathbf{V}(F)$, where

$$F = Y_{00} Y_{11} Y_{22} + Y_{00} Y_{12}^2 + Y_{11} Y_{02}^2 + Y_{22} Y_{01}^2. \tag{4.3}$$

In this case, $\mathcal{M}_4^3$ contains the plane $\mathbf{U}_3 \mathbf{U}_4 \mathbf{U}_5$. It may be noted that this plane has no point on $\mathcal{V}_2^4$.

Lemma 4.20. *If the characteristic of* $\mathbf{F}_q$ *is 2, then the Veronesean* $\mathcal{V}_2^4$ *is the intersection of the quadrics* $\mathbf{V}(F_{01}), \mathbf{V}(F_{02}), \mathbf{V}(F_{12})$, *where*

$$F_{01} = Y_{01}^2 + Y_{00} Y_{11}, \quad F_{02} = Y_{02}^2 + Y_{00} Y_{22}, \quad F_{12} = Y_{12}^2 + Y_{11} Y_{22}. \tag{4.4}$$

Proof. With $Y = (y_{00}, y_{11}, y_{22}, y_{01}, y_{02}, y_{12})$, let

$$\mathbf{P}(Y) \in \mathbf{V}(F_{01}) \cap \mathbf{V}(F_{02}) \cap \mathbf{V}(F_{12}).$$

Then

$$y_{01}^2 = y_{00} y_{11}, \quad y_{02}^2 = y_{00} y_{22};$$

so $y_{01}^2 y_{02}^2 = y_{00}^2 y_{11} y_{22}$. Since $y_{12}^2 = y_{11} y_{22}$, it follows that $y_{01}^2 y_{02}^2 = y_{00}^2 y_{12}^2$. Hence $y_{01} y_{02} = y_{00} y_{12}$; so $\mathbf{P}(Y)$ belongs to $\mathbf{V}(F_{012})$ in the notation of Lemma 4.1. Analogously, $\mathbf{P}(Y)$ belongs to $\mathbf{V}(F_{120})$ and $\mathbf{V}(F_{201})$. By Lemma 4.1,

$$\mathcal{V}_2^4 = \mathbf{V}(F_{01}) \cap \mathbf{V}(F_{02}) \cap \mathbf{V}(F_{12}). \qquad \square$$

Theorem 4.21. *The hypersurface $\mathcal{M}_4^3$ has the Veronesean $\mathcal{V}_2^4$ as double surface.*

Proof. First $\mathcal{M}_4^3 = \mathbf{V}(F)$, where

$$F = Y_{00} Y_{11} Y_{22} + 2 Y_{01} Y_{02} Y_{12} - Y_{00} Y_{12}^2 - Y_{11} Y_{02}^2 - Y_{22} Y_{01}^2. \qquad (4.5)$$

The partial derivatives of F are

$$\frac{\partial F}{\partial Y_{00}} = Y_{11} Y_{22} - Y_{12}^2 = -F_{12},$$

$$\frac{\partial F}{\partial Y_{11}} = Y_{00} Y_{22} - Y_{02}^2 = -F_{02},$$

$$\frac{\partial F}{\partial Y_{22}} = Y_{00} Y_{11} - Y_{01}^2 = -F_{01},$$

$$\frac{\partial F}{\partial Y_{01}} = 2(Y_{02} Y_{12} - Y_{22} Y_{01}) = -2F_{201},$$

$$\frac{\partial F}{\partial Y_{02}} = 2(Y_{01} Y_{12} - Y_{11} Y_{02}) = -2F_{120},$$

$$\frac{\partial F}{\partial Y_{12}} = 2(Y_{01} Y_{02} - Y_{00} Y_{12}) = -2F_{012}.$$

If the characteristic of $\mathbf{F}_q$ is not two, then the singular points of $\mathcal{M}_4^3$ are the elements of

$$\mathcal{M}_4^3 \cap \mathbf{V}(F_{12}) \cap \mathbf{V}(F_{02}) \cap \mathbf{V}(F_{01}) \cap \mathbf{V}(F_{012}) \cap \mathbf{V}(F_{120}) \cap \mathbf{V}(F_{201}) = \mathcal{M}_4^3 \cap \mathcal{V}_2^4 = \mathcal{V}_2^4.$$

If the characteristic of $\mathbf{F}_q$ is two, then the singular points of $\mathcal{M}_4^3$ are the elements of $\mathcal{M}_4^3 \cap \mathbf{V}(F_{12}) \cap \mathbf{V}(F_{02}) \cap \mathbf{V}(F_{01})$. By Lemma 4.20, this set is again $\mathcal{V}_2^4$.

Finally, it is straightforward to check that all singular points of $\mathcal{M}_4^3$ are double points. $\qquad \square$

The tangent lines of the conics of $\mathcal{V}_2^4$ are the *tangents* or *tangent lines* of $\mathcal{V}_2^4$. Since no point of the surface $\mathcal{V}_2^4$ is singular, all tangent lines of $\mathcal{V}_2^4$ at the point P of $\mathcal{V}_2^4$ are contained in a plane $\pi(P)$. This plane $\pi(P)$ is the *tangent plane* of $\mathcal{V}_2^4$ at P.

Since P is contained in exactly $q + 1$ conics of $\mathcal{V}_2^4$ and since no two conic planes through P have a line in common, the tangent plane $\pi(P)$ is the union of the $q + 1$ tangent lines of $\mathcal{V}_2^4$ through P. Also $\pi(P) \cap \mathcal{V}_2^4 = \{P\}$.

All the tangent lines and tangent planes of the surface $\mathcal{V}_2^4$ belong to the hypersurface $\mathcal{M}_4^3$. Since $\mathcal{M}_4^3$ is the union of the conic planes of $\mathcal{V}_2^4$, any point of $\mathcal{M}_4^3$ is on at least one tangent or bisecant of $\mathcal{V}_2^4$. As any two points of $\mathcal{V}_2^4$ are contained in a conic of $\mathcal{V}_2^4$, each bisecant of $\mathcal{V}_2^4$ is a line of $\mathcal{M}_4^3$. Hence $\mathcal{M}_4^3$ can also be described as the union of all tangents and bisecants of $\mathcal{V}_2^4$.

Theorem 4.22. *For any two distinct points P_1 and P_2 of $\mathcal{V}_2^4$, the tangent planes $\pi(P_1)$ and $\pi(P_2)$ have exactly one point in common.*

Proof. Let P_1 and P_2 be distinct points of $\mathcal{V}_2^4$, and let $\mathcal{C}$ be the conic of $\mathcal{V}_2^4$ through P_1 and P_2. The tangent l_i of $\mathcal{C}$ at P_i is contained in $\pi(P_i)$, $i = 1, 2$, and so $\pi(P_1)$ and $\pi(P_2)$ have the point $l_1 \cap l_2 = Q$ in common.

Suppose that $Q' \in \pi(P_1) \cap \pi(P_2)$ with $Q' \neq Q$. Then $P_i Q'$ is tangent to a conic $\mathcal{C}_i$ of $\mathcal{V}_2^4$, $i = 1, 2$. If $\mathcal{C}_1 = \mathcal{C}_2$, then $\mathcal{C}_1 = \mathcal{C}_2 = \mathcal{C}$ and so $Q = Q'$, a contradiction; so $\mathcal{C}_1 \neq \mathcal{C}_2$. If $\mathcal{C}_1 \cap \mathcal{C}_2 = \{P\}$, then the conic planes containing $\mathcal{C}_1$ and $\mathcal{C}_2$ have the distinct points P and Q' in common, contradicting Theorem 4.17. Hence it has been shown that $\pi(P_1) \cap \pi(P_2) = \{Q\}$. $\qquad\square$

Theorem 4.23. *Suppose that the characteristic of $\mathbf{F}_q$ is two. Then each tangent plane of $\mathcal{V}_2^4$ meets the plane $\mathbf{U}_3\mathbf{U}_4\mathbf{U}_5$ in a line, each conic plane meets $\mathbf{U}_3\mathbf{U}_4\mathbf{U}_5$ in a point, and $\mathbf{U}_3\mathbf{U}_4\mathbf{U}_5$ consists of the nuclei of all conics on $\mathcal{V}_2^4$.*

Proof. It was observed above that, in the case of characteristic two, the plane $\mathbf{U}_3\mathbf{U}_4\mathbf{U}_5$ belongs to the hypersurface $\mathcal{M}_4^3$. If P is any point of $\mathcal{V}_2^4$ and Q is any point of the plane $\mathbf{U}_3\mathbf{U}_4\mathbf{U}_5$, then, by Lemma 4.20, the line PQ has only the point P in common with $\mathcal{V}_2^4$.

Let π be any conic plane of $\mathcal{V}_2^4$, and let π be represented by (4.2). Then

$$\pi \cap \mathbf{U}_3\mathbf{U}_4\mathbf{U}_5 = \mathbf{V}(u_1 Y_{01} + u_2 Y_{02}, \, u_0 Y_{01} + u_2 Y_{12}, \, u_0 Y_{02} + u_1 Y_{12}).$$

Since the rank of the matrix

$$\begin{bmatrix} u_1 & u_2 & 0 \\ u_0 & 0 & u_2 \\ 0 & u_0 & u_1 \end{bmatrix}$$

is 2, so $|\pi \cap \mathbf{U}_3\mathbf{U}_4\mathbf{U}_5| = 1$. Let $\pi \cap \mathbf{U}_3\mathbf{U}_4\mathbf{U}_5 = Q$ and $\pi \cap \mathcal{V}_2^4 = \mathcal{C}$. Since $|PQ \cap \mathcal{C}| = 1$ for any point P of $\mathcal{C}$, the point Q is the nucleus of $\mathcal{C}$. Also, by Theorem 4.17, the nuclei of distinct conics of $\mathcal{V}_2^4$ are distinct. So any conic plane of $\mathcal{V}_2^4$ has exactly one point in common with $\mathbf{U}_3\mathbf{U}_4\mathbf{U}_5$, which is therefore the set of the nuclei of all conics on $\mathcal{V}_2^4$.

Finally, let $\pi(P)$ be the tangent plane of $\mathcal{V}_2^4$ at P. Since each line of $\pi(P)$ through P is tangent to some conic of $\mathcal{V}_2^4$, it contains the nucleus of this conic and hence contains a point of $\mathbf{U}_3\mathbf{U}_4\mathbf{U}_5$. Hence $\pi(P)$ and $\mathbf{U}_3\mathbf{U}_4\mathbf{U}_5$ have a line in common. $\square$

Definition 4.24. In the case of characteristic two, the plane $\mathbf{U}_3\mathbf{U}_4\mathbf{U}_5$ is the *nucleus* of the Veronesean $\mathcal{V}_2^4$.

Let $l = \mathbf{V}(F)$, where $F = u_0X_0 + u_1X_1 + u_2X_2$ be any line of $\mathrm{PG}(2,q)$. If $\mathcal{C}$ is the plane quadric whose point set coincides with l, then to $\mathcal{C}$ there corresponds the hyperplane $\Pi_4 = \mathbf{V}(F')$ of $\mathrm{PG}(5,q)$, where

$$F' = u_0^2Y_{00} + u_1^2Y_{11} + u_2^2Y_{22} + 2u_0u_1Y_{01} + 2u_0u_2Y_{02} + 2u_1u_2Y_{12}. \quad (4.6)$$

Such a hyperplane Π_4 is a *contact hyperplane* of $\mathcal{V}_2^4$. The contact hyperplanes of $\mathcal{V}_2^4$ are those which have exactly one conic on $\mathcal{V}_2^4$.

First, let the characteristic of $\mathbf{F}_q$ be 2. Then the contact hyperplane Π_4 always contains the nucleus $\mathbf{U}_3\mathbf{U}_4\mathbf{U}_5$.

Let Π_2 be the conic plane containing $l\zeta = \Pi_4 \cap \mathcal{V}_2^4$. By Theorem 4.23, the contact hyperplane Π_4 is generated by the conic plane Π_2 and the nucleus $\mathbf{U}_3\mathbf{U}_4\mathbf{U}_5$. By the same theorem, the contact hyperplane Π_4 contains the $q + 1$ planes tangent to $\mathcal{V}_2^4$ at the points of the conic $l\zeta$.

Next, let the characteristic of $\mathbf{F}_q$ be odd. Consider a point P of $\mathrm{PG}(2,q)$, and also distinct lines l, l_1, l_2 through P. Then, for $i = 1, 2$, there is a hyperplane Π_4^i of $\mathrm{PG}(5,q)$ corresponding to the plane quadric $\mathcal{C}_i = l \cup l_i$. Here Π_4^i contains the tangent lines at $P\zeta$ of the conics $l\zeta, l_1\zeta, l_2\zeta$. Hence Π_4^i contains the tangent plane $\pi(P\zeta)$ of $\mathcal{V}_2^4$ at $P\zeta$, for $i = 1, 2$. The plane quadric that is the repeated line l belongs to the pencil defined by $\mathcal{C}_1$ and $\mathcal{C}_2$; so the contact hyperplane Π_4 corresponding to $\mathcal{C}$ belongs to the pencil defined by Π_4^1 and Π_4^2. Hence Π_4 also contains $\pi(P\zeta)$. Thus, also in this case, the contact hyperplane Π_4 defined by the line l contains the $q + 1$ planes tangent to $\mathcal{V}_2^4$ at the points of the conic $l\zeta$.

From (4.6) it also follows that, if the characteristic of $\mathbf{F}_q$ is not two, the set of all contact hyperplanes of $\mathcal{V}_2^4$ is simply the dual of $\mathcal{V}_2^4$.

Theorem 4.25. *When the characteristic of* $\mathbf{F}_q$ *is not two,* $\mathrm{PG}(5,q)$ *admits a polarity that maps the set of all conic planes of* $\mathcal{V}_2^4$ *onto the set all tangent planes of* $\mathcal{V}_2^4$.

Proof. Let $l = \mathbf{V}(u_0X_0 + u_1X_1 + u_2X_2)$ be a line of $\mathrm{PG}(2,q)$. By (4.2), the conic plane defined by the line l is

$$\mathbf{V}(u_0Y_{00} + u_1Y_{01} + u_2Y_{02},\ u_0Y_{01} + u_1Y_{11} + u_2Y_{12},\ u_0Y_{02} + u_1Y_{12} + u_2Y_{22}).$$

Next, let $\mathbf{P}(A) = Q$, where $A = (a_0, a_1, a_2)$, be a point of $\mathrm{PG}(2,q)$. For any line m of $\mathrm{PG}(2,q)$ through Q, the corresponding contact hyperplane of $\mathcal{V}_2^4$ contains the tangent plane of $\mathcal{V}_2^4$ at $Q\zeta$. Hence the tangent plane $\pi(Q\zeta)$ belongs to every hyperplane

$$\mathbf{V}(u_0^2Y_{00} + u_1^2Y_{11} + u_2^2Y_{22} + 2u_0u_1Y_{01} + 2u_0u_2Y_{02} + 2u_1u_2Y_{12}),$$

where u_0, u_1, u_2 satisfy

$$u_0a_0 + u_1a_1 + u_2a_2 = 0.$$

Hence the following points are elements of $\pi(Q\zeta)$:

$$\mathbf{P}(2a_0, 0, 0, a_1, a_2, 0), \quad \mathbf{P}(0, 2a_1, 0, a_0, 0, a_2), \quad \mathbf{P}(0, 0, 2a_2, 0, a_0, a_1).$$

Let η be the polarity of $\mathrm{PG}(5, q)$ represented by

$$\rho v_{00} = \tfrac{1}{2}y_{00}, \quad \rho v_{11} = \tfrac{1}{2}y_{11}, \quad \rho v_{22} = \tfrac{1}{2}y_{22},$$
$$\rho v_{01} = y_{01}, \quad \rho v_{02} = y_{02}, \quad \rho v_{12} = y_{12},$$

where

$$\mathbf{V}(v_{00}Y_{00} + v_{11}Y_{11} + v_{22}Y_{22} + v_{01}Y_{01} + v_{02}Y_{02} + v_{12}Y_{12})$$

is a variable hyperplane of $\mathrm{PG}(5, q)$. Then $\pi(Q\zeta)\eta$ is the conic plane defined by the line l of $\mathrm{PG}(2, q)$, where

$$l = \mathbf{V}(a_0 X_0 + a_1 X_1 + a_2 X_2).$$

Hence the polarity η maps the set of all conic planes of $\mathcal{V}_2^4$ onto the set of all tangent planes of $\mathcal{V}_2^4$. □

Corollary 4.26. *Let the characteristic of $\mathbf{F}_q$ be odd. Then, for any three distinct points P_1, P_2, P_3 of $\mathcal{V}_2^4$, the intersection $\pi(P_1) \cap \pi(P_2) \cap \pi(P_3)$ is empty.*

Proof. Suppose that $\pi(P_1) \cap \pi(P_2) \cap \pi(P_3) \neq \emptyset$. By Theorem 4.22,

$$\pi(P_1) \cap \pi(P_2) \cap \pi(P_3) = \{Q\}.$$

So the hyperplane $Q\eta$ contains the three distinct conic planes $\pi(P_i)\eta$, $i = 1, 2, 3$. Hence the quadric of $\mathrm{PG}(2, q)$ which corresponds to $Q\eta$ has at least three distinct linear components, a contradiction. □

4.2 Characterisations

4.2.1 Characterisations of $\mathcal{V}_n^{2^n}$ of the first kind

First, two properties are proved which hold for the quadric Veronesean $\mathcal{V}_n$. The planes of $\mathrm{PG}(N, q)$, $N = n(n + 3)/2$, meeting $\mathcal{V}_n$ in a conic are called the *conic planes* of $\mathcal{V}_n$.

Lemma 4.27. *Two distinct conic planes of $\mathcal{V}_n$ with non-empty intersection meet in exactly one point, and this point lies in $\mathcal{V}_n$.*

Proof. Let Π_2' and Π_2'' be distinct conic planes of $\mathcal{V}_n$ and let $P \in \Pi_2' \cap \Pi_2''$. Assume that $P \in \mathcal{V}_n$.

First, let q be odd. The point P belongs to at least one bisecant l of $\Pi_2' \cap \mathcal{V}_n = \mathcal{C}'$. By Corollary 4.13, the line l is not contained in the plane Π_2''. Let $\mathcal{C}' \cap l = \{P_1, P_2\}$ and $\Pi_2'' \cap \mathcal{V}_n = \mathcal{C}''$. The plane of $\mathrm{PG}(n, q)$ containing the line $\mathcal{C}''\zeta^{-1}$ and the point $P_1\zeta^{-1}$ is denoted by Π_2, with ζ as in Section 4.1. Then $\Pi_2\zeta$ is a Veronesean $\mathcal{V}_2^4$ containing $\mathcal{C}''$ and P_1. The point P_2 belongs to the space $\mathrm{PG}(5, q)$ containing $\mathcal{V}_2^4$.

By Theorem 4.10, $\mathcal{V}_2^4$ is the complete intersection of $\mathrm{PG}(5, q)$ and $\mathcal{V}_n$, and so P_2 belongs to $\mathcal{V}_2^4$. The points P_1 and P_2 are contained in a unique conic of $\mathcal{V}_2^4$ and a unique conic of $\mathcal{V}_n$. Consequently $\mathcal{C}'$ is a conic of $\mathcal{V}_2^4$. Now, by Theorem 4.17, $P \in \mathcal{V}_2^4$, contradicting that $P \notin \mathcal{V}_n$; hence $P \in \mathcal{V}_n$. Since $\mathcal{V}_n$ is a cap it follows that P is the unique common point of Π_2' and Π_2''.

Next, let q be even. If P is not the nucleus of $\mathcal{C}' = \Pi_2' \cap \mathcal{V}_n$ or $\mathcal{C}'' = \Pi_2'' \cap \mathcal{V}_n$, then the argument of the preceding paragraph shows again that $P \in \mathcal{V}_n$. So assume that P is the nucleus of both $\mathcal{C}'$ and $\mathcal{C}''$. Let $P' \in \mathcal{C}' \backslash \mathcal{C}''$. Then the line PP' is a tangent to the conic $\mathcal{C}'$ and hence is a tangent of the algebraic variety $\mathcal{V}_n$. The plane of $\mathrm{PG}(n, q)$ containing the line $\mathcal{C}'' \zeta^{-1}$ and the point $P' \zeta^{-1}$ is denoted by Π_2. Then $\Pi_2 \zeta$ is a Veronesean $\mathcal{V}_2^4$ containing $\mathcal{C}''$ and P'. The line l belongs to the space $\mathrm{PG}(5, q)$ containing $\mathcal{V}_2^4$; hence l is a tangent of $\mathcal{V}_2^4$. By Section 4.1, the Veronesean $\mathcal{V}_2^4$ contains a unique conic $\mathcal{C}$ which is tangent to l at P'. Since P is the nucleus of $\mathcal{C}''$, by Theorem 4.23 it belongs to the nucleus of $\mathcal{V}_2^4$. Then, again by Theorem 4.23, the point P is the nucleus of the conic $\mathcal{C}$. Hence the conics $\mathcal{C}$ and $\mathcal{C}''$ have a common nucleus. This contradicts Theorem 4.23. This proves that again $P \in \mathcal{V}_n$. Since $\mathcal{V}_n$ is a cap, it now follows that P is the unique common point of Π_2' and Π_2''. □

Lemma 4.28. *Let $\mathcal{C}$ be a conic of $\mathcal{V}_n$ and let P be a point of $\mathcal{V}_n \backslash \mathcal{C}$. Then the union of the tangents at P of the conics of $\mathcal{V}_n$ which pass through P and a point of $\mathcal{C}$ is a plane.*

Proof. Let Π_2 be the plane of $\mathrm{PG}(n, q)$ which contains the line $\mathcal{C} \zeta^{-1}$ and the point $P \zeta^{-1}$. Then $\Pi_2 \zeta$ is a Veronesean $\mathcal{V}_2^4$ which contains $\mathcal{C}$ and P. By Corollary 4.13, the $q+1$ conics of $\mathcal{V}_n$ which contain P and a point of $\mathcal{C}$ are precisely the $q+1$ conics of $\mathcal{V}_n$ through P. From the remarks preceding Theorem 4.22, the union of the tangents at P to the $q+1$ conics of $\mathcal{V}_2^4$ through P is the tangent plane of $\mathcal{V}_2^4$ at P. □

Let Ω be a subspace of $\mathrm{PG}(N, q)$, with $\Omega \cap \mathcal{V}_n = \emptyset$, and let Π be a subspace of $\mathrm{PG}(N, q)$, with $\Omega \cap \Pi = \emptyset$ and where Π and Ω generate $\mathrm{PG}(N, q)$. Assume that the projection of $\mathcal{V}_n$ from Ω onto Π is bijective between $\mathcal{V}_n$ and its image Φ in Π. Then Lemmas 4.27 and 4.28 also hold for the set Φ.

In this section it is shown that a weak version of Corollary 4.13, together with Lemmas 4.27 and 4.28, characterise the (bijective) projections of the Veronesean $\mathcal{V}_n$.

From now on, let $\mathcal{K}$ be a set of k points of some projective space $\mathrm{PG}(N, q)$, with $N > 2$, where $\mathcal{K}$ generates $\mathrm{PG}(N, q)$. Further, let Γ be a set of $(q+1)$-arcs, where each element of Γ is the complete intersection of $\mathcal{K}$ and some plane. The elements of Γ are called Γ-*arcs*, and the planes of the Γ-arcs are Γ-*planes*. The tangents of the Γ-arcs are Γ-*tangents*. The Γ-tangent which is tangent to $\mathcal{C} \in \Gamma$ at $P \in \mathcal{C}$ is denoted by $t(P, \mathcal{C})$.

Suppose that $\mathcal{K}$ satisfies the following:

(a) any two distinct points of $\mathcal{K}$ belong to a Γ-plane;
(b) any two distinct intersecting Γ-planes meet on $\mathcal{K}$;
(c) if $\mathcal{C}$ is a Γ-arc and P belongs to $\mathcal{K} \backslash \mathcal{C}$, then the tangents at P of the Γ-arcs passing through P and a point of $\mathcal{C}$ are coplanar.

Note that, for $q > 2$, the set Γ is uniquely determined by $\mathcal{K}$. Indeed, since $q > 2$, any $(q + 1)$-arc $\mathcal{C}$ on $\mathcal{K}$ contains at least four different points P_1, P_2, P_3, P_4. Let $P = P_1 P_2 \cap P_3 P_4$, let Π_2 be a Γ-plane containing P_1, P_2, and let Π'_2 be a Γ-plane containing P_3, P_4. Then $P \notin \mathcal{K}$ and $P \in \Pi_2 \cap \Pi'_2$. By (b), $\Pi_2 = \Pi'_2$ and so $\mathcal{C}$ is a Γ-arc.

If $\mathcal{B}_1, \mathcal{B}_2, \ldots, \mathcal{B}_k$ are point sets of $PG(N, q)$, then $\langle \mathcal{B}_1, \mathcal{B}_2, \ldots, \mathcal{B}_k \rangle$ denotes the subspace of $PG(N, q)$ generated by $\mathcal{B}_1 \cup \mathcal{B}_2 \cup \cdots \cup \mathcal{B}_k$.

Lemma 4.29.   (i) *The set $\mathcal{K}$ is a k-cap.*
 (ii) *Any two distinct points of $\mathcal{K}$ are contained in a unique Γ-arc.*
 (iii) *A Γ-tangent $t(P, \mathcal{C})$ is tangent to a unique Γ-arc through P.*

Proof. Let $P, P' \in \mathcal{K}$ and let Π_2 be a Γ-plane containing P and P'. Since $\Pi_2 \cap \mathcal{K}$ is an arc, so $\mathcal{K} \cap PP' = \{P, P'\}$. Hence $\mathcal{K}$ is a k-cap. Let Π'_2 be another Γ-plane containing the line PP'. By (b), the line $\Pi_2 \cap \Pi'_2$ is a line of $\mathcal{K}$, contradicting the first part of the proof. Finally, consider a Γ-tangent $t(P, \mathcal{C})$, and let $t(P, \mathcal{C}) = t(P, \mathcal{C}')$, with $\mathcal{C} \neq \mathcal{C}'$. By (b), the Γ-planes containing $\mathcal{C}$ and $\mathcal{C}'$ meet on $\mathcal{K}$. Since $t(P, \mathcal{C})$ is the intersection of these Γ-planes, the set $\mathcal{K}$ contains a line, a contradiction. □

As $\mathcal{K}$ is a cap, it is called a *Veronesean cap*. The unique Γ-arc containing the points P, P', with $P \neq P'$, of $\mathcal{K}$ is denoted by $[P, P']$. For $P \in \mathcal{K}$ and $\mathcal{C}$ a Γ-arc not passing through P, the plane containing the tangents at P of the Γ-arcs passing through P and a point of $\mathcal{C}$ is denoted by $\pi(P, \mathcal{C})$. By Lemma 4.29, the number of Γ-arcs containing P and a point of $\mathcal{C}$ is exactly $q + 1$, and $\pi(P, \mathcal{C})$ is the union of the corresponding Γ-tangents.

Lemma 4.30. *The incidence structure $\mathcal{S}$ formed by the points of $\mathcal{K}$ and the $(q + 1)$-arcs of Γ is the incidence structure of points and lines of a projective space of order q and some dimension $n \geq 2$.*

Proof. By Lemma 4.29, the incidence structure $\mathcal{S}$ is a linear space in which all lines have size $q + 1$. It suffices to check the Veblen–Young (or Pasch) axiom. So let $\mathcal{C}_1$ and $\mathcal{C}_2$ be two elements of Γ meeting in a point $P \in \mathcal{K}$, and let $\mathcal{C}_3, \mathcal{C}_4 \in \Gamma$, with $P \notin \mathcal{C}_3 \cup \mathcal{C}_4$, be such that they both meet $\mathcal{C}_1$ and $\mathcal{C}_2$ in distinct points $\{P_{ij}\} = \mathcal{C}_i \cap \mathcal{C}_j$, $i \in \{1, 2\}$ and $j \in \{3, 4\}$. It must be shown that $\mathcal{C}_3$ and $\mathcal{C}_4$ are not disjoint. Both planes $\pi(P_{13}, \mathcal{C}_2)$ and $\pi(P_{13}, \mathcal{C}_4)$ contain the distinct lines $t(P_{13}, \mathcal{C}_1)$ and $t(P_{13}, [P_{13}, P_{24}])$; hence they coincide. Consequently, there is some point $P' \in \mathcal{C}_4$ such that $t(P_{13}, [P_{13}, P']) = t(P_{13}, \mathcal{C}_3)$. But then Lemma 4.29 implies that $\mathcal{C}_3 = [P_{13}, P']$. Hence $P' \in \mathcal{C}_3 \cap \mathcal{C}_4$. □

The natural number n is called the *dimension* of $\mathcal{K}$. Then $\mathcal{K}$ has $\theta(n)$ points.

Corollary 4.31. *Let $\mathcal{S}'$ be a subspace of $\mathcal{S}$ of dimension r, $2 \leq r < n$, with $\mathcal{K}'$ the set of points of $\mathcal{S}'$ and Γ' the set of Γ-arcs in $\mathcal{S}'$. Then $\mathcal{K}'$ satisfies $(a), (b), (c)$ and hence is a Veronesean sub-cap of $\mathcal{K}$ of dimension r with Γ' as the prescribed set of $(q + 1)$-arcs.*

For any point P in $\mathcal{K}$, let $T(P) = \bigcup t(P, \mathcal{C})$, with the union taken over all Γ-arcs containing P.

Lemma 4.32. *For any point P, the set $T(P)$ is an n-dimensional projective subspace of* $\mathrm{PG}(N, q)$.

Proof. Let R, R' be distinct points of $T(P)$, where R, R', P are not collinear. Let $PR = t(P, \mathcal{C})$ and $PR' = t(P, \mathcal{C}')$. For $A \in \mathcal{C} \backslash \{P\}$ and $B \in \mathcal{C}' \backslash \{P\}$, the plane $\pi(P, [A, B])$ coincides with the plane PRR'. Since $\pi(P, [A, B]) \subset T(P)$ and $\pi(P, [A, B]) = PRR'$, the line RR' belongs to $T(P)$. Hence it follows that $T(P)$ is a projective subspace of $\mathrm{PG}(N, q)$.

Since the projective space $\mathcal{K}$ has dimension n, the number of Γ-arcs through P is equal to $\theta(n-1)$. Hence the number of tangents $t(P, \mathcal{C})$ in $T(P)$ is equal to $\theta(n-1)$. Thus the projective space $T(P)$ is n-dimensional.   □

For any point P of $\mathcal{K}$ the space $T(P)$ is called the *tangent space* of $\mathcal{K}$ at P.
In the next theorem a bound on N is obtained.

Theorem 4.33. *The dimension $N \leq n(n + 3)/2$.*

Proof. Proceed by induction on n, first assuming $n > 2$. Let $\mathcal{S}'$ be a hyperplane of $\mathcal{S}$ and P a point of $\mathcal{S}$ not in $\mathcal{S}'$. Then, by induction, the point set $\mathcal{K}'$ of $\mathcal{S}'$ generates a subspace $\mathrm{PG}(N', q)$ of $\mathrm{PG}(N, q)$ with $N' \leq (n-1)(n+2)/2$. Let $P' \in \mathcal{K} \backslash \{P\}$. Then $[P, P'] \cap \mathcal{K}' = \{P''\}$ and $P' \in P''t(P, [P, P']) \subset T(P)\mathrm{PG}(N', q)$. Hence $\mathrm{PG}(N, q) = T(P)\mathrm{PG}(N', q)$, implying $N \leq 1 + n + (n-1)(n+2)/2 = n(n+3)/2$. If $n = 2$, then the same argument can be made but replacing $\mathcal{K}'$ by an element of Γ.   □

Theorem 4.34. *When $n = 2$, then*

(i) $N = 5$;
(ii) *the plane $\mathcal{S}$ is isomorphic to* $\mathrm{PG}(2, q)$;
(iii) *all Γ-arcs are conics;*
(iv) $\mathcal{K}$ *is the Veronesean $\mathcal{V}_2^4$ whose conic planes are the Γ-planes of $\mathcal{K}$.*

Proof. Let Π_2 be a fixed Γ-plane and put $\mathcal{C} = \mathcal{K} \cap \Pi_2$. First it will be shown that $M = 5$. Let $P \in \mathcal{K} \backslash \mathcal{C}$. From the proof of Theorem 4.33, $\mathrm{PG}(N, q) = T(P)\Pi_2$. Suppose by way of contradiction that there exists $P' \in T(P) \cap \Pi_2$. The line PP' is tangent to a certain Γ-arc $[P, P'']$, for some $P'' \in \mathcal{C}$. But now $P' \notin \mathcal{K}$ and yet P' belongs to Π_2 and the Γ-plane containing $[P, P'']$, contradicting (b). It now follows that $N = 5$.

Now consider a plane Π_2' skew to Π_2 and denote by ρ the projection from Π_2 onto Π_2'. It will be shown that ρ is injective on $\mathcal{K} \backslash \mathcal{C}$. Indeed, if $P\rho = R\rho$, for $P, R \in \mathcal{K} \backslash \mathcal{C}$ and $P \neq R$, then $P\Pi_2 = R\Pi_2$ and contains $[P, R]$. Hence Π_2 and the Γ-plane containing $[P, R]$ have a line in common, which by (b) belongs to $\mathcal{K}$, a contradiction. So ρ is injective on $\mathcal{K} \backslash \mathcal{C}$. The points of $\mathcal{K} \backslash \mathcal{C}$ on a Γ-arc $\mathcal{C}'$ of $\mathcal{K}$ different from $\mathcal{C}$ are mapped onto q points of a line of Π_2'; the missing point is the projection of the tangent line, minus its point on $\mathcal{C}$, of $\mathcal{C}'$ at the point $\mathcal{C} \cap \mathcal{C}'$. So a set $\mathcal{A}$ of q^2 points of Π'_2 and $q^2 + q$ lines of Π'_2 is obtained, all containing exactly q points of $\mathcal{A}$. Let l_∞ be the remaining line of Π'_2. Assume by way of contradiction that l_∞ contains a point $V \in \mathcal{A}$. Then consider two distinct lines l_1, l_2 through V, distinct from l_∞. The unique points X_1 of l_1 and X_2 of l_2 not in $\mathcal{A}$ are distinct and not contained in l_∞. But now the line $X_1 X_2$ contains at most $q-1$ points of $\mathcal{A}$, a contradiction. Hence l_∞ contains no point of $\mathcal{A}$, and so it contains the projections of all tangent lines minus their points on $\mathcal{C}$, as mentioned above. It now also follows that the projective plane $\mathcal{S}$ is isomorphic to Π'_2, and so to $\mathrm{PG}(2, q)$. Consequently, the plane $\mathcal{S}$ can be denoted by $\mathrm{PG}(2, q)$.

Consider the inverse image in $\mathcal{K}$ of the intersection of $\mathcal{A}$ and the line containing $U'\rho$ and $(T(U)\backslash\mathcal{C})\rho$. Therefore, given a point $U \in \mathcal{C}$ and a point $U' \in \mathcal{K}\backslash\mathcal{C}$, the space $T(U)U'$ meets $\mathcal{K}$ in $[U, U']$. Let $(*)$ denote this property.

Next it is shown that the Γ-arcs are conics. Consider two points $P_1, P_2 \in \mathcal{C}$. Project $\mathcal{K}\backslash\mathcal{C}$ from the line $P_1 P_2$ onto some solid Π_3 of $\mathrm{PG}(5, q)$ skew to $P_1 P_2$. Let ρ' be this projection map. Then the image of the q points not on $\mathcal{C}$ of a Γ-arc $\mathcal{C}' \neq \mathcal{C}$ containing P_1, together with the image of its tangent line (minus its point on $\mathcal{C}$) at $\mathcal{C} \cap \mathcal{C}'$ is a line of Π_3; similarly for Γ-arcs on $\mathcal{K}$ through P_2. Also, the set of images of tangent lines at P_1 to Γ-arcs different from $\mathcal{C}$ through P_1, together with the image of $\Pi_2\backslash P_1 P_2$, is also a line of Π_3 and similarly for P_2. So a set of $(q+1)^2$ points of Π_3 containing two sets of $q+1$ mutually skew lines is obtained, and lines of different sets intersect in exactly one point; hence these $(q+1)^2$ points are the points of a hyperbolic quadric $\mathcal{H}$. It follows that the image $\mathcal{D}'$ under ρ' of any Γ-arc $\mathcal{D}$ not containing P_1 nor P_2 is a conic section of $\mathcal{H}$. Hence $\mathcal{D}$ is, as the intersection of a plane and a quadratic cone $P_1 P_2 \mathcal{D}'$, itself a conic.

Now consider three points P_0, P_1, P_2 of $\mathcal{K}$ which form a triangle in $\mathrm{PG}(2, q)$. Let $\mathcal{V}_2^4$ be the quadric Veronesean in $\mathrm{PG}(5, q)$ associated with $\mathrm{PG}(2, q)$, and denote for each point or subset $\mathcal{E}$ of $\mathrm{PG}(2, q)$ the corresponding point or subset on $\mathcal{V}_2^4$ by $\mathcal{E}^*$. The set of conics of $\mathcal{V}_2^4$ will be denoted by Γ^*. Since $\mathcal{V}_2^4$ satisfies in particular (a), (b), (c), $\mathcal{V}_2^4$ may be treated as a Veronesean cap and thus appropriate notation may be used. The Γ-arcs $[P_0, P_1], [P_1, P_2], [P_2, P_0]$ generate $\mathrm{PG}(5, q)$, because the space they generate contains both $T(P_0)$ and $[P_1, P_2]$, and by the first section of the proof this space is 5-dimensional.

Now project $\mathcal{K}\backslash[P_0, P_1]$ and $\mathcal{V}_2^4\backslash[P_0^*, P_1^*]$ from the planes containing $[P_0, P_1]$ and $[P_0^*, P_1^*]$ onto the planes Π_2 and Π_2^*; then Π_2 and the Γ-plane of $[P_0, P_1]$ are skew, and Π_2^* and the Γ^*-plane of $[P_0^*, P_1^*]$ are skew. There is a collineation σ from Π_2 to Π_2^* which maps the projection of a point $P \in \mathcal{K}\backslash[P_0, P_1]$ onto the projection of P^*. Let θ be the field automorphism associated to σ. Then there is a collineation (of conics, that is, a bijection preserving the cross-ratio, up to a field automorphism) between $[P_0, P_2]$ and $[P_0^*, P_2^*]$, and between $[P_1, P_2]$ and $[P_1^*, P_2^*]$, with associated field automorphism θ, and mapping P onto P^*. By extending these collineations to the Γ-planes $\langle[P_0, P_2]\rangle$ and $\langle[P_1, P_2]\rangle$ of $[P_0, P_2]$ and $[P_1, P_2]$, and permuting the indices, it follows that there are collineations $\alpha_i : \langle[P_j, P_k]\rangle \rightarrow \langle[P_j^*, P_k^*]\rangle$, for all i, j, k, with $\{i, j, k\} = \{0, 1, 2\}$, with associated field automorphism θ, mapping P to P^*, for all $P \in [P_0, P_1] \cup [P_1, P_2] \cup [P_2, P_0]$.

Now α_0 and α_1 extend to a common collineation α' between $\langle [P_1, P_2], [P_0, P_2] \rangle$ and $\langle [P_1^*, P_2^*], [P_0^*, P_2^*] \rangle$. Consider any Γ-arc $\mathcal{C}$, with $P_0, P_1, P_2 \notin \mathcal{C}$. Let R be the intersection of the line $P_0 P_1$ and the tangent line l_2 of $[P_0, P_1]$ at $\mathcal{C} \cap [P_0, P_1]$. Consider the tangent lines l_0 and l_1 of $[P_1, P_2]$ and $[P_0, P_2]$ at $\mathcal{C} \cap [P_1, P_2]$ and $\mathcal{C} \cap [P_0, P_2]$. Letting $\langle \mathcal{C} \rangle$ play the role of Π_2 in the first part of the proof, it follows that the subspace $\langle l_0, l_1, l_2 \rangle$ is 4-dimensional and meets $\mathcal{K}$ in $\mathcal{C}$. Choose the plane Π_2' in the second section of the proof so that it contains $P_0 P_1$. As $\langle l_0, l_1 \rangle$ is a solid in the hyperplane $\langle [P_1, P_2], [P_2, P_0] \rangle$, it follows that $\langle l_0, l_1 \rangle$ intersects the line $P_0 P_1$ in R. Hence the restriction of α' to $P_0 P_1$ coincides with the restriction of α_2 to $P_0 P_1$. So there exists a collineation $\alpha : \mathrm{PG}(5, q) \to \mathrm{PG}(5, q)$ such that α induces α_i on $\langle [P_j, P_k] \rangle$, for all i, j, k, with $\{i, j, k\} = \{0, 1, 2\}$.

Now let $P \in \mathcal{K}$ be arbitrary, but not belonging to the Γ-planes containing $[P_0, P_1], [P_1, P_2], [P_2, P_0]$. Put $P_i' = [P, P_i] \cap [P_j, P_k]$, for $\{i, j, k\} = \{1, 2, 3\}$. Then, by $(*)$, the point P is the intersection of $\mathcal{K}$ with Φ, where

$$\Phi = T(P_0) P_0' \cap T(P_1) P_1' \cap T(P_2) P_2'.$$

Hence P^* is the intersection of $\mathcal{V}_2^4$ with Φ^*, where

$$\Phi^* = T(P_0^*) P_0'^* \cap T(P_1^*) P_1'^* \cap T(P_2^*) P_2'^*.$$

It is now shown that $\Phi = \{P\}$. First, $[P_1, P_2] \subset T(P_1) P_0'$; hence

$$\mathrm{PG}(5, q) = T(P_0) \langle [P_1, P_2] \rangle = T(P_0) T(P_1) P_0' P_1'.$$

Consequently, $T(P_0) P_0' \cap T(P_1) P_1'$ is a line of $\mathrm{PG}(5, q)$, which contains P and $P' = t(P_0, [P_0, P_1]) \cap t(P_1, [P_0, P_1])$. As $P' \neq P_2'$ and as $T(P_2) \cap \langle [P_0, P_1] \rangle$ is empty, the assumption $P' \in T(P_2) P_2'$ would imply that $T(P_2) P_2'$ is a hyperplane, a contradiction. Hence $\{P\} = \Phi$. However, in a similar way, $\{P^*\} = \Phi^*$. It follows that $P\alpha = P^*$, and the theorem is proved. $\square$

The projection map ρ in the second section of the proof shows the following.

Lemma 4.35. *If $n = 2$ and $\mathcal{C} \in \Gamma$, then the planes $T(P)$, with $P \in \mathcal{C}$, generate a hyperplane of $\mathrm{PG}(5, q)$ which meets $\mathcal{K}$ precisely in $\mathcal{C}$.*

Consider now the general case.

Theorem 4.36. *If $n \geq 2$ and $N \geq n(n + 3)/2$, then $N = n(n + 3)/2$ and $\mathcal{K}$ is the quadric Veronesean $\mathcal{V}_n$ of dimension n.*

Proof. The proof proceeds by induction on n, the case $n = 2$ being proved in Theorem 4.34. So suppose now that $n > 2$. By Theorem 4.33, $N = n(n + 3)/2$. Select two distinct hyperplanes $\mathcal{S}_1$ and $\mathcal{S}_2$ of $\mathcal{S}$. These correspond to two Veronesean sub-caps $\mathcal{K}_1$ and $\mathcal{K}_2$ of $\mathcal{K}$ of dimension $n - 1$. It will be shown that $\langle \mathcal{K}_1 \rangle$ and $\langle \mathcal{K}_2 \rangle$ have dimension $(n - 1)(n + 2)/2$. By Theorem 4.33, the dimension n_i of $\langle \mathcal{K}_i \rangle$ is at most $(n - 1)(n + 2)/2$, with $i = 1, 2$. From the proof of Theorem 4.33, $\mathrm{PG}(N, q) = T(P) \langle \mathcal{K}_i \rangle$ for any $P \in \mathcal{K} \backslash \mathcal{K}_i$, and hence, by Lemma 4.32,

$n(n+3)/2 \leq 1 + n + n_i \leq n(n+3)/2$, implying that $n_i = (n-1)(n+2)/2$, for $i = 1, 2$. So $\langle \mathcal{K}_1 \rangle$ and $\langle \mathcal{K}_2 \rangle$ have dimension $(n-1)(n+2)/2$. Put $\langle \mathcal{K}_i \rangle = \Omega_i$, with $i = 1, 2$. The caps $\mathcal{K}_1$ and $\mathcal{K}_2$ meet in a Veronesean cap $\mathcal{K}_3$ of dimension $n - 2$.

Let $\langle \mathcal{K}_3 \rangle = \Omega$. Considering $\mathcal{K}_3$ as a Veronesean sub-cap of $\mathcal{K}_1$, the dimension of Ω is $(n-2)(n+1)/2$. Now consider a Γ-arc $\mathcal{C}$ not meeting $\mathcal{K}_1 \cap \mathcal{K}_2$. For $q > 2$, it is immediate that $PG(N, q)$ is generated by $\mathcal{K}_1, \mathcal{K}_2$ and $\mathcal{C}$.

Now let $q = 2$. Let $P = \mathcal{C} \backslash (\mathcal{K}_1 \cup \mathcal{K}_2)$. If $P' \in \mathcal{K} \backslash (\mathcal{K}_1 \cup \mathcal{K}_2 \cup \mathcal{C})$, then consider a Veronesean sub-cap $\mathcal{K}'$ of $\mathcal{K}$ of dimension two containing P and P'. The space $\langle \mathcal{K}' \rangle$ of dimension five is generated by $\mathcal{K}_1 \cap \mathcal{K}'$, $\mathcal{K}_2 \cap \mathcal{K}'$ and P. Hence $P' \in \langle \mathcal{K}_1, \mathcal{K}_2, P \rangle$. So, also for $q = 2$, $PG(N, q)$ is generated by $\mathcal{K}_1$, $\mathcal{K}_2$ and $\mathcal{C}$. As $N = n(n+3)/2$ it follows that $\Omega_1 \cap \Omega_2 = \Omega$ and that $\Omega_1 \Omega_2 \cap \langle \mathcal{C} \rangle$ is a line. Also, by induction, the caps $\mathcal{K}_i$, for $i = 1, 2, 3$, are projectively equivalent to quadric Veroneseans and can be identified as such.

Now proceed very similarly as in the proof of Theorem 4.34. Let $\mathcal{V}_n$ be the quadric Veronesean in $PG(N, q)$ associated with $\mathcal{S} = PG(n, q)$ and denote for any point or point set $\mathcal{B}$ of $\mathcal{S}$ the corresponding point or point set on $\mathcal{V}_n$ by $\mathcal{B}^*$. It is now shown that $\mathcal{K}$ and $\mathcal{V}_n$ are projectively equivalent and that a collineation of $PG(N, q)$ can be chosen which maps any point $P \in \mathcal{K}$ to the point $P^* \in \mathcal{V}_n$. These assertions may be included in the induction hypothesis as they are valid for the case $n = 2$, by Theorem 4.34. Hence there is a collineation $\alpha_0 : \langle \mathcal{C} \rangle \rightarrow \langle \mathcal{C}^* \rangle$ with associated field automorphism θ_0, and collineations $\alpha_i : \Omega_i \rightarrow \langle \mathcal{K}_i^* \rangle$, with associated field automorphisms $\theta_i, i = 1, 2$, mapping P to P^*, for every P in $\mathcal{C}$ and $\mathcal{K}_i$, $i = 1, 2$; here α_0 is obtained by restriction to $\mathcal{C}$, after considering a Veronesean sub-cap of dimension two of $\mathcal{K}$ containing $\mathcal{C}$.

Let $\mathcal{K}'$ be a Veronesean sub-cap of dimension two of $\mathcal{K}$ containing $\mathcal{C}$, and let $\mathcal{V}_2$ be the corresponding Veronese variety on $\mathcal{V}_n$. Considering the restriction of α_i to $\mathcal{K}' \cup \mathcal{K}_i$, with $i = 1, 2$, it follows from Theorem 4.34 that $\theta_0 = \theta_1 = \theta_2$, that there exists a collineation α' from $\Omega_1 \Omega_2$ onto $\langle \mathcal{K}_1^*, \mathcal{K}_2^* \rangle$ having as restriction to Ω_1 and Ω_2 the collineations α_1 and α_2, and that α_0 and α' coincide on $\langle \mathcal{C} \rangle \cap \Omega_1 \Omega_2$. Hence there exists a collineation $\alpha : PG(N, q) \rightarrow PG(N, q)$ such that $P\alpha = P^*$, for all $P \in \mathcal{C} \cup \mathcal{K}_1 \cup \mathcal{K}_2$. Now let P be any other point of $\mathcal{K}$. Then there is a unique Veronesean cap of dimension two on $\mathcal{K}$ containing $\mathcal{C}$ and P, namely, the plane in $\mathcal{S}$ defined by the line $\mathcal{C}$ of $\mathcal{S}$ and the point P. It has a unique Γ-arc in common with both $\mathcal{K}_1$ and $\mathcal{K}_2$, and hence, as in the proof of Theorem 4.34, it follows that $P\alpha = P^*$. The theorem is thus established.   □

Now the main result of this section is proved, keeping all the previous notation.

Theorem 4.37. *If $n = 3$, then either $N = 8$ or 9. In the latter case, $\mathcal{K}$ is the quadric Veronesean of dimension three.*

Proof. Consider a quadric sub-Veronesean $\mathcal{K}_1$ of dimension two on $\mathcal{K}$. It will be shown that the 5-dimensional space $\langle \mathcal{K}_1 \rangle$ contains at most one point of $\mathcal{K} \backslash \mathcal{K}_1$. Let $P, P' \in (\mathcal{K} \backslash \mathcal{K}_1) \cap \langle \mathcal{K}_1 \rangle$, $P \neq P'$. The set of all Γ-arcs contained in $\mathcal{K}_1$ is denoted by Γ_1. The unique Γ-arc $\mathcal{C}$ containing P, P' has some point P_1 in common with $\mathcal{K}_1$ and therefore is entirely contained in $\langle \mathcal{K}_1 \rangle$. It follows that $T(P_1)$ is completely contained in $\langle \mathcal{K}_1 \rangle$. Let Π_2 be a plane containing an element of Γ_1, but not containing P_1. Then, by comparing dimensions, it follows that there exists a point U of $\mathrm{PG}(N, q)$ in $T(P_1) \cap \Pi_2$. By Lemma 4.32, there is a Γ-arc $\mathcal{C}'$ through P_1 with tangent line $P_1 U$ at P_1. Hence the plane $\langle \mathcal{C}' \rangle$ meets Π_2 in a point not belonging to $\mathcal{K}$, contradicting (b). So it has been shown that $\langle \mathcal{K}_1 \rangle$ contains at most one point of $\mathcal{K} \backslash \mathcal{K}_1$. Note that the last part of the argument shows that no space $T(P_1)$, $P_1 \in \mathcal{K}_1$, is contained in $\langle \mathcal{K}_1 \rangle$.

Assume that the point $P \in \mathcal{K} \backslash \mathcal{K}_1$ is contained in $\langle \mathcal{K}_1 \rangle$. Choose a second quadric sub-Veronesean $\mathcal{K}_2 \neq \mathcal{K}_1$ of dimension two on $\mathcal{K}$ with $P \notin \mathcal{K}_2$. The intersection $\mathcal{C}'' = \mathcal{K}_1 \cap \mathcal{K}_2$ belongs to Γ. It will be shown that $\langle \mathcal{K}_1 \rangle \cap \langle \mathcal{K}_2 \rangle = \langle \mathcal{C}'' \rangle$. Assume, by way of contradiction, that $\langle \mathcal{K}_1 \rangle \cap \langle \mathcal{K}_2 \rangle$ contains a solid Π_3 containing $\langle \mathcal{C}'' \rangle$. Let $P'' \in \mathcal{K}_2 \backslash \mathcal{K}_1$ be arbitrary. By comparing dimensions and as $P'' \neq P$, the tangent plane at P'' of $\mathcal{K}_2$ has at least one point $R \neq P''$ in common with Π_3. Hence there is a point $R' \in \mathcal{C}''$ such that RP'' is tangent to $[R', P'']$. So the line $R'R$ is contained in $\langle [P'', R'] \rangle$, implying that it must be a tangent line to $[P'', R']$ because otherwise $\langle \mathcal{K}_1 \rangle$ contains a point of $[P'', R'] \backslash \{R'\}$, contradicting that $\{P\} = (\mathcal{K} \backslash \mathcal{K}_1) \cap \langle \mathcal{K}_1 \rangle$. Consequently, $T(R')$ is generated by the tangent plane of $\mathcal{K}_1$ at R' and the line $R'R$. Hence $T(R')$ is contained in $\langle \mathcal{K}_1 \rangle$, contradicting the last remark in the previous paragraph. So it has been shown that $\langle \mathcal{K}_1 \rangle \cap \langle \mathcal{K}_2 \rangle = \langle \mathcal{C}'' \rangle$. Hence $N \geq 8$.

The assertion for $N = 9$ follows from Theorem 4.36.   □

Theorem 4.38. (i) *When $n = 3$ and $N = 8$, then there exists a projective space $\Pi_9 = \mathrm{PG}(9, q)$ containing $\mathrm{PG}(8, q)$, a point R of Π_9, and a quadric Veronesean $\overline{\mathcal{K}}$ of dimension three in Π_9, with $R \notin \overline{\mathcal{K}}$, such that $\mathcal{K}$ is the projection of $\overline{\mathcal{K}}$ from R onto $\mathrm{PG}(8, q)$.*

(ii) *The Veronesean $\overline{\mathcal{K}}$ can be chosen in such a way that $\mathcal{K} \cap \overline{\mathcal{K}}$ is the union of two quadric sub-Veroneseans of dimension two of both $\mathcal{K}$ and $\overline{\mathcal{K}}$, and $\overline{\mathcal{K}}$ is uniquely determined by this intersection, by the point R and by one point $P' \in \overline{\mathcal{K}}$ with P' not belonging to $\mathcal{K} \cap \overline{\mathcal{K}}$ with $RP' \cap \mathcal{K}$ non-empty.*

Proof. The proof of Theorem 4.37 yields the existence of two quadric sub-Veroneseans $\mathcal{K}_1$ and $\mathcal{K}_2$ of $\mathcal{K}$ such that $\langle \mathcal{K}_1, \mathcal{K}_2 \rangle = \mathrm{PG}(8, q)$ and $\langle \mathcal{K}_1 \rangle \cap \langle \mathcal{K}_2 \rangle$ is a Γ-plane. Now embed $\mathrm{PG}(8, q)$ as a hyperplane in some 9-dimensional space $\mathrm{PG}(9, q)$ and let R be any point of $\mathrm{PG}(9, q) \backslash \mathrm{PG}(8, q)$. Let $P \in \mathcal{K} \backslash (\mathcal{K}_1 \cup \mathcal{K}_2)$ and choose $P' \in PR, P \neq P' \neq R$, arbitrarily. Let $Q \in \mathcal{K} \backslash (\mathcal{K}_1 \cup \mathcal{K}_2)$ be arbitrary, $Q \neq P$. The conic $[P, Q]$ either has different points P_1, P_2 in common with $\mathcal{K}_1$ and $\mathcal{K}_2$ or has a point Z in common with $\mathcal{K}_1 \cap \mathcal{K}_2$. In the first case, define the point $Q\theta$ as the intersection of the plane $P_1 P_2 P'$ with the line RQ; this is well defined since both objects are contained in the solid $P_1 P_2 PR$. In the second case, define $Q\theta$ as the intersection of the plane $t(Z, [P, Z])P'$ with the line RQ. If $U \in \mathcal{K}_1 \cup \mathcal{K}_2$, then put $U\theta = U$. Also, $P\theta = P'$. Define $\overline{\mathcal{K}}$ as the set of points $Q\theta$ such that $Q \in \mathcal{K}$. Then θ is a well-defined map from $\mathcal{K}$ to $\overline{\mathcal{K}}$. It follows that θ is bijective and that its inverse is the restriction to $\overline{\mathcal{K}}$ of a projection mapping with centre R and image $\mathrm{PG}(8, q)$. Note that $\langle \overline{\mathcal{K}} \rangle = \mathrm{PG}(9, q)$. It is now shown that for every conic $\mathcal{C} \in \Gamma$, the set $\mathcal{C}\theta$ is a conic on $\overline{\mathcal{K}}$.

If $\mathcal{C} \subset \mathcal{K}_1 \cup \mathcal{K}_2$, then this is immediate. Also, if $\mathcal{C}$ contains P, then it follows from the construction. Now suppose that $P \notin \mathcal{C}$ and that $\mathcal{C}$ is not contained in $\mathcal{K}_1 \cup \mathcal{K}_2$. Then P and $\mathcal{C}$ are contained in the 5-dimensional space $\mathrm{PG}(5, q)$ containing the unique quadric sub-Veronesean $\mathcal{V}_2$ of dimension two which contains both P and $\mathcal{C}$. Now, $\mathcal{V}_2$ either has distinct conics $\mathcal{C}_1$ and $\mathcal{C}_2$ in common with $\mathcal{K}_1$ and $\mathcal{K}_2$ or $\mathcal{V}_2$ contains the conic $\mathcal{K}_1 \cap \mathcal{K}_2$.

Consider the first case and let $U \in \mathcal{C}$ be arbitrary. If $[U, P]$ contains distinct points of $\mathcal{K}_1$ and $\mathcal{K}_2$, then $U\theta$ is contained in the 5-dimensional space $\langle \mathcal{C}_1, \mathcal{C}_2, P' \rangle$. If $[U, P]$ contains the unique common point W of $\mathcal{C}_1$ and $\mathcal{C}_2$, then the tangents at W of $[U, P], \mathcal{C}_1, \mathcal{C}_2$ are coplanar by (c), and so again $U\theta$ is contained in $\langle \mathcal{C}_1, \mathcal{C}_2, P' \rangle$. If $R \in \langle \mathcal{C}_1, \mathcal{C}_2, P' \rangle$, then $P \in \langle \mathcal{C}_1, \mathcal{C}_2 \rangle$, so $P \in \mathcal{C}_1 \cup \mathcal{C}_2$, a contradiction. Hence, in this first case, $\mathcal{C}\theta$ is the intersection of the cone $R\mathcal{C}$ with $\langle \mathcal{C}_1, \mathcal{C}_2, P' \rangle$, implying that $\mathcal{C}\theta$ is a conic on $\overline{\mathcal{K}}$.

Now consider the second case. If $U \in \mathcal{C}$, then $U\theta$ is contained in the 5-dimensional space Π_5 generated by P' and all tangent planes of $\mathcal{V}_2$ at points of $\mathcal{K}_1 \cap \mathcal{K}_2$; see also Lemma 4.35. If R were contained in Π_5, then P would be in Π_5, so $P \in \Pi_5 \cap \mathcal{K}$; hence $P \in \mathcal{K}_1 \cap \mathcal{K}_2$ by Lemma 4.35, a contradiction. Hence, in this second case, $\mathcal{C}\theta$ is the intersection of the cone $R\mathcal{C}$ with Π_5, implying as before that $\mathcal{C}\theta$ is a conic on $\overline{\mathcal{K}}$.

Therefore it follows that every two points of $\overline{\mathcal{K}}$ are contained in a unique conic which is the image under θ of some element of Γ. Let $\overline{\Gamma}$ be the set of all these conics on $\overline{\mathcal{K}}$. Then it has been shown that $\overline{\mathcal{K}}$ satisfies (a) for $\overline{\Gamma}$. Let Π_2 and Π_2' be two planes of $\mathrm{PG}(9, q)$ containing the images under θ of distinct Γ-arcs $\mathcal{C}$ and $\mathcal{C}'$. Suppose $\Pi_2 \cap \Pi_2' \neq \emptyset$. As $\langle \mathcal{C}, \mathcal{C}' \rangle$ is at least 4-dimensional, the point R does not belong to $\Pi_2 \Pi_2'$ and $|\Pi_2 \cap \Pi_2'| = 1$. So $\langle \mathcal{C} \rangle \cap \langle \mathcal{C}' \rangle \neq \emptyset$, and consequently $\mathcal{C} \cap \mathcal{C}'$ is a point by (b). It follows that $\Pi_2 \cap \Pi_2'$ is a point of $\overline{\mathcal{K}}$. This shows that $\overline{\mathcal{K}}$ satisfies (b).

Finally, it is shown that (c) is satisfied for $\overline{\mathcal{K}}$. Therefore, let $V \in \overline{\mathcal{K}}$ and let $\overline{\mathcal{C}}$ be a conic of $\overline{\mathcal{K}}$ which is the image under θ of an element of Γ; assume also that $V \notin \overline{\mathcal{C}}$. By (c) applied to $\mathcal{K}$, the tangents at V of the elements of $\overline{\Gamma}$ which contain V and a point of $\overline{\mathcal{C}}$, are contained in a solid Π_3 containing R.

First let $q > 2$. By considering θ^{-1} it follows that all elements of $\overline{\Gamma}$ which contain V and a point of $\overline{\mathcal{C}}$ belong to the 5-dimensional space Π_5' generated by $\overline{\mathcal{C}}$ and two elements $\overline{\mathcal{C}}_1$ and $\overline{\mathcal{C}}_2$ of $\overline{\Gamma}$, defined by V and distinct points $\overline{U}_1$ and $\overline{U}_2$ of $\overline{\mathcal{C}}$. This space Π_5' does not contain R, as otherwise, by applying θ^{-1}, there arises a Veronesean sub-cap of dimension 2 on $\mathcal{K}$ contained in a 4-dimensional space. It now follows that the tangents at V of the elements of $\overline{\Gamma}$ which contain V and a point of $\overline{\mathcal{C}}$ are contained in the plane $\Pi_3 \cap \Pi_5'$. By Theorem 4.36, $\overline{\mathcal{K}}$ is a quadric Veronesean of dimension 3.

Now let $q = 2$. If it is shown that the image under θ of the point set of any Veronesean cap of dimension 2 on $\mathcal{K}$ is contained in a 5-dimensional space, then the argument of the previous paragraph applies and the theorem is proved. This is true for $\mathcal{K}_1$ and $\mathcal{K}_2$. Now consider the set $\mathcal{K}_3 = (\mathcal{K} \backslash (\mathcal{K}_1 \cup \mathcal{K}_2)) \cup (\mathcal{K}_1 \cap \mathcal{K}_2)$. In the present situation, $[P, Q]$ contains a point U_Q of $\mathcal{K}_1 \cap \mathcal{K}_2$, for all $Q \in \mathcal{K} \backslash (\mathcal{K}_1 \cup \mathcal{K}_2)$, with $P \neq Q$. All lines $t(U_Q, [P, U_Q])$ belong to a common 4-dimensional space Π_4 which intersects the Veronesean cap of dimension 2 defined by P and $\mathcal{K}_1 \cap \mathcal{K}_2$ in $\mathcal{K}_1 \cap \mathcal{K}_2$. Hence all planes $t(U_Q, [P, U_Q])P'$ belong to a common 5-dimensional space Π_5''.

This space Π_5'' does not contain R as otherwise Π_4 contains P, a contradiction. By construction, all corresponding points $Q\theta$ are contained in Π_5'', for all Q in the set $\mathcal{K}\backslash(\mathcal{K}_1 \cup \mathcal{K}_2)$, with $P \neq Q$. These points $Q\theta$ together with P' and $\mathcal{K}_1 \cap \mathcal{K}_2$ form the point set that is the image under θ of the Veronesean sub-cap $\mathcal{K}_3$ of dimension 2.

Last, consider a Veronesean cap $\mathcal{K}_4$ of dimension 2 on $\mathcal{K}$ other than $\mathcal{K}_1, \mathcal{K}_2, \mathcal{K}_3$, and hence not containing $\mathcal{K}_1 \cap \mathcal{K}_2$. Put $\{Z\} = \mathcal{K}_1 \cap \mathcal{K}_2 \cap \mathcal{K}_4$ and let Q_1, Q_2 be the other points of $\mathcal{K}_3 \cap \mathcal{K}_4$. Put $\mathcal{C}_i = \mathcal{K}_4 \cap \mathcal{K}_i$, with $i = 1, 2, 3$. By (c), the tangent l to $\mathcal{C}_3$ at Z is contained in $\langle \mathcal{C}_1, \mathcal{C}_2 \rangle$. The conic $\mathcal{C}_3\theta$ is $\{Z, Q_1\theta, Q_2\theta\}$, and, from the construction of θ, it follows that l is tangent to $\mathcal{C}_3\theta$ at Z. Hence $\langle \mathcal{K}_4\theta \rangle = \langle \mathcal{C}_1, \mathcal{C}_2, \mathcal{C}_3\theta \rangle$ is 5-dimensional and does not contain R, as otherwise $\mathcal{K}_4$ is in a 4-dimensional space.

The theorem is thus established. $\square$

Lemma 4.39. *If $N < n(n+3)/2$, then there exist two distinct Veronesean sub-caps $\mathcal{K}_1, \mathcal{K}_2$ of dimension $n-1$ such that $\langle \mathcal{K}_1, \mathcal{K}_2 \rangle = \mathrm{PG}(N, q)$.*

Proof. Suppose $M < n(n+3)/2$. Coordinatise the projective space $\mathrm{PG}(n, q)$ with respect to a basis $E_0, E_1, \ldots, E_n$ of the underlying vector space and consider the points $P_i^* = \mathbf{P}(E_i)$ and the points $P_{ij}^* = \mathbf{P}(E_i - E_j)$, with $i, j \in \{0, 1, \ldots, n\}$ and $i \neq j$. Note that $P_{ij}^* = P_{ji}^*$; so it may be assumed that $i < j$. Denote the corresponding points on $\mathcal{K}$ by P_i and P_{ij}. Let $P^* \in \mathrm{PG}(n, q)$ and let $l(P^*)$ be the minimal number of points of $\{P_0^*, P_1^*, \ldots, P_n^*\}$ needed to generate a subspace containing P^*. Put $\mathcal{S} = \{P_k, P_{ij} \mid 0 \leq k \leq n \text{ and } 0 \leq i < j \leq n\}$. For any $P \in \mathcal{K}$, let P^* be the corresponding point of $\mathrm{PG}(n, q)$. If $l(P^*) = 1$, then $P \in \langle \mathcal{S} \rangle$. If $l(P^*) = 2$, then P belongs to some plane $P_i P_j P_{ij}$, with $i < j$, and so belongs again to $\langle \mathcal{S} \rangle$.

Now, assume that $l(P^*) > 2$ and, first, take $q > 2$. Let $P^* = \mathbf{P}(\sum r_i E_i)$, with $r_i \in \mathbf{F}_q$, and $r_i \neq 0$ for $i \in \{0, 1, \ldots, l(P^*) - 1\}$, but with $r_i = 0$ otherwise. Then take a line m^* of $\mathrm{PG}(n, q)$ through P^* and the point $Q^* = \mathbf{P}(D)$, where the vector D is defined as follows. If not all r_i are equal, say $r_0 \neq r_1$, let $D = E_0 + E_1$; if all r_i are equal, let $D = E_0 + tE_1$, with $t \neq 0, 1$. Then m^* contains three distinct points Q_1^*, Q_2^*, Q_3^* such that $l(Q_i^*) \leq l(P^*) - 1$, for $i = 1, 2, 3$. By induction on $l(P^*)$, it follows that $Q_1, Q_2, Q_3 \in \langle \mathcal{S} \rangle$, and hence $P \in Q_1 Q_2 Q_3 \subset \langle \mathcal{S} \rangle$. If $q = 2$, then, without loss of generality, let $P^* = \mathbf{P}(D = \sum E_i)$, with $0 \leq i \leq l(P^*) - 1$. Hence P is contained in the 5-dimensional space Π_5 generated by the Veronesean sub-cap of dimension 2 determined by

$$\mathbf{P}(E_0), \ \mathbf{P}(E_1), \ \mathbf{P}(E_0 + E_1 + D), \ \mathbf{P}(E_0 + E_1), \ \mathbf{P}(E_0 + D), \ \mathbf{P}(E_1 + D);$$

note that these six points correspond to six points of $\mathrm{PG}(N, 2)$ which generate Π_5. By induction it now follows that $P \in \langle \mathcal{S} \rangle$. Hence $\langle \mathcal{S} \rangle = \mathrm{PG}(N, q)$.

Since $N < |\mathcal{S}| - 1 = n(n+3)/2$, some element P of $\mathcal{S}$ satisfies $P \in \langle \mathcal{S} \backslash \{P\} \rangle$. Without loss of generality, let $P = P_0$ or $P = P_{01}$. In the first case choose the two Veronesean sub-caps $\mathcal{K}_1$ and $\mathcal{K}_2$ of dimension $n-1$ as being determined by the hyperplanes $\mathbf{V}(X_0)$ and $\mathbf{V}(X_0 + X_1 + \cdots + X_n)$ of $\mathrm{PG}(n, q)$, while in the second case choose them as being determined by the hyperplanes $\mathbf{V}(X_0)$ and $\mathbf{V}(X_1)$ of $\mathrm{PG}(n, q)$. $\square$

Finally, the main result is shown.

Theorem 4.40. *There exists a projective space* $\mathrm{PG}(n(n+3)/2, q)$ *containing the space* $\mathrm{PG}(N, q)$, *a subspace* Π *of* $\mathrm{PG}(n(n+3)/2, q)$ *skew to* $\mathrm{PG}(N, q)$, *and a quadric Veronesean* $\mathcal{V}_n$ *of dimension* n *in* $\mathrm{PG}(n(n+3)/2, q)$, *with* $\Pi \cap \mathcal{V}_n = \emptyset$, *such that* $\mathcal{K}$ *is the bijective projection of* $\mathcal{V}_n$ *from* Π *onto* $\mathrm{PG}(N, q)$. *The subspace* Π *can be empty, in which case* $\mathcal{K}$ *is the quadric Veronesean* $\mathcal{V}_n$.

Proof. By Theorems 4.34, 4.36, 4.37, 4.38, the theorem is already established for $n = 2, 3$ and $N = n(n+3)/2$.

Suppose $N < n(n+3)/2$. Let $\mathcal{K}_1, \mathcal{K}_2$ be as in Lemma 4.39. Embed $\mathrm{PG}(N, q)$ as a hyperplane in a projective space $\mathrm{PG}(N+1, q)$ and let R be any point of the difference $\mathrm{PG}(N+1, q) \backslash \mathrm{PG}(N, q)$. Further, let P be any point of $\mathcal{K} \backslash (\mathcal{K}_1 \cup \mathcal{K}_2)$, and choose arbitrarily a point $P' = P\theta$ other than P and R on the line RP. Also, choose an element $\mathcal{C}$ of Γ through P which has different points P_1 and P_2 in common with $\mathcal{K}_1$ and $\mathcal{K}_2$. As in the proof of Theorem 4.38, define $Q\theta$ for $Q \in \mathcal{C}$. For $Q \in \mathcal{K}_1 \cup \mathcal{K}_2$, let $Q\theta = Q$. Now, let $Q \in \mathcal{K}$ be arbitrary, but not in $\mathcal{K}_1 \cup \mathcal{K}_2 \cup \mathcal{C}$. Then there is a Veronesean sub-cap $\mathcal{V}_2$ of dimension two of $\mathcal{K}$ containing $\mathcal{C}$ and Q; also $\mathcal{V}_2$ has different conics $\mathcal{C}_1$ and $\mathcal{C}_2$ in common with $\mathcal{K}_1$ and $\mathcal{K}_2$. Define $Q\theta$ as the intersection of the spaces $\langle \mathcal{C}_1, \mathcal{C}_2, P\theta \rangle$ and RQ. The set of all points $Q\theta$, with $Q \in \mathcal{K}$, is denoted by $\overline{\mathcal{K}}$.

Let $\mathcal{D}$ be any element in Γ. It is shown that $\mathcal{D}\theta$ is a conic. If $\mathcal{C} \cap \mathcal{D} \neq \emptyset$, then this follows immediately from the construction. Assume now that $\mathcal{C} \cap \mathcal{D} = \emptyset$. Consider the unique Veronesean sub-cap $\mathcal{K}_3$ of dimension three containing $\mathcal{C}$ and $\mathcal{D}$. Then $\mathcal{K}_3$ meets $\mathcal{K}_1$ and $\mathcal{K}_2$ in different sub-Veroneseans $\mathcal{V}_2'$ and $\mathcal{V}_2''$ of dimension two, meeting in a conic $\mathcal{D}'$ of Γ. If $P \notin \langle \mathcal{V}_2', \mathcal{V}_2'' \rangle$, then $\mathcal{D}\theta$ is the intersection of the space $\langle \mathcal{V}_2', \mathcal{V}_2'', P\theta \rangle$ with the cone RD, hence it is a conic itself. If $P \in \langle \mathcal{V}_2', \mathcal{V}_2'' \rangle$, then this follows from Theorem 4.38 and its proof. Let $\overline{\Gamma}$ be the set of all conics $\mathcal{D}\theta$ on $\overline{\mathcal{K}}$.

As in the proof of Theorem 4.38, it is shown that (a) and (b) are satisfied for $\overline{\mathcal{K}}$ and $\overline{\Gamma}$, and also (c) for $q > 2$. So let $q = 2$. With this notation, the Veronesean sub-cap $\mathcal{V}_2$ contains the points $P, Q, P_1, P_2, U \in \mathcal{C}_1 \cap \mathcal{C}_2$, the third point Q_1 of $\mathcal{C}_1$ and the third point of $\mathcal{C}_2$. Hence $U \in [P, Q]$. So $t(U, [P, U])$ is in the plane $t(U, \mathcal{C}_1) t(U, \mathcal{C}_2)$. It follows that $t(U, [P, U]) P' \cap RQ = \langle \mathcal{C}_1, \mathcal{C}_2, P' \rangle \cap RQ$. Now consider any Veronesean sub-cap $\widetilde{\mathcal{V}}_2$ of dimension two on $\mathcal{K}$ and let $\widetilde{\mathcal{K}}_3$ be a Veronesean sub-cap of dimension three on $\mathcal{K}$ containing $\widetilde{\mathcal{V}}_2$ and P. Relying on the foregoing and the case $q = 2$ of the proof of Theorem 4.38, it follows that the set $\widetilde{\mathcal{V}}_2\theta$ belongs to a 5-dimensional space which does not contain R. Hence (c) holds.

Induction on N now completes the proof of the main theorem. □

4.2.2 Characterisations of $\mathcal{V}_n^{2^n}$ of the second kind

By Theorem 4.22, for any two distinct points P_1 and P_2 of the Veronesean $\mathcal{V}_2^4$, the tangent planes $\pi(P_1)$ and $\pi(P_2)$ have exactly one point in common. By a classical theorem, the Veronesean $\mathcal{V}_2^4$ over $\mathbf{C}$ is the only surface generating $\mathrm{PG}(n, \mathbf{C})$, $n \geq 5$, which is not a cone (with non-trivial vertex) and which satisfies the property just mentioned. In this section, the aim is to prove similar characterisation theorems in the case of a Galois field $\mathbf{F}_q$ and the Veronesean $\mathcal{V}_n^{2^n}$.

Consider the quadric Veronesean $\mathcal{V}_n^{2^n}$ and the corresponding Veronesean map from $\mathrm{PG}(n,q)$ into $\mathrm{PG}(N_n,q)$, with $N_n = n(n+3)/2$. Then, by Theorem 4.10, the image of an arbitrary hyperplane of $\mathrm{PG}(n,q)$ under the Veronesean map is a quadric Veronesean $\mathcal{V}_{n-1}^{2^{n-1}}$ and the subspace of $\mathrm{PG}(N_n,q)$ generated by it has dimension $N_{n-1} = (n-1)(n+2)/2$. Such a subspace is called a $\mathcal{V}_{n-1}^{2^{n-1}}$-subspace, or, for short, a $\mathcal{V}_{n-1}$-subspace, of $\mathcal{V}_n^{2^n}$ or of $\mathrm{PG}(N_n,q)$: this is an abuse of language, since the subspace does not lie in $\mathcal{V}_n^{2^n}$. The image of a line of $\mathrm{PG}(n,q)$ is a conic. If q is even, then the intersection of all tangent lines of a conic is the *nucleus* of the conic. In the next theorem it is shown that for $n > 2$ the set of all these nuclei is a Grassmann variety. For $n = 2$, by Theorem 4.23 the set of all these nuclei is a plane, called the *nucleus of $\mathcal{V}_2^4$*.

Theorem 4.41. *If q is even, then the set of all nuclei of the conics on $\mathcal{V}_n^{2^n}$, with $n > 2$, is the Grassmann variety $\mathcal{G}_{1,n}$ of the lines of $\mathrm{PG}(n,q)$ and hence generates a subspace of dimension N_{n-1}.*

Proof. Let q be even and $n > 2$. If l is the line of $\mathrm{PG}(n,q)$ determined by the points $\mathbf{P}(x_0, x_1, \ldots, x_n)$ and $\mathbf{P}(y_0, y_1, \ldots, y_n)$, then the image of l under the Veronesean map is the set of points

$$\mathbf{P}(x_0^2 s^2 + y_0^2 t^2, \ldots, x_n^2 s^2 + y_n^2 t^2, x_0 x_1 s^2 + y_0 y_1 t^2$$
$$+ (x_0 y_1 + x_1 y_0)st, \ldots, x_0 x_n s^2 + y_0 y_n t^2 + (x_0 y_n + x_n y_0)st, \ldots,$$
$$x_{n-1} x_n s^2 + y_{n-1} y_n t^2 + (x_{n-1} y_n + x_n y_{n-1})st),$$

with $s, t \in \mathbf{F}_q$ and $(s, t) \neq (0, 0)$. This is a conic $\mathcal{C}$ in the plane generated by the three points

$$\mathbf{P}(x_0^2, \ldots, x_n^2, x_0 x_1, \ldots, x_0 x_n, \ldots, x_{n-1} x_n),$$
$$\mathbf{P}(y_0^2, \ldots, y_n^2, y_0 y_1, \ldots, y_0 y_n, \ldots, y_{n-1} y_n),$$
$$\mathbf{P}(0, \ldots, 0, x_0 y_1 + x_1 y_0, \ldots, x_0 y_n + x_n y_0, \ldots, x_{n-1} y_n + x_n y_{n-1}).$$

It can be checked that the last point is the nucleus of $\mathcal{C}$ and the result follows.   □

The subspace of dimension N_{n-1} generated by the Grassmann variety $\mathcal{G}_{1,n}$ is called the *nucleus* of the Veronesean $\mathcal{V}_n^{2^n}$.

In Section 4.1 it was mentioned that, for q even, all contact hyperplanes of $\mathcal{V}_2^4$ contain the nucleus of $\mathcal{V}_2^4$. For later reference, call this the *nucleus property* of $\mathcal{V}_2^4$. In the next theorem, this is generalised to all $n \geq 2$, but first the definition of *contact hyperplane* in the general case is given.

Let $\mathrm{PG}(n-1,q) = \mathbf{V}(F)$, where $F = u_0 X_0 + u_1 X_1 + \cdots + u_n X_n$, be any hyperplane of $\mathrm{PG}(n,q)$, with $n \geq 2$. If $\mathcal{Q}$ is the quadric whose point set coincides with $\mathrm{PG}(n-1,q)$, then to $\mathcal{Q}$ there corresponds the hyperplane $\mathbf{V}(F')$ of the space $\mathrm{PG}(n(n+3)/2,q)$, where

$$F' = u_0^2 Y_{00} + u_1^2 Y_{11} + \cdots + u_n^2 Y_{nn}$$
$$+ 2u_0 u_1 Y_{01} + \cdots + 2u_0 u_n Y_{0n} + \cdots + 2u_{n-1} u_n Y_{n-1,n}.$$

Such a hyperplane is called a *contact hyperplane* of $\mathcal{V}_n^{2^n}$. The contact hyperplanes of $\mathcal{V}_n^{2^n}$ are the hyperplanes which contain exactly one $\mathcal{V}_{n-1}$-subspace of $\mathcal{V}_n^{2^n}$.

Theorem 4.42. *If q is even, then the nucleus of $\mathcal{V}_n^{2^n}$ is the intersection of all contact hyperplanes of $\mathcal{V}_n^{2^n}$.*

Proof. From the proof of Theorem 4.41, the subspace $\mathbf{V}(Y_{00}, Y_{11}, \ldots, Y_{nn})$ of $\mathrm{PG}(n(n+3)/2, q)$ is the nucleus of $\mathcal{V}_n^{2^n}$. As $\mathbf{V}(u_0^2 Y_{00} + u_1^2 Y_{11} + \cdots + u_n^2 Y_{nn})$, with $(u_0, u_1, \ldots, u_n) \neq (0, 0, \ldots, 0)$, are the contact hyperplanes of $\mathcal{V}_n^{2^n}$, so the nucleus is the intersection of the contact hyperplanes. □

For later reference, call this the *nucleus property* of $\mathcal{V}_n^{2^n}$.

In Theorem 4.25 it was shown that $\mathrm{PG}(5, q)$, with q odd, admits a polarity which maps the set of all conic planes of $\mathcal{V}_2^4$ onto the set of all tangent planes of $\mathcal{V}_2^4$. Similarly it may be shown that $\mathrm{PG}(N_n, q)$, with q odd, admits a polarity θ which maps the set of all $\mathcal{V}_{n-1}$-subspaces of $\mathcal{V}_n^{2^n}$ onto the set of all tangent spaces of $\mathcal{V}_n^{2^n}$. This polarity is represented by the equations

$$\rho v_{00} = y_{00}/2, \ \rho v_{11} = y_{11}/2, \ldots, \rho v_{nn} = y_{nn}/2,$$
$$\rho v_{01} = y_{01}, \ \rho v_{02} = y_{02}, \ldots, \rho v_{n-1,n} = y_{n-1,n},$$

where

$$\mathbf{V}(v_{00} Y_{00} + v_{11} Y_{11} + \cdots + v_{n-1,n} Y_{n-1,n})$$

is a variable hyperplane of $\mathrm{PG}(N_n, q)$. The images of the points of $\mathcal{V}_n^{2^n}$ are the contact hyperplanes of $\mathcal{V}_n^{2^n}$.

Now let $\mathcal{S}_n$ be the set of all $\mathcal{V}_{n-1}$-subspaces of the quadric Veronesean $\mathcal{V}_n$ in $\mathrm{PG}(N_n, q)$, with $N_n = n(n+3)/2$. The set $\mathcal{S}_n$ has the following properties, which can be verified using coordinates:

(a) every two members of $\mathcal{S}_n$ generate a hyperplane of $\mathrm{PG}(N_n, q)$;
(b) every three members of $\mathcal{S}_n$ generate $\mathrm{PG}(N_n, q)$;
(c) no point is contained in every member of $\mathcal{S}_n$;
(d) the intersection of any non-empty collection of members of $\mathcal{S}_n$ is a subspace of dimension $N_i = i(i+3)/2$ for some $i \in \{-1, 0, 1, \ldots, n-1\}$;
(e) if q is even, then there exist three members π, π', π'' of $\mathcal{S}_n$ with

$$\pi \cap \pi' = \pi' \cap \pi'' = \pi'' \cap \pi.$$

For $n = 2$ and arbitrary q, property (d) follows immediately from (a), (b), (c).

From now on, let $\mathcal{S}$ be a collection of $\theta(n) = q^n + q^{n-1} + \cdots + q + 1$ subspaces of dimension $N_{n-1} = (n-1)(n+2)/2$ of $\mathrm{PG}(N_n, q)$, with $N_n = n(n+3)/2$, such that the following conditions are satisfied:

(I) every two members of $\mathcal{S}$ generate a hyperplane of $\mathrm{PG}(N_n, q)$;

(II) every three members of S generate $PG(N_n, q)$;

(III) no point is contained in every member of S.

Definition 4.43. (1) The set S is called a *Veronesean set of subspaces*.

(2) In the particular case where no three members of S meet in the same subspace, necessarily of dimension N_{n-2}, the set S is called an *ovoidal Veronesean set of subspaces*.

(3) A set of subspaces in (1) which is not ovoidal is called *proper*.

(4) If a collection S of subspaces of dimension N_{n-1} satisfies (I), (II) and (III), but no three members of S meet in the same subspace, and if S contains $\theta(n) + 1$ elements, then S is called a *hyperovoidal Veronesean set of subspaces*.

One of the purposes of this section is to classify the proper Veronesean sets of subspaces, and to show that every ovoidal Veronesean set of subspaces is contained in a unique hyperovoidal Veronesean set of subspaces.

Further it is shown that, for $q \geq n$, every Veronesean set of subspaces satisfies the following:

(IV) the intersection of any non-empty collection of members of S is a subspace of dimension $i(i + 3)/2$ for some $i \in \{-1, 0, 1, \ldots, n - 1\}$.

A further condition may be formulated:

(V) if q is even, then there exist $\pi, \pi', \pi'' \in S$ with

$$\pi \cap \pi' = \pi' \cap \pi'' = \pi'' \cap \pi.$$

If S is a proper Veronesean set of subspaces satisfying also (IV), then it will be shown that either it must be the collection of $\mathcal{V}_{n-1}$-subspaces of a quadric Veronesean $\mathcal{V}_n^{2^n}$ in $PG(N_n, q)$, or that q is even and there is a unique subspace $PG(N_{n-1}, q)$ such that $S \cup \{PG(N_{n-1}, q)\}$ is the set of $\mathcal{V}_{n-1}$-subspaces together with the nucleus of a quadric Veronesean $\mathcal{V}_n^{2^n}$ in $PG(N_n, q)$. Also, it will follow that, if S^* is a set of $\theta(n) + 1$ subspaces of dimension $N_{n-1} = (n - 1)(n + 2)/2$ of $PG(N_n, q)$ such that (I), (II), (III) hold for S^* and either also (IV) holds, or $q \geq n$, then q is even and either S^* is the set of all $\mathcal{V}_{n-1}$-subspaces together with the nucleus of a quadric Veronesean $\mathcal{V}_n^{2^n}$ in $PG(N_n, q)$, or it is a hyperovoidal Veronesean set of subspaces.

The proof proceeds by induction on n, but the smallest case $n = 2$ is handled in the course of proving the general case.

It is convenient in many situations to consider the dual projective space. The dual of an object $\mathcal{B}$ of $PG(N_n, q)$ is denoted by $\overline{\mathcal{B}}$. In particular, denote the dual of $PG(N_n, q)$ by $\overline{PG}(N_n, q)$. So consider a set $\overline{S}$ of $\theta(n)$ n-dimensional subspaces of $\overline{PG}(N_n, q)$, satisfying the following properties:

(I′) every two members of $\overline{S}$ meet in a point of $\overline{PG}(N_n, q)$;

(II′) no three members of $\overline{S}$ have a point in common;

(III′) no hyperplane of $\overline{PG}(N_n, q)$ contains all members of $\overline{S}$.

The rough idea of the strategy is to fix one member π of S and to consider all intersections of π with the other elements of S. This allows the use of induction. However, these intersections do not always satisfy (I), (II), (III); if they do not, then another member of $\mathcal{S}$ is considered. To start, the properties of the set $\mathcal{S}_\pi$ of subspaces of π of dimension $N_{n-2} = (n-2)(n+1)/2$ obtained by intersecting π with all elements of $\mathcal{S}\backslash\{\pi\}$ are collected.

Note that, for any element $\overline{\pi}$ of $\overline{\mathcal{S}}$, every point of $\overline{\pi}$ is incident with a unique element of $\overline{\mathcal{S}}\backslash\{\overline{\pi}\}$, by (I') and (II'), except for a unique point, called the *nucleus* of $\overline{\pi}$.

For sake of completeness (IV') and (V') are also formulated:

(IV') any non-empty collection of members of $\overline{\mathcal{S}}$ generates a subspace of dimension $N_n - N_i - 1$ for some $i \in \{-1, 0, 1, \ldots, n-1\}$;

(V') when q is even, then there exists a $2n$-dimensional space containing at least three elements of $\overline{\mathcal{S}}$.

Lemma 4.44. *If $q \geq n$ or if $\mathcal{S}$ satisfies* (IV), *then any two elements of $\mathcal{S}_\pi$, with $\pi \in \mathcal{S}$, generate a hyperplane of π.*

Proof. The lemma is immediate if $\mathcal{S}$ satisfies (IV). So assume that $q \geq n$.
Let $\pi^1, \pi^2 \in \mathcal{S}\backslash\{\pi\}$, with $\pi \cap \pi^1 \neq \pi \cap \pi^2$. Then

$$\langle \pi \cap \pi^1, \pi \cap \pi^2 \rangle \subset \pi \cap \langle \pi^1, \pi^2 \rangle,$$

and the last is a hyperplane of π by (II). Hence, it remains to show that $\langle \pi \cap \pi^1, \pi \cap \pi^2 \rangle$ has dimension at least $N_{n-1} - 1$. This is equivalent to showing that the dimension of $\pi \cap \pi^1 \cap \pi^2$ is at most $2N_{n-2} - N_{n-1} + 1 = n(n-3)/2 = N_{n-3}$. Suppose by way of contradiction that the dimension of $\pi \cap \pi^1 \cap \pi^2$ is larger than N_{n-3}. Then the dimension of $\langle \overline{\pi}, \overline{\pi}^1, \overline{\pi}^2 \rangle$ is at most $3n - 2$. Put

$$\overline{\gamma}^1 = \langle \overline{\pi}, \overline{\pi}^1 \rangle, \quad \overline{\gamma}^2 = \langle \overline{\gamma}^1, \overline{\pi}^2 \rangle.$$

Since $\overline{\pi}$ and $\overline{\pi}^1$ meet in a point, the subspace $\overline{\gamma}^1$ has dimension $2n$. If $\overline{\gamma}^1 = \overline{\gamma}^2$, then $\pi \cap \pi^1 = \pi \cap \pi^2$, contrary to the assumption.

Now it is shown that there is a sequence $(\overline{\pi}^3, \overline{\pi}^4, \ldots, \overline{\pi}^{n+1})$ of elements of $\overline{\mathcal{S}}$ such that, for all $i \in \{2, 3, \ldots, n\}$,

(i) the subspace $\overline{\gamma}^i$ defined inductively by $\overline{\gamma}^i = \langle \overline{\gamma}^{i-1}, \overline{\pi}^i \rangle$ has at least an i-dimensional subspace in common with $\overline{\pi}^{i+1}$,

(ii) $\overline{\gamma}^i$ does not contain π^{i+1}.

Putting $i = n$ in (i) and (ii), these two conditions give a contradiction, in view of the fact that $\overline{\pi}^{n+1}$ has dimension n.

The sequence is now constructed by an inductive argument, adding $\overline{\pi}^2$ to the sequence, putting $\overline{\pi} = \overline{\gamma}^0$, and noting that $\overline{\pi}^2$ has at least one plane in common with $\overline{\gamma}^1$. For this first step, the intersection with $\overline{\pi}^{i+1}$ is larger than asked for in (i). Suppose now that there is already a sequence $(\overline{\pi}^2, \overline{\pi}^3, \ldots, \overline{\pi}^k)$, for some integer k in $\{2, 3, \ldots, n\}$, satisfying (i) and (ii) for all $i \in \{1, 2, \ldots, k-1\}$. First, note that the dimension of $\overline{\gamma}^k$ is bounded by

$$\dim \overline{\gamma}^2 + (n-2) + (n-3) + \cdots + (n-(k-1))$$
$$\leq 3n - 2 + (n-2)(n-1)/2 = (n(n+3)/2) - 1;$$

hence $\overline{\gamma}^k$ is contained in a hyperplane of $\overline{\mathrm{PG}}(N_n, q)$. Condition (III') therefore guarantees the existence of a subspace $(\overline{\pi}^{k+1})'$ not contained in $\overline{\gamma}^k$.

Now, there are at least $q^n - 1$ elements of $\overline{S}$ meeting $(\overline{\pi}^{k+1})'$ in a point outside $\overline{\gamma}^k$, and it is shown that, for all $i \in \{2, 3, \ldots, k\}$, there are at most $(q^n - 1)/(q-1) - i$ of these meeting $\overline{\pi}^i$ in a point of $\overline{\gamma}^{i-1}$. Indeed, the i subspaces $\overline{\pi}, \overline{\pi}^1, \ldots, \overline{\pi}^{i-1}$ meet $\overline{\pi}^i$ in a point of $\overline{\gamma}^{i-1}$ and have no points outside $\overline{\gamma}^k$; therefore there still remain $(q^n - 1)/(q - 1) - i$ points of $\overline{\pi}^i \cap \overline{\gamma}^{i-1}$ that possibly could be contained in a (necessarily unique) element of $\overline{S}$ meeting $(\overline{\pi}^{k+1})'$ in a point outside $\overline{\gamma}^k$. The result follows.

A counting argument, using the fact that $k \leq n \leq q$, now shows that at least one element $\overline{\pi}^{k+1}$ of $\overline{S}$ meets $(\overline{\pi}^{k+1})'$ in a point outside $\overline{\gamma}^k$ and meets $\overline{\pi}^i$ in a point outside $\overline{\gamma}^{i-1}$, for all $i \in \{2, 3, \ldots, k\}$. Putting $k = n$, a subspace $\overline{\pi}^{n+1}$ is obtained having an n-dimensional subspace in common with $\overline{\gamma}^n$, but not contained in $\overline{\gamma}^n$, a contradiction. The lemma is thus established.   □

Now assume either that S also satisfies (IV) or that $q \geq n$.

Some more notation and terminology are required. For $\pi' \in S \backslash \{\pi\}$, define $[\pi, \pi']$ to be the set of elements of S containing $\pi \cap \pi'$. The dual of $[\pi, \pi']$ is denoted by $[\overline{\pi}, \overline{\pi}']$. The π-*number* of π' is the size of $[\pi, \pi']$. The *spectrum* $\mathrm{Spec}(\pi)$ of π is the set of all π-numbers of elements of $S \backslash \{\pi\}$. It is shown that for the π-number there are a limited number of possibilities.

Lemma 4.45.   (i) *For q even*, $\mathrm{Spec}(\pi) \subset \{2, q, q + 1\}$.
(ii) *For q odd*, $\mathrm{Spec}(\pi) \subset \{q, q + 1\}$.

Proof. First it is shown that if, for some $\pi' \in S$, there is a π-number at least three, then it is either q or $q + 1$. So, suppose that $\pi', \pi'' \in S \backslash \{\pi\}$, with $\pi' \neq \pi''$, meet π in the same subspace γ. Dualise the situation. By (III'), there is an element $\overline{\tau} \in \overline{S}$ not contained in $\overline{\gamma}$. By Lemma 4.44, $\overline{\tau}$ meets $\overline{\gamma}$ in a line $\overline{\zeta}$, which has the distinct points $\overline{\sigma}, \overline{\sigma}', \overline{\sigma}''$ in common with $\overline{\pi}, \overline{\pi}', \overline{\pi}''$. Since every element of $\overline{S}$ contained in $\overline{\gamma}$ must meet $\overline{\tau}$, by (I'), necessarily in a point of $\overline{\zeta}$, and since these points must all be distinct, it follows already that the π-number of π' is not larger than $q + 1$.

Suppose now, by way of contradiction, that the π-number of π' is strictly less than q. Then there are at least two points on $\overline{\zeta}$ that are not contained in an element of $\overline{S}$ that is entirely contained in $\overline{\gamma}$. One of these points cannot be the nucleus of $\overline{\tau}$; so there is at least one point $\overline{\delta}$ of $\overline{\zeta}$ that is contained in an element $\overline{\tau}'$ of $\overline{S} \backslash \{\overline{\tau}\}$ that does not belong to $\overline{\gamma}$. The subspace $\overline{\tau}'$ meets $\overline{\gamma}$ in a line $\overline{\zeta}'$ intersecting $\overline{\pi}, \overline{\pi}', \overline{\pi}''$ in $\overline{\theta}, \overline{\theta}', \overline{\theta}''$. Let $\{\overline{\eta}\} = \overline{\pi} \cap \overline{\pi}'$, and let $\overline{\alpha}$ be the plane spanned by $\overline{\eta}$ and $\overline{\zeta}$. Then, $\overline{\alpha} = \langle \overline{\pi}, \overline{\zeta} \rangle \cap \langle \overline{\pi}', \overline{\zeta} \rangle$. But since $\langle \overline{\pi}, \overline{\zeta} \rangle = \langle \overline{\pi}, \overline{\delta} \rangle = \langle \overline{\pi}, \overline{\zeta}' \rangle$, and similarly $\langle \overline{\pi}', \overline{\zeta} \rangle = \langle \overline{\pi}', \overline{\zeta}' \rangle$, it follows that $\overline{\zeta}'$ is contained in $\overline{\alpha}$. Hence $\overline{\pi}''$ has the two distinct points $\overline{\sigma}''$ and $\overline{\theta}''$ in common with $\overline{\alpha}$, implying that $\overline{\pi}''$ meets both $\overline{\pi}$ and $\overline{\pi}'$ in points of $\overline{\alpha}$; these intersections are on the lines $\langle \overline{\eta}, \overline{\sigma}, \overline{\theta} \rangle$ and $\langle \overline{\eta}, \overline{\sigma}', \overline{\theta}' \rangle$. So all elements of $\overline{S}$ that are contained in $\overline{\gamma}$ meet $\overline{\pi}$ in points of $\overline{\alpha}$.

Now select a point $\overline{\delta}'$ of $\overline{\pi}$, distinct from the nucleus of $\overline{\pi}$, and not lying in $\overline{\alpha}$. There is a unique element $\overline{\tau}'' \in \overline{S} \backslash \{\overline{\pi}\}$ containing $\overline{\delta}'$, and by the foregoing $\overline{\tau}''$ is not contained in $\overline{\gamma}$. Interchanging the roles of τ'' and τ, it follows that $\overline{\pi} \cap \overline{\pi}''$ is a point of the line $\langle \overline{\eta}, \overline{\delta}' \rangle$, a contradiction. Hence $\mathrm{Spec}(\pi) \subset \{2, q, q+1\}$.

Suppose now that q is odd. It is shown that the π-number of $\pi' \in \mathcal{S} \backslash \{\pi\}$ cannot be two. Assume, by way of contradiction, that the π-number of such a π' is two. Consider an arbitrary $\tau \in \mathcal{S} \backslash \{\pi, \pi'\}$. Then, again putting $\overline{\gamma} = \langle \overline{\pi}, \overline{\pi}' \rangle$, the space $\overline{\tau}$ meets $\overline{\gamma}$ in a line $\overline{\zeta}$. Let $\{\overline{\eta}\} = \overline{\pi} \cap \overline{\pi}'$. As in the previous section, the intersection of $\overline{\gamma}$ and any element $\overline{\tau}' \in \overline{\mathcal{S}} \backslash \{\overline{\pi}, \overline{\pi}', \overline{\tau}\}$ containing some point of $\overline{\zeta}$, is a line $\overline{\zeta}'$ contained in the plane $\langle \overline{\eta}, \overline{\zeta} \rangle$. If the nucleus of $\overline{\tau}$ were not on $\overline{\zeta}$, then there would be $q - 1$ choices for $\overline{\tau}'$, and since no three of the corresponding lines $\overline{\zeta}'$, together with $\overline{\zeta}, \overline{\zeta}_\pi = \overline{\alpha} \cap \overline{\pi}, \overline{\zeta}_{\pi'} = \overline{\alpha} \cap \overline{\pi}'$ meet in a common point, there arises a $(q+2)$-arc, a contradiction. Hence there is a unique point $\overline{\theta}$ on the line $\overline{\zeta}_\pi$ not contained in an element of $\overline{\mathcal{S}} \backslash \{\overline{\pi}\}$ that contains a line of $\overline{\alpha}$.

By way of contradiction, suppose that $\overline{\theta}$ is not the nucleus of $\overline{\pi}$. Then there is an element $\overline{\tau}'' \in \overline{\mathcal{S}} \backslash \{\overline{\pi}\}$ containing $\overline{\theta}$. If $\overline{\zeta}''$ is the intersection of $\overline{\tau}''$ with $\overline{\gamma}$, then a previous argument shows that there are $q - 2 > 0$ elements of $\overline{\mathcal{S}}$ different from $\overline{\tau}'', \overline{\pi}, \overline{\pi}'$ meeting $\overline{\gamma}$ in a line of the plane $\langle \overline{\eta}, \overline{\zeta}'' \rangle$. These $q - 2$ elements also contain points of the line $\overline{\zeta}_\pi$ different from $\overline{\eta}$ and $\overline{\theta}$, and so their intersections with $\overline{\gamma}$ are contained in $\overline{\alpha}$. It follows that $\overline{\alpha} = \langle \overline{\eta}, \overline{\zeta}'' \rangle$, and so $\overline{\tau}''$ contains a line of $\overline{\alpha}$, a contradiction. Hence $\overline{\theta}$ is the nucleus of $\overline{\pi}$. But τ was arbitrary in $\overline{\mathcal{S}} \backslash \{\overline{\pi}, \overline{\pi}'\}$ and this contradicts the uniqueness of the nucleus of $\overline{\pi}$. The lemma is thus proved. $\qquad \square$

Now the case $q \neq 2$ with each spectrum a subset of $\{q, q+1\}$ is considered, and also the case $q = 2$ for which each spectrum is $\{3\}$.

An extra axiom is required.

(A_n) Assume that $\mathcal{S}$ satisfies (I) to (V). Then either $\mathcal{S}$ is the set of $\mathcal{V}_{n-1}$-subspaces of a quadric Veronesean $\mathcal{V}_n^{2^n}$ in $\mathrm{PG}(N_n, q)$, or q is even, there are two elements $\pi, \pi' \in \mathcal{S}$ with the property that no other element of $\mathcal{S}$ contains $\pi \cap \pi'$, and there is a unique subspace η of dimension $(n-1)(n+2)/2$ such that $\mathcal{S} \cup \{\eta\}$ is the set of $\mathcal{V}_{n-1}$-subspaces together with the nucleus of a quadric Veronesean $\mathcal{V}_n^{2^n}$ in $\mathrm{PG}(n(n+3)/2, q)$.

Lemma 4.46. (i) *Let $q > 2$ and suppose that $\mathrm{Spec}(\pi) \subset \{q, q+1\}$. Then $\mathrm{Spec}(\pi) = \{q+1\}$. If this holds for every element of $\mathcal{S}$ and if for $n > 2$ the axiom (A_{n-1}) is satisfied, then $\mathcal{S}$ is the set of $\mathcal{V}_{n-1}$-subspaces of a quadric Veronesean $\mathcal{V}_n^{2^n}$.*

(ii) *If $q = 2$, if $\mathrm{Spec}(\pi) = \{3\}$ for every element of $\mathcal{S}$ and if for $n > 2$ the axiom (A_{n-1}) is satisfied, then the same conclusion holds.*

Proof. Assume that $q > 2$ and that $\mathrm{Spec}(\pi) \subset \{q, q+1\}$. Suppose that the π-number of some $\pi' \in \mathcal{S} \backslash \{\pi\}$ is q. Let $\tau \in \mathcal{S}$ be such that it does not contain $\pi \cap \pi'$, and let $\tau' \in [\pi, \tau]$, with $\pi \neq \tau' \neq \tau$. This means that $\langle \overline{\pi}, \overline{\tau} \rangle = \langle \overline{\pi}, \overline{\tau}' \rangle$, which implies that the lines $\langle \overline{\pi}, \overline{\tau} \rangle \cap \overline{\pi}'$ and $\langle \overline{\pi}, \overline{\tau}' \rangle \cap \overline{\pi}'$ coincide. Denoting this line by $\overline{\zeta}_\tau = \overline{\zeta}_{\tau'}$, it follows that there are two possibilities:

(a) all points on $\overline{\zeta}_\tau$ are contained in an element of $[\overline{\pi}, \overline{\tau}]$ when the π-number of τ is $q + 1$;

(b) all but exactly one point of $\overline{\zeta}_\tau$ are contained in an element of $[\overline{\pi}, \overline{\tau}]$ when the π-number of τ is q.

Also, there are exactly q points on the line $\overline{\zeta} = \overline{\tau} \cap \overline{\gamma}$, with $\overline{\gamma} = \langle \overline{\pi}, \overline{\pi}' \rangle$, contained in elements of $[\overline{\pi}, \overline{\pi}']$. So there remains a unique point $\overline{\theta}$ on $\overline{\zeta}$ which is not contained in any element of $[\overline{\pi}, \overline{\pi}']$. It is now shown that $\overline{\theta}$ is the nucleus of $\overline{\tau}$.

Assume, by way of contradiction, that $\overline{\theta}$ is not the nucleus of $\overline{\tau}$. Then, let $\overline{\beta}$ be the unique element of $\overline{S} \setminus \{\overline{\tau}\}$ containing $\overline{\theta}$. As in the proof of the previous lemma, this implies that the intersection of $\overline{\beta}$ with $\overline{\gamma}$ is a line $\overline{\zeta}'$ contained in the plane $\overline{\alpha}$ spanned by $\overline{\zeta}$ and the point $\overline{\eta}$, with $\{\overline{\eta}\} = \overline{\pi} \cap \overline{\pi}'$. Also, every element of $[\overline{\pi}, \overline{\pi}']$ meets $\overline{\zeta}'$ in a point not belonging to $\overline{\zeta}$, and hence has a line in common with $\overline{\alpha}$; this implies that every element of $[\overline{\pi}, \overline{\pi}']$ has a point in common with $\overline{\zeta}_\tau$, which is contained in $\overline{\alpha}$. This contradicts the observation made in the previous paragraph on the points of $\overline{\zeta}_\tau$. Consequently, $\overline{\theta}$ is the nucleus of $\overline{\tau}$.

Hence, as $\overline{\tau}$ was essentially arbitrary, all nuclei are contained in $\overline{\gamma}$. If the π-number of τ were also equal to q, then similarly all nuclei would be contained in $\langle \overline{\pi}, \overline{\tau} \rangle$. Assume, by way of contradiction, that this is the case. Let $\overline{\beta}'$ be an element of $\overline{S}$ containing no point of $\langle \overline{\pi}, \overline{\tau} \rangle \cap \overline{\pi}'$ and no point of $\langle \overline{\pi}, \overline{\pi}' \rangle \cap \overline{\tau}$. Then the nucleus of $\overline{\beta}'$ is on the line $\langle \overline{\pi}, \overline{\tau} \rangle \cap \overline{\beta}'$ and on the line $\langle \overline{\pi}, \overline{\pi}' \rangle \cap \overline{\beta}'$. Hence these lines coincide and so $\overline{\beta}'$ intersects $\langle \overline{\pi}, \overline{\tau} \rangle \cap \overline{\pi}'$ and $\langle \overline{\pi}, \overline{\pi}' \rangle \cap \overline{\tau}$, a contradiction.

It follows that the π-number of all elements of $S \setminus [\pi, \pi']$ is equal to $q + 1$. Counting the number of sets $[\pi, \pi'']$, with $\pi'' \in S \setminus [\pi, \pi']$, gives $(q^n + q^{n-1} + \cdots + q^2 + 1)/q$, which is not an integer. This is a contradiction, and so the spectrum of π is the singleton $\{q + 1\}$.

Suppose now that $\mathrm{Spec}(\pi) = \{q + 1\}$ for all $\pi \in S$, and let $q \geq 2$. First assume that $n = 2$. Let $\mathcal{V}$ be the set of points of $\mathrm{PG}(5, q)$ that are contained in precisely $q + 1$ elements of S. Note that

$$|\mathcal{V}| = \frac{(q^2 + q + 1)(q^2 + q)/2}{(q + 1)q/2} = q^2 + q + 1,$$

by counting the ordered triples (π', π'', P) with $\pi', \pi'' \in S$, $\pi' \neq \pi''$, and P a point of $\pi' \cap \pi''$, in two ways. Also, there are precisely $q + 1$ points of $\mathcal{V}$ in each member of S. It is shown that $\mathcal{V}$ is a cap. First it is established that, whenever a point $P \in \mathcal{V}$ is contained in a line of $\mathrm{PG}(5, q)$ intersecting $\mathcal{V}$ in at least three points, then the set of $q + 1$ points in any element of S containing P is a line of $\mathrm{PG}(5, q)$, leading to a contradiction.

Let P, P', P'' be three distinct points of $\mathcal{V}$ on a common line m. Let $\pi_P \in S$ contain P. First suppose that m is not contained in an element of S. Let $\pi_{P'}$ and $\pi_{P''}$ be two elements of S containing P' and P'', and such that the intersections with π_P, say R' and R'', are distinct; these elements exist because there are $q + 1$ elements of S through each point of $\mathcal{V}$. If P, R', R'' were not collinear, then the plane π_P would be generated by these three points; but then the planes $\pi_{P'}$ and $\pi_{P''}$ would generate a 4-dimensional space containing R', R'', P', P'', hence also containing P and thus containing π_P, contradicting (II). Fixing π_P and $\pi_{P'}$, but not $\pi_{P''}$, there arise at least q distinct points on $PR'\backslash\{P, R'\}$, a contradiction. So m is contained in an element π of $\mathcal{S}$.

Assume that π_P does not contain m. From a previous argument it follows that the points of $\mathcal{V}$ in π_P are contained in a line. Similarly, interchanging the roles of P', P'' and two arbitrary points of $\pi_P \cap \mathcal{V}$ different from P, it follows that also $\pi \cap \mathcal{V}$ is a line. But now every member of $\mathcal{S}$ has that property, since every member of $\mathcal{S}$ contains a point of π_P. Hence $\mathcal{V}$ consists of the union of $q^2 + q + 1$ lines and consequently is a projective plane Π_2 of $\mathrm{PG}(5, q)$. Every element of $\mathcal{S}$ meets Π_2 in a line, which implies that the $(q^2 + q + 1)(q^2 + q)/2$ distinct hyperplanes containing two elements of $\mathcal{S}$ all contain Π_2. But there are only $q^2 + q + 1$ hyperplanes in $\mathrm{PG}(5, q)$ through Π_2, a contradiction. Consequently, $\mathcal{V}$ is a cap.

It now follows that, for every $\pi \in \mathcal{S}$, the set $\pi \cap \mathcal{V}$ is a $(q + 1)$-arc. Hence, on $\mathcal{V}$, there is a set $\mathcal{O}$ of size $q^2 + q + 1$ of $(q + 1)$-arcs, meeting in pairs in a point, and such that every point is contained in $q + 1$ of these $(q + 1)$-arcs. It follows that every two distinct points of $\mathcal{V}$ are contained in a unique $(q + 1)$-arc. In order to apply Theorem 4.34 to conclude that $\mathcal{V}$ is the quadric Veronesean $\mathcal{V}_2^4$, it must be shown that the tangent lines at any fixed point $P \in \mathcal{V}$ to the $(q + 1)$-arcs $O \in \mathcal{O}$ containing P are coplanar. To that end, consider an arbitrary plane $\pi \in \mathcal{S}$ containing P and project $\mathcal{V}\backslash\pi$ from π onto a plane Π_2 of $\mathrm{PG}(5, q)$ skew to π; denote by θ the projection map.

First it is shown that θ is injective on $\mathcal{V}\backslash\pi$. Let $P', P'' \in \mathcal{V}\backslash\pi$ and suppose that $P'\theta = P''\theta$. Let $\pi' \in \mathcal{S}$ contain P' and P''. Then $\langle \pi, P', P'' \rangle = \langle \pi, \pi' \rangle$ is 3-dimensional, a contradiction. Now let $\pi' \in \mathcal{S}\backslash\{\pi\}$ be arbitrary. Since $\langle \pi, \pi' \rangle$ is 4-dimensional, the projection of $(\pi'\backslash\pi) \cap \mathcal{V}$ consists of q points on a line m' of Π_2. Let R' be the unique point on m' that is not an image under θ of any point of $(\pi' \cap \mathcal{V})\backslash\pi$. By way of contradiction, let R' be the image of a point $R \in \mathcal{V}\backslash\pi$; necessarily $R \notin \pi'$. The $q + 1$ planes of $\mathcal{S}$ through R, minus their intersection points with π, are mapped under θ into $q+1$ different lines of Π_2 through R', since every three distinct elements of $\mathcal{S}$ generate $\mathrm{PG}(5, q)$. Hence there is an element $\pi'' \in \mathcal{S}$ through R which yields m'. So π, π', π'' are contained in the hyperplane $\langle \pi, m' \rangle$ of $\mathrm{PG}(5, q)$, contradicting (II). It now follows that the set of planes of $\mathcal{S}\backslash\{\pi\}$ through P corresponds under θ with the set of $q + 1$ lines of Π_2 containing a fixed point P^* of Π_2, and that the 3-dimensional subspace $\Pi_3 = \langle \pi, P^* \rangle$ meets every element $\pi' \in \mathcal{S}\backslash\{\pi\}$ containing P in a line $m_{\pi'}$ through P disjoint from $\mathcal{V}\backslash\pi$. So $m_{\pi'}$ is tangent to the $(q + 1)$-arc $\pi' \cap \mathcal{V}$ at P. Now fix $\pi' \in \mathcal{S}\backslash\{\pi\}$ with $P \in \pi'$. Then, similarly, there is a solid Π_3' containing π' and the tangent lines at P to the elements of $\mathcal{O}$ containing P. As $\Pi_3 \neq \Pi_3'$, so all these tangent lines are contained in the plane $\Pi_3 \cap \Pi_3'$. This shows the lemma for $n = 2$.

Next, suppose that $n > 2$. Consider the set $\mathcal{S}_\pi = \{\pi \cap \pi' \mid \pi' \in \mathcal{S}\backslash\pi\}$, and calculate $|\mathcal{S}_\pi| = \theta(n - 1)$. In $\pi = \mathrm{PG}(N_{n-1}, q)$ the set $\mathcal{S}_\pi$ satisfies (I) and (III) for the parameter $n - 1$ instead of n. It is now shown that it also satisfies (II). Let $[\pi, \pi'] \neq [\pi, \tau]$, with $\tau \in \mathcal{S}\backslash\{\pi\}$, and let $\bar{\zeta}_\tau$ be as above. Then, all points of $\bar{\zeta}_\tau$ are contained in elements of $[\bar{\pi}, \bar{\tau}]$. Hence any $\bar{\tau}' \in \bar{\mathcal{S}}\backslash([\bar{\pi}, \bar{\pi}'] \cup [\bar{\pi}, \bar{\tau}])$ meets $\bar{\pi}'$ outside $\langle \bar{\pi}, \bar{\tau} \rangle$. This means that

$$\langle \bar{\pi}, \bar{\pi}' \rangle \cap \langle \bar{\pi}, \bar{\tau} \rangle \cap \langle \bar{\pi}, \bar{\tau}' \rangle = \bar{\pi},$$

and the dual of this is exactly (II). Also, if S satisfies (IV), then S_π satisfies (IV); if $q \geq n$, then $q \geq n - 1$. Hence, by (A_{n-1}), the set S_π is either the set of all V_{n-2}-subspaces of a quadric Veronesean $V_{n-1}^{2^{n-1}}$, or q is even and S_π is the nucleus together with all V_{n-2}-subspaces but exactly one of a quadric Veronesean $V_{n-1}^{2^{n-1}}$, or it is an ovoidal Veronesean set of subspaces.

It is now shown that the last two cases cannot occur. In both these cases, there is an element of S_π, which can be taken to be $\pi \cap \pi'$, with the property that it contains $\theta(n-1) - 1$ subspaces of dimension N_{n-2} arising as intersections of $\pi \cap \pi'$ with other elements of S_π; if S_π is not ovoidal, then take for π' the nucleus of $V_{n-1}^{2^{n-1}}$. Let δ be such a subspace of dimension N_{n-2} and let $\tau \in S \backslash [\pi, \pi']$ contain δ. Then all $q+1$ elements of $[\pi, \tau]$ contain δ, but they define $q+1$ distinct members of $S_{\pi'}$, each of which is defined by q different elements of S. As there are $\theta(n-1) - 1$ choices for δ, there are at least

$$q^{n+1} + q^n + \cdots + q^3 + (q+1) \tag{4.7}$$

elements in S, a contradiction; the last '$q+1$' in (4.7) comes from the $q+1$ elements of $[\pi, \pi']$. It follows that S_π is the set of all V_{n-2}-subspaces of a $V_{n-1}^{2^{n-1}}$.

Consider now the set V of all points of $\mathrm{PG}(N_n, q)$ that are contained in precisely $\theta(n-1)$ elements of S. From the previous section, it immediately follows that, for each element π of S, the intersection $\pi \cap V$ is a quadric Veronesean $V_{n-1}^{2^{n-1}}$. Denote by Γ the set of all conics contained in these intersections $\pi \cap V$. Now let $P', P'' \in V$, $P' \neq P''$. Then there are elements $\pi', \pi'' \in S$ with $P' \in \pi'$ and $P'' \in \pi''$. Suppose $\pi' \neq \pi''$. The $\theta(n-1)$ elements of S containing P' meet π'' in distinct subspaces, by (I) and the fact that their intersection contains P'; hence P'' is contained in at least one of them. Consequently P' and P'' are contained in a common member of S, and hence P' and P'' are contained in a conic of Γ. Assume, by way of contradiction, that $P', P'' \in V$ with $P' \neq P''$ are contained in distinct conics C' and C'' of Γ. Let $R \in C'' \backslash C'$. As before, it follows that, if $n > 2$, then C' is contained in at least one of the $\theta(n-1)$ elements of S containing R. So C' and C'' are distinct conics of a quadric Veronesean $V_{n-1}^{2^{n-1}}$ sharing two distinct points, a contradiction. So any two distinct points of V are contained in exactly one conic of Γ.

Now let C be any member of Γ and assume that $P \in V \backslash C$. As before, if $n \geq 3$, then C is contained in at least one of the $\theta(n-1)$ elements of S containing P. So P and C are contained in a common member π of S and, since $\pi \cap V$ is a Veronesean, the tangents at P of the conics through P which have a point in common with C all lie in a fixed plane. By a similar argument, it follows that two distinct elements of Γ containing P always generate a 4-dimensional space. Now assume that $C', C'' \in \Gamma$, with $C' \neq C''$, and that $P \in \langle C' \rangle \cap \langle C'' \rangle$. It will be shown that $P \in C' \cap C''$.

If $P \notin C' \cap C''$ and P is not the nucleus of at least one of C', C'', say P is not the nucleus of C', then there is an element π of S containing C'' and two distinct points P' and P'' of C'. If π also contains C', then, as C' and C'' are conics of some $V_{n-1}^{2^{n-1}}$, it follows that $P \in C' \cap C''$, a contradiction. If π does not contain C', then P' and P'' are contained in common distinct elements of Γ, again a contradiction. Now assume that P is the common nucleus of C' and C''. Let $R \in C''$. Then R and C' are contained in a common element π of $\mathcal{S}$. As P is the common nucleus of $\mathcal{C}'$ and $\mathcal{C}''$ the space π cannot contain $\mathcal{C}''$. So $\pi \cap \langle \mathcal{C}'' \rangle = PR$. Let $R' \in \mathcal{C}'' \backslash \{R\}$. If $n > 3$, then by similar arguments R, R' and $\mathcal{C}'$ are contained in a common element π' of $\mathcal{S}$. As $\pi' \cap \mathcal{V}$ is a quadric Veronesean $\mathcal{V}_{n-1}^{2^{n-1}}$, the conics $\mathcal{C}'$ and $\mathcal{C}''$ of $\pi' \cap \mathcal{V}$ cannot have a common nucleus. So $n = 3$.

If R, R', R'' are distinct points of $\mathcal{C}''$, then $\langle R, \mathcal{C}' \rangle$, $\langle R', \mathcal{C}' \rangle$, $\langle R'', \mathcal{C}' \rangle$ are contained in respective elements π, π', π'' of $\mathcal{S}$. The 5-dimensional spaces π, π', π'' are distinct and share the plane $\langle \mathcal{C}' \rangle$. Let $U \in \mathcal{C}'$. Then U and $\mathcal{C}''$ belong to a Veronesean $\mathcal{V}_2^4$, so the tangents of $\mathcal{V}_2^4$ at U are coplanar. Let $\mathcal{D}$, $\mathcal{D}'$, $\mathcal{D}''$ be the conics containing $\{U, R\}$, $\{U, R'\}$, $\{U, R''\}$. Since the tangents at U of $\mathcal{D}$, $\mathcal{D}'$, $\mathcal{D}''$ are coplanar, $\mathcal{D}''$ belongs to $\langle \pi, \pi' \rangle$. Consequently, $\pi'' \subset \langle \pi, \pi' \rangle$. So π, π', π'' are in the same hyperplane, a contradiction. Therefore $P \in \mathcal{C}' \cap \mathcal{C}''$.

By Theorem 4.36, it now follows that Γ is the set of all conics on a quadric Veronesean $\mathcal{V}_n^{2^n}$. Finally, by Theorem 4.15, $\mathcal{S}$ is the set of all $\mathcal{V}_{n-1}$-subspaces of $\mathcal{V}_n^{2^n}$. $\qquad\square$

From now on assume that there exists some member of $\mathcal{S}$ whose spectrum contains 2. Then q is even by Lemma 4.45. First consider the case where the spectrum contains 2 and has size at least two.

Lemma 4.47. *Let* $\pi \in \mathcal{S}$ *be such that* $2 \in \mathrm{Spec}(\pi)$ *and* $|\mathrm{Spec}(\pi)| \geq 2$. *Then the following hold.*

(i) $\mathrm{Spec}(\pi) = \{2, q, q+1\}$.

(ii) *If* $q > 2$, *then there exists a unique element* $\pi' \in \mathcal{S} \backslash \{\pi\}$ *such that* $|[\pi, \pi']| = 2$ *and there are precisely* $q-1$ *elements* $\pi'' \in \mathcal{S} \backslash \{\pi\}$ *such that* $|[\pi, \pi'']| = q$. *Also,* $\mathrm{Spec}(\pi') = \{2\}$ *and the spectrum of any other element of* $\mathcal{S}$ *is* $\{2, q, q+1\}$.

(iii) *If* $q = 2$, *then there are precisely two elements of* $\mathcal{S} \backslash \{\pi\}$ *with* π-*number 2, one of which has spectrum* $\{2\}$, *while any other element of* $\mathcal{S}$ *has spectrum* $\{2, 3\}$.

Proof. Let $\pi, \pi' \in \mathcal{S}$ be such that $|[\pi, \pi']| = 2$ and let $\tau \in \mathcal{S} \backslash \{\pi\}$ be such that $l = |[\pi, \tau]| > 2$. As before, let $\overline{\gamma} = \langle \overline{\pi}, \overline{\pi}' \rangle$ and let $\overline{\zeta}$ be the intersection of $\overline{\gamma}$ with $\overline{\tau}$. The elements of $[\overline{\pi}, \overline{\tau}]$ must meet $\overline{\pi}'$ in the line joining $\overline{\pi} \cap \overline{\pi}'$ and $\overline{\tau} \cap \overline{\pi}'$; on the other hand, the elements of $\overline{\mathcal{S}} \backslash \{\overline{\pi}, \overline{\pi}', \overline{\tau}\}$ containing a point of $\overline{\zeta}$ intersect $\overline{\gamma}$ in lines which are contained in the plane $\langle \overline{\eta}, \overline{\zeta} \rangle$, with $\{\overline{\eta}\} = \overline{\pi} \cap \overline{\pi}'$. It follows that there are at least $q - 1$ points on the line $\overline{\zeta}'$ joining $\overline{\eta}$ to $\overline{\pi} \cap \overline{\tau}$ contained in elements of $[\overline{\pi}, \overline{\tau}] \backslash \{\overline{\pi}\}$, if $l = q+1$, and at least $q-2$ such points if $l = q$. A similar statement holds for the line $\overline{\zeta}''$ joining $\overline{\eta}$ with $\overline{\pi} \cap \overline{\tau}'$, for every $\overline{\tau}' \in [\overline{\pi}, \overline{\tau}] \backslash \{\overline{\pi}\}$. It readily follows that, if $q > 2$, then every such line $\overline{\zeta}''$ coincides with $\overline{\zeta}'$. Suppose now that $q > 2$. Then there are precisely q points on the line $\overline{\zeta}'$ contained in elements of $[\overline{\pi}, \overline{\tau}] \backslash \{\overline{\pi}\}$, if $l = q + 1$, and precisely $q - 1$ such points if $l = q$. Since the line $\overline{\zeta}'$ is uniquely determined by $[\overline{\pi}, \overline{\tau}]$, every element $\overline{\pi}''$ of $\overline{\mathcal{S}} \backslash \{\overline{\pi}\}$ with π-number 2 must intersect $\overline{\pi}$ on the line $\overline{\zeta}'$. It follows that $\overline{\pi}' = \overline{\pi}''$ if $l = q + 1$, and there is at most one choice for $\overline{\pi}'' \neq \overline{\pi}'$ if $l = q$.

So suppose that such a space $\overline{\pi}'' \neq \overline{\pi}'$ exists. Put $\{\overline{\eta}'\} = \overline{\pi} \cap \overline{\pi}''$. Then every element $\overline{\tau}' \in \overline{\mathcal{S}} \backslash [\overline{\pi}, \overline{\tau}]$, with $\overline{\pi}' \neq \overline{\tau}' \neq \overline{\pi}''$, has π-number q, and so, by the previous arguments, all elements of $[\overline{\pi}, \overline{\tau}'] \backslash \{\overline{\pi}\}$ have a point in common with $\overline{\eta}\overline{\eta}'$, a contradiction. Hence, for $q > 2$, there is just one $\pi' \in \mathcal{S} \backslash \{\pi\}$ with $|[\pi, \pi']| = 2$.

Suppose now that $q = 2$ and put $[\overline{\pi}, \overline{\tau}] = \{\overline{\pi}, \overline{\tau}, \overline{\tau}'\}$. The only possible reason for the line $\overline{\zeta}'$ not to contain $\overline{\pi} \cap \overline{\tau}'$ is that there is no element of $\overline{\mathcal{S}} \backslash \{\overline{\pi}, \overline{\pi}', \overline{\tau}\}$ containing a point of $\overline{\zeta}$; in other words, the nucleus of $\overline{\tau}$ is contained in $\overline{\gamma}$. Assume, by way of contradiction, that there are at least three elements of $\overline{\mathcal{S}} \backslash \{\overline{\pi}\}$ with π-number 2. Then for at least two of them, say $\overline{\pi}'$ and $\overline{\pi}''$, the points $\overline{\pi} \cap \overline{\tau}, \overline{\pi}' \cap \overline{\tau}, \overline{\pi}'' \cap \overline{\tau}$ and the nucleus $\overline{\theta}$ of $\overline{\tau}$ are distinct collinear points, a contradiction. So in the case $q = 2$, there are at most two elements in ζ with π-number 2, say $\overline{\pi}'$ and $\overline{\pi}''$, and a counting argument shows that there are exactly two elements in $\overline{\mathcal{S}}$ with π-number 2. Putting $\overline{\pi} \cap \overline{\pi}' = \{\overline{\eta}\}$ and $\overline{\pi} \cap \overline{\pi}'' = \{\overline{\eta}'\}$, the same argument also shows that, if $\overline{\pi} \cap \overline{\tau} = \{\overline{\alpha}\}$ and $\overline{\pi} \cap \overline{\tau}' = \{\overline{\alpha}'\}$, then the line $\overline{\alpha}\overline{\alpha}'$ contains either $\overline{\eta}$ or $\overline{\eta}'$, and if it contains, say, $\overline{\eta}'$, then the nuclei of $\overline{\tau}$ and $\overline{\tau}'$ are contained in $\overline{\gamma}$.

With this notation, it is shown that the spectrum of $\overline{\pi}''$ is equal to $\{2\}$. For, if $[\overline{\pi}'', \overline{\tau}]$ contains an element $\overline{\tau}'' \notin \{\overline{\pi}'', \overline{\tau}\}$, then $\overline{\tau}''$ meets $\overline{\pi}$ in the point $\overline{\alpha}'$, a contradiction. Similarly, $|[\overline{\pi}'', \overline{\tau}']| = 2$ and so there are at least three elements of $\mathcal{S} \backslash \{\pi''\}$ with π''-number 2, namely π, τ and τ'. So, by previous arguments, there cannot be an element with π''-number $q + 1 = 3$. Hence $\mathrm{Spec}(\pi'') = \{2\}$.

Next, it is shown that $\mathrm{Spec}(\pi') = \{2, 3\}$. Suppose, by way of contradiction, that $\mathrm{Spec}(\pi') = \{2\}$. First note that the argument in the previous paragraph implies that the nucleus of $\overline{\pi}$ is on the line $\overline{\eta}\overline{\alpha}$, as otherwise $\overline{\tau}'$ contains the third point of $\overline{\eta}\overline{\alpha}$ as well as $\overline{\alpha}'$. Analogously, the nucleus of $\overline{\pi}$ is on the line $\overline{\eta}\overline{\alpha}'$. This yields a contradiction. It also follows that $|[\overline{\pi}', \overline{\tau}]| = |[\overline{\pi}', \overline{\tau}']| = 3$, and so the nucleus of $\overline{\pi}$ is on $\overline{\eta}\overline{\eta}'$. Taking into account all previous arguments, it follows that π'' is the unique element of $\mathcal{S}$ with spectrum $\{2\}$, and the other elements are divided in pairs $\{\varphi, \varphi'\}$ with respect to the relation "φ' has φ-number 2". Also, the nucleus of $\overline{\varphi}$ and the points $\overline{\varphi} \cap \overline{\varphi}', \overline{\varphi} \cap \overline{\pi}''$ are collinear, and the two intersection points of the elements of $[\overline{\varphi}, \overline{\beta}] \backslash \{\overline{\varphi}\}$, where $\beta \notin \{\varphi, \varphi', \pi''\}$, with $\overline{\varphi}$ are collinear with $\overline{\varphi} \cap \overline{\pi}''$.

Further, it is shown that the nucleus of any $\overline{\rho} \in \overline{\mathcal{S}}$ is contained in the space $\langle \overline{\pi}, \overline{\pi}' \rangle = \overline{\gamma}$. This is immediate if $\rho \in \{\pi, \pi'\}$. Suppose now that $\rho \notin \{\pi, \pi'\}$, and also assume that $\rho \neq \pi''$. If the nucleus of $\overline{\rho}$ were not contained in $\overline{\gamma}$, then the unique element $\overline{\rho}'$ of $[\overline{\pi}, \overline{\rho}] \backslash \{\overline{\pi}, \overline{\rho}\}$ would meet $\overline{\gamma}$ in a line of the plane spanned by $\overline{\eta}$ and $\overline{\rho} \cap \overline{\gamma}$, implying that the intersection points $\overline{\pi} \cap \overline{\rho}$ and $\overline{\pi} \cap \overline{\rho}'$ would be collinear with $\overline{\eta}$, contradicting an earlier observation. Now it has still to be proved that the nucleus of $\overline{\pi}''$ is contained in $\overline{\gamma}$. Suppose this is not the case. Then the third point of the line joining $\overline{\eta}'$ and $\overline{\pi}' \cap \overline{\pi}''$ is on an element $\overline{\beta}$ of $\overline{\mathcal{S}}$ intersecting $\overline{\pi}, \overline{\pi}', \overline{\pi}''$ in distinct points of the line $\overline{\beta} \cap \overline{\gamma}$. But the nucleus of $\overline{\beta}$ is also on that line, a contradiction.

A similar result on the nuclei of the elements of $\overline{\mathcal{S}}$ is now shown for $q > 2$. A counting argument shows the existence of at least one element $\tau \in \mathcal{S} \backslash \{\pi\}$ with π-number q. It is now shown that the nuclei of all elements of $\overline{\mathcal{S}}$ are contained in $\overline{\gamma}^* = \langle \overline{\pi}, \overline{\tau} \rangle$. Note that, similarly to the first part of the proof of Lemma 4.46, this implies that the only elements of $\mathcal{S} \backslash \{\pi\}$ with π-number q are those of $[\pi, \tau]$.

Now the assertion on the nuclei is shown. Put $\{\overline{\xi}\} = \overline{\pi} \cap \overline{\tau}$. Let φ be any element of $\mathcal{S} \backslash [\pi, \tau]$. There is a unique point $\overline{\xi}'$ on the line $\overline{\zeta}_\varphi = \overline{\gamma}^* \cap \overline{\varphi}$ not contained in an element of $[\overline{\pi}, \overline{\tau}]$. If this point would not be the nucleus of $\overline{\varphi}$, then it would be contained in an element $\overline{\varphi}' \neq \overline{\varphi}$ of $\overline{\mathcal{S}}$. By previous arguments, it follows that the elements of $[\overline{\pi}, \overline{\tau}] \backslash \{\overline{\tau}\}$ would meet $\overline{\tau}$ in points of the line $\overline{\tau} \cap \langle \overline{\xi}, \overline{\zeta}_\varphi \rangle$. But this line contains the points of intersection of $\overline{\tau}$ with any element of $[\overline{\pi}, \overline{\varphi}] \cup [\overline{\pi}, \overline{\varphi}']$. It follows that the π-number of both φ and φ' is 2, a contradiction. This proves the assertion.

It is now shown that all elements of $\mathcal{S} \backslash \{\pi'\}$ have π'-number equal to 2. If not, then, by the first section of this proof, each element of $\mathcal{S} \backslash \{\pi', \pi\}$ has π'-number q or $q + 1$, and, from above, $q - 1$ elements of $\mathcal{S} \backslash \{\pi'\}$ have π'-number q. Hence one can find $\varphi \in \mathcal{S} \backslash \{\pi, \pi'\}$ such that φ has π-number $q + 1$ and π'-number $q + 1$. It follows from previous arguments that at least $q - 2$ elements of $[\overline{\pi}, \overline{\varphi}] \backslash \{\overline{\pi}, \overline{\varphi}\}$ meet $\overline{\gamma}$ in lines belonging to the plane generated by $\overline{\eta}$, with $\{\overline{\eta}\} = \overline{\pi} \cap \overline{\pi}'$, and $\overline{\gamma} \cap \overline{\varphi}$. By symmetry, this also holds for at least $q - 2$ elements of $[\overline{\pi}', \overline{\varphi}] \backslash \{\overline{\pi}', \overline{\varphi}\}$, a contradiction.

Finally, it is shown that the spectrum of any element of $\mathcal{S} \backslash \{\pi'\}$ is $\{2, q, q + 1\}$. Assume, by way of contradiction, that the spectrum of $\varphi \in \mathcal{S} \backslash \{\pi'\}$ contains at most two elements. As $\mathrm{Spec}(\pi') = \{2\}$, the spectrum of φ contains 2. In the case that $|\mathrm{Spec}(\varphi)| = 2$, then, with π and φ interchanged, $\mathrm{Spec}(\varphi) = \{2, q, q + 1\}$, a contradiction. Hence $\mathrm{Spec}(\varphi) = \{2\}$, again a contradiction, as there is exactly one element of $\mathcal{S} \backslash \{\pi\}$ with π-number 2. Hence $\mathrm{Spec}(\varphi) = \{2, q, q + 1\}$ for any element $\varphi \in \mathcal{S} \backslash \{\pi'\}$. $\qquad\square$

Lemma 4.48. *If $\mathcal{S}$ is a proper Veronesean set of subspaces with the property that 2 is contained in the spectrum of at least one element of $\mathcal{S}$ and if, for $n > 2$, axiom (A_{n-1}) is satisfied, then $\mathcal{S}$ is the set of all $\mathcal{V}_{n-1}$-subspaces but one, together with the nucleus of a quadric Veronesean $V_n^{2^n}$.*

Proof. As $\mathcal{S}$ is proper, there is an element of $\mathcal{S}$ whose spectrum contains 2 and at least one of $q, q+1$; if $q = 2$, then there is an element of $\mathcal{S}$ whose spectrum is $\{2, 3\}$. Lemma 4.47 implies that there is a unique element π' of $\mathcal{S}$ with spectrum $\{2\}$, and all other elements of $\mathcal{S}$ have spectrum $\{2, q, q + 1\}$ for $q > 2$, and $\{2, 3\}$ for $q = 2$. Also, for $\pi \in \mathcal{S} \backslash \{\pi'\}$, there is a unique set $[\pi, \tau]$ of size q, with $\tau \in \mathcal{S} \backslash \{\pi, \pi'\}$, and all elements of $\mathcal{S} \backslash ([\pi, \tau] \cup \{\pi'\})$ have π-number $q + 1$. It also follows from the proof of Lemma 4.47 that, for each element φ with π-number $q + 1$, the set of points $\overline{\pi} \cap \overline{\varphi}'$, with $\varphi' \in [\pi, \varphi] \backslash \{\pi\}$, is contained in a line $\overline{\zeta}_\varphi$, which contains the common point $\overline{\eta}$ of $\overline{\pi}$ and $\overline{\pi}'$. The unique line $\overline{\zeta}_\tau$ of $\overline{\pi}$ through $\overline{\eta}$ that cannot be obtained in this way, contains the $q - 1$ points of intersection of $\overline{\pi}$ with the elements of $[\overline{\pi}, \overline{\tau}] \backslash \{\overline{\pi}\}$ and also the nucleus of $\overline{\pi}$.

It is now shown that the set of all nuclei is an n-dimensional subspace $\mathrm{PG}(n, q)$ of $\overline{\mathrm{PG}}(N_n, q)$. From the fact that there are exactly $\theta(n)$ nuclei, it suffices to show that all points of the line joining any two distinct nuclei are again nuclei. In other words, it is sufficient to show the following:

(1) the nuclei of all elements of $[\overline{\pi}, \overline{\varphi}]$ are collinear;
(2) the nuclei of all elements of $[\overline{\pi}, \overline{\tau}] \cup \{\overline{\pi}'\}$ are collinear.

Put $\overline{\gamma}^* = \langle \overline{\pi}, \overline{\tau} \rangle$, and for each element φ' of $[\pi, \varphi] \setminus \{\pi\}$, put $\overline{\gamma}^* \cap \overline{\varphi'} = \overline{\rho}_{\varphi'}$. Note that the unique point of such a line $\overline{\rho}_{\varphi'}$ which is not contained in any element of $[\overline{\pi}, \overline{\tau}]$ is the nucleus $\overline{\xi}'$ of $\overline{\varphi'}$, since, by the proof of Lemma 4.47, all nuclei are contained in $\langle \overline{\pi}, \overline{\tau} \rangle$. Previous arguments imply that, for each $\tau' \in [\pi, \tau] \setminus \{\pi\}$, the points $\overline{\pi} \cap \overline{\tau}'$ and $\overline{\varphi}' \cap \overline{\tau}'$, with $\varphi' \in [\pi, \varphi] \setminus \{\pi\}$, constitute a line $\overline{\delta}_{\tau'} = \langle \overline{\pi}, \overline{\varphi} \rangle \cap \overline{\tau}'$. It follows that the q disjoint lines $\overline{\xi}_\varphi, \overline{\delta}_{\tau'}$, with τ' varying in $[\pi, \tau] \setminus \{\pi\}$, all meet each of the $q + 1$ disjoint lines $\overline{\zeta}_\tau, \overline{\rho}_{\varphi'}$, with φ' varying in $[\pi, \varphi] \setminus \{\pi\}$. Hence the nuclei of all elements of $[\overline{\pi}, \overline{\varphi}]$ are contained in the unique 'missing' line of the hyperbolic quadric containing the $2q + 1$ mentioned lines. This shows (1).

Let $\overline{\zeta}_{\pi'}$ be the intersection of $\overline{\gamma}^*$ with $\overline{\pi}'$. This line contains the point $\overline{\eta}$, hence $\langle \overline{\zeta}_{\pi'}, \overline{\zeta}_\tau \rangle$ is a plane $\overline{\alpha}$. Since, for every $\tau' \in [\pi, \tau]$, the τ'-number of π' is 2, the plane $\overline{\alpha}$ contains, for every such τ', the line $\overline{\zeta}^*_{\tau'}$ consisting of all points $\overline{\tau}' \cap \overline{\tau}''$, with $\tau'' \in [\pi, \tau] \setminus \{\tau'\}$, the point $\overline{\pi}' \cap \overline{\tau}'$, and the nucleus of $\overline{\tau}'$. Note that $\overline{\zeta}^*_\pi = \overline{\zeta}_\tau$. So the set of lines $\overline{\mathcal{O}} = \{\overline{\zeta}^*_{\tau'} \mid \tau' \in [\pi, \tau]\} \cup \{\overline{\zeta}_{\pi'}\}$ is a dual $(q + 1)$-arc in $\overline{\alpha}$. As all nuclei are contained in $\overline{\gamma}^*$, the nucleus of $\overline{\pi}'$ belongs to $\overline{\zeta}_{\pi'}$. Hence, noting that q is even by Lemma 4.45, the $q + 1$ nuclei of the elements of $[\overline{\pi}, \overline{\tau}] \cup \{\overline{\pi}'\}$ form the nucleus line of the dual $(q + 1)$-arc $\overline{\mathcal{O}}$, proving (2). Consequently, the set of all nuclei is an n-dimensional subspace $\mathrm{PG}(n, q)$ of $\overline{\mathrm{PG}}(N_n, q)$.

Now it is shown that also the set $\widetilde{S} = (S \cup \{\mathrm{PG}(N_{n-1}, q)\}) \setminus \{\pi'\}$, where $\mathrm{PG}(N_{n-1}, q)$ is the dual of $\mathrm{PG}(n, q)$, is a Veronesean set of subspaces. Condition (I) follows from the fact that $\mathrm{PG}(n, q)$ meets every element of $\overline{S}$ in a unique point. Condition (II) follows from the fact that no point of $\mathrm{PG}(n, q)$ is contained in two distinct elements of $\overline{S}$. Condition (III) is also satisfied.

Assume, by way of contradiction, that the $\mathrm{PG}(N_{n-1}, q)$-number of $\varphi \in S \setminus \{\pi'\}$ equals 2. Then a subspace $\varphi' \in S \setminus \{\pi'\}$, $\varphi' \neq \varphi$ exists, for which $\|[\varphi, \varphi']\| = q$. By the foregoing, $\mathrm{PG}(n, q)$ is a subspace of $\langle \overline{\varphi}, \overline{\varphi'} \rangle$, and so $\varphi' \in [\mathrm{PG}(N_{n-1}, q), \varphi]$, a contradiction. Now, no spectrum contains 2. Next, it is shown that, if S satisfies condition (IV), then so does the set $\widetilde{S}$. This follows immediately from the fact that, for any two distinct $\varphi, \varphi' \in S \setminus \{\pi'\}$, with $\|[\varphi, \varphi']\| = q$, the equation

$$\langle \overline{\varphi}, \overline{\varphi'} \rangle = \langle \overline{\varphi}, \mathrm{PG}(n, q) \rangle = \langle \overline{\varphi'}, \mathrm{PG}(n, q) \rangle$$

holds.

As for $n > 2$ axiom (A_{n-1}) is also satisfied, it follows that S^* is the set of all V_{n-1}-subspaces of a quadric Veronesean $\mathcal{V}_n^{2^n}$. Finally, by the nucleus property of $\mathcal{V}_n^{2^n}$, the subspace π' is the nucleus of $\mathcal{V}_n^{2^n}$. $\square$

Now the main results can be stated and proved.

Theorem 4.49. *Let S be a collection of $\theta(n)$ subspaces of dimension $(n-1)(n+2)/2$ of the projective space $\mathrm{PG}(n(n+3)/2, q)$, with $n \geq 2$, satisfying (I)–(V). Then one of the following holds:*

(a) *S is the set of V_{n-1}-subspaces of a quadric Veronesean $\mathcal{V}_n^{2^n}$ in the space $\mathrm{PG}(n(n+3)/2, q)$;*

(b) *q is even, there are two elements* $\pi, \pi' \in S$ *with the property that no other element of* S *contains* $\pi \cap \pi'$, *and there is a unique subspace* $PG(N_{n-1}, q)$ *such that* $S \cup \{PG(N_{n-1}, q)\}$ *is the set of* V_{n-1}-*subspaces together with the nucleus of a quadric Veronesean* $V_n^{2^n}$ *in* $PG(n(n+3)/2, q)$.

When $n = 2$, *the statement holds under the weaker hypothesis of* S *satisfying* (I), (II), (III), (V). *In both cases, but with* $(q, n) \neq (2, 2)$ *in the latter case*, $V_n^{2^n}$ *is the set of points of* $PG(n(n+3)/2, q)$ *contained in at least* $\theta(n-1) - 1$ *elements of* S; *in the exceptional case there are* 13 *points contained in at least* 2 *elements of* S, *where* 6 *are coplanar while the others form* V_2^4.

Proof. The first part of the statement follows from Lemmas 4.46 and 4.48. For $n = 2$, condition (IV) is trivially satisfied. Now, any point of $V_n^{2^n}$ is contained in exactly $\theta(n-1)V_{n-1}$-subspaces. Conversely, let P be a point of $PG(n(n+3)/2, q)$ contained in $\theta(n-1)$ V_{n-1}-subspaces of $V_n^{2^n}$. The Veronesean $V_n^{2^n}$ is the image of some Π_n and the V_{n-1}-subspaces correspond to $\theta(n-1)$ hyperplanes of Π_n. In Π_n, there are distinct intersecting lines l and m such that l is the intersection of some of these hyperplanes, and such that m is the intersection of some of these hyperplanes. To l and m, there correspond conics C_l and C_m on $V_n^{2^n}$ such that the point $P \in \langle C_l \rangle \cap \langle C_m \rangle$; it follows that $P \in V_n^{2^n}$. Now it follows that, for $(q, n) \neq (2, 2)$, the Veronesean $V_n^{2^n}$ is the set of points of $PG(n(n+3)/2, q)$ contained in at least $\theta(n-1) - 1$ elements of S. If $(q, n) = (2, 2)$ and S is not the set of V_1-subspaces of a V_2^4, then there are 13 points contained in at least 2 elements of S, where 6 are coplanar while the others form V_2^4; here, the 6 coplanar points are contained in the nucleus of V_2^4. □

For q large enough, this set of axioms can be reduced.

Theorem 4.50. *Let* S *be a set of* $\theta(n)$ *subspaces of dimension* $(n-1)(n+2)/2$ *of the projective space* $PG(n(n+3)/2, q)$, *with* $n \geq 2$, *satisfying* (I), (II), (III). *If* $q \geq n$, *then* S *also satisfies* (IV).

Proof. For $n = 2$, condition (IV) is trivially satisfied. So let $n > 2$. Consider the set $S_\pi = \{\pi \cap \pi' \mid \pi' \in S \setminus \{\pi\}\}$. Relying on Lemma 4.44, it was shown in the proof of Lemma 4.46 that S_π satisfies (I), (II) and (III). By induction it follows that (IV) is satisfied. □

For q odd, this is a most satisfying characterisation, since conditions (I)–(IV) really characterise the set of V_{n-1}-subspaces of a quadric Veronesean $V_n^{2^n}$, and for $q \geq n$ conditions (I), (II), (III) do this.

There are two corollaries.

Corollary 4.51. *If* S^* *is a set of* $\theta(n) + 1$ *subspaces of dimension* $(n-1)(n+2)/2$ *of* $PG(n(n+3)/2, q)$ *such that* (I), (II), (III), (V) *hold for* S^* *and either* (IV) *also holds or* $q \geq n$, *then* q *is even and* S^* *is the set of all* V_{n-1}-*subspaces together with the nucleus of a quadric Veronesean* $V_n^{2^n}$ *in* $PG(N_n, q)$. *Also*, $V_n^{2^n}$ *is the set of points of* $PG(n(n+3)/2, q)$ *contained in* $\theta(n-1)$ *elements of* S^*.

Proof. There is an element π in $\mathcal{S}^*$ such that $\mathcal{S}^* \backslash \{\pi\}$ also satisfies (I)–(V). If q were odd, then by Theorem 4.49 all contact hyperplanes of some $\mathcal{V}_n^{2^n}$ would contain π, a contradiction. Hence q is even. Now, again by Theorem 4.49, $\mathcal{S}^*$ is the set of all $\mathcal{V}_{n-1}$-subspaces together with the nucleus of a quadric Veronesean $\mathcal{V}_n^{2^n}$ in $\mathrm{PG}(N_n, q)$. Finally, from the proof of Theorem 4.49, $\mathcal{V}_n^{2^n}$ is the set of points of $\mathrm{PG}(n(n+3)/2, q)$ contained in $\theta(n-1)$ elements of $\mathcal{S}^*$. $\qquad\square$

Corollary 4.52. *Let $\mathcal{S}$ be a set of $k \geq \theta(n)$ subspaces of dimension $m - n - 1$ of $\mathrm{PG}(m, q)$, with $m \geq n(n+3)/2$ and such that $q \geq n$. Suppose that*

(a) *every pair of elements of $\mathcal{S}$ is contained in some hyperplane of $\mathrm{PG}(m, q)$;*
(b) *no three elements of $\mathcal{S}$ are contained in a hyperplane of $\mathrm{PG}(m, q)$;*
(c) *no point is contained in all elements of $\mathcal{S}$;*
(d) *for q even there exist three distinct elements π, π', π'' of $\mathcal{S}$ with*

$$\pi \cap \pi' = \pi' \cap \pi'' = \pi'' \cap \pi.$$

Then

(i) $m = n(n+3)/2$;
(ii) *either $k = \theta(n)$ and $\mathcal{S}$ is the set of $\mathcal{V}_{n-1}$-subspaces of a quadric Veronesean $\mathcal{V}_n^{2^n}$ or q is even, $k \in \{\theta(n), \theta(n) + 1\}$ and $\mathcal{S}$ consists of k elements of the set of $\mathcal{V}_{n-1}$-subspaces together with the nucleus of a quadric Veronesean $\mathcal{V}_n^{2^n}$.*

In both cases, for q even but with $(q, n) \neq (2, 2)$, if $\mathcal{S}$ contains the nucleus of $\mathcal{V}_n^{2^n}$, then $\mathcal{V}_n^{2^n}$ is the set of points of $\mathrm{PG}(m, q)$ contained in at least $\theta(n-1) - 1$ elements of $\mathcal{S}$; in the exceptional case there are 13 points contained in at least 2 elements of $\mathcal{S}$, where 6 are coplanar while the other 7 form $\mathcal{V}_2^4$.

Proof. If elements π, π' of $\mathcal{S}$, with $\pi \neq \pi'$, did not generate a hyperplane, then the number of hyperplanes containing π and one element of $\mathcal{S} \backslash \{\pi\}$ is at least $(\theta(n) - 2) + (q + 1)$, a contradiction as π is contained in exactly $\theta(n)$ hyperplanes of $\mathrm{PG}(m, q)$.

Assume, by way of contradiction, that $m > n(n+3)/2$. As in the proof of Lemma 4.44, and with that notation, there is a sequence $(\bar{\pi}^3, \bar{\pi}^4, \dots, \bar{\pi}^{n+1})$ of elements of $\overline{\mathcal{S}}$ in the dual space $\overline{\mathrm{PG}}(m, q)$ of $\mathrm{PG}(m, q)$ satisfying (i) and (ii) of that proof, for all $i \in \{2, 3, \dots, n\}$. Again, as in the proof of Lemma 4.44, this gives a contradiction. Hence $m = n(n+3)/2$. Now Corollary 4.52 follows from Theorems 4.49, 4.50 and Corollary 4.51. $\qquad\square$

For q odd, relying on the polarity θ which interchanges the $\mathcal{V}_{n-1}$-subspaces and tangent spaces of $\mathcal{V}_n^{2^n}$, the following results are obtained.

Theorem 4.53. *Let $\overline{\mathcal{S}}$ be a collection of $\theta(n)$ subspaces of dimension n of the projective space $\mathrm{PG}(n(n+3)/2, q)$, with q odd and $n \geq 2$, satisfying (I′)–(IV′). Then $\overline{\mathcal{S}}$ is the set of all tangent spaces of a quadric Veronesean $\mathcal{V}_n^{2^n}$ in $\mathrm{PG}(n(n+3)/2, q)$. In particular, if $n = 2$, then the statement holds under the weaker hypotheses of $\overline{\mathcal{S}}$ satisfying (I′)–(III′). Also, $\mathcal{V}_n^{2^n}$ is the set of all points contained in exactly one element of $\overline{\mathcal{S}}$.*

Theorem 4.54. *Let $\overline{S}$ be a set of $\theta(n)$ subspaces of dimension n of the projective space $\mathrm{PG}(n(n+3)/2, q)$, with q odd and $n \geq 2$, satisfying (I')–(III'). If $q \geq 2$, then $\overline{S}$ also satisfies (IV').*

Corollary 4.55. *Let $\overline{S}$ be a set of $k \geq \theta(n)$ subspaces of dimension n of $\mathrm{PG}(m, q)$, with q odd, $m \geq n(n+3)/2$ and such that $q \geq n$. Suppose that*

(a) *every two elements of $\overline{S}$ have a non-empty intersection,*
(b) *every three distinct elements of $\overline{S}$ have an empty intersection,*
(c) *no hyperplane contains all elements of $\overline{S}$.*

Then

  (i) *$m = n(n+3)/2$, $k = \theta(n)$;*
 (ii) *$\overline{S}$ is the set of all tangent spaces of a quadric Veronesean $V_n^{2^n}$;*
(iii) *$V_n^{2^n}$ is the set of all points contained in exactly one element of $\overline{S}$.*

Something more can be said in the case that S does not satisfy (V).

Theorem 4.56. *Let S be an ovoidal Veronesean set of subspaces of the projective space $\mathrm{PG}(n(n+3)/2, q)$, with $n \geq 2$. Then q is even and S can be extended to a hyperovoidal Veronesean set of subspaces of $\mathrm{PG}(n(n+3)/2, q)$. Also, if $n = 2$, then $q \in \{2, 4\}$ and S is uniquely determined in both cases, up to a projectivity.*

Proof. It is shown that the set of all nuclei of members of $\overline{S}$ is an n-dimensional subspace of $\mathrm{PG}(N_n, q)$. Therefore it suffices to prove that all points of the line joining two arbitrary distinct nuclei are nuclei. So let $\pi', \pi'' \in S$, with $\pi' \neq \pi''$, and let $\overline{\xi}'$ be the nucleus of $\overline{\pi}'$ and $\overline{\xi}''$ the nucleus of $\overline{\pi}''$. Let $\overline{\pi}' \cap \overline{\pi}'' = \{\eta\}$. Further, let $\overline{\delta}'$ be an arbitrary point on the line $\overline{\eta}\overline{\xi}'$ different from $\overline{\eta}$ and $\overline{\xi}'$. Let $\overline{\tau} \in \overline{S} \setminus \{\overline{\pi}'\}$ contain $\overline{\delta}'$, and let $\overline{\zeta}$ be the intersection of $\overline{\tau}$ with $\langle \overline{\pi}', \overline{\pi}'' \rangle$. As before, any element of $\overline{S}$ meeting $\overline{\zeta}$ has a line in common with the plane $\langle \overline{\eta}, \overline{\zeta} \rangle$, and the set of all these lines is a dual $(q+1)$-arc if the nucleus of $\overline{\tau}$ is on $\overline{\zeta}$, or a dual $(q+2)$-arc if the nucleus of $\overline{\tau}$ is not on $\overline{\zeta}$. In the latter case, the point $\overline{\xi}'$ is contained in a unique line, different from the line $\overline{\eta}\,\overline{\xi}'$, of that dual $(q+2)$-arc, contradicting the definition of a nucleus. Hence there is a dual $(q+1)$-arc and it now follows, interchanging the roles of π'' and τ if necessary, that the unique line of $\langle \overline{\eta}, \overline{\zeta} \rangle$ extending the dual $(q+1)$-arc to a dual $(q+2)$-arc contains $q+1$ nuclei amongst which are $\overline{\xi}'$ and $\overline{\xi}''$. Hence the set of all nuclei of members of $\overline{S}$ is an n-dimensional subspace $\mathrm{PG}(n, q)$ of $\mathrm{PG}(N_n, q)$. So $S \cup \{\mathrm{PG}(n, q)\}$ is a hyperovoidal Veronesean set of subspaces of $\mathrm{PG}(n(n+3)/2, q)$.

Now take the case $n = 2$. Consider the hyperovoidal Veronesean set of subspaces $S \cup \overline{\mathrm{PG}(2, q)} = S^*$. By the first part of this proof, every three distinct elements of $\overline{S}^*$ define $q + 2$ elements of $\overline{S}^*$, which all intersect a common plane in a line. Let $\overline{B}$ be the set with as elements these sets consisting of $q + 2$ elements of $\overline{S}^*$. Now count in different ways the number of ordered pairs $(\overline{\pi}, \overline{O})$, with $\overline{\pi} \in \overline{S}^*$, $\overline{O} \in \overline{B}$ and $\overline{\pi} \in \overline{O}$. Then

$$|\overline{B}|(q+2) = (q^2 + q + 2)(q^2 + q + 1)(q^2 + q)/(q+1)q.$$

Hence $q+2$ divides 12, and so $q \in \{2, 4\}$. Also, if $q = 2$, then $\overline{S}$ and $\overline{S}^*$ are uniquely defined, up to a projectivity. For $q = 4$, there is, up to a projectivity, just one example, which is related to the simple Mathieu group M_{22}; see Section 4.7.   □

Remark 4.57. The set $\overline{S}^*$, provided with the elements of $\overline{B}$, is an extension of a projective plane of order q. Hence $q \in \{2, 4\}$. For $q = 4$, this extension is the unique 3–$(22, 6, 1)$ Witt design. This design admits M_{22} as an automorphism group; this is not the full automorphism group.

The unique example for $q = 2$ can be generalised as follows to any n, with $n \geq 2$. Let $AG(n + 1, 2)$ be an affine space in $PG(n + 1, 2)$. Consider, in the Grassmann variety of the lines of $PG(n + 1, 2)$, all subspaces corresponding to the sets of all lines with a common point in $AG(n + 1, 2)$. Then this gives a dual hyperovoidal set of subspaces in $PG(n(n + 3)/2, 2)$.

For $n \geq 3$ with $q > 2$, a classification of ovoidal and hyperovoidal sets of subspaces remains open.

4.2.3 Characterisations of $\mathcal{V}_n^{2^n}$ of the third kind

Relying on Subsection 4.2.2, a simple and elegant characterisation of the finite quadric Veronesean $\mathcal{V}_n^{2^n}$ is obtained.

Theorem 4.58. *Under the conditions that $m \geq n(n + 3)/2, n \geq 2$ and $q > 2$, let $\theta : PG(n, q) \to PG(m, q)$ be an injective map, such that the image of any line of $PG(n, q)$ under θ is a plane $(q + 1)$-arc in $PG(m, q)$, and such that the image of θ generates $PG(m, q)$. Then $m = n(n+3)/2$, the image of θ is the quadric Veronesean $\mathcal{V}_n^{2^n}$, and the images of the lines of $PG(n, q)$ are the conics on $\mathcal{V}_n^{2^n}$.*

Proof. Let $\theta : PG(n, q) \to PG(m, q)$ be an injective map from $PG(n, q)$ into $PG(m, q)$, with $n \geq 1$ and $q > 2$, such that the image of any line of $PG(n, q)$ is a plane $(q + 1)$-arc in $PG(m, q)$, and such that the image of θ generates $PG(m, q)$. Let π be the subspace of $PG(m, q)$ generated by the image under θ of any hyperplane Π_{n-1} of $PG(n, q)$. It is shown that the dimension of π is at least $m - n - 1$.

For $n = 1$, this follows since in this case $m = 2$ and the dimension of the image of a point is zero. Now let $n > 1$. Let π' be the subspace of $PG(m, q)$ generated by the image of a hyperplane Π'_{n-1} of $PG(n, q)$, with $\Pi_{n-1} \neq \Pi'_{n-1}$, and let l be a line of $PG(n, q)$ not contained in $\Pi_{n-1} \cup \Pi'_{n-1}$ for which $l \cap \Pi_{n-1} \neq l \cap \Pi'_{n-1}$. Let $\overline{\Pi}_2 = \langle l\theta \rangle$. Since $q > 2$, it follows that every point P of $PG(n, q)$ is contained in a line l' of $PG(n, q)$ meeting $l \cup \Pi_{n-1} \cup \Pi'_{n-1}$ in three distinct points. Since the images under θ of these points generate $\langle l'\theta \rangle$, the point $P\theta$ is contained in $\langle \pi, \pi', \overline{\Pi}_2 \rangle$. Hence $PG(m, q) = \langle \pi, \pi', \overline{\Pi}_2 \rangle$. If w is the dimension of π, w' the dimension of π', and u the dimension of $\pi \cap \pi'$, then this implies that $m \leq w + w' - u + 1$. By the induction hypothesis $u \geq w' - n$; hence $m - w \leq w' - u + 1 \leq n + 1$ and the result follows.

Now it is shown that, for $n \geq 2$ and $m \geq N_n = n(n+3)/2$, a direct consequence is the equality $m = N_n$. From a chain of subspaces $\Pi_1 \subset \Pi_2 \subset \cdots \subset \Pi_{n-1}$ in $PG(n, q)$, it follows that $m \leq 2 + 3 + \cdots + n + (n + 1) = N_n$. Hence $m = N_n$.

Also, in this case, the dimension of the subspace of $PG(m, q)$ generated by the image of a k-dimensional subspace Π_k of $PG(m, q)$ is equal to $N_k = k(k + 3)/2$, for $k \in \{0, 1, 2, \ldots, n\}$. It also follows that, with the notation of the previous paragraph, if $n \geq 2$ and $m \geq N_n$, then π and π' meet in a subspace of dimension N_{n-2}, and $\langle \pi, \pi', \overline{\Pi}_2 \rangle = PG(N_n, q)$. Since every hyperplane $\Pi''_{n-1} \notin \{\Pi_{n-1}, \Pi'_{n-1}\}$ of $PG(n, q)$ either contains a line meeting $\Pi_{n-1} \cup \Pi'_{n-1}$ in just two points, or else meets every such line in a unique point outside $\Pi_{n-1} \cup \Pi'_{n-1}$, the images under θ of three distinct hyperplanes generate $PG(N_n, q)$.

Hence the set

$$\mathcal{S} = \{\langle \Pi_{n-1}\theta \rangle \mid \Pi_{n-1} \text{ is a hyperplane of } PG(n, q)\}$$

satisfies (I), (II) and (IV) of Subsection 4.2.2. Assume, by way of contradiction, that there is a point P contained in all elements of $\mathcal{S}$. Then P is contained in all subspaces $\langle \Pi_{n-2}\theta \rangle$ with Π_{n-2} a subspace of dimension $n - 2$ in a given hyperplane Π_{n-1} of $PG(n, q)$. Similar arguments imply that P is contained in all planes $\langle l\theta \rangle$ with l any line of a given plane Π_2 of $PG(n, q)$. As $\langle l\theta \rangle \cap \langle l'\theta \rangle$ is $l\theta \cap l'\theta$, a contradiction is obtained. Hence, (III) is also satisfied.

Next, it is shown that (V) is satisfied. Since every subspace of dimension $n - 2$ in $PG(n, q)$ is contained in $q + 1$ hyperplanes of $PG(n, q)$, with the notation of Subsection 4.2.2, the size of $[\pi, \pi']$ is $q + 1$ for any two distinct $\pi, \pi' \in \mathcal{S}$. So, again with the same terminology, the spectrum of every element of $\mathcal{S}$ is $\{q + 1\}$.

From Theorem 4.49, it now follows that the image of θ, which is precisely the set of points of $PG(N_n, q)$ contained in $\theta(n - 1)$ elements of $\mathcal{S}$, is the quadric Veronesean $\mathcal{V}_n^{2^n}$; the images of the lines of $PG(n, q)$ are precisely the conics on $\mathcal{V}_n^{2^n}$. □

Remark 4.59. 1. For $q = 2$, every cap of size $2^{n+1} - 1, n \geq 2$, in some projective space $PG(m, 2)$, with $m \geq n(n + 3)/2$ and where the cap generates $PG(m, 2)$, can be seen as the image of a mapping θ of $PG(n, 2)$ into $PG(m, 2)$ satisfying the conditions of Theorem 4.58. Hence the condition $q > 2$ in the statement of Theorem 4.58 is necessary.

2. For $n = 2$, the plane $PG(2, q)$ in the statement of Theorem 4.58 can be replaced by any projective plane, which is not necessarily Desarguesian, with $q + 1$ points on any line.

4.2.4 Characterisations of $\mathcal{V}_n^{2^n}$ of the fourth kind

First, the Veronese surface $\mathcal{V}_2^4$ is characterised by considering its common points with the planes and hyperplanes of $PG(5, q)$. Then, without proof, recent characterisations of $\mathcal{V}_n^{2^n}$ are given, where again the common points of $\mathcal{V}_n^{2^n}$ with subspaces are considered. Since the hyperplane sections of $\mathcal{V}_2^4$ correspond to the quadrics of $PG(2, q)$, any hyperplane Π_4 of $PG(5, q)$ has $1, q + 1$ or $2q + 1$ points in common with $\mathcal{V}_2^4$. Now consider the intersections of $\mathcal{V}_2^4$ with the planes of $PG(5, q)$.

Lemma 4.60. *Any plane π of $PG(5, q)$ meets $\mathcal{V}_2^4$ in $0, 1, 2, 3,$ or $q + 1$ points.*

Proof. Suppose that the plane π contains at least four distinct points Q_1, Q_2, Q_3, Q_4 of $\mathcal{V}_2^4$. Then $q > 2$. By Corollary 4.13, the points Q_i and Q_j, $i \neq j$, are contained in a unique conic of $\mathcal{V}_2^4$. Let C' be the conic defined by Q_1 and Q_2, and let C'' be the conic defined by Q_2 and Q_3. Suppose that $C' \neq C''$. By Theorem 4.17, the conic planes π' and π'' containing C' and C'' generate a hyperplane Π_4. With the notation of Section 4.1, the set $(\Pi_4 \cap \mathcal{V}_2^4)\zeta^{-1}$ is a quadric of $PG(2, q)$; hence $|\Pi_4 \cap \mathcal{V}_2^4| \leq 2q+1$. Since $\pi \subset \Pi_4$, so $Q_4 \in \Pi_4$; since also $C' \cup C'' \subset \Pi_4$, it follows that

$$|\Pi_4 \cap \mathcal{V}_2^4| \geq |C' \cup C'' \cup \{Q_4\}| = 2q + 2,$$

a contradiction. Thus $C' = C''$, and so $C' = \pi \cap \mathcal{V}_2^4$. It follows that $|\pi \cap \mathcal{V}_2^4| = q+1$. $\square$

Now the intention is to characterise, for $q > 2$, the Veronesean $\mathcal{V}_2^4$ by the number of its common points with the hyperplanes and planes of $PG(5, q)$.

From now on, let $\mathcal{K}$ be a set of k points of $PG(m, q)$, $m \geq 5$, with the following properties:

(A) $|\Pi_4 \cap \mathcal{K}| \leq 2q + 1$ for any four-dimensional subspace Π_4 of $PG(m, q)$ with equality for some Π_4;
(B) any plane of $PG(m, q)$ meeting $\mathcal{K}$ in four points meets it in at least $q + 1$ points.

Lemma 4.61. *For any line l, either $l \subset \mathcal{K}$ or $|l \cap \mathcal{K}| \leq 3$.*

Proof. The lemma is immediate for $q \leq 3$. So, for $q \geq 4$, let l be a line of $PG(m, q)$, where $l \not\subset \mathcal{K}$ and $|l \cap \mathcal{K}| = s$. Suppose that $4 \leq s \leq q$ and let Π_4 be a 4-dimensional space containing the line l. By (B), any plane π of Π_4 containing l has at least $q + 1$ points in common with $\mathcal{K}$. Consequently,

$$|\Pi_4 \cap \mathcal{K}| \geq (q^2 + q + 1)(q + 1 - s) + s \geq q^2 + q + 1 + s \geq q^2 + q + 5.$$

By (A), $|\Pi_4 \cap \mathcal{K}| \leq 2q + 1$. So $q^2 + q + 5 \leq 2q + 1$, a contradiction. $\square$

Lemma 4.62. *For the set $\mathcal{K}$ with $q \geq 5$, there is no pair (l_1, l_2) of skew lines in the following cases:*
   (i) *l_1 and l_2 are both lines of $\mathcal{K}$;*
   (ii) *l_1 is a line of $\mathcal{K}$ and l_2 is a trisecant of $\mathcal{K}$;*
   (iii) *l_1 and l_2 are both trisecants of $\mathcal{K}$.*

Proof. (i) Let l_1 and l_2 be distinct skew lines of $\mathcal{K}$, and let Π_3 be the solid defined by them. So $|\Pi_3 \cap \mathcal{K}| \geq 2q + 2$. Hence any four-dimensional space of $PG(m, q)$ containing Π_3 has more than $2q + 1$ points in $\mathcal{K}$. This contradicts (A).

(ii), (iii) Here, l_2 is a trisecant and l_1 is a trisecant or line of $\mathcal{K}$. Then there exist distinct planes π, π', π'' containing l_2 and also a point of $l_1 \cap \mathcal{K}$. By (B), each of $|\pi \cap \mathcal{K}|, |\pi' \cap \mathcal{K}|, |\pi'' \cap \mathcal{K}|$ is at least $q + 1$. Hence the solid $\Pi_3 = l_1 l_2$ contains at least $3(q + 1 - 3) + 3 = 3q - 3$ points in $\mathcal{K}$. Consequently, any 4-dimensional subspace of $PG(m, q)$ through Π_3 has at least $3q - 3$ points in $\mathcal{K}$. Since $q \geq 5$, so $3q - 3 > 2q + 1$, contradicting (A). $\square$

Lemma 4.63. *If $\mathcal{K}$ contains distinct lines l_1 and l_2, then $l_1 \cap l_2 \neq \emptyset$ and $\mathcal{K} = l_1 \cup l_2$.*

Proof. From the first part of the proof of Lemma 4.62, $l_1 \cap l_2 \neq \emptyset$. Let $Q \in \mathcal{K}$, with $Q \notin l_1 \cup l_2$. Then any four-dimensional subspace containing l_1, l_2 and Q has at least $2q + 2$ points in $\mathcal{K}$, contradicting (A). Hence $\mathcal{K} = l_1 \cup l_2$. $\qquad\square$

Lemma 4.64. *If $\mathcal{K}$ lies in a plane π of $\mathrm{PG}(m, q)$, then $|\mathcal{K}| = 2q + 1$ and there are the following possibilities:*
(a) *$|l \cap \mathcal{K}| \leq 3$ for any line l in π, with equality for some line l;*
(b) *the set $\mathcal{K}$ consists of a line l and a q-arc $\mathcal{K}'$ of π, where $l \cap \mathcal{K}' = \emptyset$;*
(c) *the set $\mathcal{K}$ consists of two distinct lines of π.*

Proof. Any 4-dimensional subspace containing π has at most $2q + 1$ points in $\mathcal{K}$. As $\mathcal{K} \subset \pi$, it follows that $|\mathcal{K}| \leq 2q + 1$. Since at least one 4-dimensional subspace has exactly $2q + 1$ points in $\mathcal{K}$, so $|\mathcal{K}| = 2q + 1$.

First, assume that $\mathcal{K}$ does not contain a line. By Lemma 4.61, any line of π has at most three distinct points in $\mathcal{K}$. If there is no line of π with exactly three points in $\mathcal{K}$, then $\mathcal{K}$ is a $(2q + 1)$-arc, contradicting Theorem 8.5 of PGOFF2. So π contains at least one trisecant of $\mathcal{K}$.

Next, let l be a line which lies in $\mathcal{K}$. If the q points of $\mathcal{K} \backslash l$ lie on a line l', then $\mathcal{K} = l \cup l'$ by Lemma 4.63. Now assume that there is no line containing $\mathcal{K} \backslash l$. Then, by Lemma 4.61, the set $\mathcal{K} \backslash l$ is a q-arc of π. $\qquad\square$

Lemma 4.65. *Let $q \geq 5$ and suppose that $\mathcal{K}$ generates a solid Π_3. Then $\mathcal{K} = l \cup \mathcal{K}^*$, with $\mathcal{K}^*$ a q-arc of some plane π, with l a line not contained in π, and $l \cap \mathcal{K}^* = \emptyset$.*

Proof. Any 4-dimensional subspace containing Π_3 has at most $2q + 1$ points in $\mathcal{K}$. As $\mathcal{K} \subset \Pi_3$, so $|\mathcal{K}| \leq 2q + 1$. Since at least one 4-dimensional subspace has exactly $2q + 1$ points in $\mathcal{K}$, so $|\mathcal{K}| = 2q + 1$.

Two cases are distinguished.

(1) *$\mathcal{K}$ does not contain a line*

By Lemma 4.61, any line has at most three points in $\mathcal{K}$.

First, suppose that $\mathcal{K}$ has at least one trisecant l. Let $Q \in \mathcal{K} \backslash l$. By condition (B), the plane $\pi = Ql$ has at least $q + 1$ points in $\mathcal{K}$. Since $\mathcal{K}$ generates the solid Π_3, there is at least one point Q' in $\mathcal{K} \backslash \pi$. By (B), the plane $\pi' = Q'l$ has at least $q + 1$ points in $\mathcal{K}$. If $Q'' \in \mathcal{K} \backslash (\pi \cup \pi')$, then the plane $\pi'' = Q''l$ has at least $q + 1$ points in $\mathcal{K}$; so $|\mathcal{K}| \geq 3(q + 1 - 3) + 3 = 3q - 3$. Hence $2q + 1 \geq 3q - 3$, whence $q \leq 4$, a contradiction. Consequently, $\mathcal{K} \subset \pi \cup \pi'$.

Suppose now that $\mathcal{K}$ has a trisecant l', with $l' \neq l$. Then l' lies in π or π'; say, $l' \subset \pi$. The plane $\pi^* = Q'l'$ has at least $q + 1$ points in $\mathcal{K}$, at most three of which belong to π'. Hence at least $q + 1 - 6$ points of $\pi^* \cap \mathcal{K}$ do not belong to $\pi \cup \pi'$. Since $\mathcal{K} \subset \pi \cup \pi'$, so $q \leq 5$. As it was assumed that $q \geq 5$, hence $q = 5$.

By the same argument, the line $l'' = \pi^* \cap \pi'$ is a trisecant of $\mathcal{K}$ and the point $l' \cap l''$ of l is not in $\mathcal{K}$. On the lines l, l', l'' there are nine points of $\mathcal{K}$. Since $|\mathcal{K}| = 11$, there is a point $Q^* \in \mathcal{K} \backslash (l \cup l' \cup l'')$, say $Q^* \in \pi$. Since $q = 5$, at least one of the lines joining Q^* to a point of $l' \cap \mathcal{K}$ has a point in $l \cap \mathcal{K}$. Such a line l^* is a trisecant of $\mathcal{K}$. Then the plane $Q'l^*$ contains at least one point of $\mathcal{K}\backslash(\pi \cup \pi')$, a contradiction. So it has been shown that l is the only trisecant of $\mathcal{K}$.

Let $l \cap \mathcal{K} = \{Q_1, Q_2, Q_3\}$. Then $(\pi \cap \mathcal{K})\backslash\{Q_i\}$ is a k_i-arc of the plane π for $i = 1, 2, 3$. Analogously, $(\pi' \cap \mathcal{K})\backslash\{Q_i\}$ is a k_i'-arc of the plane π' for $i = 1, 2, 3$. However, since $|\mathcal{K}| = 2q + 1$ and $\mathcal{K} \subset \pi \cup \pi'$, so

$$(k_i - 2) + (k_i' - 2) + 3 = 2q + 1,$$

for $i = 1, 2, 3$. Hence $k_i + k_i' = 2q + 2$. Without loss of generality, let $k_1 \geq q+1$. From Section 8.1 of PGOFF2, the k_1-arc $(\pi \cap \mathcal{K})\backslash\{Q_1\}$ has at least $(q - 1)/2$ bisecants passing through Q_1. It follows that the plane π contains at least two trisecants of $\mathcal{K}$ through Q_1, a contradiction.

It has been shown that $\mathcal{K}$ has no trisecant; that is, $\mathcal{K}$ is a $(2q + 1)$-cap of the solid Π_3. Assume that no four points of $\mathcal{K}$ are coplanar; then $\mathcal{K}$ is a $(2q + 1)$-arc of Π_3. By Theorems 21.2.4 and 21.3.8 of FPSOTD, $|\mathcal{K}| \leq q + 1$, contradicting that $|\mathcal{K}| = 2q + 1$. Hence Π_3 contains a plane π which has at least four points in $\mathcal{K}$. By condition (B), $|\pi \cap \mathcal{K}| \geq q+1$. Since $\pi \cap \mathcal{K}$ is a k'-arc of π, so $|\pi \cap \mathcal{K}| \in \{q+1, q+2\}$. Hence $|\mathcal{K}\backslash\pi| \geq q - 1$.

Let l be a line containing at least two distinct points of $\mathcal{K}\backslash\pi$. Since $\mathcal{K}$ is a cap, the point $l \cap \pi$ is not in $\mathcal{K}$. If $\pi \cap \mathcal{K}$ is a $(q + 1)$-arc of π and q is even, then, since $q - 1 > 2$, it may be assumed that l does not contain the nucleus of $\pi \cap \mathcal{K}$. Since $q \geq 5$, the point $l \cap \pi$ is on at least two bisecants l_1 and l_2 of $\pi \cap \mathcal{K}$. By (B), the planes ll_1 and ll_2 have at least $q + 1$ points in $\mathcal{K}$. Hence

$$|\mathcal{K}| \geq 2(q + 1 - 2) + 2 + (q + 1 - 4) = 3q - 3.$$

So $2q + 1 \geq 3q - 3$, whence $q \leq 4$, a contradiction. Hence case (1) cannot occur.

(2) $\mathcal{K}$ *contains a line* l

Since $\mathcal{K}$ generates the solid Π_3, Lemma 4.63 shows that l is the only line on $\mathcal{K}$. Suppose that $\mathcal{K}\backslash l$ contains distinct collinear points Q_1, Q_2, Q_3. By Lemma 4.61, the line $Q_1 Q_2$ is a trisecant of $\mathcal{K}$; by Lemma 4.62, the line $Q_1 Q_2$ has a point on the line l. So $Q_1 Q_2$ has at least four distinct points in $\mathcal{K}$, a contradiction.

Hence $\mathcal{K}\backslash l$ is a q-cap of Π_3. On $\mathcal{K}\backslash l$, the distinct points P_1, P_2, P_3 can be chosen so that the plane $\pi = P_1 P_2 P_3$ does not contain l. Since $P_1, P_2, P_3, \pi \cap l$ are four distinct points of $\mathcal{K}$, the plane π contains at least $q + 1$ points of $\mathcal{K}$. As $|\mathcal{K}| = 2q + 1$, so $|\pi \cap \mathcal{K}| = q + 1$ and $\mathcal{K}$ consists of l and the q-arc $\mathcal{K}^* = (\pi \cap \mathcal{K})\backslash l$. $\qquad\square$

Lemma 4.66. *Let* $q \geq 5$ *and suppose that* $\mathcal{K}$ *generates an* s-*space* Π_s *with* $s \geq 4$. *Then any line* l *which is not contained in* $\mathcal{K}$ *has at most two points in* $\mathcal{K}$.

Proof. Let l be a line not contained in $\mathcal{K}$ and suppose that $|l \cap \mathcal{K}| > 2$. By Lemma 4.61, $|l \cap \mathcal{K}| = 3$. Since $\mathcal{K}$ generates Π_s and $s \geq 4$, there are points P_1, P_2, P_3 in $\mathcal{K}$ such that l, P_1, P_2, P_3 generate a Π_4. For $1, 2, 3$, let Π_2^i be the plane containing l and P_i. Since $|\Pi_2^i \cap \mathcal{K}| \geq 4$, so $|\Pi_2^i \cap \mathcal{K}| \geq q + 1$. Hence $|\Pi_4 \cap \mathcal{K}| \geq 3q - 3$. By (A), it follows that $3q - 3 \leq 2q + 1$; so $q \leq 4$, a contradiction. $\qquad\square$

Corollary 4.67. *Let $q \geq 5$ and suppose that $\mathcal{K}$ generates an s-space Π_s with $s \geq 4$. Then $\mathcal{K}$ is a k-cap or $\mathcal{K}$ contains a unique line l and any other line has at most two points in $\mathcal{K}$.*

Proof. This follows from Lemmas 4.62, 4.63, 4.66.   □

Henceforth, let $\mathcal{K}$ be a set of k points of $PG(5, q)$ satisfying the following:

(A') $|\Pi_4 \cap \mathcal{K}| \in \{1, q+1, 2q+1\}$ for any hyperplane Π_4 of $PG(5, q)$ but also that $|\Pi_4 \cap \mathcal{K}| = 2q+1$ for some hyperplane Π_4;
(B) any plane of $PG(5, q)$ with four points in $\mathcal{K}$ has at least $q+1$ points in $\mathcal{K}$.

Lemma 4.68. *For $q \geq 5$, any set satisfying* (A') *and* (B) *is a (q^2+q+1)-cap which generates $PG(5, q)$.*

Proof. By (A'), the set $\mathcal{K}$ does not generate a line. Assume that $\mathcal{K}$ generates a plane Π_2. By Lemma 4.64, there is a line l of Π_2 with $|l \cap \mathcal{K}| \in \{2, 3\}$. Let Π_4 be a hyperplane of $PG(5, q)$ which contains l but not Π_2. Then $|\Pi_4 \cap \mathcal{K}| \in \{2, 3\}$, contradicting (A'). Next, suppose that $\mathcal{K}$ generates a solid Π_3. By Lemma 4.65, $\mathcal{K} = l \cup \mathcal{K}^*$, with $\mathcal{K}^*$ a q-arc of some plane Π_2 and l a line not contained in Π_2 skew to $\mathcal{K}^*$. Let Π_2' be a plane containing two points of $\mathcal{K}^*$ and one point of $l \backslash \Pi_2$; then $|\Pi_2' \cap \mathcal{K}| = 3$. Hence any hyperplane containing Π_2' but not Π_3 has exactly three points in $\mathcal{K}$, contradicting (A').

Finally, let $\mathcal{K}$ generate a hyperplane Π_4. By (A'), $|\mathcal{K}| = 2q+1$ and each solid Π_3 of Π_4 has 1 or $q+1$ points in $\mathcal{K}$. Let l be a line with at least two points in $\mathcal{K}$ and let Π_2 be a plane of Π_4 containing l. Further, let $|l \cap \mathcal{K}| = a_1$ and $|\Pi_2 \cap \mathcal{K}| = a_2$. Counting the points of $\mathcal{K}$ in the solids of Π_4 containing Π_2,

$$(q + 1 - a_2)(q+1) + a_2 = 2q + 1.$$

Hence $a_2 = q$. Counting the points of $\mathcal{K}$ in the planes of Π_4 containing l,

$$(q - a_1)(q^2 + q + 1) + a_1 = 2q + 1.$$

Hence

$$q^3 + q^2 - a_1 q^2 - a_1 q - q = 1.$$

Consequently q divides 1, a contradiction. So it has been shown that $\mathcal{K}$ generates $PG(5, q)$.

Now, it must be shown that $\mathcal{K}$ is a k-cap. By Lemma 4.66, it is sufficient to prove that $\mathcal{K}$ does not contain a line. So assume that $\mathcal{K}$ contains some line l. By Corollary 4.67, any plane through l has at most one point in $\mathcal{K} \backslash l$. Let Π_3 be a solid skew to l. By projecting $\mathcal{K} \backslash l$ from l onto Π_3, a set $\mathcal{K}'$ of size $k - (q+1)$ is obtained. By (A'), any plane of Π_3 has 0 or q points in $\mathcal{K}'$. Let $\{\Pi_2^i \mid i = 1, \ldots, q^3 + q^2 + q + 1\}$ be the set of planes of Π_3 and let t_i be the number of points of $\mathcal{K}'$ in Π_2^i. Counting the set $\{(P, \Pi_2^i) \mid P \in \mathcal{K}', P \in \Pi_2^i\}$ in two ways gives

$$\sum t_i = (k - q - 1)(q^2 + q + 1). \qquad (4.8)$$

Now, counting the set $\{(P, P', \Pi_2^i) \mid P, P' \in \mathcal{K}'; P, P' \in \Pi_2^i; P' \neq P\}$ in two ways gives

$$\sum t_i(t_i - 1) = (k - q - 1)(k - q - 2)(q + 1). \tag{4.9}$$

Since $t_i \in \{0, q\}$ for all i, so $\sum t_i(t_i - q) = 0$. Hence

$$\sum t_i(t_i - 1) - (q - 1) \sum t_i = 0. \tag{4.10}$$

By (4.8), (4.9), (4.10),

$$(k - q - 1)(k - q - 2)(q + 1) - (q - 1)(k - q - 1)(q^2 + q + 1) = 0.$$

Since $k \neq q + 1$, so

$$(k - q - 2)(q + 1) - (q - 1)(q^2 + q + 1) = 0.$$

Thus

$$k = (q^3 + q^2 + 3q + 1)/(q + 1).$$

Therefore $q + 1$ divides $q^3 + q^2 + 3q + 1$ and so $q + 1$ divides 2, a contradiction. Hence it has been shown that $\mathcal{K}$ is a k-cap.

Finally it must be shown that $k = q^2 + q + 1$. Let $\{\Pi_4^i \mid i = 1, \ldots, \theta(5)\}$ be the set of hyperplanes of $PG(5, q)$ and let s_i be the number of points of $\mathcal{K}$ in Π_4^i. Counting the set $\{(P, \Pi_4^i) \mid P \in \mathcal{K}, P \in \Pi_4^i, i \in \theta(5)\}$ in two ways gives

$$\sum s_i = k(q^4 + q^3 + q^2 + q + 1). \tag{4.11}$$

Now, counting the set $\{(P, P', \Pi_4^i) \mid P, P' \in \mathcal{K}; P, P' \in \Pi_4^i; P' \neq P; i \in \theta(5)\}$ in two ways gives

$$\sum s_i(s_i - 1) = k(k - 1)(q^3 + q^2 + q + 1). \tag{4.12}$$

As $\mathcal{K}$ is a k-cap, so counting the number of ordered 4-tuples (P, P', P'', Π_4^i) for distinct points P, P', P'' in $\mathcal{K}$ and Π_4^i, with i varying, in two ways gives

$$\sum s_i(s_i - 1)(s_i - 2) = k(k - 1)(k - 2)(q^2 + q + 1). \tag{4.13}$$

Since $s_i \in \{1, q + 1, 2q + 1\}$ for all i,

$$\sum (s_i - 1)(s_i - q - 1)(s_i - 2q - 1) = 0, \tag{4.14}$$

which expands to

$$\sum s_i(s_i - 1)(s_i - 2) - 3q \sum s_i(s_i - 1)$$
$$+ (q + 1)(2q + 1) \sum s_i - (q + 1)(2q + 1)\theta(5) = 0. \tag{4.15}$$

From (4.11), (4.12), (4.13), (4.15),

$$k(k-1)(k-2)(q^2+q+1) - 3qk(k-1)(q^3+q^2+q+1)$$
$$+(q+1)(2q+1)k(q^4+q^3+q^2+q+1) - (q+1)(2q+1)\theta(5) = 0.$$

Hence

$$(q^2+q+1)k^3 - 3(q^4+q^3+2q^2+2q+1)k^2$$
$$+(2q^6+5q^5+9q^4+9q^3+11q^2+9q+3)k$$
$$-(2q^7+5q^6+6q^5+6q^4+6q^3+6q^2+4q+1) = 0.$$

It follows that, if $k \neq q^2+q+1$, then

$$(q^2+q+1)k^2 - (2q^4+q^3+3q^2+4q+2)k$$
$$+(2q^5+3q^4+q^3+2q^2+3q+1) = 0. \tag{4.16}$$

If $s_i = 1$ for at least one hyperplane Π_4^i, then there exists at least one solid with exactly one point in $\mathcal{K}$. Now suppose that there is at least one hyperplane Π_4^j with $s_j = q+1$. If $P \in \Pi_4^j \cap \mathcal{K}$, then there exists a line Π_1, a plane Π_2, and a solid Π_3 of Π_4^j, with $P \in \Pi_1 \subset \Pi_2 \subset \Pi_3$ and $|\Pi_1 \cap \mathcal{K}| = |\Pi_2 \cap \mathcal{K}| = |\Pi_3 \cap \mathcal{K}| = 1$. Thus, also in this case, there exists a solid with exactly one point in $\mathcal{K}$.

Next, assume that $s_i = 2q+1$ for all hyperplanes Π_4^i. Then, by (4.11),

$$\theta(5)(2q+1) = k\theta(4).$$

Hence

$$k = q(2q+1) + (2q+1)/\theta(4).$$

It follows that $\theta(4) \leq 2q+1$, a contradiction. So there is always a solid Π_3 with exactly one point P in $\mathcal{K}$. Now, counting the points of $\mathcal{K}$ in all hyperplanes containing Π_3 shows that

$$k \equiv 1 \pmod{q}. \tag{4.17}$$

Suppose that $k \neq q^2+q+1$; that is, k satisfies (4.16). Let

$$F(x) = (q^2+q+1)x^2 - (2q^4+q^3+3q^2+4q+2)x$$
$$+(2q^5+3q^4+q^3+2q^2+3q+1).$$

Then

$$F(q+1) = q^4 - q^2 > 0, \qquad F(q+2) = -q^4+q^3+q^2+q+1 < 0.$$

Consequently, $F(x)$ has a root $k' = q+c$, with $1 < c < 2$. Since $k \geq 2q+1$, so $k' \neq k$. The sum of the roots of $F(x)$ is

$$k + k' = k + q + c = (2q^4+q^3+3q^2+4q+2)/(q^2+q+1).$$

Hence

$$k = 2q^2 - 2q + 2 - c + 3q/(q^2+q+1). \tag{4.18}$$

By (4.17), q divides $2q^2 - 2q + 1 - c + 3q/(q^2 + q + 1)$. It follows that q divides $1 - c + 3q/(q^2 + q + 1)$. Since $1 < c < 2$,

$$|1 - c + 3q/(q^2 + q + 1)| < q,$$

whence

$$1 - c + 3q/(q^2 + q + 1) = 0. \tag{4.19}$$

Now, by (4.18) and (4.19),

$$k = 2q^2 - 2q + 1.$$

Hence

$$F(2q^2 - 2q + 1) = q^2(q - 2)(q - 4) = 0,$$

a contradiction, since $q \geq 5$. Therefore $k = q^2 + q + 1$. □

Lemma 4.69. *In* $\mathrm{PG}(5,3)$, *any set* $\mathcal{K}$ *satisfying* (A′) *is a 13-cap which generates* $\mathrm{PG}(5,3)$.

Proof. By (A′), the set $\mathcal{K}$ does not generate a line. So, suppose that it generates a plane Π_2. By Lemma 4.64, there is a line l of Π_2 with $|l \cap \mathcal{K}| \in \{2,3\}$. Let Π_4 be a hyperplane of $\mathrm{PG}(5,3)$ which contains l but not Π_2. Then $|\Pi_4 \cap \mathcal{K}| \in \{2,3\}$, contradicting (A′).

Next, suppose that $\mathcal{K}$ generates a solid Π_3. Then $|\mathcal{K}| = 7$ and each plane of Π_3 has one or four points in $\mathcal{K}$. Let P and P' be distinct points of $\mathcal{K}$. If the line $PP' = l$ has b points in $\mathcal{K}$, then a count of the points of $\mathcal{K}$ in the planes of Π_3 through l gives $4(4 - b) + b = 7$, whence $b = 3$. Let $l \cap \mathcal{K} = \{P, P', P''\}$ and $\Pi_2 \cap \mathcal{K} = \{P, P', P'', P'''\}$ with Π_2 some plane of Π_3 through l. Then the line PP''' has only two points in $\mathcal{K}$, a contradiction.

Finally, suppose that $\mathcal{K}$ generates a hyperplane Π_4. By (A′), $|\mathcal{K}| = 7$ and each solid Π_3 of Π_4 has one or four points in $\mathcal{K}$. Let l be a line with at least two points in $\mathcal{K}$, and let Π_2 be a plane of Π_4 containing l. Further, let $|l \cap \mathcal{K}| = a_1$ and $|\Pi_2 \cap \mathcal{K}| = a_2$. Counting the points of $\mathcal{K}$ in the solids of Π_4 containing Π_2 gives $4(4 - a_2) + a_2 = 7$; hence $a_2 = 3$. Counting the points of $\mathcal{K}$ in the planes of Π_4 containing l gives $13(3 - a_1) + a_1 = 7$; hence $a_1 = 8/3$, a contradiction. So it has been shown that $\mathcal{K}$ generates $\mathrm{PG}(5,3)$.

Now it is shown that $\mathcal{K}$ is a k-cap. First suppose that there is a line l which contains exactly three points P, P', P'' of $\mathcal{K}$. Let R_1, R_2, R_3 be points of $\mathcal{K} \backslash l$ such that the planes lR_1, lR_2, lR_3 generate a hyperplane Π_4; then $|\Pi_4 \cap \mathcal{K}| = 7$. Also $|lR_i \cap \mathcal{K}| \in \{4,5\}$ for $i = 1,2,3$. If the plane lR_1 contains five points of $\mathcal{K}$, then a count of the points of $\mathcal{K}$ in the hyperplanes through the solid lR_1R_2 shows that $|\mathcal{K}| = 4(7 - 6) + 6 = 10$. However, a count of the points of $\mathcal{K}$ in the hyperplanes through the solid lR_2R_3 gives $|\mathcal{K}| = 4(7 - 5) + 5 = 13$. This contradiction shows that the plane lR_1 contains exactly four points of $\mathcal{K}$. Analogously, the planes lR_2 and lR_3 contain exactly four points of $\mathcal{K}$.

Let $(\mathcal{K} \cap \Pi_4) \backslash l = \{R_1, R_2, R_3, R_4\}$. By a previous argument,

$$|\mathcal{K} \cap lR_1R_2| = |\mathcal{K} \cap lR_2R_3| = |\mathcal{K} \cap lR_1R_3|.$$

It follows that R_4 does not belong to any of the solids $lR_1R_2, lR_2R_3, lR_1R_3$. As above, counting the points of $\mathcal{K}$ in the hyperplanes through the solid lR_1R_2 gives $|\mathcal{K}| = 13$. Let Π_3 be a solid skew to l and let $\mathcal{K}\backslash l = \mathcal{K}'$; then $l\mathcal{K}' \cap \Pi_3 = \mathcal{K}''$ is a set of 10 points of Π_3. No three points of $\mathcal{K}''$ are collinear, and any plane containing at least three points of $\mathcal{K}''$ contains exactly four points of $\mathcal{K}''$. So $\mathcal{K}''$ is an ovaloid of Π_3. Hence $\{P, P', P''\}$ is the only set consisting of three collinear points of $\mathcal{K}$.

For a hyperplane Π_4^i of $PG(5,3)$, let s_i be the number of points of $\mathcal{K}$ in Π_4^i. Counting in two ways the number of ordered pairs (R, Π_4^i), with $R \in \mathcal{K} \cap \Pi_4^i$ and Π_4^i varying over all hyperplanes, gives

$$\sum s_i = 13.121 = 1573. \tag{4.20}$$

Counting the ordered triples (R, R', Π_4^i) with distinct R, R' in $\mathcal{K} \cap \Pi_4^i$ gives

$$\sum s_i(s_i - 1) = 13.12.40 = 6240. \tag{4.21}$$

Now a count of ordered 4-tuples (R, R', R'', Π_4^i) with distinct R, R', R'' in $\mathcal{K} \cap \Pi_4^i$ gives

$$\sum s_i(s_i - 1)(s_i - 2) = (13.12.11 - 6).13 + 6.40 = 22470. \tag{4.22}$$

From (4.20), (4.21), (4.22),

$$\sum s_i = 1573, \quad \sum s_i^2 = 7813, \quad \sum s_i^3 = 42763. \tag{4.23}$$

These equations imply that

$$\sum (s_i - 1)(s_i - 4)(s_i - 7) = 162. \tag{4.24}$$

Since (A') is satisfied, so $s_i \in \{1, 4, 7\}$, whence

$$\sum (s_i - 1)(s_i - 4)(s_i - 7) = 0,$$

contradicting (4.24). Hence there is no line containing exactly three points of $\mathcal{K}$.

Now suppose that there is a line l which is contained in $\mathcal{K}$. Let R_1, R_2, R_3 be points of $\mathcal{K}\backslash l$ such that the planes lR_1, lR_2, lR_3 generate a hyperplane Π_4; then $|\Pi_4 \cap \mathcal{K}| = 7$. Hence the planes lR_1, lR_2, lR_3 each contain exactly five points of $\mathcal{K}$, and the solids $lR_1R_2, lR_2R_3, lR_1R_3$ each contain exactly six points of $\mathcal{K}$. Counting the points of $\mathcal{K}$ in the hyperplanes through lR_1R_2 makes $|\mathcal{K}| = 4.(7 - 6) + 6 = 10$. Let Π_3 be a solid skew to l and let $\mathcal{K}\backslash l = \mathcal{K}'$. Then $l\mathcal{K}' \cap \Pi_3 = \mathcal{K}''$ is a set of six points of Π_3. As no four of these six points are coplanar, so $\mathcal{K}''$ is a 6-arc of Π_3. By Theorem 21.2.1 of FPSOTD, an arc of $PG(3,3)$ has at most five points, giving a contradiction. Thus it has been shown that $\mathcal{K}$ is a k-cap.

Finally, it must be shown that $k = 13$. For distinct points P, P', P'' in $\mathcal{K}$ and Π_4^i, count in two ways, similarly to above, the ordered pairs (P, Π_4^i), the ordered triples (P, P', Π_4^i), and the ordered 4-tuples (P, P', P'', Π_4^i):

$$\sum s_i = 121k; \tag{4.25}$$

$$\sum s_i(s_i - 1) = 40k(k - 1); \tag{4.26}$$

$$\sum s_i(s_i - 1)(s_i - 2) = 13k(k - 1)(k - 2). \tag{4.27}$$

As above, since $s_i \in \{1, 4, 7\}$ for all i, so

$$\sum (s_i - 1)(s_i - 4)(s_i - 7) = 0, \tag{4.28}$$

which implies that

$$\sum s_i(s_i - 1)(s_i - 2) - 9 \sum s_i(s_i - 1) + 28 \sum s_i - 10192 = 0. \tag{4.29}$$

Substituting from (4.25), (4.26), (4.27) into (4.29), gives

$$13k^3 - 399k^2 + 3774k - 10192 = (k - 13)(13k^2 - 230k + 784) = 0.$$

As the quadratic does not have integer roots, so $k = 13$, completing the proof. □

Lemma 4.70. *Any solid* Π_3 *of* $\mathrm{PG}(5, q)$, $q = 3$ *or* $q \geq 5$, *meets* $\mathcal{K}$ *in at most* $q + 2$ *points.*

Proof. Let $|\Pi_3 \cap \mathcal{K}| = m$ and suppose that $m \geq q + 2$. Counting the points of $\mathcal{K}$ in the hyperplanes through Π_3 gives

$$k = (q + 1)(2q + 1 - m) + m.$$

By Lemmas 4.68 and 4.69, $k = q^2 + q + 1$. Hence $m = q + 2$. □

Lemma 4.71. *When* $q = 3$ *or* $q \geq 5$, *suppose that the plane* Π_2 *meets* $\mathcal{K}$ *in more than three points. Then* $\Pi_2 \cap \mathcal{K}$ *is a* $(q + 1)$-*arc and so, for* q *odd, is a conic.*

Proof. Let $|\Pi_2 \cap \mathcal{K}| = n$. From (B), $n \geq q + 1$. Since $\mathcal{K}$ is a cap, $n \leq q + 2$ by Theorem 8.5 of PGOFF2. If $n = q + 2$, then any solid containing Π_2 and a point of $\mathcal{K} \backslash \Pi_2$ has at least $q + 3$ points in common with $\mathcal{K}$, contradicting Lemma 4.70. Hence $n = q + 1$, and $\Pi_2 \cap \mathcal{K}$ is a $(q + 1)$-arc of Π_2. By Theorem 8.14 of PGOFF2, $\Pi_2 \cap \mathcal{K}$ is a conic when q is odd. □

Lemma 4.72. *For* $q = 3$ *or* $q \geq 5$, *any two points of* $\mathcal{K}$ *are contained in a unique plane meeting* $\mathcal{K}$ *in a* $(q + 1)$-*arc.*

Proof. Let P and P' be distinct points of $\mathcal{K}$, and suppose that no plane through the line PP' meets $\mathcal{K}$ in a $(q + 1)$-arc. Then, by Lemma 4.71, any plane through PP' has at most three points in $\mathcal{K}$. Now project the set $\mathcal{K} \backslash \{P, P'\}$ from the line PP' onto a solid Π_3 skew to PP'. This gives a set $\mathcal{K}'$ of $q^2 + q - 1$ points in Π_3. Since any hyperplane of $\mathrm{PG}(5, q)$ through PP' meets $\mathcal{K}$ in $q + 1$ or $2q + 1$ points, any plane of Π_3 meets $\mathcal{K}'$ in $q - 1$ or $2q - 1$ points. For a plane Π_2^i of Π_3, let $s_i = |\mathcal{K}' \cap \Pi_2^i|$.

Again count, for distinct points R, R' in $\mathcal{K}'$ and Π_2^i, the ordered pairs (R, Π_2^i) and the ordered triples (R, R', Π_2^i):

$$\sum s_i = (q^2 + q - 1)(q^2 + q + 1); \tag{4.30}$$

$$\sum s_i(s_i - 1) = (q^2 + q - 1)(q^2 + q - 2)(q + 1). \tag{4.31}$$

Since $s_i \in \{q - 1, 2q - 1\}$ for all i, so

$$\sum (s_i - q + 1)(s_i - 2q + 1) = 0. \tag{4.32}$$

Hence

$$\sum s_i(s_i - 1) + 3(1 - q) \sum s_i + (q - 1)(2q - 1)(q^3 + q^2 + q + 1) = 0. \tag{4.33}$$

Substituting from (4.30), (4.31) into (4.33) gives

$$(q^2 + q - 1)(q^2 + q - 2)(q + 1) + 3(1 - q)(q^2 + q - 1)(q^2 + q + 1)$$
$$+ (q - 1)(2q - 1)(q^3 + q^2 + q + 1) = 0;$$

that is,

$$q^2(q - 1)(q - 2) = 0,$$

contradicting that $q > 2$. Consequently, there exists a plane through PP' which meets $\mathcal{K}$ in a $(q + 1)$-arc.

Now suppose that PP' is contained in distinct planes Π_2 and Π_2' meeting $\mathcal{K}$ in $(q + 1)$-arcs. Then the solid defined by these planes meets $\mathcal{K}$ in at least $2q$ points, contradicting Lemma 4.70. Hence the points P, P' are contained in a unique plane meeting $\mathcal{K}$ in a $(q + 1)$-arc.   □

Lemma 4.73. *Let $q = 3$ or $q \geq 5$. The number of planes meeting $\mathcal{K}$ in $(q + 1)$-arcs is $q^2 + q + 1$, and any two distinct planes meeting $\mathcal{K}$ in a $(q + 1)$-arc have exactly one point in common.*

Proof. Let b be the number of planes meeting $\mathcal{K}$ in a $(q + 1)$-arc. By Lemma 4.72, with $k = |\mathcal{K}|$,

$$k(k - 1)/\{(q + 1)q\} = b.$$

Since $k = q^2 + q + 1$, so also $b = q^2 + q + 1$. If Π_2 and Π_2' are distinct planes meeting $\mathcal{K}$ in $(q + 1)$-arcs, and if they meet in a line, then the solid containing them meets $\mathcal{K}$ in at least $2q$ points, contradicting Lemma 4.70.

Now suppose that $\Pi_2 \cap \Pi_2' = \emptyset$. For any point P in $\Pi_2 \cap \mathcal{K}$ and any point P' in $\Pi_2' \cap \mathcal{K}$, there is exactly one plane containing PP' and meeting $\mathcal{K}$ in a $(q + 1)$-arc. Hence there are at least $2 + (q + 1)^2$ planes meeting $\mathcal{K}$ in a $(q + 1)$-arc. This contradicts the first part of the proof. Therefore Π_2 and Π_2' have exactly one point in common.   □

Theorem 4.74. *If $\mathcal{K}$ is a set of k points of $\mathrm{PG}(5, q)$, $q \notin \{2, 4\}$, which satisfies (A') and (B), then $\mathcal{K}$ is a Veronesean $\mathcal{V}_2^4$.*

Proof. Let $\mathcal{L}$ be the set of all planes intersecting $\mathcal{K}$ in a $(q+1)$-arc. By Lemma 4.73, $|\mathcal{L}| = q^2 + q + 1$ and any two elements of $\mathcal{L}$ meet in exactly one point. Hence condition (I) of Subsection 4.2.2 is satisfied.

Now it is shown that condition (III) of Section 4.2.2 is also satisfied. Let $\Pi_2 \in \mathcal{L}$, let $P \in \Pi_2$, let $P' \in \Pi_2 \cap \mathcal{K}$ with $P \neq P'$, and let $P'' \in \mathcal{K} \backslash \Pi_2$. Then the element Π_2' of $\mathcal{L}$ containing P' and P'' has only the point P' in common with Π_2. Hence $P \notin \Pi_2'$. Since P was arbitrarily chosen in Π_2, this means that there is no point belonging to all elements of $\mathcal{L}$.

Let Π_2, Π_2', Π_2'' be three distinct elements of $\mathcal{L}$. If Π_2, Π_2', Π_2'' generate a hyperplane Π_4, then $|\Pi_4 \cap \mathcal{K}| \geq 3q$, contradicting that $|\Pi_4 \cap \mathcal{K}| \in \{1, q+1, 2q+1\}$. Hence Π_2, Π_2', Π_2'' generate $\mathrm{PG}(5, q)$. This is condition (II).

Now, by Theorems 4.49 and 4.56, the set $\mathcal{K}$ is a Veronesean $\mathcal{V}_2^4$. $\qquad\square$

Remark 4.75. For $q = 3$, any set $\mathcal{K}$ satisfies condition (B). Hence any set $\mathcal{K}$ of $\mathrm{PG}(5, 3)$ which satisfies (A') is a Veronesean $\mathcal{V}_2^4$.

In the next three lemmas, assume that $q = 4$ and that $\mathcal{K}$ satisfies (A') and (B).

Lemma 4.76. *The set $\mathcal{K}$ generates* $\mathrm{PG}(5, 4)$.

Proof. By (A'), the set $\mathcal{K}$ does not generate a line. Assume that $\mathcal{K}$ generates a plane Π_2. By Lemma 4.64 there is a line l of Π_2 with $|l \cap \mathcal{K}| \in \{2, 3\}$. Let Π_4 be a hyperplane of $\mathrm{PG}(5, 4)$ which contains l but not Π_2. Then $|\Pi_4 \cap \mathcal{K}| \in \{2, 3\}$, contradicting (A'). Next, assume that $\mathcal{K}$ generates a solid Π_3; then $|\mathcal{K}| = 9$. By considering the hyperplanes of $\mathrm{PG}(5, 4)$ which intersect Π_3 in a plane, each plane of Π_3 has either one or five points in $\mathcal{K}$. Let P and P' be distinct points of $\mathcal{K}$. Suppose that the line $PP' = l$ has $b \geq 2$ points in $\mathcal{K}$. Counting the points of $\mathcal{K}$ in the planes of Π_3 through the line l gives $5(5-b) + b = 9$, whence $b = 4$. Let $l \cap \mathcal{K} = \{P, P', P'', P'''\}$ and let $\Pi_2 \cap \mathcal{K} = \{P, P', P'', P''', R\}$, with Π_2 some plane of Π_3 through l. Then the line RP has only $2 \neq b$ points in $\mathcal{K}$, a contradiction. Finally, assume that $\mathcal{K}$ generates a hyperplane Π_4. By (A'), again $|\mathcal{K}| = 9$ and each solid Π_3 of Π_4 has either one or five points in $\mathcal{K}$. Let l be a line with at least two points in $\mathcal{K}$, and let Π_2 be a plane of Π_4 containing l. Further, let $|\Pi_2 \cap \mathcal{K}| = b$. Counting the points of $\mathcal{K}$ in the solids of Π_4 containing Π_2 gives $5(5-b) + b = 9$, whence $b = 4$. This contradicts (B), and the lemma is proved. $\qquad\square$

Lemma 4.77. *The set $\mathcal{K}$ is a cap.*

Proof. Let l be a line of $\mathrm{PG}(5, 4)$. By Lemma 4.61, either $l \subset \mathcal{K}$ or $|l \cap \mathcal{K}| \leq 3$.

First assume that $l \cap \mathcal{K} = \{P, P', P''\}$. Then select three points R_1, R_2, R_3 in $\mathcal{K} \backslash \{P, P', P''\}$ so that $\langle l, R_1, R_2, R_3 \rangle$ is a hyperplane Π_4. Then $|\Pi_4 \cap \mathcal{K}| = 9$. By (B), $\langle l, R_i \rangle$ necessarily contains five points of $\mathcal{K}$, $i = 1, 2, 3$. The solid $\langle l, R_1, R_2 \rangle$ contains either seven or eight points of $\mathcal{K}$. If $\langle l, R_1, R_2 \rangle$ contains eight points of $\mathcal{K}$, then it contains the three planes $\langle l, R_i \rangle, i = 1, 2, 3$, which is a contradiction. Therefore $|\mathcal{K} \cap \langle l, R_1, R_2 \rangle| = 7$. Considering the hyperplanes of $\mathrm{PG}(5, 4)$ containing $\langle l, R_1, R_2 \rangle$, it follows that $|\mathcal{K}| = 17$. Now project $\mathcal{K} \backslash l$ from l onto a solid Π_3 of

PG$(5, 4)$ skew to l. This produces a set $\mathcal{K}'$ of size seven in Π_3 which intersects each plane of Π_3 in either one or three points. However, such a set $\mathcal{K}'$ does not exist.

Next, assume that $\mathcal{K}$ contains a line l. Choose points $R_1, R_2, R_3 \in \mathcal{K} \backslash l$ such that $\langle l, R_1, R_2, R_3 \rangle$ is a hyperplane Π_4; then $|\Pi_4 \cap \mathcal{K}| = 9$. So $(\mathcal{K} \cap \Pi_4) \backslash l$ consists of four points R_1, R_2, R_3, R_4. By the previous paragraph, $R_4 \notin \langle l, R_i \rangle$, $i = 1, 2, 3$, as otherwise there is a line containing exactly three points of $\mathcal{K}$. Now project $\mathcal{K} \backslash l$ from l onto a solid Π_3 of PG$(5, 4)$ skew to l. This gives a set $\mathcal{K}'$ which intersects each plane of Π_3 in zero or four points. It follows that each line of Π_3 contains either zero or c points, with c a constant. If Π_2 is a plane of Π_3 with $|\Pi_2 \cap \mathcal{K}'| = 4$, then each line of Π_2 contains either zero or c points of this set of size four, a contradiction. $\square$

Lemma 4.78. *The cap $\mathcal{K}$ contains exactly* 21 *points.*

Proof. Put $|\mathcal{K}| = k$. Let $\{\Pi_4^i \mid i = 1, \ldots, \theta(5)\}$ be the set of hyperplanes of PG$(5, 4)$, and let $s_i = |\mathcal{K} \cap \Pi_4^i|$. Counting in two ways the number of ordered pairs (P, Π_4^i), with $P \in \mathcal{K} \cap \Pi_4^i$,

$$\sum_{i=1}^{1365} s_i = 341k. \tag{4.34}$$

Counting in two ways the number of ordered triples (P, P', Π_4^i), with $P, P' \in \mathcal{K} \cap \Pi_4^i$ and $P \neq P'$,

$$\sum_{i=1}^{1365} s_i(s_i - 1) = 85k(k - 1). \tag{4.35}$$

The set $\mathcal{K}$ is a cap; so counting in two ways the number of ordered 4-tuples (P, P', P'', Π_4^i), with distinct $P, P', P'' \in \mathcal{K} \cap \Pi_4^i$,

$$\sum_{i=1}^{1365} s_i(s_i - 1)(s_i - 2) = 21k(k - 1)(k - 2). \tag{4.36}$$

Since $s_i \in \{1, 5, 9\}$ for all i,

$$\sum_{i=1}^{1365} (s_i - 1)(s_i - 5)(s_i - 9) = 0.$$

Hence

$$\sum_{i=1}^{1365} s_i(s_i - 1)(s_i - 2) - 12 \sum_{i=1}^{1365} s_i(s_i - 1) + 45 \sum_{i=1}^{1365} s_i - 61425 = 0.$$

By (4.34), (4.35), (4.36),

$$21k(k - 1)(k - 2) - 1020k(k - 1) + 15345k - 61425 = 0.$$

Hence

$$7k^3 - 361k^2 + 5469k - 20475 = 0.$$

It follows that $k = 21$ or $k = 25$.

Assume, by way of contradiction, that $k = 25$. If Π_3 is a solid in $\mathrm{PG}(5, 4)$ which contains $c \geq 6$ points of $\mathcal{K}$, then, considering the hyperplanes of $\mathrm{PG}(5, 4)$ containing Π_3,

$$|\mathcal{K}| = 25 = c + 5(9 - c);$$

so $c = 5$, a contradiction. If Π_2 is a plane of $\mathrm{PG}(5, 4)$ which contains at least four points of $\mathcal{K}$, then, by (B), the plane Π_2 contains at least five points of $\mathcal{K}$; so there exists a solid which contains at least six points of $\mathcal{K}$, a contradiction. Hence any four points of $\mathcal{K}$ are linearly independent.

Let P be a fixed point of $\mathcal{K}$ and let t_i be the number of hyperplanes of $\mathrm{PG}(5, 4)$ which contain P and intersect $\mathcal{K}$ in i points, with $i = 1, 5, 9$. A count of pairs $\{P', \Pi_4\}$, with $P' \in \mathcal{K}, P \neq P'$, with Π_4 a hyperplane, and with $P, P' \in \Pi_4$, gives $4t_5 + 8t_9 = 2040$. Similarly, a count of triples $\{P', P'', \Pi_4\}$, with P', P'' distinct points of $\mathcal{K}$, different from P, and with $P, P', P'' \in \Pi_4$, gives $6t_5 + 28t_9 = 5796$. Finally, a count of 4-tuples $\{P', P'', P''', \Pi_4\}$, with distinct $P', P'', P''' \in \mathcal{K}$ different from P and with $P, P', P'', P''' \in \Pi_4$, gives $4t_5 + 56t_9 = 10120$; this contradicts the previous two equations. So $k = 21$ and the lemma is proved.   □

Theorem 4.79. *Let $\mathcal{K}$ be a set of points of* $\mathrm{PG}(5, q)$, *satisfying* (A') *and* (B). *Then*

(i) *for $q > 2$, the set $\mathcal{K}$ is a Veronesean V_2^4 in* $\mathrm{PG}(5, q)$;
(ii) *for $q = 2$, it is either a quadric Veronesean V_2^4 or an elliptic quadric in some subspace* $\mathrm{PG}(3, 2)$.

Proof. By Theorem 4.74 it may be assumed that $q \in \{2, 4\}$.

First, let $q = 4$. From the previous three lemmas, it follows that Lemmas 4.70 to 4.73 hold. Let $\mathcal{L}$ be the set of all planes of $\mathrm{PG}(5, 4)$ intersecting $\mathcal{K}$ in a 5-arc. As in the proof of Theorem 4.74, it is shown that $\mathcal{K}$ either is V_2^4 or $\mathcal{L}$ is the unique ovoidal Veronesean set of planes of $\mathrm{PG}(5, 4)$. Lemmas 4.72 and 4.73 imply that any point of $\mathcal{K}$ is contained in five planes of $\mathcal{L}$. Hence $\mathcal{L}$ is not an ovoidal Veronesean set of planes, and so $\mathcal{K}$ is a V_2^4.

Finally, suppose that $q = 2$. Let Π_4 be a hyperplane of $\mathrm{PG}(5, 2)$ containing five points of $\mathcal{K}$. If these five points generate Π_4, then, considering the three hyperplanes containing a common solid of Π_4 which contains four points of $\mathcal{K}$, it follows that $|\mathcal{K}| = 7$ and that any six points of $\mathcal{K}$ generate $\mathrm{PG}(5, 2)$. In this case, $\mathcal{K}$ is a quadric Veronesean V_2^4. So assume now that these five points of Π_4 do not generate Π_4. This implies that $|\mathcal{K}| = 5$. Now it can be shown that $\mathcal{K}$ generates a solid Π_3. As every plane of that solid contains either one or three points of $\mathcal{K}$, it follows that $\mathcal{K}$ is an elliptic quadric of Π_3.   □

From now on, let $\mathcal{K}$ be a set of points of $\mathrm{PG}(m, q), m \geq 5$, with the following properties :

(A'') $|\Pi_4 \cap \mathcal{K}| = 1, q+1$, or $2q+1$ for any four-dimensional subspace Π_4 of $\mathrm{PG}(m, q)$ with equality for some Π_4;

(B') any plane of $\mathrm{PG}(m, q)$ meeting $\mathcal{K}$ in four points meets it in at least $q + 1$ points.

Corollary 4.80. (i) *When $q > 2$, then $m = 5$ and $\mathcal{K}$ is a Veronesean V_2^4 in* $\mathrm{PG}(5, q)$.
  (ii) *When $q = 2$, then $m = 5$ and $\mathcal{K}$ is either a quadric Veronesean V_2^4 of $\mathrm{PG}(5, 2)$ or an elliptic quadric in some solid Π_3.*

Proof. First let $q > 2$ and assume, by way of contradiction, that $m > 5$. Consider a subspace Π_4 which intersects $\mathcal{K}$ in $2q + 1$ points, a subspace Π_5 of $\mathrm{PG}(m, q)$ which contains Π_4, and a subspace Π_6 of $\mathrm{PG}(m, q)$ which contains Π_5. By Theorem 4.79, the set $\Pi_5 \cap \mathcal{K} = \mathcal{K}'$ is a Veronesean V_2^4 in Π_5. Let $\mathcal{K}'' = \mathcal{K} \cap \Pi_6$. Considering the hyperplanes of Π_6 which contain Π_4,

$$(q + 1)(q^2 - q) + 2q + 1 = |\mathcal{K}''|.$$

Hence $|\mathcal{K}''| = q^3 + q + 1$. From Theorem 4.79, it follows that any plane containing at least four points of $\mathcal{K}''$ intersects $\mathcal{K}''$ in a $(q + 1)$-arc. Let Γ be the set of these $(q + 1)$-arcs; by Theorem 4.79, any two points of $\mathcal{K}''$ belong to exactly one element of Γ. If $P \in \mathcal{K}''$, then there are exactly $q^2 + 1$ elements of Γ containing P. Counting in two ways the number of pairs $\{P, K\}$, with $P \in \mathcal{K}''$, $K \in \Gamma$, $P \in K$,

$$(q^3 + q + 1)(q^2 + 1) = |\Gamma|(q + 1).$$

Hence $q + 1$ divides $(q^3 + q + 1)(q^2 + 1)$, a contradiction. So $m = 5$ and, by Theorem 4.79, $\mathcal{K}$ is a Veronesean V_2^4 in $\mathrm{PG}(5, q)$.

Next let $q = 2$. Consider a subspace Π_4 which intersects $\mathcal{K}$ in five points. If these five points generate a solid, then, considering the 5-dimensional subspaces of $\mathrm{PG}(m, 2)$ containing Π_4, it follows that $\mathcal{K}$ consists of these five points. Now $m = 5$ by (A'').

So from now on assume that any five points of $\mathcal{K}$ are linearly independent. Suppose, by way of contradiction, that $m > 5$. Let Π_6 be a subspace of $\mathrm{PG}(m, 2)$ containing at least five points of $\mathcal{K}$. As in the case $q > 2$, it follows that $|\mathcal{K}''| = 11$ with $|\mathcal{K} \cap \Pi_6| = \mathcal{K}''$. Also, by Theorem 4.79, any six points of $\mathcal{K}''$ are linearly independent. Any 5-dimensional subspace Π_5 of Π_6 containing at least six points of $\mathcal{K}''$ contains exactly seven points of $\mathcal{K}''$. Let Γ' be the set of all such sets $\Pi_5 \cap \mathcal{K}''$ of size seven. Counting the number of elements of Γ' containing two given points of $\mathcal{K}''$ gives $\mathbf{c}(9, 4)/\mathbf{c}(5, 4) = 126/5$, a contradiction. Hence $m = 5$ and then, by Theorem 4.79, $\mathcal{K}$ is a quadric Veronesean V_2^4 in $\mathrm{PG}(5, 2)$. □

Some recent characterisations of $V_n^{2^n}$ are now given, where again the numbers of common points of $V_n^{2^n}$ with subspaces are considered. As the proofs are technical and quite long, they are omitted.

Theorem 4.81. *With $N_n = n(n + 3)/2$, $q \geq 5$ and $n \geq 2$, a set $\mathcal{K}$ of $\theta(n)$ points generating $\mathrm{PG}(N_n, q)$ is a quadric Veronesean $V_n^{2^n}$ if and only if the following conditions are satisfied.*

(a) *If a plane of* $\mathrm{PG}(N_n, q)$ *intersects* $\mathcal{K}$ *in more than three points, then it contains exactly* $q + 1$ *points of* $\mathcal{K}$. *Also, any two distinct points are contained in a plane containing* $q + 1$ *points of* $\mathcal{K}$.

(b) *If a solid* Π_3 *of* $\mathrm{PG}(N_n, q)$ *intersects* $\mathcal{K}$ *in more than four points, then there are four points of* $\Pi_3 \cap \mathcal{K}$ *contained in a plane of* Π_3; *in particular, by* (a), *this implies that if* $|\Pi_3 \cap \mathcal{K}| > 4$, *then* $|\Pi_3 \cap \mathcal{K}| \geq q + 1$.

(c) *If a 5-dimensional subspace* Π_5 *of* $\mathrm{PG}(N_n, q)$ *intersects* $\mathcal{K}$ *in more than* $2q + 2$ *points, then it intersects* $\mathcal{K}$ *in exactly* $q^2 + q + 1$ *points.*

Remark 4.82. For $q < 5$, any quadric Veronesean $\mathcal{V}_n^{2^n}$ satisfies conditions (a), (b), (c) of Theorem 4.81.

A counterexample to the previous theorem, for $n > 2$ and $q = 2$, is given by removing one point of $\mathcal{V}_n^{2^n}$ and replacing it by a point of $\mathrm{PG}(N_n, 2)$ for which the rank of the matrix in Theorem 4.2 is maximal.

A counterexample, for $q = 3$ and $n = 2$, is given by the point set consisting of the points of an elliptic quadric $\mathcal{E}_3$ of a solid Π_3 of $\mathrm{PG}(5, 3)$ and three points on a line l of $\mathrm{PG}(5, 3)$ which does not intersect Π_3.

Theorem 4.83. *With* $N_n = n(n+3)/2$, *a set* $\mathcal{K}$ *of* $\theta(n)$ *points generating* $\mathrm{PG}(N_n, q)$, $q \geq 5$ *and* $n > 2$, *is a quadric Veronesean* $\mathcal{V}_n^{2^n}$ *if and only if the following conditions are satisfied:*

(a) *for any plane* π *of* $\mathrm{PG}(N_n, q)$, *the intersection* $\pi \cap \mathcal{K}$ *contains at most* $q + 1$ *points of* $\mathcal{K}$;

(b) *if a solid* Π_3 *intersects* $\mathcal{K}$ *in more than four points, then* $|\Pi_3 \cap \mathcal{K}| \geq q + 1$ *and* $\Pi_3 \cap \mathcal{K}$ *is not a* $(q + 1)$*-arc;*

(c) *if a 5-dimensional subspace* Π_5 *of* $\mathrm{PG}(N_n, q)$ *intersects* $\mathcal{K}$ *in more than* $2q + 2$ *points, then it intersects* $\mathcal{K}$ *in exactly* $q^2 + q + 1$ *points; also, any two distinct points of* $\mathcal{K}$ *are contained in a 5-dimensional subspace of* $\mathrm{PG}(N_n, q)$ *containing* $q^2 + q + 1$ *points of* $\mathcal{K}$.

Remark 4.84. For $q < 5$, any quadric Veronesean $\mathcal{V}_n^{2^n}$ also satisfies conditions (a), (b), (c) of Theorem 4.83.

For $n = 2$, a counterexample to the previous theorem is obtained as follows. Consider in $\mathrm{PG}(5, q)$ a point P on a $(q^2 + 1)$-cap O in $\mathrm{PG}(3, q)$, where, for $q = 2$, the cap O is assumed to be an elliptic quadric. Let l be a tangent line of O at P. Next, consider a second solid Π_3' intersecting Π_3 precisely in l and let K be a plane $(q + 1)$-arc for which $l \cap K = \{P\}$. Then $O \cup K$ satisfies (a), (b), (c) of Theorem 4.83 but is not a Veronesean $\mathcal{V}_2^4$.

4.3 Hermitian Veroneseans

In this section, Hermitian Veroneseans are introduced. Also, several properties and characterisations of these Hermitian Veroneseans are stated. There are many similarities with properties and characterisations of quadric Veroneseans. However, due to their length and technicality, the proofs are omitted.

Here x^q, with $x \in \mathbf{F}_{q^2}$, is also denoted by $\overline{x}$. An $(n+1) \times (n+1)$ matrix M over $\mathbf{F}_{q^2}$ is *Hermitian* if $M^* = \overline{M}$, where M^* is the transpose of M and $\overline{M}$ is the result of applying $x \mapsto \overline{x}$ to each entry of M. The space of all $(n+1) \times (n+1)$ Hermitian matrices over $\mathbf{F}_{q^2}$, with $n \geq 1$, is denoted by $\mathcal{H}(n+1, q^2)$; this space is a vector space of dimension $(n+1)^2$ over $\mathbf{F}_q$; the group $\mathrm{GL}(n+1, q^2)$ acts on $\mathcal{H}(n+1, q^2)$ with the action $\mathcal{I}$ given by $M\mathcal{I} = TM\overline{T}^*$.

For $1 \leq i \leq n+1$, let $\mathcal{H}_i(n+1, q^2)$ be the set of matrices in $\mathcal{H}(n+1, q^2)$ of rank i and let $\mathbf{P}\mathcal{H}_i(n+1, q^2)$, or simply $\mathbf{P}\mathcal{H}_i$, be the set of spaces of dimension one spanned by the matrices in $\mathcal{H}_i(n+1, q^2)$. Then each $\mathbf{P}\mathcal{H}_i$ is an orbit of $\mathrm{GL}(n+1, q^2)$ under the induced action on the projective space $\mathrm{PG}(\mathcal{H}(n+1, q^2))$ of dimension $n^2 + 2n$ over $\mathbf{F}_q$. Note that $\mathbf{P}\mathcal{H}_1$ is canonically in one-to-one correspondence with the projective space $\mathrm{PG}(n, q^2)$.

This can be seen as follows. Let $V = K^{n+1}$ consist of all column vectors over $\mathbf{F}_{q^2}$ of length $n+1$. For $P = \mathbf{P}(x) \in \mathrm{PG}(V)$, put $P\delta = \mathbf{P}(x\overline{x}^*)$; here, $x\overline{x}^*$ is a matrix of rank one in $\mathcal{H}(n+1, q^2)$ and so $P\delta$ is in $\mathbf{P}\mathcal{H}_1$. The linear group $\mathrm{GL}(n+1, q^2)$ preserves this action:

$$\mathbf{P}(Tx)\delta = \mathbf{P}((Tx)(\overline{Tx})^*) = \mathbf{P}(T(x\overline{x}^*)\overline{T}^*) = (\mathbf{P}(x)\delta)\mathcal{I}.$$

Since $\mathrm{GL}(n+1, q^2)$ is transitive on $\mathbf{P}\mathcal{H}_1$, it follows that $\mathrm{PG}(V)\delta = \mathbf{P}\mathcal{H}_1$. In fact, δ is an injective map from $\mathrm{PG}(V)$ onto $\mathbf{P}\mathcal{H}_1$.

Next, note that $\mathbf{P}\mathcal{H}_1$ is a cap in $\mathbf{P}\mathcal{H} = \mathrm{PG}(\mathcal{H}(n+1, q^2))$, that is, no three points are collinear. The set $\mathbf{P}\mathcal{H}_1$ is called the *Hermitian Veronesean* of $\mathrm{PG}(n, q^2)$.

Now, an alternative description of $\mathbf{P}\mathcal{H}_1$ in $\mathrm{PG}(n^2+2n, q)$ is given. This amounts to choosing an explicit basis in the vector space $\mathcal{H}(n+1, q^2)$, and then applying the map δ. Let $r \in \mathbf{F}_{q^2} \backslash \mathbf{F}_q$ be arbitrary. Then the map δ above can be given as

$$\mathbf{P}(x_0, x_1, \ldots, x_n)\delta = \mathbf{P}((y_{ij})),$$

with $x_i \in \mathbf{F}_{q^2}$, $i = 0, 1, \ldots, n$, $y_{ii} = x_i\overline{x}_i$, $y_{ij} = x_i\overline{x}_j + \overline{x}_ix_j$ for $i < j$, and $y_{ij} = rx_i\overline{x}_j + \overline{r}\,\overline{x}_ix_j$, for $i > j$. This representation is called the *r-representation*.

From the *r*-representation it follows that the inverse image with respect to δ of the intersection of $\mathbf{P}\mathcal{H}_1$ with a hyperplane of $\mathrm{PG}(n^2 + 2n, q)$ is a Hermitian variety, and conversely every Hermitian variety of $\mathrm{PG}(V)$ arises in this way. It follows that $\mathbf{P}\mathcal{H}_1$ is not contained in a hyperplane of $\mathrm{PG}(n^2+2n, q)$. The lines of $\mathrm{PG}(V)$ have a natural interpretation in terms of the geometry of $\mathbf{P}\mathcal{H}$: the span in $\mathbf{P}\mathcal{H}$ of the image $l\delta$, with l a line of $\mathrm{PG}(V)$, is a solid denoted by $\xi(l)$. Since $l\delta$ is a cap of size $q^2 + 1$ in the solid $\xi(l)$, it is an ovoid for $q > 2$; then $l\delta$ is always an elliptic quadric in $\xi(l)$ and $\xi(l) \cap \mathbf{P}\mathcal{H}_1 = l\delta$. Thus the lines of $\mathrm{PG}(V)$ can be interpreted as certain solids of $\mathbf{P}\mathcal{H}$ in which the points of $\mathbf{P}\mathcal{H}_1$ form an elliptic quadric. Denote by Σ the set of all these solids. Further, for a point $P \in \mathbf{P}\mathcal{H}_1$ and $\xi \in \Sigma$ with $P \in \xi$, let $T_P(\xi)$ be the tangent plane to $\xi \cap \mathbf{P}\mathcal{H}_1$ at P in ξ.

Now some important properties of the Hermitian Veroneseans $\mathcal{H}_{n, n^2+2n} = \mathbf{P}\mathcal{H}_1$ are listed.

Theorem 4.85. *Let $\mathcal{H}_{n, n^2+2n}$ be a Hermitian Veronesean in $\mathrm{PG}(n^2 + 2n, q)$. Then*

(i) *each elliptic quadric in some* $\mathrm{PG}(3, q) \subset \mathrm{PG}(n^2 + 2n, q)$ *contained in* $\mathcal{H}_{n,n^2+2n}$ *corresponds to a line of* $\mathrm{PG}(V)$;

(ii) *every n-dimensional subspace over* $\mathbf{F}_q$ *of* $\mathrm{PG}(V)$ *corresponds to a quadric Veronesean* $\mathcal{V}_n$ *over* $\mathbf{F}_q$ *on* $\mathcal{H}_{n,n^2+2n}$, *and* $\langle \mathcal{V}_n \rangle \cap \mathcal{H}_{n,n^2+2n} = \mathcal{V}_n$.

Definition 4.86. *A solid generated by an elliptic quadric on* $\mathcal{H}_{n,n^2+2n}$ *is an elliptic space of* $\mathcal{H}_{n,n^2+2n}$.

Every elliptic space corresponds to a line of $\mathrm{PG}(V)$ and vice versa.

Theorem 4.87. *Let* $\mathcal{H}_{n,n^2+2n}$ *be a Hermitian Veronesean in* $\mathrm{PG}(n^2 + 2n, q)$.

(i) *Any two distinct points* P, P' *of* $\mathcal{H}_{n,n^2+2n}$ *lie in a unique element of* Σ, *denoted by* $\xi[P, P']$.

(ii) *Two distinct solids of* Σ *are either disjoint or meet in a unique point of* $\mathcal{H}_{n,n^2+2n}$.

(iii) *Let* $\xi \in \Sigma$, $P \in \mathcal{H}_{n,n^2+2n}$, $P \notin \xi$ *and put* $O = \xi \cap \mathcal{H}_{n,n^2+2n}$. *Then*

$$\bigcup_{P' \in O} T_P(\xi[P, P'])$$

is a projective subspace of dimension four.

4.4 Characterisations of Hermitian Veroneseans

This section contains characterisations of Hermitian Veroneseans similar to the characterisations of the first, third and fourth kind of the quadric Veroneseans.

4.4.1 Characterisations of $\mathcal{H}_{n,n^2+2n}$ of the first kind

Let $\mathcal{K}$ be a subset of the point set of $\mathrm{PG}(N, q)$, $N > 3$, which generates $\mathrm{PG}(N, q)$ and for which there exists a set Σ of solids of $\mathrm{PG}(N, q)$, called the *elliptic spaces* of $\mathrm{PG}(N, q)$, such that for any $\xi \in \Sigma$, the set $\mathcal{K}(\xi) = \mathcal{K} \cap \xi$ is an ovoid in ξ. When $\xi \in \Sigma$ and $P \in \mathcal{K}(\xi)$, the tangent plane of $\mathcal{K}(\xi)$ at P in ξ is denoted by $T_P(\xi)$.

Suppose that $\mathcal{K}$ satisfies the following:

(a) any two distinct points P, P' of $\mathcal{K}$ lie in a unique element of Σ, denoted by $[P, P']$;

(b) if $\xi_1, \xi_2 \in \Sigma$, with $\xi_1 \neq \xi_2$ and $\xi_1 \cap \xi_2 \neq \emptyset$, then $\xi_1 \cap \xi_2 \subset \mathcal{K}$;

(c) if $P \in \mathcal{K}$ and $\xi \in \Sigma$, with $P \notin \xi$, then the planes $T_P([P, P'])$, with $P' \in \mathcal{K}(\xi)$, are contained in a common 4-dimensional subspace of $\mathrm{PG}(N, q)$, denoted by $T(P, \xi)$.

It can be shown that $\mathcal{K}$ is a cap, and subsequently $\mathcal{K}$ is called a *Hermitian cap*. By Theorem 4.87, any Hermitian Veronesean is a Hermitian cap.

Consider the Hermitian Veronesean $\mathcal{H}_{n,n^2+2n}$ in $\mathrm{PG}(n^2 + 2n, q)$, and let Π_m be a subspace of $\mathrm{PG}(n^2 + 2n, q)$ of dimension m, which does not intersect any elliptic space, nor any $T(P, \xi)$, with $P \in \mathcal{H}_{n,n^2+2n}$ and ξ an elliptic space not containing P. If $\Pi_{n^2+2n-m-1}$ is a subspace of dimension $n^2 + 2n - m - 1$ skew to Π_m, then the projection of $\mathcal{H}_{n,n^2+2n}$ from Π_m onto $\Pi_{n^2+2n-m-1}$ is also a Hermitian cap. Such a Hermitian cap is a *quotient* of the Hermitian Veronesean $\mathcal{H}_{n,n^2+2n}$.

Theorem 4.88. *Let $\mathcal{K}$ be a Hermitian cap in the projective space* $\mathrm{PG}(N, q)$, $N > 3$.

 (i) *If Σ is the corresponding set of elliptic spaces, then $\mathcal{K}$ together with the set $\Xi = \{\mathcal{K}(\xi) \mid \xi \in \Sigma\}$ is the point-line incidence structure of a projective space* $\mathrm{PG}(n, q^2)$, *with $n > 1$.*
 (ii) *The cap $\mathcal{K}$ is projectively equivalent to a quotient of the Hermitian Veronesean* $\mathcal{H}_{n,n^2+2n}$.

In order to obtain this result, some particular cases and lemmas are proved, some of which are of independent interest. In particular, the following result is significant.

Theorem 4.89. *Let*

(a) *$\mathcal{K}$ be a Hermitian cap in the projective space* $\mathrm{PG}(N, q)$, $N > 3$;
(b) *Σ be the corresponding set of elliptic spaces;*
(c) *$\Xi = \{\mathcal{K}(\xi) \mid \xi \in \Sigma\}$.*

Then the following hold:

 (i) *$\mathcal{K}$, together with the set Ξ, is the point-line structure of a projective space* $\mathrm{PG}(n, q^2)$, $n > 1$, *and $N \leq n^2 + 2n$;*
 (ii) *if $N = n^2 + 2n$, then $\mathcal{K}$ is projectively equivalent to $\mathcal{H}_{n,n^2+2n}$;*
 (iii) *if $n \in \{2, 3\}$, then $\mathcal{K}$ is projectively equivalent to $\mathcal{H}_{n,n^2+2n}$;*
 (iv) *if $\mathcal{K} \cap \langle \Pi_{n-1} \rangle = \Pi_{n-1}$ for every hyperplane Π_{n-1} of $\mathrm{PG}(n, q^2)$ and with $\langle \Pi_{n-1} \rangle$ the subspace of $\mathrm{PG}(N, q)$ generated by Π_{n-1}, then $\mathcal{K}$ is projectively equivalent to $\mathcal{H}_{n,n^2+2n}$.*

Theorem 4.88 is similar to Theorem 4.40 on Veronesean caps. Also, note that the set of elliptic spaces of a Hermitian cap $\mathcal{K}$ in $\mathrm{PG}(N, q)$ is uniquely determined if $q > 2$. This follows immediately from (b) in the definition of Hermitian cap, by considering two coplanar bisecants with no common point on $\mathcal{K}$ of a hypothetical ovoid contained in $\mathcal{K}$ and not lying in an elliptic space of $\mathcal{K}$. If $q = 2$, this is not clear.

4.4.2 Characterisation of $\mathcal{H}_{n,n^2+2n}$ of the third kind

Relying on Subsection 4.4.1, a characterisation of $\mathcal{H}_{n,n^2+2n}$, similar to the characterisation of the third kind of $V_n^{2^n}$, is obtained.

As always, a finite projective space has *order* q if $q + 1$ is the number of points on any line.

Theorem 4.90. *Let*

(a) *$\mathcal{S} = (\mathcal{P}, \mathcal{B}, I)$ be a finite projective plane of order $q^2 > 4$;*

(b) $\mathcal{P}$ be a subset of the point set of $\mathrm{PG}(d,q)$, with $d \geq 8$, not contained in a hyperplane of $\mathrm{PG}(d,q)$;

(c) the points incident with any line l of $\mathcal{S}$ form an ovoid in some solid ξ_l of $\mathrm{PG}(d,q)$.

Then

(i) $d = 8$;

(ii) the plane $\mathcal{S}$ is Desarguesian;

(iii) $\mathcal{P}$ is projectively equivalent to the Hermitian Veronesean $\mathcal{H}_{2,8}$ of $\mathrm{PG}(2,q^2)$.

A representation of a point-line incidence structure $\mathcal{S}$ as in this theorem is an *ovoidal embedding* of $\mathcal{S}$. Hence, by Theorem 4.90, all ovoidal embeddings of all finite projective planes of order $q^2 > 4$ are classified.

Theorem 4.91. *Let*

(a) $\mathcal{S} = (\mathcal{P},\mathcal{B},\mathrm{I})$ be the point-line geometry of a finite projective space of order $q^2 > 4$ and dimension $n \geq 2$;

(b) $\mathcal{P}$ be a subset of the point set of $\mathrm{PG}(d,q)$, with $d \geq n^2 + 2n$, not contained in a hyperplane of $\mathrm{PG}(d,q)$;

(c) the points incident with any line l of $\mathcal{S}$ form an ovoid in some solid ξ_l of $\mathrm{PG}(d,q)$.

Then $d = n^2 + 2n$ and $\mathcal{P}$ is projectively equivalent to the Hermitian Veronesean $\mathcal{H}_{n,n^2+2n}$ of $\mathrm{PG}(n,q^2)$.

Hence all ovoidal embeddings of all point-line geometries of the finite projective spaces of dimension $n \geq 2$ and order $q^2 > 4$ are classified. In fact, Theorem 4.90 is part of Theorem 4.91, but it is formulated separately to emphasise that, for $n = 2$, it is not assumed that $\mathcal{S}$ is Desarguesian.

4.4.3 Characterisation of $\mathcal{H}_{2,8}$ of the fourth kind

In this subsection the Hermitian Veronesean $\mathcal{H}_{2,8}$ is characterised by considering its common points with solids and hyperplanes of $\mathrm{PG}(8,q)$.

Since hyperplanes of $\mathrm{PG}(8,q)$ meet $\mathcal{H}_{2,8}$ in point sets that correspond to singular and non-singular Hermitian curves in $\mathrm{PG}(2,q^2)$, the size of such an intersection is either $q^2 + 1$, $q^3 + 1$, or $q^3 + q^2 + 1$. It is now shown that each solid of $\mathrm{PG}(8,q)$ which intersects $\mathcal{H}_{2,8}$ in at least $q + 3$ points intersects it in $q^2 + 1$ points. So, let Π_3 be such a solid. Suppose first that $\Pi_3 \cap \mathcal{H}_{2,8}$ generates Π_3 and let P_1, P_2, P_3, P_4 be four points of $\mathrm{PG}(2,q^2)$ which correspond to four points of $\Pi_3 \cap \mathcal{H}_{2,8}$ generating Π_3.

First, assume that P_1, P_2, P_3, P_4 are four points on a line l of $\mathrm{PG}(2,q^2)$. Then, to l there corresponds an ovoid on $\mathcal{H}_{2,8}$ in some solid of $\mathrm{PG}(8,q)$, and this solid coincides with Π_3. So, in this case, $\Pi_3 \cap \mathcal{H}_{2,8}$ is an ovoid of Π_3 and thus contains $q^2 + 1$ points.

Suppose next that P_1, P_2, P_3 lie on a line l and that P_4 is not on l. Then each point of $\Pi_3 \cap \mathcal{H}_{2,8}$ is also contained in every hyperplane of $\mathrm{PG}(8,q)$ containing the images of P_1, P_2, P_3, P_4. Hence every point of $\mathrm{PG}(2,q^2)$ corresponding to a point of $\Pi_3 \cap \mathcal{H}_{2,8}$ is contained in every Hermitian curve containing P_1, P_2, P_3, P_4. If P, with $P \neq P_4$, is a point of $\mathrm{PG}(2, q^2)$ not on the subline over $\mathbf{F}_q$ of l defined by P_1, P_2, P_3, then either P does not belong to the singular Hermitian curve defined by the three lines $P_4 P_1, P_4 P_2, P_4 P_3$, if P is on l, or there is a point P' on l and a line l' through P' such that P does not belong to the singular Hermitian curve defined by the three lines $l, l', P' P_4$, if P is not on l. Hence $\Pi_3 \cap \mathcal{H}_{2,8}$ contains at most $q + 2$ points, a contradiction.

Suppose now that $\{P_1, P_2, P_3, P_4\}$ is a 4-arc in $\mathrm{PG}(2, q^2)$. By considering the three, unique singular Hermitian curves containing the lines $P_1 P_2, P_1 P_3, P_1 P_4$, the lines $P_2 P_1, P_2 P_3, P_2 P_4$, and the lines $P_3 P_1, P_3 P_2, P_3 P_4$, it follows that every point P of $\Pi_3 \cap \mathcal{H}_{2,8}$ corresponds to a point of the subplane $\mathrm{PG}(2, q)$ of $\mathrm{PG}(2, q^2)$ which contains P_1, P_2, P_3, P_4. Hence P is contained in the quadric Veronesean $\mathcal{V}_2^4$ on $\mathcal{H}_{2,8}$ which is the image of $\mathrm{PG}(2, q)$. Since $\mathcal{V}_2^4$ generates a 5-dimensional subspace Π_5 of $\mathrm{PG}(8, q)$, so Π_3 is the intersection of two hyperplanes of Π_5. Consequently, the size of $\Pi_3 \cap \mathcal{H}_{2,8}$ is the size of the intersection of two distinct plane quadrics in $\mathrm{PG}(2, q)$. Since this is at most $q + 2$, a contradiction is obtained.

Finally, suppose that $\Pi_3 \cap \mathcal{H}_{2,8}$ does not generate Π_3. Then it generates a plane since $\mathcal{H}_{2,8}$ is a cap. Hence $\Pi_3 \cap \mathcal{H}_{2,8}$ is a plane k-arc, and so $k \leq q + 2$, again a contradiction.

Theorem 4.92. *Let $\mathcal{K}$ be a set of points of $\mathrm{PG}(8, q)$, $q \neq 2$, with $|\mathcal{K}| = q^4 + q^2 + 1$. Then $\mathcal{K}$ is projectively equivalent to $\mathcal{H}_{2,8}$ if and only if the following conditions are satisfied:*

(a) *every hyperplane of $\mathrm{PG}(8, q)$ intersects $\mathcal{K}$ in either $q^2 + 1$, $q^3 + 1$ or $q^3 + q^2 + 1$ points;*

(b) *if a solid of $\mathrm{PG}(8, q)$ intersects $\mathcal{K}$ in at least $q + 3$ points, then it intersects $\mathcal{K}$ in precisely $q^2 + 1$ points.*

4.5 Segre varieties

To begin with an example, let $\mathcal{L}_1$ and $\mathcal{L}_2$ be projective lines over the field K. If $\mathcal{L}_1 = \{\mathbf{P}(s_1, t_1)\}$ and $\mathcal{L}_2 = \{\mathbf{P}(s_2, t_2)\}$, then their *Segre product* or *Segre variety* is

$$\mathcal{S}_{1;1} = \{\mathbf{P}(s_1 s_2, s_1 t_2, t_1 s_2, t_1 t_2)\}$$

in $\mathrm{PG}(3, K)$. Note that $\mathcal{S}_{1;1}$ is the hyperbolic quadric $\mathcal{H}_3 = \mathbf{V}(X_0 X_3 - X_1 X_2)$.

In general, let $\mathcal{P}_1, \mathcal{P}_2, \ldots, \mathcal{P}_k$ be projective spaces with $\mathcal{P}_i = \mathrm{PG}(n_i, K)$, for $i = 1, 2, \ldots, k$, where each $n_i \geq 1$. Let

$$\mathcal{P}_i = \{\mathbf{P}(X^i)\},$$
$$\text{with} \quad X^i = (x_0^{(i)}, x_1^{(i)}, \ldots, x_{n_i}^{(i)}).$$

Let $\overline{\mathbf{N}}_r = \{0, 1, \ldots, r\}$ for any $r \geq 1$, and let

$$\eta : \overline{\mathbf{N}}_{n_1} \times \overline{\mathbf{N}}_{n_2} \times \cdots \times \overline{\mathbf{N}}_{n_k} \longrightarrow \overline{\mathbf{N}}_m$$

be a bijection, with $m + 1 = (n_1 + 1)(n_2 + 1) \cdots (n_k + 1)$.

Definition 4.93. The *Segre variety* of the k given projective spaces is

$$\mathcal{S} = \mathcal{S}_{n_1;n_2;\ldots;n_k} = \Big\{ \mathbf{P}(x_0, x_1, \ldots, x_m) \mid x_j = x_{(i_1,i_2,\ldots,i_k)\eta} = x_{i_1}^{(1)} x_{i_2}^{(2)} \cdots x_{i_k}^{(k)},$$
$$\mathbf{P}(X^i) \text{ a point of } \mathcal{P}_i \Big\}$$

in $\mathrm{PG}(m, K)$.

As $(x_0^{(i)}, x_1^{(i)}, \ldots, x_{n_i}^{(i)}) \neq (0, 0, \ldots, 0)$ all i, so $(x_0, x_1, \ldots, x_m) \neq (0, 0, \ldots, 0)$. The integers $n_1, n_2, \ldots, n_k$ are the *indices* of the variety $\mathcal{S}$, which has dimension $n_1 + n_2 + \cdots + n_k$. Also, $\mathcal{S}_{n_1;n_2;\ldots;n_k}$ is absolutely irreducible and non-singular. It has order

$$\frac{(n_1 + n_2 + \cdots + n_k)!}{n_1! \, n_2! \, \cdots \, n_k!}.$$

Any point $\mathbf{P}(x_0, x_1, \ldots, x_m)$ of the Segre variety satisfies the equations

$$x_{(i_1,i_2,\ldots,i_k)\eta} \, x_{(j_1,j_2,\ldots,j_k)\eta}$$
$$- x_{(i_1,\ldots,i_{s-1},j_s,i_{s+1},\ldots,i_k)\eta} \, x_{(j_1,j_2,\ldots,j_{s-1},i_s,j_{s+1},\ldots,j_k)\eta} = 0. \quad (4.37)$$

Theorem 4.94. *The Segre variety $\mathcal{S}_{n_1;n_2;\ldots;n_k}$ is the intersection of all quadrics of* $\mathrm{PG}(m, K)$ *defined by the equations* (4.37). *Also, any point of* $\mathrm{PG}(m, K)$ *satisfying the equations* (4.37) *corresponds to a unique element of* $\mathcal{P}_1 \times \mathcal{P}_2 \times \cdots \times \mathcal{P}_k$.

Proof. Let $\mathbf{P}(x_0', x_1', \ldots, x_m')$ be a point satisfying the equations (4.37). Without loss of generality, let $x_{(0,0,\ldots,0)\eta}' = 1$. If the points $P_i = \mathbf{P}(x_0^{(i)}, x_1^{(i)}, \ldots, x_{n_i}^{(i)})$ of $\mathcal{P}_i$ define the given point of $\mathrm{PG}(m, K)$, then $x_0^{(1)} x_0^{(2)} \cdots x_0^{(k)} \neq 0$. Hence, take $x_0^{(1)} = x_0^{(2)} = \cdots = x_0^{(k)} = 1$. Then

$$x_{(i_1,0,\ldots,0)\eta} = x_{i_1}^{(1)}, \quad x_{(0,i_2,0,\ldots,0)\eta} = x_{i_2}^{(2)}, \quad \ldots, \quad x_{(0,0,0,\ldots,0,i_k)\eta} = x_{i_k}^{(k)},$$

with $i_s = 0, 1, \ldots, n_s$. Consequently, the given point of $\mathrm{PG}(m, K)$ corresponds to at most one element of $\mathcal{P}_1 \times \mathcal{P}_2 \times \cdots \times \mathcal{P}_k$.

Consider the k points $P_i = \mathbf{P}(x_0^{(i)}, x_1^{(i)}, \ldots, x_{n_i}^{(i)})$ with

$$x_{(i_1,0,\ldots,0)\eta}' = x_{i_1}^{(1)}, \quad x_{(0,i_2,0,\ldots,0)\eta}' = x_{i_2}^{(2)}, \quad \ldots, \quad x_{(0,0,0,\ldots,0,i_k)\eta}' = x_{i_k}^{(k)}.$$

For these points,

$$
\begin{aligned}
x'_{(i_1,i_2,\ldots,i_k)\eta} &= x'_{(i_1,i_2,\ldots,i_k)\eta}\, x'_{(0,0,\ldots,0)\eta} \\
&= x'_{(i_1,i_2,\ldots,i_{k-1},0)\eta}\, x'_{(0,0,\ldots,0,i_k))\eta} \\
&= x'_{(i_1,i_2,\ldots,i_{k-1},0)\eta}\, x^{(k)}_{i_k} \\
&= x'_{(i_1,i_2,\ldots,i_{k-1},0)\eta}\, x'_{(0,0,\ldots,0)\eta} x^{(k)}_{i_k} \\
&= x'_{(i_1,i_2,\ldots,i_{k-2},0,0)\eta}\, x'_{(0,0,\ldots,i_{k-1},0)\eta} x^{(k)}_{i_k} \\
&= x'_{(i_1,i_2,\ldots,i_{k-2},0,0)\eta}\, x^{(k-1)}_{i_{k-1}} x^{(k)}_{i_k} \\
&= \cdots = x^{(1)}_{i_1} x^{(2)}_{i_2} \cdots x^{(k)}_{i_k} = x_{(i_1,i_2,\ldots,i_k)\eta}.
\end{aligned}
$$

Hence, to the element $(P_1, P_2, \ldots, P_k)$ in $\mathcal{P}_1 \times \mathcal{P}_2 \times \cdots \times \mathcal{P}_k$, there corresponds the given point of $\mathrm{PG}(m, K)$. $\qquad\square$

Let

$$
\delta : \mathcal{P}_1 \times \mathcal{P}_2 \times \cdots \times \mathcal{P}_k \to \mathcal{S}_{n_1;n_2;\ldots;n_k} \tag{4.38}
$$

be defined by

$$
(\mathbf{P}(x^{(1)}_0, x^{(1)}_1, \ldots, x^{(1)}_{n_1}), \ldots, \mathbf{P}(x^{(k)}_0, x^{(k)}_1, \ldots, x^{(k)}_{n_k})) \mapsto \mathbf{P}(x_0, x_1, \ldots, x_m),
$$

with

$$
x_j = x_{(i_1,i_2,\ldots,i_k)\eta} = x^{(1)}_{i_1} x^{(2)}_{i_2} \cdots x^{(k)}_{i_k}.
$$

By Theorem 4.94, the mapping δ is a bijection.

Theorem 4.95. *Given points $P_1, P_2, \ldots, P_{i-1}, P_{i+1}, \ldots, P_k$ of the respective spaces $\mathcal{P}_1, \mathcal{P}_2, \ldots, \mathcal{P}_{i-1}, \mathcal{P}_{i+1}, \ldots, \mathcal{P}_k$, the set of all points $(P_1, P_2, \ldots, P_k)\delta$, with P_i any point of $\mathcal{P}_i$, is an n_i-dimensional projective space.*

Proof. Up to order, the coordinates of $(P_1, P_2, \ldots, P_k)\delta$ are of the form

$$
x^{(i)}_0 r_1, x^{(i)}_1 r_1, \ldots, x^{(i)}_{n_i} r_1, x^{(i)}_0 r_2, x^{(i)}_1 r_2, \ldots, x^{(i)}_{n_i} r_2, \ldots,
$$
$$
x^{(i)}_0 r_{(m+1)/(n_i+1)}, \ldots, x^{(i)}_{n_i} r_{(m+1)/(n_i+1)},
$$

with $r_1, r_2, \ldots, r_{(m+1)/(n_i+1)}$ constants, which are not all zero. However, since $\mathbf{P}(x^{(i)}_0, x^{(i)}_1, \ldots, x^{(i)}_{n_i})$ is a variable point of $\mathcal{P}_i$, it follows that the set of all points $(P_1, P_2, \ldots, P_k)\delta$ is an n_i-dimensional projective space on $\mathcal{S}_{n_1;n_2;\ldots;n_k}$. $\qquad\square$

The variation of $(P_1, P_2, \ldots, P_{i-1}, P_{i+1}, \ldots, P_k)$ gives a system Σ_i of n_i-dimensional projective spaces on $\mathcal{S}_{n_1;n_2;\ldots;n_k}$.

Theorem 4.96. (i) *Any two distinct elements of Σ_i are skew.*
(ii) *Each point of $\mathcal{S}_{n_1;n_2;\ldots;n_k}$ is contained in exactly one element of each Σ_i.*
(iii) *For $i \neq j$, an element of Σ_i meets an element of Σ_j in at most one point.*

Proof. Let Π_{n_i} and Π'_{n_i} in Σ_i correspond to the distinct $(k-1)$-tuples

$$(P_1, P_2, \ldots, P_{i-1}, P_{i+1}, \ldots, P_k),$$
$$(P'_1, P'_2, \ldots, P'_{i-1}, P'_{i+1}, \ldots, P'_k),$$

where $P_j, P'_j \in \mathcal{P}_j$. For any points $P_i, P'_i \in \mathcal{P}_i$, the two k-tuples $(P_1, P_2, \ldots, P_k)$, $(P'_1, P'_2, \ldots, P'_k)$ are distinct; so $(P_1, P_2, \ldots, P_k)\delta \neq (P'_1, P'_2, \ldots, P'_k)\delta$. Therefore $\Pi_{n_i} \cap \Pi'_{n_i} = \emptyset$.

Let $(P_1, P_2, \ldots, P_k)\delta$ be any point of $\mathcal{S}_{n_1;n_2;\ldots;n_k}$; this point lies in the space Π_{n_i} of Σ_i corresponding to the points $P_1, P_2, \ldots, P_{i-1}, P_{i+1}, \ldots, P_k$.

Finally, let the spaces Π_{n_i} in Σ_i and Π_{n_j} in Σ_j correspond to the $(k-1)$-tuples $(P_1, P_2, \ldots, P_{i-1}, P_{i+1}, \ldots, P_k)$ and $(P'_1, P'_2, \ldots, P'_{j-1}, P'_{j+1}, \ldots, P'_k)$, with i, j distinct. If $\Pi_{n_i} \cap \Pi_{n_j} \neq \emptyset$, then $P_s = P'_s$ for all s with $s \neq i, j$. If $P_s = P'_s$ for all s with $s \neq i, j$, then $(P_1, P_2, \ldots, P_{i-1}, P'_i, P_{i+1}, \ldots, P_k)\delta$ is the unique common point of Π_{n_i} and Π_{n_j}. $\qquad\square$

From now on, it is assumed that $K = \mathbf{F}_q$, although many of the results hold in a general field.

Theorem 4.97. *The cardinalities of the Segre variety and its projective spaces Σ_i are as follows*:

(i) $|\mathcal{S}_{n_1;n_2;\ldots;n_k}| = \theta(n_1)\theta(n_2) \cdots \theta(n_k)$;
(ii) $|\Sigma_i| = \theta(n_1)\theta(n_2) \cdots \theta(n_k)/\theta(n_i)$.

Proof. Since the mapping δ of (4.38) is a bijection, so (i) follows. By Theorem 4.96, the elements of Σ_i form a partition of $\mathcal{S}_{n_1;n_2;\ldots;n_k}$, which gives (ii). $\qquad\square$

Example 4.98. (1) For $n_1 = n_2 = \cdots = n_k = 1$, the dimension $m = 2^k - 1$ and the order of $\mathcal{S}_{n_1;n_2;\ldots;n_k}$ is $k!$. Here, the elements of Σ_i are lines, $|\mathcal{S}_{1;1;\ldots;1}| = (q+1)^k$ and $|\Sigma_i| = (q+1)^{k-1}$. As at the start, $\mathcal{S}_{1;1}$ is a hyperbolic quadric of $\mathrm{PG}(3, q)$. For $k = 3$, the dimension $m = 7$, the order of $\mathcal{S}_{1;1;1}$ is 6, the size $|\mathcal{S}_{1;1;1}| = (q+1)^3$, and $|\Sigma_i| = (q+1)^2$.

(2) For $n_1 = n_2 = \cdots = n_k = n$, the dimension $m = (n+1)^k - 1$ and the order of $\mathcal{S}_{n_1;n_2;\ldots;n_k}$ is $(kn)!/(n!)^k$. Also, $|\mathcal{S}_{n;n;\ldots;n}| = \theta(n)^k$ and $|\Sigma_i| = \theta(n)^{k-1}$. When $k = n = 2$, then $m = 8$, the order of $\mathcal{S}_{2;2}$ is 6, the size $|\mathcal{S}_{2;2}| = (q^2+q+1)^2$, and $|\Sigma_i| = q^2 + q + 1$.

(3) When $k = 2$, then $m = (n_1+1)(n_2+1) - 1 = n_1 n_2 + n_1 + n_2$ and the order of $\mathcal{S}_{n_1;n_2}$ is $(n_1 + n_2)!/(n_1! n_2!)$. Also, $|\mathcal{S}_{n_1;n_2}| = \theta(n_1)\theta(n_2)$, with $|\Sigma_1| = \theta(n_2)$ and $|\Sigma_2| = \theta(n_1)$.

For $n_1 = n_2 = n$, then $m = n(n+2)$, the order of $\mathcal{S}_{n;n}$ is $(2n)!/(n!)^2$, the size $|\mathcal{S}_{n;n}| = \theta(n)^2$, and $|\Sigma_i| = \theta(n)$.

When $n_1 = 1$ and $n_2 = n$, then $m = 2n + 1$, the order of $\mathcal{S}_{1;n}$ is $n + 1$, the size $|\mathcal{S}_{1;n}| = (q+1)\theta(n)$, with $|\Sigma_1| = \theta(n)$ and $|\Sigma_2| = q + 1$. In the particular case that $n_1 = 1$ and $n_2 = 2$, then $m = 5$, the order of $\mathcal{S}_{1;2}$ is 3, with the size $|\mathcal{S}_{1;2}| = (q+1)(q^2+q+1)$; also, $|\Sigma_1| = q^2 + q + 1$ and $|\Sigma_2| = q + 1$.

Now, the variety $\mathcal{S}_{n_1;n_2}$ is considered in more detail. Several of its properties can be generalised to all Segre varieties.

Theorem 4.99. *On the Segre variety* $\mathcal{S}_{n_1;n_2}$*, each element of* Σ_1 *meets each element of* Σ_2 *in a single point.*

Proof. If Π_{n_1} corresponds to the point P_2 of $\mathcal{P}_2$ and Π_{n_2} to the point P_1 of $\mathcal{P}_1$, then $\Pi_{n_1} \cap \Pi_{n_2} = \{(P_1, P_2)\delta\}$. $\qquad\square$

Theorem 4.100. *No hyperplane of* $\mathrm{PG}(m, q)$ *contains the Segre variety* $\mathcal{S}_{n_1;n_2}$*.*

Proof. Suppose the hyperplane $\Pi_{m-1} = \mathbf{V}(F)$, with $F = \sum_j a_j X_j$, contains $\mathcal{S}_{n_1;n_2}$. Then, with the notation $a_j = a_{(i_1,i_2)\eta} = b_{i_1 i_2}$,

$$\sum_{i_1=0}^{n_1} \sum_{i_2=0}^{n_2} b_{i_1 i_2} x_{i_1}^{(1)} x_{i_2}^{(2)} = 0$$

for all $x_0^{(1)}, x_1^{(1)}, \ldots, x_{n_1}^{(1)}$ and all $x_0^{(2)}, x_1^{(2)}, \ldots, x_{n_2}^{(2)}$. Now, fix the first set of elements, namely $x_0^{(1)}, x_1^{(1)}, \ldots, x_{n_1}^{(1)}$. Since

$$\sum_{i_2=0}^{n_2} \left(\sum_{i_1=0}^{n_1} b_{i_1 i_2} x_{i_1}^{(1)} \right) x_{i_2}^{(2)} = 0$$

for all $x_0^{(2)}, x_1^{(2)}, \ldots, x_{n_2}^{(2)}$, it follows that, for any $i_2 \in \{0, 1, \ldots, n_2\}$,

$$\sum_{i_1=0}^{n_1} b_{i_1 i_2} x_{i_1}^{(1)} = 0$$

for all $x_0^{(1)}, x_1^{(1)}, \ldots, x_{n_1}^{(1)}$. Hence $b_{i_1 i_2} = 0$ for both all $i_1 \in \{0, 1, \ldots, n_1\}$ and all $i_2 \in \{0, 1, \ldots, n_2\}$. So $a_0 = a_1 = \cdots = a_m = 0$, a contradiction. $\qquad\square$

Now introduce the following notation:

$$x_0^{(1)} = y_0, \ x_1^{(1)} = y_1, \ \ldots, x_{n_1}^{(1)} = y_{n_1},$$
$$x_0^{(2)} = z_0, \ x_1^{(2)} = z_1, \ \ldots, x_{n_2}^{(2)} = z_{n_2},$$
$$x_j = x_{(i_1,i_2)\eta} = x_{i_1 i_2}.$$

Also, let $(i_1, i_2)\eta = i_1(n_2 + 1) + i_2$. The equations (4.37) become

$$x_{i_1 i_2} x_{j_1 j_2} - x_{j_1 i_2} x_{i_1 j_2} = 0. \tag{4.39}$$

Theorem 4.101. *The Segre variety* $\mathcal{S}_{n_1;n_2}$ *consists of all points* $\mathbf{P}(X)$*, with*

$$X = (x_{00}, x_{01}, \ldots, x_{0n_2}, x_{10}, \ldots, x_{1n_2}, \ldots, x_{n_1 n_2}),$$

of $\mathrm{PG}(m, q)$ *for which* rank $[x_{ij}] = 1$.

Proof. By Theorem 4.94, $\mathcal{S}_{n_1;n_2}$ consists of all points $\mathbf{P}(X)$ satisfying (4.39). This implies the result.   □

Theorem 4.102. *The intersection of* $\mathcal{S}_{n;n}$ *and the subspace*

$$\Pi_{n(n+3)/2} = \mathbf{V}(X_{01} - X_{10}, X_{02} - X_{20}, \ldots, X_{n-1,n} - X_{n,n-1})$$

is the quadric Veronesean $\mathcal{V}_n$ *of all quadrics of* $\mathrm{PG}(n, q)$.

Proof. This follows immediately from Theorems 4.2 and 4.101.   □

Let ξ_i be a projectivity of $\mathcal{P}_i$, $i = 1, 2$, and let ξ be defined by

$$(P_1, P_2)\delta\xi = (P_1\xi_1, P_2\xi_2)\delta$$

for all P_1 in $\mathcal{P}_1$ and P_2 in $\mathcal{P}_2$. Then ξ is a permutation of the Segre variety $\mathcal{S}_{n_1;n_2}$; also, ξ fixes both Σ_1 and Σ_2.

When $n_1 = n_2 = n$, let $\psi_1 : \mathcal{P}_1 \to \mathcal{P}_2$ and $\psi_2 : \mathcal{P}_2 \to \mathcal{P}_1$ be projectivities. Define ψ by $(P_1, P_2)\delta\psi = (P_2\psi_2, P_1\psi_1)\delta$ for all P_1 in $\mathcal{P}_1$ and P_2 in $\mathcal{P}_2$. Then ψ is a permutation of $\mathcal{S}_{n;n}$ that interchanges Σ_1 and Σ_2.

Let $G(\mathcal{S}_{n_1;n_2})$ be the subgroup of $\mathrm{PGL}(m + 1, q)$, with $m = n_1 n_2 + n_1 + n_2$, that fixes $\mathcal{S}_{n_1;n_2}$.

Theorem 4.103.   (i) *The permutation* ξ *of* $\mathcal{S}_{n_1;n_2}$ *is induced by a unique element* $\tilde{\xi}$ *of* $G(\mathcal{S}_{n_1;n_2})$.
 (ii) *When* $n_1 = n_2 = n$, *then the permutation* ψ *of* $\mathcal{S}_{n;n}$ *is induced by a unique element* $\tilde{\psi}$ *of* $G(\mathcal{S}_{n;n})$.

Proof. Let ξ_i be the projectivity of $\mathcal{P}_i$, $i = 1, 2$, with matrix $A_i = [a_{jk}^{(i)}]$. If

$$(x_{00}, x_{01}, \ldots, x_{n_1 n_2})\xi = (x'_{00}, x'_{01}, \ldots, x'_{n_1 n_2}),$$

then

$$x'_{i_1 i_2} = y'_{i_1} z'_{i_2} = \sum_{r=0}^{n_1} a_{r i_1}^{(1)} y_r \sum_{s=0}^{n_2} a_{s i_2}^{(2)} z_s$$

$$= \sum_{r=0}^{n_1} \sum_{s=0}^{n_2} a_{r i_1}^{(1)} a_{s i_2}^{(2)} y_r z_s = \sum_{r=0}^{n_1} \sum_{s=0}^{n_2} a_{r i_1}^{(1)} a_{s i_2}^{(2)} x_{rs}. \tag{4.40}$$

Hence ξ is induced by the element $\tilde{\xi}$ of $G(\mathcal{S}_{n_1;n_2})$ with matrix

$$A = \begin{bmatrix} a_{00}^{(1)} A_2 & a_{01}^{(1)} A_2 & \cdots & a_{0n_1}^{(1)} A_2 \\ a_{10}^{(1)} A_2 & a_{11}^{(1)} A_2 & \cdots & a_{1n_1}^{(1)} A_2 \\ \vdots & \vdots & & \vdots \\ a_{n_1 0}^{(1)} A_2 & a_{n_1 1}^{(1)} A_2 & \cdots & a_{n_1 n_1}^{(1)} A_2 \end{bmatrix}.$$

The matrix A is the Kronecker product $A_1 \otimes A_2$ of A_1 and A_2. A consequence is that $|A| = |A_1 \otimes A_2| = |A_1|^{n_2+1}|A_2|^{n_1+1}$.

Suppose that ξ is also induced by the element $\tilde{\xi}'$ of $G(\mathcal{S}_{n_1;n_2})$, where $\tilde{\xi}'$ has matrix $A' = [a'_{jk}]$. Then the projectivity $\tilde{\xi}'\tilde{\xi}^{-1}$ with matrix $A'A^{-1} = B$ induces the identity mapping on $\mathcal{S}_{n_1;n_2}$. Let $B = [b_{jk}]$ and put $b_{jk} = b_{(j_1 j_2)(k_1 k_2)}$, where $(j_1, j_2)\eta = j$ and $(k_1, k_2)\eta = k$. Then

$$ty_{k_1}z_{k_2} = \sum_{j_1=0}^{n_1}\sum_{j_2=0}^{n_2} b_{(j_1 j_2)(k_1 k_2)}y_{j_1}z_{j_2}$$

for some $t \in \mathbf{F}_q$, for all $\mathbf{P}(Y)$ of $\mathcal{P}_1$ with $Y = (y_0, y_1, \ldots, y_{n_1})$ and all $\mathbf{P}(Z)$ of $\mathcal{P}_2$ with $Z = (z_0, z_1, \ldots, z_{n_2})$. Letting $Y = (1, 0, \ldots, 0)$ and $Z = (1, 0, \ldots, 0)$ gives $b_{(00)(k_1 k_2)} = 0$ if $(k_1, k_2) \neq (0, 0)$. More generally, $b_{(j_1 j_2)(k_1 k_2)} = 0$ for $(k_1, k_2) \neq (j_1, j_2)$. So

$$ty_{k_1}z_{k_2} = b_{(k_1 k_2)(k_1 k_2)}y_{k_1}z_{k_2},$$

for all $\mathbf{P}(Y)$ of $\mathcal{P}_1$ and $\mathbf{P}(Z)$ of $\mathcal{P}_2$. Thus $b_{(k_1 k_2)(k_1 k_2)}$ is independent of k_1 and k_2. Therefore $B = tI$ and and so $\tilde{\xi}' = \tilde{\xi}$. So it has been shown that the permutation ξ of $\mathcal{S}_{n_1;n_2}$ is induced by a unique element $\tilde{\xi}$ of $G(\mathcal{S}_{n_1;n_2})$.

Now assume that $n_1 = n_2 = n$. Similarly to above, let $\psi_1 : \mathcal{P}_1 \to \mathcal{P}_2$ and $\psi_2 : \mathcal{P}_2 \to \mathcal{P}_1$ be projectivities defined by the matrices D_1 and D_2. Define ψ by $(P_1, P_2)\delta\psi = (P_2\psi_2, P_1\psi_1)\delta$ for all P_1 in $\mathcal{P}_1$ and P_2 in $\mathcal{P}_2$. Put $D_i = [d^{(i)}_{jk}]$ for $i = 1, 2$, and

$$(x_{00}, x_{01}, \ldots, x_{nn})\psi = (x'_{00}, x'_{01}, \ldots, x'_{nn}).$$

Then

$$x'_{i_1 i_2} = y'_{i_1}z'_{i_2} = \sum_{r=0}^{n} d^{(2)}_{ri_1}z_r \sum_{s=0}^{n} d^{(1)}_{si_2}y_s$$

$$= \sum_{r=0}^{n}\sum_{s=0}^{n} d^{(2)}_{ri_1}d^{(1)}_{si_2}z_r y_s = \sum_{r,s=0}^{n} d^{(2)}_{ri_1}d^{(1)}_{si_2}x_{sr}. \tag{4.41}$$

Since (4.41) represents an element $\tilde{\psi}$ of $\mathrm{PGL}(m+1, q)$, the permutation ψ is induced by the element $\tilde{\psi} \in G(\mathcal{S}_{n;n})$. Let $\tilde{\zeta} \in \mathrm{PGL}(m+1, q)$, with $m = n^2 + 2n$, be defined by $x'_{sr} = x_{rs}$ for all $r, s = 0, 1, \ldots, m$. For any point P of $\mathcal{S}_{n;n}$, the coordinates of $P\tilde{\zeta}$ satisfy (4.39); so, by Theorem 4.94, the point $P\tilde{\zeta}$ is in $\mathcal{S}_{n;n}$. Hence $\tilde{\zeta} \in G(\mathcal{S}_{n;n})$ and $\tilde{\psi}\tilde{\zeta}$ has matrix $D_1 \otimes D_2$. So, by the first part of the proof, the projectivity $\tilde{\psi}\tilde{\zeta}$ corresponds to (ξ_1, ξ_2), with ξ_i the projectivity of $\mathcal{P}_i$ with matrix D_i, $i = 1, 2$.

If ψ is also induced by the element $\tilde{\psi}'$ of $G(\mathcal{S}_{n;n})$, then $\tilde{\psi}\tilde{\zeta}$ and $\tilde{\psi}'\tilde{\zeta}$ induce the same permutation of $\mathcal{S}_{n;n}$. Now, by the preceding paragraph and the first part of the proof, it follows that $\tilde{\psi}\tilde{\zeta} = \tilde{\psi}'\tilde{\zeta}$, whence $\tilde{\psi} = \tilde{\psi}'$. The conclusion is that ψ is induced by a unique element $\tilde{\psi}$ of $G(\mathcal{S}_{n;n})$. $\qquad\square$

Since $\mathcal{S}_{n_1;n_2}$ is the intersection of the quadrics (4.39), any line l of $PG(m+1,q)$ meets $\mathcal{S}_{n_1;n_2}$ in $0,1,2$ or $q+1$ points. In the next lemma it is shown that the lines of the elements of Σ_1 and Σ_2 are the only lines which are completely contained in $\mathcal{S}_{n_1;n_2}$.

Lemma 4.104. *Any line l of $\mathcal{S}_{n_1;n_2}$ is contained in an element of Σ_1 or Σ_2.*

Proof. Let l be a line of $\mathcal{S}_{n_1;n_2}$ and let P and P' be distinct points of l. Further, let $(P_1, P_2)\delta = P$ and $(P_1', P_2')\delta = P'$. Assume that $P_1 \neq P_1'$ and $P_2 \neq P_2'$. With

$$\mathbf{U}_0' = \mathbf{P}(1,0,\ldots,0) \qquad \text{and} \qquad \mathbf{U}_0'' = \mathbf{P}(1,0,\ldots,0),$$
$$\mathbf{U}_1' = \mathbf{P}(0,1,0,\ldots,0) \qquad \text{and} \qquad \mathbf{U}_1'' = \mathbf{P}(0,1,0,\ldots,0),$$

let ξ_1 be a projectivity of $\mathcal{P}_1 = PG(n_1,q)$, for which $P_1\xi_1 = \mathbf{U}_0'$, $P_1'\xi_1 = \mathbf{U}_1'$, and ξ_2 be a projectivity of $\mathcal{P}_2 = PG(n_2,q)$, for which $P_2\xi_2 = \mathbf{U}_0''$, $P_2'\xi_2 = \mathbf{U}_1''$. The element of $G(\mathcal{S}_{n_1;n_2})$ which corresponds to (ξ_1, ξ_2) is denoted by $\tilde{\xi}$. Then $l\tilde{\xi}$ is also a line of $\mathcal{S}_{n_1;n_2}$.

The line $l\tilde{\xi}$ contains the points $(\mathbf{U}_0', \mathbf{U}_0'')\delta = \mathbf{U}_0$ and $(\mathbf{U}_1', \mathbf{U}_1'')\delta = \mathbf{U}_{n_2+2}$, where $\mathbf{U}_i = \mathbf{P}(E_i)$ with E_i the vector with one in the $(i+1)$-th place and zeros elsewhere. Hence $l\tilde{\xi}$ also contains the point $\mathbf{P}(E_0 + E_{n_2+2}) = R$. However, the coordinates of R do not satisfy (4.39); thus $R \notin \mathcal{S}_{n_1;n_2}$, giving a contradiction. Therefore, either $P_1 = P_1'$ or $P_2 = P_2'$; suppose the former.

Let P'' be any point of l and let $(P_1'', P_2'')\delta = P''$. If $P_1'' \neq P_1$, then, by the preceding paragraph, $P_2'' = P_2$ and $P_2'' = P_2'$, a contradiction. So $P_1'' = P_1$. This gives the conclusion that l is a line of the element of Σ_2 that corresponds to the point P_1 of $\mathcal{P}_1$. $\qquad\square$

An s-space Π_s which is contained in $\mathcal{S}_{n_1;n_2}$ but in no $(s+1)$-space Π_{s+1} of $\mathcal{S}_{n_1;n_2}$ is a *maximal space* or *maximal subspace* of $\mathcal{S}_{n_1;n_2}$. The next result describes what they are.

Theorem 4.105. *The maximal spaces of the Segre variety $\mathcal{S}_{n_1;n_2}$ are the elements of Σ_1 and Σ_2.*

Proof. Let Π_s be a maximal subspace of $\mathcal{S}_{n_1;n_2}$ and suppose that Π_s is not contained in an element of $\Sigma_1 \cup \Sigma_2$. Choose a point P in Π_s and also a line l of Π_s through P. By Theorem 4.96 and Lemma 4.104, the line l is contained in a unique element π' of $\Sigma_1 \cup \Sigma_2$. Since Π_s is not contained in π', there exists a line l' through P which is contained in Π_s but not in π'. Let π'' be the unique element of $\Sigma_1 \cup \Sigma_2$ which contains l'; then $\pi' \cap \pi'' = \{P\}$. Since $\Pi_s \neq (\Pi_s \cap \pi') \cup (\Pi_s \cap \pi'')$, there exists a line l'' through P not in $\pi' \cup \pi''$. The line l'' is contained in a unique element π''' of $\Sigma_1 \cup \Sigma_2$. Thus P is contained in at least three distinct elements of $\Sigma_1 \cup \Sigma_2$, a contradiction.

Hence Π_s is contained in an element of $\Sigma_1 \cup \Sigma_2$. Since Π_s is maximal, it is an element of $\Sigma_1 \cup \Sigma_2$. $\qquad\square$

Corollary 4.106. *Each s-space of $\mathcal{S}_{n_1;n_2}$, $s > 0$, is contained in a unique element of $\Sigma_1 \cup \Sigma_2$.*

Proof. Let Π_s be an s-space of $\mathcal{S}_{n_1;n_2}$, with $s > 0$. This space is contained in a maximal subspace of $\mathcal{S}_{n_1;n_2}$ and, by the theorem, in an element π of $\Sigma_1 \cup \Sigma_2$. By Theorem 4.96, the space π is uniquely determined by Π_s. $\qquad \square$

Notation 4.107. With $[r, s]_- = \prod_{i=r}^{i=s}(q^i - 1)$ for $r \leq s$ and $[r, s]_- = 1$ for $r > s$, as in PGOFF2,

$$\phi(r; n, q) = |\mathrm{PG}^{(r)}(n, q)| = [n - r + 1, n + 1]_- / [1, r + 1]_-.$$

Corollary 4.108. *Let* $n_1 \leq n_2$. *The number of s-spaces contained in* $\mathcal{S}_{n_1;n_2}$ *is*

(i) $\theta(n_1)\phi(s; n_2, q) + \theta(n_2)\phi(s; n_1, q)$, *for* $0 < s \leq n_1$;
(ii) $\theta(n_1)\phi(s; n_2, q)$, *for* $n_1 < s \leq n_2$.

Proof. This follows from Corollary 4.106. $\qquad \square$

Theorem 4.109. *Let* $P_i \in \mathcal{P}_i$ *and let* $\mathrm{PG}(d_i, q)$ *be a d_i-space of* $\mathcal{P}_i$, $i = 1, 2$. *Then*

(i) $(\{P_1\} \times \mathrm{PG}(d_2, q))\delta$ *is a d_2-space and* $(\mathrm{PG}(d_1, q) \times \{P_2\})\delta$ *is a d_1-space of* $\mathcal{S}_{n_1;n_2}$;
(ii) *all subspaces of* $\mathcal{S}_{n_1;n_2}$ *are obtained as in* (i);
(iii) *for* $d_i > 0$, $i = 1, 2$, *the set* $(\mathrm{PG}(d_1, q) \times \mathrm{PG}(d_2, q))\delta$ *is a Segre variety* $\mathcal{S}_{d_1;d_2}$ *on* $\mathcal{S}_{n_1;n_2}$;
(iv) $\mathcal{S}_{d_1;d_2} = \mathcal{S}_{n_1;n_2} \cap \Pi_{m'}$, *where* $m' = d_1 d_2 + d_1 + d_2$ *and* $\Pi_{m'}$ *is the m'-space generated by* $\mathcal{S}_{d_1;d_2}$;
(v) *all Segre varieties of* $\mathcal{S}_{n_1;n_2}$ *are obtained as in* (iii).

Proof. By Theorem 4.103, coordinates can be chosen so that $\mathrm{PG}(d_1, q)$ contains the points

$$\mathbf{U}_0' = \mathbf{P}(E_0'), \ \mathbf{U}_1' = \mathbf{P}(E_1'), \ldots, \ \mathbf{U}_{d_1}' = \mathbf{P}(E_{d_1}'),$$

and that $\mathrm{PG}(d_2, q)$ contains the points

$$\mathbf{U}_0'' = \mathbf{P}(E_0''), \ \mathbf{U}_1'' = \mathbf{P}(E_1''), \ldots, \ \mathbf{U}_{d_2}'' = \mathbf{P}(E_{d_2}''),$$

where E_i' and E_i'' are vectors with 1 in the $(i+1)$-th place and zeros elsewhere. Then $(\mathrm{PG}(d_1, q) \times \mathrm{PG}(d_2, q))\delta = \mathcal{V}$ is the set of all points

$$(y_0 z_0, y_0 z_1, \ldots, y_0 z_{n_2}, y_1 z_0, y_1 z_1, \ldots, y_1 z_{n_2}, \ldots, y_{n_1} z_{n_2}),$$

with

$$z_{d_2+1} = z_{d_2+2} = \cdots = z_{n_2} = 0 \text{ and } y_{d_1+1} = y_{d_1+2} = \cdots = y_{n_1} = 0.$$

When $d_1 = 0$, then $\mathcal{V}$ is a d_2-space of $\mathcal{S}_{n_1;n_2}$; when $d_2 = 0$, then $\mathcal{V}$ is a d_1-space of $\mathcal{S}_{n_1;n_2}$. When $d_1, d_2 > 0$, then $\mathcal{V}$ is a Segre variety $\mathcal{S}_{d_1;d_2}$. The subspace $\Pi_{m'}$, with $m' = d_1 d_2 + d_1 + d_2$, of $\mathrm{PG}(m, q)$ generated by $\mathcal{S}_{d_1;d_2}$ is the intersection of the hyperplanes $\mathbf{V}(X_{i_1 i_2})$ where $i_1 > d_1$ or $i_2 > d_2$. For any point of $\mathcal{S}_{n_1;n_2}$ in the intersection of these hyperplanes, $y_{i_1} = 0$ for $i_1 > d_1$ and $z_{i_2} = 0$ for $i_2 > d_2$. Hence $\mathcal{S}_{d_1;d_2} = \mathcal{S}_{n_1;n_2} \cap \Pi_{m'}$.

When $d_1 = 0$, then δ defines a projectivity from $PG(d_2, q)$ onto the d_2-space $(PG(d_1, q) \times PG(d_2, q))\delta$; similarly, when $d_2 = 0$, then δ defines a projectivity from Π_{d_1} onto the d_1-space $(PG(d_1, q) \times PG(d_2, q))\delta$.

Conversely, let Π_{d_2} be a d_2-space contained in $\mathcal{S}_{n_1;n_2}$. By Corollary 4.106, the space Π_{d_2} is contained in an element of $\Sigma_1 \cup \Sigma_2$. Suppose, for example, that Π_{d_2} is contained in an element Π_{n_2} of Σ_1. Let $\Pi_{n_2}\delta^{-1} = \{P_1\} \times PG(n_2, q)$. Since δ defines a projectivity from $PG(n_2, q)$ onto Π_{n_2}, so $\Pi_{d_2}\delta^{-1} = \{P_1\} \times PG(d_2, q)$, where $PG(d_2, q)$ is a d_2-space of $PG(n_2, q)$.

Next, let $\mathcal{S}_{d_1;d_2}$ be a Segre subvariety of $\mathcal{S}_{n_1;n_2}$. The systems of maximal subspaces of $\mathcal{S}_{d_1;d_2}$ are denoted by Σ'_1 and Σ'_2, where the elements of Σ'_i are contained in elements of Σ_i, $i = 1, 2$. Let $P \in \mathcal{S}_{d_1;d_2}$ and $P\delta^{-1} = (P_1, P_2)$. The elements of Σ'_1 and Σ'_2 containing P are denoted by Π_{d_2} and Π_{d_1}. Let $\Pi_{d_2}\delta^{-1} = \{P_1\} \times PG(d_2, q)$, where $PG(d_2, q)$ is a d_2-space of $PG(n_2, q)$, and let $\Pi_{d_1}\delta^{-1} = PG(d_1, q) \times \{P_2\}$, where $PG(d_1, q)$ is a d_1-space of $PG(n_1, q)$. The points of $\mathcal{S}_{d_1;d_2}$ are the points P', where $\{P'\} = \Pi'_{d_1} \cap \Pi'_{d_2}$, with $\Pi'_{d_1} \in \Sigma'_2$ and $\Pi'_{d_2} \in \Sigma'_1$. It follows that $\mathcal{S}_{d_1;d_2}$ consists of the points P', where $\{P'\} = \Pi_{n_1} \cap \Pi_{n_2}$ with Π_{n_1} any space of Σ_2 containing a point of Π_{d_2} and Π_{n_2} any space of Σ_1 containing a point of Π_{d_1}. Hence $\mathcal{S}_{d_1;d_2}\delta^{-1}$ consists of all ordered pairs (P'_1, P'_2) with $P'_1 \in PG(d_1, q)$ and $P'_2 \in PG(d_2, q)$. Therefore $\mathcal{S}_{d_1;d_2} = (PG(d_1, q) \times PG(d_2, q))\delta$. $\qquad\square$

Corollary 4.110. *Let $n_1 \leq n_2$. For given d_1, d_2, with $0 < d_1 \leq n_1, 0 < d_2 \leq n_2$ and $d_1 \leq d_2$, the number of Segre subvarieties $\mathcal{S}_{d_1;d_2}$ of $\mathcal{S}_{n_1;n_2}$ is*

(i) $\phi(d_1; n_1, q)\phi(d_2; n_2, q) + \phi(d_1; n_2, q)\phi(d_2; n_1, q)$, *for $d_1 < d_2 \leq n_1$;*
(ii) $\phi(d_1; n_1, q)\phi(d_2; n_2, q)$ *for $d_1 = d_2 \leq n_1$ and $d_1 \leq n_1 < d_2$.*

Proof. This follows from Theorem 4.109. $\qquad\square$

Corollary 4.111. *Let Π_s, with $s \geq 1$, be an s-space of $\mathcal{S}_{n_1;n_2}$ contained in an element Π_{n_1} of Σ_2. Then the elements of Σ_1 meeting Π_s in a point are the elements of a system of maximal spaces of a Segre subvariety $\mathcal{S}_{s;n_2}$ of $\mathcal{S}_{n_1;n_2}$.*

Proof. Let $\Pi_s\delta^{-1} = PG(s, q) \times \{P_2\}$, with P_2 a point of $PG(n_2, q)$ and $PG(s, q)$ an s-space of $PG(n_1, q)$. Then the elements of Σ_1 having a point in common with Π_s are the elements of a system of maximal spaces of $(PG(s, q) \times PG(n_2, q))\delta$, which is a Segre subvariety $\mathcal{S}_{s;n_2}$ of $\mathcal{S}_{n_1;n_2}$. $\qquad\square$

Let ξ_i be a projectivity of $PG(n_i, q)$, $i = 1, 2$, and let $\tilde{\xi}$ be the corresponding element of $G(\mathcal{S}_{n_1;n_2})$. By Theorem 4.103, the map $\theta : (\xi_1, \xi_2) \to \tilde{\xi}$ defines a monomorphism from $PGL(n_1 + 1, q) \times PGL(n_2 + 1, q)$ to $G(\mathcal{S}_{n_1;n_2})$. Now let $n_1 = n_2 = n$, let ζ_1 be the projectivity from $PG(n_1, q) = \Pi^1_n$ onto $PG(n_2, q) = \Pi^2_n$ with matrix I, and let ζ_2 be the projectivity from Π^2_n onto Π^1_n with matrix I. By (4.41), the element of $G(\mathcal{S}_{n;n})$ which corresponds to (ζ_1, ζ_2) is the element $\tilde{\zeta} \in PGL(m + 1, q)$, $m = n^2 + 2n$, defined by $x'_{sr} = x_{rs}$ for all $r, s = 0, 1, \ldots, m$.

Theorem 4.112. (i) *For $n_1 \neq n_2$, the mapping $\theta : (\xi_1, \xi_2) \to \tilde{\xi}$ is an isomorphism from $PGL(n_1 + 1, q) \times PGL(n_2 + 1, q)$ to $G(\mathcal{S}_{n_1;n_2})$.*

(ii) *For $n_1 = n_2 = n$,*

$$G(\mathcal{S}_{n;n}) =$$

$$(\mathrm{PGL}(n+1,q) \times \mathrm{PGL}(n+1,q))\theta \cup ((\mathrm{PGL}(n+1,q) \times \mathrm{PGL}(n+1,q))\theta)\tilde{\zeta}.$$

Proof. Let $n_1 \neq n_2$, and let $\tilde{\xi} \in G(\mathcal{S}_{n_1;n_2})$. By Theorem 4.105, $\Sigma_1\tilde{\xi} = \Sigma_1$ and $\Sigma_2\tilde{\xi} = \Sigma_2$. Hence $\tilde{\xi}$ defines permutations ξ_1 of $\mathrm{PG}(n_1,q)$ and ξ_2 of $\mathrm{PG}(n_2,q)$. To the points of a line l of $\mathrm{PG}(n_1,q)$ there correspond the elements $\Pi^0_{n_2}, \Pi^1_{n_2}, \ldots, \Pi^q_{n_2}$ of a system of maximal n_2-spaces of the Segre subvariety $\mathcal{S}_{1;n_2} = (l \times \mathrm{PG}(n_2,q))\delta$ of $\mathcal{S}_{n_1;n_2}$. From the second paragraph of the proof of Theorem 4.109, δ defines a projectivity from l onto any line $(l \times \{P_2\})\delta = m$, with $P_2 \in \mathrm{PG}(n_2,q)$. The n_2-spaces $\Pi^0_{n_2}\tilde{\xi}, \Pi^1_{n_2}\tilde{\xi}, \ldots, \Pi^q_{n_2}\tilde{\xi}$ are the elements of a system of maximal n_2-spaces of the Segre subvariety $\mathcal{S}_{1;n_2}\tilde{\xi} = \mathcal{S}'_{1;n_2}$ of $\mathcal{S}_{n_1;n_2}$. By Theorem 4.109, $\mathcal{S}'_{1;n_2} = (l' \times \mathrm{PG}(n_2,q))\delta$, with l' some line of $\mathrm{PG}(n_1,q)$. Again from the proof of Theorem 4.109, δ defines a projectivity from l' onto the line $m\tilde{\xi}$. Hence $l\xi_1 = l'$, and ξ_1 induces a projectivity from l to l'. It follows that ξ_1 is a projectivity of $\mathrm{PG}(n_1,q)$. Analogously, ξ_2 is a projectivity of $\mathrm{PG}(n_2,q)$. Hence $(\xi_1,\xi_2)\theta$ is the given $\tilde{\xi}$. Therefore $(\mathrm{PGL}(n_1+1,q) \times \mathrm{PGL}(n_2+1,q))\theta = G(\mathcal{S}_{n_1;n_2})$, which proves the first part of the theorem.

Next, let $n_1 = n_2 = n$, let $\mathrm{PG}(n_1,q) = \Pi^1_n$ and let $\mathrm{PG}(n_2,q) = \Pi^2_n$. Consider any element $\tilde{\eta}$ in $G(\mathcal{S}_{n;n})$. Then either $\Sigma_1\tilde{\eta} = \Sigma_1$ and $\Sigma_2\tilde{\eta} = \Sigma_2$ or $\Sigma_1\tilde{\eta} = \Sigma_2$ and $\Sigma_2\tilde{\eta} = \Sigma_1$. In the former case, as in the first part of the proof, there exists projectivities ξ_1 of Π^1_n and ξ_2 of Π^2_n such that $(\xi_1,\xi_2)\theta = \tilde{\eta}$. In the latter case, $\Sigma_i\tilde{\eta}\tilde{\zeta} = \Sigma_i$ for $i = 1,2$. Hence there exist projectivities η_1 of Π^1_n and η_2 of Π^2_n such that $(\eta_1,\eta_2)\theta = \tilde{\eta}\tilde{\zeta}$. Since $\tilde{\zeta}^{-1} = \tilde{\zeta}$, so $(\eta_1,\eta_2)\theta\tilde{\zeta} = \tilde{\eta}$. This gives the conclusion. $\square$

Corollary 4.113. (i) *For $n_1 \neq n_2$,*

$$|G(\mathcal{S}_{n_1;n_2})| = |\mathrm{PGL}(n_1+1,q)|\,|\mathrm{PGL}(n_2+1,q)|.$$

(ii) *For $n_1 = n_2 = n$,*

$$|G(\mathcal{S}_{n;n})| = 2\,|\mathrm{PGL}(n+1,q)|^2.$$

Proof. This follows from the theorem. $\square$

Theorem 4.114. *For $n_1 = n_2 = n$, let ψ_1 be a projectivity from $\mathrm{PG}(n_1,q) = \Pi^1_n$ onto $\mathrm{PG}(n_2,q) = \Pi^2_n$. Then the set of all points $(P_1, P_1\psi_1)\delta$, with $P_1 \in \Pi^1_n$ is a quadric Veronesean $\mathcal{V}_n$.*

Proof. Coordinates can be chosen so that ψ_1 has matrix I. Then $(P_1, P_1\psi_1)\delta$ is the set of all points

$$\mathbf{P}(y_0^2, y_0 y_1, \ldots, y_0 y_n, y_1 y_0, y_1^2, y_1 y_2, \ldots, y_1 y_n, \ldots, y_n^2),$$

with $(y_0, y_1, \ldots, y_n) \neq (0, 0, \ldots, 0)$. Hence $(P_1, P_1\psi_1)\delta$ is a quadric Veronesean $\mathcal{V}_n$. Since $(P_1, P_1\psi_1)\delta$ is the intersection of $\mathcal{S}_{n;n}$ and the space

$$\mathbf{V}(X_{01} - X_{10}, X_{02} - X_{20}, \ldots, X_{n-1,n} - X_{n,n-1}),$$

it is the quadric Veronesean $\mathcal{V}_n$ described in Theorem 4.102. $\square$

Remark 4.115. With ψ_1 as in Theorem 4.114, let $\psi_2 = \psi_1^{-1}$. Also, let $\tilde{\psi}$ be the element of $G(\mathcal{S}_{n;n})$ that corresponds to (ψ_1, ψ_2). Then the points $P = (P_1, P_2)\delta$ of $\mathcal{S}_{n;n}$ which are fixed by $\tilde{\psi}$ are determined by $(P_2\psi_2, P_1\psi_1)\delta = (P_1, P_2)\delta$. Hence these points are of the form $(P_1, P_1\psi_1)\delta$ with $P_1 \in \Pi_n^1$. By Theorem 4.114, the set of all these fixed points is a quadric Veronesean $\mathcal{V}_n$.

Finally, the coordinates of the maximal spaces merit a brief description. Consider the element $(\mathrm{PG}(n_1, q) \times \{P_2\})\delta = \Pi_{n_1}$ of Σ_1, where $P_2 = \mathbf{P}(Z)$ and $Z = (z_0, z_1, \ldots, z_{n_2})$. The space Π_{n_1} is generated by the independent points $R_i = \mathbf{P}(X_i)$, $i = 0, 1, \ldots, n$, where

$$X_0 = (z_0, z_1, \ldots, z_{n_2}, 0, \ldots, 0),$$
$$X_1 = (0, \ldots, 0, z_0, z_1, \ldots, z_{n_2}, 0, \ldots, 0)$$
$$\text{with } z_0 \text{ in the } (n_2 + 2)\text{-nd place,}$$

$$\vdots$$

$$X_{n_1} = (0, \ldots, 0, z_0, z_1, \ldots, z_{n_2}).$$

The coordinates of Π_{n_1} are denoted by $(i_0\, i_1 \cdots i_{n_1})$ with $i_0 < i_1 < \cdots < i_{n_1}$ and $\{i_0, i_1, \ldots, i_{n_1}\}$ a subset of order $n_1 + 1$ of $\{0, 1, \ldots, n_1 n_2 + n_1 + n_2 = m\}$.

Let $V_k = \{k(n_2 + 1), k(n_2 + 1) + 1, \ldots, k(n_2 + 1) + n_2\}$, $k = 0, 1, \ldots, n_1$, and $i_{ks} = k(n_2 + 1) + s$. Then

$$(i_0\, i_1 \cdots i_{n_1}) = 0 \quad \text{when} \quad (i_0\, i_1 \cdots i_{n_1}) \notin V_0 \times V_1 \times \cdots \times V_{n_1}.$$

If

$$(i_0\, i_1 \cdots i_{n_1}) \in V_0 \times V_1 \times \cdots \times V_{n_1}, \text{ with } i_0 = i_{0s_0}, i_1 = i_{1s_1}, \ldots, i_{n_1} = i_{n_1 s_{n_1}},$$

then $(i_0\, i_1 \cdots i_{n_1}) = z_{s_0} z_{s_1} \cdots z_{s_{n_1}}$.

With $m = n_1 n_2 + n_1 + n_2$, let $\Sigma_1 \mathfrak{G}$ be the image of Σ_1 on the Grassmannian $\mathcal{G}_{n_1, m}$. When $n_1 = 1$, then, from above, it follows that $\Sigma_1 \mathfrak{G}$ is the Veronesean of quadrics of $\mathrm{PG}(n_2, q)$. When $n_2 = 1$, then $\Sigma_1 \mathfrak{G}$ is a normal rational curve of order $n_1 + 1$. In particular, when $n_1 = n_2 = 1$, then $\Sigma_1 \mathfrak{G}$ and $\Sigma_2 \mathfrak{G}$ are conics of the Klein quadric $\mathcal{G}_{1,3}$.

4.6 Regular n-spreads and Segre varieties $\mathcal{S}_{1;n}$

In Section 17.1 of FPSOTD, regular spreads of lines in $\mathrm{PG}(3, q)$ were studied in detail. In particular, the reguli contained in such a spread were considered. Some of the results are extended here to regular spreads of n-spaces in $\mathrm{PG}(2n + 1, q)$.

Definition 4.116. (1) A partition of $\mathrm{PG}(2n + 1, q)$ by n-spaces is an n-*spread*.
(2) A system of maximal n-spaces of a Segre variety $\mathcal{S}_{1;n}$ is an n-*regulus*.
(3) A 1-regulus is a *regulus*.

Theorem 4.117. *If, in* $\mathrm{PG}(2n+1, q)$, $n \geq 1$, *the n-spaces* Π_n, Π_n', Π_n'' *are mutually skew, then the set of all lines having a non-empty intersection with* Π_n, Π_n', Π_n'' *is a system of maximal spaces of a Segre variety* $\mathcal{S}_{1;n}$.

Proof. Coordinates are chosen so that Π_n contains the points

$$\mathbf{U}_0 = \mathbf{P}(E_0), \ \mathbf{U}_1 = \mathbf{P}(E_1), \ldots, \ \mathbf{U}_n = \mathbf{P}(E_n),$$

where E_i is the vector with one in the $(i+1)$-th place and zeros elsewhere. Through each $\mathbf{U}_i$, $i = 0, 1, \ldots, n$, there is exactly one line l_i meeting Π_n' and Π_n'' in a point.

Suppose that intersections of $l_0, l_1, \ldots, l_n$ with Π_n' generate a space $\Pi_{n'}'$ and with Π_n'' generate a space $\Pi_{n''}''$. Then the $(n' + n'' + 1)$-space generated by $\Pi_{n'}'$ and $\Pi_{n''}''$ contains the points $\mathbf{U}_0, \mathbf{U}_1, \ldots, \mathbf{U}_n$ and hence Π_n. Since $\Pi_n \cap \Pi_{n'}' = \emptyset$, so $n = n''$; analogously, $n = n'$. Hence take $l_i \cap \Pi_n' = \{\mathbf{U}_{i+n+1}\}$, $i = 0, 1, \ldots, n$, with $\mathbf{U}_{i+n+1} = \mathbf{P}(E_{i+n+1})$. Let $l_i \cap \Pi_n'' = \{Q_i\}$, $i = 0, 1, \ldots, n$, and let U be a point of Π_n'' contained in none of the $(n-1)$-spaces generated by n of the points $Q_0, Q_1, \ldots, Q_n$. Then U may be taken as $\mathbf{P}(E)$ with $E = (1, 1, \ldots, 1)$. Then it follows that $Q_i = \mathbf{P}(E_{i,i+n+1})$ with $E_{i,i+n+1}$ the vector with one in the $(i+1)$-th and $(i+n+2)$-nd places and zeros elsewhere, $i = 0, 1, \ldots, n$.

Let $P = \mathbf{P}(Z)$, with $Z = (z_0, z_1, \ldots, z_n, 0, \ldots, 0)$, be any point of Π_n, and let l be the line through P having a non-empty intersection with Π_n' and Π_n''. Let $l \cap \Pi_n' = \{P'\}$ with $P' = \mathbf{P}(Z')$ and $Z' = (0, \ldots, 0, z_0', z_1', \ldots, z_n')$; also, let $l \cap \Pi_n'' = \{P''\}$ with $P'' = \mathbf{P}(Z'')$. Then, with $r_0, r_1 \neq 0$,

$$Z'' = (r_0 z_0, r_0 z_1, \ldots, r_0 z_n, r_1 z_0', r_1 z_1', \ldots, r_1 z_n').$$

Since Z'' is a linear combination of the vectors $E_{0,n+1}, E_{1,n+2}, \ldots, E_{n,2n+1}$, it follows that $r_0 z_i = r_1 z_i'$, $i = 0, 1, \ldots, n$. Hence take $z_i = z_i'$ all i. Then any point of the line l is of the form $\mathbf{P}(X)$, where

$$X = (y_0 z_0, y_0 z_1, \ldots, y_0 z_n, y_1 z_0, y_1 z_1, \ldots, y_1 z_n),$$

with $(y_0, y_1) \neq (0, 0)$. Therefore all lines l form a system of maximal spaces of a Segre variety $\mathcal{S}_{1;n}$. $\square$

Corollary 4.118. *If* Π_n, Π_n', Π_n'' *are mutually skew n-spaces in* $\mathrm{PG}(2n + 1, q)$, *with* $n \geq 1$, *then there is exactly one n-regulus containing all three spaces.*

Proof. A Segre variety $\mathcal{S}_{1;n}$ containing the spaces Π_n, Π_n', Π_n'' is necessarily the union of the $\theta(n)$ lines having a point in common with these three spaces. In the theorem it was shown that this union is a Segre variety $\mathcal{S}_{1;n}$. The system of maximal spaces containing Π_n, Π_n', Π_n'' of this unique Segre variety $\mathcal{S}_{1;n}$ is the unique n-regulus containing these three spaces. $\square$

Notation 4.119. The n-regulus containing Π_n, Π_n', Π_n'' is denoted $\mathcal{R}(\Pi_n, \Pi_n', \Pi_n'')$.

Theorem 4.120. *The number of n-reguli in* $\mathrm{PG}(2n + 1, q)$ *is*

$$|\mathrm{PGL}(2n + 2, q)| / \{|\mathrm{PGL}(2, q)| |\mathrm{PGL}(n + 1, q)|\}.$$

Proof. The number of n-reguli in $\mathrm{PG}(2n+1, q)$, $n > 1$, is the number of Segre varieties $\mathcal{S}_{1;n}$ in $\mathrm{PG}(2n+1, q)$, which is $|\mathrm{PGL}(2n+2, q)|/|G(\mathcal{S}_{1;n})|$. So, by Corollary 4.113, this is the required number.

For $n = 1$, the number of reguli in $\mathrm{PG}(3, q)$ is twice the number of Segre varieties $\mathcal{S}_{1;1}$ in $\mathrm{PG}(3, q)$. The number of $\mathcal{S}_{1;1}$ is $|\mathrm{PGL}(4, q)|/|G(\mathcal{S}_{1;1})|$. By Corollary 4.113, this is $|\mathrm{PGL}(4, q)|/\{2|\mathrm{PGL}(2, q)|^2\}$. Hence the number of reguli is $|\mathrm{PGL}(4, q)|/\{|\mathrm{PGL}(2, q)|^2\}$.   □

Let $l_0, l_1, \ldots, l_n$ be $n+1$ lines of $\mathrm{PG}(2n+1, q^{n+1})$ which generate the space and are conjugate in $\mathbf{F}_{q^{n+1}}$ over $\mathbf{F}_q$. Also, let P_0 be any point of l_0 and let $P_1, \ldots, P_n$ be the points conjugate to P_0. Then $P_i \in l_i$ for $i = 0, 1, \ldots, n$ and the points $P_0, P_1, \ldots, P_n$ generate an n-space $\bar{\Pi}_n$ of $\mathrm{PG}(2n + 1, q^{n+1})$. The intersection of $\bar{\Pi}_n$ and $\mathrm{PG}(2n+1, q)$ is an n-space Π_n of $\mathrm{PG}(2n+1, q)$. The set of these $q^{n+1} + 1$ spaces Π_n is denoted by $\mathcal{S}(l_0, l_1, \ldots, l_n)$. In $\mathrm{PG}(3, q)$, the set $\mathcal{S}(l_0, l_1)$ is an elliptic congruence or, equivalently, a regular spread, as in Lemma 17.1.2 of FPSOTD.

Lemma 4.121. *The set* $\mathcal{S}(l_0, l_1, \ldots, l_n)$ *is an n-spread of* $\mathrm{PG}(2n + 1, q)$.

Proof. Suppose that Π_n and Π'_n correspond to distinct points P_0 and P'_0 of l_0. Then $\bar{\Pi}_n \cap l_i \neq \bar{\Pi}'_n \cap l_i$ for $i = 0, 1, \ldots, n$, with $\bar{\Pi}_n$ the extension of Π_n and $\bar{\Pi}'_n$ the extension of Π'_n. Suppose that $\Pi_n \cap \Pi'_n \neq \emptyset$; then the space generated by Π_n and Π'_n has dimension less than $2n + 1$ and its extension contains the lines $l_0, l_1, \ldots, l_n$, a contradiction. Hence $\Pi_n \cap \Pi'_n = \emptyset$.

Since $|\mathcal{S}(l_0, l_1, \ldots, l_n)| = |l_0| = q^{n+1} + 1$, the set $\mathcal{S}(l_0, l_1, \ldots, l_n)$ is a partition of $\mathrm{PG}(2n + 1, q)$. This gives the result.   □

Definition 4.122. For $n > 1$, an n-spread of $\mathrm{PG}(2n + 1, q)$ is *regular* if there exist lines $l_0, l_1, \ldots, l_n$ of $\mathrm{PG}(2n + 1, q^{n+1})$ for which $\mathcal{S} = \mathcal{S}(l_0, l_1, \ldots, l_n)$.

Theorem 4.123.   (i) *The following are equivalent*:
   (a) *if* Π_n, Π'_n, Π''_n *are three distinct elements of the n-spread $\mathcal{S}$ of* $\mathrm{PG}(2n+1, q)$, *then the whole n-regulus* $\mathcal{R}(\Pi_n, \Pi'_n, \Pi''_n)$ *is contained in $\mathcal{S}$*;
   (b) *$\mathcal{S}$ is an n-spread of* $\mathrm{PG}(2n + 1, q)$ *such that the n-spaces of $\mathcal{S}$ meeting any line not in an element of $\mathcal{S}$ form an n-regulus.*
   (ii) *A regular n-spread satisfies* (a) *and* (b).

Proof. (a) $\Rightarrow$ (b). Suppose that $\mathcal{S}$ satisfies (a), and let l be a line not contained in an element of the n-spread $\mathcal{S}$. Also, let P, P', P'' be distinct points of l and let Π_n, Π'_n, Π''_n be the n-spaces of $\mathcal{S}$ containing these points. Then $\mathcal{R}(\Pi_n, \Pi'_n, \Pi''_n) \subset \mathcal{S}$. The elements of $\mathcal{R}(\Pi_n, \Pi'_n, \Pi''_n)$ are the n-spaces of $\mathcal{S}$ containing a point of l. Hence the n-spaces of $\mathcal{S}$ meeting l form an n-regulus.

(b) $\Rightarrow$ (a). Suppose that $\mathcal{S}$ satisfies (b), and let Π_n, Π'_n, Π''_n be distinct elements of $\mathcal{S}$. Also, let l be a line meeting them. The n-regulus consisting of the $q + 1$ elements of $\mathcal{S}$ meeting l contains Π_n, Π'_n, Π''_n and hence is $\mathcal{R}(\Pi_n, \Pi'_n, \Pi''_n)$. Therefore $\mathcal{R}(\Pi_n, \Pi'_n, \Pi''_n) \subset \mathcal{S}$.

Finally, let $\mathcal{S}$ be the regular n-spread $\mathcal{S}(l_0, l_1, \ldots, l_n)$. Consider a line l not contained in an element of $\mathcal{S}$. The elements of $\mathcal{S}$ meeting l are $\Pi_n^0, \Pi_n^1, \ldots, \Pi_n^q$. The $q+1$ spaces Π_n of $\mathcal{R}(\Pi_n^0, \Pi_n^1, \Pi_n^2)$ meet $l, l_0, l_1, \ldots, l_n$ in the extension $\mathrm{PG}(2n+1, q^{n+1})$ of $\mathrm{PG}(2n+1, q)$. Hence $\mathcal{R}(\Pi_n^0, \Pi_n^1, \Pi_n^2) = \{\Pi_n^0, \Pi_n^1, \ldots, \Pi_n^q\}$, and so $\{\Pi_n^0, \Pi_n^1, \ldots, \Pi_n^q\}$ is an n-regulus. Thus (b) is satisfied. By the previous part of the proof, (a) is also satisfied. $\qquad\square$

Theorem 4.124. *For $q > 2$, an n-spread S of $\mathrm{PG}(2n+1, q)$ satisfying (a) or (b) in the statement of Theorem 4.123 is regular.*

Proof. See Section 4.7. $\qquad\square$

Remark 4.125. For $q = 2$, conditions (a) and (b) are trivially satisfied. Many examples of non-regular n-spreads in $\mathrm{PG}(2n+1, 2)$ are known.

Theorem 4.126. (i) *The number of n-reguli contained in a regular n-spread of* $\mathrm{PG}(2n+1, q)$ *is*

$$q^n (q^{2n+2} - 1)/(q^2 - 1) = q^n (q^{2n} + q^{2n-2} + \cdots + q^2 + 1).$$

(ii) *The number of regular n-spreads of* $\mathrm{PG}(2n+1, q)$ *is*

$$q^{2n(n+1)}[1, 2n+1]_-/\{(q^{n+1} - 1)(n+1)\}.$$

Proof. (i) Let $\mathcal{S}$ be a regular n-spread of $\mathrm{PG}(2n+1, q)$. By Theorem 4.123, the number of n-reguli contained in $\mathcal{S}$ is the number of subsets of order three of $\mathcal{S}$ divided by the number of subsets of order three of an n-regulus. Hence this number is

$$(q^{n+1} + 1)q^{n+1}(q^{n+1} - 1)/\{(q+1)q(q-1)\} = q^n (q^{2n+2} - 1)/(q^2 - 1).$$

(ii) Let $\Pi_n^0, \Pi_n^1, \Pi_n^2$ be mutually skew n-spaces of $\mathrm{PG}(2n+1, q)$. For any point $P \in \Pi_n^0$, let l be the line containing P and meeting the n-spaces Π_n^1 and Π_n^2, and let $l \cap \Pi_n^1 = \{P'\}$ and $l \cap \Pi_n^2 = \{P''\}$.

To show that $\psi : P' \to P''$ is a projectivity, from the proof of Theorem 4.117 coordinates can be chosen so that the Segre variety $\mathcal{S}_{1;n}$ containing $\mathcal{R}(\Pi_n^0, \Pi_n^1, \Pi_n^2)$ consists of all points $\mathbf{P}(X)$ with

$$X = (y_0 z_0, y_0 z_1, \ldots, y_0 z_n, y_1 z_0, y_1 z_1, \ldots, y_1 z_n),$$

and with

$$\psi : P' = (0, \ldots, 0, z_0, z_1, \ldots, z_n) \to P'' = (z_0, z_1, \ldots, z_n, 0, \ldots, 0)$$

for all $(z_0, z_1, \ldots, z_n) \neq (0, \ldots, 0)$. Hence ψ is a projectivity.

Next, let $\mathcal{S}$ be a regular n-spread of $\mathrm{PG}(2n+1, q)$, where

$$\mathcal{S} = \mathcal{S}(l_0, l_1, \ldots, l_n) = \mathcal{S}(m_0, m_1, \ldots, m_n).$$

Consider elements $\Pi_n^0, \Pi_n^1, \Pi_n^2, \Pi_n^3$ of $\mathcal{S}$ not belonging to the same n-regulus. For any point P of Π_n^0, let l be the line containing P and meeting the n-spaces Π_n^1, Π_n^2, and let l' be the line containing P and meeting the n-spaces Π_n^1, Π_n^3. In addition, let $l \cap \Pi_n^1 = \{P'\}$ and $l' \cap \Pi_n^1 = \{P''\}$. Then $\psi : P \to P'$ and $\psi' : P \to P''$ are projectivities from Π_n^0 to Π_n^1. Hence $\psi^{-1}\psi' = \delta : P' \to P''$ is a projectivity of Π_n^1 to itself. Since $\Pi_n^0, \Pi_n^1, \Pi_n^2, \Pi_n^3$ do not belong to the same n-regulus, so δ is not the identity.

In the extension $\mathrm{PG}(2n+1, q^{n+1})$ of $\mathrm{PG}(2n+1, q)$, let $l_i \cap \overline{\Pi}_n^1 = \{P_i\}$ and $m_i \cap \overline{\Pi}_n^1 = \{Q_i\}$ for $i = 0, 1, \ldots, n$, where $\overline{\Pi}_n^1$ is the extension of Π_n^1. Note that, in $\mathrm{PG}(2n+1, q^{n+1})$, the points $P_0, P_1, \ldots, P_n, Q_0, Q_1, \ldots, Q_n$ are fixed by δ. Since the conjugate points $P_0, P_1, \ldots, P_n$, as well as $Q_0, Q_1, \ldots, Q_n$, are linearly independent, so $\{P_0, P_1, \ldots, P_n\} = \{Q_0, Q_1, \ldots, Q_n\}$. It follows that $\{l_0, l_1, \ldots, l_n\} = \{m_0, m_1, \ldots, m_n\}$.

Now, the number of all regular n-spreads of $\mathrm{PG}(2n+1, q)$ containing a given n-regulus $\mathcal{R}$ is calculated. Let $\Pi_n \in \mathcal{R}$ and let $\mathcal{S}$ be a regular n-spread containing $\mathcal{R}$. Then $\mathcal{S} = \mathcal{S}(l_0, l_1, \ldots, l_n)$, where $\{l_0, l_1, \ldots, l_n\}$ is uniquely defined by $\mathcal{S}$.

The points P_i, with $\{P_i\} = l_i \cap \overline{\Pi}_n$ and $\overline{\Pi}_n$ the extension of Π_n, $i = 0, 1, \ldots, n$, are linearly independent in $\overline{\Pi}_n$. Conversely, consider $n + 1$ linearly independent and conjugate points $P_0, P_1, \ldots, P_n$ of $\overline{\Pi}_n$. If $\overline{\mathcal{S}}_{1;n}$ is the extension of the Segre variety $\mathcal{S}_{1;n}$ defined by $\mathcal{R}$, then $\overline{\mathcal{S}}_{1;n}$ contains exactly one line l_i through P_i which meets all elements of $\mathcal{R}$, $i = 0, 1, \ldots, n$. The lines $l_0, l_1, \ldots, l_n$ are conjugate and generate $\mathrm{PG}(2n+1, q^{n+1})$, and so they define a unique regular n-spread $\mathcal{S} = \mathcal{S}(l_0, l_1, \ldots, l_n)$. Hence the number of regular n-spreads containing $\mathcal{R}$ is the number of sets $\{P_0, P_1, \ldots, P_n\}$. Let $\Pi_n = \mathbf{V}(X_{n+1}, X_{n+2}, \ldots, X_{2n+1})$ and let $P_0 = \mathbf{P}(Y_0)$, with

$$Y_0 = (f_0(\alpha), f_1(\alpha), \ldots, f_n(\alpha), 0, 0, \ldots, 0),$$
$$f_i(T) = a_{i0} + a_{i1}T + \cdots + a_{in}T^n, \quad a_{ij} \in \mathbf{F}_q,$$
$$\mathbf{F}_{q^{n+1}} = \{a_0 + a_1\alpha + \cdots + a_n\alpha^n \mid a_i \in \mathbf{F}_q\}.$$

The points P_i, $i = 1, 2, \ldots, n$, which are conjugate to P_0 are the points $P_i = \mathbf{P}(Y_i)$, with

$$Y_i = (f_0(\alpha^{q^i}), f_1(\alpha^{q^i}), \ldots, f_n(\alpha^{q^i}), 0, 0, \ldots, 0).$$

The points $P_0, P_1, \ldots, P_n$ are linearly independent in $\overline{\Pi}_n$ if and only if $\Delta \Delta' \neq 0$, where

$$\Delta = \begin{vmatrix} 1 & \alpha & \alpha^2 & \cdots & \alpha^n \\ 1 & \alpha^q & (\alpha^q)^2 & \cdots & (\alpha^q)^n \\ \vdots & \vdots & \vdots & \cdots & \vdots \\ 1 & \alpha^{q^n} & (\alpha^{q^n})^2 & \cdots & (\alpha^{q^n})^n \end{vmatrix}, \quad \Delta' = \begin{vmatrix} a_{00} & a_{01} & \cdots & a_{0n} \\ a_{10} & a_{11} & \cdots & a_{1n} \\ \vdots & \vdots & \cdots & \vdots \\ a_{n0} & a_{n1} & \cdots & a_{nn} \end{vmatrix}.$$

Since

$$\Delta = \prod_{i>j=0}^{n} (\alpha^{q^i} - \alpha^{q^j}),$$

so $\Delta \neq 0$. Therefore the points $P_0, P_1, \ldots, P_n$ are linearly independent if and only if $\Delta' \neq 0$. So the number of such sets $\{P_0, P_1, \ldots, P_n\}$ is equal to

$$|GL(n+1, q)|/\{(q^{n+1} - 1)(n+1)\}.$$

This is also the number of all regular n-spreads containing $\mathcal{R}$.

From Theorem 4.120 and the first part of Theorem 4.126, it now follows that the number of all regular n-spreads of $PG(2n+1, q)$ is

$$\frac{|PGL(2n+2, q)|}{|PGL(2, q)|\,|PGL(n+1, q)|} \cdot \frac{|GL(n+1, q)|}{(q^{n+1} - 1)(n+1)} \cdot \frac{q^2 - 1}{q^n(q^{2n+2} - 1)}$$

$$= q^{2n(n+1)} \prod_{i=1}^{2n+1} (q^i - 1)/\{(q^{n+1} - 1)(n+1)\}. \quad \square$$

Corollary 4.127. *The number of lines l_0 of $PG(2n+1, q^{n+1})$ for which l_0, together with its conjugates $l_1, l_2, \ldots, l_n$, generate $PG(2n+1, q^{n+1})$ is*

$$q^{2n(n+1)}[1, 2n+1]_-/(q^{n+1} - 1).$$

Proof. To each such line l_0 there corresponds one regular n-spread $\mathcal{S}(l_0, l_1, \ldots, l_n)$ of $PG(2n+1, q)$, and to each regular n-spread there correspond $n+1$ of these lines. Hence the number of such lines l_0 is equal to $n+1$ times the number of regular n-spreads. $\square$

Theorem 4.128. (i) *The group $PGL(2n+2, q)$ acts transitively on the set of all regular n-spreads.*

(ii) *The subgroup $G(\mathcal{S})$ consisting of projectivities fixing a given regular n-spread $\mathcal{S}$ has order*

$$(n+1)q^{n+1}(q^{n+1} - 1)(q^{2n+2} - 1)/(q-1).$$

Proof. Let $\mathcal{S} = \mathcal{S}(l_0, l_1, \ldots, l_n)$ and $\mathcal{S}' = \mathcal{S}(l_0', l_1', \ldots, l_n')$ be regular n-spreads of $PG(2n+1, q)$, with $\Pi_n^0, \Pi_n^1, \Pi_n^2$ distinct elements of $\mathcal{S}$ and $\Pi_n^{0'}, \Pi_n^{1'}, \Pi_n^{2'}$ distinct elements of $\mathcal{S}'$. There is an element ξ in $PGL(2n+2, q)$ for which $\Pi_n^{i'} \xi = \Pi_n^i$ for $i = 0, 1, 2$. By Theorem 4.117, $\mathcal{R}(\Pi_n^{0'}, \Pi_n^{1'}, \Pi_n^{2'})\xi = \mathcal{R}(\Pi_n^0, \Pi_n^1, \Pi_n^2)$. Also, let $l_0' \xi \cap \overline{\Pi}_n^0 = P_0'$ and $l_0 \cap \overline{\Pi}_n^0 = P_0$, with $\overline{\Pi}_n^0$ the extension of Π_n^0. Now, coordinates are chosen so that $\Pi_n^0 = \mathbf{V}(X_{n+1}, X_{n+2}, \ldots, X_{2n+1})$, $P_0 = \mathbf{P}(Y_0)$ and $P_0' = \mathbf{P}(Y_0')$, where

$$Y_0 = (f_0(\alpha), f_1(\alpha), \ldots, f_n(\alpha), 0, 0, \ldots, 0),$$
$$Y_0' = (f_0'(\alpha), f_1'(\alpha), \ldots, f_n'(\alpha), 0, 0, \ldots, 0),$$

with

$$f_i(T) = a_{i0} + a_{i1}T + \cdots + a_{in}T^n, \quad a_{ij} \in \mathbf{F}_q,$$
$$f_i'(T) = a_{i0}' + a_{i1}'T + \cdots + a_{in}'T^n, \quad a_{ij}' \in \mathbf{F}_q,$$
$$\mathbf{F}_{q^{n+1}} = \{a_0 + a_1\alpha + \cdots + a_n\alpha^n \mid a_i \in \mathbf{F}_q\}.$$

From the proof of Theorem 4.126, the matrices $A = [a_{ij}]$ and $A' = [a'_{ij}]$ are nonsingular. Also, from the proof of Theorem 4.117, it may be assumed that

$$\Pi_n^1 = \mathbf{V}(X_0, X_1, \ldots, X_n), \quad \Pi_n^2 = \mathbf{V}(X_0 - X_{n+1}, X_1 - X_{n+2}, \ldots, X_n - X_{2n+1}).$$

Given the projectivity η of $\mathrm{PG}(2n + 1, q)$ with matrix

$$\begin{bmatrix} A'A^{-1} & 0 \\ 0 & A'A^{-1} \end{bmatrix}^*,$$

then $\Pi_n^i \eta = \Pi_n^i$, $i = 0, 1, 2$, and so $\mathcal{R}(\Pi_n^0, \Pi_n^1, \Pi_n^2)\eta = \mathcal{R}(\Pi_n^0, \Pi_n^1, \Pi_n^2)$. Extended to $\mathrm{PG}(2n + 1, q^{n+1})$, this projectivity η maps P_0 to P_0'. Then

$$l_0 \eta = l_0' \xi, \quad l_0 \eta \xi^{-1} = l_0', \quad \{l_0, l_1, \ldots, l_n\} \eta \xi^{-1} = \{l_0', l_1', \ldots, l_n'\}.$$

Hence $\mathcal{S} \eta \xi^{-1} = \mathcal{S}'$.

Since $\mathrm{PGL}(2n + 2, q)$ acts transitively on the set of all regular n-spreads of $\mathrm{PG}(2n + 1, q)$, so, with $\mathcal{S}$ a regular n-spread, the order of $G(\mathcal{S})$ is equal to $|\mathrm{PGL}(2n + 2, q)|$ divided by the number of all regular n-spreads. Thus Theorem 4.126 gives the result. □

Lemma 4.129. *Let*

(a) $\mathrm{PG}(2n + 1, q^2)$ *be an extension of the projective space* $\mathrm{PG}(2n + 1, q)$;
(b) $\Pi_{n,q}$ *be an n-space over* $\mathbf{F}_q$ *in* $\mathrm{PG}(2n + 1, q^2)$ *skew to* $\mathrm{PG}(2n + 1, q)$;
(c) $P \in \Pi_{n,q}$ *and $\tilde{P}$ be the conjugate of P with respect to* $\mathbf{F}_{q^2}$ *over* $\mathbf{F}_q$.

Then the lines of $\mathrm{PG}(2n + 1, q)$ *which are intersections of* $P\tilde{P}$ *and* $\mathrm{PG}(2n + 1, q)$ *form a system of maximal spaces of a Segre variety* $\mathcal{S}_{1;n}$ *of* $\mathrm{PG}(2n + 1, q)$.

Proof. The intersection of the line $P\tilde{P}$ and the space $\mathrm{PG}(2n + 1, q)$ is a line of $\mathrm{PG}(2n + 1, q)$. Let $P_0, P_1, \ldots, P_{n+1}$ be $n + 2$ points of $\Pi_{n,q}$ such that any $n + 1$ of them are linearly independent in $\Pi_{n,q}$. If $l_0, l_1, \ldots, l_{n+1}$ are the corresponding lines of $\mathrm{PG}(2n + 1, q)$, then any $n + 1$ of them generate $\mathrm{PG}(2n + 1, q)$.

Let Q_0, Q_1, Q_2 be three distinct points of the line l_0. Through Q_i there is exactly one n-space $\Pi_{n,q}^i$ of $\mathrm{PG}(2n + 1, q)$ which has a point in common with each of the lines $l_1, l_2, \ldots, l_{n+1}$, $i = 0, 1, 2$. Let $\mathcal{S}_{1;n}$ be the Segre variety of $\mathrm{PG}(2n + 1, q)$ defined by the n-regulus $\mathcal{R}(\Pi_{n,q}^0, \Pi_{n,q}^1, \Pi_{n,q}^2) = \mathcal{R}$. The extensions of $\Pi_{n,q}, \mathcal{R}, \mathcal{S}_{1;n}$ to $\mathrm{PG}(2n + 1, q^2)$ are denoted $\overline{\Pi}_{n,q}, \overline{\mathcal{R}}, \overline{\mathcal{S}}_{1;n}$. The space $\overline{\Pi}_{n,q}$ is the unique n-space of $\mathrm{PG}(2n + 1, q^2)$ containing P_0 and meeting the lines $P_1\tilde{P}_1, P_2\tilde{P}_2, \ldots, P_{n+1}\tilde{P}_{n+1}$ in a point; hence $\overline{\Pi}_{n,q}$ belongs to $\overline{\mathcal{R}}$. If l is a line of $\mathcal{S}_{1;n}$ meeting all elements of $\mathcal{R}$, then the extension $\bar{l}$ of l has a point P in $\overline{\Pi}_{n,q}$. Since l is a line of $\mathrm{PG}(2n + 1, q)$, the line $\bar{l}$ also contains the conjugate point $\tilde{P}$. The set of all points P is projectively equivalent to $\Pi_{n,q}^0$ and hence is a projective n-space $\Pi'_{n,q}$ over $\mathbf{F}_q$. Since the points $P_0, P_1, \ldots, P_{n+1}$ are contained in a unique n-space over $\mathbf{F}_q$, so $\Pi'_{n,q} = \Pi_{n,q}$. The result is thus established. □

4.6.1 Construction method for n-spreads of $\mathrm{PG}(2n + 1, q)$

Consider a projective space $\mathrm{PG}(2m, q^2)$ and let $\mathcal{P}$ be a partition of it by projective $2m$-spaces $\Pi_{2m,q}^i$ over $\mathbf{F}_q$, $i = 1, 2, \ldots, (q^{2m+1} + 1)/(q + 1)$. By Theorem 4.29 of PGOFF2, such a partition $\mathcal{P}$ exists. Embed $\mathrm{PG}(2m, q^2)$ in the extension space $\mathrm{PG}(4m + 1, q^2)$ of $\mathrm{PG}(4m + 1, q)$, and assume that $\mathrm{PG}(2m, q^2)$ does not contain a point of $\mathrm{PG}(4m + 1, q)$. By Lemma 4.129, the $2m$-space $\Pi_{2m,q}^i$ defines a Segre variety $\mathcal{S}_{1;2m}^i$ of $\mathrm{PG}(4m + 1, q)$. These $(q^{2m+1} + 1)/(q + 1)$ Segre varieties form a partition of $\mathrm{PG}(4m+1, q)$. Hence the $q^{2m+1} + 1$ maximal $2m$-spaces of these Segre varieties form a $2m$-spread $\mathcal{S}$ of $\mathrm{PG}(4m + 1, q)$.

Next, consider a projective space $\mathrm{PG}(2m + 1, q^2)$, $m \geq 0$. Let $\mathcal{P}$ be a partition of this space consisting of α spaces $\Pi_{2m+1,q}^i$ of dimension $2m + 1$ over $\mathbf{F}_q$ and β spaces Π_{m,q^2}^j of dimension m over $\mathbf{F}_{q^2}$; then $\alpha(q+1)+\beta = q^{2m+2}+1$. By Theorem 4.1 of PGOFF2, such a partition always exists for $\alpha = 0$. Embed $\mathrm{PG}(2m + 1, q^2)$ in the extension $\mathrm{PG}(4m + 3, q^2)$ of $\mathrm{PG}(4m + 3, q)$, and assume that $\mathrm{PG}(2m + 1, q^2)$ does not contain a point of $\mathrm{PG}(4m + 3, q)$. By Lemma 4.129, the $(2m + 1)$-space $\Pi_{2m+1,q}^i$ defines a Segre variety $\mathcal{S}_{1;2m+1}^i$ of $\mathrm{PG}(4m+3, q)$. The m-space Π_{m,q^2}^j and its conjugate $\tilde{\Pi}_{m,q^2}^j$ generate a $(2m + 1)$-space Π_{2m+1,q^2}^j of $\mathrm{PG}(4m + 3, q^2)$, and $\Pi_{2m+1,q^2}^j \cap \mathrm{PG}(4m + 3, q)$ is a $(2m + 1)$-space $\Pi_{2m+1,q}^j$ of $\mathrm{PG}(4m + 3, q)$. The α Segre varieties $\mathcal{S}_{1;2m+1}^i$ and the β spaces $\Pi_{2m+1,q}^j$ form a partition of $\mathrm{PG}(4m+3, q)$. Let Σ^i be the system of maximal $(2m + 1)$-spaces of $\mathcal{S}_{1;2m+1}^i$ for $m \neq 0$, and let Σ^i be a system of lines of $\mathcal{S}_{1;2m+1}^i$ for $m = 0$. Then the elements of $\Sigma^1 \cup \Sigma^2 \cup \cdots \cup \Sigma^\alpha$ together with the β spaces $\Pi_{2m+1,q}^1, \Pi_{2m+1,q}^2, \ldots, \Pi_{2m+1,q}^\beta$ form a $(2m + 1)$-spread $\mathcal{S}$ of $\mathrm{PG}(4m + 3, q)$.

4.7 Notes and references

Section 4.1

For more details on quadric Veroneseans and their projections, see for example Bertini [17], Burau [60], Godeaux [142], Semple and Roth [283], Herzer [161].

Due to the finiteness of the field, the proofs of Theorems 4.12 and 4.15 had to be modified from the proofs for $\mathbf{C}$.

The following important result on Veroneseans over $\mathbf{C}$ is due to Kronecker and Castelnuovo.

Theorem 4.130. (i) *Any surface of* $\mathrm{PG}(m, \mathbf{C})$ *that contains* ∞^2 *plane quadrics is the Veronesean* $\mathcal{V}_2^4$ *or one of its projections.*

(ii) *Any surface of* $\mathrm{PG}(3, \mathbf{C})$ *having* ∞^2 *reducible plane sections is either the projection of* $\mathcal{V}_2^4$ *or a scroll.*

Consider the Veronesean $\mathcal{V}_2^4$ in $\mathrm{PG}(5, \mathbf{C})$ and let l be a line meeting $\mathcal{M}_4^3$ in three distinct points. By Theorem 4.21, $\mathcal{V}_2^4$ is a double surface of $\mathcal{M}_4^3$ and so l is skew to $\mathcal{V}_2^4$. The projection of $\mathcal{V}_2^4$ from l onto a solid Π_3, with $l \cap \Pi_3 = \emptyset$, is a surface $\mathcal{F}_2^4$

of order four and is a *Steiner surface*. It has three double lines which meet in a triple point of the surface. In a suitable coordinate system,

$$\mathcal{F}_2^4 = \mathbf{V}(X_0 X_1 X_2 X_3 - X_2^2 X_3^2 - X_3^2 X_1^2 - X_1^2 X_2^2).$$

Section 4.2

Mazzocca and Melone [229] formulate (a), (b) and (c), but they assume conics instead of $(q+1)$-arcs; in their paper such sets are called *Veronesean caps*. For q odd, they establish Theorem 4.36. In this paper, there is no bound on the dimension of the ambient projective space. In GGG1, there are some counterexamples; an extra condition is added to make the characterisation work. The proof of Mazzocca and Melone is modified so as to hold also in the even case. In Thas and Van Maldeghem [361], the extra condition is again deleted and conics are replaced by $(q+1)$-arcs. In particular, the original problem of Mazzocca and Melone is completely solved in the finite case. To obtain the main theorem, which is Theorem 4.40, a completely new proof is developed.

Let $\mathcal{M}$ be the algebraic variety formed by the tangent spaces of $\mathcal{V}_n$. Melone [234] gives a characterisation of $\mathcal{M}$ in terms of its points and the lines contained in the tangent spaces of $\mathcal{V}_n$.

Concerning Theorems 4.37 and 4.38, it turns out that the case $N = 8$ does not exist; so necessarily $N = 9$.

The particular case $n = 2$, with q odd, of Theorem 4.49 is due to Tallini [304]. All other results of Section 4.2.2 are taken from Thas and Van Maldeghem [360]. In Theorem 4.56, the uniqueness in the case $(n, q) = (2, 4)$ is taken from Del Fra [114].

Section 4.2.3 comes from Thas and Van Maldeghem [360].

For q odd and $q > 3$, the characterisation Theorem 4.79 is taken from Ferri [134]. The proof of Lemma 4.68 has been modified from the latter. Lemma 4.69 is essential for the characterisation of $\mathcal{V}_2^4$ in the case $q = 3$. For any q, Thas and Van Maldeghem [362] copy the proof in GGG1, except for $q \in \{2, 4\}$, for which they produce a separate argument. Theorems 4.81 and 4.83 as well as Remarks 4.82 and 4.84 are taken from Schillewaert, Thas and Van Maldeghem [272].

Section 4.3

This section is taken from Cooperstein, Thas and Van Maldeghem [76]; see also Cossidente and Siciliano [79] for the case $n = 2$.

Section 4.4

Theorems 4.88 and 4.89 are also taken from [76]. Theorems 4.90 to 4.92 are taken from Thas and Van Maldeghem [363]. For more characterisations of quadric and Hermitian Veroneseans, including the infinite cases, see Schillewaert and Van Maldeghem [273, 274], Thas and Van Maldeghem [366, 365], Akça et al. [1].

Section 4.5

For more details on Segre varieties, see, for example, C. Segre [281], Godeaux [142], Burau [60], Melone and Olanda [235]. For $n_2 = 1$, the normal rational curve $\Sigma_1\mathfrak{G}$ on the Grassmannian $\mathcal{G}_{n_1,2n_1+1}$ is also considered by Herzer [162]. Other characterisations of Segre varieties are contained in Thas and Van Maldeghem [365, 367].

Section 4.6

Theorem 4.124 is due to Bruck and Bose [51]. Its proof depends on deep theorems about translation planes. In André [2], in Segre [278], and in [51], it is shown that the study of n-spreads in $\mathrm{PG}(2n + 1, q)$ is equivalent to the study of finite translation planes. The n-spread is regular if and only if the corresponding translation plane is Desarguesian. To a translation plane of order 2^{n+1} there always corresponds an n-spread of the space $\mathrm{PG}(2n + 1, 2)$. Since there are many non-Desarguesian translation planes of order 2^{n+1}, Dembowski [116], it follows that there are many non-regular n-spreads in $\mathrm{PG}(2n + 1, 2)$. This explains Remark 4.125.

In Bruen and Thas [52] there is a construction of translation planes which is equivalent to the second construction in the last part of the section. In this connection, Corollary 2 of Lemma 17.6.6 of FPSOTD shows that $\mathrm{PG}(3, 4)$ can be partitioned into 14 lines and one $\mathrm{PG}(3, 2)$. The corresponding translation plane of order 16 is non-Desarguesian and can be shown to be isomorphic to the plane discovered by Lorimer [210]. The partitions described in the second construction are called *mixed partitions* of $\mathrm{PG}(2m + 1, q^2)$. In recent years, several papers on such partitions have been written; see, for example, Mellinger [231, 233, 232], Ebert and Mellinger [127].

Finally, interesting invariants, formulas and properties related to Segre varieties and Veroneseans are contained in Glynn [141], Havlicek, Odehnal and Saniga [160], Kantor and Shult [197].

5

Embedded geometries

5.1 Polar spaces

Definition 5.1. A *polar space* S of (finite) *rank* n or *projective index* $n - 1, n \geq 3$, is a set P of elements called *points* together with distinguished subsets called *subspaces* with the following properties.

(1) A subspace together with the subspaces it contains, is a d-dimensional projective space with $-1 \leq d \leq n - 1$.
(2) The intersection of any two subspaces is a subspace.
(3) Given a subspace π of dimension $n - 1$ and a point P in $P \backslash \pi$, there exists a unique subspace π' containing P such that the dimension of $\pi \cap \pi'$ is $n - 2$. The subspace π' contains all points of π which are joined to P by some subspace of dimension 1.
(4) There exist disjoint subspaces of dimension $n - 1$.

Definition 5.2. A polar space has *rank* 2 or *projective index* 1 if it is an incidence structure consisting of the triple $S = (P, B, I)$ in which P and B are disjoint, non-empty sets of objects called *points* and *lines*, and for which I is a symmetric point-line *incidence relation* satisfying the following axioms.

(1) Each point is incident with $1 + t$ lines, where $t \geq 1$, and two distinct points are incident with at most one line.
(2) Each line is incident with $1 + s$ points, where $s \geq 1$, and two distinct lines are incident with at most one point.
(3) If P is a point and l is a line not incident with P, then there is a unique pair $(P', l') \in P \times B$ for which $P \, \mathrm{I} \, l' \, \mathrm{I} \, P' \, \mathrm{I} \, l$; see Figure 5.1.

Polar spaces of rank 2 are usually called *generalised quadrangles*. The integers s and t are the *parameters* of the generalised quadrangle and S has *order* (s, t); when $s = t$, then S has *order* s. There is a point-line duality for generalised quadrangles of order (s, t), for which in any definition or theorem the words 'point' and 'line' are interchanged and the parameters s and t are interchanged. Normally, it is assumed

© Springer-Verlag London 2016

J.W.P. Hirschfeld, J.A. Thas, *General Galois Geometries*, Springer Monographs in Mathematics, DOI 10.1007/978-1-4471-6790-7_5

Fig. 5.1. The polar space axiom in the rank 2 case

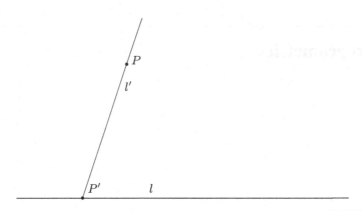

without further remark that the dual of a given theorem or definition has also been given.

The main reason for the difference between the axioms in the cases $n = 2$ and $n \geq 3$ is that, in the latter case, the axioms applied to $n = 2$ do not imply that each line contains a constant number of points and similarly that each point is on a constant number of lines.

Isomorphisms and automorphisms of polar spaces are defined in the usual way; similarly for isomorphisms (or collineations), anti-isomorphisms (or reciprocities), automorphisms, anti-automorphisms, involutions, and polarities of generalised quadrangles.

Example 5.3. (a) Let $\mathcal{Q}$ be a non-singular quadric of $\mathrm{PG}(d, q)$ of projective index $n - 1$ with $n \geq 2$. Then $\mathcal{Q}$ together with the projective subspaces lying on it is a polar space of rank n.

(b) Let $\mathcal{U}$ be a non-singular Hermitian variety of $\mathrm{PG}(d, q^2)$, $d \geq 3$. Then $\mathcal{U}$ together with the subspaces lying on it is a polar space. The projective index of this polar space is the maximum dimension of subspaces lying on $\mathcal{U}$.

(c) Let ζ be a null polarity of $\mathrm{PG}(d, q)$, with d odd. Then $\mathrm{PG}(d, q)$ together with all subspaces of the self-polar $(d - 1)/2$-dimensional spaces is a polar space of projective index $(d - 1)/2$.

(d) Let

$$\mathcal{P} = \{P_{ij} \mid i, j = 0, 1, \dots, s\}, s > 0,$$
$$\mathcal{B} = \{l_0, l_1, \dots, l_s, m_0, m_1, \dots, m_s\},$$
$$P_{ij} \mathrel{I} l_k \quad \text{if and only if } i = k,$$
$$P_{ij} \mathrel{I} m_k \quad \text{if and only if } j = k.$$

Then $(\mathcal{P}, \mathcal{B}, I)$ is a generalised quadrangle of order $(s, 1)$. Up to an isomorphism, there is only one generalised quadrangle of order $(s, 1)$, for any given $s > 0$. The generalised quadrangles with $t = 1$ are called *grids*.

(e) Let π be a plane of $PG(3, q)$, q even, and let $\mathcal{O}$ be an oval in π. Further, let $\mathcal{P} = PG(3, q)\backslash\pi$, let $\mathcal{B}$ be the set of all lines of $PG(3, q)$ not contained in π but containing a point of $\mathcal{O}$, and let I be the incidence of $PG(3, q)$. Then $(\mathcal{P}, \mathcal{B}, I)$ is a generalised quadrangle of order $(q - 1, q + 1)$ and is denoted by $\mathcal{T}_2^*(\mathcal{O})$.

A complete classification of the polar spaces of rank at least three has been obtained by Tits. This is the result, without proof, in the finite case.

Theorem 5.4. *If $\mathcal{S}$ is a finite polar space of rank at least three, then $\mathcal{S}$ is isomorphic to one of* (a), (b), (c).

The examples (d) and (e) show that this theorem is not valid in the rank 2 case. In fact, many other examples of generalised quadrangles are known.

Definition 5.5. (1) A *Shult space* $\mathcal{S}$ is a non-empty set $\mathcal{P}$ of *points* together with distinguished subsets of cardinality at least two, called *lines*, such that for each line l of $\mathcal{S}$ and each point P of $\mathcal{P}\backslash l$, the point P is collinear with (or adjacent to) either one or all points of l; two not-necessarily-distinct points P_1 and P_2 are *collinear* (or *adjacent*), with the notation $P_1 \sim P_2$, if there is at least one line of $\mathcal{S}$ containing P_1 and P_2.

(2) The space $\mathcal{S}$ is *non-degenerate* if no point of $\mathcal{S}$ is collinear with all other points.

(3) A *subspace* X of $\mathcal{S}$ is a set of pairwise collinear points such that any line meeting X in more than one point is contained in X.

(4) The space $\mathcal{S}$ has *rank n* or *projective index $n - 1$*, where $n \geq 1$, if n is the largest integer for which there is a chain $X_0 \subset X_1 \subset \cdots \subset X_n$ of distinct subspaces $X_0 = \emptyset, X_1, X_2, \ldots, X_n$.

From Theorem 5.4, it follows that, for any finite polar space $\mathcal{S}$ of rank $n \geq 3$, the point set $\mathcal{P}$ together with the subspaces of dimension 1 is a Shult space of rank n. In fact this result also holds for infinite polar spaces. Next, let $\mathcal{S}$ be a generalised quadrangle of order (s, t). If each line of $\mathcal{S}$ is identified with the set of its points, then a Shult space of rank 2 is obtained.

The following converse, stated without proof, is due to Buekenhout and Shult.

Theorem 5.6.   (i) *A non-degenerate Shult space of rank $n \geq 3$, all of whose lines have cardinality at least three, together with its subspaces, is a polar space of rank n.*

(ii) *A Shult space of rank 2, all of whose lines have cardinality at least three and all of whose points are contained in at least three lines, is a generalised quadrangle.*

Definition 5.7. If $\mathcal{S}$ is a degenerate Shult space, then the point set consisting of all points of $\mathcal{S}$ which are collinear with each point of $\mathcal{S}$ is the *radical* of $\mathcal{S}$ and is denoted by $\mathcal{R}$; it is a subspace of $\mathcal{S}$.

An equivalence relation ρ is defined on the point set $\mathcal{P}$ of $\mathcal{S}$ by putting $P \rho P'$ if and only if the set of all points collinear with P coincides with the set of all points collinear with P'. Let $\rho(P)$ denote the equivalence class containing the point P for the relation ρ; then $\rho(P) = \mathcal{R}$ for all P in $\mathcal{R}$.

Lemma 5.8. *Let $\mathcal{S}$ be a degenerate Shult space with radical $\mathcal{R}$. If the set of all points collinear with the point P is contained in the set of all points collinear with the point P', then either $P \rho P'$ or $P' \in \mathcal{R}$.*

Proof. Assume that $P' \notin \rho(P)$ and $P' \notin \mathcal{R}$. Then P and P' are distinct collinear points, and there exists a point T which is not collinear with P'. Let l be a line containing P and P', and let T' be a point of l collinear with T. Since $P' \notin \rho(P)$ there exists a point Z which is collinear with P' but not with P. Let m be a line containing Z and P'. On $m \backslash \{P'\}$ there is a point Z' which is collinear with T. Since Z is not collinear with P, also Z' is not collinear with P. On a line m' through T and Z' there is a point W which is collinear with P. Hence W is collinear with P'. Since W and Z' are collinear with P', also T is collinear with P', a contradiction. Therefore $P' \in \rho(P)$ or $P' \in \mathcal{R}$. $\qquad\square$

Let P be a point of $\mathcal{S}$ which is not contained in the radical $\mathcal{R}$. A corollary of Lemma 5.8 is that $\mathcal{R} \cup \rho(P)$ is a subspace of $\mathcal{S}$. Now a new structure $\mathcal{S}'$ is introduced with point set $\mathcal{P}'$ and line set $\mathcal{B}'$:

(1) a *point* is a class $\rho(P)$ with $P \notin \mathcal{R}$;
(2) a *line* of $\mathcal{S}'$ is a set $\{\rho(P) \mid P \in l\}$ with l a line of $\mathcal{S}$ not contained in a subspace of the form $\rho(T) \cup \mathcal{R}$.

Then the following result is readily obtained.

Theorem 5.9. *If $\mathcal{R} \neq \mathcal{P}$, then the structure $\mathcal{S}'$ is a non-degenerate Shult space.*

Finally, for a non-degenerate Shult space, the radical $\mathcal{R}$ is defined to be the empty set.

5.2 Generalised quadrangles

Only finite generalised quadrangles are considered.

A start is made by giving a brief description of three families of examples known as the *classical* generalised quadrangles, all of which are associated with classical groups.

(a) Consider a non-singular quadric $\mathcal{Q}$ of projective index 1 in the projective space $\mathrm{PG}(d, q)$, with $d = 3, 4$, or 5. Then the points of $\mathcal{Q}$ together with the lines of $\mathcal{Q}$, which are the subspaces of maximal dimension on $\mathcal{Q}$, form a generalised quadrangle $\mathcal{Q}(d, q)$ with the following parameters:

$$
\begin{aligned}
s = q, \ t = 1, &\quad \text{when } d = 3, \\
s = q, \ t = q, &\quad \text{when } d = 4, \\
s = q, \ t = q^2, &\quad \text{when } d = 5.
\end{aligned}
$$

Since $\mathcal{Q}(3,q)$ is a grid, its structure is trivial. From Section 1.1, the quadric $\mathcal{Q}$ has the following canonical form:

$$\mathcal{Q} = \mathcal{H}_3 = \mathbf{V}(X_0X_1 + X_2X_3), \qquad\qquad \text{when } d = 3;$$
$$\mathcal{Q} = \mathcal{P}_4 = \mathbf{V}(X_0^2 + X_1X_2 + X_3X_4), \qquad\quad \text{when } d = 4;$$
$$\mathcal{Q} = \mathcal{E}_5 = \mathbf{V}(f(X_0, X_1) + X_2X_3 + X_4X_5), \quad \text{when } d = 5,$$

where $f(X_0, X_1)$ is an irreducible binary quadratic form.

(b) Let $\mathcal{U}$ be a non-singular Hermitian variety in the projective space $\mathrm{PG}(d, q^2)$, with $d = 3$ or 4. Then the points of $\mathcal{U}$ together with the lines on $\mathcal{U}$ form a generalised quadrangle $\mathcal{U}(d, q^2)$ with parameters as follows:

$$s = q^2, t = q, \quad \text{when } d = 3,$$
$$s = q^2, t = q^3, \text{ when } d = 4.$$

From Section 2.1, $\mathcal{U}$ has the following canonical form:

$$\mathcal{U} = \mathbf{V}(X_0^{q+1} + X_1^{q+1} + \cdots + X_d^{q+1}).$$

(c) The points of $\mathrm{PG}(3, q)$, together with the self-polar lines of a null polarity ζ, form a generalised quadrangle $\mathcal{W}(q)$ with parameters

$$s = q, t = q.$$

From Chapter 15 of FPSOTD, the lines of $\mathcal{W}(q)$ are the elements of a general linear complex of lines of $\mathrm{PG}(3, q)$. Further, a null polarity of $\mathrm{PG}(3, q)$ has the following canonical bilinear form:

$$X_0Y_1 - X_1Y_0 + X_2Y_3 - X_3Y_2.$$

The examples (d) and (e) of Section 5.1 show that there exist generalised quadrangles other than the classical ones and their duals. The order of each known generalised quadrangle is one of the following:

$(s, 1)$	with $s \geq 1$;
$(1, t)$	with $t \geq 1$;
(q, q)	with q a prime power;
$(q, q^2), (q^2, q)$	with q a prime power;
$(q^2, q^3), (q^3, q^2)$	with q a prime power;
$(q - 1, q + 1), (q + 1, q - 1)$	with q a prime power.

Definition 5.10. Let $\mathcal{S} = (\mathcal{P}, \mathcal{B}, \mathrm{I})$ be a generalised quadrangle of order (s, t).

(1) Two, not-necessarily-distinct points P, P' of $\mathcal{S}$ are *collinear* provided that there is some line l for which $P\,\mathrm{I}\,l\,\mathrm{I}\,P'$; write $P \sim P'$. Hence $P \not\sim P'$ means that P and P' are not collinear.

(2) Dually, for $l, l' \in \mathcal{B}$, they are *concurrent* or *non-concurrent*; write $l \sim l'$ or $l \not\sim l'$.

(3) When $P \sim P'$, it is also said that P is *orthogonal* or *perpendicular* to P'; similarly for $l \sim l'$.

(4) The line incident with distinct collinear points P and P' is denoted PP', and the point incident with distinct concurrent lines l and l' is denoted $l \cap l'$.

For $P \in \mathcal{P}$, put $P^{\perp} = \{P' \in \mathcal{P} \mid P \sim P'\}$, and note that $P \in P^{\perp}$. The *trace* of a pair $\{P, P'\}$ of distinct points is defined to be the set $P^{\perp} \cap P'^{\perp}$ and is denoted $\{P, P'\}^{\perp}$; then $|\{P, P'\}^{\perp}| = s + 1$ or $t + 1$ according as $P \sim P'$ or $P \not\sim P'$. More generally, if $\mathcal{A} \subset \mathcal{P}$, the 'perp' is defined by $\mathcal{A}^{\perp} = \bigcap\{P^{\perp} \mid P \in \mathcal{A}\}$. For $P \neq P'$, the *span* of the pair $\{P, P'\}$ is

$$\mathrm{sp}(P, P') = \{P, P'\}^{\perp\perp} = \{Y \in \mathcal{P} \mid Y \in Z^{\perp} \text{ for all } Z \in P^{\perp} \cap P'^{\perp}\}.$$

When $P \not\sim P'$, then $\{P, P'\}^{\perp\perp}$ is also called the *hyperbolic line* defined by P and P', and $|\{P, P'\}^{\perp\perp}| = s + 1$ or $|\{P, P'\}^{\perp\perp}| \leq t + 1$ according as $P \sim P'$ or $P \not\sim P'$.

A *triad* (of points) is a triple of pairwise non-collinear points. Then, given a triad $\mathcal{T} = \{P, P', P''\}$, a *centre* of $\mathcal{T}$ is just a point of $\mathcal{T}^{\perp}$.

These definitions are illustrated by some examples.

Example 5.11. (a) Let $P \not\sim P'$ in $\mathcal{W}(q)$, and let ζ be the null polarity defining $\mathcal{W}(q)$. If l is the polar line of the line PP' of $\mathrm{PG}(3, q)$, then $\{P, P'\}^{\perp} = l$ and $\{P, P'\}^{\perp\perp} = PP'$. Hence each hyperbolic line of $\mathcal{W}(q)$ contains $q + 1$ points.

(b) Let $P \not\sim P'$ in $\mathcal{Q}(4, q)$; then $\{P, P'\}^{\perp}$ is a conic. For q odd, the double perp $\{P, P'\}^{\perp\perp} = \{P, P'\}$, and for q even, $\{P, P'\}^{\perp\perp}$ is the intersection of $\mathcal{Q}$ and the plane $PP'N$, where N is the nucleus of the quadric $\mathcal{Q}$. In the even case $\{P, P'\}^{\perp\perp}$ is a conic, and each hyperbolic line contains $q + 1$ points.

(c) Let $P \not\sim P'$ in $\mathcal{Q}(5, q)$, and let ζ be the polarity defined by $\mathcal{Q}$. If π is the polar solid of the line PP' of $\mathrm{PG}(5, q)$, then $\{P, P'\}^{\perp}$ is the elliptic quadric $\pi \cap \mathcal{Q}$ of π, and $\{P, P'\}^{\perp\perp} = \{P, P'\}$.

(d) Let $P \not\sim P'$ in $\mathcal{U}(3, q^2)$, and let ζ be the unitary polarity defined by $\mathcal{U}$. If l is the polar line of the line PP' of $\mathrm{PG}(3, q^2)$, then $\{P, P'\}^{\perp} = \mathcal{U} \cap l$ and the double perp $\{P, P'\}^{\perp\perp} = PP' \cap \mathcal{U}$. Hence each hyperbolic line has $q + 1$ points.

(e) Let $P \not\sim P'$ in $\mathcal{U}(4, q^2)$, and let ζ be the unitary polarity defined by $\mathcal{U}$. If π is the polar plane of the line PP' of $\mathrm{PG}(4, q^2)$, then $\{P, P'\}^{\perp}$ is the non-singular Hermitian curve $\pi \cap \mathcal{U}$, and $\{P, P'\}^{\perp\perp} = PP' \cap \mathcal{U}$. Hence each hyperbolic line has $q + 1$ points.

(f) Consider again $\mathcal{W}(q)$ and its defining polarity ζ. If $\mathcal{T} = \{P, P', P''\}$ is a triad of $\mathcal{W}(q)$ for which P, P', P'' are collinear in $\mathrm{PG}(3, q)$, then $\mathcal{T}^{\perp} = \{P, P'\}^{\perp}$ and so $|\mathcal{T}^{\perp}| = q + 1$. If $\mathcal{T} = \{P, P', P''\}$ is a triad for which P, P', P'' are not collinear, then the pole of the plane $PP'P''$ is the unique centre of $\mathcal{T}$.

(g) Finally consider again $\mathcal{Q}(5, q)$ and the corresponding polarity ζ. For the triad $\mathcal{T} = \{P, P', P''\}$ of $\mathcal{Q}(5, q)$, the perp $\mathcal{T}^{\perp}$ is the conic $\mathcal{Q} \cap \pi$, where π is the polar plane of the plane $PP'P''$. Hence, in this case, any triad has $q + 1$ centres.

Let $\mathcal{S} = (\mathcal{P}, \mathcal{B}, \mathrm{I})$ be a generalised quadrangle of order (s, t), and put $|\mathcal{P}| = v$ and $|\mathcal{B}| = b$.

Theorem 5.12. (i) $v = (s+1)(st+1)$; (ii) $b = (t+1)(st+1)$.

Proof. Let l be a fixed line of S and count in different ways the number of ordered pairs $(P, m) \in \mathcal{P} \times \mathcal{B}$ with $P \not{I} l$, $P \, I \, m$, and $l \sim m$. Then $v - s - 1 = (s+1)ts$, whence $v = (s+1)(st+1)$. Dually, $b = (t+1)(st+1)$.   □

Theorem 5.13. *The integer $s+t$ divides $st(s+1)(t+1)$.*

Proof. If $\mathcal{E} = \{\{P, P'\} \mid P, P' \in \mathcal{P}$ and $P \sim P'\}$, then it is evident that $(\mathcal{P}, \mathcal{E})$ is a strongly regular graph with parameters

$$v = (s+1)(st+1), \ k = n_1 = st + s, \ \lambda = p_{11}^1 = s - 1, \ \mu = p_{11}^2 = t + 1.$$

The graph $(\mathcal{P}, \mathcal{E})$ is called the *point graph* of the generalised quadrangle. Let the point set $\mathcal{P} = \{P_1, P_2, \ldots, P_v\}$ and let $A = [a_{ij}]$ be the $v \times v$ matrix over $\mathbf{R}$ for which $a_{ij} = 0$ if $i = j$ or $P_i \not\sim P_j$, and $a_{ij} = 1$ if $i \neq j$ and $P_i \sim P_j$; that is, A is an adjacency matrix of the graph $(\mathcal{P}, \mathcal{E})$.

If $A^2 = [c_{ij}]$, then (a) $c_{ii} = (t+1)s$; (b) $i \neq j$ and $P_i \not\sim P_j$ imply $c_{ij} = t + 1$; (c) $i \neq j$ and $P_i \sim P_j$ imply $c_{ij} = s - 1$. Consequently,

$$A^2 - (s - t - 2)A - (t+1)(s-1)I = (t+1)J;$$

here I is the $v \times v$ identity matrix and J is the $v \times v$ matrix with each entry equal to one. Evidently, $(t+1)s$ is an eigenvalue of A, and J has eigenvalues $0, v$ with respective multiplicities $v - 1, 1$. Since

$$((t+1)s)^2 - (s-t-2)(t+1)s - (t+1)(s-1)$$
$$= (t+1)(st+1)(s+1) = (t+1)v,$$

the eigenvalue $(t+1)s$ of A corresponds to the eigenvalue v of J, and so $(t+1)s$ has multiplicity 1. The other eigenvalues of A are roots of the equation

$$x^2 - (s - t - 2)x - (t+1)(s-1) = 0.$$

Denote the multiplicities of these eigenvalues θ_1, θ_2 by m_1, m_2. Then

$$\theta_1 = -t - 1, \ \theta_2 = s - 1, \ v = 1 + m_1 + m_2,$$
$$s(t+1) - m_1(t+1) + m_2(s-1) = \text{tr}(A) = 0.$$

Hence

$$m_1 = (st+1)s^2/(s+t), \qquad m_2 = st(s+1)(t+1)/(s+t).$$

Since m_1, m_2 are positive integers, $s+t$ divides both $(st+1)s^2$ and $st(s+1)(t+1)$. Note that $s+t$ divides $(st+1)s^2$ if and only if it divides $st(s+1)(t+1)$.   □

Theorem 5.14 (Higman's inequality). *If $s > 1$ and $t > 1$, then $t \le s^2$, and dually $s \le t^2$.*

Proof. Let P, P' be two non-collinear points of $\mathcal{S}$. Put

$$\mathcal{V} = \{T \in \mathcal{P} \mid P \nsim T \text{ and } P' \nsim T\};$$

so $|\mathcal{V}| = d = (s+1)(st+1) - 2 - 2(t+1)s + (t+1)$. Denote the elements of $\mathcal{V}$ by $T_1, T_2, \ldots, T_d$ and let

$$t_i = |\{Z \in \{P, P'\}^\perp \mid Z \sim T_i\}|.$$

Count in different ways the number of ordered pairs $(T_i, Z) \in \mathcal{V} \times \{P, P'\}^\perp$ with $Z \sim T_i$ to obtain

$$\sum_i t_i = (t+1)(t-1)s. \tag{5.1}$$

Next count the number of ordered triples $(T_i, Z, Z') \in \mathcal{V} \times \{P, P'\}^\perp \times \{P, P'\}^\perp$, with $Z \neq Z'$, $Z \sim T_i$, $Z' \sim T_i$, to obtain

$$\sum_i t_i(t_i - 1) = (t+1)t(t-1). \tag{5.2}$$

From (5.1) and (5.2), it follows that

$$\sum_i t_i^2 = (t+1)(t-1)(s+t).$$

With $d\bar{t} = \sum_i t_i$, the inequality $0 \le \sum_i(\bar{t} - t_i)^2$ simplifies to

$$d\sum_i t_i^2 - \left(\sum_i t_i\right)^2 \ge 0,$$

which implies

$$d(t+1)(t-1)(s+t) \ge (t+1)^2(t-1)^2 s^2,$$

or

$$t(s-1)(s^2 - t) \ge 0,$$

completing the proof. $\qquad\square$

There is an immediate corollary of the proof.

Corollary 5.15. *When $s > 1$ and $t > 1$, the following are equivalent:*

   (i) $s^2 = t$;
  (ii) $d\sum t_i^2 - (\sum t_i)^2 = 0$ *for any pair $\{P, P'\}$ of non-collinear points;*
 (iii) $t_i = \bar{t}$ *for $i = 1, 2, \ldots, d$ and any pair $\{P, P'\}$ of non-collinear points;*
 (iv) *each triad of points has a constant number of centres, in which case this number is $s + 1$.*

Theorem 5.16. *If $s \neq 1$, $t \neq 1$, $s \neq t^2$, and $t \neq s^2$, then $t \leq s^2 - s$ and dually $s \leq t^2 - t$.*

Proof. Suppose $s \neq 1$ and $t \neq s^2$. By Theorem 5.14, $t = s^2 - x$ with $x > 0$, and, by Theorem 5.13, the integer $s + s^2 - x$ divides $s(s^2 - x)(s + 1)(s^2 - x + 1)$. Hence, modulo $s + s^2 - x$,

$$0 \equiv x(-s)(-s + 1) \equiv x(x - 2s).$$

If $x < 2s$, then $s + s^2 - x \leq x(2s - x)$ forces $x \in \{s, s + 1\}$. Consequently, $x = s, x = s + 1$, or $x \geq 2s$; so $t \leq s^2 - s$.   □

The only classical generalised quadrangle which has $t = s^2$ is $\mathcal{Q}(5, q)$; the only classical example with $s = t^2$ is $\mathcal{U}(3, q^2)$. In Example 5.11 (g), it was shown that any triad of $\mathcal{Q}(5, q)$ has $q + 1$ centres.

In the next two theorems, isomorphisms and anti-isomorphisms between the classical generalised quadrangles are described.

Theorem 5.17.   (i) $\mathcal{Q}(4, q)$ *is isomorphic to the dual of* $\mathcal{W}(q)$;
 (ii) $\mathcal{Q}(4, q)$ *and* $\mathcal{W}(q)$ *are self-dual if and only if q is even.*

Proof. Let $\mathcal{H}_5 = \mathcal{G}_{1,3}$ be the Klein quadric, that is, the Grassmannian of the lines of $\mathrm{PG}(3, q)$. The image of $\mathcal{W}(q)$ on $\mathcal{H}_5$ is the intersection of $\mathcal{H}_5$ with a non-tangent hyperplane $\mathrm{PG}(4, q)$ of $\mathrm{PG}(5, q)$; see Section 15.4 of FPSOTD. The non-singular quadric $\mathcal{H}_5 \cap \mathrm{PG}(4, q)$ of $\mathrm{PG}(4, q)$ is denoted by $\mathcal{Q}$. The lines of $\mathcal{W}(q)$ which are incident with a given point form a flat pencil of lines; hence their images on $\mathcal{H}_5$ form a line of $\mathcal{Q}$. Now it follows that $\mathcal{W}(q)$ is anti-isomorphic to $\mathcal{Q}(4, q)$.

In Theorem 16.4.13 of FPSOTD, it was shown that $\mathcal{W}(q)$ is self-dual if and only if q is even. By the first part of the proof, also $\mathcal{Q}(4, q)$ is self-dual if and only if q is even.   □

In Section 16.4 of FPSOTD, it was shown that $\mathcal{W}(q)$ admits a polarity if and only if $q = 2^{2h+1}$ with $h \geq 0$.

An algebraic proof of the existence of an anti-isomorphism between $\mathcal{Q}(5, q)$ and $\mathcal{U}(3, q^2)$ can be found in Section 19.2 of FPSOTD. Here it is shown in a purely geometrical way that $\mathcal{Q}(5, q)$ and $\mathcal{U}(3, q^2)$ are anti-isomorphic.

Theorem 5.18. *The generalised quadrangle $\mathcal{Q}(5, q)$ is isomorphic to the dual of* $\mathcal{U}(3, q^2)$.

Proof. Let $\mathcal{Q}$ be an elliptic quadric in $\mathrm{PG}(5, q)$. Extend $\mathrm{PG}(5, q)$ to $\mathrm{PG}(5, q^2)$. Then the extension of $\mathcal{Q}$ is a hyperbolic quadric $\mathcal{H}_5$ in $\mathrm{PG}(5, q^2)$. Hence $\mathcal{H}_5$ is the Klein quadric of the lines of $\mathrm{PG}(3, q^2)$. So to $\mathcal{Q}$ in $\mathcal{H}_5$ there corresponds a set $\mathcal{V}$ of lines in $\mathrm{PG}(3, q^2)$. To a given line l of the generalised quadrangle $\mathcal{Q}(5, q)$ there correspond $q + 1$ lines of $\mathrm{PG}(3, q^2)$ that all lie in a plane and pass through a point P.

Let $\mathcal{U}$ be the set of points on the lines of $\mathcal{V}$. Then, to each point of $\mathcal{Q}(5, q)$, there corresponds a line of $\mathcal{V}$, and to each line l of $\mathcal{Q}(5, q)$ there corresponds a point P of

$\mathcal{U}$. To distinct lines l, l' of $\mathcal{Q}(5, q)$ correspond distinct points P, P' of $\mathcal{U}$, as a plane of $\mathcal{H}_5$ contains at most one line of $\mathcal{Q}$. Since a point T of $\mathcal{Q}(5, q)$ is on $q^2 + 1$ lines of $\mathcal{Q}(5, q)$, these $q^2 + 1$ lines are mapped onto the $q^2 + 1$ points of the image of T. Hence an anti-isomorphism is obtained from $\mathcal{Q}(5, q)$ onto the structure $(\mathcal{U}, \mathcal{V}, \mathrm{I})$, where I is the natural incidence relation. So $(\mathcal{U}, \mathcal{V}, \mathrm{I})$ is a generalised quadrangle of order (q^2, q) embedded in $\mathrm{PG}(3, q^2)$. But now, by a result of Buekenhout and Lefèvre, which is part of Theorem 5.51, the generalised quadrangle $(\mathcal{U}, \mathcal{V}, \mathrm{I})$ must be $\mathcal{U}(3, q^2)$. $\qquad\square$

5.3 Embedded Shult spaces

Definition 5.19. (1) A *projective Shult space* $\mathcal{S}$ is a Shult space for which the point set $\mathcal{P}$ is a subset of the point set of some projective space $\mathrm{PG}(d, K)$, and for which the line set $\mathcal{B}$ is a non-empty set of lines of $\mathrm{PG}(d, K)$.
(2) In this case, the Shult space $\mathcal{S}$ is *(fully) embedded* in $\mathrm{PG}(d, K)$.
(3) If $\mathrm{PG}(d', K)$ is the subspace of $\mathrm{PG}(d, K)$ generated by all points of $\mathcal{P}$, then $\mathrm{PG}(d', K)$ is the *ambient space* of $\mathcal{S}$.

Examples (a), (b), (c) of Section 5.1 are projective Shult spaces. The aim is to show that these are the only non-degenerate Shult spaces embedded in a Galois space; a direct proof is given, without relying on Theorems 5.4 and 5.6.

Theorem 5.20. *A non-degenerate Shult space $\mathcal{S}$ of rank 2 embedded in $\mathrm{PG}(d, q)$ is a generalised quadrangle.*

Proof. Let $\mathcal{P}$ be the point set of $\mathcal{S}$, and let $\mathcal{B}$ be the line set of $\mathcal{S}$. On each line of $\mathcal{B}$ there are exactly $q + 1$ points. Let $P \in \mathcal{P}, l \in \mathcal{B}$, and $P \notin l$. By the definition of Shult space, $\mathcal{B}$ contains one or $q + 1$ lines through P which are concurrent with l. If $\mathcal{B}$ contains $q + 1$ lines through P and concurrent with l, then the plane Pl is a subspace of $\mathcal{S}$. This yields a contradiction since $\mathcal{S}$ has rank 2. So there is exactly one line of $\mathcal{S}$ through P which is concurrent with l.

Let P, P' be distinct points of $\mathcal{P}$ for which PP' is not a line of $\mathcal{B}$. If l is any line of $\mathcal{B}$ through P, then there is exactly one line l' of $\mathcal{B}$ through P' which is concurrent with l. It follows that the number of lines of $\mathcal{B}$ through P is equal to the number of lines of $\mathcal{B}$ through P'.

Now it is shown that any point P of $\mathcal{P}$ is contained in at least two lines of $\mathcal{B}$. Since $\mathcal{S}$ is non-degenerate, there is a line l in $\mathcal{B}$ which does not contain P. Let l' be the line of $\mathcal{B}$ through P which is concurrent with l. The common point of l and l' is denoted by P'. Since $\mathcal{S}$ is non-degenerate, there is a point $P'' \neq P'$ in $\mathcal{P}$ such that $P'P''$ is not a line of $\mathcal{B}$. The line of $\mathcal{B}$ which contains P'' and is concurrent with l is denoted by l''. Then $P \notin l''$ and $l' \cap l'' = \emptyset$. So the line of $\mathcal{B}$ which contains P and is concurrent with l'' is distinct from l'. Consequently, P is contained in at least two distinct lines of $\mathcal{B}$.

Now consider distinct points T, T' of $\mathcal{P}$, where TT' is in $\mathcal{B}$. Let m be a line of $\mathcal{B}$ through T which is distinct from TT', and let m' be a line of $\mathcal{B}$ through T' which is distinct from TT'; then $m \cap m' = \emptyset$. Let $M \in m\backslash\{T\}$, $M' \in m'\backslash\{T'\}$, with MM' not in $\mathcal{B}$; then $T'M \notin \mathcal{B}$ and $TM' \notin \mathcal{B}$. By a previous argument, the number of lines of $\mathcal{B}$ through T is equal to the number of lines of $\mathcal{B}$ through M', which equals the number of lines of $\mathcal{B}$ through M, which in turn equals the number of lines of $\mathcal{B}$ through T'. Therefore each point of $\mathcal{P}$ is contained in a constant number $t + 1$, where $t \geq 1$, of lines of $\mathcal{B}$.

Therefore $\mathcal{S}$ is a generalised quadrangle of order (q, t). $\square$

Theorem 5.21. *Let $\mathcal{S}$ be a non-degenerate projective Shult space of rank 2 with ambient space $\mathrm{PG}(d, q)$. If some point of $\mathcal{S}$ is contained in exactly two lines of $\mathcal{S}$, then $d = 3$ and $\mathcal{S}$ is the generalised quadrangle $\mathcal{Q}(3, q)$.*

Proof. By Theorem 5.20, $\mathcal{S}$ is a generalised quadrangle of order $(q, 1)$. Since $\mathcal{S}$ is a grid, so $d = 3$ and $\mathcal{S} = \mathcal{Q}(3, q)$. $\square$

Let $\mathcal{S}$ be a Shult space embedded in $\mathrm{PG}(d, q)$. Let $\mathcal{P}$ be the point set of $\mathcal{S}$ and let $\mathcal{B}$ be the line set of $\mathcal{S}$. If $P, P' \in \mathcal{P}$ and if there is at least one line of $\mathcal{B}$ through P and P', then P and P' are *adjacent*; write $P \sim P'$. So there should be no confusion between collinearity in $\mathrm{PG}(d, q)$ and adjacency in $\mathcal{S}$.

For the subspace of $\mathrm{PG}(d, q)$ generated by the point sets or points $\mathcal{V}_1, \mathcal{V}_2, \ldots, \mathcal{V}_k$, the notation $\langle \mathcal{V}_1, \mathcal{V}_2, \ldots, \mathcal{V}_k \rangle$ is used. If π is a subspace of $\mathrm{PG}(d, q)$, then $\pi \cap \mathcal{S}$ (or $\mathcal{S} \cap \pi$) denotes the structure with point set $\pi \cap \mathcal{P}$ and line set the set of all lines of $\mathcal{S}$ contained in π. If $\mathrm{PG}(d, q)$ is the ambient space of $\mathcal{S}$, then $\langle \pi \cap \mathcal{P} \rangle$ does not necessarily coincide with π.

Assume, from now on, that $\mathcal{S}$ is a projective Shult space with point set $\mathcal{P}$, line set $\mathcal{B}$, and ambient space $\mathrm{PG}(d, q)$.

Theorem 5.22.   (i) *The radical $\mathcal{R}$ of a degenerate Shult space $\mathcal{S}$ is a subspace of $\mathrm{PG}(d, q)$.*

  (ii) *Let $\mathcal{R}$ have dimension $r \neq -1$, and let $\mathrm{PG}(d - r - 1, q) = \pi$ be a subspace of $\mathrm{PG}(d, q)$ which is skew to $\mathcal{R}$. If $\mathcal{R} \neq \mathrm{PG}(d, q)$, then*

    (a) *$\pi \cap \mathcal{S}$ is a non-degenerate Shult space;*

    (b) *$\mathcal{P}$ is the union of all lines joining every point of $\mathcal{R}$ to every point of $\pi \cap \mathcal{S}$;*

    (c) *two points of $\mathcal{P}\backslash\mathcal{R}$ are adjacent if and only if their projections from $\mathcal{R}$ onto π are adjacent.*

Proof. In Section 5.1, it was already mentioned that $\mathcal{R}$ is a subspace of the Shult space $\mathcal{S}$. Hence $\mathcal{R}$ has the property that the line joining any two distinct points of $\mathcal{R}$ is completely contained in $\mathcal{R}$. Hence $\mathcal{R}$ is a subspace of $\mathrm{PG}(d, q)$. Let $\mathcal{R}$ have dimension r, with $r < d$, and let $\mathrm{PG}(d - r - 1, q) = \pi$ be a subspace of $\mathrm{PG}(d, q)$ which is skew to $\mathcal{R}$. Also, in Section 5.1 the equivalence relation ρ on $\mathcal{P}$ was introduced: $P \rho P'$ if and only if the set of all points collinear with P coincides with the set of all points collinear with P'. Let $\rho(P)$ denote the equivalence class containing the point P. If $P \in \mathcal{P} \backslash \mathcal{R}$, then it was shown in Section 5.1 that $\mathcal{R} \cup \rho(P)$ is a subspace of $\mathcal{S}$ and hence a subspace of $\mathrm{PG}(d, q)$. Now it is shown that $\mathcal{R} \cup \rho(P)$ is the projective $(r + 1)$-space $\mathcal{R}P$.

Let $P' \in \mathcal{R}P$, with $P' \notin \mathcal{R} \cup \{P\}$, and let $P'' = PP' \cap \mathcal{R}$. If $T \sim P$, then, since $T \sim P''$, also $T \sim P'$. Analogously, $T \sim P'$ implies $T \sim P$; hence $P' \in \rho(P)$, and so $\mathcal{R}P \subset \mathcal{R} \cup \rho(P)$. Next, let $P_1 \in \rho(P) \backslash \{P\}$ and suppose that $PP_1 \cap \mathcal{R} = \emptyset$; then $PP' \subset \rho(P)$. Since $P \notin \mathcal{R}$ there is a point T not adjacent to P. On PP' there is at least one point T' adjacent to T. Since $T' \in \rho(P)$, so $T \sim P$, a contradiction. Hence $PP_1 \cap \mathcal{R} \neq \emptyset$, and so $P_1 \in \mathcal{R}P$. Consequently, $\mathcal{R} \cup \rho(P) \subset \mathcal{R}P$. It follows that $\mathcal{R} \cup \rho(P) = \mathcal{R}P$.

The set $\mathcal{P}$ is the union of all lines joining every point of $\mathcal{R}$ to every point of $\pi \cap \mathcal{S}$. Let $\rho(P) \neq \rho(P')$, with $P, P' \notin \mathcal{R}$, let $T \in \rho(P), T' \in \rho(P')$, and let $P \sim P'$. Since $T \in \rho(P)$ so $T \sim P'$ and, similarly, since $T' \in \rho(P')$ so $T \sim T'$. It now follows that two points of $\mathcal{P} \backslash \mathcal{R}$ are adjacent if and only if their projections from $\mathcal{R}$ onto π are adjacent. Finally, by Theorem 5.9, $\pi \cap \mathcal{S}$ is a non-degenerate Shult space. $\qquad\square$

The non-degenerate Shult space $\pi \cap \mathcal{S}$ is a *basis* of $\mathcal{S}$.

Lemma 5.23. *Let $\mathcal{S}$ be a non-degenerate Shult space in $\mathrm{PG}(d, q)$ and let P, P' be adjacent points of $\mathcal{S}$. Then*

(i) *there is a point T in $\mathcal{S}$ such that $T \not\sim P$ and $T \not\sim P'$;*
(ii) *each point of $\mathcal{S}$ is contained in a constant number $t + 1$ of lines of $\mathcal{B}$, where $t \geq 1$.*

Proof. Suppose (i) is false. As $\mathcal{S}$ is non-degenerate, there is a point P_1 with $P_1 \not\sim P'$; so $P_1 \sim P$. There is also a point P_2 for which $P_2 \not\sim P$ and $P_2 \sim P'$. Let l be a line through P_1 intersecting $P'P_2$ in P_2'. Then $P_2' \neq P'$ and $P \not\sim P_2'$. If T is a point on l other than P_1 and P_2', then $T \not\sim P$ and $T \not\sim P'$.

Let $M, M' \in \mathcal{P}$, with $M \not\sim M'$. If m is any line of $\mathcal{B}$ through M, then there is a unique line of $\mathcal{B}$ through M' concurrent with m, and conversely. It follows that the number of lines of $\mathcal{B}$ through M is equal to the number of lines of $\mathcal{B}$ through M'. Next, let $M, M' \in \mathcal{P}$, with $M \neq M'$ and $M \sim M'$. There is a point T in $\mathcal{S}$ such that $T \not\sim M$ and $T \not\sim M'$. The number of lines of $\mathcal{B}$ through T is equal to the number of lines of $\mathcal{B}$ through M and to the number of lines of $\mathcal{B}$ through M'. Again the number of lines of $\mathcal{B}$ through M is equal to the number of lines of $\mathcal{B}$ through M'. Hence the number of lines of $\mathcal{B}$ through the point $M \in \mathcal{P}$ is a constant $t + 1$. Let $m \in \mathcal{B}$ with $M \notin m$. On m there is a point M' which is collinear with M; so M' is contained in at least two lines of $\mathcal{B}$. Therefore $t \geq 1$. $\qquad\square$

Lemma 5.24. *If π is a subspace of $\mathrm{PG}(d, q)$ for which $\pi \cap \mathcal{P}$ is non-empty, then $\pi \cap \mathcal{S}$ is a Shult space. If $\mathcal{S}$ is non-degenerate and π is a hyperplane, then $\pi \cap \mathcal{P}$ generates π.*

Proof. Let π be a subspace of $\mathrm{PG}(d, q)$. If $\pi \cap \mathcal{P}$ is non-empty, then it is immediate that $\pi \cap \mathcal{S}$ is a Shult space. Now let $\mathcal{S}$ be non-degenerate and let π be a hyperplane.

Since $PG(d, q)$ is the ambient space of $\mathcal{S}$, there is a point P in $\mathcal{P}\backslash(\pi \cap \mathcal{P})$. It suffices to show that an arbitrary line l of $\mathcal{B}$ is in $\langle \pi \cap \mathcal{P}, P \rangle$.

Suppose that l meets π in some point P'. If $P \in l$, the result follows. So suppose $P \notin l$. If $P \sim P''$ with $P'' \in l \backslash \{P'\}$, there is a line l' of $\mathcal{B}$ through P and P'' which is in $\langle \pi \cap \mathcal{P}, P \rangle$. Hence $P'' \in \langle \pi \cap \mathcal{P}, P \rangle$, and consequently l is in $\langle \pi \cap \mathcal{P}, P \rangle$.

Finally, suppose $P \notin l$ and suppose that P' is the only point of l which is adjacent to P. Let R be a point not adjacent to P'. The line PP' contains a point R' for which $R' \sim R$. As $R' \in \langle \pi \cap \mathcal{P}, P \rangle$, so $RR' \subset \langle \pi \cap \mathcal{P}, P \rangle$. Let $T \in RR' \backslash \{R'\}$ with $T \notin \pi$. Then $T \in \langle \pi \cap \mathcal{P}, P \rangle$. Let $T' \in l$ with $T \sim T'$. Then $T' \neq P'$, since otherwise $R \sim P'$. The line TT' is contained in $\langle \pi \cap \mathcal{P}, P \rangle$; hence $T' \in \langle \pi \cap \mathcal{P}, P \rangle$. Therefore $P'T' = l$ is contained in $\langle \pi \cap \mathcal{P}, P \rangle$. □

Definition 5.25. (1) For P in $\mathcal{P}$, put $P^\perp = \{P' \in \mathcal{P} \mid P' \sim P\}$. Then $P^\perp$ is the union of all lines of $\mathcal{B}$ through P.
(2) A *tangent to $\mathcal{S}$ at $P \in \mathcal{P}$* is any line l through P such that either $l \in \mathcal{B}$ or $l \cap \mathcal{P} = \{P\}$.
(3) The union of all tangents to $\mathcal{S}$ at P is the *tangent set of $\mathcal{S}$ at P* and is denoted by $\mathcal{S}(P)$. The relation between $\mathcal{S}(P)$ and $P^\perp$ is that $P^\perp = \mathcal{P} \cap \mathcal{S}(P)$.
(4) A line l of $PG(d, q)$ is a *secant* of $\mathcal{S}$ if l intersects $\mathcal{P}$ in at least two points but is not a member of $\mathcal{B}$.

Lemma 5.26. *For each $P \in \mathcal{P}$, the set $\langle P^\perp \rangle \subset \mathcal{S}(P)$.*

Proof. It must be shown that, for each line l through P in $\langle P^\perp \rangle$, either $l \in \mathcal{B}$ or l intersects $\mathcal{P}$ exactly in P. So suppose that $P \in l \notin \mathcal{B}, l \subset \langle P^\perp \rangle$. First, let l_1 be a line of $\mathcal{B}$ through P and let l_2 be a second tangent to $\mathcal{S}$ at P for which the plane $\pi = \langle l_1, l_2 \rangle$ contains l. If l were not a tangent at P it would contain some point P', where $P \neq P' \in \mathcal{P}$. There would be a unique line $m \in \mathcal{B}$ through P' and intersecting l_1 in P_1, with $P_1 \neq P$. As m is contained in π, so m meets l_2 in a point P_2, with $P_2 \neq P$. Then $P, P_2 \in l_2$ implies $l_2 \in \mathcal{B}$, since l_2 is a tangent to $\mathcal{S}$ containing at least two points of $\mathcal{S}$. But then l_1 and l_2 are two lines of $\mathcal{S}$ through P intersecting m, contradicting $P \not\sim P'$ and the assumption that $\mathcal{S}$ is a Shult space. Hence l must be a tangent.

Now, suppose there is an integer k such that $\langle P^\perp \rangle$ is generated by k lines $l_1, l_2, \ldots, l_k$ of $\mathcal{S}$ through P. Let $\pi^{(i)} = \langle l_1 \cup l_2 \cup \cdots \cup l_i \rangle$, $i = 2, 3, \ldots, k$. From the first case, $\pi^{(2)} \in \mathcal{S}(P)$. Now use induction on i. Assume $\pi^{(i)} \subset \mathcal{S}(P)$, and let l be some line of $\pi^{(i+1)}$ through P. Take $l \neq l_{i+1}$ and $l \not\subset \pi^{(i)}$. Then the plane $\pi = \langle l, l_{i+1} \rangle$ intersects $\pi^{(i)}$ in a line l'. By the induction hypothesis, l' is tangent to $\mathcal{S}$ at P, so that $\pi = \langle l', l_{i+1} \rangle$ satisfies the hypothesis of the first case. Hence l is a tangent to $\mathcal{S}$ at P, and it follows that $\pi^{(i+1)} \subset \mathcal{S}(P)$. □

Lemma 5.27. *For any point P in $\mathcal{P}$, the set $\langle P^\perp \rangle \neq PG(d, q)$.*

Proof. If $\langle P^\perp \rangle = PG(d, q)$, then, by Lemma 5.26, $\mathcal{S}(P) = PG(d, q)$. But then P is adjacent to all points of $\mathcal{S}$, and so $\mathcal{S}$ is degenerate, a contradiction. □

Lemma 5.28. *The dimension of $\langle P^\perp \rangle$ is independent of P in $\mathcal{P}$.*

Proof. If $P \not\sim P'$, with $P, P' \in \mathcal{P}$, then it must be shown that $\langle P^\perp \rangle \cap \langle P'^\perp \rangle$ is a hyperplane in $\langle P^\perp \rangle$. As $\langle P^\perp \cap P'^\perp \rangle \subset \langle P^\perp \rangle \cap \langle P'^\perp \rangle$ and, since $\mathcal{S}$ is a Shult space, so P and $P^\perp \cap P'^\perp$ generate $\langle P^\perp \rangle$. Hence $\langle P^\perp \rangle \cap \langle P'^\perp \rangle$ is a hyperplane of $\langle P^\perp \rangle$, or $\langle P^\perp \rangle \cap \langle P'^\perp \rangle = \langle P^\perp \rangle$. If $\langle P^\perp \rangle = \langle P^\perp \rangle \cap \langle P'^\perp \rangle$, then $\langle P'^\perp \rangle \subset \langle P^\perp \rangle \subset \mathcal{S}(P)$, contradicting that $P \not\sim P'$. Consequently, $\langle P^\perp \rangle \cap \langle P'^\perp \rangle$ is a hyperplane of $\langle P^\perp \rangle$. Analogously, $\langle P^\perp \rangle \cap \langle P'^\perp \rangle$ is a hyperplane of $\langle P'^\perp \rangle$. It follows that $\langle P^\perp \rangle$ and $\langle P'^\perp \rangle$ have the same dimension.

Next, let $P \sim P'$ with P, P' distinct points of $\mathcal{P}$. By Lemma 5.23, there is a point T in $\mathcal{S}$ such that $T \not\sim P$ and $T \not\sim P'$. Now, from above, $\dim \langle T^\perp \rangle = \dim \langle P^\perp \rangle$ and $\dim \langle T^\perp \rangle = \dim \langle P'^\perp \rangle$. Hence $\dim \langle P^\perp \rangle = \dim \langle P'^\perp \rangle$. □

Lemma 5.29. *The point P in $\mathcal{P}$ is the unique point of $\mathcal{S}$ adjacent to all points of $P^\perp$.*

Proof. Let $P' \in \mathcal{P} \backslash \{P\}$ be adjacent to all points of $P^\perp$; so $P' \in P^\perp$. Since $\mathcal{S}$ is a Shult space, all points of the line PP' are adjacent to all points of $P^\perp$. Now take a point $T \in \mathcal{P} \backslash P^\perp$. Since $\langle P^\perp \rangle \subset \mathcal{S}(P)$, we have $T \notin \langle P^\perp \rangle$. There is a line m through T intersecting PP' in a point P'', and so $\langle P''^\perp \rangle$ contains $\langle P^\perp \rangle$ properly. This contradicts Lemma 5.28. □

Lemma 5.30. *For each P in $\mathcal{P}$, the subspace $\langle P^\perp \rangle$ is a hyperplane of $\mathrm{PG}(d, q)$.*

Proof. Consider a point P' in $\mathcal{P} \backslash \langle P^\perp \rangle$. By Lemma 5.24, $\langle P^\perp, P' \rangle \cap \mathcal{S}$ is a Shult space $\mathcal{S}'$. Assume that $\mathcal{S}'$ is degenerate with radical $\mathcal{R}'$. If $R \in \mathcal{R}'$, then R is adjacent to all points of $P^\perp$. By Lemma 5.29, $R = P$. Hence $P' \sim P$, a contradiction. Consequently $\mathcal{S}'$ is non-degenerate. Now assume that $\langle P^\perp, P' \rangle \neq \mathrm{PG}(d, q)$, and let $P'' \in \mathrm{PG}(d, q) \backslash \langle P^\perp, P' \rangle$. Consider a point T in $\mathcal{S}'$, with $T \sim P''$. If T is contained in $t + 1$ lines of $\mathcal{S}$ and $t' + 1$ lines of $\mathcal{S}'$, then $t \geq t' + 1$. But P is contained in exactly $t + 1$ lines of $\mathcal{S}'$. Hence $t = t'$, contradicting $t > t'$. Therefore $\langle P^\perp, P' \rangle = \mathrm{PG}(d, q)$, whence $\langle P^\perp \rangle$ is a hyperplane of $\mathrm{PG}(d, q)$. □

Lemma 5.31. *The hyperplane $\langle P^\perp \rangle$, for P in $\mathcal{P}$, is the tangent set $\mathcal{S}(P)$ of $\mathcal{S}$ at P.*

Proof. From Lemmas 5.26 and 5.30, $\langle P^\perp \rangle$ is a hyperplane contained in $\mathcal{S}(P)$. If equality did not hold, there would be some tangent l at P not in $\langle P^\perp \rangle$.

Let l_1 be a line of $\mathcal{S}$ through P and let Π_2 be the plane $\langle l, l_1 \rangle$. If there were a point $P' \in \Pi_2 \cap \mathcal{P}$ with $P' \notin l_1$, there would be a line m of $\mathcal{B}$ through P' meeting l_1 in a point other than P. But m would be in Π_2 and hence would meet l in a point of $\mathcal{P}$ other than P, an impossibility. Hence each line of Π_2 through P is a tangent of $\mathcal{S}$ at P.

Let l_2 be a line of $\mathcal{S}$ through P, with $l_2 \neq l_1$. If m, with $m \neq l_1$, is a line of Π_2 through P, then, by the previous paragraph, each line of $\langle l_2, m \rangle$ through P is a tangent of $\mathcal{S}$ at P. Hence each line of $\Pi_3 = \langle l, l_1, l_2 \rangle$ through P is a tangent of $\mathcal{S}$ at P.

Let l_3 be a line of $\mathcal{S}$ through P, with l_3 not in the plane $\langle l_1, l_2 \rangle$. If m', with m' not in the plane $\langle l_1, l_2 \rangle$, is a line of Π_3 through P, then, by the previous paragraph, each line of $\langle l_3, m' \rangle$ through P is a tangent of $\mathcal{S}$ at P. Hence each line of the space $\Pi_4 = \langle l, l_1, l_2, l_3 \rangle$ through P is a tangent of $\mathcal{S}$ at P.

Continuing in this way shows that each line of $\langle P^\perp, l \rangle = \mathrm{PG}(d, q)$ through P is a tangent of S at P. Consequently, P belongs to the radical of S, a contradiction as S is non-degenerate.

The conclusion is that $\langle P^\perp \rangle = S(P)$. □

The tangent set $S(P)$ of S at P, for P in $\mathcal{P}$, is also called the *tangent hyperplane of S at P*.

Lemma 5.32. *Let l be a secant of S containing three distinct points P, A, A' of $\mathcal{P}$. Then the perspectivity σ of $\mathrm{PG}(d, q)$ with centre P and axis $S(P)$ mapping A onto A' leaves $\mathcal{P}$ invariant.*

Proof. The map σ fixes all points of $S(P)$ and thus fixes $P^\perp$. Let $P' \in \mathcal{P} \backslash P^\perp$. First, suppose that P' is not on l and let π be the plane $\langle P, A, P' \rangle$. If $m = \langle A, P' \rangle$, then m intersects $S(P)$ at a point P'', fixed by σ. Hence $m\sigma = \langle A', P'' \rangle$.

If m is a line of S, then $P'' \in \mathcal{P}$ and so the tangent $\langle P, P'' \rangle$ is a line of S. Thus the plane $\langle P, A, P'' \rangle = \pi$ is in the tangent hyperplane $S(P'')$. Hence, since $A' \in \pi$, it follows that $A' \sim P''$, that $m\sigma$ is a line of S, and that $P'\sigma$ is a point of S.

If m is not a line of S, suppose there is a point $D \in \mathcal{P} \backslash S(P)$ with $D \in A^\perp \cap P'^\perp$. The argument above, with D in the role of P', shows that $D\sigma \in \mathcal{P}$. Then, with D and $D\sigma$ playing the roles of A and A', it follows that $P'\sigma \in \mathcal{P}$. On the other hand, suppose $A^\perp \cap P'^\perp \subset S(P)$. Consider a line l_1 of $\mathcal{B}$ containing P, and let P_1 be defined by $P_1 \in l_1$ and $A \sim P_1 \sim P'$. By Lemma 5.23, there is a point $T \in \mathcal{P}$ with $T \not\sim P$ and $T \not\sim P_1$. The line of $\mathcal{B}$ through T which is concurrent with l_1 is denoted by m_1. Let D, D' be defined by $D, D' \in m_1$ and $A \sim D, D' \sim P'$. Then D, D' are distinct points of $\mathcal{P} \backslash S(P)$. Repeated applications of the argument above show that $D\sigma, D'\sigma$, and finally $P'\sigma$ are all in $\mathcal{P}$.

Secondly, suppose P' is on l, and use the fact that, if D is any point of $\mathcal{P}$ not on l, then $D\sigma \in \mathcal{P}$. It follows readily that $P'\sigma \in \mathcal{P}$. □

Lemma 5.33. *If A, B, C are three collinear, distinct points of $\mathcal{P}$, then the intersections $S(A) \cap S(B)$, $S(B) \cap S(C)$, $S(C) \cap S(A)$ coincide.*

Proof. It is shown that $S(A) \cap S(B) \subset S(C)$.

First, suppose that A, B, C are on a line l of S. Let $P \in S(A) \cap S(B)$. If $P \in \mathcal{P} \backslash l$, then $P \sim A$ and $P \sim B$, and so $P \sim C$. If $P \in l$, then $P \in S(C)$. Now let $P \notin \mathcal{P}$. Suppose that PC is a secant of S, and let $C' \in PC \cap \mathcal{P}$ with $C \neq C'$. The plane $\langle A, P, B \rangle$ is in $S(A)$; hence $AC' \in \mathcal{B}$. Analogously, $BC' \in \mathcal{B}$. Consequently, $C \sim C'$, a contradiction. So PC is a tangent of S at C.

Secondly, suppose that A, B, C are on a secant l of S. Let $P \in S(A) \cap S(B)$. If $P \in \mathcal{P}$, then $P \sim A$ and $P \sim B$, and so the line AB is in $S(P)$. Hence $P \sim C$ and so $P \in S(C)$. Now let $P \notin \mathcal{P}$, and suppose that $P \notin S(C)$. Then there is a second point C' of $\mathcal{P}$ on the line PC. Consider the perspectivity σ with centre A and axis $S(A)$ mapping C onto B. By Lemma 5.32, $\mathcal{P}\sigma = \mathcal{P}$. This perspectivity σ fixes $P \in S(A)$ and so σ maps the line PC onto the line PB. Hence $C'\sigma$ is a point of $PB \backslash \{B\}$ on $\mathcal{P}$, a contradiction since $P \in S(B)$. Consequently, $P \in S(C)$. Hence $S(A) \cap S(B) \subset S(C)$.

Analogously, $\mathcal{S}(A) \cap \mathcal{S}(C) \subset \mathcal{S}(B)$ and $\mathcal{S}(B) \cap \mathcal{S}(C) \subset \mathcal{S}(A)$. Therefore $\mathcal{S}(A) \cap \mathcal{S}(B), \mathcal{S}(B) \cap \mathcal{S}(C)$, and $\mathcal{S}(C) \cap \mathcal{S}(A)$ coincide. $\square$

Lemma 5.34. *All secant lines contain the same number of points of $\mathcal{S}$.*

Proof. Let l and l' be secant lines of $\mathcal{S}$. First, suppose that l and l' have a point P of $\mathcal{P}$ in common, and let m be any secant line through P. If some m contains more than two points in $\mathcal{P}$, consider, by Lemma 5.32, the non-trivial group G of all perspectivities with centre P and axis $\mathcal{S}(P)$ leaving $\mathcal{P}$ invariant. The group G is regular on the set of points of m in $\mathcal{P}$ other than P. Hence each secant through P has $1 + |G|$ points of $\mathcal{P}$, so that l and l' have the same number of points of $\mathcal{S}$. If no m is incident with more than two points of $\mathcal{P}$, then l and l' contain two points of $\mathcal{S}$.

Secondly, suppose l and l' do not have any point of $\mathcal{P}$ in common, and choose points P and P' of $\mathcal{P}$ on l and l'. If $P \not\sim P'$, then PP' is a secant, and so meets $\mathcal{P}$ in the same number of points as do l and l', by the above. If $P \sim P'$, then, by Lemma 5.23, there is a point T in $\mathcal{S}$ such that $P \not\sim T \not\sim P'$. Now apply the above argument to the secants l, PT, TP', l'. $\square$

Definition 5.35. (1) For a point P of $\mathrm{PG}(d, q)$, the *collar* $\mathcal{S}_P$ of $\mathcal{S}$ *for* P is the set of all points P' of $\mathcal{S}$ such that $P = P'$ or the line $\langle P, P' \rangle$ is a tangent to $\mathcal{S}$ at P. For example, if $P \in \mathcal{P}$, then $\mathcal{S}_P$ is just $P^\perp$. If $P \notin \mathcal{P}$, the collar $\mathcal{S}_P$ is the set of points P' of $\mathcal{P}$ such that $\langle P, P' \rangle \cap \mathcal{P} = \{P'\}$.
(2) For all $P \in \mathrm{PG}(d, q)$, the *polar* $P\zeta$ of P with respect to $\mathcal{S}$ is the subspace of $\mathrm{PG}(d, q)$ generated by the collar $\mathcal{S}_P$. So, if $P \in \mathcal{P}$, then $P\zeta = \langle P^\perp \rangle = \mathcal{S}(P)$.

Lemma 5.36. *For any point P of $\mathrm{PG}(d, q)$, let P_1 and P_2 be distinct points of $\mathcal{S}_P$. Then $\mathcal{P} \cap \langle P_1, P_2 \rangle \subset \mathcal{S}_P$.*

Proof. Suppose $P' \in \mathcal{P} \cap \langle P_1, P_2 \rangle$, $P_1 \neq P' \neq P_2$. Since $P \in \mathcal{S}(P_1) \cap \mathcal{S}(P_2)$, by Lemma 5.33 the point P is also in $\mathcal{S}(P')$ and hence $P' \in \mathcal{S}_P$. $\square$

Lemma 5.37. *Each line l of $\mathcal{S}$ intersects the collar $\mathcal{S}_P$, with $P \in \mathrm{PG}(d, q)$, in exactly one point, unless each point of l is in $\mathcal{S}_P$.*

Proof. When $P \in \mathcal{P}$, the result is immediate. So suppose that $P \notin \mathcal{P}$ and put $\pi = \langle l, P \rangle$. If $\pi \cap \mathcal{P} = l$, then each point of l is in $\mathcal{S}_P$. So suppose $P' \in \pi \cap \mathcal{P}$, $P' \notin l$. Let $\langle P, P' \rangle \cap l = \{T\}$; then $T \not\sim P'$. Hence P' is adjacent to a unique point T' of l. By Lemma 5.31, each line of π through T' is tangent to $\mathcal{S}$ at T', and hence $T' \in \mathcal{S}_P$. Also, by Lemma 5.36, T' is the unique point of l in $\mathcal{S}_P$ unless each point of l is in $\mathcal{S}_P$. $\square$

Lemma 5.38. *Either $P\zeta = \langle \mathcal{S}_P \rangle$ is a hyperplane or $P\zeta = \mathrm{PG}(d, q)$.*

Proof. Again assume that $P \notin \mathcal{P}$. If the assertion is false for some point P, then $P\zeta$ is contained in some $(d-2)$-space π of $\mathrm{PG}(d, q)$. By Lemma 5.37, each line l of $\mathcal{B}$ intersects π. Therefore, if P' is a point of $\mathcal{S}$ not in π, then $\mathcal{S}_{P'} = P'^\perp$ is contained in $\langle \pi, P' \rangle$. As $\langle \mathcal{S}_{P'} \rangle$ is a hyperplane, so $\langle \pi, P' \rangle = \langle \mathcal{S}_{P'} \rangle = \mathcal{S}_{P'}$. Any line l' of $\mathcal{S}$ through P' must contain another point P'' of $\mathcal{P}$ not in π. Then it follows that $\mathcal{S}(P') = \langle \pi, P' \rangle = \langle \pi, P'' \rangle = \mathcal{S}(P'')$. This contradicts Lemma 5.29. $\square$

Lemma 5.39. *If $P\zeta$ is a hyperplane, then $\mathcal{S}_P = \mathcal{P} \cap P\zeta$.*

Proof. By definition, $\mathcal{S}_P \subset \mathcal{P} \cap P\zeta$. Suppose there were a point P' of $\mathcal{P} \cap P\zeta$ not in $\mathcal{S}_P$. Then either some line l of $\mathcal{S}$ through P' does not lie in $P\zeta$, or $P\zeta = \mathcal{S}(P')$. In the first case, l intersects $P\zeta$ exactly in P'. As $P' \notin \mathcal{S}_P$, so l is on no point of $\mathcal{S}_P$, contradicting Lemma 5.37. In the second case, as $P' \notin \mathcal{S}_P$, each line of $\mathcal{B}$ through P' has exactly one point in $\mathcal{S}_P$. So on any line of $\mathcal{B}$ through P' there is a point $P'' \ne P'$ of $\mathcal{S}(P')\backslash\mathcal{S}_P$. By Lemma 5.29, $\mathcal{S}(P') \ne \mathcal{S}(P'')$; so there is a line of $\mathcal{B}$ through P'' but not in $P\zeta = \mathcal{S}(P')$, leading back to the first case. □

Lemma 5.40. *Let P be a point of $\mathrm{PG}(d,q)$ and let A, A' be distinct points of $\mathcal{P}\backslash\{P\}$ collinear with P but not in $P\zeta$. Then the perspectivity σ of $\mathrm{PG}(d,q)$ with centre P and axis $P\zeta$ mapping A onto A' leaves $\mathcal{P}$ invariant.*

Proof. Since $A, A' \notin P\zeta$, so $P\zeta$ is a hyperplane. First, let $P \in \mathcal{P}$. Since $A, A' \notin P\zeta$, the line $\langle P, A, A' \rangle$ is a secant of $\mathcal{S}$. In this case, the result is known by Lemma 5.32. Now let $P \notin \mathcal{P}$. Note that σ fixes all points of $\mathcal{P} \cap P\zeta$. Let P' be a point of $\mathcal{P}\backslash P\zeta$ not on $\langle A, A' \rangle$. Let π be the plane $\langle P, A, P' \rangle$ and m the line $\langle A, P' \rangle$. If $m \cap P\zeta = \{P''\}$, then $m\sigma = \langle A', P'' \rangle$.

If m is a line of $\mathcal{S}$, then $P'' \in \mathcal{P} \cap P\zeta = \mathcal{S}_P$ by Lemma 5.39. So $\langle P'', P \rangle$ and m, and hence π, are in the tangent hyperplane $\mathcal{S}(P'')$. Then $A' \in \pi \subset \mathcal{S}(P'')$, showing that $m\sigma = \langle A', P'' \rangle \in \mathcal{B}$; that is, $P'\sigma \in \mathcal{P}$.

If m is not a line of $\mathcal{S}$, suppose there is a point $D \in \mathcal{P}\backslash P\zeta$ with $D \in A^{\perp} \cap P'^{\perp}$. The previous argument, with D in the role of P', shows that $D\sigma \in \mathcal{P}$. Then, with D and $D\sigma$ playing the roles of A and A', it follows that $P'\sigma \in \mathcal{P}$. On the other hand, suppose $A^{\perp} \cap P'^{\perp} \subset \mathcal{S}_P$. Let $T \in A^{\perp} \cap P'^{\perp}$. Now assume that each point of $A^{\perp} \cap P'^{\perp}$ is adjacent to T. Since each line of $\mathcal{B}$ through P' has a point in common with $\mathcal{S}_P$, it follows that T is adjacent to all points of $P'^{\perp}$, contradicting Lemma 5.29. Hence there exists a point $T' \in A^{\perp} \cap P'^{\perp}$, with $T \not\sim T'$. Let $D \in \langle A, T \rangle\backslash\{A, T\}$, and let $D' \in \langle P', T' \rangle$ with $D \sim D'$. If $D' = P'$, then $A \sim P'$, a contradiction. If $D = T'$, then $T \sim T'$, a contradiction. However, $D \ne D'$, which means that D and D' are distinct points of $\mathcal{P}\backslash P\zeta$. Repeating the argument above shows that $D\sigma, D'\sigma$, and $P'\sigma$ are all in $\mathcal{P}$.

Finally, suppose that P' is on $\langle A, A' \rangle$, and use the fact that, if D is any point of $\mathcal{P}$ not on $\langle A, A' \rangle$, then $P'\sigma \in \mathcal{P}$. It follows readily that $P'\sigma \subset \mathcal{P}$. □

By Lemma 5.34, all secant lines of $\mathcal{S}$ contain the same number α of points of $\mathcal{S}$; note that $\alpha \ge 2$.

Lemma 5.41. *If $\alpha \ne 2$, there is no point in all tangent hyperplanes of $\mathcal{S}$.*

Proof. Suppose that P belongs to all tangent hyperplanes of $\mathcal{S}$. If $P \in \mathcal{P}$, then $\mathcal{P} = P^{\perp}$, contradicting the non-degeneracy of $\mathcal{S}$; so $P \notin \mathcal{P}$. Let $P', P'' \in \mathcal{P}$, with $P' \not\sim P''$. Then the plane $\pi = \langle P, P', P'' \rangle$ contains no line of $\mathcal{S}$, since otherwise at least one of the tangents $\langle P, P' \rangle, \langle P, P'' \rangle$ contains at least two distinct points of $\mathcal{S}$. Since $P' \not\sim P''$, so $\pi \not\subset \mathcal{S}(P')$; hence $\langle P, P' \rangle$ is the only line of π which is tangent to $\mathcal{S}$ at P'. Counting the points of $\pi \cap \mathcal{P}$ on the lines of π through P' gives

$q(\alpha - 1) + 1$. For each point $T \in \pi \cap \mathcal{P}$, the line $\langle T, P \rangle$ is a tangent of $\mathcal{S}$ at T; so the number of points in $\pi \cap \mathcal{P}$ is at most the number of lines of π through P. Consequently, $q(\alpha - 1) + 1 \leq q + 1$. Hence $\alpha = 2$. $\qquad \square$

Theorem 5.42. *When* $\alpha = 2$, *then* $\mathcal{S}$ *is formed by the points and lines of a nonsingular quadric of* $\mathrm{PG}(d, q)$.

Proof. Each line of $\mathrm{PG}(d, q)$ contains $0, 1, 2$, or $q + 1$ points of $\mathcal{S}$. Since the union of all tangent lines at any point of $\mathcal{P}$ is a hyperplane, $\mathcal{P}$ is a non-singular quadratic set in the sense of Section 1.10. By Theorem 1.97, $\mathcal{P}$ is a non-singular quadric of $\mathrm{PG}(d, q)$, and all lines in $\mathcal{B}$ are lines of the quadric $\mathcal{P}$. Conversely, let l be a line of the quadric $\mathcal{P}$. Since l contains more than $\alpha = 2$ points of $\mathcal{P}$, it is a line of $\mathcal{B}$. Hence $\mathcal{B}$ is the set of all lines of the quadric $\mathcal{P}$. $\qquad \square$

From now on, it is assumed that $\alpha > 2$. By Lemma 5.41, there is no point in all tangent hyperplanes of $\mathcal{S}$.

Lemma 5.43. *If* $\mathcal{S}$ *has rank at least three, then* $P\zeta$ *is a hyperplane for any* P *in* $\mathrm{PG}(d, q)$.

Proof. Suppose that $\mathcal{S}$ contains a plane π. If $P \in \mathcal{P}$, then $P\zeta$ is the hyperplane $\langle P^{\perp} \rangle = \mathcal{S}(P)$. By way of contradiction, let $P \notin \mathcal{P}$ with $P\zeta = \mathrm{PG}(d, q)$. By Lemma 5.37, each line of π has at least one point in $\mathcal{S}_P$. Hence consider in π two points P_1 and P_2 of $\mathcal{S}_P$. By Lemma 5.36, the line $\langle P_1, P_2 \rangle$ is contained in $\mathcal{S}_P$. Hence $\mathcal{S}_P$ contains at least one line of $\mathcal{B}$. Also, the set $\mathcal{S}_P$, together with the lines of $\mathcal{S}$ in $\mathcal{S}_P$, forms a projective Shult space $\mathcal{S}'$ with ambient space $\langle \mathcal{S}_P \rangle = P\zeta = \mathrm{PG}(d, q)$. The Shult space $\mathcal{S}'$ cannot be degenerate, since otherwise $\mathcal{S}_P$ would be contained in a tangent hyperplane of $\mathcal{S}$, contradicting that $\langle \mathcal{S}_P \rangle = \mathrm{PG}(d, q)$. Consequently, the lines of $\mathcal{S}'$ through $T \in \mathcal{S}_P$ generate a hyperplane π' of $\mathrm{PG}(d, q)$. Hence $\pi' = \langle T^{\perp} \rangle$, and so the tangent hyperplanes of $\mathcal{S}'$ are tangent hyperplanes of $\mathcal{S}$.

Let $T \in \mathcal{S}_P$, and consider the secant lines of $\mathcal{S}'$ through T. As the tangent hyperplanes $\mathcal{S}'(T)$ and $\mathcal{S}(T)$ coincide, these secant lines are also the secant lines of $\mathcal{S}$ through T. Hence, by Lemma 5.36, all points of $\mathcal{P}$ non-adjacent to T are in $\mathcal{S}_P$. Next, let $T' \in T^{\perp} \backslash \{T\}$. By Lemma 5.29, there is a point T'' in $T'^{\perp}$ which is not adjacent to T. The line $\langle T', T'' \rangle$ contains q points not adjacent to T. Hence these q points are in $\mathcal{S}_P$. By Lemma 5.36, T' is also in $\mathcal{S}_P$. Consequently, $\mathcal{S}_P = \mathcal{P}$. So P is in all tangent hyperplanes of $\mathcal{S}$, contradicting Lemma 5.41. $\qquad \square$

The next few lemmas show that Lemma 5.43 also holds in the rank 2 case, that is, in the case of a generalised quadrangle.

Lemma 5.44. *If* $\mathcal{S}$ *is a generalised quadrangle, then* $\alpha = (t/q^{d-3}) + 1$ *and* $d = 3$ *or* 4.

Proof. If $\mathcal{S}$ is a generalised quadrangle, then $\mathcal{S}$ has order (q, t). Since $\mathcal{S}$ is non-degenerate, so $d \geq 3$. The secant lines through a point P of $\mathcal{P}$ are the q^{d-1} lines of $\mathrm{PG}(d, q)$ through P which do not lie in the tangent hyperplane $\mathcal{S}(P)$. Hence the total number of points of $\mathcal{S}$ is $(\alpha-1)q^{d-1}+|P^\perp|$. By Theorem 5.12, $|\mathcal{P}| = (1+q)(1+qt)$. Hence $\alpha = (t/q^{d-3})+1$. By Theorem 5.14, $t \le q^2$, so that $2 < \alpha \le (q^2/q^{d-3})+1$; this implies that $d = 3$ or $d = 4$. □

Definition 5.45. A subset $\mathcal{C}$ of $\mathcal{P}$ is called *linearly closed* in $\mathcal{P}$ or $\mathcal{S}$ if, for all points $P, P' \in \mathcal{C}$, with $P \ne P'$, the intersection $\langle P, P' \rangle \cap \mathcal{P}$ is contained in $\mathcal{C}$. Thus any subset $\mathcal{D}$ of $\mathcal{P}$ generates a *linear closure* $\overline{\mathcal{D}}$ in $\mathcal{P}$ or $\mathcal{S}$.

Lemma 5.46. *Let $\mathcal{S}$ be a generalised quadrangle with ambient space* $\mathrm{PG}(3, q)$. *If* $P_1, P_2, P_3 \in \mathcal{P}$ *are non-collinear in* $\mathrm{PG}(3, q)$ *and* $\mathcal{D} = \{P_1, P_2, P_3\}$, *then the linear closure* $\overline{\mathcal{D}} = \mathcal{P} \cap \langle P_1, P_2, P_3 \rangle$.

Proof. Let $\mathcal{S}$ have order (q, t). If the plane $\pi = \langle P_1, P_2, P_3 \rangle$ contains a line of $\mathcal{S}$, then $\mathcal{P} \cap \langle P_1, P_2, P_3 \rangle$ consists of $t+1$ distinct concurrent lines of $\mathcal{B}$. In this case, the lemma follows immediately.

Hence suppose π contains no lines of $\mathcal{B}$. As $d = 3$, by Lemma 5.44, any secant line intersects $\mathcal{P}$ in exactly $t+1$ points. Take a point P, with $P \ne P_1$, on $\langle P_1, P_2 \rangle \cap \mathcal{P}$. The $t+1$ secant lines $\langle P, T \rangle$, where T is a point of $\mathcal{P} \cap \langle P_1, P_3 \rangle$, intersect $\mathcal{P}$ in points which are in the linear closure $\overline{\mathcal{D}}$. As each of these lines $\langle P, T \rangle$ intersects $\mathcal{P}$ in $t+1$ points, there are $t(t+1)+1$ points of $\mathcal{P}$ on these lines. Hence $|\overline{\mathcal{D}}| \ge t^2 + t + 1$.

If Lemma 5.46 were false, there would be a point T' in $(\mathcal{P} \cap \pi) \backslash \overline{\mathcal{D}}$. Then every line of π through T' contains at most one point of $\overline{\mathcal{D}}$; so there are at least $t^2 + t + 1$ lines of π through T'. Hence $t^2 + t + 1 \le q + 1$, and so $t^2 < q$. By Theorem 5.14, it follows that $t = 1$. Consequently, $\alpha = 2$, a contradiction. The conclusion is that $\overline{\mathcal{D}} = \mathcal{P} \cap \langle P_1, P_2, P_3 \rangle$. □

Lemma 5.47. *Let $\mathcal{S}$ be a generalised quadrangle with ambient space* $\mathrm{PG}(4, q)$. *If* $P_1, P_2, P_3 \in \mathcal{P}$ *are non-collinear in* $\mathrm{PG}(4, q)$ *and* $\mathcal{D} = \{P_1, P_2, P_3\}$, *then the linear closure* $\overline{\mathcal{D}} = \mathcal{P} \cap \langle P_1, P_2, P_3 \rangle$.

Proof. Let $\mathcal{S}$ have order (q, t). As before, suppose that $\pi = \langle P_1, P_2, P_3 \rangle$ contains no line of $\mathcal{S}$. Fix a point $P \in \mathcal{P} \cap \pi$ and a line $l \in \mathcal{B}$ incident with P. Also, let $\pi' = \langle P_1, P_2, P_3, l \rangle$. By Lemma 5.24, $\pi' \cap \mathcal{S}$ is a Shult space with ambient space π'. For $\pi' \cap \mathcal{S}$ there are the following possibilities:

(a) $\pi' \cap \mathcal{S}$ is non-degenerate and then, by Theorem 5.20, $\pi' \cap \mathcal{S}$ is a projective subquadrangle of $\mathcal{S}$;

(b) $\pi' \cap \mathcal{S}$ is degenerate, and the lines of $\pi' \cap \mathcal{S}$ contain a distinguished point of π'.

If (a) holds, then, by Lemma 5.46, $\mathcal{P} \cap \pi$ is the linear closure of $\mathcal{D}$ in $\mathcal{S}$. If (b) holds, two cases are possible.

  (i) There exists a line l' in $\mathcal{B}$ through a point of π such that $\langle \pi, l' \rangle$ intersects $\mathcal{S}$ in a subquadrangle. Then Lemma 5.46 still applies.

  (ii) For each line l' of $\mathcal{B}$ intersecting π, the lines of $\langle \pi, l' \rangle \cap \mathcal{S}$ all contain a point T_i of l' not in π. Here the hyperplane $\langle \pi, l' \rangle$ is the tangent hyperplane of $\mathcal{S}$ at T_i. Hence π contains $1 + t$ points of $\mathcal{P}$: these are $P_1, P_2, \ldots, P_{t+1}$, and $P_j \sim T_i$ for all i and j. Further, by the definition of the points T_i, the lines of $\mathcal{S}$ through a given point P_j are the lines $\langle P_j, T_i \rangle$. Hence there are exactly $1 + t$ points T_i. This means $\mathcal{S}$ has two disjoint sets $\{P_j\}, \{T_i\}$ of $1 + t$ pairwise-non-adjacent points with $P_j \sim T_i$, $0 \leq i, j \leq t$. If $T' \in \langle P_j, T_i \rangle \cap \langle P_{j'}, T_{i'} \rangle$, with $j \neq j'$ and $i \neq i'$, then a triangle $P_j T_{i'} T'$ is obtained in $\mathcal{S}$, a contradiction, since $\mathcal{S}$ is a generalised quadrangle. Hence there are exactly $(q-1)(t+1)^2 + 2(t+1)$ points of $\mathcal{S}$ on the lines $\langle P_j, T_i \rangle$. Since

$$(q-1)(t+1)^2 + 2(t+1) \leq |\mathcal{P}| = (q+1)(qt+1),$$

so $(t - q)t(q - 1) \leq 0$, whence $t \leq q$. Then $\alpha = (t/q) + 1 \leq 2$, contradicting that $\alpha > 2$.   □

Lemma 5.48. *Let $\{P_1, P_2, \ldots, P_k\}$ be a set of points of the generalised quadrangle $\mathcal{S}$. Then the linear closure of $\{P_1, P_2, \ldots, P_k\}$ in $\mathcal{S}$ is $\mathcal{P} \cap \langle P_1, P_2, \ldots, P_k \rangle$.*

Proof. First note that if $q = 2$ then, since $\alpha > 2$, any line of $PG(d, q)$ containing at least two points of $\mathcal{P}$ is entirely contained in $\mathcal{P}$. Hence $\mathcal{P}$ is a subspace of $PG(d, q)$ and $\mathcal{P} = PG(d, q)$. Consequently, in this case, the lemma follows.

Now let $q > 2$. The result is immediate when $k = 1$ or $k = 2$. By Lemmas 5.46 and 5.47, the result also holds if $\langle P_1, P_2, \ldots, P_k \rangle$ is a plane. Further, it may be assumed that the points P_i are linearly independent in $PG(d, q)$. Now apply induction; so suppose the result is true for $k - 1$ points $P_1, P_2, \ldots, P_{k-1}$, $3 \leq k - 1$. For $\mathcal{D} = \{P_1, P_2, \ldots, P_k\}$, with $P_k \in \mathcal{P} \backslash \langle P_1, P_2, \ldots, P_{k-1} \rangle$, indices can be chosen so that $\langle P_1, P_2, \ldots, P_{k-1} \rangle \not\subset \mathcal{S}(P_1)$. Put $l_i = \langle P_1, P_i \rangle$, $i = 2, 3, \ldots, k$, and let π be any plane through l_k contained in $\langle l_2, l_3, \ldots, l_k \rangle = \langle P_1, P_2, \ldots, P_k \rangle$. Then π intersects $\langle l_2, l_3, \ldots, l_{k-1} \rangle$ in a line l.

If it is shown that $\mathcal{P} \cap \langle l_k, l \rangle \subset \overline{\mathcal{D}}$, the desired result follows immediately. Suppose l contains at least two points of $\mathcal{P}$. By the induction hypothesis, the points of $\mathcal{P}$ on l are all in $\overline{\{P_1, P_2, \ldots, P_{k-1}\}}$. Lemmas 5.46 and 5.47 then show that $\mathcal{P} \cap \langle l_k, l \rangle$ is in $\overline{\mathcal{D}}$. Now suppose that l is a tangent line whose points are not all in $\mathcal{P}$. If $\langle l_k, l \rangle$ contains no point of $\mathcal{P}$ not on l_k, there is nothing more to show.

So suppose P is a point of $\mathcal{P} \cap \langle l_k, l \rangle$ but not on l_k. Consider the plane π' generated by l and a secant through P_1 in the space $\langle l_2, l_3, \ldots, l_{k-1} \rangle$; such a secant exists since $\mathcal{S}(P_1)$ does not contain $\langle P_1, P_2, \ldots, P_{k-1} \rangle$. This plane π' is not in the tangent hyperplane $\mathcal{S}(P_1)$; so l is the unique tangent at P_1 in π'. Hence there are two secants m_1, m_2 in π' and through P_1. Each of the planes $\langle l_k, m_1 \rangle, \langle l_k, m_2 \rangle$ is not in $\mathcal{S}(P_1)$ and hence contains exactly one tangent at P_1. Consider in $\langle l_k, m_1 \rangle$ a secant m, with $m \neq l_k$, such that the plane $\langle m, P \rangle$ intersects $\langle l_k, m_2 \rangle$ in a secant m'. Note that m exists, because $\langle l_k, m_1 \rangle$ has at least four lines through P_1. By the induction hypothesis, the points of $\mathcal{P}$ on m_1 and m_2 belong to $\overline{\{P_1, P_2, \ldots, P_{k-1}\}}$. Hence, by Lemmas 5.46 and 5.47, the points of $\mathcal{P}$ on m and m' belong to $\overline{\mathcal{D}}$. But, as $P \in \langle m, m' \rangle$, again by Lemmas 5.46 and 5.47, $P \in \overline{\mathcal{D}}$.   □

Lemma 5.49. *If $\mathcal{S}$ is a generalised quadrangle, then $P\zeta$ is a hyperplane for any P in $PG(d, q)$.*

Proof. If $P \in \mathcal{P}$, then $P\zeta$ is the hyperplane $\langle P^{\perp} \rangle = \mathcal{S}(P)$. So suppose $P \notin \mathcal{P}$. Consider the intersection $P\zeta \cap \mathcal{P}$. By Lemmas 5.36 and 5.48, all points of $P\zeta \cap \mathcal{P}$

are in $\mathcal{S}_P$, implying that $\mathcal{S}_P = P\zeta \cap \mathcal{P}$. If $P\zeta$ were not a hyperplane, then, by Lemma 5.38, $P\zeta = \mathrm{PG}(d,q)$, implying $\mathcal{S}_P = \mathcal{P}$. Hence P belongs to all hyperplanes of $\mathcal{S}$, contradicting Lemma 5.41. Therefore $P\zeta$ is a hyperplane. $\qquad\square$

Theorem 5.50. (i) *The mapping* $P \mapsto P\zeta$ *is a polarity of* $\mathrm{PG}(d,q)$;
(ii) $\mathcal{P}$ *is the set of all self-conjugate points of* ζ;
(iii) $\mathcal{B}$ *consists of all lines* l *of* $\mathrm{PG}(d,q)$ *such that* $l \subset l\zeta$.

Proof. First, it is shown that $P \to P\zeta$ defines a bijection from the set of all points of $\mathrm{PG}(d,q)$ onto the set of all hyperplanes of $\mathrm{PG}(d,q)$. By Lemmas 5.43 and 5.49, $P\zeta$ is a hyperplane for any $P \in \mathrm{PG}(d,q)$. Assume that $P \neq P'$ and $P\zeta = P'\zeta$. Let $X \in \langle P, P'\rangle$ and $Y \in \mathcal{S}_P = \mathcal{P} \cap P\zeta = \mathcal{P} \cap P'\zeta = \mathcal{S}_{P'}$, with $Y \neq X, P, P'$. The lines $\langle Y, P\rangle$ and $\langle Y, P'\rangle$ are tangents of $\mathcal{S}$; hence the line $\langle Y, X\rangle$ is also a tangent of $\mathcal{S}$. If $Y = P$ and $Y \neq X$, then the line $\langle Y, X\rangle = \langle Y, P'\rangle$ is a tangent of $\mathcal{S}$. Analogously, if $Y = P'$ and $Y \neq X$, then the line $\langle Y, X\rangle = \langle Y, P\rangle$ is a tangent of $\mathcal{S}$. Hence $\mathcal{S}_P \subset \mathcal{S}_X$ for any $X \in \langle P, P'\rangle$. Consequently, $P\zeta = \langle \mathcal{S}_P\rangle \subset \langle \mathcal{S}_X\rangle = X\zeta$; so $P\zeta = X\zeta$ since $P\zeta$ and $X\zeta$ are hyperplanes. Let $P'' \in \mathcal{P}\backslash P\zeta$; then $P \neq P'' \neq P'$. Let X be the common point of $\langle P, P'\rangle$ and the hyperplane $P''\zeta$. Then $X = P''$, or $\langle X, P''\rangle$ is a tangent; so $P'' \in X\zeta$. Hence $P'' \in P\zeta = X\zeta$, a contradiction. It follows that $P \neq P'$ implies that $P\zeta \neq P'\zeta$. Since $\mathrm{PG}(d,q)$ is finite, so $P \to P\zeta$ defines a bijection of the set of all points of $\mathrm{PG}(d,q)$ onto the set of all hyperplanes of $\mathrm{PG}(d,q)$.

Next, suppose that $P' \in P\zeta$, with $P \neq P'$. Now, let $A \in \mathcal{P}\backslash P\zeta$, and also let $A' \in \langle P, A\rangle \cap \mathcal{P}$ with $A' \neq P$ and $A' \neq A$. By Lemma 5.40, the perspectivity σ with centre P and axis $P\zeta$ mapping A onto A' leaves $\mathcal{P}$ invariant. Since P' is on the axis of σ, so $P'\sigma = P'$; hence $P'\zeta\sigma = P'\zeta$. From above, $P\zeta \neq P'\zeta$; so $P'\zeta$ contains the centre P of σ. Therefore it has been shown that $P' \in P\zeta$ implies that $P \in P'\zeta$. This means that ζ is a polarity of $\mathrm{PG}(d,q)$.

Let $\mathcal{S}'$ be the Shult space defined by the polarity ζ. The point set of $\mathcal{S}'$ is denoted by $\mathcal{P}'$ and the line set of $\mathcal{S}'$ by $\mathcal{B}'$. If $P \in \mathcal{P}$, then $P \in P\zeta$ and hence $P \in \mathcal{P}'$; so $\mathcal{P} \subset \mathcal{P}'$. Also $\mathcal{B} \subset \mathcal{B}'$. Assume that $P' \in \mathcal{P}'\backslash\mathcal{P}$ and let T, T' be distinct points of $\mathcal{S}_{P'}$.

First, suppose that the plane $\langle P', T, T'\rangle = \pi$ contains no lines of $\mathcal{B}$. Then the plane $\pi \not\subset \mathcal{S}(T)$ since $T \not\sim T'$; so $\langle P', T\rangle$ is the only line of π tangent to $\mathcal{S}$ at T. A count of the points of $\pi \cap \mathcal{P}$ on the lines of π through T gives $q(\alpha - 1) + 1$. Since $P' \in \mathcal{P}'$, the lines $\langle P', T\rangle$ and $\langle P', T'\rangle$ are contained in $P'\zeta$; so, for each point $A \in \pi \cap \mathcal{P}$, the line $\langle P', A\rangle$ is contained in $P'\zeta$. Consequently, $\langle P', A\rangle$ is tangent to $\mathcal{S}$ at A. So the number of points in $\pi \cap \mathcal{P}$ is at most the number of lines of π through P'. Therefore $q(\alpha - 1) + 1 \leq q + 1$, whence $\alpha = 2$, a contradiction.

Secondly, suppose that the plane π contains a line l of $\mathcal{B}$. Since $\langle P', T\rangle$ and $\langle P', T'\rangle$ are tangents of $\mathcal{S}$, it follows that $l = \langle T, T'\rangle$. Hence any two points of $\mathcal{S}_{P'}$ are adjacent in $\mathcal{S}$. By Lemma 5.36, $T^\perp \supset \mathcal{S}_{P'}$. Hence $T\zeta \supset P'\zeta$ and so $T\zeta = P'\zeta$, giving $T = P'$, a contradiction. Thus $\mathcal{P}' = \mathcal{P}$. Next, let $l' \in \mathcal{B}'$. If D, D' are distinct points of l', then D' belongs to the tangent hyperplane $D\zeta$ of $\mathcal{S}'$ at D. As $D' \in D\zeta$ and $D, D' \in \mathcal{P}$, the line $l' = \langle D, D'\rangle$ belongs to $\mathcal{B}$. Therefore $\mathcal{B} = \mathcal{B}'$. Hence

$S = S'$, which means that $\mathcal{P}$ is the set of all self-conjugate points of the polarity ζ and that $\mathcal{B}$ consists of all lines l with $l \subset l\zeta$. □

Theorem 5.51. *Let S be a non-degenerate projective Shult space with ambient space* $\mathrm{PG}(d,q)$. *Then one of the following holds:*

(a) S *is formed by the points and lines of a non-singular quadric of* $\mathrm{PG}(d,q)$;
(b) q *is a square and S is formed by the points and lines of a non-singular Hermitian variety of* $\mathrm{PG}(d,q)$;
(c) d *is odd, the points of S are the points of* $\mathrm{PG}(d,q)$, *and the lines of S are the lines of* $\mathrm{PG}(d,q)$ *in the self-polar* $(d-1)/2$-*dimensional spaces with respect to some null polarity ζ of* $\mathrm{PG}(d,q)$.

Proof. If $\alpha = 2$, then, by Theorem 5.42, it is case (a). If $\alpha > 2$, then, by Theorem 5.50, it is one of the cases (b), (c). □

Theorem 5.52. *Let S be a projective Shult space with ambient space* $\mathrm{PG}(d,q)$. *Then S is one of the following types.*

(a) S *is formed by the points and the lines of* $\mathrm{PG}(d,q), d \geq 1$. *The radical $\mathcal{R}$ of S is* $\mathrm{PG}(d,q)$.
(b) *The point set of S is the union of k spaces* $\mathrm{PG}(r+1,q)$ *through a* $\mathrm{PG}(r,q)$, *where $k > 1$ and $r \geq 0$. The line set of S is the set of all lines in these* $(r+1)$-*spaces. The radical $\mathcal{R}$ of S is* $\mathrm{PG}(r,q)$.
(c) S *is formed by the points and lines of a quadric $\mathcal{Q}$ of projective index at least one of* $\mathrm{PG}(d,q)$, $d \geq 3$. *The radical $\mathcal{R}$ of S is the space of all singular points of $\mathcal{Q}$.*
(d) *The order q is a square and S is formed by the points and lines of a Hermitian variety $\mathcal{U}$ of* $\mathrm{PG}(d,q)$, $d \geq 3$. *The radical $\mathcal{R}$ of S is the space of all singular points of $\mathcal{U}$.*
(e) *The points of S are the points of* $\mathrm{PG}(d,q)$, $d \geq 3$. *There are skew subspaces Π_r and Π_{d-r-1}, with $r \geq -1$, and $d - r - 1$ odd and at least three; the lines of S are all the lines of* $\mathrm{PG}(d,q)$ *in the* $(r+2)$-*spaces joining Π_r to the lines of Π_{d-r-1} in the self-polar* $(d-r-2)/2$-*spaces of some null polarity ζ in Π_{d-r-1}. The radical $\mathcal{R}$ of S is Π_r.*

Proof. First, let S be non-degenerate. By Theorem 5.51, it is either case (c) with $\mathcal{R} = \emptyset$, or case (d) with $\mathcal{R} = \emptyset$, or case (e) with $\mathcal{R} = \emptyset$.

Next, let S be degenerate with radical $\mathcal{R}$ and let $\mathcal{R} = \mathrm{PG}(d,q)$. Then case (a) holds.

Now, let S be degenerate with radical $\mathcal{R} = \Pi_r$ and $-1 < r < d$. Let Π_{d-r-1} be a subspace of $\mathrm{PG}(d,q)$ skew to $\mathcal{R}$. By Theorem 5.22, $\Pi_{d-r-1} \cap S$ is a non-degenerate Shult space, the point set of S is the union of all lines joining every point of $\mathcal{R}$ to every point of $\Pi_{d-r-1} \cap S$, and two points of $\mathcal{P} \backslash \mathcal{R}$ are adjacent if and only if their projections from $\mathcal{R}$ onto Π_{d-r-1} are adjacent. If $\Pi_{d-r-1} \cap S$ contains at least one line, then $\Pi_{d-r-1} \cap S$ is a non-degenerate projective Shult space with ambient space Π_{d-r-1}. In this case, by Theorem 5.51, one of (c), (d), (e) with $\mathcal{R} \neq \emptyset$ occurs.

Finally, suppose that $\Pi_{d-r-1} \cap S$ contains no line; this gives case (b). □

Definition 5.53. Consider a pair $\mathcal{S} = (\mathcal{P}, \mathcal{B})$, where $\mathcal{P}$ is a non-empty point set of $PG(d, q)$ and $\mathcal{B}$ is a (possibly empty) line set of $PG(d, q)$. If $\mathcal{B} \neq \emptyset$, then let $\mathcal{P}$ be the union of all lines of $\mathcal{B}$.

(1) The subspace $\Pi_{d'}$ of $PG(d, q)$ generated by all points of $\mathcal{P}$ is the *ambient space* of $\mathcal{S}$.

(2) A *tangent* to $\mathcal{S}$ at $P \in \mathcal{P}$ is any line l through P such that either $l \in \mathcal{B}$ or $l \cap \mathcal{P} = \{P\}$.

(3) The union of all tangents to $\mathcal{S}$ at P is the *tangent set of $\mathcal{S}$ at P*, and is denoted $\mathcal{S}(P)$.

(4) The set $\mathcal{S}$ is a *semi-quadratic set* of $PG(d, q)$, $d \geq 2$, if $PG(d, q)$ is the ambient space of $\mathcal{S}$ and if, for each $P \in \mathcal{P}$, the tangent set $\mathcal{S}(P)$ is either a hyperplane or $PG(d, q)$.

(5) If $\mathcal{S}(P) = PG(d, q)$, then P is a *singular* point of $\mathcal{S}$.

(6) The set of all singular points of $\mathcal{S}$ is the *radical* $\mathcal{R}$ of $\mathcal{S}$.

(7) A semi-quadratic set $\mathcal{S}$ of $PG(d, q)$ is a *semi-ovaloid* if $\mathcal{B} = \emptyset$; in this case $\mathcal{R} = \emptyset$.

Theorem 5.54. *The pair $\mathcal{S} = (\mathcal{P}, \mathcal{B})$ is a semi-quadratic set of $PG(d, q)$, $d \geq 2$, if and only if one of the following holds.*

(a) *$\mathcal{S}$ is of type (a), (c), (d), (e) in the statement of Theorem 5.52. In each of these cases the radical of the Shult space $\mathcal{S}$ coincides with the radical of the semi-quadratic set $\mathcal{S}$.*

(b′) (1) *The point set of $\mathcal{S}$ is the union of k spaces Π_{r+1} through a Π_r, where $k > 1$ and $d - 3 \geq r \geq -1$.*

(2) *The line set of $\mathcal{S}$ is the set of all lines in these $(r + 1)$-spaces.*

(3) *If Π_{d-r-1} is skew to Π_r, then $\mathcal{P} \cap \Pi_{d-r-1}$ is a semi-ovaloid of Π_{d-r-1}.*

(4) *The radical $\mathcal{R}$ of $\mathcal{S}$ is Π_r.*

Proof. Let $\mathcal{S} = (\mathcal{P}, \mathcal{B})$ be a semi-quadratic set of $PG(d, q)$, $d \geq 2$. It is shown that $\mathcal{S}$ is a Shult space. Let $P \in \mathcal{P}, l \in \mathcal{B}$, and $P \notin l$. Since the tangent set $\mathcal{S}(P)$ is a hyperplane or $PG(d, q)$ itself, so $|l \cap \mathcal{S}(P)| = 1$ or $l \subset \mathcal{S}(P)$. If $P' \in l \cap \mathcal{S}(P)$, then $\langle P, P' \rangle$ is a line of $\mathcal{B}$. Hence P is adjacent to one point or to all points of l. It follows that $\mathcal{S}$ is a Shult space.

First, assume that $\mathcal{B} \neq \emptyset$. Then $\mathcal{S}$ is a projective Shult space with ambient space $PG(d, q)$. So one of the cases (a) to (e) in the statement of Theorem 5.52 occurs. Conversely, each Shult space of type (a), (c), (d), or (e) is a semi-quadratic set. Now consider a Shult space of type (b), and let Π_{d-r-1} be skew to Π_r, $r \geq 0$. Then $\mathcal{S}$ is a semi-quadratic set if and only if $\mathcal{P} \cap \Pi_{d-r-1}$ is a semi-ovaloid of Π_{d-r-1} with $d - r - 1 \geq 2$. In each of these cases the radical of the Shult space $\mathcal{S}$ coincides with the radical of the semi-quadratic set $\mathcal{S}$.

Secondly, assume that $\mathcal{B} = \emptyset$. Then the semi-quadratic set $\mathcal{S}$ is a semi-ovaloid of $PG(d, q)$. This gives case (b′) with $r = -1$. $\qquad\square$

Let $\mathcal{S} = (\mathcal{P}, \emptyset)$ be a semi-ovaloid of $PG(d, q), d \geq 2$. Any tangent of $\mathcal{S}$ has exactly one point in common with $\mathcal{S}$, and the tangent set $\mathcal{S}(P)$ of $\mathcal{S}$ at $P \in \mathcal{P}$ is always a hyperplane. In the next theorem all semi-ovaloids are classified.

Theorem 5.55. *Let* $\mathcal{S} = (\mathcal{P}, \emptyset)$ *be a semi-ovaloid of* $\mathrm{PG}(d, q)$, *then there are only two possibilities.*

(a) $d = 2$ *and* $q + 1 \leq |\mathcal{P}| \leq q\sqrt{q} + 1$. *If* $|\mathcal{P}| = q + 1$, *then* $\mathcal{P}$ *is a* $(q + 1)$-*arc of* $\mathrm{PG}(2, q)$; *if* $|\mathcal{P}| = q\sqrt{q} + 1$, *then* $\mathcal{P}$ *is a Hermitian arc of* $\mathrm{PG}(2, q)$.
(b) $d = 3$ *and* $|\mathcal{P}| = q^2 + 1$. *For* $q > 2$, *the set* $\mathcal{P}$ *is an ovaloid of* $\mathrm{PG}(3, q)$; *for* $q = 2$, *the set* $\mathcal{P}$ *is an elliptic quadric of* $\mathrm{PG}(3, 2)$.

Proof. Let $\mathcal{S} = (\mathcal{P}, \emptyset)$ be a semi-ovaloid of $\mathrm{PG}(2, q)$. Let $P \in \mathcal{P}$ and let l be the tangent of $\mathcal{S}$ at P. If the line m contains P and $m \neq l$, then $|m \cap \mathcal{P}| > 1$. Counting the points of $\mathcal{P}$ on the lines through P gives $|\mathcal{P}| \geq q + 1$. If $|\mathcal{P}| = q + 1$, then each non-tangent contains exactly zero or two points of $\mathcal{P}$, whence $\mathcal{P}$ is a $(q + 1)$-arc of $\mathrm{PG}(2, q)$. Conversely, any $(q + 1)$-arc of $\mathrm{PG}(2, q)$ is a semi-ovaloid of $\mathrm{PG}(2, q)$.

Next, let $\mathcal{S} = (\mathcal{P}, \emptyset)$ be a semi-ovaloid of $\mathrm{PG}(d, q), d > 2$. Let l be a tangent of $\mathcal{S}$ at $P \in \mathcal{P}$. If π is a plane through l which does not belong to the tangent set $\mathcal{S}(P)$, then $\pi \cap \mathcal{S}$ is a semi-ovaloid of π. Counting the points of $\mathcal{S}$ in the planes through l gives

$$|\mathcal{P}| = \sum_{\pi}(|\mathcal{P} \cap \pi| - 1) + 1 \geq q \cdot q^{d-2} + 1 = q^{d-1} + 1.$$

Let $\mathcal{S} = (\mathcal{P}, \emptyset)$ be a semi-ovaloid of $\mathrm{PG}(d, q), d \geq 2$. From the first part of the proof,

$$|\mathcal{P}| \geq q^{d-1} + 1. \tag{5.3}$$

Let $\mathcal{P} = \{P_1, P_2, \ldots, P_\alpha\}$ and $\mathrm{PG}(d, q) \backslash \mathcal{P} = \{T_1, T_2, \ldots, T_\beta\}$, with

$$\alpha + \beta = (q^{d+1} - 1)/(q - 1).$$

Further, let t_i be the number of tangents of $\mathcal{S}$ through $T_i, i = 1, 2, \ldots, \beta$. Now, count in different ways the number of ordered pairs (T_i, P_j), where $\langle T_i, P_j \rangle$ is a tangent of $\mathcal{S}$; this gives

$$\sum_{i=1}^{\beta} t_i = \alpha q \sum_{k=0}^{d-2} q^k. \tag{5.4}$$

Next, count in two ways the number of ordered triples $(T_i, P_j, P_{j'})$, with $P_j \neq P_{j'}$ and $\langle T_i, P_j \rangle, \langle T_i, P_{j'} \rangle$ tangents to $\mathcal{S}$. Hence

$$\sum_{i=1}^{\beta} t_i(t_i - 1) = \alpha(\alpha - 1) \sum_{k=0}^{d-2} q^k. \tag{5.5}$$

From (5.4) and (5.5), it follows that

$$\sum_{i=1}^{\beta} t_i^2 = \alpha(\alpha + q - 1) \sum_{k=0}^{d-2} q^k.$$

With $\beta \bar{t} = \sum_{i=1}^{\beta} t_i$, the inequality $0 \leq \sum_{i=1}^{\beta}(\bar{t} - t_i)^2$ simplifies to

$$\beta \sum_{i=1}^{\beta} t_i^2 - \left(\sum_{i=1}^{\beta} t_i\right)^2 \geq 0;$$

this implies that

$$\beta\alpha(\alpha+q-1)\sum_{k=0}^{d-2} q^k - \alpha^2 q^2 \left(\sum_{k=0}^{d-2} q^k\right)^2 \geq 0.$$

As $\beta = \sum_{k=0}^{d} q^k - \alpha$, manipulation gives

$$(\alpha - 1)^2 \leq q^{d+1}. \tag{5.6}$$

Next, let $d = 2$. From (5.3) and (5.6) it follows that

$$q + 1 \leq \alpha \leq q\sqrt{q} + 1.$$

If $q + 1 = \alpha$, then in the first part of the proof it was shown that $\mathcal{P}$ is a $(q + 1)$-arc. Let $\alpha = q\sqrt{q} + 1$. Then

$$0 = \sum_{i=1}^{\beta} (\bar{t} - t_i)^2;$$

so $t_i = \bar{t} = (\sum_{i=1}^{\beta} t_i)/\beta = \alpha q/\beta = \sqrt{q} + 1, i = 1, 2, \ldots, \beta$. From Section 12.3 of PGOFF2, the set of all tangents of $\mathcal{S}$ forms a dual Hermitian arc of $\mathrm{PG}(2,q)$, and hence $\mathcal{P}$ is a Hermitian arc of $\mathrm{PG}(2,q)$.

Next, let $d = 3$. From (5.3) and (5.6), it follows that $\alpha = q^2 + 1$. Any line l through $P_i \in \mathcal{P}$, but not in $\mathcal{S}(P_i)$, contains at least one point of $\mathcal{P}\backslash\{P_i\}$. Since there are q^2 such lines l through P_i and $|\mathcal{P}\backslash\{P_i\}| = q^2$, it follows that l contains exactly two points of $\mathcal{P}$. Hence $\mathcal{P}$ is a $(q^2 + 1)$-cap of $\mathrm{PG}(3,q)$. If $q > 2$, then $\mathcal{P}$ is an ovaloid; if $q = 2$, then the 5-cap $\mathcal{P}$ is an elliptic quadric of $\mathrm{PG}(3,2)$.

Finally, let $d > 3$; then (5.3) contradicts (5.6). □

Let $\mathcal{S} = (\mathcal{P}, \emptyset)$ be a semi-ovaloid of the plane $\mathrm{PG}(2,q)$. If $P \in \mathcal{P}$ and all non-tangents through P intersect $\mathcal{P}$ in more than two points, then $\mathcal{S}' = (\mathcal{P}\backslash\{P\}, \emptyset)$ is still a semi-ovaloid of $\mathrm{PG}(2,q)$. In $\mathrm{PG}(2,3)$, there is a class of semi-ovaloids with six points: take the vertices of a quadrangle together with two of its diagonal points.

5.4 Lax and polarised embeddings of Shult spaces

Definition 5.56. (1) A Shult space $\mathcal{S}$ with point set $\mathcal{P}$ is *laxly embedded* in $\mathrm{PG}(d, K)$, $d \geq 2$, if, for $\mathcal{P} \subset \mathrm{PG}(d, K)$, each line l of $\mathcal{S}$ is a subset of a line l' of $\mathrm{PG}(d, K)$, and distinct lines l_1, l_2 of $\mathcal{S}$ define distinct lines l_1', l_2' of $\mathrm{PG}(d, K)$.
(2) If $\Pi_{d'}$ is the subspace of $\mathrm{PG}(d, K)$ generated by all points of $\mathcal{P}$, then $\Pi_{d'}$ is the *ambient space* of $\mathcal{S}$.

(3) A lax embedding is *full* if, in (1), $l = l'$ for each line l of S. The embeddings considered in Section 5.3 are full embeddings.

(4) The embeddings described in Theorem 5.51 are the *natural embeddings* of the classical finite non-degenerate Shult spaces.

(5) A lax embedding in $PG(d, q)$ of a Shult space S, with point set P, is *weak* or *polarised* if, for any point P of S, the subspace generated by the set

$$\mathcal{A} = \{P' \in S \mid P' \text{ is collinear with } P\}$$

meets P precisely in $\mathcal{A}$.

Remark 5.57. If the non-degenerate Shult space S is isomorphic to the Shult space arising from a non-singular parabolic quadric P_n in $PG(n, q)$, n even and $n \geq 4$, then S has two natural embeddings which are not projectively equivalent: the points and lines of P_n and the points of $PG(n-1, q)$ together with the lines of $PG(n-1, q)$ in the self-polar $(n-2)/2$-dimensional spaces with respect to some null polarity ζ of $PG(n-1, q)$.

It can be shown that, for a non-degenerate Shult space S laxly embedded in $PG(d, q)$, (5) is equivalent to the following condition (5′):

(5′) the set of all points of S collinear in S with any given point of S is contained in a hyperplane of $PG(d, q)$.

All full embeddings described in Theorem 5.52 are polarised and all full embeddings described in Theorem 5.51 satisfy conditions (5) and (5′).

In what follows, 'embedded in $PG(d, K)$' means that $PG(d, K)$ is the ambient space.

The next three theorems contain the complete classification of all Shult spaces weakly embedded in $PG(d, q)$. However, proofs are not given. First, the universal weak embedding of the generalised quadrangle $W(2)$ in $PG(4, K)$ is described.

Let P_1, P_2, P_3, P_4, P_5 be consecutive vertices of a proper pentagon in $W(2)$. Let K be any field and identify P_i, $i \in \{1, 2, 3, 4, 5\}$, with the point of $PG(4, K)$ with coordinates $(0, \ldots, 0, 1, 0, \ldots, 0)$, where the 1 is in the i-th position. Identify the unique point Q_{i+3} of $W(2)$ on the line $P_i P_{i+1}$ and different from both P_i and P_{i+1}, with the point $(0, \ldots, 0, 1, 1, 0, \ldots, 0)$ of $PG(4, K)$, where the 1's are in the i-th and the $(i+1)$-th positions, and where subscripts are taken modulo 5. Finally, identify the unique point R_i of the line $P_i Q_i$ of $W(2)$ and different from both P_i and Q_i, with the point whose coordinates are all 0 except in the i-th position, where the coordinate is -1, and in the positions $i-2$ and $i+2$, where it takes the value 1; again subscripts are taken modulo 5. It is an elementary exercise to check that this defines a weak embedding of $W(2)$ in $PG(4, K)$. This embedding is the *universal weak embedding* of $W(2)$ in $PG(4, K)$.

Theorem 5.58. *Let S be a generalised quadrangle of order (s, t), $s, t \neq 1$, weakly embedded in $PG(d, q)$. Then either s is a prime power, $\mathbf{F}_s$ is a subfield of $\mathbf{F}_q$ and S is fully embedded in some subgeometry $PG(d, s)$ of $PG(d, q)$, or S is isomorphic to $W(2)$ and the weak embedding is the universal one in $PG(4, q)$ with q odd.*

Theorem 5.59. *Let S be a non-degenerate Shult space of rank at least three, all of whose lines have size at least three. If S is weakly embedded in $\mathrm{PG}(d, q)$, then S is fully embedded in some subgeometry $\mathrm{PG}(d, q')$ of $\mathrm{PG}(d, q)$, for some subfield $\mathbf{F}_{q'}$ of $\mathbf{F}_q$.*

Let S be a finite Shult space of rank at least three all of whose lines have size at least three. Then the radical $\mathcal{R}$ of S together with the lines of S in $\mathcal{R}$ is the point-line incidence structure of a projective space over some field $\mathbf{F}_{q'}$. Let the dimension of $\mathcal{R}$ be denoted by $r(S)$, or r for short.

Definition 5.60. A Shult space is *classical* if it arises, up to isomorphism, from a quadric, a Hermitian variety, or a null polarity.

Theorem 5.61. *Let S be a classical Shult space with $\mathrm{rank}(S) = R$ and satisfying $R - r \geq 4$. If S is weakly embedded in $\mathrm{PG}(d, q)$, then there is a projective space $\mathrm{PG}(\bar{d}, q)$, $\bar{d} \geq d$, containing $\mathrm{PG}(d, q)$ such that S is the projection from a $\Pi_{\bar{d}-d-1,q} \subset \mathrm{PG}(\bar{d}, q)$ into $\mathrm{PG}(d, q)$ of a classical Shult space $\overline{S}$ which is fully embedded in some subgeometry $\mathrm{PG}(\bar{d}, q')$ of $\mathrm{PG}(\bar{d}, q)$, for some subfield $\mathbf{F}_{q'}$ of $\mathbf{F}_q$.*

Surprisingly, also for the weakest form of embeddings, the lax embeddings, strong results are obtained. First, a lax embedding of the generalised quadrangle $\mathcal{U}(3, 4)$ arising from a non-singular Hermitian variety $\mathcal{U}$ in $\mathrm{PG}(3, 4)$ is described.

A *double-six* in $\mathrm{PG}(3, K)$, with K any field, is a set of 12 lines

$$a_1 \ a_2 \ a_3 \ a_4 \ a_5 \ a_6$$
$$b_1 \ b_2 \ b_3 \ b_4 \ b_5 \ b_6$$

such that each line meets only the five lines not in the same row or column. A double-six lies on a unique non-singular cubic surface $\mathcal{F}$, which contains 15 further lines. Any non-singular cubic surface $\mathcal{F}$ of $\mathrm{PG}(3, \overline{K})$, with $\overline{K}$ an algebraically closed extension of K, contains exactly 27 lines. These 27 lines form exactly 36 double-sixes. With the notation introduced above, there exists a unique polarity β of $\mathrm{PG}(3, K)$ such that $a_i \beta = b_i$, $i = 1, 2, \ldots, 6$. As the other 15 lines of the corresponding cubic surface are the lines $c_{ij} = \langle a_i, b_j \rangle \cap \langle a_j, b_i \rangle$, with $i, j = 1, 2, \ldots, 6$ and $i \neq j$, so $c_{ij} \beta = \langle a_i \cap b_j, a_j \cap b_i \rangle$. For every double-six, any line l of it together with the five lines different from l and concurrent with l, form a set of six lines every five of which are linearly independent when regarded as six points on the Klein quadric.

Conversely, given five skew lines a_1, a_2, a_3, a_4, a_5 with a transversal b_6 such that each five of the six lines are linearly independent, then the six lines belong to a unique double-six, and so belong to a unique, non-singular cubic surface. A double-six and a cubic surface with 27 lines exist in $\mathrm{PG}(3, K)$ for every field K except $K = \mathbf{F}_q$ with $q = 2, 3, 5$. Let $\mathcal{F}$ be a non-singular cubic surface of $\mathrm{PG}(3, K)$. If $P \in \mathcal{F}$ is on exactly three lines l_1, l_2, l_3 of $\mathcal{F}$, then P is an *Eckardt point* of $\mathcal{F}$; if $\mathcal{F}$ is non-singular these lines l_1, l_2, l_3 belong to the tangent plane of $\mathcal{F}$ at P. A *tritangent plane* is a plane containing three lines of $\mathcal{F}$. If $\mathcal{F}$ has 27 lines, then $\mathcal{F}$ has 45 tritangent planes. A *trihedral pair* is a set of six tritangent planes divided into two sets, each set consisting of three planes pairwise intersecting in a line not belonging to $\mathcal{F}$, such that the three planes of each set contain the same set of nine distinct lines of $\mathcal{F}$. If $\mathcal{F}$ contains 27 lines, then the 45 tritangent planes form 120 trihedral pairs.

Consider a non-singular cubic surface $\mathcal{F}$ in $\mathrm{PG}(3, K)$ and assume that $\mathcal{F}$ has 27 lines. Let $\mathcal{S}' = (\mathcal{P}', \mathcal{B}', I')$ be the following incidence structure: the elements of $\mathcal{P}'$ are the 45 tritangent planes of $\mathcal{F}$, the elements of $\mathcal{B}'$ are the 27 lines of $\mathcal{F}$, a point $\pi \in \mathcal{P}'$ is incident with a line $l \in \mathcal{B}'$ if $l \subset \pi$. Then $\mathcal{S}'$ is the unique generalised quadrangle of order $(4, 2)$; for the uniqueness, see Corollary 5.82. Let $\mathcal{D}$ be one of the double-sixes contained in $\mathcal{B}'$ and let β be the above polarity fixing $\mathcal{D}$. If $\mathcal{P} = \mathcal{P}'\beta$, $\mathcal{B} = \mathcal{B}'\beta$, and I is symmetrised containment, then $\mathcal{S} = (\mathcal{P}, \mathcal{B}, I)$ is again the unique generalised quadrangle of order $(4, 2)$. This generalised quadrangle $\mathcal{S}$ is contained in the dual surface $\widehat{\mathcal{F}}$ of $\mathcal{F}$ which also contains exactly 27 lines. If lines in $\mathcal{B}$ are identified with their set of points, $\mathcal{S}$ is laxly embedded in $\mathrm{PG}(3, K)$. If $P \in \mathcal{P}$ and the three lines of $\mathcal{S}$ incident with P are contained in a common plane π, then $\pi\beta$ is an Eckardt point of $\mathcal{F}$.

If $\mathcal{D}$ is a double-six contained in $\mathcal{B}$, then the 15 lines of $\mathcal{B}$ not contained in $\mathcal{D}$, together with the 15 points of $\mathcal{P}$ not on lines of $\mathcal{D}$, form the unique generalised quadrangle of order 2; for the uniqueness, see Theorem 5.73. In this way the 36 subquadrangles of order 2 of $\mathcal{S}$ are obtained. If P, P' are non-collinear points of $\mathcal{S}$, then let $\{P, P'\}^{\perp} = \{R, R', R''\}$, $\{R, R'\}^{\perp} = \{P, P', P''\}$ in $\mathcal{S}$. Then $\{P\beta, P'\beta, P''\beta, R\beta, R'\beta, R''\beta\}$ yields a trihedral pair of $\mathcal{B}'$. In this way the 120 trihedral pairs are obtained. If l, m, n are skew lines of $\mathcal{B}$, then $|\{l, m, n\}^{\perp}| = 3$ in $\mathcal{S}$, say $\{l, m, n\}^{\perp} = \{l', m', n'\}$. So also any three skew lines of $\mathcal{B}'$ are concurrent with three skew lines of $\mathcal{B}'$. In total, $\mathcal{B}'$ admits 360 such configurations.

As already mentioned, $\mathcal{S} = (\mathcal{P}, \mathcal{B}, I)$ is laxly embedded in $\mathrm{PG}(3, K)$. Conversely, let $\mathcal{S} = (\mathcal{P}, \mathcal{B}, I)$ be any lax embedding in $\mathrm{PG}(3, K)$ of $\mathcal{U}(3, 4)$; up to isomorphism, $\mathcal{U}(3, 4)$ is the unique generalised quadrangle of order $(4, 2)$. Let $\mathcal{D}$ be any double-six contained in $\mathcal{B}$; then $\mathcal{D}$ consists of the 12 lines not belonging to a subquadrangle of order 2. Let β be the polarity fixing $\mathcal{D}$ described above, and let $\mathcal{B}' = \mathcal{B}\beta$. The double-six $\mathcal{D}$ belongs to a unique non-singular cubic surface $\mathcal{F}$. With the notation introduced above, the other 15 lines of $\mathcal{F}$ are the lines $c_{ij} = \langle a_i, b_j \rangle \cap \langle a_j, b_i \rangle$. So $c_{ij}\beta = \langle a_i \cap b_j, a_j \cap b_i \rangle$, and, by considering a subquadrangle $\mathcal{Q}(4, 2)$ of the generalised quadrangle $\mathcal{Q}(5, 2)$, it follows that $\langle a_i \cap b_j, a_j \cap b_i \rangle$ is a line of $\mathcal{S}$. Hence $c_{ij}\beta$ is a line of $\mathcal{S}$. Consequently, $\mathcal{B}'$ is the set of the 27 lines of a unique non-singular cubic surface $\mathcal{F}$. It follows that every lax embedding of $\mathcal{U}(3, 4)$ in $\mathrm{PG}(3, K)$ is of the type described above. So, such a lax embedding is uniquely defined by five skew lines a_1, a_2, a_3, a_4, a_5 together with a transversal b_6 such that each five of the six lines are linearly independent. Such a configuration exists for every field K except for $K = \mathbf{F}_2, \mathbf{F}_3, \mathbf{F}_5$.

The embedding $\mathcal{S}$ is polarised if and only if the 45 tritangent planes of $\mathcal{F}$ define 45 Eckardt points. If $K = \mathbf{F}_q$, then necessarily $q = 4^m$, and for each such q a polarised embedding of the generalised quadrangle of order $(4, 2)$ is possible. If $\mathcal{U}(3, 4)$ is embedded in $\mathrm{PG}(3, q)$ and if the embedding $\mathcal{S}$ is polarised, then, by Theorem 5.58, $\mathcal{S}$ is a full embedding of $\mathcal{U}(3, 4)$ in a subgeometry $\mathrm{PG}(3, 4)$ of $\mathrm{PG}(3, q)$, for a subfield $\mathbf{F}_4$ of $\mathbf{F}_q$; so $\mathcal{S}$ is a Hermitian surface of $\mathrm{PG}(3, 4)$. This result can be extended to infinite fields. So, if $\mathcal{U}(3,4)$ admits a polarised embedding in $\mathrm{PG}(3,K)$, then $\mathbf{F}_4$ is a subfield of K and the embedding is full in a subgeometry $\mathrm{PG}(3,4)$ of $\mathrm{PG}(3,K)$.

Hence the following result is obtained.

Theorem 5.62. *Let* K *be any field and let* $\mathcal{S}$ *be a lax embedding of* $\mathcal{U}(3,4)$ *in* $\mathrm{PG}(3,K)$*. Then*

  (i) $|K| \neq 2,3,5$ *and* $\mathcal{S}$ *arises from a non-singular cubic surface* $\mathcal{F}$;
 (ii) *the embedding is polarised if and only if* $\mathcal{F}$ *admits* 45 *Eckardt points*;
(iii) *in the latter case, the field* $\mathbf{F}_4$ *is a subfield of* K *and* $\mathcal{S}$ *is a Hermitian surface in a subgeometry* $\mathrm{PG}(3,4)$ *of* $\mathrm{PG}(3,K)$.

Similarly to the projective case, *lax, weak* (or *polarised*) and *full embeddings* of point-line geometries in an affine space $\mathrm{AG}(d,K)$ can be defined. In the following theorem, lax embeddings of generalised quadrangles of order (s,t), $s > 1$, in $\mathrm{PG}(d,q)$, $d > 2$, are considered.

Theorem 5.63. *If the generalised quadrangle* $\mathcal{S}$ *of order* (s,t)*,* $s > 1$*, is laxly embedded in* $\mathrm{PG}(d,q)$*, then* $d \leq 5$.

  (i) *If* $d = 5$*, then* $\mathcal{S} \cong \mathcal{Q}(5,s)$ *and the full automorphism group of* $\mathcal{S}$ *is induced by* $\mathrm{PGL}(6,q)$*. Also, one of the following holds:*
   (a) *it is weakly embedded and hence, by Theorems* 5.58 *and* 5.51*, fully and naturally embedded in some subgeometry* $\mathrm{PG}(5,s)$ *of* $\mathrm{PG}(5,q)$;
   (b) $s = 2$*,* q *is odd and there exists up to an element of* $\mathrm{PSL}(6,q)$*, a unique* (*non-weak*) *lax embedding, which is a full affine embedding if* $q = 3$;
   (c) *for* $s = 2$*,* $q = 3^h$*, the* (*non-weak*) *lax embedding is a full embedding in some affine subgeometry* $\mathrm{AG}(5,3)$ *over the subfield* $\mathbf{F}_3$ *of* $\mathbf{F}_q$.
 (ii) *If* $d = 4$*, then* $s \leq t$.
   (a) *If* $s = t$*, then* $\mathcal{S} \cong \mathcal{Q}(4,s)$ *and one of the following occurs.*
     1. $s \neq 2$*,* q *is odd, and* $\mathcal{S}$ *is weakly embedded; hence, by Theorems* 5.58 *and* 5.51*, it is fully and naturally embedded in some subgeometry* $\mathrm{PG}(4,s)$ *in* $\mathrm{PG}(4,q)$.
     2. $s = 2$*,* q *is odd, and* $\mathcal{S}$ *is weakly embedded in* $\mathrm{PG}(4,q)$*, that is,* $\mathcal{S}$ *is the universal weak embedding of* $\mathcal{W}(2)$ *in* $\mathrm{PG}(4,q)$ *by Theorem* 5.58.
     3. $s = 3$*,* $q \equiv 1 \pmod 3$*, and there exists, up to an element of* $\mathrm{PSL}(5,q)$*, a unique* (*non-weak*) *lax embedding. Further,*
       I. *the case* $q = 4$ *corresponds to a full affine embedding;*
       II. *the case* q *even corresponds to a full affine embedding in an affine subgeometry over the subfield* $\mathbf{F}_4$ *of* $\mathbf{F}_q$;
       III. *the group* $\mathrm{PSp}(4,3)$ *acting naturally as an automorphism group on* $\mathcal{W}(3)$ *is induced on* $\mathcal{S}$ *by* $\mathrm{PSL}(5,q)$; *if* q *is a square and if also* $\sqrt{q} \equiv -1 \pmod 3$*, then the full automorphism group* $\mathrm{PGSp}(4,3)$ *of* $\mathcal{S}$ *is the group induced by* $\mathrm{P\Gamma L}(5,q)$; *otherwise,* $\mathrm{P\Gamma L}(5,q)$ *just induces* $\mathrm{PSp}(4,3)$.

(b) *If $s > 2$, then $t \neq s + 2$.*

(c) *If $t^2 = s^3$, then $S \cong \mathcal{U}(4, s)$ and S is weakly embedded and hence, by Theorems 5.58 and 5.51, fully and naturally embedded in some subspace $\mathrm{PG}(4, s)$ of $\mathrm{PG}(4, q)$.*

(d) *If $S \cong \mathcal{Q}(5, s)$, then there exists a $\mathrm{PG}(5, q)$ containing $\mathrm{PG}(4, q)$ and a point $P \in \mathrm{PG}(5, q) \backslash \mathrm{PG}(4, q)$ such that S is the projection from P onto $\mathrm{PG}(4, q)$ of a generalised quadrangle $\overline{S} \cong \mathcal{Q}(5, s)$ which is laxly embedded in $\mathrm{PG}(5, q)$, and hence determined by (i).*

(iii) $d = 3$.

    (a) *If $t = 1$, then S is a subquadrangle of order $(s, 1)$ of some $\mathcal{Q}(3, q)$.*

    (b) *If $s = t^2$, then $S \cong \mathcal{U}(3, s)$ and one of the following holds:*

        1. *S is weakly embedded in $\mathrm{PG}(3, q)$ and hence, by Theorems 5.58 and 5.51, fully and naturally embedded in some subgeometry $\mathrm{PG}(3, s)$ of $\mathrm{PG}(3, q)$;*

        2. *$(s, t) = (4, 2)$ with $q \notin \{2, 3, 5\}$ and S arises from a non-singular cubic surface in $\mathrm{PG}(3, q)$; see Theorem 5.62.*

    (c) *If S is classical or dual classical, but not isomorphic to $\mathcal{W}(s)$ with s odd, then the following classification is obtained.*

        1. *S is not dual to $\mathcal{U}(4, s^{2/3})$.*

        2. *If $S \cong \mathcal{U}(4, s)$, then there exists a $\mathrm{PG}(4, q)$ containing $\mathrm{PG}(3, q)$ and a point $P \in \mathrm{PG}(4, q) \backslash \mathrm{PG}(3, q)$ such that S is the projection from P onto $\mathrm{PG}(3, q)$ of a generalised quadrangle $\overline{S} \cong \mathcal{U}(4, s)$ which is fully and naturally embedded in a subgeometry $\mathrm{PG}(4, s)$ of $\mathrm{PG}(4, q)$, for some subfield $\mathbf{F}_s$ of $\mathbf{F}_q$, with s a square.*

        3. *If $S \cong \mathcal{Q}(4, s)$, then there exists a $\mathrm{PG}(4, q)$ containing $\mathrm{PG}(3, q)$ and a point $P \in \mathrm{PG}(4, q) \backslash \mathrm{PG}(3, q)$ such that S is the projection from P onto $\mathrm{PG}(3, q)$ of a generalised quadrangle $\overline{S} \cong \mathcal{Q}(4, s)$ which is laxly embedded in $\mathrm{PG}(4, q)$, and hence determined by (ii)(a).*

        4. *If $S \cong \mathcal{Q}(5, s)$, then there exists a $\mathrm{PG}(5, q)$ containing $\mathrm{PG}(3, q)$ and a line l of $\mathrm{PG}(5, q)$ skew to $\mathrm{PG}(3, q)$ such that S is the projection from l onto $\mathrm{PG}(3, q)$ of a generalised quadrangle $\overline{S} \cong \mathcal{Q}(5, s)$ which is laxly embedded in $\mathrm{PG}(5, q)$, and hence determined by (i).*

Lax embeddings of non-degenerate Shult spaces of rank at least three are considered in the next theorem.

Theorem 5.64. *Let S be a non-degenerate Shult space of rank at least three all of whose lines have size at least three and which is laxly embedded in $\mathrm{PG}(d, q)$, where $d \geq 3$.*

(i) *If S is not the Shult space arising from a null polarity of $\mathrm{PG}(2m + 1, s)$, s odd, then there exists a $\mathrm{PG}(n, q)$ containing $\Pi_d = \mathrm{PG}(d, q)$, a Π_{n-d-1} in $\mathrm{PG}(n, q)$ skew to Π_d and a non-degenerate classical Shult space $\overline{S} \cong S$ fully and naturally embedded in a subgeometry $\mathrm{PG}(n, s)$ of $\mathrm{PG}(n, q)$, such that S is the projection of $\overline{S}$ from Π_{n-d-1} onto Π_d.*

(ii) *If $d \geq 4$ and if S arises from a null polarity of* $\mathrm{PG}(2m+1, s)$, $m \geq 2$ *and s odd, then there exists a* $\mathrm{PG}(2m+1, q)$ *containing Π_d and a subspace Π_{2m-d} of* $\mathrm{PG}(2m+1, q)$ *skew to Π_d such that S is the projection from Π_{2m-d} onto Π_d of a non-degenerate Shult space* $\overline{S} \cong S$ *which is fully and naturally embedded in a subgeometry* $\mathrm{PG}(2m+1, s)$ *of* $\mathrm{PG}(2m+1, q)$.

Remark 5.65. The lax embeddings in $\mathrm{PG}(3, q)$ of the generalised quadrangles $\mathcal{W}(s)$, s odd, and of the non-degenerate Shult spaces of rank at least three arising from a null polarity of $\mathrm{PG}(2m+1, s)$, $m \geq 2$ and s odd, are not yet classified. The classification of lax embeddings of Shult spaces in $\mathrm{PG}(2, q)$, without extra conditions, seems to be hopeless.

5.5 Characterisations of the classical generalised quadrangles

In this section the most important characterisations of the classical generalised quadrangles are reviewed. Apart from a few exceptions the proofs are long, complicated, and technical. So, proofs are given only in the simpler cases. First, some new ideas are introduced.

Definition 5.66. Let $S = (\mathcal{P}, \mathcal{B}, \mathrm{I})$ be a finite generalised quadrangle of order (s, t).

(1) If $P \sim P'$, $P \neq P'$, or if $P \not\sim P'$ and $|\{P, P'\}^{\perp\perp}| = t + 1$, the pair $\{P, P'\}$ is *regular*.
(2) The point P is *regular* if $\{P, P'\}$ is regular for all $P' \in \mathcal{P}$, $P' \neq P$.
(3) A point P is *co-regular* if each line incident with P is regular.
(4) The pair $\{P, P'\}$, $P \not\sim P'$, is *anti-regular* if $|P''^{\perp} \cap \{P, P'\}^{\perp}| \leq 2$ for all $P'' \in \mathcal{P} \backslash \{P, P'\}$.
(5) A point P is *anti-regular* if $\{P, P'\}$ is anti-regular for all $P' \in \mathcal{P} \backslash P^{\perp}$.
(6) The *closure* of the pair $\{P, P'\}$ is

$$\mathrm{cl}(P, P') = \{P'' \in \mathcal{P} \mid P''^{\perp} \cap \{P, P'\}^{\perp\perp} \neq \emptyset\}.$$

Theorem 5.67. *Let $S = (\mathcal{P}, \mathcal{B}, \mathrm{I})$ be a generalised quadrangle of order $s > 1$.*

(i) *For a regular point P, the incidence structure π_P with point set $P^{\perp}$, with line set the set of spans $\{P', P''\}^{\perp\perp}$, where $P', P'' \in P^{\perp}$ with $P' \neq P''$, and with the natural incidence is a projective plane of order s.*

(ii) *For an anti-regular point P and a point P' in $P^{\perp} \backslash \{P\}$, the incidence structure $\pi(P, P')$ with point set $P^{\perp} \backslash \{P, P'\}^{\perp}$, with lines the sets $\{P, P_1\}^{\perp} \backslash \{P\}$ with $P \sim P_1 \not\sim P'$ and the sets $\{P, P_2\}^{\perp} \backslash \{P'\}$ with $P' \sim P_2 \not\sim P$, and with the natural incidence is an affine plane of order s.*

Proof. Both parts are straightforward verifications of the axioms.   □

Definition 5.68. (1) An *ovoid* of S is a set $\mathcal{O}$ of points of S such that each line of S is incident with a unique point of $\mathcal{O}$.

(2) A *spread* of S is a set $\mathcal{R}$ of lines of S such that each point of S is incident with a unique line of $\mathcal{R}$.

Remark 5.69. It follows that any ovoid or spread of S has exactly $1 + st$ elements.

Definition 5.70. (1) Let $s^2 = t > 1$; then, by Corollary 5.15, any triad $\{P, P', P''\}$ has $|\{P, P', P''\}^\perp| = s + 1$. Thus $|\{P, P', P''\}^{\perp\perp}| \le s + 1$ and $\{P, P', P''\}$ is 3-*regular* provided that $|\{P, P', P''\}^{\perp\perp}| = s + 1$.
(2) The point P is 3-*regular* if and only if each triad containing P is 3-regular.

These definitions are illustrated by some examples.

In Section 5.2, it was observed that each hyperbolic line of $\mathcal{W}(q)$ contains $q + 1$ points. Hence all points of $\mathcal{W}(q)$ are regular. Dually, all lines of $\mathcal{Q}(4, q)$ are regular. By Theorem 5.17, all lines of $\mathcal{W}(q)$, with q even, are regular. Dually, all points of $\mathcal{Q}(4, q)$, q even, are regular. From the examples in Section 5.2, it follows that each point of $\mathcal{Q}(4, q)$, q odd, is anti-regular, and, dually, that each line of $\mathcal{W}(q)$, q odd, is anti-regular. Further, each hyperbolic line of $\mathcal{U}(3, q^2)$ contains $q + 1$ points; hence each point of $\mathcal{U}(3, q^2)$ is regular. Dually, each line of $\mathcal{Q}(5, q)$ is regular.

Consider $\mathcal{Q}(5, q)$ and the corresponding polarity ζ. If $\mathcal{T} = \{P, P', P''\}$ is a triad of $\mathcal{Q}(5, q)$, then $\mathcal{T}^\perp$ is the conic $\mathcal{Q} \cap \pi$, where π is the polar of the plane $PP'P''$, and $\mathcal{T}^{\perp\perp}$ is the conic $\mathcal{Q} \cap PP'P''$. So $|\mathcal{T}^{\perp\perp}| = q + 1$, and consequently each point of $\mathcal{Q}(5, q)$ is 3-regular. Dually, each line of $\mathcal{U}(3, q^2)$ is 3-regular.

The generalised quadrangle $\mathcal{Q}(4, q)$ always has ovoids, and has spreads if and only if q is even; see Section 7.2. Further, $\mathcal{Q}(5, q)$ has spreads but no ovoids, and $\mathcal{U}(4, q^2)$ has no ovoids. For $q = 2$, the Hermitian variety $\mathcal{U}(4, q^2)$ has no spreads; for $q > 2$, the existence of a spread is an open problem.

Historically, the next result is probably the oldest combinatorial characterisation of a class of generalised quadrangles. A proof is essentially contained in a paper by Singleton, although he erroneously thought he had proved a stronger result; but the first satisfactory treatment may have been given by Benson. Undoubtedly, it was discovered independently by several authors; see Section 5.10.

Theorem 5.71. *A generalised quadrangle S of order $s \ne 1$ is isomorphic to $\mathcal{W}(s)$ if and only if all its points are regular.*

Proof. All points of $\mathcal{W}(s)$ are regular.

Conversely, let $S = (\mathcal{P}, \mathcal{B}, I)$ be a generalised quadrangle of order $s \ne 1$, for which all points are regular. Now define the incidence structure $S' = (\mathcal{P}', \mathcal{B}', I')$, with $\mathcal{P}' = \mathcal{P}$, with $\mathcal{B}'$ the set of spans of all point pairs of $\mathcal{P}$, and I' the natural incidence. Then S is isomorphic to the substructure of S' formed by all points and the spans of all pairs of points collinear in S.

Let $\mathcal{T} = \{P, P', P''\}$ be a triad of points in S. Counting all points on the lines joining a point of $\{P, P'\}^\perp$ and a point of $\{P, P'\}^{\perp\perp}$ gives

$$(s + 1)^2(s - 1) + 2(s + 1) = (s + 1)(s^2 + 1) = |\mathcal{P}|.$$

Hence P'' is on at least one line joining a point of $\{P, P'\}^\perp$ to a point of $\{P, P'\}^{\perp\perp}$, and so $\mathcal{T}^\perp \ne \emptyset$. Now, by Theorem 5.67 (i), it follows that any three non-collinear points of S' generate a projective plane of order s. Since $|P'| = s^3 + s^2 + s + 1$, so S' is the design of points and lines of $PG(3, s)$. All spans in S of collinear point pairs containing a given point P form a pencil of lines in $PG(3, s)$. By Theorem 15.2.13 of FPSOTD, the set of all spans of collinear point pairs is a general linear complex of lines of $PG(3, q)$ or, equivalently, is the set of all self-polar lines of a null polarity ζ. Thus $S \cong W(s)$. □

The next result is a slight generalisation of the preceding theorem and is stated without proof.

Theorem 5.72. *A generalised quadrangle S of order (s, t), $s \neq 1$, is isomorphic to $W(s)$ if and only if each hyperbolic line has at least $s + 1$ points.*

Theorem 5.73. *Up to isomorphism, there is only one generalised quadrangle of order 2.*

Proof. Let S be a generalised quadrangle of order 2. Consider two points P, P' with $P \not\sim P'$, and let $\{P, P'\}^{\perp} = \{Z_1, Z_2, Z_3\}$. Let $\{Z_1, Z_2\}^{\perp} = \{P, P', U\}$, let Y be the unique point of PZ_3 collinear with U, and let Y' be the unique point of $P'Z_3$ collinear with U. If Y, Z_3, Y' are distinct, then U is incident with the four distinct lines UZ_1, UZ_2, UY, UY', a contradiction since $t = 2$. Thus $Y = Z_3 = Y'$, and so $U \sim Z_3$. Hence $|\{P, P'\}^{\perp\perp}| = 3$. So every point of S is regular and now, by Theorem 5.71, $S \cong W(2)$. □

The following four characterisations are stated without proof.

Theorem 5.74. *A generalised quadrangle S of order $s \neq 1$ is isomorphic to $W(2^h)$ if and only if it has an ovoid $\mathcal{O}$ each triad of which has at least one centre.*

Theorem 5.75. *A generalised quadrangle S of order $s \neq 1$ is isomorphic to $W(2^h)$ if and only if it has an ovoid $\mathcal{O}$ each point of which is regular.*

Theorem 5.76. *A generalised quadrangle S of order $s \neq 1$ is isomorphic to $W(2^h)$ if and only if it has a regular pair $\{l_1, l_2\}$ of non-concurrent lines with the property that any triad of points lying on lines of $\{l_1, l_2\}^{\perp}$ has at least one centre.*

Theorem 5.77. *Let S be a generalised quadrangle of order $s \neq 1$ with an anti-regular point P. Then S is isomorphic to $Q(4, s)$ if and only if there is a point P' in $P^{\perp} \setminus \{P\}$, for which the associated affine plane $\pi(P, P')$ is Desarguesian.*

Corollary 5.78. *Let S be a generalised quadrangle of order $s \neq 1$ having an anti-regular point P. If $s \leq 8$, then S is isomorphic to $Q(4, s)$.*

Proof. Since each plane of order $s \leq 8$ is Desarguesian, the result follows. □

Theorem 5.79. *Let S be a generalised quadrangle of odd order $s > 1$ with co-regular point P. Then $S \cong Q(4, s)$ if and only if, for at least one line l incident with P, the projective plane π_l is Desarguesian.*

Corollary 5.80. *Let S be a generalised quadrangle of order s, with $s \in \{5, 7\}$. If S contains a co-regular point, then $S \cong Q(4, s)$.*

Proof. As the projective planes of order 5 and 7 are unique, the result follows from Theorem 5.79. □

The following characterisation theorem is very important, not only for the theory of generalised quadrangles, but also for other areas in combinatorics. The proof is again very long, and so it is not given.

Theorem 5.81. *Let S be a generalised quadrangle of order (s, s^2).*

(i) *When $s > 1$, then $S \cong Q(5, s)$ if and only if all points of S are 3-regular.*

(ii) *When s is odd and $s > 1$, then $S \cong Q(5, s)$ if and only if it has a 3-regular point.*

(iii) *When s is even, then $S \cong Q(5, s)$ if and only if it has at least one 3-regular point not incident with some regular line.*

(iv) *When s is odd and $s > 1$, then $S \cong Q(5, s)$ if and only if the following properties hold:*

  (a) *there are distinct collinear points P and P' such that each triad containing P with centre P' is 3-regular;*

  (b) *each triad containing P' with centre P is 3-regular;*

  (c) *there is at least one 3-regular triad $\{P_1, P_2, P_3\}$ with $P_1 \text{ I } PP'$ and where $\{P_1, P_2, P_3\}^{\perp}$ does not contain a point incident with PP'.*

Corollary 5.82. *Up to isomorphism there is only one generalised quadrangle of the following orders:* (i) $(2, 4)$; (ii) $(3, 9)$.

Proof. (a) Let S be a generalised quadrangle of order $(2, 4)$. If $\{P_1, P_2, P_3\}$ is a triad of points, then $\{P_1, P_2, P_3\}^{\perp\perp} = \{P_1, P_2, P_3\}$. Hence $|\{P_1, P_2, P_3\}^{\perp\perp}| = 1 + s$, every point is 3-regular, and, by Theorem 5.81 (i), $S \cong Q(5, 2)$.

(b) Let S be of order $(3, 9)$. Let $\{P_1, P_2, P_3\}$ be a triad of points, and let $\{P_1, P_2, P_3\}^{\perp} = \{U_1, U_2, U_3, U_4\}$, $\{U_1, U_2, U_3\}^{\perp} = \{P_1, P_2, P_3, P_4\}$. The number of points collinear with U_4 and also with at least two points of $\{U_1, U_2, U_3\}$ is at most six, and the number of points collinear with U_4 and incident with some line $P_4 U_i$, $i = 1, 2, 3$, is at most three. Since $3 + 6 < 10 = t + 1$, there is a line l incident with U_4, but not concurrent with $P_4 U_i$, $i = 1, 2, 3$, and not incident with an element of $\{U_i, U_j\}^{\perp}$, $i \neq j$, $1 \leq i \leq 3$, $1 \leq j \leq 3$. The point incident with l and collinear with U_i is denoted by Z_i, $i = 1, 2, 3$; the points Z_1, Z_2, Z_3 are distinct. Since S has no triangles, the point P_4 is not collinear with any Z_i, forcing $P_4 \sim U_4$. Hence the triad $\{P_1, P_2, P_3\}$ is 3-regular, and so every point of S is 3-regular. Now, by Theorem 5.81, $S \cong Q(5, 3)$. □

Definition 5.83. (1) The generalised quadrangle $S' = (\mathcal{P}', \mathcal{B}', I')$ of order (s', t') is a *subquadrangle* of the generalised quadrangle $S = (\mathcal{P}, \mathcal{B}, I)$ of order (s, t) if $\mathcal{P}' \subset \mathcal{P}$, $\mathcal{B}' \subset \mathcal{B}$, and if I' is the restriction of I to $(\mathcal{P}' \times \mathcal{B}') \cup (\mathcal{B}' \times \mathcal{P}')$.

(2) If $S' \neq S$, then S' is a *proper* subquadrangle of S.

Remark 5.84. If $|\mathcal{P}| = |\mathcal{P}'|$ it follows that $s = s'$ and $t = t'$; hence, if $\mathcal{S}'$ is a proper subquadrangle, then $\mathcal{P} \neq \mathcal{P}'$, and dually $\mathcal{B} \neq \mathcal{B}'$.

Some examples are given.

Example 5.85. (a) Consider $\mathcal{Q}(5, q)$, with $\mathcal{Q}$ a non-singular quadric of projective index 1 in $\mathrm{PG}(5, q)$. Intersect $\mathcal{Q}$ with a non-tangent hyperplane Π_4. Then the points and lines of $\mathcal{Q}' = \mathcal{Q} \cap \Pi_4$ form the generalised quadrangle $\mathcal{Q}'(4, q)$. Here $s^2 = t = q^2$, $s = s' = t'$. Since all lines of $\mathcal{Q}(5, q)$ and of $\mathcal{Q}'(4, q)$ are regular, both have subquadrangles of order (s'', t'') with $t'' = 1$ and $s'' = s' = s$, each of which is a hyperbolic quadric in some solid of $\mathrm{PG}(5, q)$.

(b) Let $\mathcal{U} = \mathcal{U}(4, q^2)$, a non-singular Hermitian variety of $\mathrm{PG}(4, q^2)$ and intersect $\mathcal{U}$ with a non-tangent hyperplane Π_3. Then the points and lines of $\mathcal{U}' = \mathcal{U} \cap \Pi_3$ form the generalised quadrangle $\mathcal{U}'(3, q^2)$. In this case, the parameters are

$$t = s^{3/2} = q^3, \; s = s', \; t' = \sqrt{s}.$$

Since all points of $\mathcal{U}'(3, q^2)$ are regular, so $\mathcal{U}'(3, q^2)$ has subquadrangles with $t'' = t' = \sqrt{s}$ and $s'' = 1$.

(c) Now consider $\mathcal{Q}(4, q)$ and extend $\mathbf{F}_q$ to $\mathbf{F}_{q^2}$. Then $\mathcal{Q}$ extends to $\overline{\mathcal{Q}}$ and $\mathcal{Q}(4, q)$ to $\overline{\mathcal{Q}}(4, q^2)$. Here, $\mathcal{Q}(4, q)$ is a subquadrangle of $\overline{\mathcal{Q}}(4, q^2)$, with $t = s = q^2$ and $t' = s' = q$.

Next consider the role of subquadrangles in characterising $\mathcal{Q}(5, q)$. Proofs are again omitted.

Theorem 5.86. *Let $\mathcal{S}$ be a generalised quadrangle of order (s, t) with $s > 1$ and $t > 1$. Then $\mathcal{S}$ is isomorphic to $\mathcal{Q}(5, s)$ if and only if either* (i) *or* (ii) *holds:*

(i) *every triad of lines with at least one centre is contained in a proper subquadrangle of order (s, t');*

(ii) *for each triad $\{P, P', P''\}$ with distinct centres Z, Z', the five points P, P', P'', Z, Z' are contained in a proper subquadrangle of order (s, t').*

For a generalised quadrangle $\mathcal{S}$ of order (s, t), let $\{l_1, l_2, l_3\}$ and $\{m_1, m_2, m_3\}$ be two triads of lines for which $l_i \not\sim m_j$ if and only if $\{i, j\} = \{1, 2\}$. Let P_i be the point defined by $l_i \, \mathrm{I} \, P_i \, \mathrm{I} \, m_i$, $i = 1, 2$. This configuration $\mathcal{T}$ of seven distinct points and six distinct lines is a *broken grid* with *carriers* P_1 and P_2; see Figure 5.2.

The broken grid $\mathcal{T}$ satisfies *axiom* (D) *with respect to the pair* $\{l_1, l_2\}$ provided the following holds: if $l_4 \in \{m_1, m_2\}^\perp$ with $l_4 \not\sim l_i, i = 1, 2, 3$, then $\{l_1, l_2, l_4\}$ has at least one centre. Interchanging l_i and m_i gives the definition of axiom (D) for $\mathcal{T}$ with respect to the pair $\{m_1, m_2\}$. Further, $\mathcal{T}$ is said to satisfy *axiom* (D) provided it satisfies axiom (D) with respect to both pairs $\{l_1, l_2\}$ and $\{m_1, m_2\}$.

Let P be any point of $\mathcal{S}$. Then $\mathcal{S}$ is said to satisfy *axiom* $(D)'_P$ if the broken grid $\mathcal{T}$ satisfies axiom (D) with respect to $\{l_1, l_2\}$ whenever $P \, \mathrm{I} \, l_1$; it satisfies *axiom* $(D)''_P$ if $\mathcal{T}$ satisfies axiom (D) with respect to $\{m_1, m_2\}$ whenever $P \, \mathrm{I} \, l_1$.

Now, another interesting characterisation of $\mathcal{Q}(5, s)$ is the following.

Fig. 5.2. Broken grid

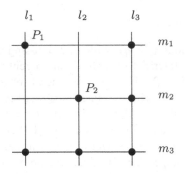

Theorem 5.87. *Let S be a generalised quadrangle of order (s,t), with $s \neq t$, and $s > 1$, $t > 1$.*

(i) *If s is odd, then $S \cong \mathcal{Q}(5,s)$ if and only if S contains a co-regular point P for which $(D)'_P$ or $(D)''_P$ is satisfied.*

(ii) *If s is even, then $S \cong \mathcal{Q}(5,s)$ if and only if all lines of S are regular and S contains a point P for which $(D)'_P$ or $(D)''_P$ is satisfied.*

In order to conclude this section dealing with characterisations of $\mathcal{Q}(5,s)$, one more basic concept is introduced.

Definition 5.88. Let $S = (\mathcal{P}, \mathcal{B}, I)$ be a generalised quadrangle of order (s,t). If $\mathcal{B}^{\perp\perp}$ is the set of all spans $\{P, P'\}^{\perp\perp}$ with $P \not\sim P'$, then let $S^{\perp\perp} = (\mathcal{P}, \mathcal{B}^{\perp\perp}, \in)$.

(1) S satisfies *property $(A)_P$* for $P \in \mathcal{P}$, if for any $m = \{P', P''\}^{\perp\perp} \in \mathcal{B}^{\perp\perp}$ with $P \in \{P', P''\}^{\perp}$, and $U \in \mathrm{cl}(P', P'') \cap (P^{\perp} \setminus \{P\})$ with $U \notin m$, the substructure of $S^{\perp\perp}$ generated by m and U is a dual affine plane.

(2) S satisfies *property (A)* if it satisfies $(A)_P$ for all $P \in \mathcal{P}$.

(3) The dual of $(A)_P$ is denoted by $(\hat{A})_l$ and of (A) by $(\hat{A})$.

So S satisfies (A) if, for any $m = \{P', P''\}^{\perp\perp} \in \mathcal{B}^{\perp\perp}$ and any U in the set $\mathrm{cl}(P', P'') \setminus (\{P', P''\}^{\perp} \cup \{P', P''\}^{\perp\perp})$, the substructure of $S^{\perp\perp}$ generated by m and U is a dual affine plane.

Again, the proof of the following theorem is not given.

Theorem 5.89. *Let S be a generalised quadrangle of order (s,t), $s \neq t$, $t > 1$.*

(i) *If $s > 1$, and s is odd, then S is isomorphic to $\mathcal{Q}(5,s)$ if and only if $(\hat{A})_l$ is satisfied for all lines l incident with some co-regular point P.*

(ii) *If s is even, then S is isomorphic to $\mathcal{Q}(5,s)$ if and only if all lines of S are regular and $(\hat{A})_l$ is satisfied for all lines l incident with some point P.*

Definition 5.90. A finite *net* of *order* $k \geq 2$ and *degree* $r \geq 2$ is an incidence structure $\mathcal{N} = (\mathcal{P}, \mathcal{B}, I)$ of points and lines satisfying the following axioms.

(1) Each point is incident with r lines and two distinct points are incident with at most one line.
(2) Each line is incident with k points and two distinct lines are incident with at most one point.
(3) If P is a point and l is a line not incident with P, then there is a unique line m incident with P and not concurrent with l.

For a net of order k and degree r, it follows that $|\mathcal{P}| = k^2$ and $|\mathcal{B}| = kr$. Also, $r \le k + 1$, with $r = k + 1$ if and only if the net is an affine plane of order k.

The following theorem gives the relation between regularity in generalised quadrangles and dual nets.

Theorem 5.91. *Let P be a regular point of the generalised quadrangle $\mathcal{S} = (\mathcal{P}, \mathcal{B}, I)$ of order (s, t), $s \ne 1$. Then the incidence structure, with*

(a) *point set $P^{\perp} \backslash \{P\}$,*
(b) *line set the set of all spans $\{Q, Q'\}^{\perp\perp}$, where $Q, Q' \in P^{\perp} \backslash \{P\}$, $Q \not\sim Q'$,*
(c) *incidence the natural one,*

is the dual of a net of order s and degree $t + 1$. If, in particular, $s = t > 1$, then a dual affine plane of order s arises.

Proof. This is a straightforward exercise.   □

Let $\mathcal{P}$ be the set of all points of $\mathrm{PG}(n, q)$ which are not contained in a fixed subspace $\mathrm{PG}(n - 2, q)$, with $n \ge 2$, let $\mathcal{B}$ be the set of all lines of $\mathrm{PG}(n, q)$ having no point in common with $\mathrm{PG}(n - 2, q)$, and let I be the natural incidence. Then $(\mathcal{P}, \mathcal{B}, I)$ is the dual of a net of order q^{n-1} and degree $q + 1$; this dual net is denoted by H_q^n.

Definition 5.92. *A point-line incidence geometry $\mathcal{S} = (\mathcal{P}, \mathcal{B}, I)$ satisfies the Veblen–Pasch axiom if and only if the following condition is satisfied:*

(VP) *if $l_1 \, I \, P \, I \, l_2$, $l_1 \ne l_2$, and $m_1 \, \not{I} \, P \, \not{I} \, m_2$, $l_i \sim m_j$, for all $i, j \in \{1, 2\}$, then $m_1 \sim m_2$.*

Theorem 5.93. *Let $\mathcal{S}$ be a dual net of order $s + 1$ and degree $t + 1$, with $s < t + 1$. If $\mathcal{S}$ satisfies (VP), then $\mathcal{S}$ is isomorphic to the dual net H_q^n, which has $s = q$ and $t = q^{n-1} - 1$.*

The following characterisation theorem of $\mathcal{Q}(5, s)$ is in terms of regularity and dual nets.

Theorem 5.94. *Let $\mathcal{S}$ be a generalised quadrangle of order (s, t) with $s \ne t$, $s > 1$ and $t > 1$.*

(i) *If s is odd, then $\mathcal{S}$ is isomorphic to the classical generalised quadrangle $\mathcal{Q}(5, s)$ if and only if it has a co-regular point P and if for each line l incident with P the corresponding dual net $\mathcal{N}_l^D$ satisfies (VP).*

(ii) *If s is even, then $\mathcal{S}$ is isomorphic to the classical generalised quadrangle $\mathcal{Q}(5, s)$ if and only if all its lines are regular and, for at least one point P and all lines l incident with P, the corresponding dual nets $\mathcal{N}_l^D$ satisfy* (VP).

The next characterisation theorem of $\mathcal{Q}(5, s)$ involves subquadrangles and ovoids.

Theorem 5.95. *Let $\mathcal{S}$ be a generalised quadrangle of order (s, s^2), $s \neq 1$, having a subquadrangle $\mathcal{S}'$ isomorphic to $\mathcal{Q}(4, s)$. If in $\mathcal{S}'$ each ovoid $\mathcal{O}_P$ consisting of all the points collinear with a given point P of $\mathcal{S} \backslash \mathcal{S}'$ is an elliptic quadric, then $\mathcal{S}$ is isomorphic to $\mathcal{Q}(5, s)$.*

Definition 5.96. (1) Let $\mathcal{S} = (\mathcal{P}, \mathcal{B}, \mathrm{I})$ be a generalised quadrangle of order (s, t), and define $\mathcal{B}^* = \{ \{P, P'\}^{\perp\perp} \mid P, P' \in \mathcal{P}, P \neq P' \}$. Then $\mathcal{S}^* = (\mathcal{P}, \mathcal{B}^*, \in)$ is a *linear space*; see Section 5.10.
(2) So as to have no confusion between collinearity in $\mathcal{S}$ and collinearity in $\mathcal{S}^*$, points $P_1, P_2, \ldots, P_r$ of $\mathcal{P}$ which are on a line of $\mathcal{S}^*$ are $\mathcal{S}^*$-*collinear*.
(3) A *linear variety* of $\mathcal{S}^*$ is a subset $\mathcal{P}' \subset \mathcal{P}$ such that $P, P' \in \mathcal{P}'$, $P \neq P'$, implies $\{P, P'\}^{\perp\perp} \subset \mathcal{P}'$.
(4) If $\mathcal{P}' \neq \mathcal{P}$ and $|\mathcal{P}'| > 1$, the linear variety is *proper*; if $\mathcal{P}'$ is generated by three points which are not $\mathcal{S}^*$-collinear, $\mathcal{P}'$ is a *plane* of $\mathcal{S}^*$.

Now a fundamental characterisation of the generalised quadrangle $\mathcal{U}(3, s)$ is stated.

Theorem 5.97. *Let $\mathcal{S} = (\mathcal{P}, \mathcal{B}, \mathrm{I})$ be a generalised quadrangle of order (s, t), with $s \neq t, s > 1$, and $t > 1$. Then $\mathcal{S}$ is isomorphic to $\mathcal{U}(3, s)$ if and only if the following hold:*

(a) *all points of $\mathcal{S}$ are regular;*
(b) *if the lines l and l' of $\mathcal{B}^*$ are contained in a proper linear variety of $\mathcal{S}^*$, then also the lines $l^\perp$ and $l'^\perp$ of $\mathcal{B}^*$ are contained in a proper linear variety of $\mathcal{S}^*$.*

A beautiful characterisation theorem, but with a long and complicated proof, is the following.

Theorem 5.98. *A generalised quadrangle $\mathcal{S}$ of order (s, t), $s^3 = t^2$ and $s \neq 1$, is isomorphic to $\mathcal{U}(4, s)$ if and only if every hyperbolic line has at least $\sqrt{s} + 1$ points.*

Relying on this result, another characterisation of $\mathcal{U}(4, s)$ can be given.

Theorem 5.99. *Let $\mathcal{S}$ have order (s, t) with $1 < s^3 \leq t^2$. Then $\mathcal{S}$ is isomorphic to $\mathcal{U}(4, s)$ if and only if each trace $\{P, P'\}^\perp$, with $P \not\sim P'$, is a plane of $\mathcal{S}^*$ which is generated by any three non-$\mathcal{S}^*$-collinear points in it.*

Next, conditions are given which simultaneously characterise several classical generalised quadrangles.

Definition 5.100. (1) A point U of $\mathcal{S}$ is *semi-regular* provided that $P'' \in \mathrm{cl}(P, P')$ whenever U is the unique centre of the triad $\{P, P', P''\}$.

(2) A point U has *property* (H) when $P'' \in \mathrm{cl}(P, P')$ if and only if $P \in \mathrm{cl}(P', P'')$, whenever $\{P, P', P''\}$ is a triad consisting of points of $U^{\perp}$. If follows that any semi-regular point has property (H).

Some examples are now given.

In $\mathcal{W}(q)$, $\mathcal{Q}(4, q)$, $\mathcal{Q}(5, q)$, and $\mathcal{U}(3, q^2)$ all points and lines are semi-regular and have property (H). In $\mathcal{U}(4, q^2)$ all points are semi-regular and have property (H); all lines have property (H). Finally, it is shown that no line of $\mathcal{U}(4, q^2)$ is semi-regular. Consider three distinct lines l, m, n of $\mathcal{U}(4, q^2)$ with $l \sim n \sim m \not\sim l$. Further, let r be a line of $\mathcal{U}(4, q^2)$ for which $r \sim n$, $l \not\sim r \not\sim m$, and which is not contained in the solid $\mathrm{PG}(3, q^2)$ defined by l and m. Then n is the unique centre of the triad $\{l, m, r\}$, but $r \notin \mathrm{cl}(l, m)$ since $\mathrm{cl}(l, m)$ consists of all lines concurrent with at least one of l, m. Hence n is not semi-regular. So property (H) does not imply semi-regularity.

The following characterisation of $\mathcal{U}(4, s^2)$ involves subquadrangles.

Theorem 5.101. *A generalised quadrangle $\mathcal{S}$ of order (s^2, s^3), with $s \neq 1$, is isomorphic to $\mathcal{U}(4, s^2)$ if and only if any two non-concurrent lines are contained in a proper subquadrangle of order (s^2, t), with $t \neq 1$.*

Now, six characterisations, most of them involving more than one classical generalised quadrangle, are stated without proof.

Theorem 5.102. *Let $\mathcal{S}$ have order (s, t) with $s \neq 1$. Then $|\{P, P'\}^{\perp\perp}| \geq (s^2/t) + 1$ for all P, P', with $P \not\sim P'$, if and only if one of the following occurs:*

(a) $t = s^2$;
(b) $\mathcal{S} \cong \mathcal{W}(s)$;
(c) $\mathcal{S} \cong \mathcal{U}(4, s)$.

Theorem 5.103. *In the generalised quadrangle $\mathcal{S}$ of order (s, t), each point has property (H) if and only if one of the following holds:*

(a) *each point is regular;*
(b) *each hyperbolic line has exactly two points;*
(c) $\mathcal{S} \cong \mathcal{U}(4, s)$.

Theorem 5.104. *Let $\mathcal{S}$ be a generalised quadrangle of order (s, t). Then each point is semi-regular if and only if one of the following occurs:*

(a) $s > t$ *and each point is regular;*
(b) $s = t$ *and $\mathcal{S} \cong \mathcal{W}(s)$;*
(c) $s = t$ *and each point is anti-regular;*
(d) $s < t$, *each hyperbolic line has exactly two points, and no triad of points has a unique centre;*
(e) $\mathcal{S} \cong \mathcal{U}(4, s)$.

Theorem 5.105. *In a generalised quadrangle $\mathcal{S}$ of order (s, t), all triads $\{P, P', P''\}$ with $P'' \notin \mathrm{cl}(P, P')$ have a constant number of centres if and only if one of the following occurs:*

(a) *all points are regular;*
(b) $s^2 = t$;
(c) $\mathcal{S} \cong \mathcal{U}(4, s)$.

Theorem 5.106. *The generalised quadrangle $\mathcal{S}$ of order $(s, t), s > 1$, is isomorphic to one of $\mathcal{W}(s), \mathcal{Q}(5, s)$, or $\mathcal{U}(4, s)$ if and only if, for each triad $\{P, P', P''\}$ with $P \notin \mathrm{cl}(P', P'')$, the set $\{P\} \cup \{P', P''\}^{\perp}$ is contained in a proper subquadrangle of order (s, t').*

Theorem 5.107. *Let $\mathcal{S}$ be a generalised quadrangle of order (s, t) with not all points regular. Then $\mathcal{S}$ is isomorphic to $\mathcal{Q}(4, s)$, with s odd, to $\mathcal{Q}(5, s)$, or to $\mathcal{U}(4, s)$ if and only if each set $\{P\} \cup \{P', P''\}^{\perp}$, where $\{P, P', P''\}$ is a triad with at least one centre and $P \notin \mathrm{cl}(P', P'')$, is contained in a proper subquadrangle of order (s, t').*

Next, a characterisation is given in terms of matroids. A finite *matroid* is an ordered pair $(\mathcal{P}, \mathfrak{M})$ where $\mathcal{P}$ is a finite set of elements called *points*, and $\mathfrak{M}$ is a closure operator which associates to each subset $\mathcal{X}$ of $\mathcal{P}$ a subset $\overline{\mathcal{X}}$, the *closure* of $\mathcal{X}$, such that the following conditions are satisfied:

(1) $\overline{\emptyset} = \emptyset$, and $\overline{\{P\}} = \{P\}$ for all $P \in \mathcal{P}$;
(2) $\mathcal{X} \subset \overline{\mathcal{X}}$ for all $\mathcal{X} \subset \mathcal{P}$;
(3) $\mathcal{X} \subset \overline{\mathcal{Y}} \Rightarrow \overline{\mathcal{X}} \subset \overline{\mathcal{Y}}$ for all $\mathcal{X}, \mathcal{Y} \subset \mathcal{P}$;
(4) $P' \in \overline{\mathcal{X} \cup \{P\}}, P' \notin \overline{\mathcal{X}} \Rightarrow P \in \overline{\mathcal{X} \cup \{P'\}}$ for all $P, P' \in \mathcal{P}$ and $\mathcal{X} \subset \mathcal{P}$.

The sets $\overline{\mathcal{X}}$ are called the *closed sets* of the matroid $(\mathcal{P}, \mathfrak{M})$. It is immediate that the intersection of closed sets is always closed. A closed set $\overline{\mathcal{X}}$ has *dimension* h if $h + 1$ is the minimum number of points in any subset of $\overline{\mathcal{X}}$ whose closure coincides with $\overline{\mathcal{X}}$. The closed sets of dimension 1 are the *lines* of the matroid.

Theorem 5.108. *Suppose that $\mathcal{S} = (\mathcal{P}, \mathcal{B}, \mathrm{I})$ is a generalised quadrangle of order (s, t), $s > 1$ and $t > 1$. Then $\mathcal{B}^* = \{\{P, P'\}^{\perp \perp} \mid P, P' \in \mathcal{P} \text{ and } P \neq P'\}$ is the line set and $\mathcal{P}$ is the point set of some matroid $(\mathcal{P}, \mathfrak{M})$ having all sets $P^{\perp}$, $P \in \mathcal{P}$, as closed sets, if and only if one of the following occurs:*

(a) $\mathcal{S} \cong \mathcal{W}(s)$;
(b) $\mathcal{S} \cong \mathcal{Q}(4, s)$;
(c) $\mathcal{S} \cong \mathcal{U}(4, s)$;
(d) $\mathcal{S} \cong \mathcal{Q}(5, s)$;
(e) *all points of $\mathcal{S}$ are regular, $s = t^2$, and every three non-$\mathcal{S}^*$-collinear points are contained in a proper linear variety of the linear space $\mathcal{S}^* = (\mathcal{P}, \mathcal{B}^*, \in)$.*

Now a characterisation of $\mathcal{Q}(4, q)$ and $\mathcal{Q}(5, q)$ is given that uses Theorem 5.123 on Moufang generalised quadrangles. The statement of the theorem, however, is purely combinatorial.

Let $\mathcal{S}$ be a generalised quadrangle of order $(s, t), s > 1$ and $t > 1$. A *quadrilateral* of $\mathcal{S}$ is just a subquadrangle of order $(1, 1)$. A quadrilateral $\mathcal{V}$ is said to be *opposite* a line l if the lines of $\mathcal{V}$ are not concurrent with l. If $\mathcal{V}$ is opposite l, the four lines incident with the points of $\mathcal{V}$ and concurrent with l are called the *lines of perspectivity of* V *from* l. Two quadrilaterals V and V' are in *perspective from* l if one of the following holds:

(1) $V = V'$ and V is opposite l;
(2) (i) $V \neq V'$,
    (ii) V and V' are both opposite l,
    (iii) the lines of perspectivity of V, and of V', from l are the same.

Theorem 5.109. *The generalised quadrangle* $S = (\mathcal{P}, \mathcal{B}, \mathrm{I})$ *of order* (s, t), $s > 1$ *and* $t > 2$, *is isomorphic to* $\mathcal{Q}(4, s)$ *or* $\mathcal{Q}(5, s)$ *if and only if, given a quadrilateral* V *opposite a line* l *and a point* P', P' I l, *incident with a line of perspectivity of* V *from* l, *there is a quadrilateral* V' *containing* P' *and in perspective with* V *from* l.

Remark 5.110. If $t = 2$ and $s > 1$, then, by Theorems 5.13 and 5.14, $s \in \{2, 4\}$. Now, by Theorem 5.73 and Corollary 5.82, $S \cong \mathcal{Q}(4, 2)$ or $S \cong \mathcal{U}(3, 4)$. It can be checked that in these two cases the quadrilateral condition of the preceding theorem is satisfied.

This section on characterisation theorems of purely combinatorial type is concluded with a fundamental characterisation of all classical and dual classical generalised quadrangles with $s > 1$ and $t > 1$. The reader is reminded of properties (A) and $(\hat{A})$ introduced in Definition 5.88.

Again, let $\mathcal{B}^{\perp\perp}$ be the set of all hyperbolic lines of the generalised quadrangle $S = (\mathcal{P}, \mathcal{B}, \mathrm{I})$, and let $S^{\perp\perp} = (\mathcal{P}, \mathcal{B}^{\perp\perp}, \in)$. Then S satisfies property (A) if, for any $m = \{P, P'\}^{\perp\perp} \in \mathcal{B}^{\perp\perp}$ and any $U \in \mathrm{cl}(P, P') \backslash (\{P, P'\}^{\perp} \cup \{P, P'\}^{\perp\perp})$, the substructure of $S^{\perp\perp}$ generated by m and U is a dual affine plane. The dual of (A) is denoted by $(\hat{A})$.

Theorem 5.111. *Let* $S = (\mathcal{P}, \mathcal{B}, \mathrm{I})$ *be a generalised quadrangle of order* (s, t), *with* $s > 1$ *and* $t > 1$. *Then* S *is a classical or a dual classical generalised quadrangle if and only if it satisfies either condition* (A) *or* $(\hat{A})$.

In the last part of this section, some important characterisations of classical generalised quadrangles, formulated in terms of automorphisms, are given without proof. The trivial cases $s = 1$ and $t = 1$ are excluded.

Definition 5.112. Let $S = (\mathcal{P}, \mathcal{B}, \mathrm{I})$ be a generalised quadrangle of order (s, t) with an automorphism τ.

(1) If τ fixes each point of $P^{\perp}$, $P \in \mathcal{P}$, then τ is a *symmetry about* P.
(2) If τ is the identity or if τ fixes each line incident with P but no point of $\mathcal{P} \backslash P^{\perp}$, then τ is an *elation about* P. It is possible to prove that any symmetry about P is automatically an elation about P.
(3) The generalised quadrangle S is an *elation generalised quadrangle* with *elation group* G and *base point* or *centre* P, if there is a group G of elations about P acting regularly on $\mathcal{P} \backslash P^{\perp}$; briefly, $(S^{(P)}, G)$ or $S^{(P)}$ is an elation generalised quadrangle.

(4) If the group G is abelian, then the elation generalised quadrangle $(\mathcal{S}^{(P)}, G)$ is a *translation generalised quadrangle* with *translation group* G and *base point* or *centre* P. It may be shown that the base point P of a translation generalised quadrangle is co-regular.

(5) If there is a group G of automorphisms fixing all lines incident with P and acting transitively, but not necessarily regularly, on $\mathcal{P} \backslash P^{\perp}$, then P is a *centre of transitivity*.

(6) If there is a group G of automorphisms fixing all points incident with l and acting transitively, but not necessarily regularly, on $\mathcal{B} \backslash l^{\perp}$, then l is an *axis of transitivity*.

(7) If the group of symmetries about P has maximum size t, then P is a *centre of symmetry*, in which case P is regular.

(8) If the group of symmetries about a line l has maximum size s, then l is an *axis of symmetry*; in this case, l is regular.

(9) Suppose that l and m are non-concurrent axes of symmetry of $\mathcal{S}$. Then it follows that every line of $\{l, m\}^{\perp \perp}$ is an axis of symmetry, and $\mathcal{S}$ is a *span-symmetric generalised quadrangle* with *base span* $\{l, m\}^{\perp \perp}$.

Definition 5.113. Let $P, P' \in \mathcal{P}$, $P \not\sim P'$.

(1) A *generalised homology* with *centres* P, P' is an automorphism τ of $\mathcal{S}$ which fixes all lines incident with P and all lines incident with P'. The group of all generalised homologies with centres P, P' is denoted $H(P, P')$.

(2) The generalised quadrangle $\mathcal{S}$ is (P, P')-*transitive* if, for each $P'' \in \{P, P'\}^{\perp}$, the group $H(P, P')$ is transitive on both the set $\{P, P''\}^{\perp} \backslash \{P, P''\}$ and the set $\{P', P''\}^{\perp} \backslash \{P', P''\}$.

Finally, the *Moufang conditions* are defined.

Definition 5.114. (1) The generalised quadrangle $\mathcal{S}$ is a *Moufang generalised quadrangle* if the following condition and its dual are satisfied:

for any point P and any two distinct lines l and m incident with P, the group of automorphisms of $\mathcal{S}$ fixing l and m point-wise and P line-wise is transitive on the lines ($\neq l$) incident with a given point P' on l, where $P' \neq P$.

(2) If one of these mutually dual conditions is satisfied, $\mathcal{S}$ is a *half Moufang generalised quadrangle*.

Definition 5.115. (1) Let $\{P, l\}$ be any incident point-line pair of $\mathcal{S}$; such a pair is called a *flag* of $\mathcal{S}$.

(2) Let $Q \,\mathrm{I}\, l$, with $Q \neq P$. The flag $\{P, l\}$ is called a *Moufang flag* if the group of automorphisms of $\mathcal{S}$ fixing l point-wise and P line-wise is transitive on the points collinear with Q which are not incident with l.

(3) If every flag of $\mathcal{S}$ is a Moufang flag, the generalised quadrangle $\mathcal{S}$ is *3-Moufang*.

(4) For the flag $\{P, l\}$ of $\mathcal{S}$, let $R \,\mathrm{I}\, m \,\mathrm{I}\, P$ and $n \,\mathrm{I}\, Q \,\mathrm{I}\, l$ such that $R \neq P$, $n \neq l$ and $Q \,\mathrm{Y}\, m$. The flag $\{P, l\}$ is *half 3-Moufang at P* if the group of automorphisms of $\mathcal{S}$ fixing P line-wise and l point-wise is transitive on the set $\{R, Q\}^{\perp} \backslash \{P\}$ for all R and Q; the flag $\{P, l\}$ is *half 3-Moufang at l* if the group of automorphisms of $\mathcal{S}$ fixing P line-wise and l point-wise, is transitive on the set $\{m, n\}^{\perp} \backslash \{l\}$ for all m and n.

(5) The generalised quadrangle S is *half* 3-*Moufang* if either every flag $\{P, l\}$ of S is half 3-Moufang at P, or every flag $\{P, l\}$ of S is half 3-Moufang at l.

(6) If all points of S are centres of transitivity, and all lines are axes of transitivity, then S is a 2-*Moufang generalised quadrangle*.

Theorem 5.116. *Let* $(S^{(P)}, G)$ *be a translation generalised quadrangle of order* (s, t), $s > 1$, $t > 1$.

(i) *If* s *is prime, then* $S \cong Q(4, s)$ *or* $S \cong Q(5, s)$.

(ii) *If all lines are regular, then* $S \cong Q(4, s)$ *or* $t = s^2$.

(iii) *Let* $t = s^2$ *with* s *odd. Then* $(S^{(P)}, G)$ *is isomorphic to* $Q(5, s)$ *if and only if for a fixed point* P', *with* $P' \not\sim P$, *the group* $H(P, P')$ *has order* $s - 1$.

(iv) *Let* $t = s^2$ *with* s *even. Then* $(S^{(P)}, G)$ *is isomorphic to* $Q(5, s)$ *if and only if all lines are regular and for a fixed point* P', *with* $P' \not\sim P$, *the group* $H(P, P')$ *has order* $s - 1$.

(v) *If* $t = s^2$, *with* $s = p^2$ *and* p *prime, and if all lines are regular, then* $(S^{(P)}, G)$ *is isomorphic to* $Q(5, s)$.

(vi) *Let* $t = s^2$ *with* s *even. Then* $(S^{(P)}, G)$ *is isomorphic to* $Q(5, s)$ *if and only if* S *has a classical subquadrangle* S' *of order* s *containing the point* P.

Theorem 5.117. *A generalised quadrangle* S *of order* (s, t), $s, t \neq 1$ *and* s *even for* $s \neq t$, *is a translation generalised quadrangle for two distinct collinear base points if and only if* S *is isomorphic to* $Q(4, s)$ *or* $Q(5, s)$.

It may be shown that $Q(4, q)$ and $Q(5, q)$ are translation and elation generalised quadrangles for any choice of the base point.

Theorem 5.118. *For* q *even, let* $S^{(P)}$ *be an elation generalised quadrangle of order* (q, q^2) *containing a classical subquadrangle* S' *of order* q *containing* P. *Then* $S^{(P)} \cong Q(5, q)$.

A characterisation of $Q(5, q)$ may also be given in terms of axes of symmetry.

Theorem 5.119. *A generalised quadrangle* S *of order* (s, s^2), $s \neq 1$, *is isomorphic to* $Q(5, s)$ *if and only if each line is an axis of symmetry.*

The following theorem concerns span-symmetric generalised quadrangles.

Theorem 5.120. *Let* S *be a span-symmetric generalised quadrangle of order* (s, t), *where* $1 < s \leq t < s^2$. *Then* $s = t$ *is a prime power and* S *is isomorphic to* $Q(4, s)$.

The following characterisations again involve more than one classical generalised quadrangle.

Theorem 5.121. *The group of symmetries about each point of the generalised quadrangle* S *of order* (s, t), $s > 1$, $t > 1$, *has even order if and only if one of the following holds:*

(a) $S \cong W(s)$;

(b) $\mathcal{S} \cong \mathcal{U}(3, s)$;
(c) $\mathcal{S} \cong \mathcal{U}(4, s)$.

Theorem 5.122. *Let* $\mathcal{S} = (\mathcal{P}, \mathcal{B}, \mathrm{I})$ *be a generalised quadrangle of order* (s, t), *with* $s \neq 1$, $t \neq 1$. *Then* $\mathcal{S}$ *is classical if and only if* $\mathcal{S}$ *is* (P, P')-*transitive for all* $P, P' \in \mathcal{P}$ *with* $P \not\sim P'$.

Theorem 5.123. *A generalised quadrangle* $\mathcal{S}$ *of order* (s, t), $s \neq 1$ *and* $t \neq 1$, *is a Moufang generalised quadrangle if and only if* $\mathcal{S}$ *is classical or dual classical.*

The next result allows the previous theorem to be considerably strengthened.

Theorem 5.124. *Let* $\mathcal{S}$ *be a generalised quadrangle of order* (s, t), $s \neq 1$ *and* $t \neq 1$. *Then* $\mathcal{S}$ *is a Moufang generalised quadrangle if and only if one of the following equivalent conditions holds:*

  (i) *it is a half Moufang generalised quadrangle;*
 (ii) *it is a 3-Moufang generalised quadrangle;*
(iii) *it is a half 3-Moufang generalised quadrangle;*
(iv) *it is a 2-Moufang generalised quadrangle.*

Remark 5.125. Many of the theorems in this section also hold in the infinite case.

5.6 Partial geometries

Definition 5.126. A (finite) *partial geometry* is an incidence structure $\mathcal{S} = (\mathcal{P}, \mathcal{B}, \mathrm{I})$ in which $\mathcal{P}$ is a set of *points*, $\mathcal{B}$ is a set of *lines* and I is a symmetric point-line incidence relation satisfying the following axioms:

(1) each point is incident with $1 + t$ lines, with $t \geq 1$, and two distinct points are incident with at most one line;
(2) each line is incident with $1 + s$ points, with $s \geq 1$, and two distinct lines are incident with at most one point;
(3) if P is a point and l is a line not incident with P, then, with $\alpha \geq 1$, there are exactly α points $P_1, P_2, \ldots, P_\alpha$ and α lines $l_1, l_2, \ldots, l_\alpha$ such that $P \mathrm{I} l_i \mathrm{I} P_i \mathrm{I} l$ for $i = 1, 2, \ldots, \alpha$.

Remark 5.127. From the axioms, a partial geometry with $\alpha = 1$ is a generalised quadrangle.

Definition 5.128. (1) The integers s, t, α are the *parameters* of the partial geometry.
(2) Given two points P, P' of $\mathcal{S}$ that are not necessarily distinct, they are *collinear*, written $P \sim P'$, if there is some line l for which $P \mathrm{I} l \mathrm{I} P'$; so $P \not\sim P'$ means that P and P' are not collinear.
(3) Dually, for $l, l' \in \mathcal{B}$, write $l \sim l'$ or $l \not\sim l'$ as they are *concurrent* or non-concurrent.

(4) The line incident with distinct collinear points P and P' is denoted PP'.
(5) The point incident with distinct concurrent lines l and l' is denoted $l \cap l'$.

Let $\mathcal{S} = (\mathcal{P}, \mathcal{B}, \mathrm{I})$ be a partial geometry with parameters s, t, α. Put $|\mathcal{P}| = v$ and $|\mathcal{B}| = b$.

Theorem 5.129.   (i) $v = (s + 1)(st + \alpha)/\alpha$;
(ii) $b = (t + 1)(st + \alpha)/\alpha$.

Proof. Let l be a fixed line of $\mathcal{S}$ and count in different ways the number of ordered pairs $(P, m) \in \mathcal{P} \times \mathcal{B}$ with $P \not\!\mathrm{I}\, l$, $P \,\mathrm{I}\, m$, and $l \sim m$. This gives

$$(v - s - 1)\alpha = (s + 1)ts,$$

whence the result. Dually, $b = (t + 1)(st + \alpha)/\alpha$.   □

Corollary 5.130. *The elements $st(s + 1)/\alpha$ and $st(t + 1)/\alpha$ are integers.*

Theorem 5.131. *The integer $\alpha(s + t + 1 - \alpha)$ divides $st(s + 1)(t + 1)$.*

Proof. This is analogous to the proof of Theorem 5.13.   □

Theorem 5.132 (The Krein inequalities). *The integers s, t, α satisfy the following inequalities*:

$$(s + 1 - 2\alpha)t \leq (s - 1)(s + 1 - \alpha)^2; \tag{5.7}$$
$$(t + 1 - 2\alpha)s \leq (t - 1)(t + 1 - \alpha)^2. \tag{5.8}$$

When equality holds in (5.7), *the number of points collinear with three points P_1, P_2, P_3 depends only on the number of collinearities in $\{P_1, P_2, P_3\}$; when equality holds in* (5.8), *the number of lines concurrent with three lines l_1, l_2, l_3 depends only on the number of concurrencies in $\{l_1, l_2, l_3\}$.*

Proof. See Section 5.10.   □

Partial geometries $\mathcal{S}$ can be divided into four, non-disjoint classes.

(I) $\mathcal{S}$ has $\alpha = s + 1$ or, dually, $\alpha = t + 1$; when $\alpha = s + 1$, then $\mathcal{S}$ is a 2-$(v, s + 1, 1)$ design.
(II) $\mathcal{S}$ has $\alpha = s$ or, dually, $\alpha = t$; when $\alpha = t$, then $\mathcal{S}$ is a net of order $s + 1$ and degree $t + 1$.
(III) When $\alpha = 1$, then $\mathcal{S}$ is a generalised quadrangle.
(IV) When $1 < \alpha < \min(s, t)$, then $\mathcal{S}$ is proper.

Example 5.133. (a) Let $\mathcal{S}$ be the design formed by the points and lines of $\mathrm{PG}(n, q)$, with $n \geq 2$; then $\mathcal{S}$ is a 2-$(\theta(n), q + 1, 1)$ design.
   (b) The points and lines of $\mathrm{AG}(n, q)$, $n \geq 2$, form a 2-$(q^n, q, 1)$ design.
   (c) Let $\mathcal{K}$ be a maximal $(qn - q + n; n)$-arc in $\mathrm{PG}(2, q)$, with $n \geq 2$. Then the points of $\mathcal{K}$ together with the non-empty intersections $l \cap \mathcal{K}$ with lines l of the plane

form a 2-$(qn - q + n, n, 1)$ design. For $n < q$, the points of $PG(2, q)\backslash\mathcal{K}$ and the lines having empty intersection with $\mathcal{K}$ form a dual design with parameters

$$s = q, \ t = q/n - 1, \ \alpha = q/n.$$

(d) When d classes of parallel lines in $AG(2, q)$ are deleted, where $0 \leq d \leq q-1$, then a net of order q and degree $q + 1 - d$ is obtained.

(e) If $\mathcal{P} = PG(n, q)\backslash\Pi_{n-2}$ and $\mathcal{B}$ is the set of lines of $PG(n, q)$ skew to Π_{n-2}, $n \geq 2$, with I the natural incidence, then $\mathcal{S} = (\mathcal{P}, \mathcal{B}, \mathrm{I})$ is a partial geometry with parameters $s = q$, $t = q^{n-1} - 1$, $\alpha = q$. This dual net is denoted by H_q^n.

(f) Let $\mathcal{K}$ be a maximal $(qn - q + n; n)$-arc in $PG(2, q)$, with $2 \leq n < q$; then, by Theorem 12.47 of PGOFF2, q is even. Let $\mathcal{P} = PG(2, q)\backslash\mathcal{K}$, let $\mathcal{B}$ be the set of all lines having non-empty intersection with $\mathcal{K}$, and let I be the natural incidence. Then $\mathcal{S}(\mathcal{K}) = (\mathcal{P}, \mathcal{B}, \mathrm{I})$ is a partial geometry with parameters

$$s = q - n, \quad t = q(n - 1)/n, \quad \alpha = (q - n)(n - 1)/n. \tag{5.9}$$

(g) Take $\mathcal{K}$ of (f) to be in the plane Π_2, which is then embedded in $PG(3, q)$. Now, let $\mathcal{P}' = PG(3, q)\backslash\Pi_2$, let $\mathcal{B}'$ be the set of all lines of $PG(3, q)$ meeting $\mathcal{K}$ in a single point, and let I' be the natural incidence. Then $T_2^*(\mathcal{K}) = (\mathcal{P}', \mathcal{B}', \mathrm{I}')$ is a partial geometry with parameters

$$s = q - 1, \quad t = (q + 1)(n - 1), \quad \alpha = n - 1. \tag{5.10}$$

(h) By Theorem 12.12 of PGOFF2, there exist maximal $(2^{m+h} - 2^h + 2^m; 2^m)$-arcs in $PG(2, 2^h)$ for any m with $1 \leq m < h$. Hence there exist partial geometries with the following parameters:

(a) $s = 2^h - 2^m, \quad t = 2^h - 2^{h-m}, \quad \alpha = (2^m - 1)(2^{h-m} - 1);$ (5.11)

(b) $s = 2^h - 1, \quad t = (2^h + 1)(2^m - 1), \quad \alpha = 2^m - 1.$ (5.12)

Such a partial geometry has $\alpha = 1$ or is proper. A partial geometry of type (a) is a generalised quadrangle if and only if $h = 2$ and $m = 1$. This gives the following model of the unique generalised quadrangle with 15 points and 15 lines: points of $\mathcal{S}$ are the 15 points of $PG(2, 4)\backslash\mathcal{K}$, with $\mathcal{K}$ a given oval; lines of $\mathcal{S}$ are the 15 bisecants of $\mathcal{K}$; incidence is the natural one. A partial geometry of type (b) is a generalised quadrangle if and only if $m = 1$. In this case, $\mathcal{K}$ is an oval of $PG(2, q)$, $q = 2^h$, and $T_2^*(\mathcal{K})$ is the generalised quadrangle described in Section 5.1.

Up to duality, the parameters of the known partial geometries are the following:

(1) $s = 2^h - 2^m, \quad t = 2^h - 2^{h-m}, \quad \alpha = (2^m - 1)(2^{h-m} - 1)$, with $h \neq 2$ and $1 \leq m < h$;

(2) $s = 2^h - 1, \quad t = (2^h + 1)(2^m - 1), \quad \alpha = 2^m - 1, \quad$ with $1 < m < h$;

(3) $s = 2^{2h-1} - 1, \quad t = 2^{2h-1}, \quad \alpha = 2^{2h-2}, \quad$ with $h > 1$;

(4) $s = 3^{2m} - 1, \quad t = (3^{4m} - 1)/2, \quad \alpha = (3^{2m} - 1)/2, \quad m \geq 1$;

(5) $s = 26, \quad t = 27, \quad \alpha = 18$;

(6) $s = t = 5, \quad \alpha = 2$;

(7) $s = 4, \quad t = 17, \quad \alpha = 2$;

(8) $s = 8, \quad t = 20, \quad \alpha = 2$.

5.7 Embedded partial geometries

Definition 5.134. (1) A *projective partial geometry* $S = (\mathcal{P}, \mathcal{B}, \mathrm{I})$ is a partial geometry for which the point set $\mathcal{P}$ is a subset of the point set of some projective space $\mathrm{PG}(n, q)$ and the line set $\mathcal{B}$ is a set of lines of $\mathrm{PG}(n, q)$.

(2) In this case, S is *embedded* in $\mathrm{PG}(n, q)$.

(3) If $\mathrm{PG}(n', q)$ is the subspace of $\mathrm{PG}(n, q)$ generated by the points of $\mathcal{P}$, then it is the *ambient space* of S.

Theorem 5.135. *If* $S = (\mathcal{P}, \mathcal{B}, \mathrm{I})$ *is a partial geometry with parameters* s, t, α *which is projective with ambient space* $\mathrm{PG}(n, s)$, *then one of the following holds:*

(a) $\alpha = s + 1$ *and* S *is the* 2-$(\theta(n), s + 1, 1)$ *design formed by the points and lines of* $\mathrm{PG}(n, s)$;

(b) $\alpha = 1$ *and* S *is a classical generalised quadrangle;*

(c) $\alpha = t + 1$, $n = 2$, $\mathrm{PG}(2, s) \backslash \mathcal{P}$ *is a maximal* $(sd - s + d; d)$*-arc* $\mathcal{K}$ *of* $\mathrm{PG}(2, s)$, s *even, with* $d = s/\alpha$ *and* $2 \leq d < s$, *and* $\mathcal{B}$ *consists of all lines of* $\mathrm{PG}(2, s)$ *not meeting* $\mathcal{K}$;

(d) $\alpha = s$, $n \geq 2$ *and* $S = H_s^n$.

Proof. If $\alpha = s + 1$, then S is a 2-$(v, s + 1, 1)$ design. Hence S consists of all points and lines of a subspace Π_m of $\mathrm{PG}(n, s)$. Since $\mathrm{PG}(n, s)$ is the ambient space of S, so $m = n$. Therefore S is the design formed by the points and lines of $\mathrm{PG}(n, s)$.

If $\alpha = 1$, then by Theorem 5.51 the partial geometry S is a classical generalised quadrangle.

Now let $\alpha = t + 1$. Since any two lines of S meet, the ambient space of S is a plane $\mathrm{PG}(2, s)$. Each line of $\mathrm{PG}(2, s)$ not in $\mathcal{B}$ has exactly s/α points in $\mathrm{PG}(2, s) \backslash \mathcal{P}$. If $\alpha = s$, then S is the dual affine plane H_s^2. If $2 \leq d < s$ with $d = s/\alpha$, then $\mathrm{PG}(2, s) \backslash \mathcal{P}$ is a maximal $(sd - s + d; d)$-arc $\mathcal{K}$; so s is even by Theorem 12.47 of PGOFF2, and $\mathcal{B}$ is the set of all lines of $\mathrm{PG}(2, s)$ not meeting $\mathcal{K}$.

Now suppose that $1 < \alpha < s$ and $\alpha \neq t + 1$. In this case, $n \geq 3$.

First, let $n = 3$. Suppose that l is a line of S and P is a point of S with $P \notin l$; let π be the plane Pl of $\mathrm{PG}(3, s)$. The points and lines of S in π constitute a partial geometry S_π with parameters

$$s_\pi = s, \quad t_\pi = \alpha - 1, \quad \alpha_\pi = \alpha.$$

Hence the points of π not in S_π form a maximal $(s(s\alpha^{-1} + \alpha^{-1} - 1); s\alpha^{-1})$-arc of π. Consequently, s is even by Theorem 12.47 of PGOFF2.

Let m be any line of $\mathrm{PG}(3, s)$ that contains at least two points P', P'' of S. Take a line $m' \in \mathcal{B}$, with $m \neq m'$ and $P' \in m'$. Considering the plane $P''m'$, the set $P''m' \backslash \mathcal{P}$ is an $(s(s\alpha^{-1} + \alpha^{-1} - 1); s\alpha^{-1})$-arc of the plane $P''m'$, and therefore $|m \cap \mathcal{P}| \in \{s + 1, s + 1 - s\alpha^{-1}\}$. Hence each line of $\mathrm{PG}(3, s)$ meets $\mathcal{P}$ in one of $0, 1, s + 1 - s\alpha^{-1}$, or $s + 1$ points; that is, $\mathcal{P}$ is a set of type $(0, 1, s + 1 - s\alpha^{-1}, s + 1)$ in $\mathrm{PG}(3, s)$. Here, the $(s + 1)$-secants of $\mathcal{P}$ are the lines of S.

Now it is shown that $\mathcal{P}$ has no 1-secants. Suppose the contrary and that l is a 1-secant with $\{P\} = l \cap \mathcal{P}$. Let the lines through P be $m_1, m_2, \ldots, m_{t+1}$. Suppose that each plane lm_i contains exactly $s+1$ points of $\mathcal{P}$. Since $n=3$, each line m of $\mathcal{B}$ contains at least one point of $\mathcal{P} \cap lm_i = m_i$ for every i. It follows that all lines of $\mathcal{S}$ lie in a common plane; so $\mathrm{PG}(3,s)$ is not the ambient space of $\mathcal{S}$, a contradiction. Consequently, $|\mathcal{P} \cap lm_i| > s+1$ for at least one index i; say, $P' \in (\mathcal{P} \cap lm_i)\backslash m_i$. From the previous paragraph, the line l contains $s+1-s\alpha^{-1}$ or $s+1$ points of the partial geometry $\mathcal{S}_{\pi_i}$, where $\pi_i = P'm_i = lm_i$. Hence $1 \in \{s+1, s+1-s\alpha^{-1}\}$, a contradiction. So $\mathcal{P}$ has no 1-secant; that is, $\mathcal{P}$ is of type $(0, s+1-s\alpha^{-1}, s+1)$.

Next, it is shown that such a set cannot exist when $1 < \alpha < s$. Counting the points of $\mathcal{P}$ on all lines of $\mathrm{PG}(3,s)$ containing a fixed point of $\mathcal{P}$ gives

$$|\mathcal{P}| = 1 + (t+1)s + (s^2+s-t)(s-s\alpha^{-1}). \tag{5.13}$$

Also, from Theorem 5.129,

$$|\mathcal{P}| = (s+1)(st+\alpha)/\alpha. \tag{5.14}$$

From (5.13) and (5.14), it follows that $t = (s+1)(\alpha-1)$. Since $\alpha \neq s+1$, so $\mathcal{P} \neq \mathrm{PG}(3,s)$. Taking $A \in \mathrm{PG}(3,s)\backslash\mathcal{P}$ and counting the points of $\mathcal{P}$ on all lines of $\mathrm{PG}(3,s)$ containing A shows that $s+1-s\alpha^{-1}$ divides $|\mathcal{P}|$. Hence $s\alpha + \alpha - s$ divides

$$(s+1)(s^2\alpha + s\alpha - s^2 - s + \alpha) = (s\alpha + \alpha - s)(s^2 + s + 1) - s^2.$$

Thus $s\alpha + \alpha - s$ divides s^2. With $s = 2^h$, $\alpha = 2^k$, where $0 < k < h$, this becomes that $2^h + 1 - 2^{h-k}$ divides 2^{2h-k}, a contradiction.

It has been shown that, for $1 < \alpha < s$ and $\alpha \neq t+1$, necessarily $n > 3$.

So, let $1 < \alpha < s$, $\alpha \neq t+1$, with $n > 3$. Let l be a line of $\mathcal{S}$, let π be the plane defined by l and a point P in $\mathcal{P}\backslash l$, and let $\mathrm{PG}(3,s)$ be the solid defined by π and a point P' in $\mathcal{P}\backslash\pi$. Let P_1, P_2 be distinct points of $\mathcal{P}$ in $\mathrm{PG}(3,s)$. Counting the number of pairs (l_1, l_2), with $l_1, l_2 \in \mathcal{B}$, and both in $\mathrm{PG}(3,s)$ with $P_1 \in l_1$, $P_2 \in l_2$, and $l_1 \sim l_2$, in different ways, it appears that the number of lines of $\mathcal{B}$ in $\mathrm{PG}(3,s)$ containing P_1 equals the number of lines of $\mathcal{B}$ in $\mathrm{PG}(3,s)$ containing P_2. It follows that the points and lines of $\mathcal{S}$ in $\mathrm{PG}(3,s)$ constitute a partial geometry $\mathcal{S}'$ with parameters t', $s' = s$, $\alpha' = \alpha$. Since $1 < \alpha' < s'$ and $\alpha' \neq t'+1$ as $\mathcal{S}'$ is not contained in a plane, such a geometry cannot exist.

So the only possibilities for α are 1, $s+1$, $t+1$, s.

Consider, therefore, the case that $\alpha = s$ with $\alpha \neq 1, t+1$; then $n \geq 3$. Let l be a line of $\mathcal{S}$, and suppose that the point P of $\mathcal{S}$ is not on l. The points and lines of $\mathcal{S}$ in the plane $\pi = Pl$ form a partial geometry with parameters

$$s = s', \; t' = \alpha - 1 = s' - 1, \; \alpha' = \alpha = s';$$

that is, it is a dual affine plane of order s. If the line m of $\mathrm{PG}(n,s)$ contains at least two points of $\mathcal{P}$, then m lies in at least one plane π in which $\mathcal{S}$ induces a dual affine plane of order s. Hence $\mathcal{P}$ is a set of type $(0, 1, s, s+1)$ in $\mathrm{PG}(n,s)$, and a line m of $\mathrm{PG}(n,s)$ contains $s+1$ points of $\mathcal{P}$ if and only if m belongs to $\mathcal{B}$. Also, if $\mathcal{P}$ has no 1-secant, then all planes of $\mathrm{PG}(n,s)$ through a fixed line l of $\mathcal{S}$ contain a point of $\mathcal{P}\backslash l$; hence the points of $\mathcal{S}$ in such a plane are the points of a dual affine plane of order s. It follows that

$$|\mathcal{P}| = s + 1 + (s^2 - 1)(s^{n-1} - 1)/(s - 1) = s^n + s^{n-1}. \tag{5.15}$$

Conversely, if $|\mathcal{P}| = s^n + s^{n-1}$, then $\mathcal{P}$ admits no 1-secant.

Now it is shown that $\mathcal{P}$ has no 1-secant. First, let $n = 3$. By an argument analogous to the above, $\mathcal{P}$ has no 1-secant; so $\mathcal{P} = s^3 + s^2$. Now, induction is used. Suppose that any projective partial geometry with $\alpha = s$ but $\alpha \neq 1, t + 1$, and ambient space $\mathrm{PG}(n-1, s)$, $n \geq 4$, has no 1-secant. Next, assume that $\mathcal{S} = (\mathcal{P}, \mathcal{B}, \mathrm{I})$ is a projective partial geometry with these parameters and ambient space $\mathrm{PG}(n, s)$, $n \geq 4$, which has at least one 1-secant l.

Let $l \cap \mathcal{P} = \{P\}$. Consider a line m of $\mathcal{B}$ containing P and $n - 2$ points $P_1, P_2, \ldots, P_{n-2}$ of $\mathcal{P}$ such that $m, P_1, \ldots, P_{n-2}$ generate a hyperplane Π_{n-1}. The geometry induced by $\mathcal{S}$ in Π_{n-1} is a partial geometry with parameters $\bar{t}, \bar{s} = s = \bar{\alpha}$ and ambient space Π_{n-1}. Here, $\bar{\alpha} \neq 1, \bar{t} + 1$ since $n - 1 > 2$. By the induction hypothesis,

$$|\bar{\mathcal{P}}| = |\mathcal{P} \cap \Pi_{n-1}| = s^{n-1} + s^{n-2}. \tag{5.16}$$

Let $m_1, m_2, \ldots, m_{t+1}$ be the lines of $\mathcal{B}$ through P. The plane lm_1 contains $s + 1$ points of $\mathcal{S}$. If the intersection of $\mathcal{P}$ and the solid lm_1m_i, with $i > 1$, generates lm_1m_i, then the set $\mathcal{P} \cap lm_1m_i$ has no 1-secant, a contradiction. Hence, for all $i > 1$, the set $\mathcal{P} \cap lm_1m_i$ is contained in the plane m_1m_i, whence $|\mathcal{P} \cap lm_1m_i| = s^2 + s$.

Let P' be any point of $\mathcal{P}\backslash m_1$. Then the plane $P'm_1$ contains s lines through P' which also belong to $\mathcal{B}$. Therefore P' belongs to at least one of the solids lm_1m_i, with $i \neq 1$. It follows that

$$|\mathcal{P}| \leq \theta(n - 3)(s^2 - 1) + s + 1 = s^{n-1} + s^{n-2}, \tag{5.17}$$

where $\theta(n - 3)$ is the number of solids containing the plane lm_1. From (5.16) and (5.17), it now follows that $\mathcal{P} = \mathcal{P} \cap \Pi_{n-1}$; hence $\mathrm{PG}(n, s)$ is not the ambient space of $\mathcal{S}$, a contradiction. So $\mathcal{P}$ has no 1-secant.

Therefore $|\mathcal{P}| = s^n + s^{n-1}$ and $\mathcal{P}$ is of type $(0, s, s+1)$. Hence $\mathrm{PG}(n, s)\backslash\mathcal{P}$ is of type $(0, 1, s + 1)$ and so $\mathrm{PG}(n, s)\backslash\mathcal{P}$ is a subspace Π_{n-2} of $\mathrm{PG}(n, s)$ of dimension $n - 2$. The lines of $\mathcal{B}$ are the lines of $\mathrm{PG}(n, s)$ skew to Π_{n-2}. Thus $\mathcal{S} = H_s^n$, and the theorem is established. $\qquad \square$

Any projective partial geometry $\mathcal{S} = (\mathcal{P}, \mathcal{B}, \mathrm{I})$ satisfies the Veblen–Pasch axiom:

(VP) if $l_1 \mathrm{I} P \mathrm{I} l_2$, $l_1 \neq l_2$, $m_1 \not{I} P \not{I} m_2$, $l_i \sim m_j$, for all $i, j \in \{1, 2\}$, then also $m_1 \sim m_2$.

The known partial geometries satisfying (VP) are as follows:

(i) all known generalised quadrangles;
(ii) all known partial geometries with $\alpha = t + 1$;
(iii) the partial geometries isomorphic to the design formed by the points and lines of some $\mathrm{PG}(n, q)$;

(iv) the partial geometries isomorphic to some H_q^n.

Theorem 5.136. *Let S be a dual net of order $s + 1$ and degree $t + 1$ with $s < t + 1$. If S satisfies* (VP), *then S is isomorphic to a partial geometry H_q^n with parameters $s = q$, $t = q^{n-1} - 1$.*

Proof. See Section 5.10. □

5.8 $(0, \alpha)$-geometries and semi-partial geometries

Definition 5.137. A (finite) $(0, \alpha)$-*geometry*, where $\alpha > 1$, is an incidence structure $S = (\mathcal{P}, \mathcal{B}, I)$ in which $\mathcal{P}$ and $\mathcal{B}$ are disjoint, non-empty sets of *points* and *lines*, and for which I is a symmetric point-line incidence relation satisfying the following axioms:

(1) two distinct points are incident with at most one line;
(2) if a point P and a line l are not incident, then there are 0 or α points which are collinear with P and incident with l;
(3) each line is incident with at least two points and each point is incident with at least two lines;
(4) S is connected; this means that, for any two elements T and T' of $\mathcal{P} \cup \mathcal{B}$, there exist elements $T_1, T_2, \ldots, T_r \in \mathcal{P} \cup \mathcal{B}$ such that $T I T_1 I T_2 I \cdots I T_r I T'$.

Terms such as 'collinear' and 'concurrent' and notation such as $\sim$ and $\not\sim$ are defined as for generalised quadrangles and partial geometries.

Theorem 5.138. *Each point is incident with a constant number $1 + t$ of lines and each line is incident with a constant number $1 + s$ of points, where $t, s \geq 1$.*

Proof. Let P and P', with $P \neq P'$, be collinear points of S; let $1 + t$ and $1 + t'$ be the respective number of lines incident with P and P'. Counting in different ways the number of ordered pairs (l, l'), with $P I l$, $P' I l'$, $l \neq l'$, $l \sim l'$, gives

$$t(\alpha - 1) = t'(\alpha - 1);$$

hence $t = t'$. By the connectedness of S, each point of S is incident with $1 + t$ lines. Dually, each line is incident with $1 + s$ points. □

Definition 5.139. The integers s, t, α are the *parameters* of the $(0, \alpha)$-geometry.

Let $|\mathcal{P}| = v$ and $|\mathcal{B}| = b$. It should be noted that v and b are not uniquely determined by s, t, α.

Definition 5.140. A (finite) *semi-partial geometry* is an incidence structure S, where $S = (\mathcal{P}, \mathcal{B}, I)$, in which $\mathcal{P}$ and $\mathcal{B}$ are disjoint, non-empty sets of *points* and *lines*, and for which I is a symmetric point-line incidence relation satisfying the following axioms:

(1) each point is incident with $1 + t$ lines, with $t \geq 1$, and two distinct points are incident with at most one line;
(2) each line is incident with $1 + s$ points, with $s \geq 1$, and two distinct lines are incident with at most one point;
(3) if a point P and a line l are not incident, then there are 0 or α points, where $\alpha \geq 1$, which are collinear with P and incident with l;
(4) if two points are not collinear, then there are μ points, where $\mu > 0$, collinear with both;
(5) The integers s, t, α, μ are the *parameters* of the semi-partial geometry.

From the definitions, a semi-partial geometry with $\alpha > 1$ is a $(0, \alpha)$-geometry. The semi-partial geometries with $\alpha = 1$ are also called *partial quadrangles*. A semi-partial geometry is a partial geometry if and only if the zero in axiom (3) does not occur; this is equivalent to the condition $\mu = (t + 1)\alpha$. This gives the following diagram, where '$\longrightarrow$' indicates 'generalises to':

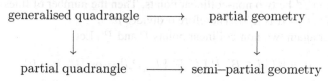

Theorem 5.141. *Let* $S = (\mathcal{P}, \mathcal{B}, I)$ *be a semi-partial geometry with parameters* $s, t, \alpha, \lambda, \mu$ *and with* $|\mathcal{P}| = v$, $|\mathcal{B}| = b$. *Then*

(i) $v(t + 1) = b(s + 1)$;
(ii) $v = 1 + (1 + t)s(1 + t(s - \alpha + 1)/\mu)$.

Proof. Counting the ordered pairs (P, l), with $P \in \mathcal{P}$, $l \in \mathcal{B}$, $P I l$, in different ways gives

$$v(t + 1) = b(s + 1).$$

Now, counting the ordered triples (P, P', P''), with $P, P', P'' \in \mathcal{P}$ and $P \sim P'$, $P \not\sim P''$, $P' \sim P''$, in different ways gives

$$v(t + 1)st(s + 1 - \alpha) = v(v - (t + 1)s - 1)\mu;$$

this implies the result.   □

Corollary 5.142. *Both the following are integers:*

$$st(t + 1)(s + 1 - \alpha)/\mu, \quad st(\mu + (t + 1)^2(s + 1 - \alpha))/(\mu(s + 1)).$$

Proof. This follows from the fact that v and b are integers.   □

Theorem 5.143. *For* $\alpha \neq s + 1$,
  (i) α *divides* $st(t + 1)$ *and* $st(s + 1)$;
 (ii) α *divides* μ;
(iii) α^2 *divides* μst;

(iv) α^2 *divides* $t(\alpha(t+1) - \mu)$;
(v) $\alpha^2 \leq \mu \leq \alpha(t+1)$.

Proof. For any non-incident point-line pair (P, l), the symbol $[P, l]$ is the number of points collinear with P and incident with l. For any line l, let v_l be the number of points P for which $[P, l] = \alpha$; for any point P, let b_P be the number of lines l for which $[P, l] = \alpha$. For a fixed line l, the following set is counted in two different ways:

$$\{(P, l') \mid P \in \mathcal{P},\ l' \in \mathcal{B},\ P \not I l,\ P I l',\ l \sim l'\}.$$

This gives $v_l\, \alpha = (s+1)ts$, and so α divides $st(s+1)$. Similarly, for a fixed point P, the following set is counted in two different ways:

$$\{(P', l) \mid P' \in \mathcal{P},\ l \in \mathcal{B},\ P \not I l,\ P' I l,\ P \sim P'\}.$$

This gives $b_P\, \alpha = (t+1)st$, and so α divides $st(t+1)$.

Let P and P' be two non-collinear points. Then the number of lines l for which $P I l$ and $[P', l] = \alpha$ is μ/α. Therefore α divides μ.

Consider again two non-collinear points P and P'. Let

$$\beta = |\{l \mid l \in \mathcal{B},\ P \not I l,\ P' \not I l,\ [P, l] = \alpha,\ [P', l] = 0\}|.$$

Now count in different ways the following set:

$$\{(l, l') \mid l, l' \in \mathcal{B},\ P \not I l,\ P' \not I l,\ P I l',\ l \sim l',\ [P', l] = 0\}.$$

Since there are μ/α lines l' with $P I l'$ and $[P', l'] = \alpha$, so

$$\begin{aligned}
\beta\alpha &= (t+1-\mu/\alpha)s(t-\mu/\alpha) + (\mu/\alpha)(s-\alpha)(t+1-\mu/\alpha)\\
&= ((t+1)\alpha - \mu)(st - \mu)/\alpha.
\end{aligned}$$

Since α divides μ and also $st(t+1)$, so α^2 divides μst.

Now consider two distinct collinear points P and P'. Then the set

$$\{l \mid l \in \mathcal{B},\ P \not I l,\ P' \not I l,\ [P, l] = \alpha,\ [P', l] = 0\}$$

has size

$$t(s+1-\alpha)(t+1-\mu/\alpha)/\alpha.$$

Since α divides all three of μ, $st(t+1)$, μst, so α^2 divides $t(\alpha(t+1) - \mu)$.

Finally, let P and P' be two non-collinear points. The number of lines l with $P I l$ and $[P', l] = \alpha$ is μ/α. Hence $\mu/\alpha \leq t+1$. Equality occurs if and only if $\mathcal{S}$ is a partial geometry. Let $P I m$, $P' I m'$, $m \sim m'$. Since $[P, m'] = \alpha$, there are at least α lines l with $P I l$ and $[P', l] = \alpha$. Therefore $\alpha \leq \mu/\alpha$. $\square$

If $\alpha = s+1$, then $\mathcal{S}$ is a 2-$(v, s+1, 1)$ design.

Theorem 5.144. *If $\mathcal{S} = (\mathcal{P}, \mathcal{B}, I)$ is a semi-partial geometry with parameters s, t, α, μ and with $\alpha \neq s+1$, then*

(i) $D = (t(\alpha - 1) + s - 1 - \mu)^2 + 4((t + 1)s - \mu)$ *is a square, except in the case* $s = t = \alpha = \mu = 1$ *where* S *is a pentagon and* $D = 5$.

(ii) $[2(t + 1)s + (v - 1)(t(\alpha - 1) + s - 1 - \mu + D^{1/2})]/(2D^{1/2})$ *is an integer.*

Proof. Let $\mathcal{E} = \{\{P, P'\} \mid P, P' \in \mathcal{P}, \ P \sim P'\}$. Then $(\mathcal{P}, \mathcal{E})$ is a strongly regular graph with the following parameters:

$$v = 1 + (1 + t)s(1 + t(s - \alpha + 1)/\mu),$$
$$k = n_1 = st + s,$$
$$\lambda = p_{11}^1 = t(\alpha - 1) + s - 1,$$
$$\mu = p_{11}^2 = \mu.$$

The graph $(\mathcal{P}, \mathcal{E})$ is the *point graph* of the semi-partial geometry.

Let $\mathcal{P} = \{P_1, P_2, \ldots, P_v\}$ and let $A = [a_{ij}]$ be the $v \times v$ matrix over $\mathbf{R}$ for which $a_{ij} = 0$ if $i = j$ or $P_i \nsim P_j$ and $a_{ij} = 1$ if $P_i \sim P_j$; that is, A is an adjacency matrix of the graph $(\mathcal{P}, \mathcal{E})$. If $A^2 = [c_{ij}]$, then

(a) $c_{ii} = (t + 1)s$;

(b) $i \neq j$ and $P_i \nsim P_j$ imply that $c_{ij} = \mu$;

(c) $i \neq j$ and $P_i \sim P_j$ imply that $c_{ij} = t(\alpha - 1) + s - 1$.

So

$$A^2 - (t(\alpha - 1) + s - 1 - \mu)A - (s(t + 1) - \mu)I = \mu J, \qquad (5.18)$$

where I is the identity matrix and J is the all-one matrix.

The matrix A has an eigenvalue $s(t + 1)$, whereas J has an eigenvalue 0 with multiplicity $v - 1$ and v with multiplicity 1. Since

$$(s(t + 1))^2 - (t(\alpha - 1) + s - 1 - \mu)s(t + 1) - (s(t + 1) - \mu) = \mu v,$$

the eigenvalue $s(t + 1)$ of A corresponds to the eigenvalue v of J, and so $s(t + 1)$ has multiplicity 1. The other eigenvalues of A are roots of the equation

$$x^2 - (t(\alpha - 1) + s - 1 - \mu)x - (s(t + 1) - \mu) = 0. \qquad (5.19)$$

Denote the multiplicities of these eigenvalues θ_1, θ_2 by m_1, m_2. The discriminant

$$D = (t(\alpha - 1) + s - 1 - \mu)^2 + 4(s(t + 1) - \mu).$$

If $D = 0$, then, as $\mu \leq (t + 1)\alpha \leq (t + 1)s$, so

$$t(\alpha - 1) + s - 1 - \mu = s(t + 1) - \mu = 0;$$

hence $t(\alpha - (s + 1)) = 1$, a contradiction. Therefore $D \neq 0$ and so $\theta_1 \neq \theta_2$. From (5.19),

$$\theta_1 = (t(\alpha - 1) + s - 1 - \mu + D^{1/2})/2,$$
$$\theta_2 = (t(\alpha - 1) + s - 1 - \mu - D^{1/2})/2.$$

Since $1 + m_1 + m_2 = v$ and $s(t+1) + m_1\theta_1 + m_2\theta_2 = \sum a_{ii} = 0$, so

$$m_2 = (2(t+1)s + (v-1)[t(\alpha-1) + s - 1 - \mu + D^{1/2}])/(2D^{1/2}).$$

It follows that

$$(2(t+1)s + (v-1)[t(\alpha-1) + s - 1 - \mu + D^{1/2}])/(2D^{1/2})$$

is an integer.

Now suppose that D is not a square. Then, as m_2 is an integer,

$$2(t+1)s + (v-1)[t(\alpha-1) + s - 1 - \mu] = 0. \tag{5.20}$$

Since $(v-1) > (t+1)s$, so

$$0 < \mu - t(\alpha-1) - s + 1 < 2.$$

Hence $\mu = s + t(\alpha-1)$. From (5.20),

$$v = 2(t+1)s + 1. \tag{5.21}$$

By Theorem 5.141,

$$v = 1 + (t+1)s \left(1 + \frac{t(s-\alpha+1)}{s+t(\alpha-1)}\right). \tag{5.22}$$

From (5.21) and (5.22), it follows that

$$\alpha = 1 + s(t-1)/(2t). \tag{5.23}$$

Hence

$$\mu = s(t+1)/2. \tag{5.24}$$

From (5.23), $2t$ divides $s(t-1)$; so t divides s. Since $\mu \leq \alpha(t+1)$, so $s \leq 2t$ by (5.23) and (5.24). Since t divides s, so $s \in \{t, 2t\}$.

First, suppose $s = 2t$. Then $\alpha = t$ and $\mu = t(t+1)$, which in turn implies that $D = (1 + 2t)^2$, a contradiction since D is not a square.

So $s = t$, $\alpha = (t+1)/2$ and $\mu = t(t+1)/2$. By Theorem 5.143, α^2 divides $\mu s t$, and so $t+1$ divides $2t^3$; hence $t = 1$. Consequently,

$$s = t = \alpha = \mu = 1, \qquad v = b = 5, \qquad D = 5.$$

This means that S is a pentagon. $\qquad\qquad\qquad\qquad\qquad\qquad\qquad\qquad \Box$

Theorem 5.145. *If S is a semi-partial geometry, but not a partial geometry, then $b \geq v$.*

Proof. Let $\mathcal{S} = (\mathcal{P}, \mathcal{B}, I)$ be a semi-partial geometry with parameters s, t, α, μ and assume that $\mathcal{S}$ is not a partial geometry; so $\mu < \alpha(t+1)$.

Let $\mathcal{P} = \{P_1, P_2, \ldots, P_v\}$ and $\mathcal{B} = \{l_1, l_2, \ldots, l_b\}$. The corresponding adjacency matrix of the point graph of $\mathcal{S}$ is denoted by $A = [a_{ij}]$. Let $M = [m_{ij}]$ be the $v \times b$ matrix over $\mathbf{R}$ for which $m_{ij} = 0$ if $P_i \not I l_j$ and $m_{ij} = 1$ if $P_i I l_j$; that is, M is an incidence matrix of the geometry $\mathcal{S}$. It follows that

$$MM^* = (t+1)I + A. \tag{5.25}$$

Suppose that $v > b$. As rank $M \le b$, so rank $MM^* \le b$. Therefore the $v \times v$ matrix MM^* is singular, whence $-(t+1)$ is an eigenvalue of A. Then, by (5.19),

$$(t+1)^2 + (t(\alpha-1) + s - 1 - \mu)(t+1) - (s(t+1) - \mu) = 0,$$

which is equivalent to $t(\mu - \alpha(t+1))$. Hence $\mu = \alpha(t+1)$, and so $\mathcal{S}$ is a partial geometry, which is a contradiction.

Hence, if $\mathcal{S}$ is not a partial geometry, then $b \ge v$. $\square$

Theorem 5.146. *The dual of a semi-partial geometry $\mathcal{S}$ is a semi-partial geometry if and only if $\mathcal{S}$ is a partial geometry or $v = b$. If $v = b$, then $\mathcal{S}$ and its dual $\overline{\mathcal{S}}$ have the same parameters.*

Proof. Let $\mathcal{S} = (\mathcal{P}, \mathcal{B}, I)$ be a semi-partial geometry with dual $\overline{\mathcal{S}}$.

Suppose that $\overline{\mathcal{S}}$ is also a semi-partial geometry. If $\mathcal{S}$ is not a partial geometry, then $\overline{\mathcal{S}}$ is not a partial geometry. By Theorem 5.145, $b \ge v$ and $v \ge b$. Hence $\mathcal{S}$ is a partial geometry or $v = b$.

If $\mathcal{S}$ is a partial geometry, then $\overline{\mathcal{S}}$ is a partial geometry and consequently also a semi-partial geometry. Now suppose that $\mathcal{S}$ is not a partial geometry, but let $b = v$. With the notation of Theorem 5.145, since $\mu \ne \alpha(t+1)$, so $-(t+1)$ is not an eigenvalue of A. Hence MM^* is non-singular, as is the $v \times v$ matrix M. By (5.25),

$$M^* = (s+1)M^{-1} + M^{-1}A,$$

as $b = v$ and so $s = t$. Hence

$$M^*M = (s+1)I + M^{-1}AM.$$

Let $M^{-1}AM = B = [b_{ij}]$. Then $b_{ii} + s + 1$ is the number of points incident with l_i; so $b_{ii} = 0$. Further, b_{ij}, for $i \ne j$, is the number of points incident with l_i and l_j. Hence B is an adjacency matrix of the point graph of $\overline{\mathcal{S}}$. Also $B^2 = M^{-1}A^2M$; hence, by (5.18),

$$B^2 = (s\alpha - 1 - \mu)B + (s(s+1) - \mu)I + \mu M^{-1}JM.$$

Since $MJ = JM = (s+1)J$, so

$$B^2 = (s\alpha - 1 - \mu)B + (s(s+1) - \mu)I + \mu J. \tag{5.26}$$

Let $B^2 = [d_{ij}]$; then, by (5.26), $d_{ij} = \mu$ for $l_i \sim l_j$. As

$$d_{ij} = \sum_{r=1}^{v} b_{ir} b_{jr},$$

so d_{ij} is the number of lines l_r with $l_i \sim l_r \sim l_j$. It follows that $\overline{S}$ is also a semi-partial geometry with the same parameters as S.   □

A list of some of the known $(0, \alpha)$-geometries and semi-partial geometries is now given.

Example 5.147. (a) Let $\mathcal{O}$ be an ovoid of $\mathrm{PG}(3, q)$, $q > 2$, or an elliptic quadric of $\mathrm{PG}(3, 2)$. Suppose that $\mathrm{PG}(3, q)$ is a hyperplane of $\mathrm{PG}(4, q)$. The points of S are those of $\mathrm{PG}(4, q) \backslash \mathrm{PG}(3, q)$. The lines of S are the lines of $\mathrm{PG}(4, q)$ that contain a point of $\mathcal{O}$ but are not contained in $\mathrm{PG}(3, q)$. Here, I is the incidence of $\mathrm{PG}(4, q)$. Then S is a partial quadrangle with parameters

$$s = q - 1, \quad t = q^2, \quad \alpha = 1, \quad \mu = q^2 - q.$$

(b) Consider a subgeometry $\mathrm{PG}(2, q)$ of the plane $\mathrm{PG}(2, q^2)$ and suppose that $\mathrm{PG}(2, q^2)$ is a plane of $\mathrm{PG}(3, q^2)$. The points of S are those of $\mathrm{PG}(3, q^2) \backslash \mathrm{PG}(2, q^2)$ and the lines of S are the lines of $\mathrm{PG}(3, q^2)$ that contain a point of $\mathrm{PG}(2, q)$ but are not contained in $\mathrm{PG}(2, q^2)$. Here, I is the incidence of $\mathrm{PG}(3, q^2)$. Then S is a semi-partial geometry with parameters

$$s = q^2 - 1, \quad t = q(q + 1), \quad \alpha = q, \quad \mu = q(q + 1).$$

(c) Let $\mathcal{U}$ be a Hermitian arc in $\mathrm{PG}(2, q^2)$. With the same construction as in (b), a semi-partial geometry is obtained; its parameters are

$$s = q^2 - 1, \quad t = q^3, \quad \alpha = q, \quad \mu = q^2(q^2 - 1).$$

(d) Let Π_{n-2} be an $(n - 2)$-dimensional subspace of $\mathrm{PG}(n, q)$, $n \geq 3$. The points of S are the lines of $\mathrm{PG}(n, q)$ skew to Π_{n-2}. The lines of S are the planes of $\mathrm{PG}(n, q)$ meeting Π_{n-2} in a point. Incidence here is inclusion. Then S is a semi-partial geometry with parameters

$$s = q^2 - 1, \quad t = q^{n-2} + q^{n-3} + \cdots + q, \quad \alpha = q, \quad \mu = q(q + 1).$$

(e) Let $\mathcal{P}$ be the set of all lines in $\mathrm{PG}(n, q)$, $n \geq 3$, let $\mathcal{B}$ be the set of all planes in $\mathrm{PG}(n, q)$, and let I be inclusion. Then $S = (\mathcal{P}, \mathcal{B}, \mathrm{I})$ is a semi-partial geometry with parameters

$$s = q(q + 1), \quad t = q^{n-2} + q^{n-3} + \cdots + q, \quad \alpha = q + 1, \quad \mu = (q + 1)^2.$$

(f) Let $\mathcal{U}$ be a non-singular Hermitian surface in $\mathrm{PG}(3, q^2)$. Points of S are the points of $\mathrm{PG}(3, q^2) \backslash \mathcal{U}$; lines of S are the 1-secants of $\mathcal{U}$ and incidence is the natural one. Then S is a semi-partial geometry with parameters

$$s = q^2 - 1, \quad t = q^3, \quad \alpha = q + 1, \quad \mu = q(q + 1)(q^2 - 1).$$

(g) Let $\mathcal{Q}$ be an elliptic quadric of $\mathrm{PG}(5, q)$. Let P be a point off $\mathcal{Q}$ and let Π_4 be a hyperplane not containing P. The projection of $\mathcal{Q}$ from P onto Π_4 is denoted by $\mathcal{Q}'$. Also, let $\mathcal{Q}''$ be the set of points P' of $\mathcal{Q}'$ for which PP' is a 1-secant of $\mathcal{Q}$. If q is odd, then $\mathcal{Q}''$ is a non-singular quadric of Π_4; if q is even, then $\mathcal{Q}''$ is a solid of Π_4. Let $\mathcal{P} = \mathcal{Q}'\backslash\mathcal{Q}''$, let $\mathcal{B}$ be the set of all lines of Π_4 that are contained in $\mathcal{Q}'$ but not in $\mathcal{Q}''$, and let I be the incidence of Π_4. Then $\mathcal{S} = (\mathcal{P}, \mathcal{B}, \mathrm{I})$ is a semi-partial geometry with parameters

$$s = q - 1, \quad t = q^2, \quad \alpha = 2, \quad \mu = 2q(q - 1).$$

For q even, see also Section 2.6.

(h) Let $\mathcal{U}$ be a non-singular Hermitian surface in $\mathrm{PG}(3, q^2)$, and let l be a fixed line of $\mathcal{U}$. The points of $\mathcal{S}$ are the points of $\mathcal{U}\backslash l$; lines of $\mathcal{S}$ are the lines of $\mathcal{U}$ not meeting l and incidence is containment. Then the dual of $\mathcal{S}$ is a partial quadrangle with parameters

$$s = q - 1, \quad t = q^2, \quad \alpha = 1, \quad \mu = q^2 - q.$$

(i) Let V be a set with h elements, $h \geq 4$, let $V_2 = \{T \subset V \mid |T| = 2\}$, and let $V_3 = \{T \subset V \mid |T| = 3\}$. If I is inclusion, then $\mathcal{S}_h = (V_2, V_3, \mathrm{I})$ is a semi-partial geometry with parameters

$$s = \alpha = 2, \quad t = m - 3, \quad \mu = 4.$$

(j) Let M be an $(n + 1) \times (n + 1)$ skew-symmetric matrix over $\mathbf{F}_q$ with $n \geq 2$; then rank $(M) = 2k$ with $k > 0$. The mapping ζ from $\mathrm{PG}(n, q)$ to its dual defined by M is a null polarity when M is non-singular. The subspace of $\mathrm{PG}(n, q)$ containing all points having no image with respect to ζ is the *radical* $\mathcal{R}$ of ζ; it has dimension $n - 2k$.

The points of $\mathcal{S}$ are the points of $\mathrm{PG}(n, q)\backslash\mathcal{R}$. The lines are the lines l of $\mathrm{PG}(n, q)\backslash\mathcal{R}$ for which $l \not\subset l\zeta$ when $n \geq 3$ and $l\zeta \not\subset l$ when $n = 2$. The incidence is the natural one. Then $\mathcal{S} = (\mathcal{P}, \mathcal{B}, \mathrm{I})$ is a $(0, \alpha)$-geometry with parameters

$$s = \alpha = q, \quad t = q^{n-1} - 1.$$

This geometry is denoted by $\overline{W(n, 2k, q)}$.

When $k = 1$, then $\overline{W(n, 2k, q)}$ is the dual net H_q^n introduced in Section 5.6; see also Section 5.7. When $2k = n + 1$ and so n is odd, then this is the case of the null polarity. Here, $\overline{W(n, n + 1, q)}$ is a semi-partial geometry with $\mu = q^{n-1}(q - 1)$, and is also denoted $\overline{W(n, q)}$. In all other cases, $\overline{W(n, 2k, q)}$ is not a semi-partial geometry; that is, it is a *proper* $(0, \alpha)$-geometry.

(k) Take a quadric $\mathcal{Q}$ in $\mathrm{PG}(n, 2)$, $n \geq 3$, and suppose that $\mathcal{Q}$ is not one of the following:

(1) Π_{n-1};
(2) $\Pi_{n-1} \cup \Pi'_{n-1}$, $\Pi_{n-1} \neq \Pi'_{n-1}$;
(3) $\mathcal{H}_3$ in $\mathrm{PG}(3, 2)$;
(4) $\Pi_{n-4}\mathcal{H}_3$, $n \geq 4$.

Let $\mathcal{B}$ be the set of lines skew to $\mathcal{Q}$, let $\mathcal{P}$ be the set of points of $PG(n,2)$ on at least one line of $\mathcal{B}$, and let I be the natural incidence. Then $\mathcal{S} = (\mathcal{P},\mathcal{B},I)$ is a $(0,2)$-geometry.

If $n = 2d - 1$ and $\mathcal{Q} = \mathcal{E}_n$, then $\mathcal{S}$ is a semi-partial geometry, denoted by $NQ^-(2d-1,2)$ with parameters

$$s = \alpha = 2, \quad t = 2^{2d-3} + 2^{d-2} - 1, \quad \mu = 2^{2d-3} + 2^{d-1}.$$

If $n = 2d - 1$ and $\mathcal{Q} = \mathcal{H}_n$, then $\mathcal{S}$ is a semi-partial geometry, denoted by $NQ^+(2d-1,2)$ with parameters

$$s = \alpha = 2, \quad t = 2^{2d-3} - 2^{d-2} - 1, \quad \mu = 2^{2d-3} - 2^{d-1}.$$

If $n = 2d$ and $\mathcal{Q} = \mathcal{P}_n$, then $\mathcal{S}$ is a semi-partial geometry, denoted by $NQ(2d,2)$ with parameters

$$s = \alpha = 2, \quad t = 2^{2d-2} - 1, \quad \mu = 2^{2d-1} - 2^{2d-2}.$$

In all other cases, $\mathcal{S}$ is a proper $(0,2)$-geometry.

(l) Let $\mathcal{Q} = \mathcal{H}_3$ in $PG(3,2^h)$, $h \geq 2$. Let $\mathcal{B}$ be the set of lines skew to $\mathcal{Q}$, let $\mathcal{P}$ be the set of points of $PG(3,2^h) \backslash \mathcal{Q}$ and let I be the incidence of $PG(3,2^h)$. Then $\mathcal{S} = (\mathcal{P},\mathcal{B},I)$ is a proper $(0,2^{h-1})$-geometry $NQ^+(3,2^h)$ with parameters

$$s = 2^h, \quad t = 2^{2h-1} - 2^{h-1} - 1.$$

.

(m) Let $\mathcal{Q} = \mathcal{E}_5$ in $PG(5,q)$, q odd. Let $\mathcal{P}$ be a subset of $\mathcal{Q}$ which meets every line of $\mathcal{Q}$ in $\frac{1}{2}(q+1)$ points. Let $\mathcal{B}$ be the set of lines of $\mathcal{Q}$ and let I be the incidence of $PG(5,q)$. Then $\mathcal{S} = (\mathcal{P},\mathcal{B},I)$ is a partial quadrangle with parameters

$$s = \tfrac{1}{2}(q-1), \quad t = q^2, \quad \mu = \tfrac{1}{2}(q-1)^2.$$

For each odd q, at least one example is known.

(n) For the parameter sets in the following table, there is at least one semi-partial geometry:

s	t	α	μ	Conditions
1	$r-1$	1	1	$r = 2, 3, 7$
6	6	6	36	
1	9	1	2	
1	15	1	4	
1	21	1	6	
2	10	1	2	
2	55	1	20	
3	77	1	14	
$q^{m+1}-1$	q^{m+2}	q^m	$q^{m+1}(q^{m+1}-1)$	$q = 2^h,\ m \geq 2$
q^m-1	q^{m+1}	$2q^{m-1}$	$2q^m(q^m-1)$	$q = p^h,\ m = 2;$
				$q = 2^h,\ m \geq 3$

5.9 Embedded $(0, \alpha)$-geometries and semi-partial geometries

A *projective* $(0, \alpha)$-geometry or semi-partial geometry $\mathcal{S} = (\mathcal{P}, \mathcal{B}, \mathrm{I})$ is a $(0, \alpha)$-geometry or semi-partial geometry whose point set $\mathcal{P}$ is a subset of the point set of some $\mathrm{PG}(n, q)$ and whose line set $\mathcal{B}$ is a set of lines of $\mathrm{PG}(n, q)$. Also, $\mathcal{S}$ is *embedded* in $\mathrm{PG}(n, q)$. If $\mathrm{PG}(n', q)$ is the subspace of $\mathrm{PG}(n, q)$ generated by the points of $\mathcal{P}$, then $\mathrm{PG}(n', q)$ is the *ambient space* of $\mathcal{S}$.

A $(0, \alpha)$-geometry or semi-partial geometry embedded in $\mathrm{PG}(2, q)$ is a partial geometry. As these were classified in Section 5.7, the dimension of the ambient space is now taken to be at least 3.

Let $\mathcal{S} = (\mathcal{P}, \mathcal{B}, \mathrm{I})$ be a projective $(0, \alpha)$-geometry or semi-partial geometry with parameters s, t, α, where $\alpha > 1$, and with ambient space $\mathrm{PG}(3, s)$. Consider a pair l, l' of intersecting lines and let $\pi = l\, l'$. The points and lines of $\mathcal{S}$ in π constitute a partial geometry $\mathcal{S}(\pi)$ with parameters $s(\pi) = s$, $t(\pi) = \alpha - 1$, $\alpha(\pi) = \alpha$, together with $m(\pi)$ isolated points, where a point P of $\mathcal{S}$ in π is *isolated* if no line of $\mathcal{S}$ through P is contained in π. By Theorem 5.135, there are the following possibilities for $\mathcal{S}(\pi)$.

(a) The parameter $\alpha = s + 1$ and $\mathcal{S}(\pi)$ is the 2-$(s^2 + s + 1, s + 1, 1)$ design formed by all points and lines of π.

(b) The points of π not in $\mathcal{S}(\pi)$ form a maximal $(sd - s + d; d)$-arc $\mathcal{K}(\pi)$ of π, with s even, $d = s/\alpha$, and $2 \leq d < s$. The lines of $\mathcal{S}(\pi)$ are the lines of π not meeting $\mathcal{K}(\pi)$.

(c) The parameter $\alpha = s$ and π contains exactly one point $P(\pi)$ which is not in $\mathcal{S}(\pi)$. The lines of $\mathcal{S}(\pi)$ are the lines of π not containing $P(\pi)$.

Lemma 5.148. (i) *The number $m = m(\pi)$ of isolated points in π is independent of the choice of π.*
  (ii) *The number of lines in $\mathcal{S}$ is*

$$b = \alpha^{-1}(s\alpha - s + \alpha)[(s + 1)(t + 1) - \alpha s] + m(t + 1), \qquad (5.27)$$

and the number of points of $\mathcal{S}$ is

$$v = [\alpha(t + 1)]^{-1}(s + 1)(s\alpha - s + \alpha)[(s + 1)(t + 1) - \alpha s] + m(s + 1). \quad (5.28)$$

Proof. The number of points of $\mathcal{S}(\pi)$ is $(s + 1)(s\alpha - s + \alpha)/\alpha$ and the number of lines is $s\alpha - s + \alpha$. The number of lines of $\mathcal{S}$ containing exactly one point of $\mathcal{S}(\pi)$ is $(t + 1 - \alpha)(s + 1)(s\alpha - s + \alpha)/\alpha$ and the number of lines of $\mathcal{S}$ containing an isolated point in π is $(t + 1)m(\pi)$. Hence

$$b = s\alpha - s + \alpha + \alpha^{-1}(t + 1 - \alpha)(s + 1)(s\alpha - s + \alpha) + m(\pi)(t + 1),$$

giving the result. So $m = m(\pi)$ is independent of the choice of π.

Counting the number of ordered pairs (P, l), with $P \in \mathcal{P}$, $l \in \mathcal{B}$, $P\,\mathrm{I}\,l$, in different ways shows that $|\mathcal{P}|(t + 1) = |\mathcal{B}|(s + 1)$, from which (5.28) follows. □

Lemma 5.149. *With respect to $\mathcal{S}$, there are three possible types of planes in the ambient space $\mathrm{PG}(3, s)$:*

(A) *those containing* $s\alpha - s + \alpha$ *lines of* S *and*

$$\rho_{\mathrm{a}} = \alpha^{-1}(s+1)(s\alpha - s + \alpha) + m \tag{5.29}$$

    points of S;
(B) *those containing exactly one line of* S *and*

$$\rho_{\mathrm{b}} = s + 1 + [\alpha(t+1)]^{-1}s(s+1)(t+1-\alpha)(\alpha-1) + m \tag{5.30}$$

    points of S;
(C) *those containing no line of* S *and*

$$\rho_{\mathrm{c}} = [\alpha(t+1)]^{-1}(s\alpha - s + \alpha)[(s+1)(t+1) - \alpha s] + m \tag{5.31}$$

    points of S.

Proof. Let π be a plane of the ambient space $\mathrm{PG}(3, s)$.

If π contains at least two lines of S, then the points and lines of S in π constitute a partial geometry $S(\pi)$ with parameters

$$s(\pi) = s, \quad t(\pi) = \alpha - 1, \quad \alpha(\pi) = \alpha,$$

together with m isolated points. Hence π contains $s\alpha - s + \alpha$ lines of S and ρ_{a} points of S.

Next, suppose that π contains exactly one line of S. Then, with ρ_{b} the number of points of S in π,
$$b = 1 + (s+1)t + (\rho_{\mathrm{b}} - s - 1)(t+1).$$

So, by Lemma 5.148, ρ_{b} is as stated.

Finally, suppose that π contains no line of S. If ρ_{c} is the number of points of S in π, then
$$b = \rho_{\mathrm{c}}(t+1).$$

Again, Lemma 5.148 gives the result. $\qquad\square$

Corollary 5.150. *If there is at least one plane of type* (B) *and at least one plane of type* (C), *then* $t + 1$ *divides* s.

Proof. From the theorem, $(t+1)(\rho_{\mathrm{b}} - \rho_{\mathrm{c}}) = s$, whence the result. $\qquad\square$

Theorem 5.151. *If* $S = (\mathcal{P}, \mathcal{B}, \mathrm{I})$ *is a projective semi-partial geometry with parameters* s, t, α, μ *and with ambient space* $\mathrm{PG}(3, s)$, *then one of the following holds:*

(a) $\alpha = s + 1$ *and* S *is the* $2\text{-}((s^2 + 1)(s+1), s+1, 1)$ *design formed by all points and lines of* $\mathrm{PG}(3, s)$;
(b) $\alpha = 1$ *and* S *is a classical generalised quadrangle;*
(c) $\alpha = s$ *and* $S = H_s^3$;
(d) $\alpha = s$ *and* $S = \overline{W}(3, s)$;
(e) $\alpha = s = 2$ *and* $S = NQ^-(3, 2)$.

Proof. If $\mathcal{S}$ is a partial geometry, then Theorem 5.135 gives one of the first three cases.

So, suppose that $\mathcal{S}$ is not a partial geometry. Then $\mu < (t + 1)\alpha$ and, from Theorems 5.143 and 5.145, $\alpha^2 \leq \mu$ and $|\mathcal{B}| = b \geq |\mathcal{P}| = v$. So $b(s + 1) = v(t + 1)$ implies that $t \geq s$. By Theorem 5.143, α divides μ, and so $\mu < (t + 1)\alpha$ implies that $\mu \leq \alpha t$.

Let $\alpha = 1$. Then, by Theorem 5.141,

$$v = 1 + (1 + t)s(1 + ts/\mu).$$

Hence $v \leq s^3 + s^2 + s + 1$ implies $\mu(s^2 + s - t) \geq st(t + 1)$. Since $\mu \leq t$, it follows that $st(t + 1) \leq t(s^2 + s - t)$ and so $t \leq s - 1 + 1/(s + 1) < s$, a contradiction. Hence $\alpha \neq 1$.

For the next step, let $\alpha = s + 1$. From the connectedness of $\mathcal{S}$, it follows that $\mathcal{S}$ is a 2-$(v, s + 1, 1)$ design, a contradiction since $\mathcal{S}$ is not a partial geometry. Hence $\alpha \neq s + 1$. As there is always a plane of type (A), either $s = \alpha$ or $s = 2^h$.

Now suppose that there is a plane of type (C). The number of points of $\mathcal{S}$ in π is $\rho_c = v/(s + 1)$. Let P be a fixed point of $\mathcal{S}$ in π, and count in different ways the set

$$\{(P', P'') \mid P', P'' \in \mathcal{P}\backslash\{P\}, \; P'' \in \pi, \; PP', P'P'' \in \mathcal{B}\}.$$

This gives

$$(\rho_c - 1)\mu = (t + 1)st, \quad \rho_c = 1 + (t + 1)st/\mu,$$

and so

$$v = s + 1 + (s + 1)(t + 1)st/\mu. \tag{5.32}$$

By Theorem 5.141,

$$v = 1 + (1 + t)s(1 + t(s - \alpha + 1)/\mu). \tag{5.33}$$

From (5.32) and (5.33), it follows that $\mu = \alpha(t + 1)$, a contradiction. Consequently, there is no plane of type (C).

From (5.28),

$$m = \frac{v}{s + 1} - \frac{(s\alpha - s + \alpha)[(s + 1)(t + 1) - \alpha s]}{\alpha(t + 1)}, \tag{5.34}$$

with v given by (5.33).

Let $P, P' \in \mathcal{P}$ with $P \neq P'$ and $PP' \notin \mathcal{B}$, and let π be a plane containing PP'. If π is of type (B), or if π is of type (A) with at least one of P, P' isolated in π, then π does not contain a point $P'' \in \mathcal{P}\backslash\{P, P'\}$ with PP'' and $P'P''$ in $\mathcal{B}$. If π is of type (A) and neither P nor P' is isolated in π, then π contains exactly α^2 points $P'' \in \mathcal{P}\backslash\{P, P'\}$ for which PP'' and $P'P''$ are lines of $\mathcal{B}$. Considering all planes through PP', it follows that α^2 divides μ.

Now suppose that all planes of $PG(3, s)$ are of type (A). Counting all points of $\mathcal{P}$ in all planes through a given line l of $\mathcal{B}$,

$$v = s + 1 + (s + 1)(\rho_{\mathrm{a}} - s - 1), \tag{5.35}$$

with ρ_{a} given by (5.29). Hence

$$m = \frac{v}{s+1} - \frac{(s+1)(s\alpha - s + \alpha)}{\alpha} + s. \tag{5.36}$$

From (5.34) and (5.36), it follows that

$$t = (s+1)(\alpha - 1). \tag{5.37}$$

Counting in different ways the set $\{(P, \pi) \mid P \in \mathcal{P}, \pi$ a plane through $P\}$,

$$v(s^2 + s + 1) = (s^3 + s^2 + s + 1)\rho_{\mathrm{a}}. \tag{5.38}$$

Eliminating ρ_{a} from (5.35) and (5.38) gives

$$v = (s^2 + 1)(s + 1),$$

and so $\mathcal{P} = \mathrm{PG}(3, s)$. Now, from Theorem 5.141 and (5.37),

$$\mu = (\alpha - 1)(s\alpha - s + \alpha).$$

Since α divides μ, it divides s. Let $s = p^h$ and $\alpha = p^r$, with p prime and $r \leq h$. Since α^2 divides μ, so p^{2r} divides $p^{h+r} - p^h + p^r$, whence p^r divides $p^h - p^{h-r} + 1$. Hence $h = r$, which means that $s = \alpha$. So

$$s = \alpha, \quad t = \alpha^2 - 1, \quad \mu = \alpha^2(\alpha - 1), \quad m = 1.$$

So any plane contains exactly one point $P(\pi)$ not in $\mathcal{P}$, and the lines of $\mathcal{B}$ in π are the lines not containing $P(\pi)$. Therefore the structure $\mathcal{S}'$ consisting of all points of $\mathrm{PG}(3, s)$ and all lines of $\mathrm{PG}(3, s)$ not in $\mathcal{B}$ is a projective generalised quadrangle. By Theorem 5.51, $\mathcal{S}'$ is a classical generalised quadrangle; so, since $\mathrm{PG}(3, s)$ is the point set of $\mathcal{S}'$, the structure $\mathcal{S}'$ is the generalised quadrangle $\mathcal{W}(s)$ arising from a null polarity of $\mathrm{PG}(3, s)$. Thus $\mathcal{S}$ is the semi-partial geometry $\overline{W}(3, s)$.

Next, suppose that there is at least one plane π of type (B). Let l be the unique line of $\mathcal{B}$ in π. Fix a point P on l and count in different ways the set

$$\{(P', P'') \mid P', P'' \in \mathcal{P} \backslash \{P\}, \ P'' \in \pi \backslash l, \ P' \notin \pi, \ PP', P'P'' \in \mathcal{B}\}.$$

This gives

$$(\rho_{\mathrm{b}} - s - 1)\mu = ts(t + 1 - \alpha). \tag{5.39}$$

Now fix a point Q in $\pi \backslash l$, and count in different ways the set

$$\{(Q', Q'') \mid Q', Q'' \in \mathcal{P} \backslash \{Q\}, \ Q'' \in \pi \backslash \{Q\}, \ Q' \notin \pi, \ QQ', Q'Q'' \in \mathcal{B}\}.$$

Hence

$$(\rho_{\mathrm{b}} - 1)\mu = ts(t + 1). \tag{5.40}$$

From (5.39) and (5.40), it follows that $\mu = \alpha t$. Hence

$$\rho_b = \alpha^{-1}s(t+1) + 1,$$

and, from Theorem 5.141,

$$v = 1 + \alpha^{-1}s(t+1)(s+1).$$

Eliminating t gives

$$v = 1 + (\rho_b - 1)(s+1). \tag{5.41}$$

Count in different ways the set

$$\{(R, r) \mid R \in \mathcal{P} \cap \pi, \; r \in \mathcal{B} \backslash \{l\}, \; R \in r\};$$

this gives

$$b - 1 = (s+1)t + (\rho_b - s - 1)(t+1).$$

Since $v(t+1) = b(s+1)$, it follows that

$$v = \frac{(s+1)(st + t + 1)}{t+1} + (\rho_b - s - 1)(s+1). \tag{5.42}$$

From (5.41) and (5.42), it follows that $s = t$. By Theorem 5.144, either

$$D = 1 + 4s(s+1-\alpha)$$

is a square or $D = 5$. Since $1 < \alpha < s+1$, so $1 + 4s(s+1-\alpha) > 5$, and so D is a square. Consequently, there exists a positive integer g for which

$$s(s+1-\alpha) = g(g+1).$$

As s is a prime power, it divides either g or $g+1$. Hence

$$g + 1 \geq s > s + 1 - \alpha \geq g.$$

It follows that $s + 1 - \alpha = g$, $s = g + 1$, $\alpha = 2$. Therefore

$$s = t, \quad \alpha = 2, \quad \mu = 2t, \quad D = (2s - 1)^2.$$

Also, since α^2 divides μ, so $t = s = 2^h$. By Theorem 5.144, $2\sqrt{D}$ divides

$$2(t+1)s + (v-1)(t(\alpha - 1) + s - 1 - \mu + \sqrt{D}),$$

and so $2^{h+1} - 1$ divides $(2^h + 1)(2^{2h} + 1)2^{h-1}$. Hence $h \in \{1, 3\}$. This gives two cases:

(I) $s = t = \alpha = 2$, $\mu = 4$;
(II) $s = t = 8$, $\alpha = 2$, $\mu = 16$.

Case (I). Here $v = 10$. Now, from (5.34), (5.29) and (5.30),

$$m = 0, \quad \rho_a = 6, \quad \rho_b = 4.$$

Since there are no planes of type (C), no three points of $Q = \mathrm{PG}(3,2)\backslash P$ are collinear. Hence Q is a 5-cap of $\mathrm{PG}(3,2)$. Since $\rho_a = 6$ and $\rho_b = 4$ so any plane of $\mathrm{PG}(3,2)$ contains one or three points of Q. Therefore Q is an elliptic quadric and B consists of the 10 external lines of Q. So it has been shown that $S = NQ^-(3,2)$.

Case (II). Here $v = 325$. Now, from (5.34), (5.29) and (5.30),

$$m = 0, \quad \rho_a = 45, \quad \rho_b = 37.$$

Let π be a plane of type (B), let l be the line of B in π, and let K be the set of the 28 points of P in $\pi\backslash l$. If P, P' are distinct points of K, then, since $\mu > 0$, it follows that there is at least one plane π' of type (A) through PP'. Since π' is of type (A) and $m = 0$, the line PP' contains five points of $\pi' \cap P$. Since PP' contains exactly one point of l, it contains four points of K. Hence K is a maximal $(28; 4)$-arc of π. Therefore each line of $\mathrm{PG}(3,8)$ has $1, 5$ or 9 points in P. In fact, from Section 19.4 of FPSOTD, P is a non-singular $28_{5,3,8}$ of $\mathrm{PG}(3,8)$. Then, from Theorems 19.4.8 and 19.4.9, $v = (8^3/2) + 8^2 + 8 + 1 = 329$, contradicting that $v = 325$. So this case cannot occur.   □

Theorem 5.152. *Let $S = (P, B, \mathrm{I})$ be a $(0, \alpha)$-geometry with parameters s, t, α, which is projective with ambient space $\mathrm{PG}(3, s)$. If $m = 0$, then one of the following holds:*

(a) *$\alpha = s + 1$ and S is the $2\text{-}((s^2 + 1)(s + 1), s + 1, 1)$ design formed by all points and lines of $\mathrm{PG}(3, s)$;*
(b) *$\alpha = s$ and $S = H_s^3$;*
(c) *$\alpha = s = 2$ and $S = NQ^-(3, 2)$.*

Proof. From the definition of a $(0, \alpha)$-geometry, $\alpha > 1$.

Suppose that all lines of some plane π of type (A) belong to B; then $\alpha = s + 1$. It follows that all lines of each plane of type (A) belong to B. Let l be a line of B not contained in π. If $l \cap \pi = P$, and if $l_1, l_2, \ldots, l_{s+1}$ are the lines of π through P, then all lines of the plane ll_i through P belong to B, for $i = 1, 2, \ldots, s + 1$; hence $t = s^2 + s$. From (5.27), $b = (s^2 + 1)(s^2 + s + 1)$ and, from (5.28), $v = (s^2 + 1)(s + 1)$. This gives (a).

Now suppose that, in each plane of type (A), there is at least one line not in B. Then either $s = \alpha$ or $s = 2^h$. Let π be a plane of type (A) and let l be a line of π which does not belong to B. Since $m = 0$, so $|l \cap P| = (s\alpha - s + \alpha)/\alpha$. Let τ be the number of planes of type (A) through l. The number of lines of B containing a point of l and contained in a plane of type (A) through l is $\tau(s\alpha - s + \alpha)$. Hence the number of planes of type (B) through l is

$$(t + 1)(s\alpha - s + \alpha)/\alpha - \tau(s\alpha - s + \alpha) = (s\alpha - s + \alpha)(t + 1 - \tau\alpha)/\alpha. \quad (5.43)$$

From (5.43), $t + 1 \geq \tau\alpha$. Since the number of planes of type (B) through l is at most $s + 1 - \tau$, so

$$(s\alpha - s + \alpha)(t + 1 - \tau\alpha)/\alpha \leq s + 1 - \tau. \qquad (5.44)$$

This inequality is equivalent to

$$t + 1 - \tau\alpha \leq 1 + \frac{1}{\alpha - 1} - \frac{t + 1}{(\alpha - 1)(s + 1)}. \qquad (5.45)$$

Hence $0 \leq t + 1 - \tau\alpha \leq 1$, and so either $t + 1 = \tau\alpha$ or $t = \tau\alpha$.

Let $t + 1 = \tau\alpha$. Then the number of planes of type (B) through l is zero. By way of contradiction, assume that S contains at least one plane π' of type (B). Let l' be the line of $\mathcal{B}$ in π'. Also, let π'' be a plane of type (A) not containing l', and let $\pi'' \cap \pi' = l''$. Since $l'' \notin \mathcal{B}$, it is contained in no plane of type (B), a contradiction since π' is of type (B). Consequently, there are no planes of type (B). Therefore S is a partial geometry, whence, by Theorem 5.135, (b) follows.

Next, let $t = \tau\alpha$. Then, by (5.44), $s \geq t$. Let π be a plane of type (A) and let P be any point of S in π. Now count in different ways the number η of ordered pairs (π', l') with π' a plane of type (B) having its line of S through P and with $l' = \pi \cap \pi'$ not in $\mathcal{B}$. Since there are $t/(\alpha - 1)$ planes of type (A) through a given line of S and thus $s + 1 - t/(\alpha - 1)$ planes of type (B) through that line, so

$$\eta = (t + 1 - \alpha)[s + 1 - t/(\alpha - 1)]. \qquad (5.46)$$

For a given line $l' \notin \mathcal{B}$ of π through P, the number of lines l'' of $\mathcal{B}$ through P, for which $l'l''$ is of type (B), is $t + 1 - \tau\alpha = 1$. Hence

$$\eta = s + 1 - \alpha. \qquad (5.47)$$

From (5.46) and (5.47),

$$s + 1 - \alpha = (t + 1 - \alpha)[s + 1 - t/(\alpha - 1)]. \qquad (5.48)$$

Since $s \geq t$,

$$s + 1 - \alpha \geq (t + 1 - \alpha)[s + 1 - s/(\alpha - 1)],$$

which is equivalent to

$$t + 1 - \alpha \leq 1 + \frac{\alpha - \alpha^2 + s}{\alpha s + \alpha - 2s - 1}. \qquad (5.49)$$

Since $t = \tau\alpha$, so $t + 1 - \alpha \geq 1$ and $\alpha - \alpha^2 + s \geq 0$. However,

$$(\alpha - \alpha^2 + s)/(\alpha s + \alpha - 2s - 1) \geq 1$$

if and only if $\alpha = 2$ and $s \geq 3$. This means that $t = \alpha$ whenever $\alpha \neq 2$. But since there are $t/(\alpha - 1)$ planes of type (A) through a given line of S, the case $\alpha \neq 2$ and $t = \alpha$ cannot occur. Therefore $\alpha = 2$. Then (5.48) becomes $(s - t)(t - 2) = 0$. Consequently $s = t$ or $t = 2$.

Now suppose that $\alpha = 2$ and $s = t$. Since $m = 0$, the dual of $\mathcal{S}$ is a semi-partial geometry $\mathcal{S}^*$ with parameters

$$s^* = t, \quad t^* = s, \quad \alpha^* = \alpha, \quad \mu^* = \alpha t/(\alpha - 1).$$

Here, $s^* = t^*$ and so, by Theorem 5.146, $\mathcal{S}$ is also a semi-partial geometry. Now, Theorem 5.151 with $s = t$ and $\alpha = 2$ gives $\mathcal{S} = NQ^-(3, 2)$.

Finally, let $\alpha = 2$ and $t = 2$; then $\tau = 1$ and $s = 2^h$. Let π be a plane of type (A) and let l be a line of π which does not belong to $\mathcal{B}$. It was shown that the number of planes of type (B) through l is $1 + s/2$. If there is at least one plane of type (C), then, by Corollary 5.150, 2^h is divisible by 3, a contradiction. So there are no planes of type (C). It follows that each of the planes through l is either of type (A) or of type (B), and so $2 + s/2 = s + 1$, whence $s = 2$. Again $s = t$ and $\alpha = 2$ and so $\mathcal{S} = NQ^-(3, 2)$. $\square$

Theorem 5.153. *Let $\mathcal{S} = (\mathcal{P}, \mathcal{B}, \mathrm{I})$ be a $(0, \alpha)$-geometry with parameters s, t, α, which is projective with ambient space $\mathrm{PG}(3, s)$. If $m \neq 0$, then there is no plane of type (B).*

Proof. Suppose that there is at least one plane of each type. The total number of planes of type (A) is

$$\frac{bt}{(\alpha - 1)(s\alpha - s + \alpha)}. \tag{5.50}$$

By (5.27), this is

$$\left[\frac{1}{\alpha}(\delta((s+1)(t+1) - \alpha s)) + m(t+1) \right] \frac{t}{(\alpha - 1)\delta}, \tag{5.51}$$

with $\delta = s\alpha - s + \alpha$. The number of planes of type (B) is

$$b(s + 1 - t(\alpha - 1)^{-1}). \tag{5.52}$$

By (5.27), this is

$$\left[\frac{1}{\alpha}(\delta((s+1)(t+1) - \alpha s)) + m(t+1) \right](s + 1 - t(\alpha - 1)^{-1}). \tag{5.53}$$

The total number of planes of type (A) and (B) is at most $(s^2 + 1)(s + 1)$. Adding (5.51) and (5.53) gives

$$\left[\alpha^{-1}(\delta((s+1)(t+1) - \alpha s)) + m(t+1) \right](s+1)(\delta - t)/\delta$$
$$\leq (s^2 + 1)(s + 1), \tag{5.54}$$

$$\alpha^{-1}((s+1)(t+1) - \alpha s)(s\alpha - s + \alpha - t)$$
$$+ m(t+1)\left(1 - \frac{t}{s\alpha - s + \alpha} \right) \leq s^2 + 1. \tag{5.55}$$

As $m \neq 0$, a line l of the plane π of type (A), which does not belong to $\mathcal{B}$, exists; so

$$|l \cap \mathcal{P}| = (s\alpha - s + \alpha)/\alpha.$$

Hence α divides s. By Corollary 5.150, $t + 1$ divides s. Since $\alpha \leq t + 1$ and s is a prime power, so α divides $t + 1$.

As, by assumption, there is at least one plane of type (B), so, by (5.52),

$$t < (\alpha - 1)(s + 1). \tag{5.56}$$

Hence

$$m(t + 1) \left(1 - \frac{t}{s\alpha - s + \alpha} \right) > \frac{m(t + 1)}{s\alpha - s + \alpha} > 0. \tag{5.57}$$

By (5.55) and (5.57),

$$\alpha^{-1}((s + 1)(t + 1) - \alpha s)(s\alpha - s + \alpha - t) < s^2 + 1,$$

which is equivalent to

$$\alpha + \alpha s - t - 1 - \frac{\alpha s(\alpha + s)}{t + s} + \frac{\alpha s(t + 1)}{(s + 1)(t + s)} < 0. \tag{5.58}$$

From this it follows that

$$t + 1 + \frac{\alpha s(\alpha + s)}{t + s} > \alpha s,$$

$$t(t + 1) + s(t + \alpha^2 - \alpha t + 1) > 0. \tag{5.59}$$

Note that $t + \alpha^2 - \alpha t + 1 < 0$ if and only if $t > \alpha + (\alpha + 1)/(\alpha - 1)$. Let l, l', l'' be distinct non-coplanar lines of $\mathcal{B}$ through P in $\mathcal{P}$. The α lines of $\mathcal{B}$ in ll' through P are $l = l_1, l' = l_2, l_3, \ldots, l_\alpha$. A count of the lines of $\mathcal{B}$ through P in the α planes $l'' l_i$ shows that $t \geq \alpha^2 - \alpha$. If $\alpha \geq 3$, then $\alpha^2 - \alpha > \alpha + (\alpha + 1)/(\alpha - 1)$ and $t + \alpha^2 - \alpha t + 1 < 0$. So, from (5.59),

$$s < \frac{t(t + 1)}{\alpha t - \alpha^2 - t - 1}. \tag{5.60}$$

Since $t + 1$ divides s, so $t + 1 < s$; then, from (5.60),

$$(\alpha - 2)t < \alpha^2 + 1. \tag{5.61}$$

Previously it was shown that α divides $t + 1$; so $t \geq \alpha^2 - \alpha$ implies that $t \geq \alpha^2 - 1$. Hence, from (5.61), $(\alpha - 2)(\alpha^2 - 1) < \alpha^2 + 1$, whence $\alpha \leq 3$ and so $\alpha \in \{2, 3\}$.

Suppose that $\alpha = 3$. From (5.61), $t < 10$; since $t \geq \alpha^2 - \alpha$, so $t \geq 6$. As α and $t + 1$ divide s, so $t = 8$. From (5.59), $s < 12$ and so $s < 9$. This gives a contradiction as $s \neq \alpha$ implies that $s = 2^h$.

Next, let $\alpha = 2$ and $t = 3$. From (5.58), $s < 3$, a contradiction since $t + 1$ divides s.

Finally, let $\alpha = 2$ and $t \neq 3$. Since α divides $t + 1$, so $t > 3$. As α divides s, so $s = 2^h$. As $t + 1$ divides s, so $t \geq 7$. With $s = (t + 1)r$, (5.59) implies that

$(r-1)(t-5) < 5$. Since $t \geq 7$ and $r = 2^k$, so $r \in \{1,2\}$. If $r = 1$, then $s = t+1$ and (5.58) gives $t(t-1) < 4$, contradicting that $t \geq 7$. If $r = 2$, then $s = 2(t+1)$ and (5.59) gives $t < 10$. Hence $s = 16$ and $t = 7$, contradicting (5.58). Thus planes of type (B) and (C) cannot both occur.

Suppose that there are no planes of type (C), but at least one plane of type (B). Then, by (5.52),

$$t < (\alpha - 1)(s+1). \tag{5.62}$$

The total number of planes of types (A) and (C) is $(s^2 + 1)(s + 1)$. Hence, from (5.50) and (5.52),

$$b\left(1 - \frac{t}{s\alpha - s + \alpha}\right) = s^2 + 1. \tag{5.63}$$

From (5.63) and (5.27),

$$m(t+1) = \frac{t(s^2+1)}{s\alpha - s + \alpha - t} + 1 - (s+1)s(t+1-\alpha)$$
$$+ (s+1)(t+1)(s/\alpha - 1). \tag{5.64}$$

If all the points of $PG(3,s)$ are elements of $\mathcal{P}$, then $\rho_a = s^2 + s + 1$ and so, by (5.29), $m = (s+1-\alpha)s/\alpha$. Since $v = (s^2+1)(s+1)$, so (5.28) implies that $t = (\alpha - 1)(s+1)$, contradicting (5.62). Hence $PG(3,s)$ contains at least one point which does not belong to $\mathcal{P}$.

Let $P \in PG(3,s)\backslash\mathcal{P}$, and let τ_b denote the number of planes of type (B) through P. Counting the pairs (π, l), with $l \in \mathcal{B}$ and $\pi = lP$ shows that

$$b = \tau_b + (s^2 + s + 1 - \tau_b)(s\alpha - s + \alpha),$$

and so, using (5.27),

$$m(t+1) = -\tau_b(s+1)(\alpha-1) + (s\alpha - s + \alpha)(s+1) \times$$
$$(s - \alpha^{-1}(t+1) + 1). \tag{5.65}$$

From (5.65), $s + 1$ divides $m(t+1)$; therefore, by (5.64), $s + 1$ divides

$$1 + [t(s^2+1)/(s\alpha - s + \alpha - t)].$$

Hence $s + 1$ divides $t + 1$. Since $t + 1 \leq (s+1)(\alpha - 1)$, so

$$\text{either} \quad t+1 = (s+1)(\alpha - 1) \quad \text{or} \quad t+1 \leq (s+1)(\alpha - 2). \tag{5.66}$$

Let $P' \in \mathcal{P}$. The number of planes of type (A) containing at least one line of $\mathcal{B}$ through P' is $t(t+1)/[\alpha(\alpha-1)]$. The number of planes of type (B) having its line of $\mathcal{B}$ through P' is

$$(t+1)[s+1-t/(\alpha-1)].$$

Since P' is in exactly $s^2 + s + 1$ planes, so

$$t(t+1)/[\alpha(\alpha-1)] + (t+1)[s+1-t/(\alpha-1)] \leq s^2 + s + 1,$$

which can be written as

$$t^2 - t(\alpha s + \alpha - 1) + s^2 \alpha \geq 0. \tag{5.67}$$

The corresponding discriminant is $D = (\alpha s + \alpha - 1)^2 - 4s^2\alpha$. So $D < 0$ if either $\alpha = 3$ and $s > 4$ or $\alpha = 2$. When $\alpha = s = 3$ and $\alpha \geq 4$, then $D > 0$. Since $m \neq 0$, it cannot be that $\alpha = s + 1 = 3$; since $\alpha < s + 1$, so α divides s and it cannot be that $\alpha = 3$ and $s = 4$.

There are now six cases to consider.

Case 1 : $t + 1 \leq (s + 1)(\alpha - 2)$, $D > 0$, $t \geq \frac{1}{2}(\alpha s + \alpha - 1 + \sqrt{D})$
From these inequalities,

$$s\alpha - 2s + \alpha - 2 \geq t + 1 \geq \frac{1}{2}(\alpha s + \alpha + 1 + \sqrt{D}); \tag{5.68}$$
$$s^2(4 - \alpha) + 2s(5 - 2\alpha) + 2(3 - \alpha) \geq 0. \tag{5.69}$$

If $\alpha \geq 4$, then $s^2(4 - \alpha) \leq 0$, $2s(5 - 2\alpha) < 0$, $2(3 - \alpha) < 0$, a contradiction. For $\alpha = s = 3$, (5.68) becomes $4 \geq t + 1 \geq 9$, again a contradiction.

Consequently, $D > 0$ implies one of the following:

(a) $t + 1 > (s + 1)(\alpha - 2)$;
(b) $t < \frac{1}{2}(\alpha s + \alpha - 1 + \sqrt{D})$.

By (5.66), (a) is equivalent to $t + 1 = (s + 1)(\alpha - 1)$; by (5.67), $D > 0$ and (b) imply that $t \leq \frac{1}{2}(\alpha s + \alpha - 1 - \sqrt{D})$.

Case 2 : $t \leq \frac{1}{2}(\alpha s + \alpha - 1 - \sqrt{D})$, $D > 0$, $t \geq 2s + 1$
From these inequalities,

$$4s + 2 \leq 2t \leq \alpha s + \alpha - 1 - \sqrt{D}; \tag{5.70}$$
$$s^2(4 - \alpha) + 3s(2 - \alpha) + 2 - \alpha \geq 0. \tag{5.71}$$

If $\alpha \geq 4$, then $s^2(4 - \alpha) \leq 0$, $3s(2 - \alpha) < 0$, $2 - \alpha < 0$, a contradiction. If $\alpha = s = 3$, (5.71) also gives a contradiction.

Thus $D > 0$ and $t \leq \frac{1}{2}(\alpha s + \alpha - 1 - \sqrt{D})$ imply that $t < 2s + 1$. Since $s + 1$ divides $t + 1$, so $t = s$.

Case 3 : $t \leq \frac{1}{2}(\alpha s + \alpha - 1 - \sqrt{D})$, $D > 0$, $t = s$
From (5.64),

$$m(s + 1) = \frac{s(s^2 + 1)}{s\alpha - 2s + \alpha} + 1 - (s + 1)s(s + 1 - \alpha)$$
$$+ (s + 1)^2(s/\alpha - 1); \tag{5.72}$$
$$m\alpha(s\alpha - 2s + \alpha) = -s(\alpha - 2)[(\alpha - 1)(s - \alpha)(s + 1) + s\alpha]. \tag{5.73}$$

Since $D > 0$, so $\alpha > 2$. Hence the left-hand side of (5.73) is positive and the right-hand side is negative, a contradiction.

Case 4 : $t + 1 = (s + 1)(\alpha - 1)$, $D > 0$

Counting the planes of type (A) through a line of $\mathcal{B}$ shows that $\alpha - 1$ divides t; so $\alpha - 1$ also divides $t + 1$. Hence $\alpha = 2$, contradicting that $D > 0$.

Case 5 : $\alpha = 2$, $D < 0$

From (5.62), $t < s + 1$. Since $s + 1$ divides $t + 1$, so $t = s$. Then (5.64) implies that $m = 0$, a contradiction.

Case 6 : $\alpha = 3$, $s > 4$, $D < 0$

From (5.62), $t < 2s + 2$. Since $s + 1$ divides $t + 1$, so $t = s$ or $t = 2s + 1$. If $t = s$, then (5.64) is equivalent to (5.73), giving $3m(s + 3) = -s(2(s - 3)(s + 1) + 3s)$, a contradiction. If $t = 2s + 1$, then (5.64) becomes $2m = -(2s^2 + 5s + 3)/6$, again a contradiction.

The conclusion is that there are no planes of type (B).   □

Corollary 5.154. *Let* $\mathcal{S} = (\mathcal{P}, \mathcal{B}, I)$ *be a* $(0, \alpha)$*-geometry with parameters* s, t, α, *which is projective with ambient space* $PG(3, s)$*. If* $m \neq 0$*, then* $t = (s + 1)(\alpha - 1)$*; if also there is no plane of type* (C)*, then* $m = s(s - \alpha + 1)/\alpha$ *and* $\mathcal{P} = PG(3, s)$.

Proof. Suppose that $m \neq 0$. By Theorem 5.153, there is no plane of type (B). Now, from (5.52), it follows that $t = (s + 1)(\alpha - 1)$.

If there is also no plane of type (C), then every plane is of type (A); so, from (5.50),

$$\frac{bt}{(\alpha - 1)(s\alpha - s + \alpha)} = (s^2 + 1)(s + 1).$$

So $b = (s^2 + 1)(s\alpha - s + \alpha)$ and $v = (s^2 + 1)(s + 1)$. Now, by (5.28), the result follows.   □

Corollary 5.155. *Let* $\mathcal{S} = (\mathcal{P}, \mathcal{B}, I)$ *be a* $(0, \alpha)$*-geometry with parameters* s, t, α, *which is projective with ambient space* $PG(3, s)$*. If there is at least one plane of type* (B)*, then* $s = \alpha = 2$ *and* $\mathcal{S} = NQ^-(3, 2)$.

Proof. By Theorem 5.153, $m = 0$. Now, by Theorem 5.152, $\mathcal{S}$ is either the design formed by all points and all lines of $PG(3, s)$ or $\mathcal{S} = H_s^3$ or $\mathcal{S} = NQ^-(3, 2)$. However, only $NQ^-(3, 2)$ admits planes of type (B).   □

Theorem 5.156. *Let* $\mathcal{S} = (\mathcal{P}, \mathcal{B}, I)$ *be a projective* $(0, \alpha)$*-geometry with parameters* s, t, α, *with ambient space* $PG(3, s)$ *and with* $m = 1$*. Then one of the following holds:*

(a) $\alpha = s$ *and* $\mathcal{S} = \overline{W(3, s)}$;
(b) $\alpha = s/2$, $s = 2^h$, $h > 1$ *and* $\mathcal{S} = NQ^+(3, s)$.

Proof. Suppose $\alpha \neq s$; then $s = 2^h$. Now, count the number τ of lines of $PG(3, s)$ containing exactly $(s\alpha - s + \alpha)/\alpha$ points of $\mathcal{P}$. In a plane of type (A) there are

$$s^2 + s + 1 - (s\alpha - s + \alpha) - (s + 1) = (s - \alpha)(s + 1)$$

such lines. By (5.50) and Corollary 5.154, there are $b(s+1)/(s\alpha - s + \alpha)$ planes of type (A). So, by (5.27), the number of planes of type (A) is

$$(s + 1)[(s + 1)(s - s/\alpha) + 2].$$

If a line $l \notin \mathcal{B}$ of PG$(3, s)$ contains ρ points of $\mathcal{P}$, then the number of planes of type (A) containing l is $\rho(t + 1)/(s\alpha - s + \alpha) = \rho$. It follows that

$$\tau = (s - \alpha)(s + 1)^2[(s + 1)(s - s/\alpha) + 2]\alpha/(s\alpha - s + \alpha). \qquad (5.74)$$

Hence $s\alpha - s + \alpha$ divides the following:

$$(s - \alpha)(s + 1)^2[(s + 1)s(\alpha - 1) + 2\alpha],$$
$$(s - \alpha)(s + 1)^2\alpha(s - 1) = (s^2 - \alpha s + s - \alpha)(\alpha s + \alpha)(s - 1),$$
$$s^3(s - 1).$$

As $\alpha = 2^k$, with $0 < k < h$, so $2^{h+k} - 2^h + 2^k$ divides $2^{3h}(2^h - 1)$, and so $2^h - 2^{h-k} + 1$ divides $2^{3h-k}(2^h - 1)$. Since $(2^h - 2^{h-k} + 1, 2^{3h-k}) = 1$, so $2^h - 2^{h-k} + 1$ divides $(2^h - 1)$; therefore $2^h - 2^{h-k} + 1$ divides $2^{h-k} - 2$. If $2^{h-k} \neq 2$, then $2^h - 2^{h-k} + 1 \leq 2^{h-k} - 2$, implying that $2^h - 2^{h-k+1} < 0$; so $\alpha = 2^k < 2$, a contradiction. Hence $2^{h-k} = 2$, whence $\alpha = s/2$. Thus either $\alpha = s$ or $\alpha = s/2$.

Assume that $\alpha = s$; then $t = s^2 - 1$. By (5.27) and (5.28), $b = s^4 + s^2$ and $v = (s^2 + 1)(s + 1)$; hence $\mathcal{P} = \text{PG}(3, s)$. By (5.50), the number of planes of type (A) is $(s^2 + 1)(s + 1)$. Consequently, there are no planes of type (C). In any plane π of type (A) there is exactly one point P not in $\mathcal{P}$, and the lines of π not in $\mathcal{B}$ form the pencil of π through P. Now, by Theorem 15.2.13 of FPSOTD, the lines of PG$(3, s)$ which are not in $\mathcal{B}$ are the lines of a general linear complex. Hence $\mathcal{S} = W(3, s)$.

Next, assume that $\alpha = s/2$; then $t = (s - 2)(s + 1)/2$. By (5.27) and (5.28), $b = s^2(s-1)^2/2$ and $v = s(s^2 - 1)$. A plane of type (A) contains s^2 points of $\mathcal{P}$ and $s(s - 1)/2$ lines of $\mathcal{B}$; a plane of type (C) contains $s(s - 1)$ points of $\mathcal{P}$. By (5.50), there are $s(s^2 - 1)$ planes of type (A), and so $(s + 1)^2$ planes of type (C).

A line $l \notin \mathcal{B}$ containing ρ points of $\mathcal{P}$ is in ρ planes of type (A). Hence, if such a line l contains at least one point of $\mathcal{P}$, then l is in at least one plane π of type (A) and $|l \cap \mathcal{P}| \in \{s - 1, s, s + 1\}$. Therefore, for any line l of PG$(3, s)$, necessarily $|l \cap \mathcal{P}| \in \{0, s - 1, s, s + 1\}$.

Let $\mathcal{P}' = \text{PG}(3, s)\backslash\mathcal{P}$; then $|\mathcal{P}'| = (s+1)^2$, and any line with at least three points in $\mathcal{P}'$ lies entirely in it. Now, by Theorem 16.2.2 of FPSOTD, $\mathcal{P}'$ is a hyperbolic quadric or consists of a plane and a line or is a cone joining an oval to a vertex. By definition, $\mathcal{P}'$ does not contain a plane. If it is a cone, then there are planes through the vertex which contain exactly $s^2 + s$ points of $\mathcal{P}$. Such planes cannot be of type (A) nor of type (C), a contradiction. Hence $\mathcal{P}'$ is a hyperbolic quadric. As no line of $\mathcal{B}$ has a point in $\mathcal{P}'$ and as $b = s^2(s - 1)^2/2$, so $\mathcal{B}$ consists of all lines having empty intersection with the quadric $\mathcal{P}'$. Thus $\mathcal{S} = NQ^+(3, s)$. Also, $s \neq 2$, as for $s = 2$ the geometry $NQ^+(3, s)$ is not connected.   □

Theorem 5.157. *In a* $(0, \alpha)$*-geometry* $\mathcal{S} = (\mathcal{P}, \mathcal{B}, I)$ *with ambient space* $\mathrm{PG}(3, s)$, *the number* m *of isolated points satisfies* $m \neq 2$.

Proof. Suppose that $m = 2$. Now count the number τ of lines of $\mathrm{PG}(3, s)$ containing exactly $(s\alpha - s + 3\alpha)/\alpha$ points of $\mathcal{P}$. In any plane of type (A) there is exactly one such line. By (5.50) and Corollary 5.154, there are $b(s + 1)/(s\alpha - s + \alpha)$ planes of type (A). By (5.27), the number of planes of type (A) is $(s + 1)[s(s + 1)(\alpha - 1) + 3\alpha]/\alpha$. However, if a line l not in $\mathcal{B}$ contains ρ points of $\mathcal{P}$, then l is in ρ planes of type (A). Hence

$$\tau = (s + 1)[s(s + 1)(\alpha - 1) + 3\alpha]/(s\alpha - s + 3\alpha).$$

Consequently, $s\alpha - s + 3\alpha$ divides $(s + 1)[s(s + 1)(\alpha - 1) + 3\alpha]$, and so $s - s/\alpha + 3$ divides

$$(s - s/\alpha + 3 + s/\alpha - 2)[(s - s/\alpha + 3)(s + 1) - 3(s + 1) + 3];$$

so $s - s/\alpha + 3$ divides $3s(s/\alpha - 2)$.

Let $s = p^h$ with p prime; then $\alpha = p^k$ with $0 < k \leq h$. If $\alpha = s$, then for any plane of type (A) the union of its $s\alpha - s + \alpha = s^2$ lines of $\mathcal{B}$ is a set of order $s^2 + s$; so $m \leq 1$, a contradiction. Hence $k < h$ and so $p = 2$. It has therefore been shown that $2^h - 2^{h-k} + 3$ divides $3.2^h(2^{h-k} - 2)$.

Since $(2^h - 2^{h-k} + 3, 2^h) = 1$, the integer $2^h - 2^{h-k} + 3$ divides $3(2^{h-k} - 2)$. Let $2^{h-k} \neq 2$. Then $2^h - 2^{h-k} + 3 \leq 3(2^{h-k} - 2)$, and so $2^h - 2^{h-k+2} + 9 \leq 0$, whence $2^h - 2^{h-k+2} < 0$. Hence $2^k < 4$, and so $k = 1$. Consequently, $2^{h-1} + 3$ divides $3(2^{h-1} - 2)$; so $2^{h-1} + 3$ divides $2^{h-1} - 2$. Hence $h = 2$ and $2^{h-k} = 2$, a contradiction. Therefore $h = k + 1$.

Let π be a plane of type (A) and let P be a point of π not in $\mathcal{P}$. Since $s/\alpha = 2$ and $m = 2$, the set $\pi \backslash \mathcal{P}$ is an s-arc $\mathcal{K}$ of π. However, $\mathcal{K}$ contains P, and $\mathcal{K}$ together with the two isolated points of π forms an oval of π. It follows that π contains exactly two lines of $\mathcal{B}$ through P having s points in common with $\mathcal{P}$. The number of planes of type (A) through P is $b/(s\alpha - s + \alpha) = s^2 - s + 1$. Therefore the number of lines through P having exactly s points in $\mathcal{P}$ is $2(s^2 - s + 1)/s$. Hence s divides 2; so $s = 2$ and $\alpha = 1$, a contradiction. $\square$

Theorem 5.158. *Let* $\mathcal{S} = (\mathcal{P}, \mathcal{B}, I)$ *be a* $(0, \alpha)$*-geometry with parameters* s, t, α *which is projective with ambient space* $\mathrm{PG}(3, s)$. *If* s *is odd, then* $\mathcal{S}$ *is a semi-partial geometry and hence is known.*

Proof. By the first part of Section 5.9, either $\alpha = s + 1$ or $\alpha = s$. If $\alpha = s + 1$, then $m = 0$, and, by Theorem 5.152, $\mathcal{S}$ is a partial geometry; so $\alpha = s$. By (5.29), $\rho_a = s^2 + s + m$. So $m \in \{0, 1\}$. If $m = 0$, then, by Theorem 5.152, $\mathcal{S}$ is a semi-partial geometry; if $m = 1$, then, by Theorem 5.156, $\mathcal{S}$ is a semi-partial geometry. $\square$

Theorem 5.159. *If* $\mathcal{S}$ *is a* $(0, \alpha)$*-geometry which is projective with ambient space* $\mathrm{PG}(3, 2)$, *then* $\mathcal{S}$ *is a semi-partial geometry and hence is known.*

Proof. As $m < 3$, the result follows from Theorems 5.152, 5.156 and 5.157. $\square$

By Theorems 5.153, 5.156, 5.157, 5.158, 5.159 all projective $(0, \alpha)$-geometries with ambient space $\mathrm{PG}(3, s)$ are known if either $m \leq 2$, s is odd, or $s = 2$.

In the first edition of General Galois Geometries it was wrongly conjectured that any projective $(0, \alpha)$-geometry with ambient space $\mathrm{PG}(3, s)$ is one of the following:

(a) $m = 0$ and S is either the design formed by all points and all lines of $\mathrm{PG}(3, s)$, or $S = H_s^3$, or $S = NQ^-(3, 2)$;
(b) $m = 1$ and either $S = \overline{W}(3, s)$, or $s \neq 2$ is even and $S = NQ^+(3, s)$.

Let $S = (\mathcal{P}, \mathcal{B}, \mathrm{I})$ be a geometry whose point set $\mathcal{P}$ is a subset of the point set of some $\mathrm{PG}(3, q)$, $q > 2$, and whose line set $\mathcal{B}$ is a non-empty set of lines of $\mathrm{PG}(3, q)$. If $\mathcal{P}$ generates $\mathrm{PG}(3, q)$, then $\mathrm{PG}(3, q)$ is the *ambient space* of S. From Theorems 5.152 and 5.153 it follows that S is a projective $(0, \alpha)$-geometry with ambient space $\mathrm{PG}(3, q)$ if and only if every pencil of lines of $\mathrm{PG}(3, q)$ contains either 0 or α, with $\alpha > 1$, lines of $\mathcal{B}$.

A point set $\mathcal{K}$ of the non-singular hyperbolic quadric $\mathcal{H}_5$ of $\mathrm{PG}(5, q)$ is a $(0, \alpha)$-*set* if every line of $\mathcal{H}_5$ contains either 0 or α points of $\mathcal{K}$. By considering $\mathcal{H}_5$ as the Klein quadric $\mathcal{G}_{1,3}$, that is, as the image of the line set of $\mathrm{PG}(3, q)$ under the Klein correspondence, it is seen that any projective $(0, \alpha)$-geometry with ambient space $\mathrm{PG}(3, q)$, $q > 2$, is equivalent to a $(0, \alpha)$-set, with $\alpha > 1$, of $\mathcal{H}_5$, and conversely.

For $q > 2$, the *deficiency* of a projective $(0, \alpha)$-geometry S with ambient space $\mathrm{PG}(3, q)$, and also of the corresponding $(0, \alpha)$-set of $\mathcal{H}_5$, is $\delta = q(q+1-\alpha)/\alpha - m$. Four special cases are the following:

(1) S consists of all points and all lines of $\mathrm{PG}(3, q)$, with $\alpha = q + 1$, $\delta = 0$;
(2) $S = \overline{W}(3, q)$, with $\alpha = q$, $\delta = 0$;
(3) $S = H_q^3$, with $\alpha = q$, $\delta = 1$;
(4) $S = NQ^+(3, 2^h)$, with $\alpha = q/2$, $\delta = q + 1$.

For $\alpha < q$, with $q > 2$ and $\alpha > 1$, the $(0, \alpha)$-sets of $\mathcal{H}_5$, and so the projective $(0, \alpha)$-geometries with ambient space $\mathrm{PG}(3, q)$, are known in the following cases:

(a) $q = 2^h$, $h > 1$, for any $\alpha \in \{2, 2^2, \ldots, 2^{h-1}\}$, and any $\delta \in \{1, q + 1\}$;
(b) $q = 2^{2e+1}$, $e > 0$, $\alpha = 2$, for any $\delta \in \{q + \sqrt{2q} + 1, q - \sqrt{2q} + 1\}$.

See Section 5.10.

Lemma 5.160. *Let* $S = (\mathcal{P}, \mathcal{B}, \mathrm{I})$ *be a projective* $(0, \alpha)$-*geometry with ambient space* $\mathrm{PG}(n, s)$, $n \geq 3$. *If* P *is any point of* $\mathcal{P}$, *then the* $t + 1$ *lines of* $\mathcal{B}$ *through* P *do not lie in the same hyperplane of* $\mathrm{PG}(n, s)$.

Proof. Suppose that the $t + 1$ lines of $\mathcal{B}$ through P are contained in a hyperplane Π_{n-1}. If l is one of these lines, then a point $P' \in l$, with $P' \neq P$, lies on $t + 1$ lines $l, l_1, l_2, l_3, \ldots, l_t$ of $\mathcal{B}$. On $l_i \backslash \{P'\}$ there are $\alpha - 1$ (≥ 1) points which are joined to P by a line of $\mathcal{B}$. It follows that l_i is contained in Π_{n-1} for $i = 1, 2, \ldots, t$. Since S is connected, repeated application of this argument shows that S is contained in Π_{n-1}, a contradiction. $\square$

Suppose that $S = (\mathcal{P}, \mathcal{B}, \mathrm{I})$ is a projective $(0, \alpha)$-geometry with ambient space $\mathrm{PG}(n, s)$, $n \geq 3$. Let $P \in \mathcal{P}$ and let $l_1, l_2, \ldots, l_r$, $r \geq 2$, be lines of $\mathcal{B}$ through P that generate a $\mathrm{PG}(n', s)$, with $2 \leq n' \leq n$. Further, let $S \cap \mathrm{PG}(n', s)$ denote the structure formed by all points and lines of S in $\mathrm{PG}(n', s)$. By definition, the *connected component* of $S \cap \mathrm{PG}(n', s)$ through P is the structure S' formed by all elements T of $S \cap \mathrm{PG}(n', s)$ for which there exist elements $T_1, T_2, \ldots, T_u$ of $S \cap \mathrm{PG}(n', s)$ with $T \, \mathrm{I} \, T_1 \, \mathrm{I} \cdots \mathrm{I} \, T_u \, \mathrm{I} \, P$. Then S' is a projective $(0, \alpha)$-geometry with ambient space $\mathrm{PG}(n', s)$. In particular, each point of S' is incident with the same number $t' + 1$, $t' \geq 1$, of lines of S'.

Theorem 5.161. *Let $S = (\mathcal{P}, \mathcal{B}, \mathrm{I})$ be a projective $(0, \alpha)$-geometry with ambient space $\mathrm{PG}(n, s)$, $n \geq 4$, $s > 2$. Then S is either the design formed by all points and lines of $\mathrm{PG}(n, s)$ or $S = \overline{W}(n, 2k, s)$ with $2k \in \{2, 4, \ldots, n+1\}$.*

Proof. It is shown that $\alpha = s$ or $\alpha = s + 1$.

(I) $n = 4$

Let P be a point of S. By Lemma 5.160, there exist lines l_1, l_2, l_3, l_4 of $\mathcal{B}$ through P that are not contained in a solid. Let $\mathrm{PG}(3, s)$ be the solid containing l_1, l_2, l_3. The connected component S' of $S \cap \mathrm{PG}(3, s)$ through P is a projective $(0, \alpha)$-geometry with ambient space $\mathrm{PG}(3, s)$. Since $s > 2$, Corollary 5.155 implies that no plane of $\mathrm{PG}(3, s)$ contains exactly one line of S'. Considering all planes of $\mathrm{PG}(3, s)$ through l_1, it follows that the parameter t' of S' is $(\alpha - 1)(s + 1)$. Hence the parameter t of S satisfies $t > (\alpha - 1)(s + 1)$.

Let l be any line of S through P, and assume that l is contained in a plane π such that l is the only line of S in π. The other lines of S through P are denoted by $l'_1, l'_2, \ldots, l'_t$. The solid defined by π and l'_i is denoted by π_i for $i = 1, 2, \ldots, t$. Then P belongs to at least $\alpha - 2$ lines l'_j, $j \neq i$, of π_i. Now suppose that P is on exactly α lines of $S \cap \pi_i$, for all $i = 1, 2, \ldots, t$; that is, the lines of $S \cap \pi_i$ through P lie in a plane. Considering all solids of $\mathrm{PG}(4, s)$ through π shows that $t \leq (\alpha - 1)(s + 1)$, a contradiction. Hence there exist lines l'_i, l'_j, $j \neq i$, such that $\pi_i = \pi_j$ and with l'_i, l'_j, l not coplanar. Since the plane π of π_i contains exactly one line of the connected component of $S \cap \pi_i$ through P, so Corollary 5.155 gives a contradiction. Therefore each plane π through l contains exactly $s\alpha - s + \alpha$ lines of S.

Now, let $\mathrm{PG}(3, s)$ be a solid in $\mathrm{PG}(4, s)$ not containing P. The $t + 1$ lines of $\mathcal{B}$ through P meet $\mathrm{PG}(3, s)$ in the $t+1$ points $P_0, P_1, \ldots, P_t$. From above, each line of $\mathrm{PG}(3, s)$ contains either 0 or α points of the set $\mathcal{V} = \{P_0, P_1, \ldots, P_t\}$. Considering all lines of $\mathrm{PG}(3, s)$ through P_0 gives that $t = (s^2 + s + 1)(\alpha - 1)$. If π is any plane of $\mathrm{PG}(3, s)$ through P_0, then $\pi \cap \mathcal{V}$ is a maximal $(s\alpha - s + \alpha; \alpha)$-arc of π. Let $\alpha \neq s + 1$ and let l' be a line of π not meeting this arc. From the planes of $\mathrm{PG}(3, s)$ through l', it follows that $s\alpha - s + \alpha$ divides $t + 1$. Hence $s\alpha - s + \alpha$ divides $s(s\alpha - s + \alpha) - s + \alpha$, and so $s\alpha - s + \alpha$ divides $s - \alpha$. Since $s\alpha - s + \alpha > s - \alpha$, so $s = \alpha$. Therefore either $\alpha = s + 1$ or $\alpha = s$.

(II) $n > 4$

Let P be a point of S. By Lemma 5.160, there exist lines l_1, l_2, l_3, l_4 of $\mathcal{B}$ through P which generate a subspace $\mathrm{PG}(4, s)$ of $\mathrm{PG}(n, s)$. The connected component S' of $S \cap \mathrm{PG}(4, s)$ through P is a projective $(0, \alpha)$-geometry with ambient space $\mathrm{PG}(4, s)$. Hence, from (I), either $\alpha = s + 1$ or $\alpha = s$.

Suppose that S is a projective $(0, \alpha)$-geometry with ambient space $\mathrm{PG}(n, s)$, with $s > 2$, $\alpha = s + 1$, $n \geq 4$. Let P be a point of S, and let l, l' be two lines of S through P. Since $\alpha = s + 1$, every line of the plane ll' is a line of S. It follows that the union of all lines of S through P is a subspace $\mathrm{PG}(n', s)$ of $\mathrm{PG}(n, s)$. Now, by Lemma 5.160, $n' = n$. Hence S is the design formed by all points and all lines of $\mathrm{PG}(n, s)$.

Next, let S be a projective $(0, \alpha)$-geometry with ambient space $\mathrm{PG}(n, s)$, with $s > 2$, $\alpha = s$, $n \geq 4$. It is shown that $t = s^{n-1} - 1$. Considering all planes containing a given line l of $\mathcal{B}$ indicates that this is equivalent to proving that no plane through l contains exactly one line of $\mathcal{B}$. By (I), this holds for $n = 4$; in this case, $t = s^3 - 1$.

Now proceed by induction on n. For any line l of S, assume that it is contained in a plane π and that no other line of S is in π. Let $P \in l$ and let the other lines of $\mathcal{B}$ through P be $l_1, l_2, \ldots, l_t$. The solid through π and l_i is denoted by π_i, $i = 1, \ldots, t$. Then P belongs to at least $\alpha - 2$ lines l_j, $j \neq i$, of π_i. Now suppose that P is on exactly α lines of $S \cap \pi_i$, for all $i = 1, \ldots, t$. Considering all solids of $\mathrm{PG}(n, s)$ through π shows that $t \leq s^{n-2} - 1$. By Lemma 5.160, it may be assumed that $l, l_1, l_2, \ldots, l_{n-1}$ generate $\mathrm{PG}(n, s)$. Let $\mathrm{PG}(n-1, s) = ll_1 \cdots l_{n-2}$ and let S' be the connected component of $S \cap \mathrm{PG}(n - 1, s)$ through P. By the induction hypothesis, the parameter t' of S' is $s^{n-2} - 1$. Since l_{n-1} is not in $\mathrm{PG}(n-1, s)$, so $t > s^{n-2} - 1$, contradicting that $t \leq s^{n-2} - 1$. Therefore, each plane through l contains exactly s^2 lines of $\mathcal{B}$; this is equivalent to the relation $t = s^{n-1} - 1$.

Let P be any point of S and let $l_0, l_1, \ldots, l_t$ be the lines of $\mathcal{B}$ through P. Further, let Π_{n-1} be a hyperplane not through P and let $l_i \cap \Pi_{n-1} = P_i$ for $i = 0, 1, \ldots, t$. From above, any line of Π_{n-1} meets $\mathcal{V} = \{P_0, P_1, \ldots, P_t\}$ in 0 or s points. Hence $\mathcal{V}$ is the complement of a hyperplane of Π_{n-1}. Therefore, the union of $l_0, l_1, \ldots, l_t$ is the complement of a hyperplane of $\mathrm{PG}(n, s)$.

Now consider the incidence structure S' formed by all points of $\mathrm{PG}(n, s)$ and its lines not in $\mathcal{B}$. By the above, the union of all lines of S' through any point of $\mathrm{PG}(n, s)$ is a hyperplane. Hence S' is a projective Shult space. Since $\mathrm{PG}(n, s)$ is the point set of S', so either (b) or (e) of Theorem 5.52 occurs. In case (b), with the notation of Theorem 5.52, $r = n - 2$ and $k = s + 1$; then $S = H_s^n = \overline{W}(n, 2, s)$. In case (e), $S = \overline{W}(n, 2k, s)$ with $2k \in \{4, 6, \ldots, n + 1\}$. □

Theorem 5.162. *Let* $S = (\mathcal{P}, \mathcal{B}, \mathrm{I})$ *be a projective semi-partial geometry with ambient space* $\mathrm{PG}(n, s)$, $n \geq 4$. *If* $\alpha > 1$ *and* $s > 2$, *then it is one of the following types:*

(a) $\alpha = s + 1$ *and* S *is the design formed by all points and all lines of* $\mathrm{PG}(n, s)$;
(b) $\alpha = s$ *and* $S = H_s^n$;

(c) $\alpha = s$, n is odd and $S = \overline{W}(n,s)$.

Proof. First, S is a projective $(0,\alpha)$-geometry with ambient space $\mathrm{PG}(n,s)$, where $s > 2$, $n \geq 4$. By Theorem 5.161, S is the design formed by all points and lines of $\mathrm{PG}(n,s)$ or $S = \overline{W}(n,2k,s)$ with $2k \in \{2,4,\ldots,n+1\}$. In Section 5.8, it was observed that $\overline{W}(n,2k,s)$ is a semi-partial geometry if and only if either $2k = n+1$, in which case n is odd, or $k = 1$. When $k = 1$, then $S = \overline{W}(n,2,s)$; however, when $2k = n+1$, then $S = \overline{W}(n,n+1,s) = \overline{W}(n,s)$. □

Theorem 5.163. *Let* $S = (\mathcal{P},\mathcal{B},\mathrm{I})$ *be a projective semi-partial geometry with ambient space* $\mathrm{PG}(n,s)$, $n \geq 3$ *and* $\alpha > 1$. *Then it is one of the following types:*

(a) $\alpha = s+1$ *and* S *is the design formed by all points and all lines of* $\mathrm{PG}(n,s)$;
(b) $\alpha = s$ *and* $S = H_s^n$;
(c) $\alpha = s = 2$ *and* $S = NQ^-(3,2)$.

Proof. For any plane π containing at least two lines of $\mathcal{B}$, let $\bar{\pi}$ be the set of all lines of π in $\mathcal{B}$; then $|\bar{\pi}| = s\alpha - s + \alpha$. Let $\bar{\mathcal{P}}$ denote the set of all these sets $\bar{\pi}$. Now consider the structure $\bar{S} = (\bar{\mathcal{P}},\mathcal{B},\bar{\mathrm{I}})$, where, for $\bar{\pi} \in \bar{\mathcal{P}}$ and $l \in \mathcal{B}$, the relation $\bar{\pi} \bar{\mathrm{I}} l$ holds when $l \in \bar{\pi}$. Then $\bar{S}$ is a $(0,\alpha)$-geometry with $\bar{\alpha} = \alpha$, $\bar{t} = (\alpha-1)(s+1)$, $\bar{s} = t/(\alpha-1)-1$. Two lines of $\mathcal{B}$ are concurrent in $\bar{S}$ if and only if they are concurrent in S. Hence $\bar{S}$ is also a dual semi-partial geometry.

First suppose that S is a partial geometry. Then, by Theorem 5.135, either (a) or (b) holds.

Now suppose that S is not a partial geometry. Then $s \geq t$ by Theorem 5.145; hence $\bar{s} < \bar{t}$. Again, by Theorem 5.145, $\bar{S}$ is a partial geometry. Consequently, for any line $l \in \mathcal{B}$ and any element $\bar{\pi} \in \bar{\mathcal{P}}$ with $l \notin \bar{\pi}$, the line l is concurrent with α lines of $\mathcal{B}$ in π. So, for any point $P \in \mathcal{P}$ with $P \notin \pi$, the lines of $\mathcal{B}$ through P are contained in the solid $P\pi$. Hence $n = 3$ by Lemma 5.160. Now suppose that $m \neq 0$ and let P' be a point of $\mathcal{P}$ in π which is on none of the $s\alpha - s + \alpha$ lines of $\mathcal{B}$ in π. If $P' \in l'$, with $l' \in \mathcal{B}$, then l' does not meet any line of $\mathcal{B}$ in π, a contradiction. Therefore $m = 0$. Since S is not a partial geometry, so $S = NQ^-(3,2)$ by Theorem 5.152. □

Remark 5.164. The determination of all projective $(0,2)$-geometries with ambient space $\mathrm{PG}(n,2)$ is complicated. However, this problem has been completely solved, and the main references are given in Section 5.10.

Open problems

Concerning the determination of all projective $(0,\alpha)$-geometries with ambient space $\mathrm{PG}(n,s)$, for given α,n,s, the following problems are still open.

(a) Determine all projective partial quadrangles, that is, $\alpha = 1$, with ambient space $\mathrm{PG}(n,s)$, $n \geq 4$.
(b) Determine all projective dual partial quadrangles, that is, $\alpha = 1$, with ambient space $\mathrm{PG}(n,s)$, $n \geq 3$. An infinite class of projective dual partial quadrangles, which are not generalised quadrangles, is given by (h) of Example 5.147.

(c) Determine all projective $(0, \alpha)$-geometries with ambient space $PG(3, s)$ when $s = 2^h$, $h > 1$.

5.10 Notes and references

Section 5.1

Theorem 5.4 is due to Tits [380], and Theorem 5.6 is due to Buekenhout and Shult [59]. Example (e) for a generalised quadrangle is taken from Hall [148]. Lemma 5.8 and Theorem 5.9 are from [59].

Now two important characterisations of Grassmann varieties are stated, in which generalised quadrangles play a central role. Let $\mathcal{S} = (\mathcal{P}, \mathcal{B})$ be a pair consisting of a non-empty finite set $\mathcal{P}$ of *points* and a set $\mathcal{B}$ of distinguished subsets of cardinality at least three of $\mathcal{P}$ called *lines*. For any point P let $P^\perp$ be the set of all points collinear with P, and for any line l let $l^\perp$ be the set of all points collinear with each point of l. Suppose that $\mathcal{S}$ is connected, that $\mathcal{P}$ contains at least two non-collinear points, and that $P^\perp \backslash \{P\}$ is connected for each point P. A *subspace* of $\mathcal{S}$ is a set $\mathcal{X}$ of pairwise collinear points such that any line meeting $\mathcal{X}$ in more than one point is contained in $\mathcal{X}$. A subspace that is not properly contained in a larger subspace is called a *max space*. Now consider the following conditions:

(a) for any point P and any line l, the point P is collinear with $0, 1$ or all points of l;
(b) if P and P' are non-collinear points such that $|P^\perp \cap P'^\perp| \geq 2$, then $P^\perp \cap P'^\perp$
   with the lines it contains is a generalised quadrangle;
(c) if $P \in \mathcal{P}$ and $l \in \mathcal{B}$ such that $P^\perp \cap l = \emptyset$ but $P^\perp \cap l^\perp \neq \emptyset$, then $P^\perp \cap l^\perp$ is a line;
(c') each line is contained in exactly two max spaces.

Theorem 5.165. *If $\mathcal{S}$ satisfies (a), (b), (c), then either $\mathcal{S}$ is a non-degenerate Shult space of rank 3 or $\mathcal{S}$ is the incidence structure formed by all points and all lines of a Grassmann variety $\mathcal{G}_{r,n}$, with $n \geq 3$ and $1 < r \leq n - 2$.*

Theorem 5.166. *If $\mathcal{S}$ satisfies (a), (b), (c'), then $\mathcal{S}$ is the incidence structure formed by all points and all lines of a Grassmann variety $\mathcal{G}_{r,n}$, $n \geq 3$, $1 \leq r \leq n - 2$.*

In the case of these Grassmann varieties, the generalised quadrangle of (b) is always the classical grid $\mathcal{Q}(3, q)$. The corresponding versions of these theorems have also been proved in the infinite case. Theorem 5.165 is due to Cooperstein [75] and Cohen [72], and Theorem 5.165 is due to Hanssens [154] and Hanssens and Thas [155].

Section 5.2

This is taken from Payne and Thas [259].

Generalised quadrangles were introduced by Tits [379]. The classical generalised quadrangles, all of which are associated with classical groups, were first recognised as generalised quadrangles also by Tits. Higman [163, 164] first proved Theorem 5.14 by a complicated matrix-theoretic method. The argument given here was used by Bose and Shrikhande [38] to show that, when $t = s^2 > 1$, then each triad has $1 + s$ centres; Cameron [63] first observed that the above technique also provides the inequality.

Concerning proper subquadrangles, the following theorem is due to Payne [254]; see also Thas [319], and Payne and Thas [259].

Theorem 5.167. Let $S' = (\mathcal{P}', \mathcal{B}', I')$ be a proper subquadrangle of the generalised quadrangle $S = (\mathcal{P}, \mathcal{B}, I)$. Then the following hold:
  (i) either $s = s'$ or $s \geq s't'$;
 (ii) if $s = s'$, then each external point is collinear with $1 + st'$ points of S';
(iii) if $s = s't'$, then each external point is collinear with $1 + s'$ points of S'.
The dual holds similarly.

Now, generalised quadrangles with small parameters are briefly described. The detailed proofs of all these results can be found in Payne and Thas [259]. Here, let $S = (\mathcal{P}, \mathcal{B}, I)$ be a generalised quadrangle of order (s, t), with $2 \leq s \leq 4$ and $s \leq t$. By Theorems 5.13 and 5.14, (s, t) is one of the following:

$$(2, 2), \ (2, 4), \ (3, 3), \ (3, 5), \ (3, 6), \ (3, 9),$$
$$(4, 4), \ (4, 6), \ (4, 8), (4, 11), (4, 12), (4, 16).$$

A short proof shows that up to isomorphism there is only one generalised quadrangle of order 2. The uniqueness of the generalised quadrangle of order $(2, 4)$ was proved independently at least five times, by Seidel [282], Shult [284], Thas [317], Freudenthal [138] and Dixmier and Zara [123]. Payne [255] and independently Dixmier and Zara [123] showed that a generalised quadrangle of order 3 is isomorphic to $\mathcal{W}(3)$ or its dual $\mathcal{Q}(4, 3)$. The uniqueness of the case $(3, 5)$ was proved by Dixmier and Zara [123], who also proved the uniqueness of the case $(3, 9)$; the latter was independently done by Cameron, for which see Payne and Thas [258]. The non-existence in the case of order $(3,6)$ was also shown in [123]. Payne [256] proved the uniqueness of the generalised quadrangle of order 4. The long proof required a correction by Tits; see Payne and Thas [259]. Single examples are known in the cases $(4, 6), (4, 8), (4, 16)$, but nothing is known about the cases $(4, 11)$ and $(4, 12)$.

Section 5.3

Theorems 5.51, 5.52, 5.54 are taken from Buekenhout and Lefèvre [57, 58] and Lefèvre-Percsy [201]. Theorem 5.55 is due to Thas [318].

All finite projective generalised quadrangles were first determined by Buekenhout and Lefèvre [57] by a proof most of which is valid in the infinite case. Independently, Olanda [251, 252] gave a typically finite proof; Thas and De Winne [347] gave a different combinatorial proof under the assumption that the 3-dimensional case is already settled.

The infinite case was settled by Dienst [121, 120]. For projective Shult spaces of rank at least three, the infinite case was completely solved in [58] and [201]. Because the generalised quadrangles, and more generally the Shult spaces, in this book are finite, the presentation of Buekenhout and Lefèvre has been modified.

All finite generalised quadrangles fully embedded in the affine space $AG(d, q)$, for $d \geq 2$, were determined by Thas [325]. The 3-dimensional case was settled independently by Bichara [21].

Section 5.4

For $d = 3$, Theorem 5.58 is due to Lefèvre-Percsy [207], and, for $d > 3$, to Thas and Van Maldeghem [357], although the former used a stronger definition for 'weak embedding', proved by the latter to be equivalent to the notion in Section 5.4. Weak embeddings of Shult spaces in $PG(d, q)$, $d \geq 3$, were introduced by Lefèvre-Percsy [204, 206, 207], but only in the case $d = 3$ was a complete classification obtained. Theorem 5.59, which contains the complete classification in the case where the Shult space has rank at least three and is non-degenerate, is due to Thas and Van Maldeghem [354]; the degenerate case is handled in Theorem 5.61 and is also taken from Thas and Van Maldeghem [356]. For details on double-sixes, cubic surfaces, tritangent planes, trihedral pairs and the case $q = 4$, see Chapter 20 of FPSOTD. Theorems 5.62 and 5.63 on lax embeddings of generalised quadrangles are due to Thas and Van Maldeghem [359, 364]; Theorem 5.64 on lax embeddings of Shult spaces is also taken from Thas and Van Maldeghem [358].

Section 5.5

The detailed proofs of most of Theorems 5.67 to 5.116 can be found in Payne and Thas [259, 260]. Theorem 5.71, which is probably the oldest combinatorial characterisation of a class of generalised quadrangles, was discovered independently by several authors; for example, Singleton [293], Benson [16], and Tallini [306]. Theorem 5.72 is due to Thas [321], Theorems 5.74 and 5.75 to Thas [315], Theorems 5.76 and 5.77 to Payne and Thas [258], and Theorem 5.77 independently to Mazzocca [228].

Theorem 5.79 is taken from Thas [339]. The first three parts of Theorem 5.81 are due to Thas [324], part (iv) to Brown and Thas [49, 50], and Theorem 5.81(i) independently to Mazzocca [227]. Theorems 5.86 and 5.87 are taken from Thas [324] and Theorem 5.89 from Thas [327]. Theorem 5.91 on generalised quadrangles and dual nets is due to Payne and Thas [259, 260], and Theorem 5.93 on dual nets to Thas and De Clerck [346].

Theorem 5.94 was proved by Thas and Van Maldeghem [355], Theorem 5.95 in the even case by Thas and Payne [350], in the odd case by Brown [43], and a proof for both cases may be found in Brouns, Thas and Van Maldeghem [39] and in Brown [46]. Theorem 5.97 is taken from Tallini [306], Theorem 5.98 from Thas

[320], Theorem 5.99 from Thas and Payne [349], and Theorem 5.101 from Thas and Van Maldeghem [359]. Theorem 5.102 is due to Thas [321], Theorems 5.103 and 5.104 to Thas and Payne [349] and Thas [321], and Theorems 5.105, 5.106, 5.107 to Thas [321]. Theorem 5.108, which is a characterisation in terms of matroids, is taken from Mazzocca and Olanda [230]. A considerable shortening of the original proof was given by Payne and Thas [259, 260].

Theorem 5.109 is due to Ronan [269]. This approach includes infinite generalised quadrangles and relies on topological methods. Payne and Thas [259, 260] offer an 'elementary' treatment which is more combinatorial than topological, and which corrects a slight oversight in the case $t = 2$. Theorem 5.111 is taken from Thas [327].

Theorem 5.116(i)–(v) is due to Payne and Thas [259, 260], while (vi) is due to Brown and Lavrauw [48]. The proof of Theorem 5.117 can be found in K. Thas [370, 375, 376], Theorem 5.118 in K. Thas [377], and Theorem 5.119 in Kantor [194] and K. Thas [368]. Theorem 5.120 on span-symmetric generalised quadrangles is due independently to Kantor [196] and K. Thas [371]. Theorem 5.121 was proved by Ealy [126], Theorem 5.122 by Thas [330, 331].

Using the language of BN pairs, Fong and Seitz [136, 137] obtained a characterisation of the finite generalised polygons, in particular the finite classical generalised quadrangles. Tits [381, 382, 383, 384] and Tits and Weiss [385] determined all finite and infinite Moufang generalised polygons; the finite case, in particular Theorem 5.123, is essentially the theorem of Fong and Seitz. De Medts [99, 100] unified and shortened the proof of Tits and Weiss for generalised quadrangles. The proofs of the different parts of Theorem 5.124 can be found in Thas, Payne and Van Maldeghem [351], Van Maldeghem [391], Tent [309], Haot and Van Maldeghem [156, 157] and K. Thas and Van Maldeghem [378].

In order to state Theorem 5.97 the notion of *linear space* was used. Here, a linear space is an incidence structure $\mathcal{S}^* = (\mathcal{P}, \mathcal{B}^*, \in)$ with $\mathcal{P}$ a non-empty set, $\mathcal{B}^*$ a non-empty set of subsets of $\mathcal{P}$, where each of the subsets has cardinality at least two, and having the property that any two distinct elements (points) of $\mathcal{P}$ are contained in a unique element (line) of $\mathcal{B}^*$.

Section 5.6

Partial geometries were introduced by Bose [36]. Theorem 5.132 is taken from Cameron, Goethals and Seidel [64]. The proper partial geometries $\mathcal{S}(\mathcal{K})$ are due to Thas [313, 316] and independently to Wallis [394]; the proper partial geometries $\mathcal{T}_2^*(\mathcal{K})$ are due to Thas [313, 316]. For surveys on partial geometries, see De Clerck [85], Thas [322], Brouwer and van Lint [42], De Clerck and Van Maldeghem [93], and De Clerck [87]; constructions of partial geometries are also contained in Mathon [226], De Clerck [86], De Clerck, Delanote, Hamilton and Mathon [90], and Hamilton and Mathon [151].

Section 5.7

Theorem 5.135 is taken from De Clerck and Thas [91] and Theorem 5.136 is taken from Thas and De Clerck [346]. All finite partial geometries embedded in the affine space $AG(d, q)$, $d \geq 2$, were determined by Thas [325].

Section 5.8

Partial quadrangles were introduced by Cameron [63], semi-partial geometries by Debroey and Thas [111, 112, 113] and $(0, \alpha)$-geometries by De Clerck and Thas [92]. Theorems 5.138 to 5.145 are taken from [111]. Theorem 5.146 is due to Debroey [109, 110], but here a new and simpler proof is given. Most of the examples of semi-partial geometries can be found in Brouwer and van Lint [42], Cameron [63], Debroey and Thas [111], Thas [328], De Clerck and Thas [92], Hall [147], De Clerck and Van Maldeghem [93], Thas [342], De Clerck [87], De Winter [102], De Winter and Thas [106], Cossidente and Penttila [77], Devillers and Van Maldeghem [118], and De Winter and Van Maldeghem [108].

Section 5.9

Lemmas 5.148, 5.149 and Theorems 5.152, 5.153, 5.163 are taken from De Clerck and Thas [92]. Theorem 5.151 is due to Debroey and Thas [113]; Theorems 5.156 and 5.157 are unpublished results of Thas. Several constructions of $(0, \alpha)$-sets of $\mathcal{H}_5$, or, equivalently, of projective $(0, \alpha)$-geometries with ambient space $PG(3, q)$, are due to De Clerck, De Feyter, and Durante [88]. Lemma 5.160 and Theorems 5.161, 5.162 are taken from Thas, Debroey, and De Clerck [348].

Farmer and Hale [130] proved that any projective $(0, s)$-geometry with ambient space $PG(n, q)$, $s > 2, n \geq 3$, is a $\overline{W}(n, 2k, s)$ with $2k \in \{2, 4, \ldots, n + 1\}$. For the results on $(0, 2)$-geometries, see Shult [285], Hall [147], Thas, Debroey, and De Clerck [348], and GGG1, Section 26.9.

All finite semi-partial geometries embedded in the affine space $AG(d, q)$, with $d \in \{2, 3\}$, were determined by Debrocy and Thas [112]. Many results on $(0, \alpha)$-geometries and semi-partial geometries embedded in $AG(d, q)$ were obtained by De Clerck and Delanote [89], Brown, De Clerck, and Delanote [47], De Winter [103], and De Feyter [94, 95, 96, 97, 98]. In this series of papers, De Feyter determines all $(0, 2)$-geometries and semi-partial geometries with $\alpha = 2$ embedded in $AG(d, 2^h)$, up to the classification of all sets of type $(0, 1, k)$, $k > 2$, in $PG(d - 1, 2^h)$.

6

Arcs and caps

6.1 Introduction

A $(k; r, s; n, q)$-*set* $\mathcal{K}$ is defined to be a set of k points in $PG(n, q)$ with at most r points in any s-space such that $\mathcal{K}$ is not contained in a proper subspace. This is a slight variation on the definition of Section 3.3 of PGOFF2, where the last condition is not present. The large question is to describe all such sets. Four questions particularly are of interest. The set $\mathcal{K}$ is *complete* if it is not contained in a $(k+1; r, s; n, q)$-set.

   I. Find the maximum value $m(r, s; n, q)$ of k.
  II. Characterise the sets, the *maximum sets,* with this value of k.
 III. Find the size $m'(r, s; n, q)$ of the second-largest, complete $(k; r, s; n, q)$-set.
 IV. Characterise the complete $(k; r, s; n, q)$-sets.

Question IV includes I, II, and III. The importance of III is that, if $\mathcal{K}$ is a $(k; r, s; n, q)$-set with $k > m'(r, s; n, q)$, then $\mathcal{K}$ is contained in a maximum set. So upper bounds on $m'(r, s; n, q)$ permit inductive arguments.

In fact, these questions are examined only when $r = s + 1$, and then only in the cases $s = 1$ and $s = n - 1$.

A $(k; 2, 1; n, q)$-set is a k-set with at most two points on any line of $PG(n, q)$ and is also called a k-*cap*. The number $m(2, 1; n, q)$ is written as $m_2(n, q)$. The only precise values known are the following:

$$m_2(n, 2) = 2^n; \tag{6.1}$$

$$m_2(2, q) = \begin{cases} q+1, & q \text{ odd}, \\ q+2, & q \text{ even}; \end{cases} \tag{6.2}$$

$$m_2(3, q) = q^2 + 1, \quad q > 2; \tag{6.3}$$

$$m_2(4, 3) = 20; \tag{6.4}$$

$$m_2(5, 3) = 56; \tag{6.5}$$

$$m_2(4, 4) = 41. \tag{6.6}$$

© Springer-Verlag London 2016
J.W.P. Hirschfeld, J.A. Thas, *General Galois Geometries*, Springer Monographs
in Mathematics, DOI 10.1007/978-1-4471-6790-7_6

Upper bounds for $m_2(n, q)$ are determined in Sections 6.2 to 6.4. In the case that $s = n - 1$, a $(k; n, n - 1; n, q)$-set is a k-set not contained in a hyperplane with at most n points in any hyperplane of $\mathrm{PG}(n, q)$ and is also called a k-arc; for $n = 2$, a k-cap and a k-arc are equivalent. The number $m(n, n-1; n, q)$ is written $m(n, q)$; by definition $m(2, q) = m_2(2, q)$. Values obtained in previous chapters are as follows:

$$m(n, q) = n + 2 \quad \text{for } q \le n + 1; \tag{6.7}$$

$$m(2, q) = \begin{cases} q + 1, & q \text{ odd}, \\ q + 2, & q \text{ even}; \end{cases} \tag{6.8}$$

$$m(3, q) = q + 1, \quad q > 3. \tag{6.9}$$

All other values known and all the values determined in Sections 6.5 to 6.7 are $m(n, q) = q + 1$ or $q + 2$. This leads naturally to the following.

Conjecture 6.1 (The Main Conjecture for Arcs).

(1) If $q > n + 1$ with q odd, then $m(n, q) = q + 1$.
(2) If $q > n + 1$, with q even and $n \notin \{2, q - 2\}$, then $m(n, q) = q + 1$.

In deciding the value of $m(n, q)$, the value of $m'(2, q)$, the size of the second largest complete arc in $\mathrm{PG}(2, q)$, is crucial. It is also useful to write, for $q \ge 5$,

$$f(q) = q - m'(2, q). \tag{6.10}$$

For $q \le 5$, there is only one complete plane arc and $m'(2, q)$ is not defined. For other small values of q the results are as in Table 6.1.

q	7	8	9	11	13	16	17	19	23	25	27	29	31	32
$m'(2, q)$	6	6	8	10	12	13	14	14	17	21	22	24	22	24
$f(q)$	1	2	1	1	1	3	3	5	6	4	5	5	9	8

Table 6.1. Complete plane arcs

Also

$$m'(2, q) \le q - \tfrac{1}{4}\sqrt{q} + \tfrac{25}{16}, \quad q \text{ odd}; \tag{6.11}$$

the slightly weaker result

$$m'(2, q) \le q - \tfrac{1}{4}\sqrt{q} + \tfrac{7}{4}, \quad q \text{ odd}, \tag{6.12}$$

has been frequently used. Further,

$$m'(2,q) \leq q - \sqrt{q} + 1, \quad q \text{ even}, \quad q > 2; \qquad (6.13)$$

$$m'(2,q) = q - \sqrt{q} + 1, q = 2^{2m}, \quad m > 1; \qquad (6.14)$$

$$m'(2,q) \leq \tfrac{44}{45}q + \tfrac{8}{9}, \qquad q \text{ prime}; \qquad (6.15)$$

$$m'(2,q) \leq q - \tfrac{1}{4}\sqrt{pq} + \tfrac{29}{16}p + 1, \ q = p^{2m+1}, \ m \geq 1, \ p \text{ odd}; \qquad (6.16)$$

$$m'(2,q) \leq q - \sqrt{2q} + 2, \ q = 2^{2m+1}, \ m \geq 1; \qquad (6.17)$$

$$m'(2,q) \leq q - \tfrac{1}{2}\sqrt{q} + 5, \qquad q = p^h \text{ for } p \geq 5; \qquad (6.18)$$

$$m'(2,q) \leq q - \tfrac{1}{2}\sqrt{q} + 3, \qquad q = p^h, \ p \geq 3; \quad q = 3^{2e} \text{ when } p = 3;$$
$$q \geq 23^2 \text{ and } q \neq 3^6 \text{ or } 5^5; \qquad (6.19)$$

$$m'(2,q) \leq q - 22 \quad \text{when } q = 5^5; \qquad (6.20)$$

$$m'(2,q) \leq q - 9 \quad \text{when } q = 3^6; \qquad (6.21)$$

$$m'(2,q) \leq q - 9 \quad \text{when } q = 23^2; \qquad (6.22)$$

$$m'(2,q) \leq q - 5 \quad \text{when } q = 19^2; \qquad (6.23)$$

$$m'(2,q) < q - 1 \quad \text{when } q > 13 \text{ except possibly for } q = 37, 41, 43,$$
$$47, 49, 53, 59, 61, 67, 71, 73, 79, 81, 83. \qquad (6.24)$$

It is convenient to record an elementary result which is subsequently applied in Section 6.3.

Lemma 6.2. *If* A, B, C *are three sets such that* $C \supset A \cup B$, *then*

$$|A \cap B| \geq |A| + |B| - |C|.$$

Proof. $|A \cap B| = |A| + |B| - |A \cup B| \geq |A| + |B| - |C|.$ □

6.2 Caps and codes

Let $\mathcal{K}$ be the k-cap $\{P_1, \ldots, P_k\}$ where $P_i = \mathbf{P}(a_{i0}, a_{i1}, \ldots, a_{in})$. This gives the $k \times (n+1)$ matrix

$$A = [a_{ij}] \qquad i = 1, \ldots, k; \ j = 0, \ldots, n,$$

which is a *matrix* of $\mathcal{K}$. Any permutation of the rows of A or multiplication of a row of A by an element of $\mathbf{F}_q{}^*$ gives another matrix of $\mathcal{K}$. For any projectivity $\mathfrak{T}$, write

$$\mathcal{K}\mathfrak{T} = \{P_1\mathfrak{T}, \ldots, P_k\mathfrak{T}\}.$$

So the caps $\mathcal{K}_1$ and $\mathcal{K}_2$ are (projectively) equivalent if $\mathcal{K}_1\mathfrak{T} = \mathcal{K}_2$ for some projectivity $\mathfrak{T}$.

Lemma 6.3. *Let $\mathcal{K}_1$ and $\mathcal{K}_2$ be caps with matrices A_1 and A_2. Then $\mathcal{K}_1$ and $\mathcal{K}_2$ are equivalent if and only if A_2 can be obtained from A_1 via a sequence of operations of the following type:*

(C1) *multiplication of a column by a non-zero scalar;*
(C2) *interchange of two columns;*
(C3) *addition of a scalar multiple of one column to another;*
(R1) *multiplication of a row by a non-zero scalar;*
(R2) *interchange of two rows.*

Proof. The operation of $\mathfrak{T}$ on $\mathcal{K}_1$ corresponds to a sequence of operations of types (C1), (C2), and (C3). The operations (R1) and (R2) leave $\mathcal{K}_1$ fixed. $\square$

If A_2 can be obtained from A_1 as in the lemma, then A_1 and A_2 are *C-equivalent*. An $[N, d]$-code over $\mathbf{F}_q$ is usually defined to be a d-dimensional subspace of the N-dimensional vector space $V(n, q)$. In this section, it is more convenient to replace a vector by its non-zero multiples. So an $[N, d]$-*code* is a Π_{d-1} of $\mathrm{PG}(N-1, q)$. It is also convenient to represent the points of an $[N, d]$-code by column vectors rather than row vectors. A *generator matrix* A for an $[N, d]$-code C is thus an $N \times d$ matrix whose columns generate C. Other generator matrices of C are obtained from A via operations of types (C1), (C2), (C3).

Two codes are *equivalent* if one can be obtained from the other by a permutation of the coordinate indices combined with multiplication of some coordinates by a non-zero scalar.

Lemma 6.4. *Let C_1 and C_2 be codes with generator matrices A_1 and A_2. Then C_1 and C_2 are equivalent if and only if A_1 and A_2 are C-equivalent.*

Proof. From the definition of equivalent codes, a generator matrix of C_2 is obtained from a generator matrix of C_1 by operations (R1) and (R2). $\square$

If $\mathcal{K}$ is a k-cap in $\mathrm{PG}(n, q)$ with matrix A, the *code C of $\mathcal{K}$* is the $[k, n+1]$-code with generator matrix A; such a code is a *cap-code*. It is assumed that $\mathcal{K}$ is not contained in a subspace of lower dimension than n.

Theorem 6.5. *Let $\mathcal{K}_1$ and $\mathcal{K}_2$ be caps with codes C_1 and C_2. Then $\mathcal{K}_1$ is equivalent to $\mathcal{K}_2$ if and only if C_1 is equivalent to C_2.*

Proof. This follows from Lemmas 6.3 and 6.4. $\square$

An $[N, d]$-code is *projective* if the rows of a generator matrix are distinct points of $\mathrm{PG}(N-1, q)$. Any cap-code is projective.

With, as usual, $\theta(n) = |\mathrm{PG}(n, q)|$, given a $[k, n+1]$-code C, denote by $M(C)$ a $k \times \theta(n)$ matrix whose columns are the points of C. Given a linearly independent set $\{X_0, \ldots, X_t\}$ of $\mathrm{PG}(k-1, q)$, let $M(X_0, \ldots, X_t)$ denote a $k \times \theta(t)$ matrix whose columns are the points X_j.

Lemma 6.6. *The number of zeros in the i-th row of the matrix $M(X_0, \ldots, X_t)$ is $\theta(t)$ if all the X_j have zero in the i-th row and is $\theta(t-1)$ otherwise.*

Proof. Let Π_t be the subspace spanned by $X_0, \ldots, X_t$. Intersect Π_t with $\mathbf{V}(X_i)$. Then $\Pi_t \cap \mathbf{V}(X_i) = \Pi_t$ if $\Pi_t \subset \mathbf{V}(X_i)$ and $\Pi_t \cap \mathbf{V}(X_i) = \Pi_{t-1}$ if $\Pi_t \not\subset \mathbf{V}(X_i)$. $\square$

Given a projective $[k, n+1]$-code C, let M_1 be the matrix obtained from $M(C)$ by omitting one row and also those columns having a non-zero entry in that row. By Lemma 6.6, M_1 is a $(k-1) \times \theta(n-1)$ matrix whose columns are the vectors of a $[k-1, n]$-code C_1. The code C_1 is a *residual* code of C. The code C has k residuals, one for each row of C; some or all of these residuals may be equivalent codes. By identifying the vectors of C_1 with those of C from which one zero entry has been omitted, C_1 can be regarded as a $[k, n]$-subcode of C. If A_1 is a generator matrix for a residual code C_1 of C, then C is equivalent to a code with generator matrix

$$A = \begin{bmatrix} 0 & 1 \\ A_1 & * \end{bmatrix}.$$

The matrix A_1 is a *residual* of A. If C_1 and A_1 are residuals of C and A, then C and A are *extensions* of C_1 and A_1.

Theorem 6.7. *A projective code C is a cap-code if and only if every residual code of C is projective.*

Proof. Suppose C_1 with matrix A_1 is a non-projective residual of C with matrix A an extension of A_1. Then two rows, the i-th and j-th, say, of A_1 are the same, up to a scalar multiple; so the first, $(i+1)$-th and $(j+1)$-th rows of A are collinear. So C is not a cap-code.

Conversely, suppose C is not a cap-code and let A be any generator matrix of C. Then three rows, say the first, second and third of A, are collinear. However, using suitable column operations, A is C-equivalent to a matrix whose first row is $(0, 0, \ldots, 0, 1)$. Since column operations preserve the collinearity of rows 1, 2 and 3, the residual obtained by omitting the first row is non-projective. $\square$

To fix ideas, consider the small example of the 4-arc ($=$ 4-cap) $\mathcal{K}$ in $PG(2, 2)$ with points $\mathbf{P}(1, 0, 0), \mathbf{P}(0, 1, 0), \mathbf{P}(0, 0, 1), \mathbf{P}(1, 1, 1)$. A matrix for $\mathcal{K}$ is

$$A = \begin{bmatrix} 1 & 0 & 0 \\ 0 & 1 & 0 \\ 0 & 0 & 1 \\ 1 & 1 & 1 \end{bmatrix}.$$

It is a generator matrix for a $[4, 3]$-code C for which a suitable $M(C)$ is

$$M(C) = \begin{bmatrix} 1 & 0 & 0 & 1 & 1 & 0 & 1 \\ 0 & 1 & 0 & 1 & 0 & 1 & 1 \\ 0 & 0 & 1 & 0 & 1 & 1 & 1 \\ 1 & 1 & 1 & 0 & 0 & 0 & 1 \end{bmatrix}.$$

The residuals with respect to the four rows of $M(C)$ are

$$\begin{bmatrix} 1 & 0 & 1 \\ 0 & 1 & 1 \\ 1 & 1 & 0 \end{bmatrix}, \quad \begin{bmatrix} 1 & 0 & 1 \\ 0 & 1 & 1 \\ 1 & 1 & 0 \end{bmatrix}, \quad \begin{bmatrix} 1 & 0 & 1 \\ 0 & 1 & 1 \\ 1 & 1 & 0 \end{bmatrix}, \quad \begin{bmatrix} 1 & 1 & 0 \\ 1 & 0 & 1 \\ 0 & 1 & 1 \end{bmatrix}.$$

Now, the weight distribution of a cap-code is considered. Let X be a vector, that is, a column of the code C.

The *weight* of X, denoted $w(X)$, is the number of non-zero entries in X. Denote the vectors of a $[k, n + 1]$-code C by $X_1, X_2, \ldots, X_{\theta(n)}$ and let $w_i = w(X_i)$, for $i \in \mathbf{N}_{\theta(n)}$. Order the X_i so that $w_1 \geq w_2 \geq \cdots \geq w_{\theta(n)}$. The ordered $\theta(n)$-tuple $(w_1, \ldots, w_{\theta(n)})$ is the *weight distribution* of C. Equivalent codes have the same weight distribution. It turns out that cap-codes have large minimum weight. For a $[k, n + 1]$-code C, let $C^\perp$ be the *dual code*; that is, $C^\perp$ is the $[k, k - n - 1]$-code consisting of points $\mathbf{P}(Y)$ in $\mathrm{PG}(k - 1, q)$ such that $XY^* = 0$ for all X in C. If C is a code of the cap $\mathcal{K}$, then $C^\perp$ has minimum weight at least four; for, if any vector in $C^\perp$ had three or less non-zero entries, its orthogonality with C would force C to be non-projective or would induce a collinearity of the corresponding three rows of C. Conversely, if $C^\perp$ has minimum weight at least four, then C is a cap-code. Thus $m_2(n, q)$ is the maximum value of N for which one error can be corrected and three detected with certainty by an $[N, N - n - 1]$-code.

Theorem 6.8. *Let $\mathcal{K}$ be a k-cap in $\mathrm{PG}(n, q)$ with code C. Then the minimum weight of C, as well as that of any residual, is at least $k - m_2(n - 1, q)$.*

Proof. Let X be any vector in C and suppose X has t zeros. Let A be a generator matrix of C with X as first column. Since the rows of A form a cap and since any subset of a cap is also a cap, it follows that those rows having a zero as first coordinate form a t-cap in a hyperplane of $\mathrm{PG}(n, q)$. Hence $t \leq m_2(n - 1, q)$ and therefore $w(X) = k - t \geq k - m_2(n - 1, q)$. The result holds for a residual C_1, since any vector in C_1 is obtained from a vector in C by omitting one zero and so leaving the weight unchanged. □

Lemma 6.9. *Let $(w_1, \ldots, w_{\theta(n)})$ be the weight distribution of a projective $[k, n+1]$-code C. Then*

(i)
$$\sum w_i = kq^n; \tag{6.25}$$

(ii)
$$\sum w_i^2 = kq^{n-1}\{k(q - 1) + 1\}. \tag{6.26}$$

Proof. (i) By summing over columns, the number of non-zero entries in $M(C)$ is $\sum w_i$ and, by Lemma 6.6, is also kq^n by summing over rows.

(ii) Let $Z_1, Z_2, \ldots, Z_k$ be the rows of the matrix $M(C)$ and also let B be the $k(k - 1)(q - 1) \times \theta(n)$ matrix with rows $Z_i + \lambda Z_j$, $(i, j) \in \mathbf{N}_k^2$, $i \neq j$, $\lambda \in \mathbf{F}_q^*$. Since C is projective, the rows of B are all non-zero; so, by Lemma 6.6, each row has $\theta(n - 1)$ zeros. The i-th column of B has $w_i(w_i - 1) + (k - w_i)(k - w_i - 1)(q - 1)$ zeros. Counting the zero entries of B via columns and rows thus gives

$$q \sum w_i^2 - \{1 + (2k-1)(q-1)\} \sum w_i + \theta(n)k(k-1)(q-1)$$
$$= k(k-1)(q-1)\theta(n-1). \quad (6.27)$$

Substituting $\sum w_i$ from (6.25) gives the result. ☐

Now write $m_1 = m_2(n-1, q)$. Then Theorem 6.8 says that, for a cap-code, $w_i \geq k - m_1$. So consider the *amended weights* u_i given by

$$u_i = w_i - (k - m_1). \quad (6.28)$$

Then $u_1 \geq u_2 \geq \cdots \geq u_{\theta(n)} \geq 0$ and $(u_1, \ldots, u_{\theta(n)})$ is the *amended weight distribution* of C. For a residual code C_1 of C, let $(v_1, \ldots, v_{\theta(n-1)})$ consist of the amended weights of the corresponding columns of $M(C)$ with

$$v_1 \geq v_2 \geq \cdots \geq v_{\theta(n-1)} \geq 0;$$

each v_i is some u_j. By abuse of language, $(v_1, \ldots, v_{\theta(n-1)})$ is called the amended weight distribution of C_1.

Lemma 6.10. *Let C be a projective $[k, n+1]$-cap-code with amended weight distribution $(u_1, u_2, \ldots, u_{\theta(n)})$. Then*

(i)
$$\sum u_i = m_1 \theta(n) - k\theta(n-1); \quad (6.29)$$

(ii)
$$\sum u_i^2 = k^2 \theta(n-2) + k(q^{n-1} - 2m_1\theta(n-1)) + m_1^2\theta(n). \quad (6.30)$$

For a residual code C_1 of C with amended weight distribution $(v_1, v_2, \ldots, v_{\theta(n-1)})$,

(iii)
$$\sum v_i = (m_1 - 1)\theta(n-1) - (k-1)\theta(n-2); \quad (6.31)$$

(iv)
$$\sum v_i^2 = (k-1)^2\theta(n-3) + (k-1)\{q^{n-2} - 2(m_1 - 1)\theta(n-2)\}$$
$$+ (m_1 - 1)^2\theta(n-1). \quad (6.32)$$

Proof. Equations (6.29)–(6.32) are just restatements of the previous lemma. ☐

It should be noted that (6.31) and (6.32) only hold because a residual of a cap-code is projective, which is not true for general codes.

Theorem 6.11. *Let C be a $[k, n+1]$-cap-code with weight distribution and amended weight distribution $(w_1, w_2, \ldots, w_{\theta(n)})$ and $(u_1, u_2, \ldots, u_{\theta(n)})$. Then*

(i)
$$w_1 + w_2 \leq m_1(q-1) + k; \quad (6.33)$$

(ii)
$$u_1 + u_2 \leq m_1(q+1) - k. \quad (6.34)$$

For a residual of C with amended weight distribution $(v_1, v_2, \ldots, v_{\theta(n-1)})$,

(iii)
$$v_1 + v_2 \leq m_1(q+1) - q - k. \quad (6.35)$$

Proof. (i) By Lemma 6.6, each row of the $k \times (q+1)$ matrix $M(X_1, X_2)$ has at least one zero. Hence, counting the zeros of $M(X_1, X_2)$ gives

$$(k - w_1) + (k - w_2) + \sum_{\lambda \in \mathbf{F}_q^*} \{k - w(X_1 + \lambda X_2)\} \geq k.$$

By Theorem 6.8, $w(X_1 + \lambda X_2) \geq k - m_1$. So

$$w_1 + w_2 \leq k + m_1(q - 1).$$

(ii), (iii) These follow immediately. $\square$

Now the bounds of Theorems 6.8 and 6.11 together with the identities of Lemma 6.10 give restrictions on $m_2(n, q)$.

Theorem 6.12. *For $n \geq 4$ and $q \neq 2$,*

$$m_2(n, q) \leq q m_2(n - 1, q) - q + 1. \tag{6.36}$$

Proof. Induction on n is used. Suppose $\mathcal{K}$ is a k-cap in $\mathrm{PG}(n, q)$ with code C. A residual code C_1 is a projective $[k - 1, n]$-code. So, if C_1 has weight distribution $w_1', \ldots, w_{\theta(n-1)}'$, then Lemma 6.9 gives

$$\sum w_i' = (k - 1)q^{n-1} \tag{6.37}$$

and Theorem 6.8 gives

$$\sum w_i' \geq (k - m_1)\theta(n - 1). \tag{6.38}$$

Hence (6.37) and (6.38) imply that

$$k \leq m_1 q + (m_1 - q^{n-1})(q - 1)/(q^{n-1} - 1). \tag{6.39}$$

Next, it is deduced by induction that $m_2(n, q) \leq q^{n-1} + 1$. First, by Theorem 16.1.5 of FPSOTD, $m_2(3, q) = q^2 + 1$. However, the induction hypothesis is that $m_1 = m_2(n - 1, q) \leq q^{n-2} + 1$. Then (6.39) gives

$$k \leq q^{n-1} + 1 - (q^{n-1} - q^{n-2})/(q^{n-1} - 1)$$

and so $k \leq q^{n-1} + 1$. Hence $m_1 \leq q^{n-2} + 1$. Substituting this in the second occurrence of m_1 in (6.39) implies that

$$k \leq m_1 q + (q^{n-2} + 1 - q^{n-1})(q - 1)/(q^{n-1} - 1) \leq m_1 q - q + 1.   \square$$

Now (6.36) is improved by showing that equality cannot hold.

Theorem 6.13. *For $n \geq 4$ and $q \neq 2$,*

$$m_2(n, q) \leq q m_2(n - 1, q) - q. \tag{6.40}$$

Proof. Suppose $\mathcal{K}$ is a k-cap with $k = qm_1 - q + 1$. Let C_1 be any residual of the code C of $\mathcal{K}$ and let C_1 have amended weight distribution $(v_1, \ldots, v_{\theta(n-1)})$. Then (6.31) and (6.32) give

$$\sum v_i = m_1 - 1, \tag{6.41}$$

$$\sum v_i^2 = (m_1 - 1)\{q^{n-1} - (q-1)(m_1 - 1)\}. \tag{6.42}$$

Since $\sum v_i^2 \le (\sum v_i)^2$,

$$q^{n-1} - (q-1)(m_1 - 1) \le m_1 - 1;$$

this gives $m_1 \ge q^{n-2} + 1$. However, in Theorem 6.12 it was shown inductively that $m_1 \le q^{n-2} + 1$; that is, $m_1 = q^{n-2} + 1$. By (6.40) and since

$$m_2(n - 2, q) \le q^{n-3} + 1,$$

it now follows that $m_2(n - 2, q) = q^{n-3} + 1$. Proceeding in this way,

$$m_2(s, q) = q^{s-1} + 1$$

for all $3 \le s < n$. To prove the theorem, it therefore suffices to show that

$$m_2(4, q) < q^3 + 1.$$

This is shown geometrically in Theorem 6.16.

Alternatively, suppose $\mathcal{K}$ is a $(q^3 + 1)$-cap in $PG(4, q)$. Then (6.41) and (6.42) give

$$\left(\sum v_i \right)^2 = \sum v_i^2 = q^4.$$

So

$$(v_1, v_2, \ldots, v_{\theta(n-1)}) = (q^2, 0, \ldots, 0),$$
$$(w_1', w_2', \ldots, w_{\theta(n-1)}') = (q^3, q^3 - q^2, \ldots, q^3 - q^2).$$

Thus, each of the k residuals of C contains a vector of weight q^3. Since any vector of weight q^3 is contained in $(q^3 + 1) - q^3 = 1$ residual, there are k distinct vectors of weight q^3 in C. In particular,

$$w_1 + w_2 = 2q^3.$$

However, from (6.33),

$$w_1 + w_2 \le (q^2 + 1)(q - 1) + q^3 + 1 < 2q^3. \qquad \Box$$

The next result is an improvement.

Theorem 6.14. *In* $\mathrm{PG}(n, q)$, $q > 2$,

(i) $$m_2(n, q) \leq q m_2(n - 1, q) - (q + 1) \, \textit{for } n \geq 4;$$ (6.43)

(ii) $$m_2(n, q) \leq q^{n-4} m_2(4, q) - q^{n-4} - 2\theta(n - 5) + 1 \ \textit{for } n \geq 5.$$ (6.44)

Proof. (i) See Section 6.8.
(ii) This follows from (i) by induction. $\qquad\qquad\qquad\qquad\qquad\qquad\square$

For q odd, there is a major improvement; the larger n is, the better the improvement.

Theorem 6.15. *In* $\mathrm{PG}(n, q)$, $n \geq 4$, $q = p^h$ *with* p *odd*,

$$m_2(n, q) \leq \frac{nh + 1}{(nh)^2} q^n + m_2(n - 1, q).$$

6.3 The maximum size of a cap for q odd

In this section some upper bounds for $m_2(n, q)$ are proved. First, a general bound is given that is useful in that it holds for all $q > 2$. The first result is also implicit in Theorem 6.13.

Theorem 6.16. *For* $q > 2$ *and* $n \geq 4$,

$$m_2(n, q) \leq q^{n-1}.$$ (6.45)

Proof. Suppose there exists a k-cap $\mathcal{K}$ with $k = q^{n-1} + 1$. First, it is shown that, for q even, any plane meets $\mathcal{K}$ in at most $q + 1$ points. Suppose therefore that $\pi \cap \mathcal{K}$ is a $(q + 2)$-arc for some plane π, and consider the $\theta(n - 3)$ solids $\Pi_3^{(i)}$ through π, $i \in \mathbf{N}_{\theta(n-3)}$. If $|\Pi_3^{(i)} \cap (\mathcal{K} \backslash \pi)| = d_i$, then

$$\sum d_i = q^{n-1} + 1 - (q + 2) = q^{n-1} - q - 1.$$ (6.46)

Since $m_2(3, q) = q^2 + 1$,

$$d_i + q + 2 \leq q^2 + 1,$$

whence

$$d_i \leq q^2 - q - 1.$$ (6.47)

So

$$\sum d_i \leq (q^2 - q - 1)\theta(n - 3)$$
$$= q^{n-1} - q - \theta(n - 3)$$
$$< q^{n-1} - q - 1,$$

for $n \geq 4$. This contradiction implies that $\mathcal{K}$ has no $(q + 2)$-arc as a plane section.

Now, for all $q > 2$, let P_1 and P_2 be points of $\mathcal{K}$ and consider the $\theta(n - 2)$ planes through the line $P_1 P_2$. They contain at most

$$(q - 1)\theta(n - 2) + 2 = q^{n-1} + 1 = k$$

points. So each of these planes must meet $\mathcal{K}$ in a $(q + 1)$-arc.

Consider a hyperplane Π_{n-1} through $P_1 P_2$. The $\theta(n - 3)$ planes through $P_1 P_2$ in Π_{n-1} all meet $\mathcal{K}\backslash\{P_1, P_2\}$ in $q - 1$ points, whence

$$|\Pi_{n-1} \cap \mathcal{K}| = (q - 1)\theta(n - 3) + 2 = q^{n-2} + 1.$$

So $\mathcal{K}' = \Pi_{n-1} \cap \mathcal{K}$ is a $(q^{n-2} + 1)$-cap. It follows from this argument that any hyperplane meeting $\mathcal{K}$ meets it either in a single point or in a $(q^{n-2} + 1)$-cap. Let Q be a point of $\mathcal{K}$. There are q^{n-1} bisecants of $\mathcal{K}$ through Q and so the number of unisecants through Q is $\theta(n - 1) - q^{n-1} = \theta(n - 2)$. Let ℓ_1 and ℓ_2 be two of them. Then every line through Q in the plane $\ell_1 \ell_2$ is also a unisecant to $\mathcal{K}$, as otherwise the plane $\ell_1 \ell_2$ would meet $\mathcal{K}$ in a $(q + 1)$-arc with two unisecants at Q, a contradiction. So the set S of points on the unisecants through Q has the property that the line joining two points of S is in S. Thus S is a subspace and hence a hyperplane, which is called the *tangent hyperplane* to $\mathcal{K}$ at Q.

Let P_1 and P_2 be two points of $\mathcal{K}$ with tangent hyperplanes T_1 and T_2. As T_1 and T_2 are distinct, they meet in a Π_{n-2} skew to $\mathcal{K}$. Let r be the number of hyperplanes through Π_{n-2} other than T_1 and T_2 which are tangent to $\mathcal{K}$ and s the number of other hyperplanes through Π_{n-2} meeting $\mathcal{K}$. So

$$r + s \leq q - 1. \tag{6.48}$$

Counting the points of $\mathcal{K}\backslash\{P_1, P_2\}$ gives

$$r + s(q^{n-2} + 1) = q^{n-1} - 1. \tag{6.49}$$

However, from (6.48),

$$\begin{aligned} r + s(q^{n-2} + 1) &\leq q - 1 - s + s(q^{n-2} + 1) \\ &= q - 1 + sq^{n-2} \\ &\leq q - 1 + (q - 1)q^{n-2} \\ &= (q - 1)(q^{n-2} + 1). \end{aligned} \tag{6.50}$$

But, $(q - 1)(q^{n-2} + 1) < q^{n-1} - 1$. So (6.49) and (6.50) cannot both hold. This proves the theorem. $\qquad\square$

Now, Theorem 6.16 is improved first for q odd and then for q even. In both cases, however, it is necessary for q to be sufficiently large. The main result for q odd is a consequence of the following result, which appeared as the corollary of Theorem 18.4.8 of FPSOTD.

Theorem 6.17. *In* $PG(3, q)$, *q odd and* $q \geq 67$, *if* $\mathcal{K}$ *is a complete k-cap which is not an elliptic quadric, then*

$$k < q^2 - \tfrac{1}{4}q^{3/2} + 2q.$$

More precisely,

$$k \leq q^2 - \tfrac{1}{4}q^{3/2} + R(q),$$

where

$$R(q) = (31q + 14\sqrt{q} - 53)/16.$$

To obtain a similar result in $PG(4, q)$, consider a k-cap $\mathcal{K}$ and examine the sections of $\mathcal{K}$ by three solids through a plane π which has a sufficiently large intersection with $\mathcal{K}$.

Lemma 6.18. *In* $PG(4, q)$, $q \geq 67$ *and odd, let* $\mathcal{K}$ *be a k-cap and* π *a plane such that* $\pi \cap \mathcal{K}$ *is an s-arc with* $s > q - \tfrac{1}{4}\sqrt{q} + \tfrac{7}{4}$. *Then there do not exist three distinct solids* $\alpha_1, \alpha_2, \alpha_3$ *containing* π *such that* $\mathcal{K}_i = \alpha_i \cap \mathcal{K}$ *is a* k_i-*cap with* $k_i > q^2 - \tfrac{1}{4}q^{3/2} + R(q)$ *for* $i = 1, 2, 3$.

Proof. Suppose that the lemma is false. Then, by Theorem 6.17, each $\mathcal{K}_i$ is contained in a (unique) elliptic quadric $\mathcal{Q}_i$. So $\mathcal{Q}_1 \cap \mathcal{Q}_2 \cap \mathcal{Q}_3$ is the unique conic in π containing $\pi \cap \mathcal{K}$.

(I) *There exists a quadric* $\mathcal{Q}$ *meeting* α_i *in* $\mathcal{Q}_i$, $i = 1, 2, 3$

The set $\mathcal{M} = \mathcal{K}_1 \cup \mathcal{K}_2 \cup \mathcal{K}_3$ is an m-cap contained in $\mathcal{K}$ with

$$m = s + \sum(k_i - s) = k_1 + k_2 + k_3 - 2s.$$

As $s \leq q + 1$, so

$$\begin{aligned} m &> 3(q^2 - \tfrac{1}{4}q^{3/2} + R(q)) - 2(q + 1) \\ &= 3(q^2 - \tfrac{1}{4}q^{3/2}) + \tfrac{1}{16}(61q + 42\sqrt{q} - 191). \end{aligned} \tag{6.51}$$

There are two possibilities for $\mathcal{Q}$. Either $\mathcal{Q} = \mathcal{P}_4$, the non-singular quadric, or else $\mathcal{Q} = \Pi_0 \mathcal{E}_3$, the singular quadric with vertex Π_0 and base $\mathcal{E}_3$.
(a) $\mathcal{Q} = \mathcal{P}_4$

First, $\mathcal{P}_4$ comprises $(q^2 + 1)(q + 1)$ points on the same number of lines with $q + 1$ lines through a point. Through each point of a line ℓ on $\mathcal{P}_4$ there pass q other lines, whence $q(q + 1)$ lines ℓ' on $\mathcal{P}_4$ meet ℓ. No two of these lines ℓ' meet off ℓ, as otherwise their plane would meet $\mathcal{P}_4$ in a cubic curve. Also, $\mathcal{P}_4$ contains $q^2(q + 1)$ points off ℓ. So, through each point of $\mathcal{P}_4 \backslash \ell$ there is exactly one line ℓ'. The m-cap $\mathcal{M}$ has at most two points on ℓ and on each ℓ', and every point of $\mathcal{M}$ lies on ℓ or some ℓ'. Hence

$$m \leq 2 + 2q(q + 1) = 2(q^2 + q + 1). \tag{6.52}$$

From (6.51) and (6.52),

$$3(q^2 - \tfrac{1}{4}q^{3/2}) + \tfrac{1}{16}(61q + 42\sqrt{q} - 191) < m \leq 2(q^2 + q + 1).$$

Hence

$$q^2 - \tfrac{3}{4}q^{3/2} + \tfrac{3}{4}(29q + 42\sqrt{q} - 223) < 0,$$

a contradiction.

(b) $\mathcal{Q} = \Pi_0 \mathcal{E}_3$

Through Π_0 there are $q^2 + 1$ generators of $\mathcal{Q}$ each containing at most two points of $\mathcal{M}$. So

$$m \leq 2(q^2 + 1). \tag{6.53}$$

From (6.51) and (6.53),

$$3(q^2 - \tfrac{1}{4}q^{3/2}) + \tfrac{1}{16}(61q + 42\sqrt{q} - 191) < m \leq 2(q^2 + 1).$$

Hence

$$q^2 - \tfrac{3}{4}q^{3/2} + \tfrac{1}{16}(61q + 42\sqrt{q} - 223) < 0,$$

a contradiction.

(II) *There is a pencil Φ of quadric hypersurfaces through $\mathcal{Q}_1$ and $\mathcal{Q}_2$ none of which contains $\mathcal{Q}_3$*

The members of Φ cut out on α_3 a pencil Φ' of quadric surfaces all containing the conic $\mathcal{C}$, the unique conic through $\pi \cap \mathcal{K}$. One member of Φ' is π repeated, and $\mathcal{Q}_3 \notin \Phi'$. So Φ' cuts out on $\mathcal{Q}_3$ a set of quartic curves $\mathcal{C} \cup \mathcal{C}'$ with $\mathcal{C}'$ quadric curves in planes π' of a pencil in α_3; each quadric $\mathcal{C}'$ is either a conic or a point. Denote the set of quadrics $\mathcal{C}'$ by Ψ. Then $\mathcal{C} \in \Psi$ and the planes π' have a common line ℓ in π.

As $k_3 - (q+1) > 2(q+1) + (q-2)(q - \tfrac{1}{4}\sqrt{q})$, there are at least three planes π' other than π meeting $\mathcal{K}_3$ in a k'-arc $\mathcal{K}'$ with $k' > q - \tfrac{1}{4}\sqrt{q}$. Since $\mathcal{K}' \subset \mathcal{Q}_3$, each of these $\mathcal{K}'$ is contained in a conic $\mathcal{C}' = \pi' \cap \mathcal{Q}_3$. For at least one of these planes the quadric $\mathcal{V}$ of Φ meeting $\mathcal{Q}_3$ in $\mathcal{C} \cup \mathcal{C}'$ is non-singular. It is now shown that, for such a $\mathcal{K}'$, there exists a line $P'P_1P_2$, where $P' \in \mathcal{K}'$, $P_1 \in \mathcal{K}_1$, $P_2 \in \mathcal{K}_2$.

Take a point $P' \in \mathcal{C}'\backslash\mathcal{C}$. Since it is simple for $\mathcal{V}$, the tangent space $T_{P'}(\mathcal{V})$ to $\mathcal{V}$ at P' meets $\mathcal{V}$ in a cone $P'P_2$. So there are $q+1$ lines ℓ' of $\mathcal{V}$ in $T_{P'}(\mathcal{V})$. As $\mathcal{C}'$ is non-singular, the space $T_{P'}(\mathcal{V})$ does not contain $\mathcal{C}$. Consequently, $T_{P'}(\mathcal{V})$ meets $\mathcal{C}$ in at most two points whence at most two lines ℓ' meet π. The others, in number at least $q - 1$, all meet α_1 in a point P_1 of $\mathcal{Q}_1$ and α_2 in a point P_2 of $\mathcal{Q}_2$, with P_1, P_2 not in $\mathcal{C}$. Also $P_1 \neq P_2$ since $\alpha_1 \cap \alpha_2 = \pi$ and $P_1, P_2 \notin \pi$. Further, $P_i \neq P'$ for $i = 1, 2$, since every point of $\alpha_i \cap \mathcal{C}'$ lies on $\mathcal{C}$.

Let $P' \in \mathcal{K}'\backslash\mathcal{C}$ and note that $|\mathcal{K}'\backslash\mathcal{C}| > q - \tfrac{1}{4}\sqrt{q} - 2$. For each such P', there are at least $q - 1$ points P_1. Conversely, each P_1 is derived from at most two P', namely $\mathcal{K}' \cap T_{P_1}(\mathcal{V})$, unless $T_{P_1}(\mathcal{V})$ contains π' and hence $\mathcal{K}'$. The exceptional case can only occur twice, when P_1 lies on the polar line of the plane π'. Thus each P' gives at least $q - 3$ points P_1, apart from the exceptions; each P_1 comes from at most two P'. Thus, with $A = \{P_1 \in \mathcal{Q}_1$ obtainable from some $P' \in \mathcal{K}'\}$,

$$|A| > \tfrac{1}{2}(q - \tfrac{1}{4}\sqrt{q} - 2)(q - 3)$$
$$= \tfrac{1}{2}q^2 - \tfrac{1}{8}q^{3/2} - \tfrac{5}{2}q + \tfrac{3}{8}\sqrt{q} + 3.$$

Let $B = \mathcal{K}_1$, $C = \mathcal{Q}_1$ and $\mathcal{K}_1^* = \{P_1 \in \mathcal{K}_1 \text{ obtainable from some } P' \in \mathcal{K}'\}$. Then $\mathcal{K}_1^* = A \cap B$. So, by Lemma 6.2,

$$|\mathcal{K}_1^*| > \tfrac{1}{2}q^2 - \tfrac{1}{8}q^{3/2} - \tfrac{5}{2}q + \tfrac{3}{8}\sqrt{q} + 3 + (q^2 - \tfrac{1}{4}q^{3/2} + R(q)) - (q^2 + 1)$$
$$= \tfrac{1}{2}q^2 - \tfrac{3}{8}q^{3/2} - \tfrac{1}{16}(9q - 20\sqrt{q} + 21).$$

The line $P'P_1$ with P' in $\mathcal{K}'$ and P_1 in $\mathcal{K}_1^*$ meets α_2 in a point P_2 of $\mathcal{Q}_2 \backslash \mathcal{C}$. Such a P_2 is obtained at most twice when $|T_{P_2}(\mathcal{V}) \cap \mathcal{K}'| \leq 2$, unless $T_{P_2}(\mathcal{V}) \supset \pi'$, which can occur for at most two points P_2, where the polar line of π' meets $\mathcal{Q}_2$. Thus, with $A = \{P_2 \in \mathcal{Q}_2 \text{ obtainable from some } P_1P'\}$,

$$|A| \geq 2 + \tfrac{1}{2}(|\mathcal{K}_1^*| - 2|\mathcal{K}'\backslash\mathcal{C}|) \geq 2 + \tfrac{1}{2}|\mathcal{K}_1^*| - (q + 1)$$
$$= \tfrac{1}{4}q^2 - \tfrac{3}{16}q^{3/2} - \tfrac{1}{32}(41q - 20\sqrt{q} - 11).$$

Now, let $B = \mathcal{K}_2$ and $C = \mathcal{Q}_2$. Therefore, if

$$\mathcal{K}_2^* = \{P_2 \in \mathcal{K}_2 \mid P'P_1P_2 \text{ is a line with } P' \in \mathcal{K}', \ P_1 \in \mathcal{K}_1^*\},$$

Lemma 6.2 gives

$$|\mathcal{K}_2^*| > \tfrac{1}{4}q^2 - \tfrac{3}{16}q^{3/2} - \tfrac{1}{32}(41q - 20\sqrt{q} - 11)$$
$$+ (q^2 - \tfrac{1}{4}q^{3/2} + R(q)) - (q^2 + 1)$$
$$= \tfrac{1}{4}q^2 - \tfrac{7}{16}q^{3/2} + \tfrac{1}{32}(21q + 48\sqrt{q} - 127)$$
$$> 0.$$

So there is a line meeting $\mathcal{K}', \mathcal{K}_1, \mathcal{K}_2$ in distinct points. Therefore $\mathcal{K}$ is not a cap, which provides the desired contradiction. □

Theorem 6.19. *In* $\mathrm{PG}(n, q)$, $n \geq 4$, $q \geq 197$ *and odd,*

$$m_2(n, q) < q^{n-1} - \tfrac{1}{4}q^{n-3/2} + 2q^{n-2}.$$

In fact, for $q \geq 67$ *and odd,*

$$m_2(n, q) < q^{n-1} - \tfrac{1}{4}q^{n-3/2}$$
$$+ \tfrac{1}{16}(31q^{n-2} + 22q^{n-5/2} - 112q^{n-3} - 14q^{n-7/2} + 69q^{n-4})$$
$$- 2(q^{n-5} + q^{n-6} + \cdots + q + 1) + 1,$$

where there is no term $-2(q^{n-5} + \cdots + 1)$ *for* $n = 4$.

Proof. Let $\mathcal{K}$ be a k-cap in $\mathrm{PG}(n, q)$.

(I) $n = 4$

(a) There is no plane π such that $\pi \cap \mathcal{K}$ is an s-arc with

$$s > q - \tfrac{1}{4}\sqrt{q} + \tfrac{7}{4}.$$

Take a line ℓ meeting $\mathcal{K}$ in two points. There are $q^2 + q + 1$ planes π through ℓ each meeting $\mathcal{K}$ in an m-arc with $m \le q - \tfrac{1}{4}\sqrt{q} + \tfrac{7}{4}$. So

$$
\begin{aligned}
k &\le 2 + (q^2 + q + 1)(q - \tfrac{1}{4}\sqrt{q} - \tfrac{1}{4}) \\
&= q^3 - \tfrac{1}{4}q^{5/2} + \tfrac{1}{4}(3q^2 - q^{3/2} + 3q - \sqrt{q} - 1) \\
&< q^3 - \tfrac{1}{4}q^{5/2} + \tfrac{1}{16}(31q^2 + 22q^{3/2} - 112q - 14q^{1/2} + 85).
\end{aligned}
$$

(b) There is a plane π such that $\pi \cap \mathcal{K}$ is an s-arc with $s > q - \tfrac{1}{4}\sqrt{q} + \tfrac{7}{4}$.

Then, by Lemma 6.18, for $q \ge 67$, there are at most two solids through π meeting $\mathcal{K}$ in an elliptic quadric, and, for the other $q - 1$ solids α through π,

$$|\alpha \cap \mathcal{K}| \le q^2 - \tfrac{1}{4}q^{3/2} + R(q).$$

So

$$
\begin{aligned}
k &\le s + 2(q^2 + 1 - s) + (q - 1)[q^2 - \tfrac{1}{4}q^{3/2} + \tfrac{1}{16}(31q + 14\sqrt{q} - 53) - s] \\
&= q^3 - \tfrac{1}{4}q^{5/2} + \tfrac{1}{16}(47q^2 + 18q^{3/2} - 84q - 14q^{1/2} + 85) - sq \\
&< q^3 - \tfrac{1}{4}q^{5/2} + \tfrac{1}{16}(47q^2 + 18q^{3/2} - 84q - 14q^{1/2} + 85) - q(q - \tfrac{1}{4}\sqrt{q} + \tfrac{7}{4}) \\
&= q^3 - \tfrac{1}{4}q^{5/2} + \tfrac{1}{16}(31q^2 + 22q^{3/2} - 112q - 14q^{1/2} + 85) \\
&< q^3 - \tfrac{1}{4}q^{5/2} + 2q^2 \text{ for } q \ge 197.
\end{aligned}
$$

(II) $n > 4$

The induction formula of Theorem 6.14 gives that

$$
\begin{aligned}
m_2(n, q) &\le q^{n-4}m_2(4, q) - q^{n-4} - 2(q^{n-5} + \cdots + 1) + 1 \\
&< q^{n-1} - \tfrac{1}{4}q^{n-3/2} + \tfrac{1}{16}(31q^{n-2} + 22q^{n-5/2} \\
&\qquad - 112q^{n-3} - 14q^{n-7/2} + 69q^{n-4}) \\
&\qquad - 2(q^{n-5} + \cdots + 1) + 1 \text{ for } q \ge 67 \\
&< q^{n-1} - \tfrac{1}{4}q^{n-3/2} + 2q^{n-2} - q^{n-4} \\
&\qquad - 2(q^{n-5} + \cdots + 1) + 1 \text{ for } q \ge 197 \\
&< q^{n-1} - \tfrac{1}{4}q^{n-3/2} + 2q^{n-2} \text{ for } q \ge 197.
\end{aligned}
$$

$\square$

6.4 The maximum size of a cap for q even

Before looking at an upper bound for $m_2(n, q)$ for q even and $q > 2$, it is necessary to improve the upper bound for $m_2'(3, q)$, the size of the second largest cap in PG$(3, q)$;

alternatively, if a k-cap of $\mathrm{PG}(3, q)$ has $k > m_2'(3, q)$, then it is contained in a $(q^2 + 1)$-cap. In Theorem 18.3.2 of FPSOTD it was shown that, for q even with $q > 2$,

$$m_2'(3, q) \leq q^2 - \tfrac{1}{2}\sqrt{q} + 1.$$

For any k-cap $\mathcal{K}$ in $\mathrm{PG}(n, q)$, as above, a 1-secant line is called a *tangent* or a *unisecant*, a 2-secant line is a *bisecant*, and a 0-secant line is an *external* line. Also, let t be the number of tangents through a point P of $\mathcal{K}$; for a point Q not in $\mathcal{K}$, let $\sigma_1(Q)$ be the number of tangents through Q and let $\sigma_2(Q)$ be the number of bisecants through Q.

Lemma 6.20. *For a k-cap $\mathcal{K}$ in $\mathrm{PG}(n, q)$,*

   (i) $t + k = \theta(n - 1) + 1$;
   (ii) $\sigma_1(Q) + 2\sigma_2(Q) = k$.

Lemma 6.21. *In $\mathrm{PG}(n, q)$ with q even, if Q is a point not on the k-cap $\mathcal{K}$ such that $\sigma_2(Q) \geq 1$, then $\sigma_1(Q) \leq t$.*

Proof. See Lemma 18.3.1 of FPSOTD, where the proof is given for $\mathrm{PG}(3, q)$, but it extends immediately to $\mathrm{PG}(n, q)$. □

Corollary 6.22. *If $\mathcal{K}$ is a complete k-cap of $\mathrm{PG}(n, q)$, with $n \geq 3$ and q even, then $\sigma_1(Q) \leq t$ for all points Q off $\mathcal{K}$.*

Lemma 6.23. *Let $\mathcal{K}$ be a complete k-cap in $\mathrm{PG}(3, q)$ with q even. If Π is a plane such that $|\Pi \cap \mathcal{K}| = x$, then $t(t - 1) \geq q(q + 2 - x)x$.*

Proof. As each of the $q^2 + q + 1$ lines through a point $P \in \mathrm{PG}(3, q) \backslash \Pi$ meets Π in a point, each of the $t(k - x)$ tangents through the points of $\mathcal{K} \backslash \Pi$ meets Π in a point of $\Pi \backslash \mathcal{K}$. As there are $x(q + 2 - x)$ tangents to $\mathcal{K}$ on Π, so, counting the pairs (Q, ℓ) with ℓ a tangent of $\mathcal{K}$, $Q \in \Pi \backslash \mathcal{K}$ and $Q \in \ell$,

$$\sum_{Q \in \Pi \backslash \mathcal{K}} \sigma_1(Q) = t(k - x) + x(q + 2 - x)q$$

$$\leq t(q^2 + q + 1 - x) = t(t + k - 1 - x).$$

Hence $t(t - 1) \geq q(q + 2 - x)x$. □

Theorem 6.24. *For q even, $q \geq 8$,*

$$m_2'(3, q) \leq q^2 - q + 5.$$

Proof. Let $\mathcal{K}$ be a complete k-cap in $\mathrm{PG}(3, q)$, with q even, $q \geq 8$ and $k < q^2 + 1$. Suppose there exists a plane Π_0 such that $4 \leq |\Pi_0 \cap \mathcal{K}| \leq q - 2$. Let $x_0 = |\Pi_0 \cap \mathcal{K}|$ and $f(y) = q(q + 2 - y)y$. Then, by Lemma 6.23,

$$t(t - 1) \geq f(4) = f(q - 2) = 4q(q - 2).$$

So

$$t \geq \tfrac{1}{2}\left\{1 + \sqrt{1 + 16q(q-2)}\right\} \geq 2q - 2.$$

Therefore $k \leq q^2 - q + 5$.

Suppose that either $|\Pi \cap \mathcal{K}| \leq 3$ or $|\Pi \cap \mathcal{K}| \geq q-1$ for any plane Π of $PG(3,q)$. Let $l_1, \ldots, l_t$ be the t tangents to $\mathcal{K}$ through a point $P \in \mathcal{K}$. There are three cases to consider.

(I) *There exists exactly one plane Π_{l_i} through any l_i such that $|\Pi_{l_i} \cap \mathcal{K}| \leq 3$ for $1 \leq i \leq t$*

Suppose that there is exactly one plane Π through P with $|\Pi \cap \mathcal{K}| \leq 3$. Then $\Pi_{l_i} = \Pi$ for all i. Hence all tangents to $\mathcal{K}$ through P are in Π. Therefore $t \leq q + 1$ and so $k \geq q^2 + 1$, a contradiction. So there are at least two planes Π_j, $j = 1, 2$, through P such that $|\Pi_j \cap \mathcal{K}| \leq 3$. Each plane Π_j contains at least $q - 1$ tangents to $\mathcal{K}$ through P. Thus $t \geq 2(q-1)$ and so $k \leq q^2 - q + 4$.

(II) *There exist two planes through some tangent l_c, $1 \leq c \leq t$, with at most three points in $\mathcal{K}$*

Any plane through l_c meets $\mathcal{K}$ in at most $q + 1$ points since l_c is a tangent of $\mathcal{K}$. Counting the points of $\mathcal{K}$ on the $q + 1$ planes through l_c shows that

$$k - 1 \leq 2 \times 2 + (q-1)q.$$

So $k \leq q^2 - q + 5$.

(III) *There is no plane through some tangent l_d, $1 \leq d \leq t$, with at most three points in $\mathcal{K}$*

Then $|\Pi_i \cap \mathcal{K}| \geq q - 1$ for any plane Π_i through l_d. By (6.13), $\Pi_i \cap \mathcal{K}$ can be completed to a $(q + 2)$-arc, which meets l_d in a point P_i other than P. As there are $q + 1$ points P_i and only q points on l_d for them to occupy, so two of the P_i coincide; say $P_1 = P_2$. Thus the number of tangents to $\mathcal{K}$ through P includes the joins of P_1 to the points of $\Pi_1 \cap \mathcal{K}$ and $\Pi_2 \cap \mathcal{K}$. Hence $t \geq \sigma_1(P_1) \geq 2(q-2) + 1$, whence $k \leq q^2 - q + 5$. $\qquad\square$

Theorem 6.25. *In* $PG(3,4)$,

  (i) $m_2'(3,4) = 14$;

  (ii) *a complete 14-cap consists of the points on the generators of a cone $P\beta$, where P is the vertex and β is a $PG(2,2)$, outside a $PG(3,2)$ containing P and β.*

Proof. See Section 6.8. $\qquad\square$

Theorem 6.26. *Let q be even.*

  (i) *For $q \geq 8$,*

$$m_2'(3,q) < q^2 - (\sqrt{5} - 1)q + 5.$$

(ii) *For $q \geq 16$,*
$$m_2'(3, q) \leq q^2 - q + 2.$$

(iii) *For $q \geq 128$,*
$$m_2'(3, q) \leq q^2 - 2q + 8.$$

Proof. See Section 6.8. □

Now the maximum number of points on a cap in $PG(4, q)$ is considered.

Theorem 6.27. *For q even, and $q \geq 8$,*
$$m_2(4, q) \leq q^3 - q^2 + 6q - 3.$$

Proof. Suppose there exists a complete k-cap $\mathcal{K}$ in $PG(4, q)$ with
$$k > q^3 - q^2 + 6q - 3. \tag{6.54}$$

Then, with t the number of tangents through a point of $\mathcal{K}$,
$$t < 2q^2 - 5q + 5, \tag{6.55}$$

by Lemma 6.20(i). A contradiction is obtained in several stages.

(I) $\mathcal{K}$ *contains no plane q-arc*

Suppose that π is a plane such that $\pi \cap \mathcal{K}$ is a q-arc $\mathcal{Q}$. Consider two subcases.
   (a) Suppose there exist three solids $\delta_1, \delta_2, \delta_3$ containing the plane π such that, for $i = 1, 2, 3$,
$$|\delta_i \cap \mathcal{K}| > q^2 - q + 5.$$
Then, by Theorem 6.24, $\delta_i \cap \mathcal{K}$ can be completed to an ovoid $\mathcal{O}_i$. So $\mathcal{O}_i \cap \pi$ is a $(q + 1)$-arc $\mathcal{Q} \cup \{N_i\}$. However, since $\mathcal{Q}$ can be contained in no more than two $(q + 1)$-arcs, at least two of the N_i coincide; say $N_1 = N_2$. The joins of N_1 to the points of $\delta_1 \cap \mathcal{K}$ and $\delta_2 \cap \mathcal{K}$ are all tangents to $\mathcal{K}$. Hence
$$\sigma_1(N_1) > 2(q^2 - 2q + 5) + q. \tag{6.56}$$

Since $\mathcal{K}$ is complete, $\sigma_1(N_1) \leq t$ by Lemma 6.21. So (6.55) and (6.56) imply that
$$2q + 5 < 0,$$

a contradiction.
   (b) If there are at most two solids δ_1 and δ_2 through π such that, for $i = 1, 2$,
$$|\delta_i \cap \mathcal{K}| > q^2 - q + 5,$$

then, counting the points of $\mathcal{K}$ on the solids through π,
$$\begin{aligned} k &\leq (q - 1)(q^2 - 2q + 5) + 2(q^2 + 1 - q) + q \\ &= q^3 - q^2 + 6q - 3, \end{aligned}$$

in contradiction to (6.54).

(II) *There exists no solid δ such that $q^2 + 1 > |\delta \cap \mathcal{K}| > q^2 - q + 5$*

Suppose δ exists. Let $\delta \cap \mathcal{K} = \mathcal{K}'$. Then $\mathcal{K}'$ can be completed to an ovoid $\mathcal{O}$ by Theorem 6.24. Let $N \in \mathcal{O} \backslash \mathcal{K}'$ and let $N' \in \mathcal{K}'$. Consider the $q + 1$ planes of δ through NN'. Since each of these planes meets $\mathcal{O}$ in a $(q+1)$-arc, each plane meets $\mathcal{K}'$ in at most a q-arc. By (I), there is no q-arc on $\mathcal{K}$; so each of these planes meets $\mathcal{K}'$ in at most a $(q - 1)$-arc. Therefore a count on the points of $\mathcal{K}'$ gives

$$q^2 - q + 5 < |\mathcal{K}'| \leq (q + 1)(q - 2) + 1,$$

whence $6 < 0$, a contradiction.

(III) *For a point N not in $\mathcal{K}$, there do not exist planes π_1, π_2 with $\pi_1 \cap \pi_2 = \{N\}$ and such that $\pi_i \cap \mathcal{K}$ is a $(q + 1)$-arc with nucleus N for $i = 1, 2$*

Suppose π_1 and π_2 exist. Let δ be a solid containing π_1. Then $\delta \cap \mathcal{K}$ contains at least $q + 2$ tangents through N, of which $q + 1$ are in π_1 and one of which is in π_2; so $|\delta \cap \mathcal{K}| < q^2 + 1$. Suppose now that

$$|\delta \cap \mathcal{K}| \leq q^2 - q + 5$$

for any such solid δ. Then a count on the points of $\mathcal{K}$ in the solids through π_1 gives

$$k \leq (q + 1)(q^2 - 2q + 5) + (q + 1);$$

that is,

$$k \leq q^3 - q^2 + 4q + 6. \tag{6.57}$$

But (6.54) and (6.57) imply that $2q - 9 < 0$, a contradiction; thus there exists a solid δ such that

$$q^2 + 1 > |\delta \cap \mathcal{K}| > q^2 - q + 5.$$

But this contradicts (II). So π_1 and π_2 do not exist.

(IV) *The tangents through any point Q off $\mathcal{K}$ lie in a solid*

Let δ be a solid not containing Q and let $\mathcal{V}$ be the set of intersections of tangents to $\mathcal{K}$ through Q with δ. It is shown that each point of $\mathcal{V}$ is on at least two lines of $\mathcal{V}$.

Let $R \in \mathcal{V}$ and let r be the corresponding tangent. Suppose, for at most one plane π through r, that $|\pi \cap \mathcal{K}| = q + 1$. Then, since there is no q-arc on $\mathcal{K}$, counting the points of $\mathcal{K}$ on the planes through r gives

$$k \leq (q^2 + q)(q - 2) + q + 1,$$

a contradiction to (6.54).

Now let π_1 and π_2 be planes through r meeting $\mathcal{K}$ in $(q + 1)$-arcs. If Q is the nucleus of both $\pi_1 \cap \mathcal{K}$ and $\pi_2 \cap \mathcal{K}$, then there are two lines of $\mathcal{V}$ through R, namely $\pi_1 \cap \delta$ and $\pi_2 \cap \delta$. Therefore, suppose that Q is not the nucleus of $\pi_1 \cap \mathcal{K}$. If, for at most one solid δ' through π_1, the equality $|\delta' \cap \mathcal{K}| = q^2 + 1$ holds, then, by (II),

$$k \leq q^2 + 1 + q(q^2 - q + 5 - q - 1),$$

whence

$$k \leq q^3 - q^2 + 4q + 1. \tag{6.58}$$

But (6.58) contradicts (6.54). Thus there are two solids δ_1 and δ_2 through π_1 for which $\delta_i \cap \mathcal{K} = \mathcal{O}_i$ is an ovoid. Then Q is the nucleus of a $(q+1)$-arc $\mathcal{M}_i$ on $\mathcal{O}_i$ for $i = 1, 2$. The tangents of $\mathcal{M}_i$ meet δ in a line ℓ_i through R, and the lines ℓ_1 and ℓ_2 are distinct since Q is not the nucleus of $\pi_1 \cap \mathcal{K}$. Thus R always lies on at least two lines of $\mathcal{V}$.

If there existed two skew lines in $\mathcal{V}$, then there would be two planes π_1 and π_2 through Q with $\pi_1 \cap \pi_2 = \{Q\}$ and Q the nucleus of both $\pi_1 \cap \mathcal{K}$ and $\pi_2 \cap \mathcal{K}$, in contradiction to (III). Thus the lines of $\mathcal{V}$ either all have a common point or all lie in a plane. Since each point of $\mathcal{V}$ is on at least two lines of $\mathcal{V}$, all lines of $\mathcal{V}$ lie in plane. Hence $\mathcal{V}$ is a subset of a plane and the tangents to $\mathcal{K}$ through Q lie in a solid.

(V) *The final contradiction is obtained by counting the tangents of $\mathcal{K}$*

Consider the function

$$G(x) = x(q^3 + q^2 + q + 2 - x).$$

It attains its maximum value for $x = \frac{1}{2}(q^3 + q^2 + q + 2)$. Since, by Theorem 6.16 and (6.54),

$$q^3 \geq k > q^3 - q^2 + 6q - 3 > \tfrac{1}{2}(q^3 + q^2 + q + 2),$$

so

$$kt = k(q^3 + q^2 + q + 2 - k) \geq G(q^3) = q^3(q^2 + q + 2). \tag{6.59}$$

By (IV), all tangents through a point Q off $\mathcal{K}$ lie in a solid, which contains at most $q^2 + 1$ points of $\mathcal{K}$. However, an ovoid has exactly $q + 1$ tangents through an external point. So, through Q, there are at most q^2 tangents of $\mathcal{K}$. A count of the pairs (R, r) where R is a point off $\mathcal{K}$ and r a tangent to $\mathcal{K}$ through R gives

$$(q^4 + q^3 + q^2 + q + 1 - k)q^2 \geq ktq. \tag{6.60}$$

From (6.54), (6.59) and (6.60),

$$(q^4 + q^3 + q^2 + q + 1 - q^3 + q^2 - 6q + 3)q$$
$$> kt \geq q^3(q^2 + q + 2).$$

Hence

$$q^4 + 2q^2 - 5q + 4 > q^4 + q^3 + 2q^2,$$

and

$$q^3 + 5q - 4 < 0,$$

the final contradiction.   □

Similar methods give an improvement to this result.

Theorem 6.28.   (i) $m_2(4,8) \leq 479$;
(ii) *for q even, $q > 8$,*

$$m_2(4,q) < q^3 - q^2 + 2\sqrt{5}q - 8;$$

(iii) *for q even, $q \geq 128$,*

$$m_2(4,q) \leq q^3 - 2q^2 + 14q - 20.$$

Proof. See Section 6.8. □

Finally, upper bounds for the size of a k-cap in $PG(n,q)$ can be obtained when q is even, $q > 2$ and $n \geq 5$.

Theorem 6.29. *Let q be even, with $q > 2$ and $n \geq 5$.*

(i) $m_2(n,4) \leq \frac{118}{3} \cdot 4^{n-4} + \frac{5}{3}$;

(ii) $m_2(n,8) \leq 478 \cdot 8^{n-4} - 2(8^{n-5} + \cdots + 8 + 1) + 1$;

(iii) $m_2(n,q) \leq q^{n-1} - (n-4)q^{n-2} + (n-3)^2 q^{n-3}$ *for $q \geq 8$ and $4 \leq n \leq 2q/3$;*

(iv) $m_2(5,q) < q^4 - q^3 + 5q^2 + 3q - 1$ *for $q \geq 16$;*

(v) $m_2(n,q) < q^{n-1} - q^{n-2} + 5q^{n-3} + 2q^{n-4} + 2(q^{n-5} + q^{n-6} + \cdots + q) + q - 1$
for $q \geq 16$ and $n > 5$;

(vi) $m_2(n,q) < q^{n-1} - q^{n-2} + 2\sqrt{5}q^{n-3} - 9q^{n-4} - 2(q^{n-5} + q^{n-6} + \cdots + q + 1) + 1$
for $q \geq 16$;

(vii) $m_2(n,q) \leq q^{n-1} - 2q^{n-2} + 14q^{n-3} - 21q^{n-4} - 2(q^{n-5} + q^{n-6} + \cdots + q + 1) + 1$
for $q \geq 128$.

Proof. Parts (i), (ii), (vi), (vii) follow from (6.6), Theorem 6.28 and Theorem 6.14(ii). For (iii), (iv) and (v), see Section 6.8. □

6.5 General properties of k-arcs and normal rational curves

As in Section 21.1 of FPSOTD, a *rational curve* C_d of order d in $PG(n,q)$ is the set of points

$$\{P(t_0,t_1) = \mathbf{P}(g_0(t_0,t_1), \ldots, g_n(t_0,t_1)) \mid t_0, t_1 \in \mathbf{F}_q\}, \tag{6.61}$$

where each g_i is a binary form of degree d and they have no non-trivial common factor. The curve C_d may also be written

$$\{P(t) = \mathbf{P}(f_0(t), f_1(t), \ldots, f_n(t)) \mid t \in \mathbf{F}_q{}^+\}, \tag{6.62}$$

where $f_i(t) = g_i(1, t)$. As the g_i have no non-trivial common factor, at least one f_i has degree d. Also C_d is *normal* if it is not the projection of a rational C'_d in $PG(n + 1, q)$, where C'_d is not contained in a hyperplane.

Theorem 6.30. *Let C_d be a normal rational curve in $PG(n, q)$ not contained in a hyperplane. Then*

(i) $q \geq n$;
(ii) $d = n$;
(iii) C_n *is projectively equivalent to*

$$\{P(t) = \mathbf{P}(t^n, t^{n-1}, \ldots, t, 1) \mid t \in \mathbf{F}_q{}^+\}; \tag{6.63}$$

(iv) C_n *consists of $q + 1$ points no $n + 1$ in a hyperplane;*
(v) *if $q \geq n+2$, there is a unique C_n through any $n+3$ points of $PG(n, q)$ no $n+1$ of which lie in a hyperplane;*
(vi) *there is a subgroup H of $PGL(n + 1, q)$ isomorphic to $PGL(2, q)$ that acts 3-transitively on C_n.*

Proof. (i)–(v) See Theorem 21.1.1 of FPSOTD.
(vi) With C_n as in (6.63), the transformation τ given by $t \mapsto (at + b)/(ct + d)$, with $ad - bc \neq 0$ induces the transformation

$$(t^n, t^{n-1}, \ldots, t, 1)$$
$$\mapsto ((at + b)^n, (at + b)^{n-1}(ct + d), \ldots, (at + b)(ct + d)^{n-1}, (ct + d)^n)$$
$$= (t^n, t^{n-1}, \ldots, t, 1)T$$

for a suitable non-singular matrix T. Hence $H = \{T \mid \tau \in PGL(2, q)\}$.   □

Now, further properties of C_n are considered. With C_n as in (6.61), write the derivative $\partial g_j / \partial t_i = g_j^i$. If, for a given i in $\{0, 1\}$, not all $g_j^i(t_0, t_1)$ are zero, then the point with $g_j^i(t_0, t_1)$ as $(j + 1)$th coordinate is denoted by $P^i(t_0, t_1)$. If such is the case for both i and if $P(t_0, t_1) \neq P^i(t_0, t_1)$ also for both i, then

$$P(t_0, t_1)P^0(t_0, t_1) = P(t_0, t_1)P^1(t_0, t_1)$$

since

$$t_0 g_j^0(t_0, t_1) + t_1 g_j^1(t_0, t_1) = n g_j(t_0, t_1).$$

For at least one i in $\{0, 1\}$, the point $P^i(t_0, t_1)$ exists and is distinct from $P(t_0, t_1)$; for such an i, the line $P(t_0, t_1)P^i(t_0, t_1)$ is the *tangent* of C_n at P and is denoted by ℓ_p.

Lemma 6.31. *Let $q \geq n \geq 3$, let C_n be a normal rational curve, and let P be a point of C_n.*

(i) *The image of the projection map $\mathcal{G}$ of $\mathcal{C}_n \backslash \{P\}$ from P onto a hyperplane Π not containing P together with $P' = \ell_P \cap \Pi$ is a normal rational curve in Π, and is denoted by $\mathcal{C}_n \mathcal{G}$.*

(ii) *No two tangents to $\mathcal{C}_n$ intersect.*

(iii) *If ℓ_P lies in a hyperplane Π', then $|\Pi' \cap \mathcal{C}_n| \leq n - 1$.*

(iv) *$\ell_P \cap \mathcal{C}_n = \{P\}$.*

(v) *If $Q, R \in \mathcal{C}_n \backslash \{P\}$, then QR does not meet ℓ_P.*

Proof. (i) Take $\mathcal{C}_n$ in canonical form

$$\{P(t) = \mathbf{P}(t^n, t^{n-1}, \ldots, t, 1) \mid t \in \mathbf{F}_q{}^+\}.$$

By Theorem 6.30 (vi), choose $P = \mathbf{U}_0$. Let $\mathcal{G}'$ be the projection of Π from P onto the hyperplane $\mathbf{u}_0$. Then, for $t \in \mathbf{F}_q$,

$$P\mathcal{G}\mathcal{G}' = \mathbf{P}(0, t^{n-1}, t^{n-2}, \ldots, t, 1).$$

Also the tangent $\ell_P = \mathbf{U}_0 \mathbf{U}_1$, which meets $\mathbf{u}_0$ in $\mathbf{U}_1$. So $\{P\mathcal{G}\mathcal{G}' \mid t \in \mathbf{F}_q\} \cup \{\mathbf{U}_1\}$ is the normal rational curve

$$\{\mathbf{P}(0, t^{n-1}, t^{n-2}, \ldots, t, 1) \mid t \in \mathbf{F}_q{}^+\}.$$

Thus $\mathcal{C}_n \mathcal{G}$ is a normal rational curve of degree $n - 1$ in Π.

    (ii) Let $P = P(t)$ and $Q = P(s)$, $s, t \in \mathbf{F}_q$, $s \neq t$. To show that $\ell_P \cap \ell_Q = \emptyset$, consider the matrix

$$\begin{bmatrix} t^n & t^{n-1} & \cdots & t^3 & t^2 & t & 1 \\ nt^{n-1} & (n-1)t^{n-2} & \cdots & 3t^2 & 2t & 1 & 0 \\ s^n & s^{n-1} & \cdots & s^3 & s^2 & s & 1 \\ ns^{n-1} & (n-1)s^{n-2} & \cdots & 3s^2 & 2s & 1 & 0 \end{bmatrix}.$$

It has rank 4, since the submatrix formed by the last four columns has determinant $(t - s)^4$. So, when $P, Q \neq \mathbf{U}_0$, the lines ℓ_P and ℓ_Q do not meet. By the transitivity of the group, this is also true when $P = \mathbf{U}_0$.

    (iii) Let $\Pi' = \mathbf{V}(a_0 X_0 + a_1 X_1 + \cdots + a_n X_n)$ and take $P = \mathbf{U}_0$. From (i), $\ell_P = \mathbf{U}_0 \mathbf{U}_1$; so ℓ_P lies in Π' if and only if $a_0 = a_1 = 0$. Hence, apart from P, a point $P(t)$ of $\mathcal{C}_n$ lies in Π' if and only if

$$a_2 t^{n-2} + \cdots + a_n = 0.$$

This has at most $n - 2$ solutions.

    (iv) $\mathbf{U}_0 \mathbf{U}_1 \cap \mathcal{C}_n = \{\mathbf{U}_0\}$.

    (v) Take $P = \mathbf{U}_0$, $Q = \mathbf{U}_n$, and $R = \mathbf{U}$. $\square$

Theorem 6.32. *If $q \geq n + 2$, then*

(i) *the group $G(\mathcal{C}_n)$ of projectivities in $\mathrm{PG}(n, q)$ fixing $\mathcal{C}_n$ is isomorphic to the projective linear group $\mathrm{PGL}(2, q)$, given by the transformations*

$$t \mapsto (at + b)/(ct + d),$$

with $ad - bc \neq 0$, acting on (6.63);

(ii) *the number of normal rational curves in* $\mathrm{PG}(n, q)$ *is*

$$\nu_n = q^{(n-1)(n+2)/2}[3, n+1]_-$$

$$= \prod_{i=0}^{n}(q^{n+1} - q^i)/\{q(q^2 - 1)(q - 1)\}.$$

Proof. (i) From Theorem 6.30(vi), there is a subgroup H of $G(\mathcal{C}_n)$ isomorphic to $\mathrm{PGL}(2, q)$. It must be shown that $H = G(\mathcal{C}_n)$.

Now, suppose that an element $\mathcal{U}$ of $G(\mathcal{C}_n)$ is given by the matrix $A = [a_{ij}]$, with $0 \le i, j \le n$, and $\mathcal{C}_n$ is taken in the form (6.63). Since

$$(t^n, t^{n-1}, \ldots, t, 1)A = (\sum_{i=0}^{n} a_{i0}t^{n-i}, \sum_{i=0}^{n} a_{i1}t^{n-i}, \ldots, \sum_{i=0}^{n} a_{in}t^{n-i})$$

$$= (s^n, s^{n-1}, \ldots, s, 1),$$

there exists a permutation ρ_n of $\mathbf{F}_q{}^+$ such that $t\rho_n = s$, whence

$$t\rho_n = \sum_{i=0}^{n} a_{i0}t^{n-i} / \sum_{i=0}^{n} a_{i1}t^{n-i}.$$

It is now shown by induction on n that there exist a, b, c, d in $\mathbf{F}_q$ with $ad - bc \ne 0$ such that

$$t\rho_n = (at + b)/(ct + d).$$

For $n = 1$,

$$t\rho_n = (a_{00}t + a_{10})/(a_{01}t + a_{11});$$

so the result is proved.

Assume that the result is true for

$$\mathcal{C}_{n-1} = \{\mathbf{P}(t^{n-1}, t^{n-2}, \ldots, t, 1, 0) \mid t \in \mathbf{F}_q{}^+\}, \ n \ge 2.$$

By the transitivity of H, take $\mathbf{U}_n\mathcal{U} = \mathbf{U}_n$; hence $a_{ni} = 0$, $0 \le i \le n - 1$, and $a_{n0} = a_{n1} = 0$ in particular. So

$$t\rho_n = \sum_{i=0}^{n-1} a_{i0}t^{n-i} / \sum_{i=0}^{n-1} a_{i1}t^{n-i}$$

$$= \sum_{i=0}^{n-1} a_{i0}t^{n-1-i} / \sum_{i=0}^{n-1} a_{i1}t^{n-1-i}.$$

Since the tangent at $\mathbf{U}_n$ is fixed, also $a_{n-1,i} = 0$ for $0 \le i \le n - 2$.

Let $\mathcal{G}$ be the projection map from $\mathbf{U}_n$ onto $\mathbf{u}_n$. Then, as in Lemma 6.31(i), let $\mathcal{C}_{n-1} = \mathcal{C}_n\mathcal{G}$. Let $A' = [a'_{ij}]$, $0 \le i, j \le n - 1$ with $a'_{ij} = a_{ij}$. Also let $\mathcal{U}'$ be the projectivity on $\mathbf{u}_n$ corresponding to the matrix A'. Then, for $t \in \mathbf{F}_q{}^+\backslash\{0\}$,

$$\mathbf{P}(t^n, \ldots, t, 1)\mathcal{G}\mathcal{U}' = \mathbf{P}(t^{n-1}, \ldots, t, 1, 0)\mathcal{U}' = \mathbf{P}(t^n, \ldots, t, 1)\mathcal{U}\mathcal{G} \in \mathcal{C}_{n-1}.$$

Also $\mathbf{U}_{n-1}\mathcal{U}' = \mathbf{U}_{n-1} \in \mathcal{C}_{n-1}$. Since $\mathcal{C}_{n-1}\mathcal{U}'$ is also a normal rational curve of order $n-1$, so $\mathcal{C}_{n-1}\mathcal{U}' = \mathcal{C}_{n-1}$ by Theorem 6.30(v). So $\mathcal{U}'$ is a projectivity of $\mathbf{u}_n$ fixing $\mathcal{C}_{n-1}$. However,

$$t\rho_{n-1} = \sum_{i=0}^{n-1} a'_{i0}t^{n-i-1} \Big/ \sum_{i=0}^{n-1} a'_{i1}t^{n-1-i}$$

$$= \sum_{i=0}^{n-1} a_{i0}t^{n-1-i} \Big/ \sum_{i=0}^{n-1} a_{i1}t^{n-1-i}$$

$$= t\rho_n.$$

By the induction hypothesis, ρ_n has the required form.

(ii) Since $\mathcal{C}_n$ is projectively unique and since $G(\mathcal{C}_n) = \mathrm{PGL}(2, q)$, so

$$\nu_n = |\mathrm{PGL}(n+1, q)| / |\mathrm{PGL}(2, q)|. \qquad \square$$

Now consider the existence of k-arcs in $\mathrm{PG}(n, q)$.

Theorem 6.33. *A k-arc in $\mathrm{PG}(n, q)$, $k \geq n+4$, exists if and only if a k-arc exists in $\mathrm{PG}(k - n - 2, q)$.*

Proof. Choose $n+1$ points of a k-arc $\mathcal{K}$ as the simplex of reference. Consider the $(k - n - 1) \times (n + 1)$ matrix M whose rows are the vectors of the other $k - n - 1$ points of $\mathcal{K}$. Since no $n+1$ points of $\mathcal{K}$ lie in a hyperplane, taking $n - s + 1$ points, where $0 \leq s \leq \min(k - n - 1, n + 1)$, of the simplex of reference and s other points shows that all $s \times s$ minors of M are non-zero. So now take the rows of the transpose M^* as vectors of points in $\mathrm{PG}(k - n - 2, q)$ and add the simplex of reference in this space. This gives a k-arc $\mathcal{K}'$ in $\mathrm{PG}(k - n - 2, q)$. The process is reversible. $\quad\square$

As in Chapter 3, let $\mathcal{G}_{r,n}$ denote the Grassmannian of r-spaces in $\mathrm{PG}(n, q)$. Let $\mathcal{A}_{k,n}$ be the set of all k-arcs in $\mathrm{PG}(n, q)$. Now consider a relation between $\mathcal{G}_{n,k-1}$ and $\mathcal{A}_{k,n}$.

Let $\mathcal{K} = \{P_1, \ldots, P_k\}$ be a k-arc in $\mathrm{PG}(n, q)$ with $k \geq n+3$, let $G(\mathcal{K})$ be the group of projectivities fixing $\mathcal{K}$, and let $g(\mathcal{K}) = |G(\mathcal{K})|$. Let $P_i = \mathbf{P}(X_i)$. Then to $\mathcal{K}$ there correspond $(q - 1)^k k!$ matrices each with k rows and $n + 1$ columns:

$$\begin{bmatrix} \rho_1 X_{i1} \\ \rho_2 X_{i2} \\ \vdots \\ \rho_k X_{ik} \end{bmatrix}. \qquad (6.64)$$

Here $\{i_1, \ldots, i_k\} = \{1, \ldots, k\}$ and $\rho_i \in \mathbf{F}_q{}^*$. Every subdeterminant of order $n+1$ is non-zero, since $\mathcal{K}$ is a k-arc. This matrix is denoted by

$$M = M_{R,\sigma}, \qquad (6.65)$$

where $R = (\rho_1, \ldots, \rho_k)$ and $\sigma = (i_1, \ldots, i_k)$. Now, take the columns of $M_{R,\sigma}$ as the vectors of $n+1$ points in $PG(k-1, q)$. These points define a $PG(n, q)$ which is denoted $\Pi_n(M)$. From the $(q-1)^k k!$ matrices M, the $(q-1)^k k!$ subspaces $\Pi_n(M)$ of $PG(k-1, q)$ are obtained; these are not necessarily distinct. Suppose that for two matrices M and M' the equation $\Pi_n(M) = \Pi_n(M')$ holds. Then there exists a unique non-singular $(n+1) \times (n+1)$ matrix A such that

$$MA = M'. \tag{6.66}$$

However (6.65) also defines a unique projectivity of $PG(n, q)$ fixing $\mathcal{K}$.

Conversely, if B is the matrix of a projectivity fixing $\mathcal{K}$, then

$$\Pi_n(\rho M B) = \Pi_n(M)$$

for all $\rho \neq 0$; that is, $(q-1)g(\mathcal{K})$ matrices M give the same $\Pi_n(M)$. So, to $\mathcal{K}$, there correspond

$$\chi(\mathcal{K}) = (q-1)^{k-1} k! / g(\mathcal{K}) \tag{6.67}$$

distinct subspaces Π_n of $PG(k-1, q)$.

The Grassmannian $\mathcal{G}_{n,k-1}$ is embedded in $PG(N, q)$, with $N = c(k, n+1) - 1$, and contains

$$[k-n, k]_- / [1, n+1]_-$$

points, Section 3.2. From above, it follows that to the k-arc $\mathcal{K}$ correspond $\chi(\mathcal{K})$ points of $\mathcal{G}_{n,k-1}$ lying in no face of the simplex of reference of $PG(N, q)$. Now consider how many k-arcs correspond to one of these $\chi(\mathcal{K})$ points Q of $\mathcal{G}_{n,k-1}$. To Q corresponds one Π_n' of $PG(k-1, q)$. The number of ordered $(n+1)$-tuples $(Q_1, \ldots, Q_{n+1})$ of linearly independent points of Π_n' is

$$\phi = \prod_{i=0}^{n} (q^{n+1} - q^i) / (q-1)^{n+1}.$$

So, to Π_n' there correspond

$$(q-1)^{n+1} \phi = \prod_{i=0}^{n} (q^{n+1} - q^i)$$

matrices M.

Suppose now that the two $k \times (n+1)$ matrices Y and Z have, as their columns, vectors of $n+1$ linearly independent points of Π_n' and give the same k-arc $\mathcal{K}'$ of $PG(n, q)$. Then

$$YA = Z$$

for a unique non-singular $(n+1) \times (n+1)$ matrix A, which consequently defines a projectivity of $PG(n, q)$ fixing $\mathcal{K}'$. Conversely, a projectivity of $PG(n, q)$ fixing $\mathcal{K}'$ gives $q-1$ matrices corresponding to Π_n'. So $\mathcal{K}'$ comes from $(q-1)g(\mathcal{K}')$ matrices corresponding to Π_n'.

Since the k-arcs $\mathcal{K}$ and $\mathcal{K}'$ come from two ordered $(n+1)$-tuples of linearly independent points of Π'_n, they are projectively equivalent; hence $g(\mathcal{K}) = g(\mathcal{K}')$. So, to Q of $\mathcal{G}_{n,k-1}$ there correspond

$$\prod_{i=0}^{n}(q^{n+1} - q^i)/[(q-1)g(\mathcal{K})]$$

k-arcs of $\mathrm{PG}(n,q)$ and these are precisely the ones projectively equivalent to $\mathcal{K}$.

Let $\mathcal{V}_{n,k-1}$ denote the set of points of $\mathcal{G}_{n,k-1}$ on no face of the simplex of reference and let $\mathcal{V}_{n,k-1}(\mathcal{K})$ denote the set of $\chi(\mathcal{K})$ points corresponding to the k-arc $\mathcal{K}$. Hence $\mathcal{V}_{n,k-1}$ is partitioned by the sets $\mathcal{V}_{n,k-1}(\mathcal{K})$. In $\mathcal{A}_{k,n}$, which is the set of k-arcs of $\mathrm{PG}(n,q)$, let $\mathcal{A}_{k,n}(\mathcal{K})$ denote the set of k-arcs corresponding to a point of $\mathcal{V}_{n,k-1}(\mathcal{K})$. Then the sets $\mathcal{A}_{k,n}(\mathcal{K})$ partition $\mathcal{A}_{k,n}$. If $\mathcal{A}_{k,n}(\mathcal{K})$ is mapped onto $\mathcal{V}_{n,k-1}(\mathcal{K})$, then a bijection of the quotient set corresponding to the partition of $\mathcal{A}_{k,n}$ and the quotient set corresponding to the partition of $\mathcal{V}_{n,k-1}$ is obtained. This discussion gives the following results.

Theorem 6.34.   (i)

$$|\mathcal{A}_{k,n}(\mathcal{K})| = \prod_{i=0}^{n}(q^{n+1} - q^i)/[(q-1)g(\mathcal{K})]. \tag{6.68}$$

(ii)

$$|\mathcal{V}_{n,k-1}(\mathcal{K})|/|\mathcal{A}_{k,n}(\mathcal{K})| = (q-1)^k k!/\prod_{i=0}^{n}(q^{n+1} - q^i). \tag{6.69}$$

(iii)

$$|\mathcal{V}_{n,k-1}|/|\mathcal{A}_{k,n}| = (q-1)^k k!/\prod_{i=0}^{n}(q^{n+1} - q^i). \tag{6.70}$$

Theorem 6.35. *For* $q \geq \max(n+2, k-n)$, $n \geq 2$, $k \geq n+4$,

$$|\mathcal{A}_{k,k-2-n}|/|\mathcal{A}_{k,n}| = \nu_{k-2-n}/\nu_n$$

$$= \prod_{i=0}^{k-2-n} (q^{k-n-1} - q^i)/\prod_{i=0}^{n}(q^{n+1} - q^i). \tag{6.71}$$

Proof. To a point of $\mathcal{V}_{n,k-1}$ there corresponds a Π_n of $\mathrm{PG}(k-1,q)$ skew to every $(k-2-n)$-dimensional edge of the simplex of reference of $\mathrm{PG}(k-1,q)$, and conversely. By the principle of duality, the number of such Π_n is the same as the number of Π_{k-2-n} of $\mathrm{PG}(k-1,q)$ skew to every n-dimensional edge of the simplex of reference. Hence

$$|\mathcal{V}_{n,k-1}| = |\mathcal{V}_{k-2-n,k-1}|. \tag{6.72}$$

The result now follows from (6.70), (6.72) and Theorem 6.30(vii).   □

There are numerous consequences that can be drawn from Theorems 6.33 and 6.35. For the moment, only results that follow from properties of $\mathrm{PG}(2,q)$ and $\mathrm{PG}(3,q)$ are considered.

Corollary 6.36.   (i) $m(q-3,q) = q+1$ *for* $q \geq 5$;
   (ii) $m(q-2,q) = q+1$ *for* q *odd with* $q \geq 5$;
   (iii) $m(q-2,q) = q+2$ *for* q *even with* $q \geq 4$.

Proof.   (i) For $q \geq 5$, $|\mathcal{A}_{q+2,3}| = 0 \Rightarrow |\mathcal{A}_{q+2,q-3}| = 0$.
   (ii) For q odd with $q \geq 5$, $|\mathcal{A}_{q+2,2}| = 0 \Rightarrow |\mathcal{A}_{q+2,q-2}| = 0$.
   (iii) (a) For q even with $q \geq 4$, $|\mathcal{A}_{q+2,2}| > 0 \Rightarrow |\mathcal{A}_{q+2,q-2}| > 0$;
      (b) also, for q even, $|\mathcal{A}_{q+3,3}| = 0 \Rightarrow |\mathcal{A}_{q+3,q-2}| = 0$.   □

Corollary 6.37. *For* $q \geq n+3$, $n \geq 2$, *if every* $(q+1)$*-arc of* $\mathrm{PG}(n,q)$ *is a normal rational curve, then every* $(q+1)$*-arc of* $\mathrm{PG}(q-n-1,q)$ *is a normal rational curve.*

Proof. For $q \geq n+3$, $n \geq 2$,

$$|\mathcal{A}_{q+1,q-n-1}| = \nu_{q-n-1} \iff |\mathcal{A}_{q+1,n}| = \nu_n.$$ □

6.6 The maximum size of an arc and the characterisation of such arcs

In Sections 21.2 and 21.3 of FPSOTD it is shown that $m(3,q) = q+1$ for $q > 3$; also a $(q+1)$-arc is a twisted cubic for q odd, while, for q even, it is of the form

$$\{\mathbf{P}(t^{2^m+1}, t^{2^m}, t, 1) \mid t \in \mathbf{F}_{2^h} \cup \{\infty\}\}$$

for some m coprime to h. This result is now generalised to higher dimensions.

Theorem 6.38. *Let* $\mathcal{K}$ *be a* k*-arc in* $\mathrm{PG}(n,q)$ *with* $k \geq n+3 \geq 6$. *If there exist points* P_0 *and* P_1 *in* $\mathcal{K}$ *such that the projections* $\mathcal{K}_0$ *of* $\mathcal{K}\backslash\{P_0\}$ *from* P_0 *and* $\mathcal{K}_1$ *of* $\mathcal{K}\backslash\{P_1\}$ *from* P_1 *onto a hyperplane* Π_{n-1} *are both contained in normal rational curves in* Π_{n-1}, *then* $\mathcal{K}$ *is contained in a unique normal rational curve of* $\mathrm{PG}(n,q)$.

Proof. Let $\mathcal{L} = \{P_0, \ldots, P_{n+2}\}$ be an $(n+3)$-arc in $\mathcal{K}$. For $i = 0, 1$, let $\mathcal{L}_i$ and $\mathcal{K}_i$ be the projections from P_i of $\mathcal{L}\backslash\{P_i\}$ and $\mathcal{K}\backslash\{P_i\}$ onto Π_{n-1}. By Theorem 6.30(v), there exist unique normal rational curves $\mathcal{C}$ in $\mathrm{PG}(n,q)$ and $\mathcal{C}^{(0)}$, $\mathcal{C}^{(1)}$ in Π_{n-1} such that $\mathcal{L} \subset \mathcal{C}, \mathcal{L}_0 \subset \mathcal{C}^{(0)}, \mathcal{L}_1 \subset \mathcal{C}^{(1)}$. Since $\mathcal{K}_0$ and $\mathcal{K}_1$ are assumed to be in normal rational curves, so $\mathcal{K}_0 \subset \mathcal{C}^{(0)}$ and $\mathcal{K}_1 \subset \mathcal{C}^{(1)}$. As $\mathcal{K}$ is contained in $P_0\mathcal{K}_0$ and in $P_1\mathcal{K}_1$, so

$$\mathcal{K}\backslash\{P_0, P_1\} \subset (P_0\mathcal{C}^{(0)} \cap P_1\mathcal{C}^{(1)})\backslash P_0P_1; \tag{6.73}$$

the right-hand side of (6.73) is now shown to lie in $\mathcal{C}$.

   Let $\mathcal{C} = \{Q_j \mid j \in \overline{\mathbf{N}}_q\}$ with $Q_j = P_j$ for $j \in \overline{\mathbf{N}}_{n+2}$. Also let $\mathcal{D}^{(i)}$ be the projection of $\mathcal{C}\backslash\{P_i\}$ onto Π_{n-1} from P_i for $i = 0, 1$. Since $\mathcal{D}^{(i)} \cup \{\ell_{P_i} \cap \Pi_{n-1}\}$

is a normal rational curve in Π_{n-1} containing $\mathcal{L}_i$ by Lemma 6.31(i), so it coincides with $\mathcal{C}^{(i)}$ by Theorem 6.30(v). Thus a line on the cone $P_i\mathcal{C}^{(i)}$ other than P_0P_1 is either the tangent ℓ_{P_i} or a line P_iQ_j, where $j \neq 0, 1$. Let ℓ_0 be a line on $P_0\mathcal{C}^{(0)}$ and ℓ_1 a line on $P_1\mathcal{C}^{(1)}$ such that neither line is P_0P_1 but with ℓ_0 and ℓ_1 intersecting. If $\ell_0 = \ell_{P_0}$, then $\ell_1 \neq \ell_{P_1}$ by Lemma 6.31(ii); thus $\ell_1 = P_1Q_j$ for some $j \neq 0, 1$. Since the plane $\pi = \ell_0\ell_1$ contains P_0, P_1, Q_j, there exists a hyperplane containing n points of $\mathcal{C}_n$ as well as ℓ_0, contradicting Lemma 6.31(iii). Thus $\ell_0 = P_0Q_u$ and $\ell_1 = P_1Q_v$ for $u, v \neq 0, 1$. Since π contains at most three points of $\mathcal{C}$, so $Q_u = Q_v$. Hence $\ell_0 \cap \ell_1$ is a point of $\mathcal{C}$. Thus $\mathcal{K}\backslash\{P_0, P_1\} \subset \mathcal{C}\backslash\{P_0, P_1\}$, whence $\mathcal{K} \subset \mathcal{C}$.   □

Theorem 6.39.   (i) *Let $\mathcal{K}$ be a $(q+2)$-arc in $PG(n, q)$ with $q + 1 \geq n + 3 \geq 6$. If P_0 and P_1 are points of $\mathcal{K}$ and Π_{n-1} is a hyperplane containing neither P_0 nor P_1, then the projections $\mathcal{K}_0$ of $\mathcal{K}\backslash\{P_0\}$ and $\mathcal{K}_1$ of $\mathcal{K}\backslash\{P_1\}$ from P_0 and P_1 onto Π_{n-1} cannot both be normal rational curves.*
(ii) *If every $(q+1)$-arc in $PG(n-1, q)$, with $q + 1 \geq n + 3 \geq 6$, is a normal rational curve, then $m(n, q) = q + 1$.*

Proof. (i) Suppose $\mathcal{K}_0$ and $\mathcal{K}_1$ are normal rational curves in Π_{n-1}. For a point P in $\mathcal{K}\backslash\{P_0, P_1\}$, let $\mathcal{K}'$ be the $(q+1)$-arc $\mathcal{K}\backslash\{P\}$. Then, by Theorem 6.38, $\mathcal{K}'$ is a normal rational curve in $PG(n, q)$. Let ℓ_{P_i} be the tangent of $\mathcal{K}'$ at P_i for $i = 0$ and 1, and let $\mathcal{K}'_i$ be the projection of $\mathcal{K}'\backslash\{P_i\}$ from P_i onto Π_{n-1}. Then $\mathcal{K}' \cup \{\ell_{P_i} \cap \Pi_{n-1}\}$ is a normal rational curve in Π_{n-1} by Lemma 6.31(i). So $\mathcal{K}_i = \mathcal{K}'_i \cup \{\ell_{P_i} \cap \Pi_{n-1}\}$ since both curves have q points in common and $q \geq n - 1 + 3$. Thus $\ell_{P_i} = P_iP$ contradicting Lemma 6.31(ii).
(ii) If there is a $(q+2)$-arc $\mathcal{K}$ in $PG(n, q)$, then $\mathcal{K}_0$ and $\mathcal{K}_1$ are $(q+1)$-arcs and so normal rational curves, contradicting (i).   □

Theorem 6.40. *In $PG(4, q)$, $q \geq 5$,*

$$m(4, q) = q + 1.$$

Proof. (i) For q odd, since every $(q+1)$-arc in $PG(3, q)$ is a twisted cubic, the result follows by Theorem 6.39(ii).
(ii) For q even, suppose there exists a $(q+2)$-arc $\mathcal{K} = \{P, Q, R_1, R_2, \ldots, R_q\}$. Take a solid Π_3 in $PG(4, q)$ containing neither P nor Q. Let $\mathcal{K}_1$ and $\mathcal{K}_2$ be the projections of $\mathcal{K}\backslash\{P\}$ and $\mathcal{K}\backslash\{Q\}$ onto Π_3 from P and Q. In Theorem 21.3.10 of FPSOTD, it is shown that, at any point L of a $(q+1)$-arc $\mathcal{L}$, there are precisely two lines, called special unisecants, such that every plane through such a unisecant meets $\mathcal{L}$ in at most one point other than L; further, the special unisecants to $\mathcal{L}$ are the generators of a hyperbolic quadric $\mathcal{H}_3$. Let $\mathcal{H}_3^{(1)}$ and $\mathcal{H}_3^{(2)}$ be the corresponding quadrics containing $\mathcal{K}_1$ and $\mathcal{K}_2$. Also, let $PQ \cap \Pi_3 = S$, let $PR_i \cap \Pi_3 = P_i$, and let $QR_i \cap \Pi_3 = Q_i$, $i \in \mathbf{N}_q$. Then $\mathcal{K}_1 = \{S, P_1, \ldots, P_q\}$ and $\mathcal{K}_2 = \{S, Q_1, \ldots, Q_q\}$. Since the plane PQR_j meets Π_3 in a line, so S, P_j, Q_j are collinear, for $j \in \mathbf{N}_q$, on the line ℓ_j. Also let ℓ and m be the special unisecants at S to $\mathcal{K}_1$. Then, each of the q planes $\ell\ell_j$ meets $\mathcal{K}_1$ in precisely two points S and P_j, and also meets $\mathcal{K}_2$ in precisely two points S and Q_j. So ℓ, and similarly m, is a special unisecant to $\mathcal{K}_2$

at S. Thus the two quadric cones $P\mathcal{H}_3^{(1)}$ and $Q\mathcal{H}_3^{(2)}$ contain the planes $\ell(PQ)$ and $m(PQ)$. They therefore intersect residually in a quadric surface $\mathcal{W}_3$ which either (a) lies in a solid Π_3', or (b) lies in no solid. However, $\mathcal{W}_3$ contains $R_1, \ldots, R_q$. In case (a), Π_3' contains at most four points of $\mathcal{K}$ and so $q \leq 4$. In case (b), $\mathcal{W}_3$ is a pair of planes with just one common point, which can contain at most six points of $\mathcal{K}$, whence $q \leq 6$. Thus both (a) and (b) are impossible.   □

Theorem 6.41. *In* $\mathrm{PG}(n, q)$, *q odd,* $n \geq 3$,

  (i) *if* $\mathcal{K}$ *is a* k-arc with $k > q - \frac{1}{4}\sqrt{q} + n - \frac{1}{4}$, *then* $\mathcal{K}$ *lies on a unique normal rational curve;*

  (ii) *if* $q > (4n - 5)^2$, *every* $(q + 1)$-arc is is a normal rational curve;

  (iii) *if* $q > (4n - 9)^2$,

$$m(n, q) = q + 1.$$

Proof. (i) This follows by induction from Theorem 6.38 and Theorem 10.25 of PGOFF2.

  (ii) $q + 1 > q - \frac{1}{4}\sqrt{q} + n - \frac{1}{4} \Leftrightarrow q > (4n - 5)^2$.

  (iii) This follows from Theorem 6.39(ii) and part (ii).   □

The next result shows that, in part (ii) of this theorem, some restriction on q is necessary.

Theorem 6.42. *In* $\mathrm{PG}(4, q)$, *q odd,*

  (i) *for* $q \leq 7$, *a* $(q + 1)$-arc is a normal rational curve;

  (ii) *for* $q = 9$, *there exist precisely two projectively distinct* 10-arcs, *the normal rational curve and one other.*

Proof. (i) For $q = 3$ and 5, the result is immediate. For $q = 7$, Corollary 6.37 with $n = 2$ can be applied.

  (ii) With $\mathbf{F}_9 \backslash \{0\} = \{\sigma^i \mid i \in \overline{\mathbf{N}}_7, \sigma^2 = \sigma + 1\}$, every 10-arc in $\mathrm{PG}(4, 9)$ other than a normal rational curve is projectively equivalent to

$$\mathcal{K} = \{\mathbf{P}(1, t, t^2 + \sigma t^6, t^3, t^4) \mid t \in \mathbf{F}_9\} \cup \{\mathbf{P}(0, 0, 0, 0, 1)\};$$

see Section 6.8. The 10-arc $\mathcal{K}$ projects to the unique complete 8-arc in $\mathrm{PG}(2, 9)$, as in Section 14.7 of PGOFF2.   □

Remark 6.43. Theorem 6.33 and Corollary 6.36 can be applied to the previous results.

The situation is surprisingly different for $(q + 1)$-arcs in $\mathrm{PG}(4, q)$ with q even, as is now demonstrated.

Let $\mathcal{K} = \{P_0, P_1, \ldots, P_q\}$ be a $(q + 1)$-arc in $\mathrm{PG}(4, q)$, $q = 2^h$, $h \geq 3$. At each point P_i of $\mathcal{K}$ there is an induced incidence structure $\mathcal{S}(P_i)$ isomorphic to $\mathrm{PG}(3, q)$, whose points, lines, and planes are the lines, planes and solids of $\mathrm{PG}(4, q)$ through P_i; the incidence is that induced by $\mathrm{PG}(4, q)$. As usual, a subspace of dimension r

is denoted Π_r; however, Π_s^i also denotes an s-dimensional subspace of $\mathcal{S}(P_i)$. Thus a Π_r through P_i is also a Π_{r-1}^i. This notation is only used for the remainder of this section.

From the definition of $\mathcal{K}$ any solid through P_i contains at most three other points of $\mathcal{K}$. Thus the set of q lines $P_i P_j, j \in \overline{\mathbf{N}}_q \backslash \{i\}$, is a q-arc $\mathcal{K}_i$ of $\mathcal{S}(P_i)$. By Theorem 6.71 for $q > 16$, this arc $\mathcal{K}_i$ can be completed to a $(q+1)$-arc $\mathcal{K}_i'$ of $\mathcal{S}(P_i)$ by adding a unique Π_1 through P_i; for $q \leq 16$, see Section 6.8. Let this line be denoted ℓ_i and called the *tangent line* to $\mathcal{K}$ at P_i. From Theorem 21.3.10 of FPSOTD, the points of $\mathcal{K}_i'$ lie on a hyperbolic quadric, denoted $\mathcal{Q}_i$. Let $\mathcal{S}_i$ denote the quadric cone of $PG(4, q)$ whose points lie on the Π_0^i of $\mathcal{Q}_i$; that is, $\mathcal{S}_i = P_i \mathcal{H}_3$, where $\mathcal{H}_3$ is a solid section of $\mathcal{Q}_i$ regarded as a set of Π_0.

Lemma 6.44.   (i) *The tangent lines ℓ_i of $\mathcal{K}$ are pairwise skew.*
(ii) *For $i \neq j$, the solid $\ell_i \ell_j$ meets $\mathcal{K}$ in $\{P_i, P_j\}$.*
(iii) *There is a unique plane α_{ij} through P_i and P_j which is both a Π_1^i of $\mathcal{Q}_i$ and a Π_1^j of $\mathcal{Q}_j$. Further, $P_i \ell_j$ is a Π_1^i of $\mathcal{Q}_i$ and $P_j \ell_i$ is a Π_1^j of $\mathcal{Q}_j$.*

Proof. (i) By construction, $\ell_i \cap \mathcal{K} = \{P_i\}$. Suppose $\ell_i \cap \ell_j$ is a Π_0; then $\ell_i \ell_j$ is a Π_2 as well as a Π_1^i and a Π_1^j. By Theorem 21.3.10 of FPSOTD, through $P_i P_j$ there are exactly two special unisecants Π_1^i of $\mathcal{K}_i'$, and these are generators of $\mathcal{Q}_i$. Any Π_3 containing a special unisecant Π_1^i of $\mathcal{K}_i'$ through $P_i P_j$ meets $\mathcal{K}_i'$ in at most one Π_0^i other than $P_i P_j$. Hence, for $\ell_i \not\subset \Pi_3$, any such Π_3 meets $\mathcal{K}$ in P_i, P_j, and at most one further point; for $\ell_i \subset \Pi_3$, it meets $\mathcal{K}$ in P_i, P_j. Since $\ell_i \subset \Pi_3$ if and only if $\ell_j \subset \Pi_3$, these two Π_1^i of $\mathcal{Q}_i$ are also Π_1^j of $\mathcal{Q}_j$ through $P_i P_j$. Thus the two cones $\mathcal{S}_i$ and $\mathcal{S}_j$ intersect in these Π_2 and hence residually in a quadric surface $\mathcal{Q}$. Since $\mathcal{K} \backslash \{P_i, P_j\}$ is in $\mathcal{Q}$ and $q + 1 \geq 9$, the surface $\mathcal{Q}$ does not contain a plane. So $\mathcal{Q}$ lies in a solid and also contains the $q - 1 \geq 7$ points of $\mathcal{K} \backslash \{P_i, P_j\}$, a contradiction.

(ii), (iii) Project $\mathcal{K} \backslash \{P_i, P_j\}$ from $P_i P_j$ onto a plane Π_2 skew to $P_i P_j$; the projection of $\mathcal{K} \backslash \{P_i, P_j\}$ is a $(q - 1)$-arc $\mathcal{K}'$ of Π_2. Then both $P_j \ell_i \cap \Pi_2 = \{Q\}$ and $P_i \ell_j \cap \Pi_2 = \{Q'\}$ extend $\mathcal{K}'$ to a q-arc. From Section 10.3 of PGOFF2, $\mathcal{K}' \cup \{Q, Q'\}$ is a $(q + 1)$-arc. Hence $\ell_i \ell_j \cap \mathcal{K} = \{P_i, P_j\}$, $P_i \ell_j$ is a Π_1^i of $\mathcal{Q}_i$ and $P_j \ell_i$ is a Π_1^j of $\mathcal{Q}_j$. Let Q'' be the unique point of Π_2 which extends $\mathcal{K}' \cup \{Q, Q'\}$ to a $(q + 2)$-arc. Then the plane $Q'' P_i P_j = \alpha_{ij}$ is both a Π_1^i of $\mathcal{Q}_i$ and a Π_1^j of $\mathcal{Q}_j$. $\square$

Lemma 6.45. *For a given i, the planes $P_i \ell_j$, for j in $\overline{\mathbf{N}}_q \backslash \{i\}$, are q of the Π_1^i of a regulus $\mathcal{R}_i$ of $\mathcal{Q}_i$.*

Proof. Let ℓ_j and ℓ_k be distinct lines of $\mathcal{K}$ with $j, k \neq i$. By Lemma 6.44(ii), the point $P_i \notin \ell_j \ell_k$, and hence $P_i \ell_j$ and $P_i \ell_k$ are skew Π_1^i; they are generators of $\mathcal{Q}_i$, by Lemma 6.44(iii), and so belong to a regulus $\mathcal{R}_i$. $\square$

Lemma 6.46. *Let g_i and g_i' be the two Π_1^i of $\mathcal{Q}_i$ through ℓ_i and let $g_i \in \mathcal{R}_i$. Then $g_i' \cap \ell_j$ is a Π_0, for $j \neq i$.*

Proof. Let g_i' belong to the regulus $\mathcal{R}_i'$ complementary to $\mathcal{R}_i$. By the previous lemma, $P_i \ell_j \in \mathcal{R}_i$ for $j \neq i$. Since lines of complementary reguli meet, so $g_i' \cap P_i \ell_j$ is a Π_0^i. Thus $g_i' \cap \ell_j$ is a Π_0. $\square$

Corollary 6.47. *Through each ℓ_i there are two Π_1^i of Q_i; they are g_i, which is skew to all ℓ_j for $j \neq i$, and g_i', which meets all ℓ_j for $j \neq i$.*

Lemma 6.48. *The $q + 1$ generators g_i' of Q_i contain a unique Π_1, denoted by ℓ, which is disjoint from K and meets the ℓ_i in distinct points.*

Proof. For $i \neq j$, let $g_i' \cap \ell_j = \{Q\}$ and $g_j' \cap \ell_i = \{R\}$. Since ℓ_i and ℓ_j are skew, so Q and R are distinct points. Since $\ell_i \subset g_i'$ and $\ell_j \subset g_j'$, so $g_i' \cap g_j' = QR$. Thus $g_0', \ldots, g_q'$ are $q + 1$ planes Π_2, meeting in pairs in a Π_1; so either they pass through a common Π_1 or they lie in a common Π_3. The latter is impossible since K lies in the space they generate. So they meet in a Π_1, denoted by ℓ. Now, ℓ is disjoint from K, since otherwise every g_i' would contain $\ell \cap K$. Since $\ell_i \subset g_i'$ and $\ell_i \not\subset g_j', j \neq i$, so $|\ell_i \cap \ell| = 1$. Since the ℓ_i are skew, each point of ℓ lies on a unique tangent line ℓ_i of K. $\qquad\square$

Theorem 6.49. *In* $\mathrm{PG}(4, q)$, $q = 2^h$, *every* $(q+1)$-*arc* K *is a normal rational curve.*

Proof. For $h = 1$ and 2, the result is immediate. So, let $h \geq 3$ and use the above notation. For $K = \{P_0, \ldots, P_q\}$, it is possible to choose coordinates so that

$$P_0 = \mathbf{U}_0, P_1 = \mathbf{U}_4, A = \alpha_{01} \cap g_0 \cap g_1 = \mathbf{U}_2, \ B = \ell \cap \ell_1 = \mathbf{U}_3, \ C = \ell \cap \ell_0 = \mathbf{U}_1,$$

and $\mathbf{U}$ is any point of $K \backslash \{P_0, P_1\}$. Note that

$$\ell \ell_0 = g_0', \ \ell \ell_1 = g_1', \ A\ell_0 = g_0, \ A\ell_1 = g_1$$

and $P_0 P_1 A = \alpha_{01}$.

Let $\beta_1 = g_1 g_1' = \mathbf{u}_0$ and consider the $(q + 1)$-arc $\beta_1 \cap K_0'$. This $(q + 1)$-arc (in a Π_3) contains the points $C = \mathbf{U}_1, P_1 = \mathbf{U}_4$, and $P_0 \mathbf{U} \cap \beta_1 = \mathbf{P}(0, 1, 1, 1, 1)$. The special unisecants of $\beta_1 \cap K_0'$ at C and P_1 are the intersections of β_1 with the Π_1^0 of Q_0 containing ℓ_0 and $P_0 P_1$; these unisecants are therefore

$$CA = g_0 \cap \beta_1, \ CB = \ell = g_0' \cap \beta_1, \ P_1 B = P_0 \ell_1 \cap \beta_1, \ P_1 A = \alpha_{01} \cap \beta_1.$$

Thus these unisecants intersect at $A = \mathbf{U}_2$ and $B = \mathbf{U}_3$.

It now follows from Theorem 21.3.15 of FPSOTD and its proof that

$$K_0 = \beta_1 \cap K_0' = \{\mathbf{P}(0, 1, \mu, \mu^\sigma, \mu^{\sigma+1}) \mid \mu \in \mathbf{F}_q{}^+\},$$

where σ is a generator of the automorphism group of $\mathbf{F}_q$; hence $x^\sigma = x^{2^n}$ for some n coprime to h. Since, by definition, K_0 is $\ell_0 \cap \beta_1$ together with a projection of $K \backslash \{\mathbf{U}_0\}$ from $\mathbf{U}_0$,

$$K = \{\mathbf{P}(1, f(\mu), \mu f(\mu), \mu^\sigma f(\mu), \mu^{\sigma+1} f(\mu)) \mid \mu \in \mathbf{F}_q{}^+\}$$

for some function $f : \mathbf{F}_q \to \mathbf{F}_q$ with $f(0) = 0, f(1) = 1$. $\qquad\square$

Next consider the $(q+1)$-arc $\beta_0 \cap \mathcal{K}_1'$, where $\beta_0 = g_0 g_0' = \mathbf{u}_4$. Similarly to the above,

$$\mathcal{K}_1 = \beta_0 \cap \mathcal{K}_1' = \{\mathbf{P}(1, \lambda, \lambda^\tau, \lambda^{\tau+1}, 0) \mid \lambda \in \mathbf{F}_q{}^+\},$$

where τ is an automorphism of $\mathbf{F}_q$ such that $x^\tau = x^{2^m}$ with m coprime to h. Since $\mathcal{K}_1$ is $\ell_1 \cap \beta_0$ together with a projection of $\mathcal{K} \backslash \{\mathbf{U}_4\}$ from $\mathbf{U}_4$, so

$$\mathcal{K} = \{\mathbf{P}(1, \lambda, \lambda^\tau, \lambda^{\tau+1}, f'(\lambda)) \mid \lambda \in \mathbf{F}_q\} \cup \{\mathbf{U}_4\},$$

where f' is a function on $\mathbf{F}_q$ with $f'(0) = 0, f'(1) = 1$. The two forms for $\mathcal{K}$ are the same if, for all λ and μ in $\mathbf{F}_q$,

$$f(\mu) = \lambda, \ \mu f(\mu) = \lambda^\tau, \ \mu^\sigma f(\mu) = \lambda^{\tau+1}, \ \mu^{\sigma+1} f(\mu) = f'(\lambda).$$

Hence

$$\mu = \lambda^{\tau-1}, \ \mu^{\sigma-1} = \lambda, \text{ and } \ \lambda = (\lambda^{\tau-1})^{\sigma-1}.$$

From the definitions of τ and σ,

$$\lambda^\tau = \lambda^{2^m}, \quad \lambda^\sigma = \lambda^{2^n};$$

Let $1 \leq n \leq m < h$; so, $\mod (2^h - 1)$,

$$(2^m - 1)(2^n - 1) \equiv 1,$$
$$2^{m+n} - 2^m - 2^n \equiv 0,$$
$$2^m - 2^{m-n} - 1 \equiv 0.$$

Since $0 \leq 2^m - 2^{m-n} - 1 < 2^h - 1$, so

$$2^m - 2^{m-n} - 1 = 0.$$

Therefore $m = n = 1$. Thus

$$\mathcal{K} = \{\mathbf{P}(1, \lambda, \lambda^2, \lambda^3, \lambda^4) \mid \lambda \in \mathbf{F}_q{}^+\},$$

where $\lambda = \infty$ gives the point $\mathbf{U}_4$.

Theorem 6.50. For $q \geq 5$, $n \geq 5$,

$$m(n, q) \leq \begin{cases} q + n - 3, & q \text{ odd}, \\ q + n - 4, & q \text{ even}. \end{cases}$$

Proof. For q odd, the result follows from Theorem 6.39 and induction, using the fact that a $(q+1)$-arc in $PG(3, q)$ is a normal rational curve. For q even, a similar argument applies, but now the fact that a $(q+1)$-arc in $PG(4, q)$ is a normal rational curve must be used. □

Theorem 6.51. (i) If $n \leq 2p - 3$, $q = p^h$, with p prime, then $m(n, q) = q + 1$.
(ii) If $q \geq n+1 \geq p+1 \geq 4$, $q = p^h$, with p prime, then $m(n, q) \leq q - p + n + 1$.
(iii) For $n \leq p - 1$, all $(q+1)$-arcs are normal rational curves.

Remark 6.52. Theorem 6.33 and Corollary 6.36 can be applied to Theorem 6.51.

Theorem 6.53. If p is a prime with $p > n + 1$, then $m(n, p) = p + 1$.

Proof. See Section 6.8. □

6.7 Arcs and hypersurfaces

In this section, a connection is obtained between arcs and hypersurfaces. The main aim is to obtain an upper bound for $m(n, q)$ with q even. To do this, a more sophisticated notion of algebraic variety than in the previous chapters is required.

Let $H, H_1, \ldots, H_r$ be forms in $\Omega = \mathbf{F}_q[X_0, \ldots, X_n]$; in fact, $H, H_1, \ldots, H_r$ are always linear. The variety $\mathcal{A}$ in $\mathrm{PG}(n, q)$ defined by H and $H_1, \ldots, H_r$ is denoted

$$\mathcal{A} = \mathbf{A}(H, H_1, \ldots, H_r)$$

and consists of the pair $(\mathbf{V}(\mathcal{A}), \mathbf{I}(\mathcal{A}))$, where $\mathbf{V}(\mathcal{A}) = \mathbf{V}(H, H_1, \ldots, H_r)$ is the set of zeros of $H, H_1, \ldots, H_r$ in $\mathrm{PG}(n, q)$ and $\mathbf{I}(\mathcal{A}) = \mathbf{I}(H, H_1, \ldots, H_r)$ is the ideal generated by $H, H_1, \ldots, H_r$ in Ω; that is,

$$\mathbf{V}(\mathcal{A}) = \{\mathbf{P}(X) \in \mathrm{PG}(n, q) \mid H(X) = H_1(X) = \cdots = H_r(X) = 0\},$$
$$\mathbf{I}(\mathcal{A}) = \{F \in \Omega \mid F = GH + G_1 H_1 + \cdots + G_r H_r$$
$$\text{for some } G, G_1, \ldots, G_r \in \Omega\}.$$

The number of points in $\mathbf{V}(\mathcal{A})$ is denoted by $|\mathcal{A}|$.

If $\mathcal{A}$ and $\mathcal{B}$ are varieties in $\mathrm{PG}(n, q)$, then $\mathcal{A}$ is *algebraically contained* in $\mathcal{B}$, denoted $\mathcal{A} \subset \mathcal{B}$, if $\mathbf{I}(\mathcal{A}) \supset \mathbf{I}(\mathcal{B})$. The varieties $\mathcal{A}$ and $\mathcal{B}$ are *(algebraically) equal* if $\mathbf{I}(\mathcal{A}) = \mathbf{I}(\mathcal{B})$. A variety $\mathcal{A} = \mathbf{A}(H)$ with $\deg H = d$ is a *hypersurface of degree d*.

If $\mathcal{A} = \mathbf{A}(H, H_1, \ldots, H_r)$ is a variety with H, $H_i \neq 0$, all i, and $H_{r+1}, \ldots, H_u$ are other linear forms in $\Omega \backslash \{0\}$, then

$$\mathcal{A} \cap \pi_1 \cap \cdots \cap \pi_u = \mathcal{A} \cap \pi_{r+1} \cap \cdots \cap \pi_u$$

is the variety $\mathbf{A}(H, H_1, \ldots, H_u)$, where π_j is the hyperplane $\mathbf{V}(H_j)$, $j = 1, \ldots, u$. As $H_1, \ldots, H_r$ are linear, the terms $\mathbf{V}(H_j)$ and $\mathbf{A}(H_j)$ are used interchangeably.

Theorem 6.54. *In* $\Sigma = \mathrm{PG}(n, q)$, $n \geq 3$, *let* $\mathcal{K} = \{\pi_1, \pi_2, \ldots, \pi_k\}$ *be a set of hyperplanes, any three of which are linearly independent, and such that to* π_i *is associated a hypersurface* Φ_i *of* Σ *of degree d with the following properties:*

(a) $\Phi_i \cap \pi_i \cap \pi_j = \Phi_j \cap \pi_i \cap \pi_j$ *for all distinct i, j;*
(b) $|\Phi_i \cap \pi_i \cap \pi_j| < \theta(n - 2)$ *for all distinct i, j;*
(c) $|\Phi_i \cap \pi_i \cap \pi_j \cap \pi_u| < \theta(n - 3)$ *for all distinct i, j, u.*

Then there exists a hypersurface Φ *in* Σ *of degree d such that, for all i,*

$$\Phi \cap \pi_i = \Phi_i \cap \pi_i. \tag{6.74}$$

Proof. The proof is by induction on k. For $k = 1$, there is nothing to prove. Suppose that $k \geq 2$ and that the statement holds for $k - 1$. Let Φ' be a hypersurface of degree d such that, for $1 \leq i \leq k - 1$,

$$\Phi' \cap \pi_i = \Phi_i \cap \pi_i. \tag{6.75}$$

Then, for $1 \le i \le k - 1$,

$$\Phi' \cap \pi_k \cap \pi_i = \Phi_i \cap \pi_i \cap \pi_k = \Phi_k \cap \pi_k \cap \pi_i. \tag{6.76}$$

Let $\Phi' = \mathbf{A}(D')$, $\Phi_k = \mathbf{A}(D_k)$ and $\pi_j = \mathbf{A}(H_j)$, $j = 1, \ldots, k$. By (6.76), for $1 \le i \le k - 1$,

$$\mathbf{I}(D', H_k, H_i) = \mathbf{I}(D_k, H_k, H_i).$$

So

$$D' = u_i D_k + r_i H_k + s_i H_i,$$

where $u_i, r_i, s_i \in \mathbf{F}_q[X_0, \ldots, X_n]$. Comparing terms of degree d shows that

$$D' + t_i D_k \in \mathbf{I}(H_k, H_i), \tag{6.77}$$

with $t_i \in \mathbf{F}_q$; here $t_i \ne 0$, since otherwise $|\Phi' \cap \pi_k \cap \pi_i| \ge \theta(n - 2)$, whence $|\Phi_i \cap \pi_i \cap \pi_k| \ge \theta(n - 2)$, a contradiction.

It is now shown that $t_i = t_j$. From (6.77),

$$(t_i - t_j)D_k = (D' + t_i D_k) - (D' + t_j D_k) \in \mathbf{I}(H_k, H_i, H_j). \tag{6.78}$$

Since $D_k \notin \mathbf{I}(H_k, H_i, H_j)$ by (c), so (6.78) implies that $t_i - t_j = 0$. Write $t_i = \lambda$ and note that $\lambda \ne 0$. Next, choose coordinates so that $\pi_k = \mathbf{A}(X_0)$ and

$$\pi_i = \mathbf{A}(a_{i0}X_0 + \cdots + a_{in}X_n) \text{ for } 1 \le i \le k - 1.$$

For $1 \le i \le k - 1$, put

$$D' + \lambda D_k - G H_k + G_i(H_i - a_{in}H_k),$$

where G_i is chosen so that it contains no terms in X_0. Thus

$$D' + \lambda D_k - G H_k = G_i(H_i - a_{i0}H_k) = G_j(H_j - a_{j0}H_k).$$

Hence

$$G_j(H_j - a_{j0}H_k) = F \prod_{i=1}^{k-1} (H_i - a_{i0}H_k), \tag{6.79}$$

since π_k, π_i, π_j are linearly independent for distinct k, i, j.

Finally, it is shown that $\Phi = \mathbf{A}(D)$, with

$$D = D' - F \prod_{i=1}^{k-1} H_i,$$

has the required properties. The only thing to check is that $\Phi \cap \pi_k = \Phi_k \cap \pi_k$; that is, $\mathbf{I}(D, H_k) = \mathbf{I}(D_k, H_k)$. This can be shown as follows:

$$\mathbf{I}(D, H_k) = \mathbf{I}(D' - F \prod_{i=1}^{k-1} H_i, H_k)$$

$$= \mathbf{I}(D' - F \prod_{i=1}^{k-1} (H_i - a_{i0} H_k), H_k)$$

$$= \mathbf{I}(GH_k - \lambda D_k, H_k)$$

$$= \mathbf{I}(D_k, H_k). \qquad \qquad \square$$

Remark 6.55. For $k \geq 3$, hypothesis (b) follows from the others. This is because $|\Phi_i \cap \pi_i \cap \pi_j| = \theta(n - 2)$ implies $|\Phi_i \cap \pi_i \cap \pi_j \cap \pi_u| = \theta(n - 3)$, contradicting (c).

Theorem 6.56. *In* $\Sigma = \mathrm{PG}(n, q)$, $n \geq 3$, *let* $\mathcal{K} = \{\pi_1, \pi_2, \ldots, \pi_k\}$, $k \geq n$, *be a set of hyperplanes, any* n *of which are linearly independent, such that, for each plane* $\pi_{i_1} \cap \pi_{i_2} \cap \cdots \cap \pi_{i_{n-2}}$, *there is an associated hypersurface* $\mathcal{C}_{\{i_1,\ldots,i_{n-2}\}}$ *of degree d. Suppose*

(a) $\mathcal{C}_{\{i_1,\ldots,i_{n-2}\}} \cap \pi_{i_1} \cap \pi_{i_2} \cap \cdots \cap \pi_{i_{n-1}} = \mathcal{C}_{\{j_1,\ldots,j_{n-2}\}} \cap \pi_{j_1} \cap \cdots \cap \pi_{j_{n-1}}$ *for all subsets* $\{i_1,\ldots,i_{n-1}\} = \{j_1,\ldots,j_{n-1}\}$ *of size* $n - 1$ *of* $\{1, 2, \ldots, k\}$;
(b) $|\mathcal{C}_{\{i_1,\ldots,i_{n-2}\}} \cap \pi_{i_1} \cap \pi_{i_2} \cap \cdots \cap \pi_{i_n}| = 0$ *for any subset* $\{i_1,\ldots,i_n\}$ *of size* n *of* $\{1,\ldots,k\}$.

Then there exist hypersurfaces $\Phi, \Phi_1, \ldots, \Phi_k$ *in* Σ *of degree d such that*

(i) $\Phi_{i_1} \cap \pi_{i_1} \cap \cdots \cap \pi_{i_{n-2}} = \mathcal{C}_{\{i_1,\ldots,i_{n-2}\}} \cap \pi_{i_1} \cap \cdots \cap \pi_{i_{n-2}}$ *for all distinct* $i_1, \ldots, i_{n-2}$;
(ii) $\Phi \cap \pi_i = \Phi_i \cap \pi_i$ *for* $1 \leq i \leq k$.

Proof. For $n = 3$, the statement holds by the previous theorem and the subsequent remark. Consider the 3-space $\pi_{i_1} \cap \cdots \cap \pi_{i_{n-3}}$. Again, by the previous theorem and remark, in this 3-space and so in Σ, there exists a hypersurface $\Phi_{\{i_1,\ldots,i_{n-3}\}}$ of degree d with

$$\Phi_{\{i_1,\ldots,i_{n-3}\}} \cap \pi_{i_1} \cap \cdots \cap \pi_{i_{n-2}} = \mathcal{C}_{\{i_1,\ldots,i_{n-2}\}} \cap \pi_{i_1} \cap \cdots \cap \pi_{i_{n-2}}$$

for any $i_{n-2} \in \mathbf{N}_k \backslash \{i_1, \ldots, i_{n-3}\}$. Also, in each 4-space $\pi_{i_1} \cap \cdots \cap \pi_{i_{n-4}}$ and so in Σ, there exists a hypersurface $\Phi_{\{i_1,\ldots,i_{n-4}\}}$ of degree d with

$$\Phi_{\{i_1,\ldots,i_{n-4}\}} \cap \pi_{i_1} \cap \cdots \cap \pi_{i_{n-3}} = \Phi_{\{i_1,\ldots,i_{n-3}\}} \cap \pi_{i_1} \cap \cdots \cap \pi_{i_{n-3}}$$

for any $i_{n-3} \in \mathbf{N}_k \backslash \{i_1, \ldots, i_{n-4}\}$. For distinct i_{n-3}, i_{n-2} in $\mathbf{N}_k \backslash \{i_1, \ldots, i_{n-4}\}$,

$$\Phi_{\{i_1,\ldots,i_{n-4}\}} \cap \pi_{i_1} \cap \cdots \cap \pi_{i_{n-2}} = \Phi_{\{i_1,\ldots,i_{n-3}\}} \cap \pi_{i_1} \cap \cdots \cap \pi_{i_{n-2}}$$

$$= \mathcal{C}_{\{i_1,\ldots,i_{n-2}\}} \cap \pi_{i_1} \cap \cdots \cap \pi_{i_{n-2}}.$$

Continuing in this way,

$$\Phi_{i_1} \cap \pi_{i_1} \cap \pi_{i_2} = \Phi_{\{i_1,i_2\}} \cap \pi_{i_1} \cap \pi_{i_2}$$

for hypersurfaces Φ_{i_1}, $\Phi_{\{i_1,i_2\}}$ of degree d and any i_2 in $\mathbf{N}_k \backslash \{i_1\}$. Hence, for distinct $i_2, \ldots, i_{n-2}$ in $\mathbf{N}_k \backslash \{i_1\}$,

$$
\begin{aligned}
\Phi_{i_1} \cap \pi_{i_1} \cap \cdots \cap \pi_{i_{n-2}} &= \Phi_{\{i_1,i_2\}} \cap \pi_{i_1} \cap \cdots \cap \pi_{i_{n-2}} \\
&= \mathcal{C}_{\{i_1,\ldots,i_{n-2}\}} \cap \pi_{i_1} \cap \cdots \cap \pi_{i_{n-2}}.
\end{aligned}
$$

Finally, a hypersurface Φ of degree d in Σ is obtained such that, for any i_1 in $\mathbf{N}_k$ and for any $i_2, \ldots, i_{n-2}$ in $\mathbf{N}_k \backslash \{i_1\}$,

$$
\begin{aligned}
\Phi \cap \pi_{i_1} &= \Phi_{i_1} \cap \pi_{i_1}, \\
\Phi \cap \pi_{i_1} \cap \cdots \cap \pi_{i_{n-2}} &= \Phi_{i_1} \cap \pi_{i_1} \cap \cdots \cap \pi_{i_{n-2}} \\
&= \mathcal{C}_{\{i_1,\ldots,i_{n-2}\}} \cap \pi_{i_1} \cap \cdots \cap \pi_{i_{n-2}}. \qquad \square
\end{aligned}
$$

The essential construction

In $\mathrm{PG}(n,q)$, $n \geq 3$, $q = 2^h$, let $\mathcal{K} = \{\pi_1, \pi_2, \ldots, \pi_k\}$, $k \geq n+1$, be an arc of hyperplanes; that is, every $n+1$ hyperplanes in $\mathcal{K}$ are linearly independent or, equivalently, no $n+1$ hyperplanes in $\mathcal{K}$ have a point in common. For distinct $i_1, \ldots, i_{n-1}$, there are exactly $t = q + n - k$ points on the line $\pi_{i_1} \cap \cdots \cap \pi_{i_{n-1}}$ contained in no other hyperplane of $\mathcal{K}$. With $S = \{i_1, \ldots, i_{n-1}\}$, denote this set of t points by

$$
\mathbf{Z}_S = \{\mathbf{Z}_S^{(1)}, \ldots, \mathbf{Z}_S^{(t)}\}.
$$

In the plane $\Pi_2 = \pi_{i_1} \cap \cdots \cap \pi_{i_{n-2}}$, the other points of $\mathcal{K}$ cut out a $(k-n+2)$-arc of lines. As $q + 2 - (k - n + 2) = t$, so by Theorem 10.1 of PGOFF2, the points in $\mathbf{Z}_S$ lie on an algebraic curve $\mathcal{C}_{\{i_1,\ldots,i_{n-2}\}}$ of degree t in Π_2 with

$$
\mathcal{C}_{\{i_1,\ldots,i_{n-2}\}} \cap \pi_{i_{n-1}} = \mathbf{Z}_S.
$$

Also

$$
\mathcal{C}_{\{i_1,\ldots,i_{n-2}\}} \cap \pi_{i_{n-1}} = \mathcal{C}_{\{j_1,\ldots,j_{n-2}\}} \cap \pi_{j_{n-1}}
$$

for all equal subsets $\{i_1, \ldots, i_{n-1}\}$ and $\{j_1, \ldots, j_{n-1}\}$ of size $n-1$ in $\mathbf{N}_k$. Further,

$$
|\mathcal{C}_{\{i_1,\ldots,i_{n-2}\}} \cap \pi_{i_{n-1}} \cap \pi_{i_n}| = 0
$$

for any subset $\{i_1, \ldots, i_n\}$ of size n in $\mathbf{N}_k$.

By Theorem 6.56, the curves $\mathcal{C}_{\{i_1,\ldots,i_{n-2}\}}$ for a fixed $\imath$ are algebraically contained in a hypersurface Φ_{i_1} in π_{i_1} of degree t with

$$
\Phi_{i_1} \cap \pi_{i_2} \cap \cdots \cap \pi_{i_{n-2}} = \mathcal{C}_{\{i_1,\ldots,i_{n-2}\}};
$$

the varieties $\Phi_1, \ldots, \Phi_k$ are algebraically contained in a hypersurface $\Phi = \Phi(\mathcal{K})$ of $\mathrm{PG}(n,q)$ of degree t with $\Phi \cap \pi_i = \Phi_i$ for $1 \leq i \leq k$. The hypersurface $\Phi = \Phi(\mathcal{K})$ is the *hypersurface associated to* $\mathcal{K}$.

Theorem 6.57. *In* $\mathrm{PG}(n,q)$ *with* $n \geq 3$, *and* $q = 2^h$, *let* $\mathcal{K} = \{\pi_1, \ldots, \pi_k\}$, *where* $k \geq n+1$, *be a* k-*arc of hyperplanes. For distinct* $i_1, i_2, \ldots, i_{n-1}$ *in* $\mathbf{N}_k$, *let* $Z_{\{i_1,\ldots,i_{n-1}\}}$ *denote the set of* $t = q + n - k$ *points on the line* $\pi_{i_1} \cap \cdots \cap \pi_{i_{n-1}}$ *that lie on no other hyperplane of* $\mathcal{K}$. *Then*

(i) *there exists a curve* $\mathcal{C}_{\{i_1,\ldots,i_{n-2}\}}$ *of degree t in the plane* $\pi_{i_1} \cap \cdots \cap \pi_{i_{n-2}}$ *such that*

$$\mathcal{C}_{\{i_1,\ldots,i_{n-2}\}} \cap \pi_{i_{n-1}} = Z_{\{i_1,\ldots,i_{n-1}\}};$$

(ii) *for fixed i, the curves* $\mathcal{C}_{\{i_1,\ldots,i_{n-2}\}}$ *are algebraically contained in a hypersurface* Φ_{i_1} *of* π_{i_1} *of degree t with* $\Phi_{i_1} \cap \pi_{i_2} \cap \cdots \cap \pi_{i_{n-2}} = \mathcal{C}_{\{i_1,\ldots,i_{n-2}\}}$ *and each variety* Φ_{i_1} *is algebraically contained in a hypersurface* $\Phi = \Phi(\mathcal{K})$ *of* $\mathrm{PG}(n,q)$ *of degree t with* $\Phi(\mathcal{K}) \cap \pi_i = \Phi_i$;

(iii) *if* $k > \frac{1}{2}q + n - 1$, *the hypersurface* $\Phi(\mathcal{K})$ *is unique*;

(iv) *with* $k > \frac{1}{2}q + n - 1$, *if* $\mathcal{L} = \{\pi_1, \ldots, \pi_k, \ldots, \pi_u\}$ *is an arc of hyperplanes containing* $\mathcal{K}$, *the hypersurface* $\Phi(\mathcal{K})$ *has components* $\Phi(\mathcal{L}), \pi_{k+1}, \ldots, \pi_u$;

(v) *if* $k > \frac{1}{2}q + n - 1$, *there is a bijection between hyperplanes of* $\mathrm{PG}(n,q)$ *extending* $\mathcal{K}$ *to a* $(q+1)$-*arc and linear components over* $\mathbf{F}_q$ *of* $\Phi(\mathcal{K})$.

Proof. (i), (ii) These were proved in the previous theorem and the subsequent remarks.

(iii) Since $k - n + 2 > t = q - k + n$, it follows from Theorem 10.1 of PGOFF2 that the curve $\mathcal{C}_{\{i_1,\ldots,i_{n-2}\}}$ is unique. Suppose Φ and Φ' are distinct hypersurfaces of degree t for which

$$\Phi \cap \pi_{i_1} \cap \cdots \cap \pi_{i_{n-2}} = \Phi' \cap \pi_{i_1} \cap \cdots \cap \pi_{i_{n-2}} = \mathcal{C}_{\{i_1,\ldots,i_{n-2}\}}.$$

Let $\Phi = \mathbf{A}(D)$ and $\Phi' = \mathbf{A}(D')$.

First, let $n = 3$ and fix an index i_1. As in Theorem 6.54, there exists λ in $\mathbf{F}_q^*$ for which $D + \lambda D'$ vanishes at all points of the $k - 1$ lines $\pi_{i_1} \cap \pi_{i_2}$ with $i_2 \neq i_1$. The surfaces Φ and Φ' both have degree $t = q + 3 - k$. Since $k > \frac{1}{2}q + 2$, so $k - 1 > t = q + 3 - k$. Therefore $D + \lambda D'$ vanishes at all points of π_{i_1}. Since the surface $\Phi'' = \mathbf{A}(D + \lambda D')$ contains all points of the lines $\pi_{i_1} \cap \pi_{i_2}$ it follows that Φ'' has the k planes $\pi_1, \ldots, \pi_k$ as components. However, $k > t = \deg(\Phi'')$, a contradiction. So $\Phi' = \Phi$.

Next, let $n > 3$ and proceed by induction on n. Since $k - 1 > \frac{1}{2}q + (n-1) - 1$, assume that the varieties $\Phi_{i_1} = \Phi'_{i_1}$ are unique for i_1 in $\mathbf{N}_k$. Fix an index i_1. Again, as in Theorem 6.54, there exists λ in $\mathbf{F}_q^*$ such that $D + \lambda D'$ vanishes at all points of the $(n-2)$-spaces $\pi_{i_1} \cap \pi_{i_2}$, of which there are $k - 1$. Now, both Φ and Φ' have degree $t = q + n - k$. Since $k > \frac{1}{2}q + n - 1$, so $k - 1 > t$. Hence $D + \lambda D'$ vanishes at all points of π_{i_1}. Since $\Phi'' = \mathbf{A}(D + \lambda D')$ contains all points of $\pi_{i_1} \cap \pi_{i_2}$ the hypersurface Φ'' has the k hyperplanes as components. As $k > t = \deg(\Phi'')$, a contradiction is obtained and $\Phi = \Phi'$.

(iv) $\Phi(\mathcal{K}) \cap \pi_{i_1} \cap \cdots \cap \pi_{i_{n-1}}$ consists of the set $\Phi(\mathcal{L}) \cap \pi_{i_1} \cap \cdots \cap \pi_{i_{n-1}}$ together with the points

$$\pi_{i_1} \cap \cdots \cap \pi_{i_{n-1}} \cap \pi_{k+1}, \quad \ldots, \quad \pi_{i_1} \cap \cdots \cap \pi_{i_{n-1}} \cap \pi_u.$$

Since $\Phi(\mathcal{K})$ is unique, the required factorisation is obtained.

(v) Suppose $\mathcal{K} \cup \{\pi\}$ is a $(k+1)$-arc of hyperplanes. By (iv), π is a linear component of $\Phi(\mathcal{K})$. Conversely, let σ be a linear component over $\mathbf{F}_q$ of $\mathcal{K}$. Let $\pi_{i_1}, \ldots, \pi_{i_n}$ be any n hyperplanes in $\mathcal{K}$. Then these n hyperplanes and σ have no point in common, since such a point would lie on $\Phi(\mathcal{K}) \cap \pi_{i_1} \cap \cdots \cap \pi_{i_n}$, contradicting the defining property of $\Phi(\mathcal{K})$. So no $n + 1$ hyperplanes in $\mathcal{K} \cup \{\sigma\}$ have a point in common, whence $\mathcal{K} \cup \{\sigma\}$ is a $(k + 1)$-arc of hyperplanes. $\qquad\square$

Theorem 6.58. *Let* $\mathcal{K} = \{\pi_1, \ldots, \pi_k\}$ *be a* k-*arc of hyperplanes in* $\mathrm{PG}(n, q), n \geq 3$ *and* $q = 2^s$. *If* $k > \frac{1}{2}q + n - 1$, *then* $\mathcal{K}$ *is contained in a unique complete arc.*

Proof. Let $\mathcal{K}'$ and $\mathcal{K}''$ be distinct complete arcs of hyperplanes containing $\mathcal{K}$, and assume that $\pi \in \mathcal{K}' \backslash \mathcal{K}''$. By Theorem 6.57(v), π is a component of $\Phi(\mathcal{K})$. Since $\pi \notin \mathcal{K}''$, by part (iv), the hyperplane π is a component of $\Phi(\mathcal{K}'')$. Again, by part (v) the arc $\mathcal{K}''$ can be extended to an arc $\mathcal{K}'' \cup \{\pi\}$ where $\pi \notin \mathcal{K}''$. This contradicts the completeness of $\mathcal{K}''$. $\qquad\square$

Now these results are applied in $\mathrm{PG}(3, q)$, $q = 2^h$. First, the necessary results for $n = 3$ are restated.

Theorem 6.59. *Let* $\mathcal{K} = \{\pi_1, \ldots, \pi_k\}$ *be a* k-*arc of planes in* $\Sigma = \mathrm{PG}(3, q)$, *with* $q = 2^h$. *For any two distinct planes* π_i *and* π_j, *let* Z_{ij} *be the set of points of* $\pi_i \cap \pi_j$ *in exactly two planes of* $\mathcal{K}$. *Then*

(i) *there exists an algebraic curve* C_i *in* π_i *containing all sets* Z_{ij} *and such that* $C_i \cap \pi_j = Z_{ij}$;
(ii) *there exists an algebraic surface* $\Phi = \Phi(\mathcal{K})$ *of degree* $t = q + 3 - k$ *algebraically containing the curves* C_i *and with* $\Phi \cap \pi_i = C_i$.

Suppose further that $k > \frac{1}{2}q + 2$. *Then*

(iii) *the surface* Φ *is unique;*
(iv) *if* $\mathcal{L} = \mathcal{K} \cup \{\pi_{k+1}, \ldots, \pi_u\}$ *is a* u-*arc of planes,* $u > k$, *the surface* $\Phi(\mathcal{K})$ *factors into* $\Phi(\mathcal{L}), \pi_{k+1}, \ldots, \pi_u$;
(v) *there is a bijection between planes of* Σ *extending* $\mathcal{K}$ *to a* $(k+1)$-*arc and linear components over* $\mathbf{F}_q$ *of* $\Phi(\mathcal{K})$;
(vi) $\mathcal{K}$ *is contained in a unique complete arc of* Σ.

Lemma 6.60. *Let* $k > q - \sqrt{q} + 2$, $q = 2^h$. *With notation as in Theorem* 6.59,

(i) *the curve* C_i *of degree* $t = q + 3 - k$ *factors into* t *lines forming an arc of lines in* π_i;
(ii) *these* t *lines* $\ell_{i1}, \ldots, \ell_{it}$, *called* S-*lines, together with the* $k - 1$ *lines* $\pi_i \cap \pi_j$, $j \in \mathbf{N}_k \backslash \{i\}$, *form a* $(q + 2)$-*arc of lines in* π_i;
(iii) *each* S-*line lies in a unique plane of* $\mathcal{K}$;
(iv) *each point* P *on an* S-*line lies on exactly one other* S-*line.*

Proof. (i) This follows from Section 10.3 of PGOFF2.

(ii),(iii) These follow from Theorem 6.59.

(iv) Each point $Z_{ij}^{(u)}$ in Z_{ij} lies on one S-line ℓ_{ia} in π_i and on one S-line ℓ_{jb} in π_j. If P is on ℓ_{ia} and is not of type $Z_{ij}^{(u)}$, then, by (ii), P lies on exactly one other S-line, which will be of type ℓ_{jb} with $b \neq a$. $\qquad\square$

Theorem 6.61. *Let* $\mathcal{K} = \{\pi_1, \pi_2, \ldots, \pi_k\}$ *be an arc of planes in* $\Sigma = \mathrm{PG}(3, q)$, *with* $q = 2^h$ *and* $k > \frac{1}{2}q + 2$. *Assume that* $\mathcal{K}$ *is contained in a* $(q + 1)$-*arc* $\mathcal{L}$ *of planes in* Σ. *Then*

   (i) $\Phi(\mathcal{K})$ *factors into* $t - 2$ *linear components over* $\mathbf{F}_q$ *and one quadratic component* $\mathcal{H}$, *where* $t = q + 3 - k$ *and* $\mathcal{H}$ *is a hyperbolic quadric* $\mathcal{H}_3$;

   (ii) $\Phi(\mathcal{L}) = \mathcal{H}$;

   (iii) *each plane of* $\mathcal{L}$ *is a tangent plane to* $\mathcal{H}$;

   (iv) *for any plane* π *in* $\mathcal{L}$ *the planes of* $\mathcal{L}\backslash\mathcal{H}$ *together with* $\mathcal{H}$ *cut out a* $(q + 2)$-*arc of lines in* π.

Proof. (i), (ii) Let $\mathcal{L} = \{\pi_1, \pi_2, \ldots, \pi_{q+1}\}$. By the definition of an arc, $q > 2$. From Theorem 6.59(iv), $\Phi(\mathcal{K})$ factors into $\Phi(\mathcal{L}), \pi_{k+1}, \ldots, \pi_{q+1}$ with $\Phi(\mathcal{L}) = \mathcal{H}$ of degree 2. Since $q > 2$, Theorem 21.3.8 of FPSOTD says that $\mathcal{L}$ is complete. From Theorem 6.59(v), the quadric $\mathcal{H}$ is irreducible.

Let $\pi_{k+1} = \pi_1', \ldots, \pi_{q+1} = \pi_{t-2}'$. Each plane π_j' in $\mathcal{L}\backslash\mathcal{K}$ intersects each plane π_i of $\mathcal{K}$ in a line. Also, if π_j', π_u', π_v' are distinct, then $\pi_j' \cap \pi_u' \cap \pi_v' \cap \pi_i = \emptyset$ since these four planes belong to an arc. Therefore the $t - 2$ planes $\pi_1', \ldots, \pi_{t-2}'$ cut out a $(t - 2)$-arc of lines in π_i.

Let $\mathcal{C}_i$ be the curve of degree t in π_i corresponding to the arc $\mathcal{K}$. Then $\mathcal{C}_i$ has $t - 2$ linear components $\pi_1' \cap \pi_i, \ldots, \pi_{t-2}' \cap \pi_i$ and one component $\mathcal{H}^{(i)} = \mathcal{H} \cap \pi_i$ of degree 2. For each $j \neq i$ with π_j in $\mathcal{K}$, each of the lines $\pi_u' \cap \pi_i$ contains exactly one of the t points $Z_{ij}^{(u)}$. If $a \neq b$, then $\pi_a' \cap \pi_i \cap Z_{ij} \neq \pi_b' \cap \pi_i \cap Z_{ij}$ since otherwise π_a', π_b', π_i and π_j have a point in common, contradicting that these four planes are part of an arc of planes of Σ. Therefore $\mathcal{H}^{(i)}$ contains exactly two points of Z_{ij}, say $Z_{ij}^{(1)}$ and $Z_{ij}^{(2)}$, $i \neq j$. Then $|\mathcal{H}^{(i)}| \geq 2(k - 1) > q + 2$. Since $|\mathcal{H}^{(i)}| > q + 2$, so $\mathcal{H} \cap \pi_i$ cannot be a conic. It follows that $\mathcal{H} \cap \pi_i$ factors into a pair of distinct lines x_i, y_i with $x_i \cap Z_{ij} = Z_{ij}^{(1)}$, $y_i \cap Z_{ij} = Z_{ij}^{(2)}$, and $Z_{ij}^{(1)} \neq Z_{ij}^{(2)}$. Hence each plane of $\mathcal{K}$ contains exactly two lines of $\mathcal{H}$. Further, each line of $\mathcal{H}$ is on at most one plane of $\mathcal{K}$. So $\mathcal{H}$ contains at least $2k > q + 4$ lines. It follows that $\mathcal{H}$ cannot be a cone and is in fact a hyperbolic quadric of Σ.

(iii), (iv) Let π be any plane of $\mathcal{L}$. The planes of $\mathcal{L}\backslash\{\pi\}$ cut out a q-arc of lines in π since $\mathcal{L}$ is an arc of planes in Σ. Hence $\Phi(\mathcal{L}) \cap \pi$ is a curve $\mathcal{C}$ of degree 2. Also, since $q + 1 > q - \sqrt{q} + 2$, it follows from Section 10.3 of PGOFF2 that $\mathcal{C}$ factors into two lines ℓ, m, which are therefore lines of $\Phi(\mathcal{L})$. Since $\Phi(\mathcal{L})$ is a hyperbolic quadric, (iii) follows. By Lemma 6.60(ii), part (iv) also follows. $\qquad\square$

For the remainder of this section assume that $q > 2$.

Lemma 6.62. *Let* $\mathcal{K}$ *be a* k-*arc of planes in* $\mathrm{PG}(3, q)$, $q = 2^h$. *If*

$$(q - 1)/t^2 + 2/t > t$$

with $t = q + 3 - k$, *then* $k > q - \sqrt{q} + 2$.

Proof. Suppose that $k \leq q - \sqrt{q} + 2$; that is, $\sqrt{q} \leq t - 1$. Therefore

$$(q-1)/(q+1+2\sqrt{q}) \ge (q-1)t^2.$$

This implies that

$$t - 2/t \ge \sqrt{q}+1-2/t \ge \sqrt{q}+1-2/(\sqrt{q}+1) > 1 > (q-1)/t^2.$$

So $t - 2/t > (q-1)/t^2$, a contradiction. $\qquad\square$

Theorem 6.63. *Let $\mathcal{K} = \{\pi_1, \pi_2, \ldots, \pi_k\}$ be a complete k-arc of planes in the space $\Sigma = \mathrm{PG}(3,q)$, $q = 2^h$. If $(q-1)/t^2 + 2/t > t$, then each $\mathcal{C}_i$ factors into t S-lines and each S-line belongs to a hyperbolic quadric algebraically contained in $\Phi(\mathcal{K})$.*

Proof. From Lemmas 6.60(i) and 6.62, the curve $\mathcal{C}_i$ in π_i factors into t S-lines. Let ℓ_{iu} be a fixed S-line in π_i. By Lemma 6.60(iv), there are exactly $q+1$ S-lines distinct from ℓ_{iu} having a point in common with ℓ_{iu}; denote these by $\ell_1, \ell_2, \ldots, \ell_{q+1}$. Let i, j, g be distinct. The S-lines in π_j are the lines ℓ_{jv} and the S-lines in π_g are denoted by ℓ_{gw}. Let f_{vw} be the number of lines ℓ_r having a point in common with ℓ_{jv} and ℓ_{gw}. If ℓ_r is not in π_j nor in π_g, then ℓ_r is concurrent with one line of type ℓ_{jv} and one of type ℓ_{gw}. If ℓ_r is in the plane π_j, then it is concurrent with t lines ℓ_{jv} and one line ℓ_{gw}. If ℓ_r is in π_g, it is concurrent with t lines ℓ_{gw} and one line ℓ_{jv}. Consequently,

$$\sum_{v,w=1}^{t} f_{vw} = q - 1 + 2t.$$

Averaging gives $\bar{f} = (q-1)/t^2 + 2/t$. Therefore there exist two S-lines, say $\ell_{jv} = \ell'$ and $\ell_{gw} = \ell''$, for which $f_{vw} \ge \bar{f} = (q-1)/t^2 + 2/t$. Now it is shown that $\ell_{iu} = \ell$, ℓ' and ℓ'' are mutually skew.

Suppose for example that ℓ' and ℓ'' meet in a point P. Since

$$f_{vw} \ge (q-1)/t^2 + 2/t$$

and since, by Lemma 6.60(iv), P lies on exactly two S-lines, the plane $\ell'\ell''$ contains at least $(q-1)/t^2 + 2/t$ S-lines. Because $\ell'\ell'' \cap \Phi(\mathcal{K})$ is an algebraic curve of degree t or $\ell'\ell''$ is a component of $\Phi = \Phi(\mathcal{K})$, so $(q-1)/t^2 + 2/t \le t$ or else the plane $\ell'\ell''$ is a component of Φ. By assumption $(q-1)/t^2 + 2/t > t$; so the plane $\ell'\ell''$ is a component of Φ. But then, by Theorem 6.59(v), the plane $\ell'\ell''$ extends $\mathcal{K}$, contradicting the fact that $\mathcal{K}$ is complete. It follows that the lines ℓ, ℓ', ℓ'' are mutually skew.

If λ is the integer defined by $(q-1)/t^2 + 2/t + 1 > \lambda \ge (q-1)/t^2 + 2/t$, then there are at least λ S-lines of the form ℓ_i, say $\ell_1, \ldots, \ell_\lambda$, which meet ℓ, ℓ', ℓ''. Hence $\ell_1, \ell_2, \ldots, \ell_\lambda$ belong to a regulus $\mathcal{R}$. Let m be a line of the complementary regulus $\mathcal{R}'$. Then m has at least $\lambda > t$ points in common with $\Phi = \Phi(\mathcal{K})$. Consequently, m is a line of Φ. So the hyperbolic quadric $\mathcal{H}_3$ with reguli $\mathcal{R}, \mathcal{R}'$ is a component of Φ. Therefore, each S-line ℓ_{iu} belongs to a hyperbolic quadric which is algebraically contained in Φ. $\qquad\square$

Lemma 6.64. *If $q - \frac{1}{2}\sqrt{q} + \frac{9}{4} < k < q + 1$, then, for any plane π of $\mathrm{PG}(3,q)$, the curve $\Phi \cap \pi$ is reducible over $\overline{\mathbf{F}}_q$.*

Proof. As $k > q - \frac{1}{2}\sqrt{q} + 2$ and since there is an integer between $q - \frac{1}{2}\sqrt{q} + \frac{9}{4}$ and $q + 1$, so $q \geq 32$. It may be assumed that no S-line is contained in π. Then any S-line has exactly one point in common with π. Since the number of S-lines is equal to $k(q + 3 - k) = kt$ and each point is on either zero or two S-lines, so

$$|\Phi \cap \pi| \geq kt/2.$$

Suppose that $\Phi \cap \pi$ is absolutely irreducible, that is, irreducible over $\overline{\mathbf{F}}_q$. By Corollary 2.29 of PGOFF2,

$$(q + 3 - k)k/2 \leq q + 1 + (q + 2 - k)(q + 1 - k)\sqrt{q}.$$

Consequently, either $k \geq q + 1$ or $k \leq q - \frac{1}{2}\sqrt{q} + \frac{9}{4} - 1/(4 + 8\sqrt{q}) < q - \frac{1}{2}\sqrt{q} + \frac{9}{4}$, a contradiction. Therefore $\Phi \cap \pi$ is reducible over $\overline{\mathbf{F}}_q$. $\qquad\square$

Lemma 6.65. *If $q - \frac{1}{2}\sqrt{q} + \frac{9}{4} < k < q + 1$, and k is even, then, for each plane π of PG$(3, q)$, the curve $\Phi \cap \pi = C$ contains a line as a component over $\mathbf{F}_q$.*

Proof. It may be assumed that no S-line is contained in π. By Lemma 6.64, the curve C is reducible over $\overline{\mathbf{F}}_q$. Also $2 < t < \frac{1}{2}\sqrt{q} + \frac{3}{4}$ and $q \geq 32$. If C' is an absolutely irreducible component of C of degree m, with $m \geq 4$, then it can now be shown that $|C'| < (q + 1)m/2$.

If C' is not defined over $\mathbf{F}_q$ then $|C'| \leq m^2$, Lemma 2.24(i) of PGOFF2; if C' is defined over $\mathbf{F}_q$ then, by the Hasse–Weil bound, $|C'| \leq q + 1 + (m - 1)(m - 2)\sqrt{q}$. Since $q + 1 + (m - 1)(m - 2)\sqrt{q} \geq m^2$, it follows that $|C'| \leq q + 1 + (m - 1)(m - 2)\sqrt{q}$.

Suppose that

$$q + 1 + (m - 1)(m - 2)\sqrt{q} \geq (q + 1)m/2.$$

Then, either

$$\tfrac{1}{4}\sqrt{q} + \tfrac{3}{2} + 1/(4\sqrt{q}) + \tfrac{1}{4}\{(\sqrt{q} - 2)^2 + 2 - 4/\sqrt{q} + 1/q\}^{1/2} \leq m$$

or

$$\tfrac{1}{4}\sqrt{q} + \tfrac{3}{2} + 1/(4\sqrt{q}) - \tfrac{1}{4}\{(\sqrt{q} - 2)^2 + 2 - 4/\sqrt{q} + 1/q\}^2 \geq m.$$

Hence, either

$$m > \tfrac{1}{4}\sqrt{q} + \tfrac{3}{2} + \tfrac{1}{4}(\sqrt{q} - 2) = \tfrac{1}{2}\sqrt{q} + 1$$

or

$$m < \tfrac{1}{4}\sqrt{q} + \tfrac{3}{2} + 1/(4\sqrt{q}) - \tfrac{1}{4}(\sqrt{q} - 2) = 2 + 1/(4\sqrt{q}).$$

This contradicts that $4 \leq m < t < \frac{1}{2}\sqrt{q} + \frac{3}{2}$. Hence $|C'| < (q + 1)m/2$.

If C'' is an absolutely irreducible component of C of odd degree m, with $m \geq 5$, then it is now shown that $|C''| < \frac{1}{2}(m - 3)(q + 1) + q + 1 + 2\sqrt{q}$. Note that, by the Hasse–Weil bound, $q + 1 + 2\sqrt{q}$ is the maximum number of points of an absolutely irreducible plane cubic curve over $\mathbf{F}_q$.

As for C',

$$|C''| \leq q + 1 + (m - 1)(m - 2)\sqrt{q}.$$

Since $5 \le m \le q + 1 - k < \frac{1}{2}\sqrt{q} - \frac{5}{4}$, so $q \ge 256$. Assume that

$$\tfrac{1}{2}(m-3)(q+1) + q + 1 + 2\sqrt{q} \le q + 1 + (m-1)(m-2)\sqrt{q}.$$

Then, either

$$m \le \tfrac{1}{4}\sqrt{q} + 1/(4\sqrt{q}) + \tfrac{3}{2} - \tfrac{1}{4}\{(\sqrt{q} - 6)^2 + 2 - 12/\sqrt{q} + 1/q\}^{1/2}$$

or

$$m \ge \tfrac{1}{4}\sqrt{q} + 1/(4\sqrt{q}) + \tfrac{3}{2} + \tfrac{1}{4}\{(\sqrt{q} - 6)^2 + 2 - 12/\sqrt{q} + 1/q\}^{1/2}.$$

Since $q \ge 256$, so $2 - 12/(\sqrt{q}) + 1/q > 0$. Hence, either

$$m < \tfrac{1}{4}\sqrt{q} + \tfrac{3}{2} + 1/(4\sqrt{q}) - \tfrac{1}{4}(\sqrt{q} - 6) = 3 + 1/(4\sqrt{q}),$$

or

$$m > \tfrac{1}{4}\sqrt{q} + \tfrac{3}{2} + \tfrac{1}{4}(\sqrt{q} - 6) = \tfrac{1}{2}\sqrt{q}.$$

This contradicts that $5 \le m < \frac{1}{2}\sqrt{q} - \frac{5}{4}$, and so $|\mathcal{C}''| < \frac{1}{2}(m-3)(q+1) + q + 1 + 2\sqrt{q}$.

Now suppose that $\mathcal{C}$ contains no linear component over $\mathbf{F}_q$, but contains $\beta \ge 0$ linear components over $\overline{\mathbf{F}}_q$. Since $\mathcal{C}$ has odd degree, the number of components over $\mathbf{F}_q$ of $\mathcal{C}$ of odd degree is odd. By the preceding sections and using the fact that $2(q + 1 + 2\sqrt{q}) < 3(q + 1)$,

  (i) for β even,

$$\begin{aligned}|\mathcal{C}| &\le \tfrac{1}{2}(t - \beta - 3)(q + 1) + \beta + (q + 1 + 2\sqrt{q}) \\ &\le \tfrac{1}{2}(t - 3)(q + 1) + q + 1 + 2\sqrt{q};\end{aligned}$$

  (ii) for β odd,

$$|\mathcal{C}| \le \tfrac{1}{2}(t - \beta)(q + 1) + \beta < \tfrac{1}{2}(t - 3)(q + 1) + q + 1 + 2\sqrt{q}.$$

Hence, in each case,

$$|\mathcal{C}| \le \tfrac{1}{2}(t - 3)(q + 1) + q + 1 + 2\sqrt{q}.$$

Consequently,

$$\tfrac{1}{2}k(q + 3 - k) \le |\mathcal{C}| \le \tfrac{1}{2}(q - k)(q + 1) + q + 1 + 2\sqrt{q}.$$

So, either

$$k \le q + 2 - \{(\sqrt{q} - 3)^2 + 2\sqrt{q} - 7\}^{1/2}$$

or

$$k \ge q + 2 + \{(\sqrt{q} - 3)^2 + 2\sqrt{q} - 7\}^{1/2}.$$

Hence, either

$$k < q + 2 - (\sqrt{q} - 3) = q - \sqrt{q} + 5$$

or

$$k > q + 2 + (\sqrt{q} - 3) = q + \sqrt{q} - 1.$$

This contradicts that $q - \frac{1}{2}\sqrt{q} + \frac{9}{4} < k < q + 1$. Therefore $\mathcal{C}$ contains a line as a component over $\mathbf{F}_q$. $\qquad\square$

Theorem 6.66. *If* $q - \frac{1}{2}\sqrt{q} + \frac{9}{4} < k < q + 1$ *and* k *is even, then* Φ *contains a plane as a component over* $\mathbf{F}_q$.

Proof. Let π_i be a plane of the k-arc. In π_i there are $q + 3 - k = t < \frac{1}{2}\sqrt{q} + \frac{3}{4}$ S-lines which form an arc of lines in π_i. Since $q \geq t(t-3)/2 + 2$, this arc is incomplete by Theorem 9.12 of PGOFF2; so there is a line ℓ which intersects the t lines of the arc at t different points. Hence $|\ell \cap \Phi| = t$ and the t points of $\Phi \cap \ell$ are simple for Φ. Considering the $q + 1$ planes of PG$(3, q)$ through ℓ and using Lemma 6.65, at least one point P of $\Phi \cap \ell$ is contained in at least $(q+1)/t$ lines of Φ. Hence P is contained in at least $2\sqrt{q} - 4 + (\frac{1}{2}\sqrt{q} + 4)/(\frac{1}{2}\sqrt{q} + \frac{3}{4})$ lines of Φ. It follows that the tangent plane π_P of Φ at P contains more than $2\sqrt{q} - 4$ lines of Φ. Since $2\sqrt{q} - 4 > t$, the plane π_P is a component of Φ. $\square$

Theorem 6.67. *Any* k-arc $\mathcal{K}$ *of* PG$(3, q)$, *with* q *even,* k *even, and with the bound* $q - \frac{1}{2}\sqrt{q} + \frac{9}{4} < k < q + 1$, *can be extended to a* $(k+1)$-arc.

Proof. This follows immediately from Theorems 6.59(v) and 6.66. $\square$

Lemma 6.68. *If* $q - \frac{1}{2}\sqrt{q} + \frac{9}{4} < k < q + 1$ *and* k *is odd, then, for each plane* π *of* PG$(3, q)$, *the curve* $\Phi \cap \pi = \mathcal{C}$ *either contains a line as component over* $\mathbf{F}_q$ *or consists of* $t/2$ *conics defined over* $\mathbf{F}_q$.

Proof. It may be assumed that no S-line is contained in π. If $\mathcal{C}'$ is an absolutely irreducible component of $\mathcal{C}$ of degree m, with $q + 2 - k > m > 4$ and $q > 512$, then it is now shown that $|\mathcal{C}'| < \frac{1}{2}(m-4)(q+1) + q + 1 + 2\sqrt{q}$. Note, that by the Hasse–Weil bound, $q + 1 + 2\sqrt{q}$ is the maximum number of points of an absolutely irreducible plane cubic curve over $\mathbf{F}_q$. Ignore for the moment the condition $q > 512$. If $\mathcal{C}'$ is not defined over $\mathbf{F}_q$, then $|\mathcal{C}'| \leq m^2$; if $\mathcal{C}'$ is defined over $\mathbf{F}_q$, then

$$|\mathcal{C}'| \leq q + 1 + (m-1)(m-2)\sqrt{q}.$$

Hence $|\mathcal{C}'| \leq q + 1 + (m-1)(m-2)\sqrt{q}$. Since $5 \leq m \leq q + 1 - k < \frac{1}{2}\sqrt{q} - \frac{5}{4}$, so $q \geq 256$. Suppose that

$$\tfrac{1}{2}(m-4)(q+1) + q + 1 + 2\sqrt{q} \leq q + 1 + (m-1)(m-2)\sqrt{q}. \tag{6.80}$$

Then, either

$$m \leq \tfrac{1}{4}\sqrt{q} + \tfrac{3}{2} + 1/(4\sqrt{q}) - \tfrac{1}{4}\{(\sqrt{q} - 11)^2 + 2\sqrt{q} - 20/\sqrt{q} + 1/q - 83\}^{1/2}$$

or

$$m \geq \tfrac{1}{4}\sqrt{q} + \tfrac{3}{2} + 1/(4\sqrt{q}) + \tfrac{1}{4}\{(\sqrt{q} - 11)^2 + 2\sqrt{q} - 20/\sqrt{q} + 1/q - 83\}^{1/2}.$$

For $q > 1024$, the inequality $2\sqrt{q} - 20/\sqrt{q} + 1/q - 83 > 0$ holds. Hence, for $q > 1024$, either

$$m < \tfrac{1}{4}\sqrt{q} + \tfrac{3}{2} + 1/(4\sqrt{q}) - \tfrac{1}{4}(\sqrt{q} - 11) = \tfrac{17}{4} + 1/(4\sqrt{q})$$

or

$$m > \tfrac{1}{4}\sqrt{q} + \tfrac{3}{2} + \tfrac{1}{4}(\sqrt{q} - 11) = \tfrac{1}{2}\sqrt{q} - \tfrac{5}{4}.$$

This contradicts that $5 \leq m < \tfrac{1}{2}\sqrt{q} - \tfrac{5}{2}$. For $q = 256$, the inequality (6.80) is satisfied; for $q = 512$, (6.80) and the inequality $5 \leq m \leq q + 1 - k < \tfrac{1}{2}\sqrt{q} - \tfrac{5}{4}$ give $m = 10$ and $k = 503$; for $q = 1024$, (6.80) is in contradiction to $5 \leq m < \tfrac{1}{2}\sqrt{q} - \tfrac{5}{4}$.

Let $\mathcal{C}''$ be an absolutely irreducible component of degree four of $\mathcal{C}$. If $\mathcal{C}''$ is not defined over $\mathbf{F}_q$, then $|\mathcal{C}''| \leq 16 < 2(q+1)$. If $\mathcal{C}''$ is defined over $\mathbf{F}_q$, then the bound $|\mathcal{C}''| \leq q + 1 + 6\sqrt{q}$ holds; so $|\mathcal{C}''| \leq 2(q+1)$, for $q \geq 32$. Hence $|\mathcal{C}''| \leq 2(q+1)$ is always true.

Now suppose that, over $\overline{\mathbf{F}}_q$, the curve $\mathcal{C}$ neither contains a line nor consists entirely of conics and irreducible quartic curves. Suppose also that $q \notin \{256, 512\}$. Let β be the number of absolutely irreducible components of degree three and let α be the number of absolutely irreducible components of degree at least five. If $\alpha = 0$, then, since $q + 3 - k = t$ is even, β is even and so $\alpha + \beta$ is even. Also $\alpha + \beta > 0$. By the preceding sections,

$$|\mathcal{C}| \leq \tfrac{1}{2}(t - 3\beta - 4\alpha)(q+1) + (\alpha + \beta)(q + 1 + 2\sqrt{q}).$$

Further, note that $2(q + 1 + 2\sqrt{q}) < 3(q+1)$. If $\alpha + \beta$ is odd, so $\alpha \neq 0$; then

$$|\mathcal{C}| \leq \tfrac{1}{2}(t - \alpha - 3)(q+1) + q + 1 + 2\sqrt{q} \leq \tfrac{1}{2}(t - 4)(q+1) + q + 1 + 2\sqrt{q}.$$

If $\alpha + \beta$ is even, so $\alpha + \beta \geq 2$; then

$$|\mathcal{C}| \leq \tfrac{1}{2}(t - \alpha - 6)(q+1) + 2(q + 1 + 2\sqrt{q})$$
$$\leq \tfrac{1}{2}(t - 6)(q+1) + 2(q + 1 + 2\sqrt{q}).$$

Thus, in both cases,

$$|\mathcal{C}| \leq \tfrac{1}{2}(t - 6)(q+1) + 2(q + 1 + 2\sqrt{q})$$
$$= \tfrac{1}{2}(q - k - 3)(q+1) + 2(q + 1 + 2\sqrt{q}).$$

Consequently,

$$\tfrac{1}{2}k(q + 3 - k) \leq \tfrac{1}{2}(t - 6)(q+1) + 2(q + 1 + 2\sqrt{q}).$$

So, either

$$k \leq q + 2 - \{2q - 8\sqrt{q} + 3\}^{1/2}$$

or

$$k \geq q + 2 + \{2q - 8\sqrt{q} + 3\}^{1/2}.$$

This contradicts $q - \tfrac{1}{2}\sqrt{q} + \tfrac{9}{4} < k < q + 1$. Hence, for $q \notin \{256, 512\}$, $\mathcal{C}$ contains over $\overline{\mathbf{F}}_q$ either a linear component or consists entirely of conics and absolutely irreducible quartic curves. Let $\mathcal{C}$ consist of ϵ conics and δ absolutely irreducible quartic curves, with $\delta \geq 1$. If $q = 32$, then $t = 3$ and so k is even; hence $q \geq 64$. Since

$$\tfrac{1}{2}(t-4\delta)(q+1)+\delta(q+1+6\sqrt{q}) = \tfrac{1}{2}(q+3-k)(q+1)+\delta(-q-1+6\sqrt{q})$$
$$\leq \tfrac{1}{2}(q+3-k)(q+1)-q-1+6\sqrt{q},$$

so

$$\tfrac{1}{2}k(q+3-k) \leq |\mathcal{C}| \leq \tfrac{1}{2}(q+3-k)(q+1)-q-1+6\sqrt{q}.$$

Consequently, either

$$k \leq q+2-\{2q-12\sqrt{q}+3\}^{1/2}$$

or

$$k \geq q+2+\{2q-12\sqrt{q}+3\}^{1/2}.$$

This contradicts

$$q-\tfrac{1}{2}\sqrt{q}+\tfrac{9}{4} < k < q+1.$$

So over $\overline{\mathbf{F}}_q$, and with $q \notin \{256,512\}$, the curve $\mathcal{C}$ either contains a line or consists entirely of conics.

Let ℓ be a line of $\mathcal{C}$, and suppose that ℓ is not defined over $\mathbf{F}_q$. Then $|\ell| \leq 1$. Let π_i be a plane of $\mathcal{K}$ not passing through a point of ℓ over $\mathbf{F}_q$. The line $\ell' = \pi \cap \pi_i$ intersects $\pi_i \cap \Phi$ and so Φ, only in points over $\mathbf{F}_q$. Hence the intersection of ℓ and ℓ' is a point over $\mathbf{F}_q$, a contradiction.

Now suppose that $\mathcal{C}$ consists of $t/2$ conics, ρ of which are not defined over $\mathbf{F}_q$, with $\rho \geq 1$. Then

$$\tfrac{1}{2}k(q+3-k) \leq |\mathcal{C}| \leq \tfrac{1}{2}(q+3-k-2\rho)(q+1)+4\rho$$
$$\leq \tfrac{1}{2}(q+1-k)(q+1)+4.$$

Hence, either

$$k \leq q+2-(2q-5)^{1/2}$$

or

$$k \geq q+2+(2q-5)^{1/2},$$

a contradiction. Consequently, for $q \notin \{256,512\}$, the curve $\mathcal{C}$ either contains a line over $\mathbf{F}_q$ or consists of $t/2$ conics over $\mathbf{F}_q$.

Let $q = 512$. Then (6.80) together with $5 \leq m \leq q+1-k < \tfrac{1}{2}\sqrt{q}-\tfrac{5}{4}$ gives $m = 10$ and $k = 503$. So suppose that $m = 10$ and $t = 12$. If $\mathcal{C}$ does not contain a line, then

$$\tfrac{1}{2}kt = \tfrac{1}{2}(503.12) \leq |\mathcal{C}| \leq (q+1)+(q+1+72\sqrt{q}) = 1026+72\sqrt{512},$$

a contradiction. Now, in the same way as for $q \notin \{256,512\}$, the curve $\mathcal{C}$ contains a linear component over the ground field $\mathbf{F}_q$. If $q = 512$ and if (6.80) is not satisfied for $m > 4$, then the procedure is as in the case $q \notin \{256,512\}$.

Finally, let $q = 256$. Since $2 < t < \tfrac{1}{2}\sqrt{q}+\tfrac{3}{4}$ with t even, then, from the bound $4 < m < t-1$, it follows that $t = 8$ and $m \in \{5,6\}$. If $t = 8, m = 5$, and $\mathcal{C}$ does not contain a line, then

$$\tfrac{1}{2}kt = \tfrac{1}{2}(251.8) \le |\mathcal{C}| \le (q+1+12\sqrt{q}) + (q+1+2\sqrt{q}) = 738,$$

a contradiction; if $t = 8$, $m = 6$, and $\mathcal{C}$ does not contain a line, then

$$\tfrac{1}{2}kt = \tfrac{1}{2}(251.8) \le |\mathcal{C}| \le (q+1+20\sqrt{q}) + (q+1) = 834,$$

also a contradiction. As for $q \notin \{256, 512\}$, the curve $\mathcal{C}$ contains a linear component over the ground field $\mathbf{F}_q$. If $q = 256$ and if there is no absolutely irreducible component with $m > 4$, then proceed as in the case $q \notin \{256, 512\}$ with $\alpha = 0$ and $\alpha + \beta = \beta \ge 2$. □

Theorem 6.69. *If* $q - \tfrac{1}{2}\sqrt{q} + \tfrac{9}{4} < k < q + 1$ *and* k *is odd, then* Φ *contains a plane as component over* $\mathbf{F}_q$ *or consists of* $(q + 3 - k)/2$ *hyperbolic quadrics over* $\mathbf{F}_q$.

Proof. If Φ contains a plane ξ as component, then, for each plane π_i of $\mathcal{K}$, the line $\xi \cap \pi_i$ is an S-line; so ξ contains at least k lines over $\mathbf{F}_q$ and consequently ξ is defined over $\mathbf{F}_q$. Suppose therefore that Φ does not contain a linear component. From the proof of Theorem 6.66, there is at least one plane which does not contain a line of Φ. Let π be a plane for which $\Phi \cap \pi = \mathcal{C}$ consists of $t/2$ conics over $\mathbf{F}_q$. First it is shown that no two of these conics coincide. Therefore suppose that at least two of the conics do coincide. Then

$$\tfrac{1}{2}k(q + 3 - k) \le |\mathcal{C}| \le \tfrac{1}{2}(q + 1 - k)(q + 1).$$

So, either

$$k \le q + 2 - (2q + 3)^{1/2}$$

or

$$k \ge q + 2 + (2q + 3)^{1/2}.$$

This contradicts $q - \tfrac{1}{2}\sqrt{q} + \tfrac{9}{4} < k < q + 1$. Hence the $t/2$ conics are distinct. The number of points common to at least two of these conics is at most

$$4(t/2)(t/2 - 1)/2 = \tfrac{1}{2}t(t - 2) < \tfrac{1}{2}(\tfrac{1}{2}\sqrt{q} + \tfrac{3}{4})(\tfrac{1}{2}\sqrt{q} - \tfrac{5}{4}) < q/8.$$

Let $\mathcal{C}_1$ be one of the conics and let P be a point of π, with $P \notin \mathcal{C}_1$, and P distinct from the nucleus of $\mathcal{C}_1$. Then there is at least one line ℓ of π through P which neither contains a point over $\mathbf{F}_q$ of $\mathcal{C}_1$ nor contains a point over $\overline{\mathbf{F}}_q$ common to at least two conics. For this line ℓ, each point of $\ell \cap \Phi$ is a simple point for Φ. Over $\mathbf{F}_q$,

$$|\ell \cap \Phi| \le t - 2 < \tfrac{1}{2}\sqrt{q} - \tfrac{5}{4}.$$

Since Φ does not contain a linear component, at each point of $\ell \cap \Phi$ the tangent plane of Φ contains at most t different lines of Φ. Hence each point of $\ell \cap \Phi$ is contained in at most t lines of Φ. Consequently, the number of planes π' through ℓ for which $\pi' \cap \Phi$ contains a line as component over $\mathbf{F}_q$ is at most

$$(t - 2)t < (\tfrac{1}{2}\sqrt{q} - \tfrac{5}{4})(\tfrac{1}{2}\sqrt{q} + \tfrac{3}{4}) = \tfrac{1}{4}q - \tfrac{1}{4}\sqrt{q} - \tfrac{15}{16} < \tfrac{1}{4}q - \tfrac{1}{4}\sqrt{q}.$$

It follows that, for more than $q+1-\frac{1}{4}q+\frac{1}{4}\sqrt{q} > \frac{3}{4}q+\frac{1}{4}\sqrt{q}$ planes π' through ℓ, the curve $\Phi \cap \pi'$ consists over $\mathbf{F}_q$ of $t/2$ conics. Hence, over $\overline{\mathbf{F}}_q$, the two conjugate points in the set $\ell \cap C_1$ are contained in more than $(3q+\sqrt{q})/4$ conics of Φ, all defined over $\mathbf{F}_q$, lying in different planes through ℓ. Let $C_1, C_2, \ldots, C_r$ be these conics, and let ξ_j be the plane of C_j.

For any S-line ℓ_i, let t_i be the number of conics of $\{C_1, C_2, \ldots, C_r\} = V$ containing at least (and so exactly one) point of ℓ_i. The number of points of $\xi_j \cap \Phi$ not belonging to an S-line is at most

$$\tfrac{1}{2}t(q+1) - tk/2 = t(t-2)/2 < \tfrac{1}{2}(\tfrac{1}{2}\sqrt{q} + \tfrac{3}{4})(\tfrac{1}{2}\sqrt{q} - \tfrac{5}{4}) < \tfrac{1}{8}q - \tfrac{1}{8}\sqrt{q}.$$

Hence the number N_j of points of C_j belonging to an S-line satisfies

$$N_j > q+1 - \tfrac{1}{8}q + \tfrac{1}{8}\sqrt{q} > \tfrac{7}{8}q + \tfrac{1}{8}\sqrt{q}.$$

As each point of Φ is on zero or two S-lines, it now follows that

$$\sum_i t_i > \tfrac{1}{4}(3q + \sqrt{q}) \cdot \tfrac{1}{8}(7q + \sqrt{q}) \cdot 2 = \tfrac{1}{16}(21q^2 + 10q\sqrt{q} + q). \tag{6.81}$$

The number of S-lines is equal to $(q+3-k)k$. Since the function $f(x) = (q+3-x)x$ is strictly decreasing for $x \geq (q+3)/2$, so

$$(q+3-k)k < (\tfrac{1}{2}\sqrt{q} + \tfrac{3}{4})(q - \tfrac{1}{2}\sqrt{q} + \tfrac{9}{4})$$
$$= \tfrac{1}{2}q\sqrt{q} + \tfrac{1}{2}q + \tfrac{3}{4}\sqrt{q} + \tfrac{27}{16}. \tag{6.82}$$

From (6.81) and (6.82), it now follows that

$$\bar{t} = (\textstyle\sum t_i)/\{(q+3-k)k\}$$
$$> \{21q^2 + 10q\sqrt{q} + q\}/\{8q\sqrt{q} + 8q + 12\sqrt{q} + 27\}$$
$$> (21\sqrt{q})/8 - 2.$$

Hence there is an S-line ℓ' which has a point in common with more than $21\sqrt{q}/8 - 2$ conics of the set V, say with $C_1, C_2, \ldots, C_s$.

The common points of $C_1, C_2, \ldots, C_s$ are denoted by Q and Q'. Recall that Q, Q' are conjugate points over $\overline{\mathbf{F}}_q$. For $i = 2, 3$, let R_i be the common point of ℓ' and C_i and let m_2, m'_2 be the tangent lines of C_2 at the respective points Q, Q'. The absolutely irreducible quadric (over $\mathbf{F}_q$) containing C_1, R_2, R_3 and having m_2, m'_2 as tangent lines is denoted by $\mathcal{Q}$. Since $R_2 \in \mathcal{Q}$ and since the tangent lines m_2, m'_2 of C_2 are tangent lines of $\mathcal{Q}$, the conic C_2 belongs to $\mathcal{Q}$. Since ℓ' has at least three points in common with $\mathcal{Q}$, it also belongs to $\mathcal{Q}$.

The common point of ℓ' and C_i is denoted by R_i, and the tangent lines of C_i at the points Q, Q' are denoted by m_i, m'_i, with $i = 1, 2, \ldots, s$. The tangent plane of Φ at Q is $m_1 m_2$ and the tangent plane of $\mathcal{Q}$ at Q' is $m'_1 m'_2$. Therefore, the tangent lines $m_i = m_1 m_2 \cap \xi_i$ and $m'_i = m'_1 m'_2 \cap \xi_i$ of C_i, are also tangent lines of $\mathcal{Q}, i = 3, 4, \ldots, s$. Since also $R_i \in \mathcal{Q}$, the conic C_i belongs to $\mathcal{Q}$, $i = 3, 4, \ldots, s$.

Consequently, the s conics $C_1, C_2, \ldots, C_s$ belong to Q. As $2s > 21\sqrt{q}/4 - 4 > 2t$, so $Q \subset \Phi$ by Bézout's theorem.

Instead of C_1 take any other conic of $\Phi \cap \pi$. It then follows that Φ consists of $t/2$ absolutely irreducible quadrics over $\mathbf{F}_q$. For any plane $\pi_i \in \mathcal{K}$, the curve $\pi_i \cap \Phi$ consists of t different S-lines and so necessarily π_i contains exactly two different lines of any of the $t/2$ quadrics. It follows that any of the quadrics contains at least $2k > 2q - \sqrt{q} + \frac{9}{2}$ lines, and hence is hyperbolic. Therefore Φ either contains a plane as component over $\mathbf{F}_q$ or consists of $(q + 3 - k)/2$ hyperbolic quadrics over $\mathbf{F}_q$. $\square$

Theorem 6.70. *Any k-arc $\mathcal{K}$ of $\Sigma = \mathrm{PG}(3, q)$, with q even, k odd, and such that $q - \frac{1}{2}\sqrt{q} + \frac{9}{4} < k < q + 1$, can be extended to a $(k + 1)$-arc.*

Proof. The hypotheses imply that $q \geq 64$. By Theorem 6.69, Φ contains a plane as component over $\mathbf{F}_q$ or consists of $(q + 3 - k)/2$ hyperbolic quadrics over $\mathbf{F}_q$.

If Φ contains a plane as component over $\mathbf{F}_q$, then, by Theorem 6.59, $\mathcal{K}$ can be extended to a $(k + 1)$-arc. Now suppose that Φ consists of $t/2$ hyperbolic quadrics. By Theorem 6.59(v), the arc $\mathcal{K}$ is complete. Consider a k-arc of planes $\mathcal{K} = \{\pi_1, \ldots, \pi_k\}$.

Let Δ_1, Δ_2 be distinct hyperbolic quadrics algebraically contained in Φ. The k planes π_i are tangent planes of Δ_1 and Δ_2. Using any correlation θ of Σ, consider the situation that a k-arc $\mathcal{K}\theta$ of points of $\mathrm{PG}(3, q)$ lies on the intersection $\Psi_1 \cap \Psi_2 = C$ of the two quadrics Ψ_1, Ψ_2, where $\Psi_i = \Delta_i\theta$, $i = 1, 2$. The extension of the curve C to the algebraic closure $\overline{\mathbf{F}}_q$ is denoted by $\bar{C}$.

There are three possible cases.

(I) *$\bar{C}$ contains as a component a line or a conic but not an irreducible cubic curve*

In this case, for any k-arc in C the result is that $k \leq 8$. However, since

$$q + 1 > k > q - \tfrac{1}{2}\sqrt{q} + \tfrac{9}{4}$$

and k is odd, so $k \geq 63$ and a contradiction is obtained.

(II) *$\bar{C}$ factors into a twisted cubic curve C' and a line ℓ*

In this case $|C' \cap \mathcal{K}\theta| \geq 61$; so the points of C' in $\mathrm{PG}(3, q)$ form a $(q + 1)$-arc $\mathcal{K}'$. put $\mathcal{K}'' = \mathcal{K}\theta \cap C'$; then $|\mathcal{K}''| \geq k - 2 > q - \frac{1}{2}\sqrt{q} + \frac{1}{4}$. Since $q \geq 64$ so $|\mathcal{K}''| > q - \frac{1}{2}\sqrt{q} + \frac{1}{4} > (q + 4)/2$. By the dual of Theorem 6.59(vi), all points of $\mathcal{K}\theta$ lie on the $(q + 1)$-arc $\mathcal{K}'$. By duality, $\mathcal{K}$ itself lies on a $(q + 1)$-arc and is not complete, a contradiction.

(I) *$\bar{C}$ is an irreducible quartic*

Let π be any plane of Σ not containing P_1 where $\pi_i\theta = P_i$, $1 \leq i \leq k$, and where P_1 is non-singular for $\bar{C}$, noting that $\bar{C}$ has at most one singular point. Projecting C from P_1 onto π gives an irreducible cubic C' over $\mathbf{F}_q$ in π. If, for $i > 1$, the meet $P_1 P_i \cap \pi = \{Q_i\}$, then $\{Q_2, \ldots, Q_k\} = \mathcal{L}$ is a $(k - 1)$-arc of points of π contained in the curve C'.

First suppose that C' has genus 1. Then, from the Hasse–Weil formula, Corollary 2.27 of PGOFF2, $|C'| \leq q + 1 + 2\sqrt{q}$. For a given $i > 1$, at least

$$(k - 2) - \{|C'| - (k - 1)\}$$

lines $Q_i Q_j$, $1 < j \neq i$, contain exactly two points of C'. So, at points of $\mathcal{L}$ the curve C' has at least $(2k - 3 - |C'|)(k - 1)/2 = F(k)$ distinct tangents. From the Hasse–Weil formula, $F(k) \geq (2k - 3 - q - 2\sqrt{q} - 1)(k - 1)/2 = G(k)$. Therefore $G(k) \leq q + 1 + 2\sqrt{q}$. Since $k > q - \frac{1}{2}\sqrt{q} + \frac{9}{4}$, so $8q^2 - 28q\sqrt{q} + 10q - 64\sqrt{q} - 11 < 0$, a contradiction as $q \geq 64$.

Next suppose that C' has genus 0. Then, as for cubics in Section 11.4 of PGOFF2, $|C'| \leq q + 2$. For a given $i > 1$, at least $(k-2) - \{|C'| - (k-1)\}$ lines $Q_i Q_j$, $1 < j \neq i$, contain exactly two points of C'. Since at most one point of $\mathcal{L}$ is singular for C', the curve C' has at least $(k - 2)(2k - 4 - |C'|)/2 = F(k)$ distinct tangents at points of $\mathcal{L}$ which are simple for C'. As $|C'| \leq q + 2$ so $F(k) \geq (k - 2)(2k - 4 - q - 2)/2 = G(k)$. Since C' has at most $q + 1$ simple points, $G(k) \leq q + 1$; hence

$$(k - 2)(2k - q - 6) \leq 2q + 2.$$

Since $k > q - \frac{1}{2}\sqrt{q} + \frac{9}{4}$, it follows that $8q^2 - 12q\sqrt{q} - 22q + 4\sqrt{q} - 19 < 0$, a contradiction.

From (I), (II), (III), Φ does not consist of $t/2$ hyperbolic quadrics; so $\mathcal{K}$ extends to a $(k + 1)$-arc. □

Theorem 6.71. *Let $\mathcal{K}$ be any k-arc of points or planes in $\mathrm{PG}(3, q)$, q even and $q \neq 2$. If $k > q - \frac{1}{2}\sqrt{q} + \frac{9}{4}$, then $\mathcal{K}$ can be completed to a $(q + 1)$-arc $\mathcal{L}$, which is uniquely determined by $\mathcal{K}$.*

Proof. Assume that $q - \frac{1}{2}\sqrt{q} + \frac{9}{4} < k < q + 1$. By Theorems 6.67 and 6.70, the k-arc $\mathcal{K}$ is not complete and so it extends to a $(k + 1)$-arc $\mathcal{K}'$. If $k + 1 = q + 1$, the result is proved. If $k + 1 < q + 1$ then, since $k + 1 > q - \frac{1}{2}\sqrt{q} + \frac{9}{4}$, the arc $\mathcal{K}'$ extends to a $(k+2)$-arc $\mathcal{K}''$. Proceeding in this way, $\mathcal{K}$ can be extended to a $(q+1)$-arc $\mathcal{L}$. By Theorem 6.59(vi), $\mathcal{L}$ is uniquely determined by $\mathcal{K}$ since $q - \frac{1}{2}\sqrt{q} + \frac{9}{4} > (q + 4)/2$. □

Before proceeding to n dimensions, it is necessary to consider the analogue of Theorem 6.59 in $\mathrm{PG}(4, q)$. Then, using the result of Theorem 6.49 that a $(q + 1)$-arc in $\mathrm{PG}(4, q)$ is a normal rational curve, Theorem 6.38 can be applied.

First, Theorems 6.57 and 6.58 are restated for $n = 4$.

Theorem 6.72. *Let $\mathcal{K} = \{\pi_1, \ldots, \pi_k\}$ be a k-arc of solids in $\Sigma = \mathrm{PG}(4, q)$, with $q = 2^s$. For i, j, m distinct, let Z_{ijm} denote the set of $t = q + 4 - k$ points on the line $\pi_i \cap \pi_j \cap \pi_m$ that lie on no other solid of $\mathcal{K}$. Then*

(i) *there exists a curve $C_{ij} = C_{ji}$ of degree t in the plane $\pi_i \cap \pi_j$ such that the intersection $C_{ij} \cap \pi_m = Z_{ijm}$;*

(ii) *for fixed i, the algebraic curves C_{ij} are algebraically contained in an algebraic surface Φ_i of degree t in π_i with $\Phi_i \cap \pi_j = C_{ij}$;*

(iii) *all the algebraic surfaces* Φ_i *are algebraically contained in a hypersurface* $\Phi = \Phi(\mathcal{K})$ *for which* $\Phi(\mathcal{K}) \cap \pi_i = \Phi_i$;

(iv) *if* $k > (q+6)/2$, *the hypersurface* $\Phi(\mathcal{K})$ *is unique*;

(v) *if* $\mathcal{L} = \mathcal{K} \cup \{\pi_{k+1}, \ldots, \pi_u\}$ *is an arc of solids with* $u > k$ *and if* $k > (q+6)/2$, *the hypersurface* $\Phi(\mathcal{K})$ *factors into* $\Phi(\mathcal{L}), \pi_{k+1}, \ldots, \pi_u$;

(vi) *if* $k > (q+6)/2$, *there is a bijection between solids of* Σ *extending* $\mathcal{K}$ *to a* $(k+1)$-*arc and linear components over* $\mathbf{F}_q$ *of* Φ;

(vii) *if* $k > (q+6)/2$, *the arc* $\mathcal{K}$ *is contained in a unique complete arc of* $\mathrm{PG}(4,q)$.

Theorem 6.73. *Let* $\mathcal{K}$ *be a* k-*arc of solids in* $\Sigma = \mathrm{PG}(4,q)$, $q = 2^h$, *with cardinality* $k > q - \frac{1}{2}\sqrt{q} + \frac{13}{4}$ *and* $t = q + 4 - k$. *Then*

(i) Φ_i *factors over* $\mathbf{F}_q$ *into* $t - 2$ *planes* $\alpha_{i1}, \alpha_{i2}, \ldots, \alpha_{i,t-2}$, *called* S-*planes, and a hyperbolic quadric* Ψ_i, *called an* S-*quadric*;

(ii) *the* $t - 2$ S-*planes in* π_i *form an arc of planes*;

(iii) *in* $\pi_i \cap \pi_j$, $i \neq j$, *there are exactly two lines* ℓ_{ij} *and* m_{ij} *which are lines of* Ψ_i;

(iv) *also, in* $\pi_i \cap \pi_j$, $i \neq j$, *the lines* ℓ_{ij} *and* m_{ij} *together with the* $t - 2$ *lines* $\pi_j \cap \alpha_{i1}, \ldots, \pi_j \cap \alpha_{i,t-2}$ *and the* $k - 2$ *lines* $\pi_i \cap \pi_j \cap \pi_u$, $u \in \mathbf{N}_k \backslash \{i, j\}$ *form a* $(q+2)$-*arc of lines*;

(v) *each plane* α_{ij} *contains two lines of* Ψ_i;

(vi) *in an* S-*plane* α_{is} *of* π_i, *the lines* $\alpha_{is} \cap \alpha_{iu}$, $s \neq u$, *the lines* $\alpha_{is} \cap \pi_j$, $i \neq j$, *and the two lines of* Ψ_i *in* α_{is} *form a* $(q+2)$-*arc of lines*;

(vii) *the* $2(q+1)$ *generators of* Ψ_i *are the* $2(k-1)$ *lines of* $\Psi_i \cap \pi_j$, $j \neq i$, *and the* $2(t-2)$ *lines of* $\Psi_i \cap \alpha_{is}$;

(viii) *at most two members of the set* $\mathcal{V}$ *of the* $(t-2)k + k$ *planes* α_{ij} *and surfaces* Ψ_i *contain any line of* Σ;

(ix) *for any* S-*plane* α_{is} *of* π_i *and any point* P *of* α_{is}, *there are at most two* S-*planes containing* P *and meeting* α_{is} *in a line*.

Proof. By Theorem 6.40, $k \leq q + 1$ for $q > 4$; for $q = 4$, also $k \leq 6$ from Section 6.1. Since $k > q - \frac{1}{2}\sqrt{q} + \frac{13}{4}$, it follows that $q \geq 32$; also $k > (q+6)/2$.

(i), (ii) Since $\mathcal{K}$ is an arc, for a fixed i, the $k - 1$ planes $\pi_i \cap \pi_j, i \neq j$, form an arc $\mathcal{M}$ of planes in π_i. Since $k > q - \frac{1}{2}\sqrt{q} + \frac{13}{4}$, so $k - 1 = |\mathcal{M}| > k > q - \frac{1}{2}\sqrt{q} + \frac{9}{4}$. By Theorem 6.71 $\mathcal{M}$ is embedded in a $(q+1)$-arc $\mathcal{L}$ of planes in π_i. The planes of $\mathcal{L}\backslash\mathcal{M}$ are the S-planes in π_i. Since $\mathcal{L}$ is a $(q+1)$ arc, then from the structure of $\Phi(\mathcal{L})$ the S-quadric in π_i as in Theorem 6.61(ii) is obtained.

(iii) This follows from Theorem 6.61(iii).

(iv)–(vii) These all follow from (iii) and (iv) of Theorem 6.61.

(viii) Since the S-planes in a given solid π_i form an arc of planes in π_i, no three S-planes of a given solid have a common line. From (vi), it cannot be that two S-planes in π_i and the S-quadric Ψ_i of π_i all contain a common line.

Let m be any line of Σ lying in an S-plane σ_1. Then σ_1 lies in a unique solid of $\mathcal{K}$, say π_i; so σ_1 is one of the S-planes α_{is} in π_i. Suppose that m lies in some other S-plane σ_2 not in π_i. Then σ_2 is in π_j, say, with $j \neq i$. Now $\pi_i \cap \pi_j$ is a plane π_{ij} containing m. In π_i, the plane α_{is} meets π_{ij} in exactly one line: this must be m. From (iv), there are no other S-planes or S-quadrics containing m. A similar argument handles the case that m lies in α_{is} and in Ψ_j, $j \neq i$, or lies in Ψ_i and in Ψ_j, $j \neq i$.

In summary, if m lies in an element of $\mathcal{V}$ from π_i and in an element of $\mathcal{V}$ from π_j, $j \neq i$, then m lies in exactly two elements of $\mathcal{V}$. If the only elements of $\mathcal{V}$ containing m lie in a given arc solid, say π_i, then again m lies in at most two elements from $\mathcal{V}$.

(ix) An argument similar to that proving (viii) applies.   □

Theorem 6.74. Let $\mathcal{K} = \{\pi_1, \ldots, \pi_k\}$ be a k-arc of solids in $\Sigma = \mathrm{PG}(4, q)$, for $q = 2^h$, with $q > k > q - \frac{1}{2}\sqrt{q} + \frac{13}{4}$. Then $\mathcal{K}$ can be extended to a q-arc.

Proof. Since $q > k > q - \frac{1}{2}\sqrt{q} + \frac{13}{4}$, so $q \geq 128$ and $k > (q + 6)/2$. By way of contradiction, assume that $\mathcal{K}$ is complete. Since $k > q - \frac{1}{2}\sqrt{q} + \frac{13}{4}$, the results of the previous theorem apply. Any S-plane of π_j, $j > 1$, meets π_1 in a line of $\pi_1 \cap \pi_j$. This line lies either in an S-plane of π_1 or in the quadric Ψ_1. Now, there are exactly two lines of Ψ_1 in $\pi_1 \cap \pi_j$. Therefore, putting $t = q + 4 - k$, the number of S-planes not in π_1 and having a line in common with some S-plane in π_1 is at least $(k-1)(t-2) - 2(k-1) = (k-1)(t-4)$. Recall that each solid contains exactly $t - 2$ S-planes. Hence there exists an S-plane $\alpha = \alpha_{1s}$ in π_1 having a line in common with at least $(k - 1)(t - 4)/(t - 2)$ S-planes not in π_1. Thus the total number of S-planes having at least one line in common with α is at least $(k - 1)(t - 4)/(t - 2) + (t - 2)$.

Denote the set of such planes by $V = \{\beta_1, \beta_2, \ldots\}$ with

$$\{\beta_1, \ldots, \beta_{t-2}\} = \{\alpha_{11}, \ldots, \alpha_{1,t-2}\}, \quad \beta_1 = \alpha_{11} = \alpha,$$

and $\beta_i \cap \alpha = \ell_i$ for $\beta_i \neq \alpha$. Since $k < q$, so $t > 4$; hence $(k - 1)(t - 4)/(t - 2) \geq 1$. Let the line ℓ_{t-1} of α lie in π_j, say, $j \neq 1$. Put

$$V_j = \{\alpha_{j1}, \ldots, \alpha_{j,t-2}, \Psi_j\} = \{\delta_{j1}, \ldots, \delta_{j,t-1}\}.$$

Let ρ_n be the number of planes of V containing a line of δ_{jn}. Since any two S-planes of V_j meet in a line and any S-plane of V_j contains a line (actually two lines) of Ψ_j by Theorem 6.73(v),

$$\sum \rho_n \geq (k - 1)(t - 4)/(t - 2) + (t - 2) - 1 + (t - 1).$$

Averaging gives $\bar{\rho} \geq [(k - 1)(t - 4)/(t - 2) + 2t - 4]/(t - 1)$. So there exists an element δ_{jn} for which

$$\rho_n \geq [(k - 1)(t - 4)/(t - 2) + 2t - 4]/(t - 1).$$

Note that, in obtaining $\sum \rho_n$, the fact that each plane of V not in π_j meets exactly one element of V_j in a line is used.

Now two cases are considered.

Case 1: $t < [(k - 1)(t - 4)/(t - 2) + 2t - 4]/(t - 1)$.

Then $\delta_{jn} = \Psi_j$ or an S-plane of π_j. Assume that $\delta_{jn} = \Psi_j$. The S-plane α in π_1 meets $\pi_1 \cap \pi_j$ in a unique line ℓ_{t-1} and ℓ_{t-1} lies on a unique S-plane η_j

of π_j. Now, Ψ_j meets $\pi_1 \cap \pi_j$ in exactly two lines ℓ, ℓ' both different from ℓ_{t-1}. The lines ℓ, ℓ', ℓ_{t-1} are part of an arc of lines in $\pi_1 \cap \pi_j$ by Theorem 6.73(iv). So $\Psi_j \cap \alpha = \{\ell_{t-1} \cap \ell, \ell_{t-1} \cap \ell'\}$. Any S-plane containing a line of Ψ_j and a line of α must pass through $\ell_{t-1} \cap \ell$ or $\ell_{t-1} \cap \ell'$. One such S-plane is η_j. So, by Theorem 6.73(ix), there are at most three S-planes altogether meeting Ψ_j and α each in lines. Therefore $3 \geq (k-1)(t-4)/(t-2) + (2t-4) > t$; so $3 > t$, contradicting the fact that $t \geq 5$ since $t = q + 4 - k$ and $k < q$ by hypothesis.

Next assume that δ_{jn} is an S-plane, say α_{j1}. If α_{j1} is the unique S-plane η_j of π_j containing ℓ_{t-1} then the planes α and α_{j1} lie in a Π_3. From Theorem 6.73(viii), ℓ_{t-1} lies in no other S-plane. Then any other S-plane containing a line of α and α_{j1} lies in this Π_3. Therefore this Π_3 contains more than t S-planes and so is a linear component of Φ. From Theorem 6.72(vi), this Π_3 is a solid of Σ extending $\mathcal{K}$. This contradicts the initial assumption that $\mathcal{K}$ is complete. Therefore α_{j1} is distinct from η_j; so $\alpha_{j1} \cap \alpha$ is a point P. Any S-plane containing a line of α_{j1} and α contains P. It follows that there are more than t S-planes containing P and intersecting α in a line, contradicting Theorem 6.73(ix) as $t > 2$.

Case 2: $t \geq [(k-1)(t-4)/(t-2) + 2t - 4]/(t-1)$.

This means that

$$t^3 - 4t^2 - (q-3)t + 4q + 4 \geq 0. \tag{6.83}$$

From the hypothesis, since $t = q + 4 - k$,

$$4 < t < \tfrac{1}{2}\sqrt{q} + \tfrac{3}{4}. \tag{6.84}$$

For $t = 5, 6, 7$, the inequality (6.83) implies that $q \leq 32$, a contradiction in each case. So

$$8 \leq t < \tfrac{1}{2}\sqrt{q} + \tfrac{3}{4}. \tag{6.85}$$

Rewriting (6.83) gives

$$(t^2 - q)t + 4q \geq 4t^2 - 3t - 4. \tag{6.86}$$

For $t \geq 2$, the right-hand side of (6.86) is positive. However, from (6.85),

$$\begin{aligned}
(t^2 - q)t + 4q &< ((\tfrac{1}{2}\sqrt{q} + \tfrac{3}{4})^2 - q)t + 4q \\
&= \tfrac{3}{4}(\sqrt{q} + \tfrac{3}{4} - q)t + 4q \\
&\leq 6(\sqrt{q} + \tfrac{3}{4} - q) + 4q \\
&= -2q + 6\sqrt{q} + \tfrac{9}{2} \\
&< 0.
\end{aligned}$$

This gives the desired contradiction.

It has thus been shown that $\mathcal{K}$ extends to a $(k+1)$-arc $\mathcal{K}'$. If $k + 1 < q$, then since $k + 1 > q - \tfrac{1}{2}\sqrt{q} + \tfrac{13}{4}$, the arc $\mathcal{K}'$ extends to a $(k+2)$-arc $\mathcal{K}''$. This process shows that $\mathcal{K}$ can be extended to a q-arc. $\qquad\square$

Using the notation above, the following result is shown.

Theorem 6.75. *Any q-arc of solids in $\Sigma = \mathrm{PG}(4,q)$, $q = 2^h$, $q \geq 64$, can be extended to a $(q+1)$-arc.*

Proof. Let $\mathcal{K} = \{\pi_1, \ldots, \pi_q\}$ be a q-arc of solids in Σ. By way of contradiction, assume that $\mathcal{K}$ is complete. The number of S-planes not in π_1 and having a line in common with some S-plane in π_1, or with Ψ_1, is exactly $(q-1)(t-2) = 2(q-1)$. So one of α_{11}, α_{12} or Ψ_1 has a line in common with at least $2(q-1)/3$ of these S-planes.

There are two possibilities:

(a) α_{11} or α_{12} has a line in common with at least $2(q-1)/3$ of these S-planes;
(b) Ψ_1 has a line in common with at least $2(q-1)/3$ of these S-planes.

In case (b), there are two further possibilities.

(I) *For each solid $\pi_i, i > 1$, at most one of the planes α_{i1}, α_{i2} contains a line of Ψ_1*

Then at least $q-1$ of the planes α_{i1}, α_{i2}, $i > 1$, contain a line of one of α_{i1}, α_{i2}. So at least one of α_{i1}, α_{i2} contains a line of at least $\frac{1}{2}(q-1)$ S-planes not contained in π_1.

(II) *There is a solid $\pi_i, i > 1$, for which the two S-planes α_{i1} and α_{i2} contain a line of Ψ_1*

Then at least $(2(q-1)/3 - 2)/3$ S-planes not in π_1 and π_i contain a line of Ψ_1 and one of $\alpha_{i1}, \alpha_{i2}, \Psi_i$. From Theorem 6.73, in $\pi_1 \cap \pi_i$, the intersection $\Phi(\mathcal{K}) \cap \pi_1 \cap \pi_i$ contains exactly four lines $\ell_1, \ell_2, \ell_3, \ell_4$ with no three concurrent. Also, suppose that ℓ_1 and ℓ_2 lie in Ψ_1 and that ℓ_3 and ℓ_4 lie in Ψ_2. So Ψ_1 meets Ψ_2 in four points no three of which are collinear. An analysis shows that through each of these four points there is at most one S-plane not in π_1 or π_i and having a line in common with Ψ_1 and Ψ_i. Since $(2(q-1)/3 - 2)/3 = (2q-8)/9$ is larger than 4, at least $(2q-8)/9$ S-planes not in π_1 and π_i contain a line of Ψ_1 and one of α_{i1}, α_{i2}.

Each solid of $\mathcal{K}$ contains two S-planes. Since

$$\min\{2(q-1)/3, (q-1)/2, (2q-8)/9\} + 2 = (2q+10)/9,$$

it follows from (a) and (b) that there exists an S-plane $\alpha = \alpha_{j1}$ having a line in common with at least $(2q+10)/9$ S-planes. Let the set of these S-planes be denoted by $V = \{\beta_1, \beta_2, \ldots\}$, with $\{\beta_1, \beta_2\} = \{\alpha_{j1}, \alpha_{j2}\}$, $\beta_1 = \alpha_{j1} = \alpha$; also denote $\beta_i \cap \alpha$ by ℓ_i, when $\beta_i \neq \alpha$. Note that $(2q-8)/9 > 0$. Then let ℓ_3 lie in π_g, $g \neq j$. Put $V_g = \{\alpha_{g1}, \alpha_{g2}, \Psi_g\} = \{\delta_{g1}, \delta_{g2}, \delta_{g3}\}$. Let ρ_n be the number of planes of V containing a line of δ_{gn}. Then, as in the previous theorem,

$$\sum \rho_n \geq (2q+10)/9 - 1 + 3 = (2q+28)/9.$$

Averaging gives $\bar{\rho} \geq (2q+28)/27$. Therefore there is an element δ_{jn} for which $\rho_n \geq (2q+28)/27$. Note that $t = 4 < (2q+28)/27$.

Suppose, for example, that $\delta_{gn} = \Psi_g$. Recall that α meets $\pi_j \cap \pi_g$ in just one line ℓ_3 which, by assumption, lies in exactly one of the planes δ_{g1} or δ_{g2}. Therefore Ψ_g has no line in common with α but meets α in exactly two points lying on ℓ_3. Then, from Theorem 6.73(ix), there are at most three S-planes intersecting Ψ_g and α in a line one of which is the S-plane of π_g through ℓ_3. So $(2q + 28)/27 \le 3$ and $q \le 16$, contradicting $q \ge 64$. Therefore δ_{gn} is an S-plane, say α_{g1}. If α_{g1} is the (unique) S-plane of π_g containing ℓ_3 then the planes α_{g1} and α are contained in a solid Π_3. Since, from Theorem 6.73(viii), ℓ_3 is in just two S-planes, this solid contains at least $(2q + 28)/27$ S-planes. Since $(2q + 28)/27 > 4 = t$ and $\Phi(\mathcal{K})$ has degree $t = 4$, the solid Π_3 is a linear component of $\Phi(\mathcal{K})$. From Theorem 6.72(vi), this solid extends $\mathcal{K}$, contradicting that $\mathcal{K}$ is complete. Therefore α_{g1} does not contain the line ℓ_3; so $\alpha_{j1} \cap \alpha$ is a point P. Any S-plane containing a line of α_{j1} and α contains P. It follows that there are at least $(2q + 28)/27$ S-planes containing P and intersecting α in a line, contradicting Theorem 6.73(ix).   □

Theorem 6.76. *Let $\mathcal{K}$ be a k-arc of points in $\mathrm{PG}(4, q)$, $q = 2^h$, $q \ne 2$. If also $k > q - \frac{1}{2}\sqrt{q} + \frac{13}{4}$, then $\mathcal{K}$ can be completed uniquely to a $(q + 1)$-arc that is a normal rational curve.*

Proof. Since $k > q - \frac{1}{2}\sqrt{q} + \frac{13}{4}$, so $q \ne 4$. By Theorem 6.40, $k \le q + 1$ for $q > 4$; so $k > q - \frac{1}{2}\sqrt{q} + \frac{13}{4}$ implies that $q \ge 32$ and $k \ge 33$. If $q > q - \frac{1}{2}\sqrt{q} + \frac{13}{4}$, then $q \ge 64$. From the previous two theorems, $\mathcal{K}$ lies in a $(q + 1)$-arc, which is complete by Theorem 6.40 and which is a normal rational curve by Theorem 6.49.   □

This gives the climactic result of this section.

Theorem 6.77. *In $\mathrm{PG}(n, q)$, $q = 2^h$, $q \ne 2$, $n \ge 4$,*

(i) *if $\mathcal{K}$ is a k-arc with $k > q - \frac{1}{2}\sqrt{q} + n - \frac{3}{4}$, then $\mathcal{K}$ lies on a unique normal rational curve;*

(ii) *if $q > (2n - \frac{7}{2})^2$, every $(q + 1)$-arc is a normal rational curve;*

(iii) *if $q > (2n - \frac{11}{2})^2$, then $m(n, q) = q + 1$.*

Proof. (i) This follows by induction from Theorems 6.38 and 6.76.

(ii) $q + 1 > q - \frac{1}{2}\sqrt{q} + n - \frac{3}{4} \Leftrightarrow q > (2n - \frac{7}{2})^2$.

(iii) This follows from Theorem 6.39(ii) and part (ii).   □

Corollary 6.78. *In $\mathrm{PG}(n, q)$, $q = 2^h$, $q \ne 2$, $n > q - \frac{1}{2}\sqrt{q} - \frac{11}{4}$,*

(i) *if $\mathcal{K}$ is a k-arc with $k \ge n + 6$, then $\mathcal{K}$ lies on a unique normal rational curve;*

(ii) *if $n \le q - 5$, then every $(q + 1)$-arc is a normal rational curve;*

(iii) *if $n \le q - 4$, then $m(n, q) = q + 1$.*

Proof. These follow from Theorem 6.77, Theorems 6.33 and 6.35, and Corollary 6.37.   □

6.8 Notes and References

Section 6.1

The bound (6.1) is due to Bose [35]; it follows from Lemma 6.20. It is also immediate that a 2^n-cap in $PG(n, 2)$ is the complement of a hyperplane. The bound (6.2) is also due to Bose [35] and is discussed in Sections 8.1–8.2 of PGOFF2. The bound (6.3) is due to Bose [35] for q odd and to Qvist [268] for q even; see Section 16.1 of FPSOTD for the bound, and for the characterisation when q is odd or $q = 4$, and Section 16.4 of FPSOTD for another example of a $(q^2 + 1)$-cap when $q = 2^h$, h odd and $h \geq 3$. For $q = 8$, every $(q^2 + 1)$-cap is one of these two types, Fellagara [131].

For $q = 16$, every $(q^2 + 1)$-cap is an elliptic quadric, O'Keefe and Penttila [248]. The bound (6.4) is due to Pellegrino [261] and the bound (6.5) is due to Hill [166]; the classification in $PG(4, 3)$ is due to Hill [169] and in $PG(5, 3)$ is due to Hill [168]. For the bound in $PG(4, 4)$, see Tallini [305], Edel and Bierbrauer [128, 129], Bierbrauer and Edel [27].

The bounds (6.12) and (6.13) for $m'(2, q)$ are due to Segre [280] and proved in Chapter 10 of PGOFF2. The improvement from (6.12) to (6.11) is due to Thas [332]. The exact value for $m(2, q)$, q an even square, is due to Fisher, Hirschfeld and Thas [135] and to Boros and Szőnyi [34] independently.

The bound (6.15) is due to Voloch [392], and by similar methods (6.16) and (6.17) are due to Voloch [393]. The results (6.15)–(6.17) depend on an improvement to the Hasse–Weil theorem as in Section 10.2 of PGOFF2, which gives an upper bound on the number of points on a non-singular, irreducible, projective, algebraic curve with a fixed-point-free linear series. This result, due to Stöhr and Voloch [297], depends on q, on the genus g, on the order and dimension of the linear series, and on the Frobenius order sequence. For (6.18)–(6.23), see Hirschfeld, Korchmáros and Torres [176, Chapter 13].

Section 6.2

This is entirely based on Hill [168], apart from Theorem 6.15, which is due to Meshulam [236].

Section 6.3

The proof of Theorem 6.16 is taken from Tallini [302]. The remainder of the section is taken from Hirschfeld [172]; the essence of the argument is found in Segre [280]. The argument used to obtain the final result, Theorem 6.19, depends intricately on the upper bound used for $m'(2, q)$. This is both explicit in the proof of Theorem 6.19 and implicit in Theorem 6.17, on which the result heavily depends. Throughout, the bound (6.12) is used. The nature of the argument given precludes a formula or a bound for $m_2(n, q)$ in terms of $m'(2, q)$; a change in the bound for $m'(2, q)$ means a complete reworking of the argument. This can be done separately for the bounds

(6.11), (6.15), (6.16), (6.18)–(6.23). For example, if q is a sufficiently large prime and $n \geq 4$, then J. F. Voloch (personal communication) has shown that

$$m_2(n, q) \leq \tfrac{1983}{2025} q^{n-1} + O(q^{n-2}).$$

For many other bounds, see Hirschfeld and Storme [177].

Section 6.4

Lemma 6.23 and Theorem 6.24 are taken from Chao [71].

Hirschfeld and Thas [181] contains the details of Theorem 6.25. For Theorem 6.26, part (ii) is taken from Ferret and Storme [133], part (iii) from Cao and Ou [66] and part (i) from Thas (2015, unpublished). The proof of Theorem 6.27 proceeds as in the proof of Theorem 27.4.5 in GGG1 or Theorem 4.1 in Hirschfeld and Thas [181], using Chao's bound for $m_2'(3, q)$. Theorem 6.28(i) and (ii) is taken from Thas (2015, unpublished); Theorem 6.28(iii) comes from Cao and Ou [66]. Theorem 6.29(iii) is taken from Ferret and Storme [133]; parts (iv)–(v) come from Storme, Thas and Vereecke [301].

Section 6.5

The first part is an amalgam of Section 21.1 of FPSOTD and Segre [275]. The proof of Theorem 6.32(i) is based on Kaneta and Maruta [190]. The remainder of the section is taken from Thas [311], although this proof of Theorem 6.33 is taken from Halder and Heise [146].

Section 6.6

Theorem 6.40 is due to Segre [275] for q odd and to Casse [69] for q even. Theorem 6.41 is due to Thas [310], although the treatment here and hence the necessary Theorems 6.38 and 6.39 follow Kaneta and Maruta [190]; part (iii) is an improvement of Thas's result from $q > (4n - 5)^2$ to $(4n - 9)^2$. Theorem 6.42(ii) is due to Glynn [140]. Lemmas 6.44 to 6.48, Corollary 6.47, and Theorem 6.49 follow Casse and Glynn [70]. Included in this paper is an elementary proof of the result that a q-arc in $\mathrm{PG}(3, q)$, q even, is contained in a $(q + 1)$-arc; see also Kaneta and Maruta [189]. Theorem 6.50 implies that $m(5, q) = q + 1$ for q even and $q \geq 8$. Maruta and Kaneta [223] have shown, for q even and $q \geq 16$, that (i) in $\mathrm{PG}(3, q)$, a $(q - 1)$-arc is contained in a unique $(q + 1)$-arc; (ii) in $\mathrm{PG}(4, q)$, a q-arc is contained in a unique $(q + 1)$-arc; (iii) in $\mathrm{PG}(5, q)$, a $(q + 1)$-arc is a normal rational curve; (iv) $m(6, q) = q + 1$. These results are dependent on those in Section 6.7. Theorem 6.51(i) is due to Ball and De Beule [9], parts (ii) and (iii) to Ball [8].

Section 6.7

This is based on three papers: Bruen, Thas and Blokhuis [53], Blokhuis, Bruen and Thas [31], and Storme and Thas [300]. In [31], Theorems 6.54 and 6.56 are also applied to the case of q odd. This gives the following result analogous to Theorem 6.58.

Theorem 6.79. *Let $\mathcal{K}$ be a k-arc in $\mathrm{PG}(n, q)$ with $n \geq 3$ and q odd. If the cardinality $k > \frac{2}{3}(q - 1) + n$, then $\mathcal{K}$ is contained in a unique complete arc.*

Using other bounds for $m'(2, q)$ from Section 6.1, improvements of parts of Theorems 6.41 and 6.77 should be obtainable.

7

Ovoids, spreads and m-systems of finite classical polar spaces

7.1 Finite classical polar spaces

In this chapter, ovoids, spreads and m-systems of finite classical polar spaces are introduced. Also SPG-reguli, SPG-systems, BLT-sets and sets with the BLT-property are defined. The main results on these topics are given, all without proof.

There are five types of finite *classical polar spaces* $S = (\mathcal{P}, \mathcal{B})$.

(1) $\mathcal{W}_n(q)$: the elements of $\mathcal{P}$ are the points of $\mathrm{PG}(n, q)$, n odd and $n \geq 3$; the elements of $\mathcal{B}$ are the subspaces of the self-polar $(n-1)/2$-dimensional spaces of a null polarity of $\mathrm{PG}(n, q)$; the rank $r = (n+1)/2$.

(2) $\mathcal{P}(2n, q)$: the elements of $\mathcal{P}$ are the points of a non-singular quadric $\mathcal{P}_{2n}$ of $\mathrm{PG}(2n, q)$, $n \geq 2$; the elements of $\mathcal{B}$ are the subspaces of the $(n-1)$-dimensional spaces on $\mathcal{P}_{2n}$; the rank $r = n$.

(3) $\mathcal{H}(2n+1, q)$: the elements of $\mathcal{P}$ are the points of a non-singular hyperbolic quadric $\mathcal{H}_{2n+1}$ of $\mathrm{PG}(2n+1, q)$, $n \geq 1$; the elements of $\mathcal{B}$ are the subspaces of the n-dimensional spaces on $\mathcal{H}_{2n+1}$; the rank $r = n+1$.

(4) $\mathcal{E}(2n+1, q)$: the elements of $\mathcal{P}$ are the points of a non-singular elliptic quadric $\mathcal{E}_{2n+1}$ of $\mathrm{PG}(2n+1, q)$, $n \geq 2$; the elements of $\mathcal{B}$ are the subspaces of the $(n-1)$-dimensional spaces on $\mathcal{E}_{2n+1}$; the rank $r = n$.

(5) $\mathcal{U}(n, q^2)$: the elements of $\mathcal{P}$ are the points of a non-singular Hermitian variety $\mathcal{U}_n$ of $\mathrm{PG}(n, q^2)$, $n \geq 3$; when n is odd, the elements of $\mathcal{B}$ are the subspaces of the $\frac{1}{2}(n-1)$-dimensional spaces on $\mathcal{U}_n$ and the rank $r = \frac{1}{2}(n+1)$; when n is even, the elements of $\mathcal{B}$ are the subspaces of the $(\frac{1}{2}n - 1)$-dimensional spaces on $\mathcal{U}_n$ and the rank $r = n/2$.

Definition 7.1. For a polar space S of rank r, the subspaces of dimension $r-1$ are the *generators* of S.

Theorem 7.2. *With $\mathcal{G}(S)$ the set of generators of the finite classical polar space S, the numbers of points and generators are given in* Table 7.1.

© Springer-Verlag London 2016
J.W.P. Hirschfeld, J.A. Thas, *General Galois Geometries*, Springer Monographs in Mathematics, DOI 10.1007/978-1-4471-6790-7_7

Table 7.1. Classical polar spaces

$\mathcal{S}$	$|\mathcal{P}|$	$|\mathcal{G}(\mathcal{S})|$
$\mathcal{W}_n(q)$	$(q^{n+1} - 1)/(q - 1)$	$(q + 1)(q^2 + 1) \cdots (q^{(n+1)/2} + 1)$
$\mathcal{P}(2n, q)$	$(q^{2n} - 1)/(q - 1)$	$(q + 1)(q^2 + 1) \cdots (q^n + 1)$
$\mathcal{H}(2n + 1, q)$	$(q^n + 1)(q^{n+1} - 1)/(q - 1)$	$2(q + 1)(q^2 + 1) \cdots (q^n + 1)$
$\mathcal{E}(2n + 1, q)$	$(q^n - 1)(q^{n+1} + 1)/(q - 1)$	$(q^2 + 1)(q^3 + 1) \cdots (q^{n+1} + 1)$
$\mathcal{U}(2n, q^2)$	$(q^{2n+1} + 1)(q^{2n} - 1)/(q^2 - 1)$	$(q^3 + 1)(q^5 + 1) \cdots (q^{2n+1} + 1)$
$\mathcal{U}(2n + 1, q^2)$	$(q^{2n+2} - 1)(q^{2n+1} + 1)/(q^2 - 1)$	$(q + 1)(q^3 + 1) \cdots (q^{2n+1} + 1)$

7.2 Ovoids and spreads of finite classical polar spaces

Definition 7.3. Let $\mathcal{S}$ be a finite classical polar space of rank $r \geq 2$.

(1) An *ovoid* $\mathcal{O}$ of $\mathcal{S}$ is a point set that meets every generator in exactly one point.
(2) A *spread* $\mathcal{T}$ of $\mathcal{S}$ is a set of generators that partitions the point set of $\mathcal{S}$.

Theorem 7.4. *The sizes of ovoids $\mathcal{O}$ and spreads $\mathcal{T}$ are given in* Table 7.2.

Table 7.2. Sizes of ovoids and spreads

$\mathcal{S}$	$|\mathcal{O}| = |\mathcal{T}|$
$\mathcal{W}_n(q)$	$q^{(n+1)/2} + 1$
$\mathcal{P}(2n, q)$	$q^n + 1$
$\mathcal{H}(2n + 1, q)$	$q^n + 1$
$\mathcal{E}(2n + 1, q)$	$q^{n+1} + 1$
$\mathcal{U}(2n, q^2)$	$q^{2n+1} + 1$
$\mathcal{U}(2n + 1, q^2)$	$q^{2n+1} + 1$

Definition 7.5. The number of points of a hypothetical ovoid $\mathcal{O}$ of $\mathcal{S}$ is the *ovoid number* of $\mathcal{S}$.

7.3 Existence of ovoids

The existence and non-existence of ovoids $\mathcal{O}$ in a finite classical polar space $\mathcal{S}$ is described in Table 7.3.

Any ovoid of $\mathrm{PG}(3, q)$, q even, is an ovoid of some $\mathcal{W}_3(q)$ and, conversely.

Table 7.3. Existence of ovoids

$\mathcal{S}$	Existence of $\mathcal{O}$	References
$\mathcal{W}_3(q)$, q even	Yes	[312]
$\mathcal{W}_3(q)$, q odd	No	[312]
$\mathcal{W}_n(q)$, $n = 2t + 1$ and $t > 1$	No	[328]
$\mathcal{P}(4, q)$	Yes	[213], [191], [259, 260], [350], [263]
$\mathcal{P}(6, q)$, q prime, $q > 3$	No	[10]
$\mathcal{P}(6, q)$, $q = 3^h$	Yes	[326, 328], [191]
$\mathcal{P}(2n, q)$, $n > 2$ and q even	No	[328]
$\mathcal{P}(2n, q)$, $n > 3$ and q odd	No	[145]
$\mathcal{H}(3, q)$	Yes	
$\mathcal{H}(5, q)$	Yes	Table 15.10 of FPSOTD
$\mathcal{H}(7, q)$, q odd, with q prime	Yes	[125], [326, 328]
or $q \equiv 0$ or $2 \pmod 3$		[192], [193], [191], [287], [74], [238] (∗)
$\mathcal{H}(7, q)$, q even	Yes	[326, 328]
$\mathcal{H}(2n + 1, q)$, $n > 3$, $q = p^h$, p prime and $p^n > \binom{2n+p}{2n+1} - \binom{2n+p-2}{2n+1}$	No	[33]
$\mathcal{E}(2n + 1, q)$, $n > 1$	No	[328]
$\mathcal{U}(3, q^2)$	Yes	[329], [259, 260], [350]
$\mathcal{U}(2n, q^2)$, $n \geq 2$	No	[328]
$\mathcal{U}(2n + 1, q^2)$, $n > 1$, $q = p^h$, p prime and $p^{2n+1} > \binom{2n+p}{2n+1}^2 - \binom{2n+p-1}{2n+1}^2$	No	[239]
$\mathcal{U}(5, 4)$	No	[83]

7.4 Existence of spreads

The existence and non-existence of spreads $\mathcal{T}$ in a finite classical polar space $\mathcal{S}$ is described in Table 7.4.

Table 7.4. Existence of spreads

S	Existence of T	References
$\mathcal{W}_n(q)$, $n = 2t + 1$ and $t \geq 1$	Yes	[213], [323], [191], [5], [350], [263]
$\mathcal{P}(2n, q)$, $n \geq 2$ and q even	Yes	[125], [326, 328], [350]
$\mathcal{P}(6, q)$, q odd, with q prime or $q \equiv 0$ or $2 \pmod{3}$	Yes	See ($*$) in Table 7.3
$\mathcal{P}(4n, q)$, q odd	No	[312], [334]
$\mathcal{H}(3, q)$	Yes	
$\mathcal{H}(7, q)$, q odd, with q prime or $q \equiv 0$ or $2 \pmod{3}$	Yes	See ($*$) in Table 7.3
$\mathcal{H}(4n + 3, q)$, q even	Yes	[125], [326, 328]
$\mathcal{H}(4n + 1, q)$	No	
$\mathcal{E}(5, q)$	Yes	[329], [259, 260], [350]
$\mathcal{E}(2n + 1, q)$, $n > 2$, q even	Yes	[125], [326, 328]
$\mathcal{U}(4, 4)$	No	
$\mathcal{U}(2n + 1, q^2)$	No	[328], [334]

A spread of $\mathcal{W}_n(q)$, $n = 2t + 1$, is also a t-spread of $\mathrm{PG}(n, q)$. For every $n = 2t + 1$, the polar space $\mathcal{W}_n(q)$ has a spread that is also a regular t-spread of $\mathrm{PG}(n, q)$.

Any non-singular hyperbolic quadric of $\mathrm{PG}(2n + 1, q)$, $n \geq 1$, has two families of generators; see Section 1.4. If π, π' are generators, then they belong to the same family if and only if the dimension of their intersection has the same parity as n. It follows that $\mathcal{H}(4n + 1, q)$ has no spread.

7.5 Open problems

For ovoids, establish the existence or non-existence in the following cases:

(a) $\mathcal{P}(6, q)$, q odd, $q \neq 3^h$ and q not prime;
(b) $\mathcal{H}(7, q)$, q odd, $q \equiv 1 \pmod{3}$ and q not prime;
(c) $\mathcal{H}(2n + 1, q)$, $n > 3$, $q = p^h$, p prime and

$$p^n \leq \binom{2n + p}{2n + 1} - \binom{2n + p - 2}{2n + 1};$$

(d) $\mathcal{U}(2n + 1, q^2)$, $(n, q) \neq (2, 2)$, $q = p^h$, p prime and

$$p^{2n+1} \leq \binom{2n + p}{2n + 1}^2 - \binom{2n + p - 1}{2n + 1}^2.$$

For spreads, establish the existence or non-existence in the following cases:

(a) $\mathcal{P}(6, q)$, q odd with $q \equiv 1 \pmod{3}$ and q not prime;
(b) $\mathcal{P}(4n + 2, q)$, q odd, for $n > 1$;
(c) $\mathcal{H}(7, q)$, q odd with $q \equiv 1 \pmod{3}$ and q not prime;
(d) $\mathcal{H}(4n + 3, q)$, q odd, for $n > 1$;
(e) $\mathcal{E}(2n + 1, q)$, $n > 2$, and q odd;
(f) $\mathcal{U}(4, q^2)$, for $q > 2$;
(g) $\mathcal{U}(2n, q^2)$, for $n > 2$.

7.6 m-systems and partial m-systems of finite classical polar spaces

Definition 7.6. Let S be a finite classical polar space of rank r, with $r \geq 2$.

(1) A *partial m-system* of S, with $0 \leq m \leq r - 1$, is any set $\{\pi_1, \pi_2, \ldots, \pi_k\}$ of m-dimensional subspaces of S such that no generator containing π_i has a point in common with $(\pi_1 \cup \pi_2 \cup \cdots \cup \pi_k) \backslash \pi_i$, with $i = 1, 2, \ldots, k$.
(2) A partial 0-system of size k is also called a *partial ovoid*, or a *cap*, or a *k-cap*.
(3) A partial $(r - 1)$-system is also called a *partial spread*.

Theorem 7.7. *An upper bound for the size of a partial m-system $\mathcal{M}$ of the classical polar space S is given by the following table:*

S	$|\mathcal{M}| \leq$
$\mathcal{W}_{2n+1}(q)$	$q^{n+1} + 1$
$\mathcal{P}(2n, q)$	$q^n + 1$
$\mathcal{H}(2n + 1, q)$	$q^n + 1$
$\mathcal{E}(2n + 1, q)$	$q^{n+1} + 1$
$\mathcal{U}(2n, q^2)$	$q^{2n+1} + 1$
$\mathcal{U}(2n + 1, q^2)$	$q^{2n+1} + 1$

Definition 7.8. A partial m-system $\mathcal{M}$ of the finite classical polar space S is an *m-system* if the upper bound in Theorem 7.7 is attained.

For $m = 0$, the m-system is an ovoid of S; for $m = r - 1$, with r the rank of S, the m-system $\mathcal{M}$ is a spread of S. The fact that $|\mathcal{M}|$ is independent of m explains why an ovoid and a spread of a finite classical polar space S have the same size.

Theorem 7.9. *Let $\mathcal{M}$ be a partial m-system of the finite classical polar space S of rank r with $m < r - 1$. Then the number $N_\mathcal{M}$ of $(m + 1)$-dimensional subspaces of S containing an element of $\mathcal{M}$ and a given point P of S not in an element of $\mathcal{M}$ is independent of the choice of P; it is given by the following table:*

$\mathcal{S}$	$N_{\mathcal{M}}$
$\mathcal{W}_{2n+1}(q)$	$q^{n-m}+1$
$\mathcal{P}(2n,q)$	$q^{n-m-1}+1$
$\mathcal{H}(2n+1,q)$	$q^{n-m-1}+1$
$\mathcal{E}(2n+1,q)$	$q^{n-m}+1$
$\mathcal{U}(2n,q^2)$	$q^{2n-2m-1}+1$
$\mathcal{U}(2n+1,q^2)$	$q^{2n-2m-1}+1$

Remark 7.10. If P is a point of $\mathcal{S}$ not in an element of the m-system $\mathcal{M}$, then, for $m < r - 1$ and $\mathcal{S} \neq \mathcal{W}_{2n+1}(q)$, Theorem 7.9 says that the tangent hyperplane of $\mathcal{S}$ at P contains exactly $N_{\mathcal{M}}$ elements of $\mathcal{M}$; for $m < r - 1$ and $\mathcal{S} = \mathcal{W}_{2n+1}(q)$, the hyperplane $P^{\perp}$, that is, $P\mathcal{I}$ with $\mathcal{I}$ the null polarity defining $\mathcal{S}$, contains exactly $N_{\mathcal{M}}$ elements of $\mathcal{M}$.

7.7 Intersections with hyperplanes and generators

In this section, $\mathcal{S}$ is a finite classical polar space of rank r, with $r \geq 2$, and $\mathcal{M}$ is an m-system of $\mathcal{S}$.

Theorem 7.11. *For $\mathcal{S} \neq \mathcal{W}_{2n+1}(q)$, let $R_{\mathcal{M}}$ be the number of elements of $\mathcal{M}$ contained in a hyperplane π which is not tangent to $\mathcal{S}$; for $\mathcal{S} = \mathcal{W}_{2n+1}(q)$, let $R_{\mathcal{M}}$ be the number of elements of $\mathcal{M}$ contained in a hyperplane $P^{\perp}$, with P a point not in an element of $\mathcal{M}$. This number is given in the following table:*

$\mathcal{S}$	$N_{\mathcal{M}} = R_{\mathcal{M}}$
$\mathcal{W}_{2n+1}(q)$	$q^{n-m}+1$
$\mathcal{P}(2n,q)$	$q^{n-m-1}+1$
and $\pi \cap \mathcal{P}(2n,q) = \mathcal{H}(2n-1,q)$	
$\mathcal{E}(2n+1,q)$	$q^{n-m}+1$
$\mathcal{U}(2n,q^2)$	$q^{2n-2m-1}+1$

Remark 7.12. 1. When $\mathcal{S} = \mathcal{P}(2n,q)$ and also $\pi \cap \mathcal{P}(2n,q) = \mathcal{E}(2n-1,q)$, then $R_{\mathcal{M}} = 0$ for $m = n - 1$, and $R_{\mathcal{M}}$ depends on the choice of π for $m < n - 1$.
2. When $\mathcal{S} = \mathcal{H}(2n+1,q)$, then $R_{\mathcal{M}} = 0$ for $m = n$, and $R_{\mathcal{M}}$ depends on the choice of π for $m < n$.
3. When $\mathcal{S} = \mathcal{U}(2n+1,q^2)$, then $R_{\mathcal{M}} = 0$ for $m = n$, and $R_{\mathcal{M}}$ depends on the choice of π for $m < n$.

Theorem 7.13. *For $\mathcal{S} = \mathcal{W}_{2n+1}(q), \mathcal{E}(2n+1,q), \mathcal{U}(2n,q^2)$, again $N_{\mathcal{M}} = R_{\mathcal{M}}$; that is, any hyperplane contains either one or $N_{\mathcal{M}}$ elements of $\mathcal{M}$. Hence the union $\widetilde{\mathcal{M}}$ of the elements of $\mathcal{M}$ has two intersection numbers β_1, β_2 with respect to hyperplanes:*

(i) *for $S = W_{2n+1}(q)$,*

$$\beta_1 = \frac{(q^{m+1} - 1)(q^n + 1)}{q - 1} - q^n, \quad \beta_2 = \frac{(q^{m+1} - 1)(q^n + 1)}{q - 1};$$

(ii) *for $S = \mathcal{E}(2n + 1, q)$,*

$$\beta_1 = \frac{(q^{m+1} - 1)(q^n + 1)}{q - 1} - q^n, \quad \beta_2 = \frac{(q^{m+1} - 1)(q^n + 1)}{q - 1};$$

(iii) *for $S = \mathcal{U}(2n, q^2)$,*

$$\beta_1 = \frac{(q^{2m+2} - 1)(q^{2n-1} + 1)}{q^2 - 1} - q^{2n-1}, \quad \beta_2 = \frac{(q^{2m+2} - 1)(q^{2n-1} + 1)}{q^2 - 1}.$$

Corollary 7.14. *For $S = W_{2n+1}(q), \mathcal{E}(2n + 1, q), \mathcal{U}(2n, q^2)$, any m-system defines a strongly regular graph and a linear projective two-weight code.*

Remark 7.15. For more details on these graphs and codes, see Section 7.10.

Theorem 7.16. *Let $\mathcal{M}$ be an m-system of the finite classical polar space S over $\mathbf{F}_q$, and let $\widetilde{\mathcal{M}}$ be the union of the elements of $\mathcal{M}$. Then, for any generator γ of S,*

$$|\gamma \cap \widetilde{\mathcal{M}}| = (q^{m+1} - 1)/(q - 1).$$

Definition 7.17. Let S be a finite, not necessarily classical, polar space of rank r, with $r \geq 2$. Hence S may be a non-classical generalised quadrangle. A point set $\mathcal{K}$ of $\mathcal{E}$ is a k-ovoid of S if each generator of S contains exactly k points of $\mathcal{K}$. A 1-ovoid is just an ovoid.

Corollary 7.18. *The union of all elements of an m-system of S over $\mathbf{F}_q$ is a k-ovoid with $k = (q^{m+1} - 1)/(q - 1)$.*

7.8 Bounds on partial m-systems and non-existence of m-systems

Notation 7.19. Write

(a) $\mathbf{c}(n, r) = \binom{n}{r}$;
(b) $\mathbf{b}(r_1, r_2, r_3, r_4) = \mathbf{c}(\mathbf{c}(r_1, r_2) + r_3, r_4)$;
(c) $\mathbf{b}'(m_1, m_2, m_3, m_4, m_5, m_6) = \mathbf{c}(\mathbf{c}(m_1, m_2) - \mathbf{c}(m_3, m_4) + m_5, m_6)$.

Theorem 7.20. *If $\mathcal{K}$ is a partial ovoid of size k of the finite classical polar space S in $\mathrm{PG}(n, q)$, with $q = p^h$ and p prime, then*

$$k \leq \mathbf{c}(p + n - 1, n)^h + 1. \tag{7.1}$$

(i) *If S comes from a quadric in $\mathrm{PG}(n,q)$, then (7.1) can be improved to*

$$k \le [\mathbf{c}(p+n-1,n) - \mathbf{c}(p+n-3,n)]^h + 1. \tag{7.2}$$

(ii) *If S arises from a quadric in $\mathrm{PG}(n,q)$ and if n and q are both even, then (7.2) can be improved to*

$$k \le n^h + 1.$$

(iii) *If $S = \mathcal{U}(n,q^2)$, with $q = p^h$, p prime, then (7.1) can be improved to*

$$k \le \left[\mathbf{c}(p+n-1,n)^2 - \mathbf{c}(p+n-2,n)^2\right]^h + 1.$$

Remark 7.21. Some results of Table 7.3 are deduced from Theorem 7.20.

Theorem 7.20 can be extended to m-systems of finite classical polar spaces. First, a useful theorem on subspaces of $\mathrm{PG}(n,q)$ is stated.

Theorem 7.22. *Consider in $\mathrm{PG}(n,q)$, $n \ge 2$, with $q = p^h$, p prime, a set of m-dimensional subspaces $\pi_1, \pi_2, \ldots, \pi_k$ and a set of $(n-m-1)$-dimensional subspaces $\pi_1', \pi_2', \ldots, \pi_k'$, with $m \le (n-1)/2$, where $\pi_i \cap \pi_i' \ne \emptyset$ and $\pi_j \cap \pi_i' = \emptyset$ for all $i,j = 1,2,\ldots,k$ with $i \ne j$. Then*

$$k \le \mathbf{b}(n+1, m+1, p-2, p-1)^h + 1. \tag{7.3}$$

Theorem 7.23.   (i) *If $\mathcal{M}$ is a partial m-system of size k of the finite classical polar space S in $\mathrm{PG}(n,q)$, with $q = p^h$, p prime, then*

$$k \le \mathbf{b}(n+1, m+1, p-2, p-1)^h + 1. \tag{7.4}$$

(a) *For $S = \mathcal{P}(n,q)$, $\mathcal{H}(n,q)$, $\mathcal{E}(n,q)$, with q odd, and for $S = \mathcal{W}_n(q)$, with q and m odd, the inequality (7.4) can be improved to*

$$k \le (\mathbf{b}(n+1, m+1, p-2, p-1) - \mathbf{b}(n+1, m+1, p-4, p-3))^h + 1.$$

(b) *When $S = \mathcal{U}(n,q^2)$, then (7.4) can be improved to*

$$k \le (\mathbf{b}(n+1, m+1, p-2, p-1)^2 - \mathbf{b}(n+1, m+1, p-3, p-2)^2)^h + 1.$$

(ii) *If S admits an m-system, then the following hold:*
    (a) *for $S = \mathcal{W}_{2n+1}(q)$, $q = p^h$ and with m even if p is odd,*

$$p^{n+1} \le \mathbf{b}(2n+2, m+1, p-2, p-1);$$

(b) *for $S = \mathcal{W}_{2n+1}(q)$, $q = p^h$ odd and m odd,*

$$p^{n+1} \le \mathbf{b}(2n+2, m+1, p-2, p-1) - \mathbf{b}(2n+2, m+1, p-4, p-3);$$

(c) *for $S = \mathcal{P}(2n,q)$, $q = 2^h$,*

$$2^n \le \mathbf{c}(2n+1, m+1);$$

(d) *for* $\mathcal{S} = \mathcal{H}(2n+1, q)$, $q = 2^h$,

$$2^n \le \mathbf{c}(2n+2, m+1);$$

(e) *for* $\mathcal{S} = \mathcal{E}(2n+1, q)$, $q = 2^h$,

$$2^{n+1} \le \mathbf{c}(2n+2, m+1);$$

(f) *for* $\mathcal{S} = \mathcal{P}(2n, q)$, $q = p^h$ *and* q *odd*,

$$p^n \le \mathbf{b}(2n+1, m+1, p-2, p-1) - \mathbf{b}(2n+1, m+1, p-4, p-3);$$

(g) *for* $\mathcal{S} = \mathcal{H}(2n+1, q)$, $q = p^h$ *and* q *odd*,

$$p^n \le \mathbf{b}(2n+2, m+1, p-2, p-1) - \mathbf{b}(2n+2, m+1, p-4, p-3);$$

(h) *for* $\mathcal{S} = \mathcal{E}(2n+1, q)$, $q = p^h$ *and* q *odd*,

$$p^{n+1} \le \mathbf{b}(2n+2, m+1, p-2, p-1) - \mathbf{b}(2n+2, m+1, p-4, p-3);$$

(i) *for* $\mathcal{S} = \mathcal{U}(2n, q^2)$, $q = p^h$,

$$p^{2n+1} \le \mathbf{b}(2n+1, m+1, p-2, p-1)^2 - \mathbf{b}(2n+1, m+1, p-3, p-2)^2;$$

(j) *for* $\mathcal{S} = \mathcal{U}(2n+1, q^2)$, $q = p^h$,

$$p^{2n+1} \le \mathbf{b}(2n+2, m+1, p-2, p-1)^2 - \mathbf{b}(2n+2, m+1, p-3, p-2)^2.$$

Theorem 7.24.   (i) *Let* $\mathcal{M}$ *be a partial m-system of size k of* $\mathcal{W}_{2n+1}(q)$, $q = p^h$ *and* $m > 0$.

(a) *For p odd with m even and for $p = 2$,*

$$k \le \mathbf{b}'(2n+2, m+1, 2n+2, m-1, p-2, p-1)^h + 1.$$

(b) *For p odd with m odd,*

$$k \le [\mathbf{b}'(2n+2, m+1, 2n+2, m-1, p-2, p-1)$$
$$- \mathbf{b}'(2n+2, m+1, 2n+2, m-1, p-4, p-3)]^h + 1.$$

(c) *If $\mathcal{M}$ is a partial m-system of size k of $\mathcal{P}(2n, q)$, with $m > 0$ and $q = 2^h$, then*

$$k \le [\mathbf{c}(2n, m+1) - \mathbf{c}(2n, m-1)]^h + 1.$$

(ii) *If $\mathcal{W}_{2n+1}(q)$ admits an m-system, $q = p^h$ and $m > 0$, then*
(a) *for p odd with m even and for $p = 2$,*

$$p^{n+1} \le \mathbf{b}'(2n+2, m+1, 2n+2, m-1, p-2, p-1);$$

(b) *for p odd with m odd,*

$$p^{n+1} \le \mathbf{b}'(2n+2, m+1, 2n+2, m-1, p-2, p-1)$$
$$-\mathbf{b}'(2n+2, m+1, 2n+2, m-1, p-4, p-3).$$

(iii) *If $P(2n,q)$ with q even admits an m-system, with $m > 0$, then*

$$2^n \le \mathbf{c}(2n, m+1) - \mathbf{c}(2n, m-1).$$

The bound (7.3) has been improved and, as a corollary, a better bound for partial m-systems has been obtained. However, this formula is more complicated than that of (7.4).

Theorem 7.25. *If $\mathcal{M}$ is a partial m-system of size k of the finite classical polar space S in $\mathrm{PG}(n, q)$, with $q = p^h$ and p prime, then*

$$k \le 1 + \prod_{j=0}^{h-1} \sum_{i=0}^{K} (-1)^i \binom{n+1}{i} \binom{n + (p-1)(m+1) - ip}{n}, \qquad (7.5)$$

where $K = \left\lfloor \frac{(m+1)(p-1)}{p} \right\rfloor$ and $\lfloor s \rfloor$ is the integer part of s.

Remark 7.26. For $m = 0$, the inequality (7.5) becomes (7.1).

7.9 m'-systems arising from a given m-system

Here the constructions of m'-systems arising from a given m-system are surveyed.

Definition 7.27. The *0-system* (or *spread, ovoid*) of

$$\mathcal{P}_2, \ \mathcal{E}_3, \ \mathcal{H}_1, \ \mathcal{U}_2, \ \mathcal{W}_1 = \mathrm{PG}(1, q)$$

is the set of all their points.

Notation 7.28. In the context of polar spaces,

$$\mathcal{P}(2, q) = \mathcal{P}_2,$$
$$\mathcal{E}(3, q) = \mathcal{E}_3,$$
$$\mathcal{H}(1, q) = \mathcal{H}_1,$$
$$\mathcal{U}(2, q) = \mathcal{U}_2.$$

The sets $\mathcal{P}_2, \mathcal{E}_3, \mathcal{H}_1, \mathcal{U}_2, \mathcal{W}_1$ are also *polar spaces of rank* 1 or *projective index* 0.

Theorem 7.29. (i) *If $\mathcal{E}(2n+1, q)$, $n \ge 1$, has an m-system, then also $\mathcal{P}(2n+2, q)$ and $\mathcal{H}(2n+3, q)$ have m-systems; if $\mathcal{P}(2n+2, q)$, $n \ge 0$, has an m-system, then $\mathcal{H}(2n+3, q)$ also has an m-system.*

(ii) *If the polar space $\mathcal{S}_n$, $n \geq 4$, in $\mathrm{PG}(n, q)$ admits an ovoid, then the polar space $\mathcal{S}_{n-2}$ in $\mathrm{PG}(n-2, q)$, of the same type as $\mathcal{S}_n$, admits an ovoid.*

(iii) *If $\mathcal{H}(4n-1, q)$, $n \geq 1$, admits a spread, then $\mathcal{P}(4n-2, q)$ admits a spread; if $\mathcal{P}(2n, q)$, $n \geq 2$, admits a spread, then $\mathcal{E}(2n-1, q)$ admits a spread.*

Corollary 7.30. *The spaces $\mathcal{P}(4, q)$, $\mathcal{H}(5, q)$, $\mathcal{H}(3, q)$ each admit an ovoid.*

Proof. Put $(n, m) = (1, 0)$, $(0, 0)$ in Theorem 7.29(i). $\qquad\qquad\square$

Theorem 7.31. *For q even, the polar space $\mathcal{P}(2n, q)$, $n \geq 1$, has an m-system if and only if $\mathcal{W}_{2n-1}(q)$ has an m-system.*

Theorem 7.32. *Let S_1 and S_2 be spreads of $\mathcal{H}(7, q)$, where the generators of S_1 and the generators of S_2 belong to different systems of generators of $\mathcal{H}(7, q)$. Then,*

(i) *for each $\pi \in S_1$, there is exactly one $\pi' \in S_2$ with $\pi \cap \pi'$ a plane;*

(ii) *these $q^3 + 1$ planes $\pi \cap \pi'$ form a 2-system of $\mathcal{H}(7, q)$.*

Theorem 7.33. *If $\mathcal{H}(4n+3, q)$ admits a 2n-system, $n \geq 0$, then it admits a spread. This spread is obtained by considering all generators of a given system of generators of $\mathcal{H}(4n+3, q)$ containing an element of the 2n-system.*

Theorem 7.34. (i) *If $\mathcal{P}(2n, q^2)$, with $n \geq 1$ and q odd, admits an m-system $\mathcal{M}$, then $\mathcal{H}(4n+1, q)$ admits a $(2m+1)$-system $\mathcal{M}'$.*

(ii) *If $\mathcal{P}(2n, q^2)$, with $n \geq 1$ and q even, admits an m-system $\mathcal{M}$, then $\mathcal{P}(4n, q)$, and hence also $\mathcal{H}(4n+1, q)$, admits a $(2m+1)$-system $\mathcal{M}'$.*

(iii) *If $\mathcal{E}(2n+1, q^e)$, with $n \geq 1$, admits an m-system $\mathcal{M}$, then $\mathcal{E}(2e(n+1)-1, q)$ admits an $(me+e-1)$-system $\mathcal{M}'$.*

Corollary 7.35. (i) *There exists a 1-system in $\mathcal{H}(5, q)$ and $\mathcal{H}(9, q)$.*

(ii) *For q even, there is a spread in $\mathcal{P}(4, q)$.*

(iii) *There is an $(e-1)$-system in $\mathcal{E}(4e-1, q)$.*

Proof. (i), (ii) Put $m = 0$ and $n \in \{1, 2\}$ in Theorem 7.34(i).
  (iii) Put $m = 0$ and $n = 1$ in Theorem 7.34(ii). $\qquad\qquad\square$

Theorem 7.36. *If $\mathcal{W}_{2n-1}(q^e)$, $n \geq 1$, admits an m-system $\mathcal{M}$, then $\mathcal{W}_{2ne-1}(q)$ admits an $(me+e-1)$-system $\mathcal{M}'$.*

Corollary 7.37. *The space $\mathcal{W}_{2e-1}(q)$ admits an $(e-1)$-system.*

Proof. Put $(n, m) = (1, 0)$ in Theorem 7.36. $\qquad\qquad\square$

Theorem 7.38. *If $\mathcal{U}(2n, q^{2e})$, with $n \geq 1$ and e odd, admits an m-system $\mathcal{M}$, then $\mathcal{U}(2ne+e-1, q^2)$ admits an $(me+e-1)$-system $\mathcal{M}'$.*

Corollary 7.39. *The space $\mathcal{U}(3e-1, q^2)$, with e odd, admits an $(e-1)$-system.*

Proof. Put $(n, m) = (1, 0)$ in Theorem 7.38. $\qquad\qquad\square$

Theorem 7.40. (i) *If $\mathcal{U}(2n, q^2)$, $n \geq 1$, admits an m-system $\mathcal{M}$, then $\mathcal{E}(4n+1, q)$ admits a $(2m + 1)$-system $\mathcal{M}'$.*
(ii) *If $\mathcal{U}(2n+1, q^2)$ admits an m-system $\mathcal{M}$, then $\mathcal{H}(4n+3, q)$ admits a $(2m+1)$-system $\mathcal{M}'$.*

Corollary 7.41. *Both $\mathcal{E}(5, q)$ and $\mathcal{H}(7, q)$ admit 1-systems.*

Proof. Put $(n, m) = (1, 0)$ in Theorem 7.40.   □

Theorem 7.42. *If $\mathcal{U}(2n, q^2)$, $n \geq 1$, admits an m-system $\mathcal{M}$, then $\mathcal{W}_{4n+1}(q)$ admits a $(2m + 1)$-system $\mathcal{M}'$.*

Corollary 7.43. *$\mathcal{W}_5(q)$ admits a 1-system.*

Proof. Put $(n, m) = (1, 0)$ in Theorem 7.42.   □

Remark 7.44. Many infinite classes of examples can be constructed using the results of this section.

7.10 m-systems, strongly regular graphs and linear projective two-weight codes

Let $\mathcal{V} \subset \mathrm{PG}(n, q)$, with $n \geq 2$, such that, for any hyperplane π,

$$|\pi \cap \mathcal{V}| \in \{\beta_1, \beta_2\},$$

with $\beta_1 \neq \beta_2$.

Let $\mathrm{PG}(n, q)$ be embedded in $\mathrm{PG}(n + 1, q)$; then two distinct points P_1, P_2 of $\mathrm{PG}(n+1, q) \backslash \mathrm{PG}(n, q)$ are *adjacent* if the line $P_1 P_2$ contains a point of $\mathcal{V}$. With this adjacency, $\mathrm{PG}(n + 1, q) \backslash \mathrm{PG}(n, q)$ becomes a strongly regular graph.

Let $\mathcal{V} = \{P_1, P_2, \ldots, P_s\}$, with $P_i = \mathbf{P}(x_i)$, $x_i = (x_{i0}, x_{i1}, \ldots, x_{in})$ and $i = 1, 2, \ldots, s$. Then the matrix

$$G = [x_1^* \ x_2^* \ \cdots \ x_s^*]$$

generates a linear projective $[s, n + 1]_q$ code C whose words can only have weights $s - \beta_1$ and $s - \beta_2$. Recall that a linear code C is projective if and only if any two columns of a generator matrix are linearly independent, that is, if and only if the minimum weight of the dual code $C^{\perp}$ is at least 3. Conversely, any linear projective two-weight $[s, n+1]_q$ code C defines a set $\mathcal{V}$ of s points in $\mathrm{PG}(n, q)$, which has two intersection numbers with respect to hyperplanes.

By Theorem 7.13 for any m-system $\mathcal{M}$ of the polar space $\mathcal{S}$, where $\mathcal{S}$ is one of $\mathcal{W}_{2n+1}(q)$, $\mathcal{E}(2n + 1, q)$, $\mathcal{U}(2n, q^2)$, $n \geq 1$, the union $\widetilde{\mathcal{M}}$ of the elements of $\mathcal{M}$ is a set with two intersection numbers β_1, β_2 with respect to hyperplanes.

Expressing that the strongly regular graph arising from an m-system of these polar spaces has $\lambda \geq 0$ gives the following result.

Theorem 7.45. *An m-system of $\mathcal{W}_{2n+1}(q)$, $\mathcal{E}(2n+1, q)$, $\mathcal{U}(2n, q^2)$, $n \geq 1$, satisfies $n \leq 2m + 1$.*

Corollary 7.46. *An m-system of $\mathcal{P}(2n + 2, q)$, with q even, satisfies $n \leq 2m + 1$.*

From m-systems, other sets with two intersection numbers can be constructed.

Theorem 7.47. *Let $\mathcal{M}_i$ be an m_i-system of $\mathcal{W}_{2n+1}(q)$, $i = 1, 2, \ldots, k$, for some integer $k > 1$. For $i = 1, 2, \ldots, k$, let*

$$a_i = \frac{(q^{m_i+1} - 1)(q^n + 1)}{q - 1}.$$

(i) *If, for all $i \neq j$, $\mathcal{M}_i$ and $\mathcal{M}_j$ are disjoint, that is, $\widetilde{\mathcal{M}}_i \cap \widetilde{\mathcal{M}}_j = \emptyset$, then the set $\widetilde{\mathcal{M}}_1 \cup \widetilde{\mathcal{M}}_2 \cup \cdots \cup \widetilde{\mathcal{M}}_k$ has two intersection numbers $a_1 + a_2 + \cdots + a_k$ and $a_1 + a_2 + \cdots + a_k - q^n$ with respect to hyperplanes in $\mathrm{PG}(2n + 1, q)$.*

(ii) *If $\mathcal{M}_i$ is covered by $\mathcal{M}_{i+1}$, that is, every element of $\mathcal{M}_i$ is a subspace of a unique element of $\mathcal{M}_{i+1}$, $i = 1, 2, \ldots, k - 1$, then*

(a) *if k is even, the set*

$$\mathcal{K} = (\widetilde{\mathcal{M}}_k \backslash \widetilde{\mathcal{M}}_{k-1}) \cup (\widetilde{\mathcal{M}}_{k-2} \backslash \widetilde{\mathcal{M}}_{k-3}) \cup \cdots \cup (\widetilde{\mathcal{M}}_2 \backslash \widetilde{\mathcal{M}}_1)$$

has two intersection numbers $a_k - a_{k-1} + a_{k-2} - a_{k-3} + \cdots + a_2 - a_1$ and $a_k - a_{k-1} + a_{k-2} - a_{k-3} + \cdots + a_2 - a_1 - q^n$ with respect to hyperplanes of $\mathrm{PG}(2n + 1, q)$;

(b) *if k is odd, the set*

$$\mathcal{K} = (\widetilde{\mathcal{M}}_k \backslash \widetilde{\mathcal{M}}_{k-1}) \cup (\widetilde{\mathcal{M}}_{k-2} \backslash \widetilde{\mathcal{M}}_{k-3}) \cup \cdots \cup (\widetilde{\mathcal{M}}_3 \backslash \widetilde{\mathcal{M}}_2) \cup \widetilde{\mathcal{M}}_1$$

has two intersection numbers $a_k - a_{k-1} + a_{k-2} - a_{k-3} + \cdots + a_3 - a_2 + a_1$ and $a_k - a_{k-1} + a_{k-2} - a_{k-3} + \cdots + a_3 - a_2 + a_1 - q^n$ with respect to hyperplanes of $\mathrm{PG}(2n + 1, q)$.

Theorem 7.48. *Let $\mathcal{M}_i$ be an m_i-system of $\mathcal{E}(2n + 1, q)$, $i = 1, 2, \ldots, k$, for some integer $k > 1$. For $i = 1, 2, \ldots, k$, let a_i be as in Theorem 7.47. Then the same conclusions hold.*

Theorem 7.49. *Let $\mathcal{M}_i$ be an m_i-system of $\mathcal{U}(2n, q^2)$, $i = 1, 2, \ldots, k$, for some integer $k > 1$. For $i = 1, 2, \ldots, k$, define*

$$a_i = \frac{(q^{2m_i+2} - 1)(q^{2n-1} + 1)}{q^2 - 1}.$$

(i) *If $\mathcal{M}_i$ and $\mathcal{M}_j$ are disjoint for all $i \neq j$, then the set $\widetilde{\mathcal{M}}_1 \cup \widetilde{\mathcal{M}}_2 \cup \cdots \cup \widetilde{\mathcal{M}}_k$ has two intersection numbers $a_1 + a_2 + \cdots + a_k$ and $a_1 + a_2 + \cdots + a_k - q^{2n-1}$ with respect to hyperplanes of $\mathrm{PG}(2n, q^2)$.*

(ii) *If $\mathcal{M}_i$ is covered by $\mathcal{M}_{i+1}$, $i = 1, 2, \ldots, k - 1$, then*

(a) *if k is even, the set*

$$\mathcal{K} = (\widetilde{\mathcal{M}}_k \setminus \widetilde{\mathcal{M}}_{k-1}) \cup (\widetilde{\mathcal{M}}_{k-2} \setminus \widetilde{\mathcal{M}}_{k-3}) \cup \cdots \cup (\widetilde{\mathcal{M}}_2 \setminus \widetilde{\mathcal{M}}_1)$$

has two intersection numbers $a_k - a_{k-1} + a_{k-2} - a_{k-3} + \cdots + a_2 - a_1$ and $a_k - a_{k-1} + a_{k-2} - a_{k-3} + \cdots + a_2 - a_1 - q^{2n-1}$ with respect to hyperplanes of $\mathrm{PG}(2n, q^2)$;
(b) *if k is odd, the set*

$$\mathcal{K} = (\widetilde{\mathcal{M}}_k \setminus \widetilde{\mathcal{M}}_{k-1}) \cup (\widetilde{\mathcal{M}}_{k-2} \setminus \widetilde{\mathcal{M}}_{k-3}) \cup \cdots \cup (\widetilde{\mathcal{M}}_3 \setminus \widetilde{\mathcal{M}}_2) \cup \widetilde{\mathcal{M}}_1$$

has two intersection numbers $a_k - a_{k-1} + a_{k-2} - a_{k-3} + \cdots + a_3 - a_2 + a_1$ and $a_k - a_{k-1} + a_{k-2} - a_{k-3} + \cdots + a_3 - a_2 + a_1 - q^{2n-1}$ with respect to hyperplanes of $\mathrm{PG}(2n, q^2)$.

Example 7.50. (a) There are no examples of an m_1-system $\mathcal{M}_1$ and an m_2-system $\mathcal{M}_2$ of the finite classical polar space $\mathcal{S} \in \{\mathcal{W}_{2n+1}(q), \mathcal{E}(2n+1, q), \mathcal{U}(2n, q^2)\}$ with $\widetilde{\mathcal{M}}_1 \cap \widetilde{\mathcal{M}}_2 = \emptyset$.

(b) For q even and for $s_1, s_2, \ldots, s_k, t$, with $s_1 < s_2 < \cdots < s_k$ where s_i divides s_{i+1} for $i = 1, 2, \ldots, k-1$, $s_k \le s$ and $st \ge s_k + 1$, there exists a chain of $(st - s_i - 1)$-systems $\mathcal{M}_i$ of $\mathcal{E}(2st - 1, q)$, $i = 1, 2, \ldots, k$, where each element $\mathcal{M}_i$ is covered by $\mathcal{M}_{i+1}$, $i = 1, 2, \ldots, k-1$.

7.11 m-systems and maximal arcs

Maximal arcs in $\mathrm{PG}(2, q)$ were defined in Section 12.1 of PGOFF2. For general planes of order q, not necessarily Desarguesian, *maximal arcs* are defined in a similar way. Translation planes of order q can be constructed from n-spreads of the space $\mathrm{PG}(2n + 1, q)$; see Section 4.7. Such a plane is Desarguesian if and only if the n-spread is regular.

Theorem 7.51. *Let $\overline{S}$ be a spread of the polar space $\mathcal{E}(2n+1, q)$ of $\mathrm{PG}(2n+1, q)$, with $n > 0$. Suppose there exists an n-spread $S = \{\pi_1, \pi_2, \ldots, \pi_{q^{n+1}+1}\}$ of $\mathrm{PG}(2n+1, q)$ such that*

$$\overline{S} = \{\mathcal{E}_{2n+1} \cap \pi_i \mid i = 1, 2, \ldots, q^{n+1} + 1\}.$$

Embed $\mathrm{PG}(2n+1, q)$ as a hyperplane in $\mathrm{PG}(2n+2, q)$ and choose some point P in $\mathrm{PG}(2n+2, q) \setminus \mathrm{PG}(2n+1, q)$. Let $\mathcal{K}$ be the set of all points not in $\mathrm{PG}(2n+1, q)$ of the cone $P\mathcal{E}_{2n+1}$. Then $\mathcal{K}$ is a maximal $(q^{2n+1} - q^{n+1} + q^n; q^n)$-arc in the projective translation plane $\pi(S)$ of order q^{n+1} determined by the n-spread S.

Remark 7.52. 1. Theorem 7.51 also holds if, for $n = 1$, the quadric $\mathcal{E}_3$ is replaced by any ovoid of $\mathrm{PG}(3, q)$.
2. For q odd, a pair $(S, \overline{S})$ as in Theorem 7.51 does not exist.

3. For S regular, the maximal arc $\mathcal{K}$ of the Desarguesian plane $\pi(S)$ can also be constructed as in Theorem 12.12 of PGOFF2.
4. For q even, the n-spread S is always a spread of the polar space $\mathcal{W}_{2n+1}(q)$ defined by the polar space $\mathcal{E}(2n+1,q)$. Also, $\overline{S}$ is an $(n-1)$-system of $\mathcal{W}_{2n+1}(q)$.
5. Let q be even, let $\mathcal{P}_{2n+2}$ be a non-singular quadric of $\mathrm{PG}(2n+2,q)$ and let $\mathcal{E}_{2n+1}$ be contained in $\mathcal{P}_{2n+2}$. If S^* is any spread of $\mathcal{P}(2n+2,q)$, then S^* induces a spread $\overline{S}$ of $\mathcal{E}(2n+1,q)$. By projection from the nucleus of $\mathcal{P}_{2n+2}$ onto the hyperplane $\mathrm{PG}(2n+1,q)$ containing $\mathcal{E}(2n+1,q)$, the spread S^* yields an n-spread S of $\mathrm{PG}(2n+1,q)$ with the desired property. All possible pairs $(S,\overline{S})$ are obtained in this way.

Theorem 7.53. (a) *Let $\mathcal{M}$ be an m-system of the polar space $\mathcal{W}_{2n+1}(q)$ in the space* $\mathrm{PG}(2n+1,q)$, $n > 0$.
(b) *Suppose there exists a spread S of $\mathcal{W}_{2n+1}(q)$ such that $\mathcal{M}$ is covered by S. Embed the space $\mathrm{PG}(2n+1,q)$ in $\mathrm{PG}(2n+2,q)$ and choose some point P in the difference $\mathrm{PG}(2n+2,q)\backslash\mathrm{PG}(2n+1,q)$.*
(c) *Let $\mathcal{K}$ be the set of all points not in $\mathrm{PG}(2n+1,q)$ of the cone $P\widetilde{\mathcal{M}}$, with $\widetilde{\mathcal{M}}$ the set of all points contained in elements of $\mathcal{M}$.*
  Then $\mathcal{K}$ is a maximal $(q^{n+m+2} - q^{n+1} + q^{m+1}; q^{m+1})$-arc in the projective translation plane $\pi(S)$ of order q^{n+1} determined by the n-spread S.

Remark 7.54. Consider an $(n-1)$-system of $\mathcal{E}(2n+1,q^s)$, q even, covered by a spread of the associated $\mathcal{W}_{2n+1}(q^s)$, as in Theorem 7.51. By Remark 7.52(4), this $(n-1)$-system is also an $(n-1)$-system of $\mathcal{W}_{2n+1}(q)$. It follows that there exists an $(ns-1)$-system of $\mathcal{W}_{2ns+2s-1}(q)$ covered by a spread of $\mathcal{W}_{2ns+2s-1}(q)$; see Theorem 7.36. However, the translation plane and maximal arc thus obtained are isomorphic to the original translation plane and maximal arc. So nothing new is constructed.

Consider a non-singular quadric $\mathcal{H}_{4n-1}$ of $\mathrm{PG}(4n-1,q)$, $n \geq 2$ and q even, and let $\mathcal{P}_{4n-2}$ be a non-singular parabolic quadric on $\mathcal{H}_{4n-1}$.

1. Let N be the nucleus of $\mathcal{P}_{4n-2}$. Project $\mathcal{P}_{4n-2}$ from N onto a hyperplane $\mathrm{PG}(4n-3,q)$ of the subspace $\mathrm{PG}(4n-2,q)$ containing $\mathcal{P}_{4n-2}$, with N not in $\mathrm{PG}(4n-3,q)$. Then the subspaces on $\mathcal{P}_{4n-2}$ are projected onto the subspaces of a polar space $\mathcal{W}_{4n-3}(q)$.
2. Let S be a spread of $\mathcal{W}_{4n-3}(q)$. To S there corresponds a spread S' of the quadric $\mathcal{P}(4n-2,q)$. The generators of a chosen system of generators of $\mathcal{H}_{4n-1}$, which contain the elements of S', constitute a spread $\hat{S}$ of $\mathcal{H}(4n-1,q)$.
3. Now considering a non-singular parabolic quadric $\mathcal{P}^*_{4n-2}$ on $\mathcal{H}_{4n-1}$ and intersecting it with the elements of $\hat{S}$ gives a spread $S^{*\prime}$ of $\mathcal{P}^*(4n-2,q)$.
4. Projecting again from the nucleus N^* of $\mathcal{P}^*_{4n-2}$ gives a spread S^* of some $\mathcal{W}^*_{4n-3}(q)$. Such a spread S^* is a *cousin* of S.
5. The projective translation plane $\pi(S^*)$ of order q^{2n-1} arising from S^* is a *cousin* of the projective translation plane $\pi(S)$.

Let S be a spread of $\mathcal{W}_{4n-3}(q)$, $n \geq 2$ and q even. Then four cases are distinguished:

(I) $N^* = N$;
(II) $N^* \neq N$, $N^* \in N\mathcal{I}$, with $\mathcal{I}$ the null polarity defined by $\mathcal{H}_{4n-1}$;
(III) $N^* \neq N$, and the line NN^* meets $\mathcal{H}_{4n-1}$ in two distinct points;
(IV) $N^* \neq N$ and $NN^* \cap \mathcal{H}_{4n-1} = \emptyset$.

For S regular, spreads S^* corresponding to different classes yield non-isomorphic translation planes $\pi(S^*)$ of order q^{2n-1}; spreads corresponding to the same class do not necessarily yield isomorphic translation planes. It follows that, for S regular and so $\pi(S)$ Desarguesian, and $N^* \neq N$, the plane $\pi(S^*)$ is always non-Desarguesian.

Theorem 7.55. *Let $\mathcal{M}$ be an m-system of a polar space $\mathcal{E}(4n - 3, q)$ in the space* $\mathrm{PG}(4n-3, q)$, $n \geq 2$ *and q even. Suppose that the associated polar space $\mathcal{W}_{4n-3}(q)$ admits a spread S such that $\mathcal{M}$ is covered by S. Then the m-system gives rise to maximal $(q^{2n+m} - q^{2n-1} + q^{m+1}; q^{m+1})$-arcs in at least q of the projective planes arising from the cousins of class* (IV) *of S.*

Corollary 7.56. *Let s, t be positive integers, with $t > 1$, such that st is odd. Then, for q even, there exist maximal $(q^{2st-s} - q^{st} + q^{st-s}; q^{st-s})$-arcs in at least q of the cousins of class* (IV) *of $\mathrm{PG}(2, q^{st})$.*

Remark 7.57. When $s = 1$, the maximal arcs of Corollary 7.56 are the maximal arcs described in Theorem 7.51. However, for $s > 1$, the maximal arcs are new.

This procedure can also be applied to non-Desarguesian planes of order q^{st}. But the isomorphism problem for cousins of type (IV) has been solved only in the Desarguesian case, so that only in this case the maximal arcs can be identified as new.

7.12 Partial m-systems, BLT-sets and sets with the BLT-property

Definition 7.58. (1) A BLT-*set* is a non-empty set $\mathcal{B}$ of disjoint lines of $\mathcal{W}_3(q)$, with the property that every line of $\mathcal{W}_3(q)$ which is not a member of $\mathcal{B}$ meets non-trivially exactly either two or none of the lines of $\mathcal{B}$.
(2) The dual concept in the generalised quadrangle $\mathcal{P}(4, q)$, which is the dual of the generalised quadrangle $\mathcal{W}_3(q)$, is a *dual* BLT-*set*.
(3) More generally, a BLT-*set* of $\mathrm{PG}(m, q)$'s is a non-empty collection $\mathcal{B}$ of disjoint subspaces of a polar space S of rank $r \geq 2$, having the property that each line of S not contained in a member of $\mathcal{B}$ meets non-trivially exactly either zero or two members of $\mathcal{B}$.

Remark 7.59. 1. BLT-sets play a key role in the theory of translation planes and in the theory of generalised quadrangles; see Section 7.15.

2. A BLT-set of points of S is a subset of the points of S with each line of S containing either zero or two points of this subset. Examples are the 56-cap on $\mathcal{E}(5,3)$ and the union of two disjoint ovoids of $\mathcal{U}(3, q^2)$.

Theorem 7.60. *A BLT- set $\mathcal{B}$ of* $\mathrm{PG}(m, q)$*'s, with $m > 0$, of a polar space S exists only in the following cases*:

(a) $m = 1$ *and* $S = \mathcal{W}_3(q)$, q *odd, with* $|\mathcal{B}| = q + 1$;
(b) $m = 1$ *and* $S = \mathcal{E}(5, q)$, q *odd, with* $|\mathcal{B}| = q^2 + 1$.

Remark 7.61. 1. For $\mathcal{W}_3(q)$, q odd, many non-isomorphic BLT-sets are known. They lead to new generalised quadrangles, new projective planes, new ovoids of $\mathcal{H}(5, q)$ and new ovoids of $\mathcal{P}(4, q)$.
2. For $\mathcal{E}(5, q)$, q odd, a unique example is known. Let π and π' be two disjoint planes of $\mathrm{PG}(5, q^2)$ which are conjugate with respect to $\mathbf{F}_{q^2}$ over $\mathbf{F}_q$; that is, $\{\pi, \pi'\}$ is an orbit of the Galois group with respect to this extension. Let C be a conic of π and let C' be the conic of π' consisting of the points conjugate to those of C. Joining the points of C to their conjugates gives $q^2 + 1$ lines of $\mathrm{PG}(5, q)$. For q odd, these lines are contained in a $\mathcal{E}(5, q)$ and form a BLT-set; these lines are also contained in a $\mathcal{H}(5, q)$, and the union of the $q^2 + 1$ lines is the intersection of $\mathcal{E}_5$ and $\mathcal{H}_5$. Under the Klein correspondence with the Klein quadric $\mathcal{H}_5$, the points on these lines are the images of all tangent lines of some $\mathcal{E}_3$.

Definition 7.62. A non-empty set $\mathcal{B}$ of disjoint m-dimensional subspaces of a polar space S of rank $r \geq 2$ possesses the BLT-*property* if there is no line of S meeting three distinct members of $\mathcal{B}$ non-trivially.

From partial m-systems of suitable size possessing the BLT-property, generalised quadrangles can be constructed. A summary is given of all possible m-systems, having the BLT-property, which yield a generalised quadrangle; there are some interesting open problems.

1. $S = \mathcal{W}_5(q)$, q odd, and $\mathcal{B}$ is a 1-system of $q^3 + 1$ lines. The corresponding generalised quadrangle has order (q^2, q^3). It was shown that there is exactly one such $\mathcal{B}$, and the corresponding generalised quadrangle is isomorphic to $\mathcal{U}(4, q^2)$.
2. $S = \mathcal{H}(5, q)$, q odd, and $\mathcal{B}$ is a 1-system of $q^2 + 1$ lines. The corresponding generalised quadrangle has order (q^2, q^2). It was shown that there is exactly one such $\mathcal{B}$, and the corresponding generalised quadrangle is isomorphic to $\mathcal{P}(4, q^2)$.
3. $S = \mathcal{H}(9, q)$, q odd, and $\mathcal{B}$ is a 2-system of $q^4 + 1$ planes. The corresponding generalised quadrangle has order (q^3, q^4). No such generalised quadrangle is known.
4. $S = \mathcal{E}(4r - 5, q)$, $r \geq 3$, and $\mathcal{B}$ is an $(r-2)$-system of $(r-2)$-dimensional spaces in number $q^{2r-2} + 1$. For each value of r one such $\mathcal{B}$ is known; the corresponding generalised quadrangle is isomorphic to $\mathcal{E}(5, q^{r-1})$. It has been established that, for $r = 3$, the 1-system $\mathcal{B}$ of $\mathcal{E}(7, q)$ is unique.
5. $S = \mathcal{U}(9, q^2)$ and $\mathcal{B}$ is a 2-system of $q^9 + 1$ planes. The corresponding generalised quadrangle has order (q^6, q^9). No such $\mathcal{B}$ is known.

7.13 m-systems and SPG-reguli

Definition 7.63. An SPG-*regulus* is a set $\mathcal{R}$ of subspaces $\pi_1, \pi_2, \ldots, \pi_r$, $r > 1$, all m-dimensional, of $\mathrm{PG}(n, q)$, $n > 1$, satisfying the following conditions:

(a) $\pi_i \cap \pi_j = \emptyset$ for all $i \neq j$ with $i, j \in \{1, 2, \ldots, r\}$;
(b) if $\mathrm{PG}(m+1, q)$ contains π_i, with $i \in \{1, 2, \ldots, r\}$, then it has a point in common with either 0 or $\alpha > 0$ spaces in $\mathcal{R}\backslash\{\pi_i\}$;
(c) if such a $\mathrm{PG}(m+1, q)$ has no point in common with π_j for all $j \neq i$, then it is a *tangent $(m+1)$-space* of $\mathcal{R}$ at π_i;
(d) if the point P of $\mathrm{PG}(n, q)$ is not contained in an element of $\mathcal{R}$, then it is contained in a constant number $\theta \geq 0$ of tangent $(m+1)$-spaces of $\mathcal{R}$.

By (a), $n \geq 2m + 1$; if $n = 2m + 1$ then there are no tangent $(m+1)$-spaces, and so $\alpha = r - 1$.

It can be shown that

$$\alpha(q - 1) \text{ divides } (r - 1)(q^{m+1} - 1)$$

and

$$\theta = \frac{(\alpha(q^{n-m} - 1) - (r - 1)(q^{m+1} - 1))rq^{m+1}}{\alpha((q^{n+1} - 1) - r(q^{m+1} - 1))}.$$

Hence $\theta = 0$ if and only if $\alpha(q^{n-m} - 1) = (r - 1)(q^{m+1} - 1)$.

From an SPG-regulus a semi-partial geometry $\mathcal{S}$ with parameters

$$s = q^{m+1} - 1, \quad t = r - 1, \quad \mu = (r - \theta)\alpha$$

can be constructed. Then $\mathcal{S}$ is a partial geometry if and only if $\theta = 0$; if $\mathcal{S}$ is not a partial geometry, that is, if $\theta \neq 0$, or, equivalently, $\alpha(q^{n-m}-1) \neq (r-1)(q^{m+1}-1)$, then, by Theorem 5.145, $t \geq s$, and so $r \geq q^{m+1}$.

Example 7.64. (a) $n = 2m + 1$. Then the SPG-regulus has no tangent $(m+1)$-spaces; hence $\alpha = r - 1$ and $\theta = 0$. In this case, the semi-partial geometry is a net of order q^{m+1} and degree r; see Section 5.6.
(b) $n = 2m + 2$. If $\theta \neq 0$, then the corresponding semi-partial geometry has parameters

$$s = q^{m+1} - 1, \quad t = r - 1,$$
$$\alpha = \frac{r^2(q^{m+1} - 1) - r(q^{m+1} - 1)(q^{m+1} + 2) + q^{2m+3} - 1}{r(q^{m+2} - 1) - (q^{2m+3} - 1)},$$
$$\mu = \frac{r(r - 1)(q^{m+1} - 1)(r - (q^{m+1} + 1))}{r(q^{m+2} - 1) - (q^{2m+3} - 1)}.$$

Theorem 7.65. *Let $\mathcal{R}$ be a set of m-dimensional subspaces $\pi_1, \pi_2, \ldots, \pi_r$, with r at least 2, of $\mathrm{PG}(2m + 2, q)$ satisfying* (a) *and* (b) *of Definition 7.63. If $\mathcal{R}$ admits tangent $(m+1)$-spaces, then*

$$\alpha(r(q^{m+2} - 1) - (q^{2m+3} - 1))$$
$$\leq r^2(q^{m+1} - 1) - r(q^{m+1} - 1)(q^{m+1} + 2) + (q^{2m+3} - 1),$$

with equality if and only if $\mathcal{R}$ is an SPG-regulus.

Remark 7.66. Theorem 7.65 can be generalised to $PG(n, q)$ by assuming that, for all i, each tangent $(m + 1)$-space at π_i intersects at most ν tangent $(m + 1)$-spaces at the other $r - 1$ elements of $\mathcal{R}$. In particular, this applies if each tangent $(m+1)$-space at π_i intersects exactly $\tilde{\nu}$ tangent $(m + 1)$-spaces at π_j, for all $j \neq i$. One case is if two tangent $(m + 1)$-spaces at different elements of $\mathcal{R}$ intersect.

Definition 7.67. (1) An SPG-regulus $\mathcal{R}$ satisfies the *polar property* if
  (a) it has tangent $(m + 1)$-spaces,
  (b) $n > 2m + 1$,
  (c) the union of the tangent $(m + 1)$-spaces at each element π_i of $\mathcal{R}$ is an $(n - m - 1)$-dimensional subspace τ_i, with $i = 1, 2, \ldots, r$.
(2) The subspace τ_i is the *tangent* $(n - m - 1)$-*space* of $\mathcal{R}$ at π_i.

Remark 7.68. 1. If an SPG-regulus $\mathcal{R}$ satisfies the polar property, then

$$r = \alpha q^{n-2m-1} + 1.$$

2. Let $\mathcal{R}$ be a set of m-dimensional subspaces $\pi_1, \pi_2, \ldots, \pi_r$, $r > 1$, of $PG(n, q)$ satisfying (a) and (b) of Definition 7.63. Assume also that $\mathcal{R}$ has tangent $(m+1)$-spaces and that for all $i = 1, 2, \ldots, r$, the union of all tangent $(m + 1)$-spaces at π_i is an $(n - m - 1)$-dimensional subspace. Then $r \leq 1 + q^{(n+1)/2}$ with equality if and only if $\mathcal{R}$ is an SPG-regulus.
3. If an SPG-regulus $\mathcal{R}$ has the polar property, then the corresponding semi-partial geometry $\mathcal{S}$ has parameters

$$s = q^{m+1} - 1, \quad t = q^{(n+1)/2}, \quad \alpha = q^{2m-(n/2)+3/2}, \quad \mu = q^{m+1}(q^{m+1} - 1).$$

It follows that $4m \geq n - 3$.

Theorem 7.69. (a) *Let* $\mathcal{R} = \{\pi_1, \pi_2, \ldots, \pi_r\}$, *with* $r > 1$, *be a set of* r *disjoint* m-*dimensional subspaces in* $PG(n, q)$, *with* $n > 2m + 1$, *such that, for each* $i = 1, 2, \ldots, r$, *there is an* $(m + 1)$-*dimensional subspace containing* π_i *and disjoint from all* π_j, $j \neq i$.
(b) *Assume that, for each* i, *the union of these* $(m + 1)$-*dimensional subspaces containing* π_i *and disjoint from all* π_j, $j \neq i$, *contains an* $(n - m - 1)$-*dimensional space* τ_i.
Then the following hold when $|\mathcal{R}| = 1 + q^{(n+1)/2}$:

  (i) *the set* $\mathcal{R}$ *is an SPG-regulus satisfying the polar property;*
  (ii) *the* r *subspaces* τ_i *form an SPG-regulus* $\mathcal{R}^*$ *satisfying the polar property in the dual space of* $PG(n, q)$.

Theorem 7.70. *Let*

(a) $\mathcal{R} = \{\pi_1, \pi_2, \ldots, \pi_r\}$, *with* $r > 1$, *be an SPG-regulus satisfying the polar property, with* $\pi_1, \pi_2, \ldots, \pi_r$ *all* m-*dimensional subspaces of* $\mathrm{PG}(n, q)$ *and* $r = 1 + q^{(n+1)/2}$;

(b) τ_i *be the tangent* $(n - m - 1)$-*space of* $\mathcal{R}$ *at* π_i;

(c) *the tangent* $(n - m - 1)$-*spaces* $\tau_1, \tau_2, \ldots, \tau_s$ *have a* $\mathrm{PG}(n - 2m - 2, q) = \pi$ *in common*;

(d) $\{\bar{\pi}_1, \bar{\pi}_2, \ldots, \bar{\pi}_s\}$ *be a set of disjoint* m-*dimensional subspaces covering the same point set as* $\pi_1, \pi_2, \ldots, \pi_s$.

Then

$$\mathcal{R}' = (\mathcal{R} \cup \{\bar{\pi}_1, \bar{\pi}_2, \ldots, \bar{\pi}_s\}) \setminus \{\pi_1, \pi_2, \ldots, \pi_s\}$$

is also an SPG-regulus satisfying the polar property, and is said to be derived *from* $\mathcal{R}$.

The relation between m-systems and SPG-reguli is given by the following theorem.

Theorem 7.71. *An* m-*system* $\mathcal{M}$ *of a polar space* $\mathcal{S}$ *in* $\mathrm{PG}(N, q)$ *is an SPG-regulus in the following cases:*

(i) $\mathcal{S} = \mathcal{E}(2n + 1, q) \subset \mathrm{PG}(2n + 1, q)$, $n > 0$;

(ii) $\mathcal{S} = \mathcal{W}_{2n+1}(q) \subset \mathrm{PG}(2n + 1, q)$, $n > 0$;

(iii) $\mathcal{S} = \mathcal{U}(2n, q^2) \subset \mathrm{PG}(2n, q^2)$, $n > 0$.

Remark 7.72. It follows from Section 7.7 that the m-systems in Theorem 7.71 are the only ones which are also SPG-reguli.

7.14 Small cases

Theorem 7.73.   (i) *Up to isomorphism, the polar space* $\mathcal{W}_5(2)$ *admits a unique 1-system and a unique spread. Hence each symplectic 2-spread of* $\mathrm{PG}(5, 2)$, *that is, each 2-spread whose elements are self-polar for some null polarity, is regular.*

(ii) *Up to isomorphism, the polar space* $\mathcal{W}_7(2)$ *admits a unique 1-system, a unique 2-system and a unique spread. Hence all symplectic 3-spreads of* $\mathrm{PG}(7, 2)$ *are regular.*

(iii) *The polar space* $\mathcal{W}_9(2)$ *admits no 1-systems and no 2-systems. Up to isomorphism* $\mathcal{W}_9(2)$ *admits exactly two spreads and ten 3-systems.*

Theorem 7.74. *The only cases that a* k-*cap* $\mathcal{K}$ *is an SPG-regulus are as follows:*

(i) $\mathcal{K}$ *is a* $(q + 2)$-*arc of* $\mathrm{PG}(2, q)$, q *even;*

(ii) $\mathcal{K}$ *is an ovoid of* $\mathrm{PG}(3, q)$;

(iii) $\mathcal{K}$ *is the projectively unique 11-cap in* $\mathrm{PG}(4, 3)$;

(iv) $\mathcal{K}$ *is the projectively unique* 56-*cap in* $\mathrm{PG}(5,3)$;

(v) $\mathcal{K}$ *is a particular* 78-*cap in* $\mathrm{PG}(5,4)$, *an example of which has been constructed*;

(vi) $\mathcal{K}$ *is a particular* 430-*cap in* $\mathrm{PG}(6,4)$, *whose existence is unknown.*

Remark 7.75. The semi-partial geometries corresponding to the SPG-reguli (i)–(vi) have respective parameters as follows:

$$(q-1, q+1, 1, q+2), \quad (q-1, q^2, 1, q^2-q), \quad (2, 10, 1, 2),$$
$$(2, 55, 1, 20), \quad (3, 77, 1, 14), \quad (3, 429, 1, 110).$$

Theorem 7.76. *There exists an* SPG-*regulus consisting of* 21 *lines in* $\mathrm{PG}(5,3)$, *with* $\alpha = 2$, $\theta = 0$. *Its partial geometry has parameters* $s = 8$, $t = 20$, $\alpha = 2$.

7.15 Notes and references

Section 7.1

For the size of orbits of subspaces under the symplectic group, see Wan [395]. In particular these give $|\mathcal{G}(\mathcal{W}_n(q))|$.

For similar results on orbits under the pseudo-symplectic group, see Liu and Wan [208], Pless [265, 266].

Sections 7.2–7.5

Apart from the references in Tables 7.3 and 7.4, see also the surveys by Thas [336, 341, 312, 323] and De Beule, Klein and Metsch [82]. The non-existence of spreads in $\mathcal{U}(4,4)$ is a computer result of A. E. Brouwer (unpublished, 1981).

Section 7.6

Partial m-systems and m-systems were introduced by Shult and Thas [289], who proved Theorems 7.7 and 7.9.

Section 7.7

This is taken from Shult and Thas [289]. For rank $r = 2$, the concept of a k-ovoid was introduced by Thas [333]. Results on k-ovoids are also contained in Bamberg, Kelly, Law and Penttila [13], and Bamberg, Law, Penttila [14].

Section 7.8

Theorem 7.20 is due to Blokhuis and Moorhouse [33] and Moorhouse [239]. They rely on a classical result of Hamada [149] on the rank of the incidence matrix of points and m-dimensional subspaces of a $\mathrm{PG}(n,q)$; see also Goethals and Delsarte [143], MacWilliams and Mann [221], and Smith [294]. From [33] and [239], Shult and Thas [291] obtain Theorems 7.22, 7.23 and 7.24. Theorem 7.25 is due to Sin [292].

Section 7.9

This is taken from Shult and Thas [289, 290]; they also show that partial m-systems can be constructed from m-systems.

Section 7.10

The relation between strongly regular graphs, linear projective two-weight codes and sets of points in $\mathrm{PG}(n, q)$ with two intersection numbers with respect to hyperplanes is due to Delsarte [115]; see also Calderbank and Kantor [62]. Theorem 7.45 is due to Hamilton and Mathon [150]. Theorems 7.47 to 7.49 and Examples 7.50 are taken from Hamilton and Quinn [153].

Section 7.11

Theorem 7.51 and Remarks 1, 4, 5 of 7.52 are due to Thas [316, 326]. Remark 2 of 7.52 is taken from Blokhuis, Hamilton and Wilbrink [32] and Remark 3 of 7.52 from Hamilton and Penttila [152]. Theorems 7.53 and 7.55, Corollary 7.56 and Remarks 7.54 and 7.57 are due to Hamilton and Quinn [153]. The description of the construction of cousins of spreads of $\mathcal{W}_{4n-3}(q)$ follows that given in Kantor's Kerdock set papers [192, 193].

Material related to Theorem 7.51 is also contained in Maschietti [224, 225], and Bader and Lunardon [6].

Section 7.12

BLT-sets were introduced by Bader, Lunardon and Thas [7]; the name is due to Kantor [195]. In [7] it is shown that, from any *flock* of a cone $P\mathcal{P}_2$ with vertex P in $\mathrm{PG}(3, q)$, q odd, that is, a partition of $P\mathcal{P}_2 \backslash \{P\}$ into q disjoint conics, q *derived* flocks can be constructed. Crucial to the construction is a dual BLT-set coming from the given flock. BLT-sets and flocks gave rise to many new translation planes and generalised quadrangles. In Shult and Thas [290], BLT-sets of m-dimensional subspaces were defined; BLT-sets of lines in polar spaces of rank 2 were previously introduced by Knarr [198]. In a paper on characterisations of generalised quadrangles of order $(s, s + 2)$, De Soete and Thas [101] introduced dual BLT-sets, which they called $\{0, 2\}$-*sets*, in generalised quadrangles of order (s, t), but their definition is conceptually distinct from the definition given here.

For a description of the Hill's 56-cap, see Hill [166] or Section 19.3 of FPSOTD. For disjoint ovoids of $\mathcal{U}(3, q^2)$, see Hamilton and Quinn [153]. For $m = 1$, Theorem 7.60 is due to Thas: the proof is contained in Knarr [198]. For $m > 1$, the proof is in Shult and Thas [290]. For the relationships between BLT-sets, projective planes and ovoids, see Bader, Lunardon and Thas [7], Kantor [195], Thas, K. Thas and Van Maldeghem [352], and Payne and Thas [260]; many results on BLT-sets of lines of $\mathcal{E}(5, q)$, q odd, are contained in Thas [345]. Definition 7.62 and the list of m-systems, having the BLT-property, which yield a generalised quadrangle, is due to Shult and

Thas [290]. The uniqueness of the 1-system of lines of $\mathcal{W}_5(q)$ satisfying the BLT-property, with q odd, is due to Thas [343]; the uniqueness of the 1-system of lines of $\mathcal{H}(5, q)$, q odd, is due to Shult and Thas [290]; the uniqueness of the 1-system of lines of $\mathcal{E}(7, q)$ is due to Luyckx and Thas [216, 220].

Section 7.13

Definition 7.63 and Examples 7.64 are taken from Thas [329], and Theorem 7.65 is taken from Thas [342]. The results on SPG-reguli satisfying the polar property are also taken from [329]. Theorems 7.69 and 7.70 are due to De Winter and Thas [107], and Theorem 7.71 to Luyckx [214].

Section 7.14

Theorem 7.73 is due to Hamilton and Mathon [150] and Theorem 7.74 is taken from Cameron [63], Calderbank [61], Coxeter [80], Hill [166, 167], Pellegrino [262], and Tzanakis and Wolfskill [387]. Finally, Theorem 7.76 is taken from De Clerck, Delanote, Hamilton and Mathon [90].

Other papers relevant to this chapter are Bamberg and Penttila [15], Bloemen, Thas and Van Maldeghem [30], Cardinali, Lunardon, Polverino and Trombetti [67], Cardinali and Trombetti [68], De Winter and Thas [106], Luyckx and Thas [215, 217, 219, 218, 220], Lunardon and Polverino [212], Offer [243, 244, 245], Offer, K. Thas and Van Maldeghem [246], Offer and Van Maldeghem [247], O'Keefe and Thas [250], and Thas [338, 341, 344]. In particular, related material is contained in Thas [342] and Thas, K. Thas and Van Maldeghem [352], where SPG-systems are introduced and eggs are studied in great detail.

REFERENCES

This bibliography is confined mainly to works cited in the text. Abbreviations for periodicals almost always follow those used by *Mathematical Reviews*.

See the bibliography of the second edition of *Projective Geometries over Finite Fields* for a comprehensive bibliography of the topic up to 1998.

An online bibliography of Finite Geometry is maintained at

http://www.maths.susx.ac.uk/Staff/JWPH/RESEARCH

[1] Z. Akça, A. Bayar, S. Ekmekci, R. Kaya, and J. A. Thas. Generalized veronesean embeddings of projective spaces, Part II. the lax case. *Ars Combin.*, 103:65–80, 2012.

[2] J. André. Uber nicht-Desarguessche Ebenen mit transitiver Translationsgruppe. *Math. Z.*, 60:156–186, 1954.

[3] E. Artin. *Geometric Algebra*. Interscience, New York, 1957, 214 pp.

[4] E. F. Assmus and J. D. Key. *Designs and their Codes*. Cambridge University Press, Cambridge, 1992, 352 pp.

[5] L. Bader, W. M. Kantor, and G. Lunardon. Symplectic spreads from twisted fields. *Boll. Un. Mat. Ital. A*, 8:383–389, 1994.

[6] L. Bader and G. Lunardon. Desarguesian spreads. *Ric. Mat.*, 60:15–37, 2011.

[7] L. Bader, G. Lunardon, and J. A. Thas. Derivation of flocks of quadratic cones. *Forum Math.*, 2:163–174, 1990.

[8] S. Ball. On sets of vectors of a finite vector space in which every subset of basis size is a basis. *J. Eur. Math. Soc.*, 14:733–748, 2012.

[9] S. Ball and J. De Beule. On sets of vectors of a finite vector space in which every subset of basis size is a basis II. *Des. Codes Cryptogr.*, 65:5–14, 2012.

[10] S. Ball, P. Govaerts, and L. Storme. On ovoids of parabolic quadrics. *Des. Codes Cryptogr.*, 38:131–145, 2006.

[11] S. Ball and M. Lavrauw. How to use Rédei polynomials in higher dimensional spaces. *Matematiche (Catania)*, 59:39–52, 2004.

[12] J. Bamberg, M. Giudici, and G. F. Royle. Every flock generalized quadrangle has a hemisystem. *Bull. London Math. Soc.*, 42:795–810, 2010.

[13] J. Bamberg, S. Kelly, M. Law, and T. Penttila. Tight sets and m-ovoids of finite polar spaces. *J. Combin. Theory Ser. A*, 114:1293–1314, 2007.

[14] J. Bamberg, M. Law, and T. Penttila. Tight sets and m-ovoids of generalised quadrangles. *Combinatorica*, 29:1–17, 2009.

© Springer-Verlag London 2016

J.W.P. Hirschfeld, J.A. Thas, *General Galois Geometries*, Springer Monographs
in Mathematics, DOI 10.1007/978-1-4471-6790-7

[15] J. Bamberg and T. Penttila. A classification of transitive ovoids, spreads, and m-systems of polar spaces. *Forum Math.*, 21:181–216, 2009.

[16] C. J. Benson. On the structure of generalized quadrangles. *J. Algebra*, 15:443–454, 1970.

[17] E. Bertini. *Introduzione alla Geometria Proiettiva degli Iperspazi*. Casa Editrice Giuseppe Principato, Messina, 1923, 517 pp.

[18] T. Beth, D. Jungnickel, and H. Lenz. *Design Theory*. Bibliographisches Institut, Mannheim, 1984, 688 pp.

[19] T. Beth, D. Jungnickel, and H. Lenz. *Design Theory, Second Edition, Volume 2*. Cambridge University Press, Cambridge, 1999, 493 pp.

[20] T. Beth, D. Jungnickel, and H. Lenz. *Design Theory, Second Edition, Volume 1*. Cambridge University Press, Cambridge, 1999, 607 pp.

[21] A. Bichara. Caratterizzazione dei sistemi rigati immersi in $A_{3,q}$. *Riv. Mat. Univ. Parma*, 4:277–290, 1978.

[22] A. Bichara and F. Mazzocca. On a characterization of Grassmann space representing the lines in an affine space. *Simon Stevin*, 56:129–141, 1982.

[23] A. Bichara and F. Mazzocca. On the independence of the axioms defining the affine and projective Grassmann spaces. In *Combinatorial and Geometric Structures and their Applications*, volume 14 of *Ann. Discrete Math.*, pages 123–128. North-Holland, Amsterdam, 1982. (Trento, 1980).

[24] A. Bichara and F. Mazzocca. On a characterization of the Grassmann spaces associated with an affine space. In *Combinatorics '81*, volume 18 of *Ann. Discrete Math.*, pages 95–112. North-Holland, Amsterdam, 1983. (Rome, 1981).

[25] A. Bichara and G. Tallini. On a characterization of the Grassmann manifold representing the planes in a projective space. In *Combinatorial and Geometric Structures and their Applications*, volume 14 of *Ann. Discrete Math.*, pages 129–150. North-Holland, Amsterdam, 1982. (Trento, 1980).

[26] A. Bichara and G. Tallini. On a characterization of Grassmann space representing the h-dimensional subspaces in a projective space. In *Combinatorics '81*, volume 18 of *Ann. Discrete Math.*, pages 113–132. North-Holland, Amsterdam, 1983. (Rome, 1981).

[27] J. Bierbrauer and Y. Edel. Blocking sets in projective spaces. In *Current Research Topics in Galois Geometry*, pages 87–104. Nova Science Publishers, New York, 2012.

[28] M. Biliotti, V. Jha, and N. L. Johnson. *Foundations of Translation Planes*. Dekker, New York, 2001, 542 pp.

[29] P. Biondi. A characterisation of the Grassmann space representing the 2-dimensional subspaces of projective space. *Boll. Un. Mat. Ital.*, 7:713–727, 1987.

[30] I. Bloemen, J. A. Thas, and H. Van Maldeghem. Translation ovoids of generalized quadrangles and hexagons. *Geom. Dedicata*, 72:19–62, 1998.

[31] A. Blokhuis, A. A. Bruen, and J. A. Thas. Arcs in $PG(n, q)$, MDS-codes and three fundamental problems of B. Segre - some extensions. *Geom. Dedicata*, 35:1–11, 1990.

[32] A. Blokhuis, N. Hamilton, and H. Wilbrink. On the non-existence of Thas maximal arcs in odd order projective planes. *European J. Combin.*, 19:413–417, 1998.

[33] A. Blokhuis and G. E. Moorhouse. Some p-ranks related to orthogonal spaces. *J. Algebraic Combin.*, 4:295–316, 1995.

[34] E. Boros and T. Szőnyi. On the sharpness of a theorem of B. Segre. *Combinatorica*, 6:261–268, 1986.

[35] R. C. Bose. Mathematical theory of the symmetrical factorial design. *Sankhyā*, 8:107–166, 1947.

[36] R. C. Bose. Strongly regular graphs, partial geometries and partially balanced designs. *Pacific J. Math.*, 13:389–419, 1963.

[37] R. C. Bose and I. M. Chakravarti. Hermitian varieties in a finite projective space $PG(N, q^2)$. *Canad. J. Math.*, 18:1161–1182, 1966.

[38] R. C. Bose and S. S. Shrikhande. Geometric and pseudo-geometric graphs $(q^2 + 1, q + 1, 1)$. *J. Geom.*, 2:75–94, 1972.

[39] L. Brouns, J. A. Thas, and H. Van Maldeghem. A characterization of $Q(5, q)$ using one subquadrangle $Q(4, q)$. *European J. Combin.*, 23:163–177, 2002.

[40] A. E. Brouwer. Some unitals on 28 points and their embeddings in projective planes of order 9. In *Geometries and Groups*, volume 893 of *Lecture Notes in Math.*, pages 183–188. Springer, Berlin, 1981. (Berlin, 1981).

[41] A. E. Brouwer, A. M. Cohen, and A. Neumaier. *Distance-Regular Graphs*. Springer, Berlin, 1989, 495 pp.

[42] A. E. Brouwer and J. H. van Lint. Strongly regular graphs and partial geometries. In *Enumeration and Design*, pages 85–122. Academic Press, Toronto, 1984. (Waterloo, 1982).

[43] M. R. Brown. *Generalised quadrangles and associated structures*. PhD thesis, University of Adelaide, 1999.

[44] M. R. Brown. The determination of ovoids of $PG(3, q)$ containing a pointed conic. *J. Geom.*, 67:61–72, 2000. (Samos, 1999).

[45] M. R. Brown. Ovoids of $PG(3, q)$, q even, with a conic section. *J. London Math. Soc.*, 62:569–582, 2000.

[46] M. R. Brown. A characterisation of the generalized quadrangle $Q(5, q)$ using cohomology. *J. Algebraic Combin.*, 15:107–125, 2002.

[47] M. R. Brown, F. De Clerck, and M. Delanote. Affine semipartial geometries and projections of quadrics. *J. Combin. Theory Ser. A*, 103:281–289, 2003.

[48] M. R. Brown and M. Lavrauw. Eggs in $PG(4n - 1, q)$, q even, containing a pseudo-conic. *Bull. London Math. Soc.*, 36:633–639, 2004.

[49] M. R. Brown and J. A. Thas. Subquadrangles of order s of generalized quadrangles of order (s, s^2), Part I. *J. Combin. Theory Ser. A*, 106:15–32, 2004.

[50] M. R. Brown and J. A. Thas. Subquadrangles of order s of generalized quadrangles of order (s, s^2), Part II. *J. Combin. Theory Ser. A*, 106:33–48, 2004.

[51] R. H. Bruck and R. C. Bose. Linear representations of projective planes in projective spaces. *J. Algebra*, 4:117–172, 1966.

[52] A. A. Bruen and J. A. Thas. Partial spreads, packings and Hermitian manifolds in $PG(3, q)$. *Math. Z.*, 151:207–214, 1976.

[53] A. A. Bruen, J. A. Thas, and A. Blokhuis. On M.D.S. codes, arcs in $PG(n, q)$ with q even, and a solution of three fundamental problems of B. Segre. *Invent. Math.*, 92:441–459, 1988.

[54] F. Buekenhout. Ensembles quadratiques des espaces projectifs. *Math. Z.*, 110:306–318, 1969.

[55] F. Buekenhout, editor. *Handbook of Incidence Geometry: Buildings and Foundations*. North-Holland, Amsterdam, 1995, 1420 pp.

[56] F. Buekenhout and A. M. Cohen. *Diagram geometry*, volume 57 of *Ergebnisse der Mathematik*. Springer, Heidelberg, 2013, xiv+594 pp.

[57] F. Buekenhout and C. Lefèvre. Generalized quadrangles in projective spaces. *Arch. Math.*, 25:540–552, 1974.

[58] F. Buekenhout and C. Lefèvre. Semi-quadratic sets in projective spaces. *J. Geom.*, 7:17–24, 1976.

[59] F. Buekenhout and E. E. Shult. On the foundations of polar geometry. *Geom. Dedicata*, 3:155–170, 1974.

[60] W. Burau. *Mehrdimensionale Projektive und Höhere Geometrie*. VEB Deutscher Verlag der Wissenschaften, Berlin, 1961, 436 pp.

[61] A. R. Calderbank. On uniformly packed $[n, n - k, 4]$ codes over $GF(q)$ and a class of caps in $PG(k - 1, q)$. *J. London Math. Soc.*, 26:365–384, 1982.

[62] A. R. Calderbank and W. M. Kantor. The geometry of two-weight codes. *Bull. London Math. Soc.*, 18:97–122, 1986.

[63] P. J. Cameron. Partial quadrangles. *Quart. J. Math. Oxford*, 26:61–74, 1975.

[64] P. J. Cameron, J.-M. Goethals, and J. J. Seidel. Strongly regular graphs having strongly regular subconstituents. *J. Algebra*, 55:257–280, 1978.

[65] P. J. Cameron and J. H. van Lint. *Graphs, Codes and Designs*, volume 43 of *London Math. Soc. Lecture Note Series*. Cambridge University Press, Cambridge, 1980, 147 pp.

[66] J. M. Cao and L. Ou. Caps in $PG(n, q)$ with q even and $n \geq 3$. *Discrete Math.*, 326:61–65, 2014.

[67] I. Cardinali, G. Lunardon, O. Polverino, and R. Trombetti. Spreads in $H(q)$ and 1-systems of $Q(6, q)$. *European J. Combin.*, 23:367–376, 2002.

[68] I. Cardinali and R. Trombetti. On Hermitian spreads. *Bull. Belg. Math. Soc. Simon Stevin*, 11:63–67, 2004.

[69] L. R. A. Casse. A solution to Beniamino Segre's 'Problem $I_{r,q}$' for q even. *Atti Accad. Naz. Lincei Rend.*, 46:13–20, 1969.

[70] L. R. A. Casse and D. G. Glynn. On the uniqueness of $(q + 1)_4$-arcs of $PG(4, q)$, $q = 2^h, h \geq 3$. *Discrete Math.*, 48:173–186, 1984.

[71] J. M. Chao. On the size of a cap in $PG(n, q)$ with q even and $n \geq 3$. *Geom. Dedicata*, 74:91–94, 1999.

[72] A. M. Cohen. On a theorem of Cooperstein. *European J. Combin.*, 4:107–126, 1983.

[73] C. J. Colbourn and J. H. Dinitz, editors. *Handbook of Combinatorial Designs*. Chapman and Hall/CRC Press, Boca Raton, second edition, 2006, 1016 pp.

[74] J. H. Conway, P. B. Kleidman, and R. A. Wilson. New families of ovoids in O_8^+. *Geom. Dedicata*, 26:157–170, 1988.

[75] B. N. Cooperstein. A characterization of some Lie incidence structures. *Geom. Dedicata*, 6:205–258, 1977.

[76] B. N. Cooperstein, J. A. Thas, and H. Van Maldeghem. Hermitian Veroneseans over finite fields. *Forum Math.*, 16:365–381, 2004.

[77] A. Cossidente and T. Penttila. Hemisystems on the Hermitian surface. *J. London Math. Soc.*, 72:731–741, 2005.

[78] A. Cossidente and T. Penttila. A new hemisystem on $\mathcal{H}(3, 49)$. *Ars Combin.*, 119:257–262, 2015.

[79] A. Cossidente and A. Siciliano. On the geometry of Hermitian matrices of order three over finite fields. *European J. Combin.*, 22:1047–1058, 2001.

[80] H. S. M. Coxeter. Twelve points in $PG(5, 3)$ with 95040 self-transformations. *Proc. Roy. Soc. London Ser. A*, 427:279–293, 1958.

[81] Z. D. Dai and X. N. Feng. Studies in finite geometries and the construction of incomplete block designs. IV. Some 'Anzahl' theorems in orthogonal geometry over finite fields of characteristic $\neq 2$. *Chinese Math.*, 7:265–279, 1965.

[82] J. De Beule, A. Klein, and K. Metsch. Substructures of finite classical polar spaces. In *Current Research Topics in Galois Geometry*, pages 35–61. Nova Science Publishers, New York, 2012.

[83] J. De Beule and K. Metsch. The Hermitian variety $H(5,4)$ has no ovoid. *Bull. Belg. Math. Soc. Simon Stevin*, 12:727–733, 2005.

[84] B. De Bruyn. Canonical equations for nonsingular quadrics and Hermitian varieties of Witt index at least $\frac{n-1}{2}$ of $\mathrm{PG}(n, \mathbb{K})$, n odd. *Int. Electron. J. Geom.*, 4:48–69, 2011.

[85] F. De Clerck. Partial geometries – a combinatorial survey. *Bull. Soc. Math. Belg. Sér. B*, 31:135–145, 1979.

[86] F. De Clerck. New partial geometries constructed from old ones. *Bull. Belg. Math. Soc. Simon Stevin*, 5:255–263, 1998. (Deinze, 1997).

[87] F. De Clerck. Partial and semipartial geometries: an update. *Discrete Math.*, 267:75–86, 2003. (Gaeta, 2000).

[88] F. De Clerck, N. De Feyter, and N. Durante. Two-intersection sets with respect to lines on the Klein quadric. *Bull. Belg. Math. Soc. Simon Stevin*, 12:743–750, 2005.

[89] F. De Clerck and M. Delanote. On $(0, \alpha)$-geometries and dual semipartial geometries fully embedded in an affine space. *Des. Codes Cryptogr.*, 32:103–110, 2004.

[90] F. De Clerck, M. Delanote, N. Hamilton, and R. Mathon. Perp-systems and partial geometries. *Adv. Geom.*, 2:1–12, 2002.

[91] F. De Clerck and J. A. Thas. Partial geometries in finite projective spaces. *Arch. Math.*, 30:537–540, 1978.

[92] F. De Clerck and J. A. Thas. The embedding of $(0, \alpha)$-geometries in $PG(n, q)$. Part I. In *Combinatorics '81*, volume 18 of *Ann. Discrete Math.*, pages 229–240. North-Holland, Amsterdam, 1983. (Rome, 1981).

[93] F. De Clerck and H. Van Maldeghem. Some classes of rank 2 geometries. In *Handbook of Incidence Geometry: Buildings and Foundations*, pages 433–475. North-Holland, Amsterdam, 1995.

[94] N. de Feyter. Planar oval sets in Desarguesian planes of even order. *Des. Codes Cryptogr.*, 32:111–119, 2004.

[95] N. de Feyter. The embedding in $AG(3, q)$ of $(0, 2)$-geometries containing a planar net. *Discrete Math.*, 292:45–54, 2005.

[96] N. de Feyter. The embedding in $AG(3, q)$ of $(0, 2)$-geometries with no planar nets. *J. Combin. Theory Ser. A*, 109.1–23, 2005.

[97] N. de Feyter. The embedding of $(0, 2)$-geometries and semipartial geometries in $AG(n, q)$. *Adv. Geom.*, 5:279–292, 2005.

[98] N. De Feyter. Classification of $(0, 2)$-geometries embedded in $AG(3, q)$. *Des. Codes Cryptogr.*, 43:21–32, 2007.

[99] T. De Medts. *Moufang quadrangles: a unifying algebraic structure, and some results on exceptional quadrangles*. PhD thesis, Ghent University, 2003.

[100] T. De Medts. *An Algebraic Structure for Moufang Quadrangles*, volume 173 of *Mem. Amer. Math. Soc.* Amer. Math. Soc., Providence, 2005, 99 pp.

[101] M. De Soete and J. A. Thas. A characterization theorem for the generalized quadrangle $T_2^*(O)$ of order $(s, s + 2)$. *Ars Combin.*, 17:225–242, 1984.

[102] S. De Winter. Elation and translation semipartial geometries. *J. Combin. Theory Ser. A*, 108:313–330, 2004.

[103] S. De Winter. Linear representations of semipartial geometries. *Bull. Belg. Math. Soc. Simon Stevin*, 12:767–780, 2005.

[104] S. De Winter and J. Schillewaert. Characterizations of finite classical polar spaces by intersections with hyperplanes and subspaces of codimension 2. *Combinatorica*, 30:25–45, 2010.

[105] S. De Winter and J. Schillewaert. A note on quasi-Hermitian varieties and singular quasi-quadrics. *Bull. Belg. Math. Soc. Simon Stevin*, 17:911–918, 2010.

[106] S. De Winter and J. A. Thas. SPG-reguli satisfying the polar property and a new semi-partial geometry. *Des. Codes Cryptogr.*, 32:153–166, 2004.

[107] S. De Winter and J. A. Thas. On semi-pseudo-ovoids. *J. Algebraic Combin.*, 22:139–149, 2005.

[108] S. De Winter and H. Van Maldeghem. The automorphism group of a class of strongly regular graphs related to $Q(6, q)$. *European J. Combin.*, 29:617–621, 2008.

[109] I. Debroey. *Semi-partiële meetkunden.* PhD thesis, Ghent University, 1978.

[110] I. Debroey. Semi partial geometries. *Bull. Soc. Math. Belg. Sér. B*, 31:183–190, 1979.

[111] I. Debroey and J. A. Thas. On semipartial geometries. *J. Combin. Theory Ser. A*, 25:242–250, 1978.

[112] I. Debroey and J. A. Thas. Semi partial geometries in $AG(2, q)$ and $AG(3, q)$. *Simon Stevin*, 51:195–209, 1978.

[113] I. Debroey and J. A. Thas. Semi partial geometries in $PG(2, q)$ and $PG(3, q)$. *Atti Accad. Naz. Lincei Rend.*, 64:147–151, 1978.

[114] A. Del Fra. On d-dimensional dual hyperovals. *Geom. Dedicata*, 79:157–178, 2000.

[115] P. Delsarte. Weights of linear codes and strongly regular normed spaces. *Discrete Math.*, 3:47–64, 1972.

[116] P. Dembowski. *Finite Geometries.* Springer, New York, 1968, 375 pp.

[117] J. B. Derr. Stabilizers of isotropic subspaces in classical groups. *Arch. Math.*, 34:100–107, 1980.

[118] A. Devillers and H. Van Maldeghem. Partial linear spaces built on hexagons. *European J. Combin.*, 28:901–915, 2007.

[119] L. E. Dickson. *Linear Groups with an Exposition of the Galois Field Theory.* Teubner, Leipzig, 1901, 312 pp. (Dover, 1958).

[120] K. J. Dienst. Verallgemeinerte Vierecke in pappusschen projektiven Räumen. *Geom. Dedicata*, 9:199–206, 1980.

[121] K. J. Dienst. Verallgemeinerte Vierecke in projektiven Räumen. *Arch. Math.*, 35:177–186, 1980.

[122] J. A. Dieudonné. *La Géometrie des Groupes Classiques.* Springer, Berlin, third edition, 1971, 129 pp.

[123] S. Dixmier and F. Zara. Étude d'un quadrangle généralisé autour de deux de ses points non-liés. preprint, 1976.

[124] R. H. Dye. On the transitivity of the orthogonal and symplectic groups in projective space. *Proc. Cambridge Philos. Soc.*, 68:33–43, 1970.

[125] R. H. Dye. Partitions and their stabilizers for line complexes and quadrics. *Ann. Mat. Pura Appl.*, 114:173–194, 1977.

[126] C. E. Ealy. *Generalized quadrangles and odd transpositions.* PhD thesis, University of Chicago, 1977.

[127] G. L. Ebert and K. E. Mellinger. Mixed partitions and related designs. *Des. Codes Cryptogr.*, 44:15–23, 2007.

[128] Y. Edel and J. Bierbrauer. 41 is the largest size of a cap in $PG(4, 4)$. *Des. Codes Cryptogr.*, 16:151–160, 1999.

[129] Y. Edel and J. Bierbrauer. The largest cap in $AG(4, 4)$ and its uniqueness. *Des. Codes Cryptogr.*, 29:99–104, 2003. (Oberwolfach, 2001).

[130] K. B. Farmer and M. P. Hale. Dual affine geometries and alternative bilinear forms. *Linear Algebra Appl.*, 30:183–199, 1980.

[131] G. Fellegara. Gli ovaloidi di uno spazio tridimensionale di Galois di ordine 8. *Atti Accad. Naz. Lincei Rend.*, 32:170–176, 1962.

[132] X. N. Feng and Z. D. Dai. Studies in finite geometries and the construction of incomplete block designs. V. Some 'Anzahl' theorems in orthogonal geometry over finite fields of characteristic 2. *Chinese Math.*, 7:392–410, 1965.

[133] S. Ferret and L. Storme. On the size of complete caps in $PG(3, 2^h)$. *Finite Fields Appl.*, 10:306–314, 2004.

[134] O. Ferri. Su di una caratterizzazione grafica della superficie di Veronese di un $S_{5,q}$. *Atti Accad. Naz. Lincei Rend.*, 61:603–610, 1976.

[135] J. C. Fisher, J. W. P. Hirschfeld, and J. A. Thas. Complete arcs in planes of square order. In *Combinatorics '84*, volume 30 of *Ann. Discrete Math.*, pages 243–250. North-Holland, Amsterdam, 1986. (Bari, 1984).

[136] P. Fong and G. Seitz. Groups with a BN-pair of rank 2, I. *Invent. Math.*, 21:1–57, 1973.

[137] P. Fong and G. Seitz. Groups with a BN-pair of rank 2, II. *Invent. Math.*, 24:191–239, 1974.

[138] H. Freudenthal. Une étude de quelques quadrangles généralisés. *Ann. Mat. Pura Appl.*, 102:109–133, 1975.

[139] D. G. Glynn. On the characterization of certain sets of points in finite projective geometry of dimension three. *Bull. London Math. Soc.*, 15:31–34, 1983.

[140] D. G. Glynn. The non-classical 10-arc of $PG(4, q)$, q even. *Discrete Math.*, 59:43–51, 1986.

[141] D. G. Glynn. Permanent formulae from the Veronesean. *Des. Codes Cryptogr.*, 68:39–47, 2013.

[142] L. Godeaux. *Géométrie Algébrique I, II*. Sciences et Lettres, Liège, 1948, 1949.

[143] J.-M. Goethals and P. Delsarte. On a class of majority-logic decodable cyclic codes. *IEEE Trans. Inform. Theory*, 14:182–188, 1968.

[144] V. D. Goppa. *Geometry and Codes*. Kluwer, Dordrecht, 1988, 157 pp. (translated by N. G. Shartse).

[145] A. Gunawardena and G. E. Moorhouse. The non-existence of ovoids in $O_9(q)$. *European J. Combin.*, 18:171–173, 1997.

[146] H.-R. Halder and W. Heise. On the existence of finite chain-m-structures and k-arcs in finite projective space. *Geom. Dedicata*, 3:483–486, 1974.

[147] J. I. Hall. Linear representations of cotriangular spaces. *Linear Algebra Appl.*, 49:257–273, 1983.

[148] M. Hall. Affine generalized quadrilaterals. In *Studies in Pure Mathematics*, pages 113–116. Academic Press, London, 1971.

[149] N. Hamada. The rank of the incidence matrix of points and d-flats in finite geometries. *J. Sci. Hiroshima Univ.*, 32:381–396, 1968.

[150] N. Hamilton and R. Mathon. Existence and non-existence of m-systems of polar spaces. *European J. Combin.*, 22:51–61, 2001.

[151] N. Hamilton and R. Mathon. On the spectrum of non-Denniston maximal arcs in $PG(2, 2^h)$. *European J. Combin.*, 25:415–421, 2004.

[152] N. Hamilton and T. Penttila. Groups of maximal arcs. *J. Combin. Theory Ser. A*, 94:63–86, 2001.

[153] N. Hamilton and C. T. Quinn. m-systems of polar spaces and maximal arcs in projective planes. *Bull. Belg. Math. Soc. Simon Stevin*, 7:237–248, 2000.

[154] G. Hanssens. *Punt-rechte meetkunden van sferische gebouwen*. PhD thesis, State University of Ghent, 1984.

[155] G. Hanssens and J. A. Thas. Pseudopolar spaces of polar rank three. *Geom. Dedicata*, 22:117–135, 1987.

[156] F. Haot and H. Van Maldeghem. Some characterizations of Moufang generalized quad-
       rangles. *Glasgow Math. J.*, 46:335–343, 2004.

[157] F. Haot and H. Van Maldeghem. A half 3-Moufang quadrangle is Moufang. *Bull. Belg.
       Math. Soc. Simon Stevin*, 12:805–811, 2005.

[158] H. Havlicek. Zur Theorie linearer Abbildungen I. *J. Geom.*, 16:152–167, 1981.

[159] H. Havlicek. Zur Theorie linearer Abbildungen II. *J. Geom.*, 16:168–180, 1981.

[160] H. Havlicek, B. Odehnal, and M. Saniga. On invariant notions of Segre varieties in
       binary projective spaces. *Des. Codes Cryptogr.*, 62:343–356, 2012.

[161] A. Herzer. Die Schmieghyperebenen an die Veronese-Mannigfaltigkeit bei beliebiger
       Charakteristik. *J. Geom.*, 18:140–154, 1982.

[162] A. Herzer. On a projective representation of chain geometries. *J. Geom.*, 22:83–99,
       1984.

[163] D. G. Higman. Partial geometries, generalized quadrangles and strongly regular graphs.
       In *Geometria Combinatoria e Sue Applicazioni*, pages 263–293. Università di Perugia,
       Perugia, 1971. (Perugia, 1970).

[164] D. G. Higman. Invariant relations, coherent configurations and generalized polygons.
       In *Combinatorics, Part 3*, pages 27–43. Math. Centre Tracts 57, Amsterdam, 1974.

[165] D. G. Higman. Classical groups. Department of Mathematics, Technological Univer-
       sity, Eindhoven, 1978, 85 pp. (appendix by D. E. Taylor).

[166] R. Hill. On the largest size of cap in $S_{5,3}$. *Atti Accad. Naz. Lincei Rend.*, 54:378–384,
       1973.

[167] R. Hill. Caps and groups. In *Teorie Combinatorie*, volume II, pages 389–394. Accad.
       Naz. dei Lincei, Rome, 1976. (Rome, 1973).

[168] R. Hill. Caps and codes. *Discrete Math.*, 22:111–137, 1978.

[169] R. Hill. On Pellegrino's 20-caps in $S_{4,3}$. In *Combinatorics '81*, volume 18 of *Ann.
       Discrete Math.*, pages 433–447. North-Holland, Amsterdam, 1983. (Rome, 1981).

[170] R. Hill. *A First Course in Coding Theory*. Oxford University Press, Oxford, 1986, 251
       pp.

[171] J. W. P. Hirschfeld. *Projective Geometries over Finite Fields*. Oxford University Press,
       Oxford, 1979, xii+474 pp.

[172] J. W. P. Hirschfeld. Caps in elliptic quadrics. In *Combinatorics '81*, volume 18 of *Ann.
       Discrete Math.*, pages 449–466. North-Holland, Amsterdam, 1983. (Rome, 1981).

[173] J. W. P. Hirschfeld. *Finite Projective Spaces of Three Dimensions*. Oxford University
       Press, Oxford, 1985, x+316 pp.

[174] J. W. P. Hirschfeld. Quadrics over finite fields. In *Combinatorica*, volume 28 of *Sym-
       posia Mathematica*, pages 53–87. Academic Press, London, 1986. (Rome, 1983).

[175] J. W. P. Hirschfeld. *Projective Geometries over Finite Fields*. Oxford University Press,
       Oxford, second edition, 1998, xiv+555 pp.

[176] J. W. P. Hirschfeld, G. Korchmáros, and F. Torres. *Algebraic Curves over a Finite Field*.
       Princeton University Press, Princeton, 2008, xxii+696 pp.

[177] J. W. P. Hirschfeld and L. Storme. The packing problem in statistics, coding theory and
       finite projective spaces: update 2001. In *Finite Geometries*, pages 201–246. Kluwer,
       Dordrecht, 2001. (Chelwood Gate, 2000).

[178] J. W. P. Hirschfeld and J. A. Thas. The characterization of projections of quadrics over
       finite fields of even order. *J. London Math. Soc.*, 22:226–238, 1980.

[179] J. W. P. Hirschfeld and J. A. Thas. Sets of type $(1, n, q+1)$ in $PG(d, q)$. *Proc. London
       Math. Soc.*, 41:254–278, 1980.

[180] J. W. P. Hirschfeld and J. A. Thas. The generalized hexagon $H(q)$ and the associated
       generalized quadrangle $K(q)$. *Simon Stevin*, 59:407–435, 1985.

[181] J. W. P. Hirschfeld and J. A. Thas. Linear independence in finite spaces. *Geom. Dedicata*, 23:15–31, 1987.

[182] J. W. P. Hirschfeld and J. A. Thas. *General Galois Geometries*. Oxford University Press, Oxford, 1991, x+407 pp.

[183] W. V. D. Hodge and D. Pedoe. *Methods of Algebraic Geometry*, volume I. Cambridge University Press, Cambridge, 1947, 440 pp.

[184] W. V. D. Hodge and D. Pedoe. *Methods of Algebraic Geometry*, volume II. Cambridge University Press, Cambridge, 1953, 394 pp.

[185] W. V. D. Hodge and D. Pedoe. *Methods of Algebraic Geometry*, volume III. Cambridge University Press, Cambridge, 1954, 336 pp.

[186] D. R. Hughes and F. C. Piper. *Design Theory*. Cambridge University Press, Cambridge, 1985, 240 pp.

[187] N. L. Johnson, V. Jha, and M. Biliotti. *Handbook of Finite Translation Planes*. Chapman and Hall/CRC, Boca Raton, 2007, xxii+861 pp.

[188] C. Jordan. *Traité des Substitutions et des Équations Algébriques*. Gauthier-Villars, Paris, 1870, 667 pp.

[189] H. Kaneta and T. Maruta. An algebraic geometrical proof of the extendability of q-arcs in $PG(3, q)$ with q even. *Simon Stevin*, 64:363–366, 1989.

[190] H. Kaneta and T. Maruta. An elementary proof and extension of Thas' theorem on k-arcs. *Math. Proc. Cambridge Philos Soc.*, 105:459–462, 1989.

[191] W. M. Kantor. Ovoids and translation planes. *Canad. J. Math.*, 34:1195–1203, 1982.

[192] W. M. Kantor. Spreads, translation planes and Kerdock sets. I. *SIAM J. Algebraic Discrete Methods*, 3:151–165, 1982.

[193] W. M. Kantor. Spreads, translation planes and Kerdock sets. II. *SIAM J. Algebraic Discrete Methods*, 3:308–318, 1982.

[194] W. M. Kantor. Automorphism groups of some generalized quadrangles. In *Advances in Finite Geometries and Designs*, pages 251–256. Oxford University Press, Oxford, 1991. (Isle of Thorns, 1990).

[195] W. M. Kantor. Generalized quadrangles, flocks, and BLT sets. *J. Combin. Theory Ser. A*, 58.153–157, 1991.

[196] W. M. Kantor. Note on span-symmetric generalized quadrangles. *Adv. Geom.*, 2:197–200, 2002.

[197] W. M. Kantor and E. E. Shult. Veroneseans, power subspaces and independence. *Adv. Geom.*, 13:511–531, 2013.

[198] N. Knarr. A geometric construction of generalized quadrangles from polar spaces of rank three. *Results Math.*, 21:332–344, 1992.

[199] O. Krauss, J. Schillewaert, and H. Van Maldeghem. Veronesean representations of projective planes over quadratic alternative division algebras. *Mich. Math. J.*, to appear.

[200] C. Lefèvre. Tallini sets in projective spaces. *Atti Accad. Naz. Lincei Rend.*, 59:392–400, 1975.

[201] C. Lefèvre-Percsy. Sur les semi-quadriques en tant qu'espaces de Shult projectifs. *Acad. Roy. Belg. Bull. Cl. Sci.*, 63:160–164, 1977.

[202] C. Lefèvre-Percsy. An extension of a theorem of G. Tallini. *J. Combin. Theory Ser. A*, 29:297–305, 1980.

[203] C. Lefèvre-Percsy. Classification d'une famille d'ensembles de classe $(0, 1, n, q + 1)$ de $PG(d, q)$. *J. Combin. Theory Ser. A*, 31:270–276, 1981.

[204] C. Lefèvre-Percsy. Espaces polaires faiblement plongés dans un espace projectif. *J. Geom.*, 16:126–137, 1981.

[205] C. Lefèvre-Percsy. Polar spaces embedded in a projective space. In *Finite Geometries and Designs*, volume 49 of *London Math. Soc. Lecture Note Series*, pages 216–220. Cambridge University Press, Cambridge, 1981. (Isle of Thorns, 1980).

[206] C. Lefèvre-Percsy. Projectivités conservant un espace polaire faiblement plongé. *Acad. Roy. Belg. Bull. Cl. Sci.*, 67:45–50, 1981.

[207] C. Lefèvre-Percsy. Quadrilatères généralisés faiblement plongés dans $PG(3, q)$. *European J. Combin.*, 2:249–255, 1981.

[208] Y. Liu and Z. X. Wan. Pseudo-symplectic geometries over finite fields of characteristic two. In *Advances in Finite Geometries and Designs*, pages 265–288. Oxford University Press, Oxford, 1991. (Isle of Thorns, 1990).

[209] P. M. Lo Re and D. Olanda. Grassmann spaces. *J. Geom.*, 17:50–60, 1981.

[210] P. Lorimer. A projective plane of order 16. *J. Combin. Theory Ser. A*, 16:334–347, 1974.

[211] G. Lunardon and O. Polverino. On the twisted cubic of $PG(3, q)$. *J. Algebraic Combin.*, 18:255–262, 2004.

[212] G. Lunardon and O. Polverino. Translation ovoids of orthogonal polar spaces. *Forum Math.*, 16:663–669, 2004.

[213] H. Lüneburg. *Die Suzukigruppen und ihre Geometrien*, volume 10 of *Lecture Notes in Math.* Springer, Berlin, 1965, 111 pp.

[214] D. Luyckx. m-systems of polar spaces and SPG reguli. *Bull. Belg. Math. Soc. Simon Stevin*, 9:177–183, 2002.

[215] D. Luyckx and J. A. Thas. Flocks and locally hermitian 1-systems of $Q(6, q)$. In *Finite Geometries*, pages 257–275. Kluwer, Dordrecht, 2001. (Chelwood Gate, 2000).

[216] D. Luyckx and J. A. Thas. The uniqueness of the 1-system of $Q^-(7, q)$, q odd. *J. Combin. Theory Ser. A*, 98:253–267, 2002.

[217] D. Luyckx and J. A. Thas. On 1-systems of $Q(6, q)$, q even. *Des. Codes Cryptogr.*, 29:179–197, 2003. (Oberwolfach, 2001).

[218] D. Luyckx and J. A. Thas. Locally Hermitian 1-systems of $Q^+(7, q)$. *Discrete Math.*, 282:223–231, 2004.

[219] D. Luyckx and J. A. Thas. Semipartial geometries, arising from locally Hermitian 1-systems of $W_5(q)$. *Bull. Belg. Math. Soc. Simon Stevin*, 11:69–76, 2004.

[220] D. Luyckx and J. A. Thas. The uniqueness of the 1-system of $Q^-(7, q)$, q even. *Discrete Math.*, 294:133–138, 2005.

[221] F. J. MacWilliams and H. B. Mann. On the p-rank of the design matrix of a difference set. *Information and Control*, 12:474–488, 1968.

[222] F. J. MacWilliams and N. J. A. Sloane. *The Theory of Error-Correcting Codes*. North-Holland, Amsterdam, 1977, 762 pp.

[223] T. Maruta and H. Kaneta. On the uniqueness of $(q + 1)$-arcs of $PG(5, q)$, $q = 2^h$, $h \geq 4$. *Math. Proc. Cambridge Philos. Soc.*, 110:91–94, 1991.

[224] A. Maschietti. Symplectic translation planes and line ovals. *Adv. Geom.*, 3:123–143, 2003.

[225] A. Maschietti. Two-transitive ovals. *Adv. Geom.*, 6:323–332, 2006.

[226] R. Mathon. A new family of partial geometries. *Geom. Dedicata*, 73:11–19, 1998.

[227] F. Mazzocca. Caratterizzazione dei sistemi rigati isomorfi ad una quadrica ellittica dello $S_{5,q}$, con q dispari. *Atti Accad. Naz. Lincei Rend.*, 57:360–368, 1974.

[228] F. Mazzocca. Immergibilità in $S_{4,q}$ di certi sistemi rigati di seconda specie. *Atti Accad. Naz. Lincei Rend.*, 56:189–196, 1974.

[229] F. Mazzocca and N. Melone. Caps and Veronese varieties in projective Galois spaces. *Discrete Math.*, 48:243–252, 1984.

[230] F. Mazzocca and D. Olanda. Sistemi rigati in spazi combinatori. *Rend. Mat.*, 12:221–229, 1979.

[231] K. E. Mellinger. A note on line-Baer subspace partitions of PG(3, 4). *J. Geom.*, 72:128–131, 2001.

[232] K. E. Mellinger. Classical mixed partitions. *Discrete Math.*, 283:267–271, 2004.

[233] K. E. Mellinger. Mixed partitions of PG(3, q^2). *Finite Fields Appl.*, 10:626–635, 2004.

[234] N. Melone. Veronese spaces. *J. Geom.*, 20:169–180, 1983.

[235] N. Melone and D. Olanda. A characteristic property of the Grassmann manifold representing the lines of a projective space. *European J. Combin.*, 5:323–330, 1984.

[236] R. Meshulam. On subsets of finite abelian groups with no 3-term arithmetic progressions. *J. Combin. Theory Ser. A*, 71:168–172, 1995.

[237] G. E. Moorhouse. On codes of Bruck nets and projective planes. In *Coding Theory, Design Theory, Group Theory*, pages 237–242. Wiley, New York, 1993.

[238] G. E. Moorhouse. Root lattice constructions of ovoids. In *Finite Geometry and Combinatorics*, volume 191 of *London Math. Soc. Lecture Note Series*, pages 269–275. Cambridge University Press, Cambridge, 1993. (Deinze, 1992).

[239] G. E. Moorhouse. Some p-ranks related to Hermitian varieties. *J. Statist. Plann. Inference*, 56:229–241, 1996.

[240] C. J. Moreno. *Algebraic Curves over Finite Fields*. Cambridge University Press, Cambridge, 1991, 246 pp.

[241] G. L. Mullen and D. Panario, editors. *Handbook of Finite Fields*. Chapman and Hall/CRC Press, Boca Raton, 2013, xxxv+1033 pp.

[242] H. Niederreiter and C. P. Xing. *Rational Points on Curves over Finite Fields*. Cambridge University Press, Cambridge, 2001, 245 pp.

[243] A. Offer. Translation ovoids and spreads of the generalized hexagon $H(q)$. *Geom. Dedicata*, 85:135–145, 2001.

[244] A. Offer. On the order of a generalized hexagon admitting an ovoid or spread. *J. Combin. Theory Ser. A*, 97:184–186, 2002.

[245] A. Offer. Translation spreads of the split Cayley hexagon. *Adv. Geom.*, 3:105–121, 2003.

[246] A. Offer, K. Thas, and H. Van Maldeghem. Generalized quadrangles with an ovoid that is translation with respect to opposite flags. *Arch. Math.*, 84:375–384, 2005.

[247] A. Offer and H. Van Maldeghem. Spreads and ovoids translation with respect to disjoint flags. *Des. Codes Cryptogr.*, 32:351–367, 2004.

[248] C. M. O'Keefe and T. Penttila. Ovoids of $PG(3, 16)$ are elliptic quadrics. *J. Geom.*, 38:95–106, 1990.

[249] C. M. O'Keefe and T. Penttila. Polynomials for hyperovals of Desarguesian planes. *J. Austral. Math. Soc. Ser. A*, 51:436–447, 1991.

[250] C. M. O'Keefe and J. A. Thas. Ovoids of the quadric $Q(2n, q)$. *European J. Combin.*, 16:87–92, 1995.

[251] D. Olanda. Sistemi rigati immersi in uno spazio proiettivo. Relazione 26, Università di Napoli, 1973.

[252] D. Olanda. Sistemi rigati immersi in uno spazio proiettivo. *Atti. Accad. Naz. Lincei Rend.*, 62:489–499, 1977.

[253] A. Pasini. *Diagram Geometries*. Oxford University Press, New York, 1994, 488 pp.

[254] S. E. Payne. A restriction on the parameters of a subquadrangle. *Bull. Amer. Math. Soc.*, 79:747–748, 1973.

[255] S. E. Payne. All generalized quadrangles of order 3 are known. *J. Combin. Theory Ser. A*, 18:203–206, 1975.

[256] S. E. Payne. Generalized quadrangles of order 4, I; II. *J. Combin. Theory Ser. A*, 22:267–279; 280–288, 1977.

[257] S. E. Payne. Collineations of the Subiaco generalized quadrangles. *Bull. Belg. Math. Soc. Simon Stevin*, 1:427–438, 1994.

[258] S. E. Payne and J. A. Thas. Generalized quadrangles with symmetry, Part II. *Simon Stevin*, 49:81–103, 1976.

[259] S. E. Payne and J. A. Thas. *Finite Generalized Quadrangles*. Pitman, London, 1984, 312 pp.

[260] S. E. Payne and J. A. Thas. *Finite Generalized Quadrangles*. European Mathematical Society, Zurich, second edition, 2009, xii+287 pp.

[261] G. Pellegrino. Sul massimo ordine delle calotte in $S_{4,3}$. *Matematiche (Catania)*, 25:1–9, 1970.

[262] G. Pellegrino. Sulle calotte massime dello spazio $S_{4,3}$. *Atti Accad. Sci. Lett. Arti Palermo*, 34:297–328, 1976.

[263] T. Penttila and B. Williams. Ovoids of parabolic spaces. *Geom. Dedicata*, 82:1–19, 2000.

[264] W. W. Peterson and E. J. Weldon. *Error-Correcting Codes*. MIT Press, Cambridge, Mass., 1972, 560 pp.

[265] V. Pless. On Witt's theorem for nonalternating symmetric bilinear forms over a field of characteristic 2. *Proc. Amer. Math. Soc.*, 15:979–983, 1964.

[266] V. Pless. The number of isotropic subspaces in a finite geometry. *Atti Accad. Naz. Lincei Rend.*, 39:418–421, 1965.

[267] O. Pretzel. *Error-Correcting Codes and Finite Fields*. Oxford University Press, Oxford, 1992, 398 pp.

[268] B. Qvist. Some remarks concerning curves of the second degree in a finite plane. *Ann. Acad. Sci. Fenn. Ser. A*, 134, 1952.

[269] M. A. Ronan. Semiregular graph automorphisms and generalized quadrangles. *J. Combin. Theory Ser. A*, 29:319–328, 1980.

[270] J. Schillewaert. A characterization of quadrics by intersection numbers. *Des. Codes Cryptogr.*, 47:165–175, 2008.

[271] J. Schillewaert and J. A. Thas. Characterizations of Hermitian varieties by intersection numbers. *Des. Codes Cryptogr.*, 50:41–60, 2009.

[272] J. Schillewaert, J. A. Thas, and H. Van Maldeghem. A characterization of the finite Veronesean by intersection properties. *Ann. Comb.*, 16:331–348, 2012.

[273] J. Schillewaert and H. Van Maldeghem. Hermitian Veronesean caps. In *Buildings, Finite Geometries and Groups*, volume 10 of *Springer Proc. Math.*, pages 175–191. Springer, New York, 2012.

[274] J. Schillewaert and H. Van Maldeghem. Quadric Veronesean caps. *Bull. Belg. Math. Soc. Simon Stevin*, 20:19–25, 2013.

[275] B. Segre. Curve razionali normali e k-archi negli spazi finiti. *Ann. Mat. Pura Appl.*, 39:357–379, 1955.

[276] B. Segre. Le geometrie di Galois. *Ann. Mat. Pura Appl.*, 48:1–97, 1959.

[277] B. Segre. *Lectures on Modern Geometry*. Cremonese, Rome, 1961, 479 pp. (with an appendix by L. Lombardo-Radice).

[278] B. Segre. Teoria di Galois, fibrazioni proiettive e geometrie non desarguesiane. *Ann. Mat. Pura Appl.*, 64:1–76, 1964.

[279] B. Segre. Forme e geometrie hermitiane, con particolare riguardo al caso finito. *Ann. Mat. Pura Appl.*, 70:1–201, 1965.

[280] B. Segre. Introduction to Galois geometries. *Atti Accad. Naz. Lincei Mem.*, 8:133–236, 1967. (edited by J. W. P. Hirschfeld).

[281] C. Segre. Sulle varietà che rappresentano le coppie di punti di due piani o spazi. *Rend. Circ. Mat. Palermo*, 5:192–204, 1891.

[282] J. J. Seidel. Strongly regular graphs with $(-1, 1, 0)$ adjacency matrix having eigenvalue 3. *Linear Algebra Appl.*, 1:281–298, 1968.

[283] J. G. Semple and L. Roth. *Introduction to Algebraic Geometry*. Oxford University Press, Oxford, 1949, 446 pp.

[284] E. E. Shult. Characterizations of certain classes of graphs. *J. Combin. Theory Ser. B*, 13:142–167, 1972.

[285] E. E. Shult. Groups, polar spaces and related structures. In *Combinatorics, Part 3*, pages 130–161. Math. Centre Tracts 57, Amsterdam, 1975.

[286] E. E. Shult. Characterizations of the Lie incidence geometries. In *Surveys in Combinatorics*, volume 82 of *London Math. Soc. Lecture Note Series*, pages 157–186. Cambridge University Press, Cambridge, 1983. (Southampton, 1983).

[287] E. E. Shult. A sporadic ovoid in $\Omega^+(8, 7)$. *Algebras, Groups and Geometries*, 2:495–513, 1985.

[288] E. E. Shult. *Points and Lines. Characterizing the Classical Geometries*. Universitext. Springer, Heidelberg, 2011, xxii+676 pp.

[289] E. E. Shult and J. A. Thas. m-systems of polar spaces. *J. Combin. Theory Ser. A*, 68:184–204, 1994.

[290] E. E. Shult and J. A. Thas. Constructions of polygons from buildings. *Proc. London Math. Soc.*, 71:397–440, 1995.

[291] E. E. Shult and J. A. Thas. m-systems and partial m-systems of polar spaces. *Des. Codes Cryptogr.*, 8:229–238, 1996.

[292] P. Sin. The p-rank of the incidence matrix of intersecting linear subspaces. *Des. Codes Cryptogr.*, 31:213–220, 2004.

[293] R. R. Singleton. Minimal regular graphs of maximal even girth. *J. Combin. Theory*, 1:306–332, 1966.

[294] K. J. C. Smith. On the p-rank of the incidence matrix of points and hyperplanes in a finite projective geometry. *J. Combin. Theory*, 7:122–129, 1969.

[295] A. P. Sprague. Pasch's axiom and projective spaces. *Discrete Math.*, 33:79–87, 1981.

[296] A. P. Sprague. Rank 3 incidence structures admitting dual-linear, linear diagram. *J. Combin. Theory Ser. A*, 38:254–259, 1985.

[297] K. O. Stöhr and J. F. Voloch. Weierstrass points and curves over finite fields. *Proc. London Math. Soc.*, 52:1–19, 1986.

[298] L. Storme and J. De Beule, editors. *Current Research Topics in Galois Geometry*. Nova Science Publishers, New York, 2012, 276 pp.

[299] L. Storme and J. A. Thas. Complete k-arcs in $PG(n, q)$, q even. *Discrete Math.*, 106/107:455–469, 1992.

[300] L. Storme and J. A. Thas. M.D.S. codes and arcs in $PG(n, q)$ with q even: an improvement of the bounds of Bruen, Thas, and Blokhuis. *J. Combin. Theory Ser. A*, 62:139–154, 1993.

[301] L. Storme, J. A. Thas, and S. K. J. Vereecke. New upper bounds for the sizes of caps in finite projective spaces. *J. Geom.*, 73:176–193, 2002.

[302] G. Tallini. Sulle k-calotte di uno spazio lineare finito. *Ann. Mat. Pura Appl.*, 42:119–164, 1956.

[303] G. Tallini. Caratterizzazione grafica delle quadriche ellittiche negli spazi finiti. *Rend. Mat. e Appl.*, 16:328–351, 1957.

[304] G. Tallini. Una proprietà grafica caratteristica delle superficie di Veronese negli spazi finiti (Note I; II). *Atti Accad. Naz. Lincei Rend.*, 24:19–23; 135–138, 1958.

[305] G. Tallini. Calotte complete di $S_{4,q}$ contenenti due quadriche ellittiche quali sezioni iperpiane. *Rend. Mat. e Appl.*, 23:108–123, 1964.

[306] G. Tallini. Strutture di incidenza dotate di polarità. *Rend. Sem. Mat. Fis. Milano*, 41:3–42, 1971.

[307] G. Tallini. On the characterization of the Grassmann manifold representing the lines in a projective space. In *Finite Geometries and Designs*, volume 49 of *London Math. Soc. Lecture Note Series*, pages 354–358. Cambridge University Press, Cambridge, 1981. (Isle of Thorns, 1980).

[308] M. Tallini Scafati. Caratterizzazione grafica delle forme hermitiane di un $S_{r,q}$. *Rend. Mat. e Appl.*, 26:273–303, 1967.

[309] K. Tent. Half-Moufang implies Moufang for generalized quadrangles. *J. Reine Angew. Math.*, 566:231–236, 2004.

[310] J. A. Thas. Normal rational curves and k-arcs in Galois spaces. *Rend. Mat.*, 1:331–334, 1968.

[311] J. A. Thas. Connection between the Grassmannian $G_{k-1;n}$ and the set of the k-arcs of the Galois space $S_{n,q}$. *Rend. Mat.*, 2:121–134, 1969.

[312] J. A. Thas. Ovoidal translation planes. *Arch. Math.*, 23:110–112, 1972.

[313] J. A. Thas. Construction of partial geometries. *Simon Stevin*, 46:95–98, 1973.

[314] J. A. Thas. Deduction of properties, valid in the projective space $S_{3n-1}(K)$, using the projective plane over the total matrix algebra $M_n(K)$. *Simon Stevin*, 46:3–16, 1973.

[315] J. A. Thas. On 4-gonal configurations. *Geom. Dedicata*, 2:317–326, 1973.

[316] J. A. Thas. Construction of maximal arcs and partial geometries. *Geom. Dedicata*, 3:61–64, 1974.

[317] J. A. Thas. On 4-gonal configurations with parameters $r = q^2 + 1$ and $k = q + 1$, Part I. *Geom. Dedicata*, 3:365–375, 1974.

[318] J. A. Thas. On semi ovals and semi ovoids. *Geom. Dedicata*, 3:229–231, 1974.

[319] J. A. Thas. A remark concerning the restriction on the parameters of a 4-gonal subconfiguration. *Simon Stevin*, 48:65–68, 1974.

[320] J. A. Thas. On generalized quadrangles with parameters $s = q^2$ and $t = q^3$. *Geom. Dedicata*, 5:485–496, 1976.

[321] J. A. Thas. Combinatorial characterizations of the classical generalized quadrangles. *Geom. Dedicata*, 6:339–351, 1977.

[322] J. A. Thas. Combinatorics of partial geometries and generalized quadrangles. In *Higher Combinatorics*, pages 183–199. Reidel, Dordrecht, 1977. (Berlin, 1976).

[323] J. A. Thas. Two infinite classes of perfect codes in metrically regular graphs. *J. Combin. Theory Ser. B*, 23:236–238, 1977.

[324] J. A. Thas. Combinatorial characterizations of generalized quadrangles with parameters $s = q$ and $t = q^2$. *Geom. Dedicata*, 7:223–232, 1978.

[325] J. A. Thas. Partial geometries in finite affine spaces. *Math. Z.*, 158:1–13, 1978.

[326] J. A. Thas. Polar spaces, generalized hexagons and perfect codes. *J. Combin. Theory Ser. A*, 29:87–93, 1980.

[327] J. A. Thas. New combinatorial characterizations of generalized quadrangles. *European J. Combin.*, 2:299–303, 1981.

[328] J. A. Thas. Ovoids and spreads of finite classical polar spaces. *Geom. Dedicata*, 10:135–144, 1981.

[329] J. A. Thas. Semi-partial geometries and spreads of classical polar spaces. *J. Combin. Theory Ser. A*, 35:58–66, 1983.

[330] J. A. Thas. Characterizations of generalized quadrangles by generalized homologies. *J. Combin. Theory Ser. A*, 40:331–341, 1985.

[331] J. A. Thas. The classification of all (x, y)-transitive generalized quadrangles. *J. Combin. Theory Ser. A*, 42:154–157, 1986.

[332] J. A. Thas. Complete arcs and algebraic curves in $PG(2, q)$. *J. Algebra*, 106:451–464, 1987.

[333] J. A. Thas. Interesting pointsets in generalized quadrangles and partial geometries. *Linear Algebra Appl.*, 114/115:103–131, 1989.

[334] J. A. Thas. Old and new results on spreads and ovoids of finite classical polar spaces. In *Combinatorics '90*, pages 529–544. North-Holland, Amsterdam, 1992. (Gaeta, 1990).

[335] J. A. Thas. Generalized quadrangles of order (s, s^2). I. *J. Combin. Theory Ser. A*, 67:140–160, 1994.

[336] J. A. Thas. Projective geometry over a finite field. In *Handbook of Incidence Geometry: Buildings and Foundations*, pages 295–347. North-Holland, Amsterdam, 1995.

[337] J. A. Thas. Generalized quadrangles of order (s, s^2), II. *J. Combin. Theory Ser. A*, 79:223–254, 1997.

[338] J. A. Thas. Symplectic spreads in $PG(3, q)$, inversive planes and projective planes. *Discrete Math.*, 174:329–336, 1997. (Montesilvano, 1994).

[339] J. A. Thas. Characterizations of translation generalised quadrangles. *Des. Codes Cryptogr.*, 23:249–257, 2001.

[340] J. A. Thas. Flocks and partial flocks of quadrics: a survey. *J. Statist. Plann. Inference*, 94:335–348, 2001.

[341] J. A. Thas. Ovoids, spreads and m-systems of finite classical polar spaces. In *Surveys in Combinatorics, 2001*, pages 241–267. Cambridge University Press, Cambridge, 2001. (Brighton, 2001).

[342] J. A. Thas. SPG systems and semipartial geometries. *Adv. Geom.*, 1:229–244, 2001.

[343] J. A. Thas. The uniqueness of 1-systems of $W_5(q)$ satisfying the BLT-property, with q odd. *Des. Codes Cryptogr.*, 44:3–10, 2007.

[344] J. A. Thas. SPG-reguli, SPG-systems, BLT-sets and sets with the BLT-property. *Discrete Math.*, 309:462–474, 2009.

[345] J. A. Thas. Generalized ovals in $PG(3n - 1, q)$, with q odd. *Pure Appl. Math. Q.*, 7:1007–1035, 2011. (Special Issue: In honor of Jacques Tits).

[346] J. A. Thas and F. De Clerck. Partial geometries satisfying the axiom of Pasch. *Simon Stevin*, 51:123–137, 1977.

[347] J. A. Thas and P. De Winne. Generalized quadrangles in finite projective spaces. *J. Geom.*, 10:126–137, 1977.

[348] J. A. Thas, I. Debroey, and F. De Clerck. The embedding of $(0, \alpha)$-geometries in $PG(n, q)$: Part II. *Discrete Math.*, 51:283–292, 1984.

[349] J. A. Thas and S. E. Payne. Classical finite generalized quadrangles, a combinatorial study. *Ars Combin.*, 2:57–110, 1976.

[350] J. A. Thas and S. E. Payne. Spreads and ovoids in finite generalized quadrangles. *Geom. Dedicata*, 52:227–253, 1994.

[351] J. A. Thas, S. E. Payne, and H. Van Maldeghem. Half Moufang implies Moufang for finite generalized quadrangles. *Invent. Math.*, 105:153–156, 1991.

[352] J. A. Thas, K. Thas, and H. Van Maldeghem. *Translation Generalized Quadrangles*, volume 6 of *Series in Pure Mathematics*. World Sci. Publ., Hackensack, 2006, xxx+345 pp.

[353] J. A. Thas and H. Van Maldeghem. Embedded thick finite generalized hexagons in projective space. *J. London Math. Soc.*, 54:566–580, 1996.

[354] J. A. Thas and H. Van Maldeghem. Orthogonal, symplectic and unitary polar spaces sub-weakly embedded in projective space. *Compositio Math.*, 103:75–93, 1996.

[355] J. A. Thas and H. Van Maldeghem. Generalized quadrangles and the Axiom of Veblen. In *Geometry, Combinatorial Designs and Related Structures*, volume 245 of *London Math. Soc. Lecture Note Series*, pages 241–253. Cambridge University Press, Cambridge, 1997. (Spetses, 1996).

[356] J. A. Thas and H. Van Maldeghem. Sub-weakly embedded singular and degenerate polar spaces. *Geom. Dedicata*, 65:291–298, 1997.

[357] J. A. Thas and H. Van Maldeghem. Generalized quadrangles weakly embedded in finite projective space. *J. Statist. Plann. Inference*, 73:353–361, 1998. (Fort Collins, 1995).

[358] J. A. Thas and H. Van Maldeghem. Lax embeddings of polar spaces in finite projective spaces. *Forum Math.*, 11:349–367, 1999.

[359] J. A. Thas and H. Van Maldeghem. Lax embeddings of generalized quadrangles in finite projective spaces. *Proc. London Math. Soc.*, 82:402–440, 2001.

[360] J. A. Thas and H. Van Maldeghem. Characterizations of the finite quadric Veroneseans $\mathcal{V}_n^{2^n}$. *Quart. J. Math. Oxford*, 55:99–113, 2004.

[361] J. A. Thas and H. Van Maldeghem. Classification of finite Veronesean caps. *European J. Combin.*, 25:275–285, 2004.

[362] J. A. Thas and H. Van Maldeghem. On Ferri's characterization of the finite quadric Veronesean $\mathcal{V}_2^4$. *J. Combin. Theory Ser. A*, 110:217–221, 2005.

[363] J. A. Thas and H. Van Maldeghem. Some characterizations of finite Hermitian Veroneseans. *Des. Codes Cryptogr.*, 34:283–293, 2005.

[364] J. A. Thas and H. Van Maldeghem. Embeddings of small generalized polygons. *Finite Fields Appl.*, 12:565–594, 2006.

[365] J. A. Thas and H. Van Maldeghem. Characterizations of Veronese and Segre varieties. *J. Geom.*, 101:211–222, 2011.

[366] J. A. Thas and H. Van Maldeghem. Generalized Veronesean embeddings of projective spaces. *Combinatorica*, 31:615–629, 2011.

[367] J. A. Thas and H. Van Maldeghem. Characterizations of Segre varieties. *J. Combin. Theory Ser. A*, 120:795–802, 2013.

[368] K. Thas. Automorphisms and characterizations of finite generalized quadrangles. In *Generalized Polygons*, pages 111–172. Vlaams Kennis- en Cultuurforum, Brussels, 2001. (Brussels, 2000).

[369] K. Thas. On symmetries and translation generalized quadrangles. In *Finite Geometries*, pages 333–345. Kluwer, Dordrecht, 2001. (Chelwood Gate, 2000).

[370] K. Thas. The classification of generalized quadrangles with two translation points. *Beiträge Algebra Geom.*, 43:365–398, 2002.

[371] K. Thas. Classification of span-symmetric generalized quadrangles of order s. *Adv. Geom.*, 2:189–196, 2002.

[372] K. Thas. Nonexistence of complete $(st - t/s)$-arcs in generalized quadrangles of order (s, t), I. *J. Combin. Theory Ser. A*, 97:394–402, 2002.

[373] K. Thas. A theorem concerning nets arising from generalized quadrangles with a regular point. *Des. Codes Cryptogr.*, 25:247–253, 2002.

[374] K. Thas. On semi quadrangles. *Ars Combin.*, 67:65–87, 2003.

[375] K. Thas. Translation generalized quadrangles for which the translation dual arises from a flock. *Glasg. Math. J.*, 45:457–474, 2003.

[376] K. Thas. *Symmetry in Finite Generalized Quadrangles*. Birkhäuser, Basel, 2004, xxi+214 pp.

[377] K. Thas. Elation generalized quadrangles of order (q, q^2), q even, with a classical subquadrangle of order q. *Adv. Geom.*, 6:265–273, 2006.

[378] K. Thas and H. Van Maldeghem. Geometric characterizations of finite Chevalley groups of type B_2. *Trans. Amer. Math. Soc.*, 360:2327–2357, 2008.

[379] J. Tits. Sur la trialité et certains groupes qui s'en déduisent. *Inst. Hautes Études Sci. Publ. Math.*, 2:13–60, 1959.

[380] J. Tits. *Buildings of Spherical Type and Finite BN-pairs*, volume 386 of *Lecture Notes in Math.* Springer, Berlin, 1974, 299 pp.

[381] J. Tits. Classification of buildings of spherical type and Moufang polygons: a survey. In *Teorie Combinatorie*, volume I, pages 229–246. Accad. Naz. dei Lincei, Rome, 1976. (Rome, 1973).

[382] J. Tits. Non-existence de certains polygones généralisés. I. *Invent. Math.*, 36:275–284, 1976.

[383] J. Tits. Moufang octagons and the Ree groups of type $^2F_4$. *Amer. J. Math.*, 105:539–594, 1983.

[384] J. Tits. Spheres of radius 2 in triangle buildings. I. In *Finite geometries, buildings, and related topics (Pingree Park, CO, 1988)*, Oxford Sci. Publ., pages 17–28. Oxford Univ. Press, New York, 1990.

[385] J. Tits and R. M. Weiss. *Moufang Polygons*. Springer Monographs in Mathematics. Springer-Verlag, Berlin, 2002.

[386] V. D. Tonchev. *Combinatorial Configurations: Designs, Codes, Graphs*. Wiley, New York, 1988, 189 pp.

[387] N. Tzanakis and J. Wolfskill. The diophantine equation $x^2 = 4q^{a/2} + 4q + 1$ with an application to coding theory. *J. Number Theory*, 26:96–116, 1987.

[388] J. H. van Lint. *Introduction to Coding Theory*. Springer, New York, 1982, 171 pp.

[389] J. H. van Lint and G. van der Geer. *Introduction to Coding Theory and Algebraic Geometry*. Birkhäuser, Basel, 1988, 83 pp.

[390] H. Van Maldeghem. A note on finite self-polar generalized hexagons and partial quadrangles. *J. Combin. Theory Ser. A*, 81:119–120, 1998.

[391] H. Van Maldeghem. *Generalized Polygons*. Birkhäuser, Basel, 1998, xv+502 pp.

[392] J. F. Voloch. Arcs in projective planes over prime fields. *J. Geom.*, 38:198–200, 1990.

[393] J. F. Voloch. Complete arcs in Galois planes of nonsquare order. In *Advances in Finite Geometries and Designs*, pages 401–406. Oxford University Press, Oxford, 1991. (Isle of Thorns, 1990).

[394] W. D. Wallis. Configurations arising from maximal arcs. *J. Combin. Theory*, 15:115–119, 1973.

[395] Z. X. Wan. Studies in finite geometries and the construction of incomplete block designs. I. Some 'Anzahl' theorems in symplectic geometry over finite fields. *Chinese Math.*, 7:55–62, 1965.

[396] Z. X. Wan and B. F. Yang. Studies in finite geometries and the construction of incomplete block designs. III. Some 'Anzahl' theorems in unitary geometry over finite fields and their applications. *Chinese Math.*, 7:252–264, 1965.

[397] A. L. Wells. Universal projective embeddings of the Grassmannian, half spinor, and dual orthogonal geometries. *Quart. J. Math. Oxford*, 34:375–386, 1983.

Index

© Springer-Verlag London 2016
J.W.P. Hirschfeld, J.A. Thas, *General Galois Geometries*, Springer Monographs
in Mathematics, DOI 10.1007/978-1-4471-6790-7

Printed in the United States
by Bookmasters

Printed in the United States
By Bookmasters